Managing Agricultural Greenhouse Gases

Coordinated Agricultural Research through GRACEnet to Address our Changing Climate

Managing Agricultural Greenhouse Gases

Coordinated Agricultural Research through GRACEnet to Address our Changing Climate

Mark A. Liebig

USDA-ARS
Northern Great Plains Research Laboratory
Mandan, ND

Alan J. Franzluebbers

USDA-ARS
Watkinsville, GA

Ronald F. Follett

USDA-ARS
Soil-Plant-Nutrient Research Unit
Fort Collins, CO

ELSEVIER

Amsterdam • Boston • Heidelberg • London • New York • Oxford
Paris • San Diego • San Francisco • Singapore • Sydney • Tokyo
Academic Press is an imprint of Elsevier

Academic Press is an imprint of Elsevier
32 Jamestown Road, London NW1 7BY, UK
225 Wyman Street, Waltham, MA 02451, USA
525 B Street, Suite 1800, San Diego, CA 92101-4495, USA

First edition 2012

Photograph and Figure Credits:

- *Background: Marty Schmer, USDA-ARS, Lincoln, NE*
- *Soil profile: Elizabeth Pruessner, USDA-ARS, Ft. Collins, CO*
- *Greenhouse gas graph: Adapted from Climate change 2007: The physical science basis. 4th Assessment Report. Intergovernmental Panel on Climate Change.*
- *Agricultural greenhouse gas summary: Amy Swan, Colorado State University. Reprinted by permission from Intergovernmental Panel on Climate Change 2006 National Greenhouse Gas Inventory Methods, Vol. 4.*

British Library Cataloguing-in-Publication Data
A catalogue record for this book is available from the British Library

Library of Congress Cataloging-in-Publication Data
A catalog record for this book is available from the Library of Congress

ISBN : 978-0-12-386897-8

For information on all Academic Press publications
visit our website at elsevierdirect.com

Typeset by TNQ Books and Journals Pvt Ltd.
www.tnq.co.in

Printed and bound in United States of America

Transferred to Digital Printing, 2012

CONTENTS

FOREWORD

This book is a tangible illustration of what a scientific community can accomplish with dedication to a vision, even when scarce resources seem to be an overwhelming limitation. The researchers whose work is described herein have set a trajectory for a conceptually simple project that continues the tradition of the most highly effective and useful agricultural research programs: those that can *explain* the natural workings behind production systems and their foundation of natural resources; *predict* how those systems will respond to changes in environmental conditions and management scenarios; and *support decisions* wherever and whenever hard choices must be made to balance environmental goals with feeding the world's people, now and in the future.

The Greenhouse gas Reduction through Agricultural Carbon Enhancement network (GRACEnet) is a multi-location project of the Agricultural Research Service (ARS), the primary in-house research agency for the U.S. Department of Agriculture (USDA). Initiated during 2002, GRACEnet was an outcome of the process ARS uses to establish and plan the centrally coordinated research of its National Programs, which number around 20 at any given time. The intent of those who had the original idea was to link ARS's long-standing research on soil organic matter with the increasing attention and need for information on how to use working lands to sequester carbon from the atmosphere in an attempt to mitigate climate change. GRACEnet is an example of how an idea born in the midst of a procedural requirement can quickly take on a very productive life of its own, if the basic idea is a valid one and a community of creative people takes ownership.

As an experiment, GRACEnet was developed to satisfy some key criteria. First, the research plan had to recognize that there are many kinds and sizes of land-based agricultural systems across the United States, and we need to know about carbon sequestration and greenhouse gas emissions in as many of them as possible. Second, we had to keep the carbon and emissions information in the context of the production systems' main purpose, namely, agricultural yield: what producer would sacrifice substantial production and market value in favor of carbon sequestration alone? Third, we wanted many different scientists to participate, so the plan had to be scalable and applicable to many different places for potential implementation across a range of research funding and personnel resources. Fourth, we needed comparability and sharing of the data across locations if we wanted to make broad conclusions and recommendations. And fifth, we needed a name for the project that the research community could recognize; one that would provide an easy shorthand for discussion.

To meet these criteria yet provide useful information, the overall approach to GRACEnet was reduced to four experimental management scenarios, based on the well-established principle that many of the best management practices to improve soil quality would also increase soil organic matter and thus sequester carbon. If the scientists would continue to build upon research successes while using the data generated to inform climate change and carbon sequestration needs, it would have value to more than one user community, and a broad cross-section of ARS's research community could participate.

The scientists and National Program Leaders thinking about this had to face a hard fact, namely, that not all laboratories had the same resources to focus on this. Thus, some might not be able to undertake research addressing all four of the management scenarios. We accepted that a location's experiment consisting of the baseline scenario and at least one of the other three would be better than no experiment at all, as long as the measurement and data sharing agreements were met. Moreover, we hoped that a participating researcher could set up multiple experiments, with different practices or in different environments. Most optimistically, we hoped that, if the idea was successful, other researchers would want to join in, perhaps non-ARS researchers—maybe even in other countries. We envisioned the effort growing in numbers of locations, management practices, researchers—and usefulness to the science, policy, and agricultural production communities.

Amidst charging forward with the proposed project and listing words and phrases that were central to describing the effort, in a mess of multi-colored ink on the whiteboard in Shafer's office, we came up with the acronym, GRACEnet.

Once the word got out and the scientists got the ball rolling, it was hard to stop. The first GRACEnet scientists' workshop was held later in 2002, the researchers held a methodology workshop in early 2003, and the research began almost immediately thereafter. Since then, GRACEnet has grown to more than 30 ARS locations and has resulted in more than 250 publications and presentations.

Not bad for a group effort without a specific line item in the budget for ARS or anyone else. GRACEnet continues in the newest 5-year iteration of climate and soils research in ARS, and given the challenges we face in agriculture, we see no end in sight for the project.

Over the years, we have been asked occasionally (usually tongue-in-cheek) if there is a hidden meaning of some sort behind the name GRACEnet; perhaps disappointingly, the answer is no, it's just an acronym that works and is pronounceable. However, the first definition for *grace* in the on-line Encarta Dictionary is "elegance, beauty, and smoothness of form or movement." That certainly describes the function and spirit of GRACEnet's scientific community. In naming GRACEnet, perhaps we were graced with more foresight and inspiration than we realized.

Steven R. Shafer
Deputy Administrator, Natural Resources and Sustainable Agricultural Systems,
Agricultural Research Service, U.S. Department of Agriculture, Beltsville, MD

Michael D. Jawson
Director, Upper Midwest Environmental Sciences Center, U.S. Geological Survey,
U.S. Department of the Interior, La Crosse, WI

PREFACE

Atmospheric greenhouse gases (GHGs) absorb and emit radiation within the thermal infrared range, a natural process that regulates the temperature of the earth. Long-term changes in GHG emissions could negatively or positively affect global surface temperature (USGCRP, 2009). The abatement of climate change is one of the most pressing modern-day environmental issues (IPCC, 2007) and raises dire concerns about global warming and its associated impacts on agriculture (CAST, 2011) and key ecosystem functions. Although a view still exists among some that global climate change associated with increases in atmospheric GHGs from human activity is not occurring, evidence to the affirmative has been documented for decades.

Jean Baptiste Joseph Fourier (1768–1830) was the first person to propose what is now referred to as the greenhouse effect (Cowie, 2007). Subsequently in 1859, John Tyndall determined in laboratory studies that several gases, including simple water vapor (H_2O) and carbon dioxide (CO_2), could trap heat rays (AIP, 2011). Svante Arrhenius in Stockholm in 1896 completed a laborious numerical computation which suggested that cutting the amount of CO_2 in the atmosphere by half could lower the temperature in Europe by 4–5°C, to roughly that during the ice age. But the question existed of whether such large changes in atmospheric composition were really possible. For that question Arrhenius turned to a colleague, Arvid Högbom, who compiled estimates of CO_2 fluxes from natural geochemical processes, including emission from volcanoes and ocean uptake, as well as from human-engineered processes from factories and industrial sources. These calculations showed that human activities were adding CO_2 to the atmosphere at rates nearly equal to those of natural geochemical processes (AIP, 2011).

In 1938, Guy Stewart Callendar estimated that a doubling of carbon dioxide (CO_2) could gradually bring a future temperature rise of 2°C (AIP, 2011). Most scientists at the time held the complacent view that increases in atmospheric CO_2 from human activity could never become a problem and that global climate change was implausible. This view changed with the consequences of World War II and the Cold War. New urgency was brought to many fields of science during the 1950s and government funding was notably increased, especially by military agencies. Government officials were aiming to answer questions associated with pressing military needs and almost any information about the atmosphere and oceans could be considered important to national security. Data to quantify absorbed infrared radiation (a topic of more interest to weapons engineers than meteorologists) were one of the first outcomes of this research (Weart, 1997). Other major advances during this time included the evolution of computers, and extensive numerical computations, including the 1952 calculations by Gilbert N. Plass. Compared to Callendar's estimate, Plass indicated that doubling of atmospheric CO_2 would cause a 3–4°C rise in global temperature and warned that climate change could be "a serious problem to future generations" (AIP, 2011). Fossil-derived CO_2 was detected in the atmosphere during the 1950s and no longer could it be assumed that emissions of CO_2 by industry would remain constant. In 1957, Roger Revelle at the Scripps Institution of Oceanography noted that the greenhouse warming effect "may become significant during future decades if industrial fuel combustion continues to rise exponentially." He also wrote "Human beings are now carrying out a large scale geophysical experiment of a kind that could not have happened in the past nor be reproduced in the future" (AIP, 2011).

By the late 1970s, methane (CH_4), nitrous oxide (N_2O), and other GHGs emitted by human activities were found to have global warming potentials tens to hundreds of times greater than that of CO_2 (AIP, 2011). By this time, it was increasingly apparent how rapidly human population and GHG emissions from industry were rising. The global human population was 1 billion in 1800, doubling to 2 billion by 1927, doubling again to 4 billion by 1974, and currently is at 7 billion (Nova, 2011). Shi (2001), using a data set from 93 countries for the period 1975–1996, provided the following observations: (1) rising income levels can be associated with a monotonically upward shift in emissions; (2) population growth was one of the major driving forces behind increasing CO_2 emissions worldwide; and (3) half of the increase in GHG emissions by 2025 could be contributed by future population growth alone.

With recognition that global human population growth and changing climate were threats to sustainability, countries around the world joined together in 1992 to produce an international environmental treaty at the United Nations Conference on Environment and Development. The conference, held in Rio de Janeiro, Brazil, established the United Nations Framework Convention on Climate Change (UNFCCC). The treaty did not set mandatory limits on GHG emissions for individual countries and contained no enforcement mechanisms. Instead, it provided updates (called "protocols") that would set mandatory emission limits. By 1995 voluntary emission reductions through the UNFCCC were realized to be inadequate, negotiations were launched to strengthen the global response to climate change, and the resulting Kyoto Protocol was adopted and signed in 1997 by most countries. The U.S. did not ratify the Protocol. The Kyoto Protocol's first commitment period extended from 2008 to 2012 and it legally bound developed countries to emission reduction targets. In anticipation of the end of the Kyoto commitment period, representatives from 194 countries met in Durban, South Africa, in 2011 and worked towards a new treaty with "legal force" under which emission targets would be adopted by both developed and developing countries. Even with the new treaty, current emission targets are still considered inadequate to prevent average global temperatures from continuing to rise. Furthermore, prolonging the Kyoto Protocol beyond 2012 appears uncertain with some countries already choosing to "opt out" of their emission reduction commitments.

The complexity of climate is widely recognized by countries around the world. Climate change has consequences for all spheres of existence on our planet. It either impacts or is impacted by global issues, including human population growth, poverty, economic development, sustainable development, protection of ecosystems, and resource management. Future solutions must come from all disciplines and fields of research and development. One of the more important impacts of climate change will be on agriculture. Rising food and energy prices and possible water shortages have led many scientists to predict a global food crisis by 2050 when the world population is expected to reach 9 billion. Accordingly, questions related to agricultural production and future food security are currently being seriously addressed. If the definition of "food insecurity" (Webb and Rogers, 2003) is stated positively, then "food security" exists when people are adequately nourished as a result of the physical availability of food, social or economic access to adequate food, and/or adequate food utilization. Attaining future food security will require proactive implementation of adaptive responses to conditions of anticipated climate change. In this regard, it is increasingly important to define "adaptation to what," "who or what is adapting," and "how adaptation occurs" (Follett, 2012). Some of the issues that agriculture will need to respond to include: (1) changes in long-term mean temperature or precipitation conditions; (2) year-to-year variability of climate; (3) increasing challenges of invasive or better adapted pests and diseases; (4) increasing competition for natural resources among different sectors of society; and (5) greater demands for food, feed, fiber, and biofuel. Adaptation to climate change will be an ongoing process, whereby improvements are already being realized in cost reductions, improved varieties, better adapted livestock, improved management, and enhanced marketing strategies.

Considerable literature exists about how agriculture contributes to climate change through its own exchanges of GHG with the atmosphere. Management of agricultural systems to sequester atmospheric CO_2 as soil organic C and to minimize GHG emissions has potential to help mitigate climate change (CAST, 2011). For example, the C sink capacity of the world's agricultural and degraded soils can be considered 50–66% of the historic C loss of 42–78 Pg (10^{15} g) (Lal, 2004). Implementing management strategies to increase the soil C pool size will serve to concurrently enhance food security while offsetting fossil fuel emissions by 0.4–1.2 Pg C yr^{-1}, or 5–15% of current global fossil-fuel emissions. As new strategies and technologies emerge to mitigate GHGs, it will remain important for agriculture to continue to develop and implement successful conservation practices that maintain or enhance soil C pools, improve N use-efficiency to minimize emissions of N_2O, minimize CH_4 emissions from animal and crop production systems, and enhance the resilience of agriculture to adapt to future climate.

This book is envisioned to provide the reader with a new appreciation of the central role of agriculture to address human needs for food, feed, fiber, and fuel, as well as its important role in both mitigating and adapting to future climate scenarios. This book provides recent developments in: (1) the evaluation of agricultural C sequestration and GHG management, measurement, and modeling; (2) economic and policy considerations for the near future; and (3) a look forward to future opportunities and a need for research collaborations.

Ronald F. Follett

Alan J. Franzluebbers

Mark A. Liebig

References

AIP (American Institute of Physics), 2011. The discovery of global warming: the carbon dioxide greenhouse effect. Available online at: <http://www.aip.org/history/climate/co2.htm#M_63_> (verified December 15, 2011).

CAST (Council for Agricultural Science and Technology), 2011. Carbon sequestration and greenhouse gas fluxes in agriculture: challenges and opportunities. Task Force Report No. 142, p. 106.

Cowie, J., 2007. Climate change: biological and human aspects. P. 3. Cambridge University Press, United Kingdom. ISBN 978–0521696197.

Follett, R.F., 2012. Beyond mitigation: adaptation of agricultural strategies to overcome projected climate change. In: Liebig, M.A., Franzluebbers, A.J., Follett, R.F. (Eds.), Managing Agricultural Greenhouse Gases: Coordinated Agricultural Research through GRACEnet to Address our Changing Climate. Academic Press.

IPCC (Intergovernmental Panel on Climate Change), 2007. Climate Change 2007: Working Group III: Mitigation of Climate Change. In: Metz, B., Davidson, O.R., Bosch, P.R., Dave, R., Meyer, L.A. (Eds.)., Cambridge University Press, Cambridge, U.K.

Lal, R., 2004. Soil carbon sequestration impacts on global climate change and food security. Science 304, 1623–1627.

Nova, 2011. World in the Balance. PBS Online by WGBH Educational Foundation Available online at: <http://www.pbs.org/wgbh/nova/worldbalance/numb-nf.htmlw> and <http://www.pbs.org/wgbh/nova/worldbalance/numb-nf.html> (verified on December 15, 2011).

Shi, A., 2001. Population growth and global carbon dioxide emissions. Proc. Int. Union Sci. Stud. Population Conf. in Brazil/session-s09. Development Research Group, The World Bank, Washington, DC.

USGCRP (U.S. Global Change Research Program), 2009. Global Climate Change Impacts in the US. Available online at: <http://www.globalchange.gov/what-we-do/assessment/previous-assessments/global-climate-change-impacts-in-the-us-2009> (verified June 3, 2011).

Webb, P., Rogers, B., 2003. Addressing the "In" in food insecurity. United States Agency for International Development, Office of Food for Peace, Occasional Paper No. 1, Washington, DC. Available online at: <http://www.usaid.gov/our_work/humanitarian_assistance/ffp/ffpop1.pdf> (verified December 19, 2011).

Weart, S.R., 1997. Global warming, Cold War, and the evolution of research plans. Hist. Stud. Phys. Biol. Sci. 27 (2), 319–356.

ACKNOWLEDGMENTS

Efforts culminating toward the publication of this book benefited from the support of many people. First and foremost, we acknowledge those who wrote, either as lead- or co-authors, the 29 chapters comprising this book. These individuals accepted their writing assignments with enthusiasm and zeal and produced superlative contributions within a relatively short timeline. Their contributions—done amid the normal chaos of academic life—are an inspiration to all who take on "one more task" for the benefit of large, synthesis efforts such as this.

We are grateful for the many individuals who served as reviewers. Whether their input was directed to the prospectus or individual chapters, their perspectives, technical insights, and keen editorial skills led to significant improvements throughout the book in both content and readability. Thanks to Mahdi Al-Kaisi, Mark Alley, Francesco Alluvione, Francisco Arriaga, Carl Bernacchi, Eliav Bitan, Humberto Blanco, Signe Kynding Borgen, Larry Cihacek, Sjoerd Duiker, Robert Dungan, Alison Eagle, Richard Farrell, Ron Follett, Alan Franzluebbers, Andrew French, Barry Glaz, Robert Gleason, Ardell Halvorson, Gary Heathman, Gerrit Hoogenboom, Holly Johnson, Jane Johnson, Rob Kallenbach, Doug Karlen, John Kimble, Mark Liebig, Shuguang Liu, Randy Miles, Jack Morgan, Michael Popp, Chuck Rice, Tom Sauer, Marty Schmer, Silvia Secchi, Narasinha Shurpali, Patrick Splichal, Kurt Spokas, Peter Stahl, Bo Stenberg, Jonathon Straube, Gary Varvel, Larry West, Tris West, David Whitehead, and Lewis Ziska. We also thank the seven anonymous reviewers of the prospectus prior to writing the book.

We appreciate the contributions for cover photos from Marty Schmer, USDA-ARS, Lincoln, NE (background photo), Elizabeth Pruessner, USDA-ARS, Ft. Collins, CO (soil profile), Climate Change 2007: The Physical Science Basis, 4th Assessment Report, Intergovernmental Panel on Climate Change (greenhouse gas graph), and Amy Swan, Colorado State University (agricultural greenhouse gas flow diagram; reprinted by permission from Intergovernmental Panel on Climate Change, 2006 National Greenhouse Gas Inventory Methods, Vol. 4).

We thank the wonderful staff at Elsevier. We were fortunate to work with two pairs of Acquisition and Development Editors while preparing this book. Kristi Gomez and Pat Gonzalez were instrumental in getting the book approved and provided key editing guidance through April 2011, while Nancy Maragioglio, Carrie Bolger and Caroline Johnson continued the excellent editorial support to complete the book on schedule. Their helpfulness, encouragement, and patience made the process much less daunting than anticipated.

We acknowledge the scientific and technical contributions of the GRACEnet team throughout the years. The contributions were numerous at all levels of activity. We hope that no one's work was omitted from this acknowledgment. The team included:

- Akron CO—Francisco Calderon, Maysoon Mikha, and Merle Vigil
- Ames IA—Cindy Cambardella, Jerry Hatfield, Kevin Jensen, Doug Karlen, Tom Kaspar, Ben Knutson, Jody Ohmacht, Timothy Parkin, Tom Sauer, Mark Tomer, and Otis Smith
- Auburn AL—Kip Balkcom, Catherine Gacengo, Stephen Prior, Brett Runion, Allen Torbert, Dexter Watts, and Wes Wood
- Beaver WV—Jay Halvorson, Jennifer Harrah, Katherine O'Neill, and Tammy Robertson
- Beltsville MD—Bakhyt Akhmedov, Joe Alfeiri, Gwen Bagley, Prasad Bandaru, Peter Beeson, Jonathan Brevard, James Bunce, Michel Cavigelli, Rachel Chapman-Steele, Anne Conklin, Gary Cooke, Craig Daughtry, Carlos deFreitas, Shaina Denise, Milutin Djurickovic, Paul Doraiswamy, Farnoosh Farahi, Barry Francis, Erwin Fuentes, Grace Garst, Steven Green,

Ariela Haber, Dean Hively, Raymond Hunt, Cesar Izaurralde, Nate Jackson, Linda Jawson, Mike Jawson, Sophie Jin, Abie Johnson, Jessica Kelly, Emily Kmitch, Mikhail Kozlov, Bill Kustas, Victoria Lake, Megan Lang, Jill Lau, Anh Le, Richard Maduka, Betty Marose, Bruno Martins, Greg McCarty, Patrick McLaughlin, Sushil Milak, Oluwaremilekun Ojurongbe, Olubukola Oriola, Melody Poland, Chris Rasmann, James Reeves, Vanessa Rivera, Andrew Russ, Ali Sadeghi, Guy Serbin, Steven Shafer, John Spargo, Alan Stern, Catherine Swoboda, Claire Teitelbaum, Crysta Tran, Charlie Walthall, Tris West, Robert Wright, Lewis Ziska, and Nate Zuk

- Bowling Green KY—Kim Cook, John Loughrin, Nanh Lovanh, and Karamat Sistani
- Brookings SD—Amy Christie, Kurt Dagel, Michael Lehman, Sharon Nichols, Shannon Osborne, and Ann Qualm
- Canal Point FL—Barry Glaz and Dolen Morris
- Cheyenne WY—Justin Derner, Matt Mortenson, and Jerry Schuman
- Corvallis OR—Stephen Griffith and Jeff Steiner
- Davis CA—Kerri Steenwerth
- Florence SC—Warren Busscher, Keri Cantrell, Dean Evans, Patrick Hunt, Jeffrey Novak, Kyoung Ro, and Don Watts
- Fort Collins CO—Laj Ahuja, Edward Buenger, Megan Busick, Stephen Del Grosso, Feike Dijkstra, Ally Eden, Ronald Follett, Ardell Halvorson, Anna Hansen, Dan LeCain, Liwang Ma, Jack Morgan, Elizabeth Pruessner, Melissa Reyes-Fox, David Smith, Mary Smith, Catherine Stewart, and Katie Tylka
- Gainesville FL—Leon Hartwell Allen Jr.
- Kimberly ID—Dave Bjorneberg, Shanna Breeding, James Ippolito, April Leytem, Robert Sojka, David Tarkalson, and Dale Westerman
- Lincoln NE—Chris Bauer, Virginia Jin, Paul Koerner, Susan Siragusa, Steve Swanson, Gary Varvel, and Wally Wilhelm
- Lubbock TX—Veronica Acosta-Martinez, Jon Cotton, Dean Holder, and Ted M. Zobeck
- Mandan ND—Dave Archer, Scott Bylin, Justin Feld, Jason Gross, Holly Johnson, Daniel Kilzer, Scott Kronberg, Mark Liebig, Rebecca Phillips, Jonathan Rolfson, and Sarah Waldron
- Maricopa AZ—Kevin Bronson, Bruce Kimball, Jean McLain, Gerard W. Wall, and Jeff White
- Morris MN—Nancy Barbour, Beth Burmeister, Jim Eklund, Kathy Eystad, Frank Forcella, Sandy Groneberg, Jay Hanson, Charles Hennen, Abdullah Jaradat, Jane Johnson, Hila Kludze, Sharon Lachnicht-Weyers, Scott Larson, Dean Peterson, Don Reicosky, Jana Rinke, Chad Rollofson, Steve VanKempen, Steve Wagner, Chris Wente, and Alan Wilts
- Orono ME—Dwight Cowperthwaite, Tim Griffin, Wayne Honeycutt, Peggy Pinette, and Lawrence Sikora
- Pendleton OR—Stephen Albrecht, Amy Baker, Robert Correa, Hero Gollany, Christopher Iacoboni, Shujiang Kang, Yi Liang, Wayne Polumsky, Chris Roager, and Joe St. Claire
- Prosser WA—Ashok Alva, Rick Boydston, Rebecca Cochran, Harold Collins, Shawel Haile-Mariam, Monica Silva, and Jason Streubel
- Pullman WA—Dave Huggins and Jeff Smith
- Sidney MT—Joy Barsotti, TheCan Ceasar, Mark Gaffri, Michael Johnson, Andy Lenssen, Johnny Rieger, Chris Russell, Upendra Sainju, and Lyn Solberg
- St. Paul MN—John Baker, William Breiter, Michael Dolan, Martin DuSaire, Jason Leonard, Tyson Ochsner, Kurt Spokas, and Rod Venterea
- Temple TX—Jeff Arnold and Ken Potter
- Tifton GA—Janzen Brewer, Dakota Garrick, Nicole Jennings, Christopher Larsen, Lorine Lewis, Anthony O'Day, Coby Smith, Andrew Stinson, Tim Strickland, Dana Sullivan, Brock Ward, and DeeAnn Webb
- Tucson AZ—Dean Martens
- University Park PA—Paul Adler, Denny Bookhamer, Curtis Dell, Steve Lamar, Bart Moyer, Matt Myers, Bill Priddy, and Howard Skinner

- Watkinsville GA—Andrea Banks, Deborah Beese, Devin Berry, Faye Black, Josh Cown, Anthony Dillard, Eric Elsner, Alan Franzluebbers, Heather Hart, Steven Knapp, Kelley Lyness, Kim Lyness, Robert Martin, Carson Pruitt, Zack Schroer, Ron Sharpe, Stephanie Steed, and Stephen Williams
- West Lafayette IN—Adam Beck, Dennis Bucholtz, Daniel Burgin, Rachel Clayton, Rhonda Graef, Amanda Lecklitner, Ryan Lewis, Whitney Pavalko, Kasey Pickett, Emily Pohlman, Sarah Rutkowski, Brian Schleck, Jeremy Schulman, Chris Smith, Doug Smith, and Diane Stott
- Wyndmoor PA—David Douds, Paul Hepperly, Joe Lee, Gerald Nagahashi, and Rita Seidel

Finally, readers are encouraged to visit the GRACEnet website for more information on protocols, methods, database, and recent developments in the project at: www.ars.usda.gov/gracenet.

CONTRIBUTORS

Paul R. Adler
USDA-Agricultural Research Service, Pasture Systems and Watershed Management Research Unit, University Park, PA

Joseph G. Alfieri
USDA-Agricultural Research Service, Hydrology and Remote Sensing Laboratory, Beltsville, MD

Ralph Alig
USDA Forest Service, Pacific Northwest Research Station, Corvallis, OR

Leon Hartwell Allen Jr.
USDA-Agricultural Research Service, Center for Medical, Agricultural, and Veterinary Entomology (CMAVE), Chemistry Research Unit, Gainesville, FL

Jorge Alvaro-Fuentes
Natural Resource Ecology Laboratory, Colorado State University, Ft. Collins, CO

David W. Archer
USDA-Agricultural Research Service, Northern Great Plains Research Laboratory, Mandan, ND

John M. Baker
USDA-Agricultural Research Service, Soil & Water Management Unit, St. Paul MN

Peter C. Beeson
USDA-Agricultural Research Service, Hydrology and Remote Sensing Laboratory, Beltsville, MD

Kevin F. Bronson
USDA-Agricultural Research Service, U.S. Arid-Land Agricultural Research Center, Maricopa, AZ

Tabitha T. Brown
Dept. Crop & Soil Sciences, Washington State University, Pullman, WA

Francisco Calderon
USDA-Agricultural Research Service, Central Great Plains Research Station, Akron, CO

Cynthia A. Cambardella
USDA-Agricultural Research Service, National Laboratory for Agriculture and the Environment, Ames, IA

Michel A. Cavigelli
USDA-Agricultural Research Service, Sustainable Agricultural Systems Laboratory, Henry A. Wallace Beltsville Agricultural Research Center, Beltsville, MD

Adam Chambers
USDA-Natural Resources Conservation Service, National Air Quality and Atmospheric Change Team, Portland, OR

Al S.K. Chan
Collaborator, USDA-Agricultural Research Service, National Laboratory for Agriculture and the Environment, Ames, IA

Mingshi Chen
ASRC Research and Technology Solutions, Contractor to USGS EROS Center, Sioux Falls, SD

Harold P. Collins
USDA-Agricultural Research Service, Vegetable and Forage Crops Research Unit, Prosser, WA

Craig S.T. Daughtry
USDA-Agricultural Research Service, Hydrology and Remote Sensing Laboratory, Beltsville, MD

Sarah C. Davis
Energy Biosciences Institute, University of Illinois at Urbana-Champaign, Urbana, IL

Stephen J. Del Grosso
USDA-Agricultural Research Service, Soil Plant Nutrient Research Unit, Fort Collins, CO

Curtis J. Dell
USDA-Agricultural Research Service, Pasture Systems and Watershed Management Research Unit, University Park, PA

Justin D. Derner
USDA-Agricultural Research Service, Rangeland Resources Research Unit, Cheyenne, WY and Fort Collins, CO

Feike A. Dijkstra
Faculty of Agriculture, Food and Natural Resources, the University of Sydney, Eveleigh NSW, Australia

Xuejun Dong
North Dakota State University Central Grasslands Research Extension Center, Streeter, ND

Mark Easter
Natural Resource Ecology Laboratory, Colorado State University, Ft. Collins, CO

Dale Enerson
North Dakota Farmers Union, Jamestown, ND

Stephen Faulkner
Aquatic Ecology Laboratory, USGS Leetown Science Center, Kearneysville, WV

Ronald F. Follett
USDA-Agricultural Research Service, Soil Plant Nutrient Research Unit, Ft. Collins, CO

Alan J. Franzluebbers
USDA-Agricultural Research Service, J. Phil Campbell Sr., Natural Resource Conservation Center, Watkinsville, GA

Tom W. Goddard
Alberta Agriculture & Rural Development, Edmonton, Alberta, Canada

Hero T. Gollany
USDA-Agricultural Research Service, Columbia Plateau Conservation Research Center, Pendleton, OR

Ram Gurung
Natural Resource Ecology Laboratory, Colorado State University, Ft. Collins, CO

Ardell D. Halvorson
USDA-Agricultural Research Service, Fort Collins, CO

Richard L. Haney
USDA-Agricultural Research Service, Grassland Soil and Water Research Laboratory, Temple TX

Jerry L. Hatfield
USDA-Agricultural Research Service, National Laboratory for Agriculture and the Environment, Ames, IA

W. Dean Hively
Eastern Geographic Science Center, USGS, Reston, VA

Shengli Huang
ASRC Research and Technology Solutions, Contractor to USGS EROS Center, Sioux Falls, SD

David Huggins
USDA-Agricultural Research Service, Washington State University, Pullman, WA

E. Raymond Hunt Jr.
USDA-Agricultural Research Service, Hydrology and Remote Sensing Laboratory, Beltsville, MD

Daniel Inman
National Bioenergy Center, National Renewable Energy Laboratory, Golden, CO

R. César Izaurralde
Joint Global Change Research Institute, Pacific Northwest National Laboratory/University of Maryland, College Park, MD

Michael D. Jawson
Director, Upper Midwest Environmental Sciences Center, U.S. Geological Survey, U.S. Department of the Interior, LaCrosse, WI

Robin E. Jenkins
DuPont Engineering Research and Technology, Wilmington, DE

Virginia L. Jin
USDA-Agricultural Research Service, Agroecosystem Management Research Unit, Lincoln, NE

Robert Johansson
USDA Office of the Chief Economist, Climate Change Program Office, Washington, DC

Greg Johnson
USDA-Natural Resources Conservation Service, National Air Quality and Atmospheric Change Team, Portland, OR

Jane M.F. Johnson
USDA-Agricultural Research Service, North Central Soil Conservation Research Laboratory, Morris, MN

Cindy Keough
Natural Resource Ecology Laboratory, Colorado State University, Fort Collins, CO

Kendrick Killian
Natural Resource Ecology Laboratory, Colorado State University, Ft. Collins, CO

Lyubov A. Kurkalova
North Carolina A&T State University, Greensboro, NC

Megan W. Lang
USDA Forest Service, Northern Research Station, Beltsville, MD

Greg Latta
Department of Forest Engineering, Resources and Management, Oregon State University Corvallis, OR

Jan Lewandrowski
USDA Office of the Chief Economist, Climate Change Program Office, Washington, DC

Zhengpeng Li
ASRC Research and Technology Solutions, Contractor to USGS EROS Center, Sioux Falls, SD

Yi Liang
Department of Biological & Agricultural Engineering, University of Arkansas Division of Agriculture, Fayetteville, AR

Mark A. Liebig
USDA-Agricultural Research Service, Northern Great Plains Research Laboratory, Mandan, ND

Jinxun Liu
Stinger Ghaffarian Technologies (SGT) Inc., Contractor to the USGS EROS Center, Sioux Falls, SD

Shuguang Liu
U.S. Geological Survey (USGS) Earth Resources Observation and Science (EROS) Center, Sioux Falls, SD

Gregory W. McCarty
USDA-Agricultural Research Service, Hydrology and Remote Sensing Laboratory, Beltsville, MD

William B. McGill
Ecosystem Science and Management, University of Northern British Columbia, Prince George, BC, Canada

Jean E.T. McLain
USDA-Agricultural Research Service, U.S. Arid-Land Agricultural Research Center, Maricopa, AZ

Ernest Marx
Natural Resource Ecology Laboratory, Colorado State University, Fort Collins, CO

Miles Merwin
Ridgeback Tree Farm, Portland, OR

Maysoon M. Mikha
USDA-Agricultural Research Service, Central Great Plains Research Station, Akron, CO

Sushil Milak
Science Systems and Applications, Inc., Lanham, MD

Jack A. Morgan
USDA-Agricultural Research Service, Rangeland Resources Research Unit, Fort Collins, CO

Jeffrey M. Novak
USDA-Agricultural Research Service, Coastal Plains Soil, Water, and Plant Research Center, Florence, SC

Jennifer Oeding
Stinger Ghaffarian Technologies (SGT) Inc., Contractor to the USGS EROS Center, Sioux Falls, SD

Stephen Ogle
Natural Resource Ecology Laboratory, Colorado State University, Ft. Collins, CO

Carolyn Olson
USDA-Natural Resources Conservation Service, Office of the Deputy Chief, Washington, DC

Lloyd B. Owens
USDA-Agricultural Research Service (retired), Coshocton OH

Timothy B. Parkin
USDA-Agricultural Research Service, National Laboratory for Agriculture and the Environment, Ames, IA

William J. Parton
Natural Resource Ecology Laboratory, Colorado State University, Fort Collins, CO

Keith Paustian
Department of Soil and Crop Sciences and Natural Resource Ecology Laboratory, Colorado State University, Ft. Collins, CO

John H. Prueger
USDA-Agricultural Research Service, National Laboratory for Agriculture and the Environment, Ames, IA

James B. Reeves III
USDA-Agricultural Research Service (retired), Environmental Management and Byproduct Utilization Laboratory, Beltsville, MD

Don C. Reicosky
USDA-Agricultural Research Service (retired), Morris, MN

Ali M. Sadeghi
USDA-Agricultural Research Service, Hydrology and Remote Sensing Laboratory, Beltsville, MD

Upendra M. Sainju
USDA-Agricultural Research Service, Northern Plains Agricultural Research Laboratory, Sidney, MT

Thomas J. Sauer
USDA-Agricultural Research Service, National Laboratory for Agriculture and the Environment, Ames, IA

Jill Schuler
Natural Resource Ecology Laboratory, Colorado State University, Ft. Collins, CO

Guy Serbin
InuTeq, LLC, Washington, DC

Steven R. Shafer
Deputy Administrator, Natural Resources and Sustainable Agricultural Systems, Agricultural Research Service, U.S. Department of Agriculture, Beltsville, MD

Gilbert C. Sigua
USDA-Agricultural Research Service (retired), Brooksville FL

R. Howard Skinner
USDA-Agricultural Research Service, Pasture Systems & Watershed Management Research Unit, University Park, PA

Jeffrey L. Smith
USDA-Agricultural Research Service, Washington State University, Pullman, WA

Sabrina Spatari
Civil, Architectural & Environmental Engineering, Drexel University, Philadelphia, PA

Kerri L. Steenwerth
USDA-Agricultural Research Service, RMI North, Davis, CA

Emma C. Suddick
University of California, Dept. of Plant Sciences, Davis, CA

Andrew E. Suyker
School of Natural Resources, University of Nebraska-Lincoln, Lincoln, NE

Amy Swan
Natural Resource Ecology Laboratory, Colorado State University, Ft. Collins, CO

Zhengxi Tan
ASRC Research and Technology Solutions, Contractor to USGS EROS Center, Sioux Falls, SD

Gary E. Varvel
USDA-Agricultural Research Service, Agroecosystem Management Research Unit, Lincoln, NE

Rodney T. Venterea
USDA-Agricultural Research Service, Soil and Water Management Research Unit, St. Paul, MN

Shashi B. Verma
School of Natural Resources, University of Nebraska-Lincoln, Lincoln, NE

Roel Vining
USDA-Natural Resources Conservation Service, Information Technology Center, Fort Collins, CO

Claudia Wagner-Riddle
School of Environmental Sciences, University of Guelph, Guelph, Ontario, Canada

Charles L. Walthall
USDA-Agricultural Research Service, Natural Resources and Sustainable Agricultural Systems, Beltsville, MD

Anne Wein
U.S. Geological Survey, Menlo Park, CA

Eric White
Department of Forest Engineering, Resources and Management, Oregon State University Corvallis, OR

Jimmy R. Williams
Texas Agri-Life Extension Service, Texas A&M University, Temple, TX

Steve Williams
Natural Resource Ecology Laboratory, Colorado State University, Ft. Collins, CO

Claudia Young
Earth Resources Technology (ERT), Inc., Contractor to the USGS EROS Center, Sioux Falls, SD

Yimin Zhang
Strategic Energy Analysis Center, National Renewable Energy Laboratory, Golden, CO

EXECUTIVE SUMMARY

Mark A. Liebig[1], Alan J. Franzluebbers[2], Ronald F. Follett[3]
[1]USDA-ARS, Northern Great Plains Research Laboratory, Mandan, ND
[2]USDA-ARS, J. Phil Campbell Sr., Natural Resource Conservation Center, Watkinsville, GA
[3]USDA-ARS, Soil Plant Nutrient Research Unit, Ft. Collins, CO

Global climate change presents numerous challenges to agriculture. Concurrent efforts to mitigate agricultural contributions to climate change while adapting to its projected consequences will be essential to ensure long-term sustainability and food security. To facilitate successful responses to climate change, relevant and timely research will be critical to ensure appropriate application of novel management practices and technologies on agricultural lands throughout the U.S. Such research should provide a mechanistic understanding of underlying processes affecting natural resources, be scalable to provide useful predictions for select management scenarios, and be translated in such a way that it effectively supports informed decision making, whether on a farm or ranch or for national policy. It is under this broad rubric of research goals that GRACEnet (Greenhouse gas Reduction through Agricultural Carbon Enhancement Network) is based.

Initiated by the U.S. Department of Agriculture-Agricultural Research Service (USDA-ARS) in 2002, GRACEnet seeks to (1) determine the effects of agricultural management on soil carbon (C) sequestration and storage, greenhouse gas (GHG) fluxes, and environmental quality; (2) provide land managers with strategies to help mitigate GHG emissions, improve soil quality, and adapt to climate change; and (3) provide policy- and decision-makers with timely and relevant information about management practices and strategies that can be used to mitigate and adapt to climate change (Walthall et al., 2012). As a national project, GRACEnet possesses significant geographical scope, cross-location consistency in measurement protocols and data handling, strong and effective leadership, and a clear plan for disseminating information (USDA-ARS, 2011a).

Incumbent to disseminating GRACEnet information is the need to provide integrative syntheses to foster the communication of research accomplishments in a broad context. Previous efforts were successful in understanding the "state of the science" regarding agricultural management effects on GHG flux and mitigation potential in different regions of North America (Franzluebbers and Follett, 2005; Franzluebbers et al., 2006). To extend these efforts further, this volume—*Managing Agricultural Greenhouse Gases: Coordinated Agricultural Research through GRACEnet to Address our Changing Climate*—provides an inclusive update to the previous synthesis effort. Organized into seven sections, this volume provides regional syntheses of soil organic C (SOC) and GHG dynamics across a broad portfolio of agricultural land uses, as well as additional chapters addressing activities that are central to GRACEnet (e.g. modeling, method development, economic outcomes of GHG mitigation options, adaptation research, and international collaboration). By drawing on the expertise from a broad base of ARS and university scientists, we envision that this volume will support ARS' goal of providing knowledge and information on soil C sequestration, GHG emissions, and environmental benefits to better implement scientifically-based agricultural management practices from field to national policy scales.

Each section in this volume addresses a distinct theme. A brief introduction is provided here with supplementary "Key Points" for each chapter.

SECTION ONE: AGRICULTURAL RESEARCH FOR A CARBON-CONSTRAINED WORLD

This section establishes context for the contributions that follow. Chapter 1 reviews current trends in GHG emissions, agricultural contributions to those emissions, and anticipated risks associated with global climate change. Opportunities to mitigate GHG emissions from agriculture, along with strategies for adapting to impacts of climate change, are also reviewed. Chapter 2 explores the vision behind the development of GRACEnet as a successful national, cross-location research project. Background on the organization, objectives, and treatment structure of GRACEnet is reviewed. Historical milestones and major accomplishments of GRACEnet complete the second chapter.

Chapter 1 (Agriculture and Climate Change: Mitigation Opportunities and Adaptation Imperatives) key points:

- Proactive responses to climate change are needed to ensure maintenance of agroecosystem functions in the future;
- Management practices that mitigate GHG emissions from agriculture and/or adapt to anticipated climate change can provide significant production and environmental co-benefits; and
- Communication of GRACEnet research accomplishments to an international audience will be essential.

Chapter 2 (GRACEnet: Addressing Policy Needs through Coordinated Cross-location Research) key points:

- Increasing interest in terrestrial C sequestration, both nationally and internationally, was a driving force in the conception of GRACEnet. Since the establishment of GRACEnet in 2002, participating researchers have significantly expanded GHG mitigation science;
- GRACEnet has been successful because the vision, leadership structure, communication, and collaboration mechanisms were addressed at the beginning of the project; and
- Common sampling protocols, organized data management, scalable management scenarios that accommodate different production systems while maintaining a focus on agricultural yield, and a commitment to a broad portfolio of deliverables (e.g. producer guidelines, policy-maker friendly documents, refereed science papers, and synthesis publications) make GRACEnet a model for future science collaborations.

SECTION TWO: AGRICULTURAL MANAGEMENT AND SOIL CARBON DYNAMICS

Chapters 3 through 8 provide literature syntheses on SOC dynamics for prevalent agro-ecosystems in the U.S. (cropland, rangeland, pasture, and biofeedstock production systems). Similar to the regional compendia published in 2005 (Franzluebbers and Follett, 2005), chapters review and synthesize literature on SOC change under a broad range of management practices, but with a focus on published research since 2004. Because of the abundance of SOC data for cropland treatments, the continental U.S. was delineated into three regions based on differences in climate and soil type (Figure 1). Accordingly, cropland data were synthesized within a region (east, central, and west reported in Chapters 3, 4, and 5, respectively).

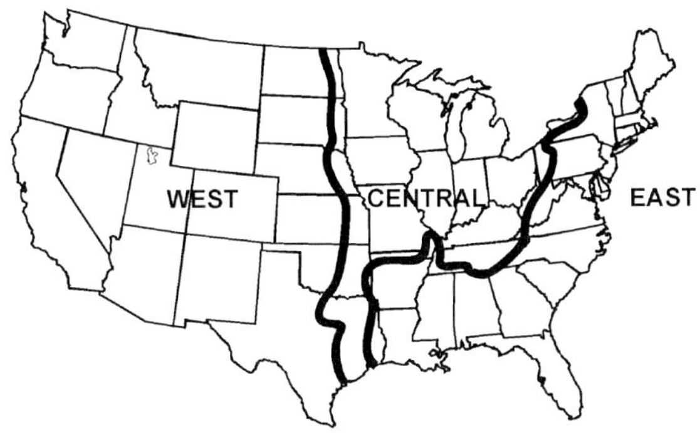

FIGURE 1
Approximate regional delineations for syntheses used in this volume.

Treatment effects on SOC in rangeland and pasture in the U.S. are included in Chapters 6 and 7, while SOC dynamics and related environmental responses to biofeedstock production are included in Chapter 8. Serendipitously, the synthesis provided in Chapter 8 also contributes to another USDA-ARS cross-location research effort, the Renewable Energy Assessment Project (REAP) (USDA-ARS, 2011b).

Chapter 3 (Cropland Management in the Eastern United States for Improved Soil Organic Carbon Sequestration) key points:

- Adoption of no-till crop production in the eastern U.S. typically results in SOC sequestration (~0.5 Mg C ha^{-1} yr^{-1});
- Impact of conservation tillage on SOC below 30 cm is unclear; and
- While conservation tillage, cover crops, and manure application are generally expected to increase SOC, data are needed for a broader range of practices and soils within the region.

Chapter 4 (Soil Carbon Sequestration in Central U.S. Agroecosystems) key points:

- No-tillage management increases surface SOC compared to plow tillage, but in some studies, subsoil C content is greater under plow tillage;
- Enhanced crop rotation complexity fosters SOC sequestration; and
- Nitrogen fertilization effects on SOC are variable due to competing processes of increased C inputs and enhanced decomposition.

Chapter 5 (Agricultural Management and Soil Carbon Dynamics: Western U.S. Croplands) key points:

- Accrual of SOC in dryland cropping systems is generally slow due to low biomass yields and C inputs;
- Adopting no-till in the region increases residue accumulation and surface SOC due to less soil disturbance, less residue incorporation and oxidation, and decreased soil erosion; and
- Intensively managed irrigated lands have the potential for significant SOC gains through the use of improved grazing regimes, fertilization practices, and irrigation management.

Chapter 6 (Soil Carbon Dynamics and Rangeland Management) key points:

- Rangelands are typically characterized by short periods of high C uptake during the growing season followed by long periods of C balance or small losses during the remainder of the year;
- Rangeland SOC response to stocking rate and grazing intensity is variable with differences attributable to differential grazing impacts across environmental gradients and management by climate interactions; and
- Adaptive management practices that optimize rangeland vegetation and soil responses to changing environmental conditions will play a critical role in determining whether rangelands serve as C sources or sinks.

Chapter 7 (Soil Organic Carbon under Pasture Management) key points:

- Significant potential exists to increase SOC in pastures with improved management;
- Additional SOC can be sequestered in pastures with attention to soil fertility, botanical composition, stocking rate, and stocking method; and
- Long-term sequestration of SOC under pastures is possible, but research is needed to establish potential limitations under different soil types and climatic conditions.

Chapter 8 (Sustainable Bioenergy Feedstock Production Systems: Integrating Carbon Dynamics, Erosion, Water Quality and Greenhouse Gas Production) key points:

- Overzealous biomass feedstock harvest may exacerbate a decline in SOC and water quality, and accelerate soil erosion;
- Management options such as reducing or eliminating tillage, adding cover, and including or expanding the presence of perennial crops are strategies to compensate for potential losses of SOC resulting from biomass feedstock harvest; and
- Changing from row crops to perennial energy grasses can sequester 0.49 to 0.75 $Mg\,C\,ha^{-1}\,yr^{-1}$.

SECTION THREE: AGRICULTURAL MANAGEMENT AND GREENHOUSE GAS FLUX

Chapters 9 through 13 mirror the regional format outlined in Section Two, but with a focus on responses of CO_2, CH_4, and N_2O fluxes to management. Chapter 9 compiles recently published GHG research for croplands in the central and eastern U.S., while Chapter 10 addresses the same topic for the western U.S. Limited data for rangeland and pasture treatments justified an aggregated chapter on GHG flux for U.S. grasslands (Chapter 11). Emerging findings on GHG flux and life-cycle analyses for biofeedstock production systems are addressed in Chapter 12. Chapter 13 synthesizes research on GHG flux from organic and flooded soils used for agriculture, both major sources of GHGs that require special management considerations.

Chapter 9 (Cropland Management Contributions to Greenhouse Gas Flux: Eastern and Central U.S.) key points:

- Based on recent findings, reducing N fertilizer rates to economically optimum levels could decrease soil N_2O emissions by 50%;
- Reducing acres in high N-demanding crops could provide even more N_2O mitigation as N_2O emissions from corn are—on average—2.7 times greater than from soybean; and
- Long-term (>20 year) no-till may reduce soil N_2O emissions by 50% compared to conventional tillage.

Chapter 10 (Management to Reduce Greenhouse Gas Emissions in Western U.S. Croplands) key points:

- N_2O emissions tend to be lower under dryland than irrigated production systems;
- Reducing N fertilization rates and selecting appropriate N sources and irrigation systems can reduce N_2O emissions by as much as 50%; and
- CH_4 flux is generally not influenced by crop management practices, except in rice production systems and where manure is applied to cropland.

Chapter 11 (Greenhouse Gas Flux from Managed Grasslands in the U.S.) key points:

- Most rangeland ecosystems in the U.S. are net CO_2 sinks, with net assimilation rates ranging from 0.2 to 1.1 $Mg\,C\,ha^{-1}\,yr^{-1}$;
- Both rangelands and pasturelands have capacity to serve as minor CH_4 sinks, while wetland landscapes with hydrophytic vegetation are significant CH_4 sources; and
- Emissions of N_2O from U.S. grasslands are moderate, and increase with woody plant encroachment, invasion of non-native grasses, and N fertilization, but not grazing intensity.

Chapter 12 (Mitigation Opportunities for Life-Cycle Greenhouse Gas Emissions during Feedstock Production across Heterogeneous Landscapes) key points:

- Global warming intensity (GWI) of ethanol feedstock production can vary significantly from 50 to 500 kg CO_2e Mg^{-1} feedstock produced;
- Factors with the greatest potential to reduce feedstock GWI include N fertilizer use, N_2O emissions, and tillage impacts on SOC; and
- Including entity-level measurements of GWI could incentivize feedstock producers to use practices with lower GHG emissions.

Chapter 13 (Greenhouse Gas Fluxes of Drained Organic and Flooded Mineral Agricultural Soils in the United States) key points:

- Emission of CO_2 from drained organic soils is significant, offsetting about 40% of the C sequestered in U.S. cropland agriculture and grazed lands;
- Growth of adaptable crops in shallow water table of organic soils can mitigate CO_2 and N_2O emissions; and
- Reducing soil incorporation of crop residues, improving N management, selection of cultivars with low CH_4 emissions, and periodically drawing down water tables can reduce CH_4 and N_2O emissions in rice production systems.

SECTION FOUR: MODELING TO ESTIMATE SOIL CARBON DYNAMICS AND GREENHOUSE GAS FLUX FROM AGRICULTURAL PRODUCTION SYSTEMS

Section Four provides reviews of five commonly used ecosystem models for estimating SOC dynamics and GHG flux (i.e. DayCent, COMET2.0, CQESTR, EPIC, and GEMS). Each chapter provides a description of the respective model and includes an application of its use for select management scenarios. The models, each capable of process-based simulations (e.g. IPCC Tier 3 method), vary considerably in their modeled output in both content and spatial scale. Accordingly, readers can compare and contrast models for making informed decisions regarding potential use.

Chapter 14 (DayCent Model Simulations for Estimating Soil Carbon Dynamics and Greenhouse Gas Fluxes from Agricultural Production Systems) key points:

- DayCent is a moderately complex biogeochemical model that simulates exchanges of C, nutrients, and trace gases among the atmosphere, soil, and vegetation;
- DayCent is used to estimate N_2O emissions from cropped and grazed soils for the U.S. National GHG inventory; and
- The latest version of DayCent accounts for management events and plant production on a daily time step.

Chapter 15 (COMET2.0—Decision Support System for Agricultural Greenhouse Gas Accounting) key points:

- COMET2.0 is a web-based decision support system for estimating the GHG footprint of agricultural production systems;
- Using process-based models, COMET2.0 incorporates user-specific information related to current and previous land management, climate, and soil properties; and
- Recent enhancements to COMET2.0 include an expanded set of crop management systems, addition of orchard and vineyards, new agroforestry options, and an N_2O emissions estimator.

Chapter 16 (CQESTR Simulations of Soil Organic Carbon Dynamics) key points:

- CQESTR is a process-based model that simulates SOC dynamics in agricultural management systems;
- CQESTR operates on a daily time-step and performs long-term (100-year) simulations; and
- Model outputs from CQESTR effectively simulated temporal and spatial changes in SOC from crop residue removal, conservation tillage, organic amendment application, and cropping intensification in long-term experiments from three diverse regions of the U.S.

Chapter 17 (Development and Application of the EPIC Model for Carbon Cycle, Greenhouse Gas Mitigation, and Biofuel Studies) key points:

- Originally developed to quantify impacts of erosion on soil productivity, EPIC has evolved into a comprehensive process-based model for simulating numerous ecosystem processes, including SOC dynamics and GHG emissions;
- EPIC allows for simulation of a wide range of management practices, including tillage, harvest, fertilization, irrigation, drainage, liming, burning, pesticide application; and
- Future improvements in EPIC will include addition of simulation modules for biochar application, sorption of soluble C in subsoil horizons, nitrification-induced N_2O emissions, and CH_4 flux.

Chapter 18 (The General Ensemble Biogeochemical Modeling System (GEMS) and its Applications to Agricultural Systems in the United States) key points:

- GEMS integrates well-established process-based models to simulate biogeochemical cycles over large areas;
- GEMS is used for the U.S. Geological Survey's assessment of national potentials for biological C sequestration and reduction of CH_4 and N_2O emissions; and
- GEMS quantifies the uncertainty of model outputs in time and space.

SECTION FIVE: MEASUREMENTS AND MONITORING: IMPROVING ESTIMATES OF SOIL CARBON DYNAMICS AND GREENHOUSE GAS FLUX

An inherent strength of GRACEnet includes the consistent use of similar methods across multiple locations, thereby facilitating data comparisons and scalability assurance for modeling efforts. This strength is shared in Section Five through the contributions of four chapters (Chapters 19–22) addressing key attributes of analytical methods used to estimate C change in soil and GHG flux. The chapters are organized to reflect an increasing assessment scale: <1 m (chamber-based measurements of GHG flux; Chapter 19), 1 to 100 m (spectroscopic methods for quantifying soil C; Chapter 20), 100 to 300 m (micrometeorological methods for assessing GHG flux; Chapter 21), and 5 to >1000 m (remote sensing SOC and GHG dynamics; Chapter 22). Each chapter addresses strengths and limitations of the respective methods, and—where applicable—outlines adjustments to raw data to minimize measurement bias and improve accuracy.

Chapter 19 (Quantifying Biases in Non-Steady-State Chamber Measurements of Soil–Atmosphere Gas Exchange) key points:

- Non-steady-state (NSS) chambers offer the best available tool for statistically evaluating effects of management practices on soil GHG emissions;
- Despite their benefits, NSS chambers tend to produce biased estimates of the actual pre-deployment flux, which can be important for estimating N_2O flux. Numerical methods have been developed to minimize these biases; and
- Advances in instrumentation have the potential to further reduce errors associated with GHG measurements.

Chapter 20 (Advances in Spectroscopic Methods for Quantifying Soil Carbon) key points:

- Expedient methods for assessing soil C are possible through diffuse reflectance spectroscopy in the visible, near-infrared, and mid-infrared spectral ranges;
- Advantages in spectroscopic methods include decreased analytical requirements, multiple-analyte prediction capability, and application of results to large spatial scales using remote sensing imagery; and
- Local calibrations are required to establish relationships between spectral data and reference analytical data.

Chapter 21 (Micrometeorological Methods for Assessing Greenhouse Gas Flux) key points:

- Micrometeorological methods provide opportunities for large-scale, long-term monitoring of GHG flux;
- Micrometeorological methods are especially valuable for capturing periodic high-flux events that have disproportional effects on total annual flux, but whose occurrence is difficult to predict; and
- Requirements of large plots and highly trained staff, coupled with high equipment costs, currently limit the use of micrometeorological methods.

Chapter 22 (Remote Sensing of Soil Carbon and Greenhouse Gas Dynamics across Agricultural Landscapes) key points:

- Remote sensing offers a practical method to account for spatial and temporal variability inherent across agricultural landscapes;
- Spatial scaling of soil C sequestration and GHG emissions is difficult because mechanisms controlling exchange of nutrients, water, and energy interact and are nonlinear; and
- Recently developed data fusion and data assimilation techniques have shown considerable promise for merging data acquired at differing spatial and temporal resolutions and creating synthesized datasets with high spatial and temporal resolution.

SECTION SIX: ECONOMIC AND POLICY CONSIDERATIONS ASSOCIATED WITH REDUCING NET GREENHOUSE GAS EMISSIONS FROM AGRICULTURE

Section Six explores economic outcomes, incentive programs, and policy scenarios associated with reducing GHG emissions from U.S. agriculture. Chapter 23 provides a review of economic outcomes from employing GHG mitigation practices, including reductions in tillage intensity, diversifying crop rotations, and N fertilizer management. Chapter 24 discusses market incentives for GHG mitigation (e.g. cap-and-trade programs), and provides a historical perspective on GHG offset trading in North America. Chapter 25 reviews outcomes from policy scenarios associated with incentivizing GHG mitigation on agricultural and forested land.

Chapter 23 (Economic Outcomes of Greenhouse Gas Mitigation Options) key points:

- Complex interactions among tillage, rotation, fertilizer, and residue management practices affect economic performance and their impacts on GHG mitigation;
- Reducing tillage is a cost-effective GHG mitigation practice, although outcomes vary among and within regions depending on differences in profitability and SOC sequestration rates; and
- As markets and demands for agricultural products evolve, research is needed to anticipate changes in profitability of GHG mitigation practices in response to changes in input and crop prices, as well as conservation and energy policies.

Chapter 24 (Agricultural Greenhouse Gas Trading Markets in North America) key points:

- Reducing GHG emissions across entire economies may be possible through market mechanisms, such as a cap-and-trade system;
- Voluntary GHG markets in the U.S. and Canada have had episodic success, but are currently stymied by uncertainty over future policies; and
- To be successful, future GHG trading/accounting systems must be transparent, consistent, comparable, accurate, and verifiable.

Chapter 25 (Eligibility Criteria Affecting Landowner Participation in Greenhouse Gas Programs) key points:

- Policy-makers recognize the benefits of incentivizing GHG mitigation in agriculture and forestry sectors in the context of pursuing a broad set of environmental goals;

- Simulations indicate that when faced with a modest C payment of $5 MT^{-1} of GHG mitigation, U.S. agriculture and forest landowners could increase income and reduce emissions on average by $>100\ Tg\ yr^{-1}$ over the next 40 years; and
- Restrictive GHG mitigation criteria, such as disallowing bioenergy inclusion or land-use change, could lead to lower emission reductions at any C price.

SECTION SEVEN: LOOKING AHEAD: OPPORTUNITIES FOR FUTURE COLLABORATION AND RESEARCH

Section Seven addresses critical issues facing agriculture in the future. The section begins with a review of networks throughout the world involved in climate change-related research (Chapter 26). The inclusion of this chapter in the final section of this volume is fitting, as future agricultural challenges associated with global climate change will likely be most effectively addressed through broad, multinational collaborative research efforts. Remaining chapters in this section also point to issues of the future, by addressing responses of agro-ecosystems to elevated CO_2 and temperature (Chapters 27 and 28), and agricultural challenges associated with—and options to adapt to—global climate change (Chapter 29).

Chapter 26 (Potential GRACEnet Linkages with Other Greenhouse Gas and Soil Carbon Research and Monitoring Programs) key points:

- Subject complexity, regional variability, and the need for method standardization have contributed to the development of GHG and SOC research networks throughout the world;
- Greater communication and collaboration among research networks would serve to benefit participating members; and
- A long-term commitment to funding is needed for research networks to effectively address climate change related questions.

Chapter 27 (Elevated CO_2 and Warming Effects on Soil Carbon Sequestration and Greenhouse Gas Exchange in Agroecosystems: A Review) key points:

- Climate change will continue to have profound effects on agroecosystems, but impacts on SOC and GHG exchange remain unclear;
- On average, SOC increases under elevated CO_2, particularly in combination with N fertilization; however, N_2O emissions also increase under elevated CO_2 with N fertilization, such that when expressed in terms of global warming potential, N_2O emissions almost completely offset gains in SOC; and
- Exchange of CO_2, CH_4, and N_2O from clayey soils is more sensitive to elevated CO_2 than from sandy soils.

Chapter 28 (Mitigation Opportunities from Land Management Practices in a Warming World: Increasing Potential Sinks) key points:

- Anticipated climate change will not be uniform in time or space, which increases uncertainty about the magnitude of impacts on agricultural systems;
- Because CO_2 emission is closely related to soil temperature, increasing global temperatures may stimulate soil heterotrophic processes and increase SOC decomposition; and
- Improved N management can mitigate N_2O emissions, but net effects will vary depending upon soil water regime.

Chapter 29 (Beyond Mitigation: Adaptation of Agricultural Strategies to Overcome Projected Climate Change) key points:

- Global climate change will impose human, economic, and environmental costs that are country or region specific with climate change impacts being both *biophysical* (e.g. water and temperature) and *socio-economic* (e.g. GDP contribution from agriculture and dietary energy supply per capita);

- Effective adaptation to climate change will require planners to evaluate risk, inventory the availability of needed resources, determine the most effective strategies, and recognize that agriculture is only one component of the total ecosystem; and
- Successful adaptation to global climate change can be facilitated across a range of scales depending upon need by efficiently utilizing resources within an asset portfolio (i.e. land, water, energy, physical infrastructure, genetic resources, research capacity, information technology, human resources, political institutions, and access to markets).

References

Franzluebbers, A.J., Follett, R.F., 2005. Greenhouse gas contributions and mitigation potential in agricultural regions of North America: introduction. Soil Tillage Res. 83, 1–8.

Franzluebbers, A.J., Follett, R.F., Johnson, J.M.F., Liebig, M.A., Gregorich, E.G., Parkin, T.B., Smith, J.L., Del Grosso, S.J., Jawson, M.D., Martens, D.A., 2006. Agricultural exhaust: A reason to invest in soil. J. Soil Water Conserv 61 (3), 98–101.

USDA-ARS (United States Department of Agriculture-Agricultural Research Service), 2011a. National Program 212. Climate change, soils, and emissions: GRACEnet. Available online at: <http://www.ars.usda.gov/research/GRACEnet> (verified August 31, 2011).

USDA-ARS, 2011b. National Program 212. Climate change, soils, and emissions: REAP. Available online at: <http://www.ars.usda.gov/research/programs/programs.htm?np_code=212&docid=21224> (verified September 1, 2011).

Walthall, C.D., Shafer, S.R., Jawson, M.D., 2012. GRACEnet: addressing policy needs through coordinated cross-location research. In: Liebig, M.A., Franzluebbers, A.J., Follett, R.F. (Eds.), Managing Agricultural Greenhouse Gases: Coordinated Agricultural Research through GRACEnet to Address Our Changing Climate. Academic Press, San Diego, CA.

SECTION 1

Agricultural Research for a Carbon-Constrained World

CHAPTER 1

Agriculture and Climate Change: Mitigation Opportunities and Adaptation Imperatives

Mark A. Liebig[1], Alan J. Franzluebbers[2], Ronald F. Follett[3]
[1]USDA-ARS, Northern Great Plains Research Laboratory, Mandan, ND
[2]USDA-ARS, J. Phil Campbell Sr., Natural Resource Conservation Center, Watkinsville, GA
[3]USDA-ARS, Soil Plant Nutrient Research Unit, Ft. Collins, CO

CHAPTER OUTLINE

NB: The U.S. Department of Agriculture, Agricultural Research Service is an equal opportunity/affirmative action employer and all agency services are available without discrimination.

Abbreviations: C, carbon; CO_2, carbon dioxide; CO_{2e}, carbon dioxide equivalent; GWP, global warming potential; GRACEnet, Greenhouse gas Reduction through Agricultural Carbon Enhancement Network; GHG, greenhouse gas; IPCC, Intergovernmental Panel on Climate Change; CH4, methane; N, nitrogen; N_2O, nitrous oxide.

INTRODUCTION

Carbon (C) and nitrogen (N) are critically important elements for sustaining life on earth. The balance of photosynthesis and respiration, along with methanotrophy and methanogenesis,

Managing Agricultural Greenhouse Gases. DOI: 10.1016/B978-0-12-386897-8.00001-2

regulate the presence of C among the atmosphere, biomass, and soil. Nitrogen, as an integral part of nucleotides and proteins, often limits net primary production (Schlesinger, 1997). Accordingly, C and N—and the key metabolic processes that regulate their transfer between compartments in the biosphere—affect the production of food, feed, fiber, and fuel needed for our daily lives.

Carbon and N also play important roles in regulating environmental quality. Reactive forms of both elements—when present in excess of biological requirements—can adversely impact environmental quality across a range of spatial scales (Vitousek et al., 1997; Janzen, 2005). Balancing concurrent needs of food security and a healthy environment is a crucial challenge given projections for human population growth (Godfrey et al., 2010). As such, documenting C and N dynamics within the biosphere will be essential to assess our relative success in achieving these concurrent goals.

Agricultural production contributes to C and N dynamics through the flux of carbon dioxide (CO_2), methane (CH_4,), and nitrous oxide (N_2O), which represent the three greenhouse gases (GHG) principally associated with agricultural activities (Paustian et al., 2006). These three GHGs differ considerably in their atmospheric concentration, residence time in the atmosphere, global warming potential, and radiative forcing (Table 1.1). Carbon dioxide, the most abundant of the three GHGs, is fixed by plants and a portion of it is respired back to the atmosphere. Destruction of plant material through harvesting, natural decay, or burning also contributes to CO_2 emissions through microbial respiration and/or direct combustion. Agricultural-induced fluxes of CH_4 include emissions from ruminant livestock, flooded rice paddies, wetlands, livestock manure, and burned biomass, and, conversely, uptake by methanotrophic bacteria in soil under aerobic conditions. Fluxes of N_2O from agriculture are typically unidirectional through processes of nitrification or denitrification, with emissions most prevalent from cultivated soils, livestock manure, and biomass burning (Schlesinger, 1997; Greenhouse Gas Working Group, 2010; Climate Change Position Statement Working Group, 2011).

Agricultural contributions to total GHG emissions in the U.S. are relatively small, accounting for approximately 6.3% of total emissions in 2009, or 419 of 6633 Tg $CO_{2e}\,yr^{-1}$ (U.S.-EPA, 2011). Of the three agricultural GHGs, emissions of CH_4 and N_2O are dominant, and considered in the U.S.-EPA Agriculture report exclusive of CO_2 emissions and removals. Methane emissions from enteric fermentation and manure management account for 96% of the total CH_4 emissions from agriculture (189 Tg $CO_{2e}\,yr^{-1}$), and are the second and fifth largest anthropogenic sources of CH_4 emissions in the U.S., respectively. Nitrous oxide emissions from soil management practices make up 92% of agricultural N_2O emissions (205 Tg $CO_{2e}\,yr^{-1}$), and are by far the largest source of anthropogenic N_2O emissions in the U.S., accounting for 69% of the total. Emissions of CO_2 from agriculture are largely constrained to fossil fuel combustion, land conversion to cropland, lime application, and urea fertilization ($\Sigma = 83.1$ Tg $CO_2\,yr^{-1}$). However, agricultural practices in the U.S. sequester approximately 49.3 Tg $CO_2\,yr^{-1}$ through conversion of cropland to grassland, increased use

TABLE 1.1 Attributes of atmospheric CO_2, CH_4, and N_2O (IPCC, 2007; NOAA, 2011)

Species	Atmospheric concentration, 2009 (ppm; ppb)†	Residence time in atmosphere (yr)	Global warming potential‡	Global radiative forcing 2009 ($W\,m^{-2}$)	Global radiative forcing Increase since 1979 ($W\,m^{-2}$; %)
CO_2	387	5	1	1.760	0.734; 72
CH_4	1818	9	25	0.502	0.083; 20
N_2O	323	120	298	0.173	0.074; 75

†CO_2 (ppm); CH_4 and N_2O (ppb).
‡Capacity to trap heat in the atmosphere over a 100-year time horizon relative to CO_2.

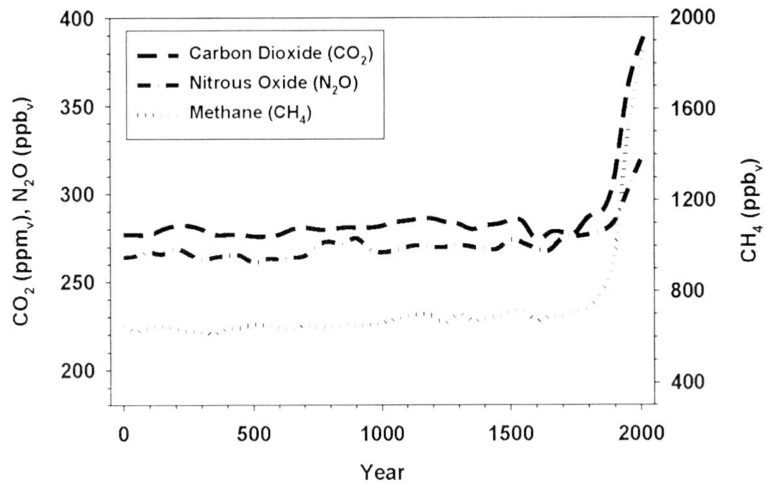

FIGURE 1.1
Concentrations of atmospheric CO_2, CH_4, and N_2O during the previous two millennia (after IPCC, 2007).

of conservation tillage and continuous cropping, and improved management of organic fertilizers (U.S.-EPA, 2011).

Atmospheric concentrations of GHGs have increased significantly since the mid-1700s (Figure 1.1). This increase has been driven mainly by fossil fuel combustion and land-use change resulting from human activities. The capacity of GHGs to trap outgoing long-wave radiation and emit it back to the earth's surface as heat has contributed to global-scale climate change (Paustian et al., 2006). Direct effects of climate change are significant and long-lasting, and include an increase in global average surface temperature, altered precipitation patterns, reduced snow cover, increased sea level rise, and ocean acidification (IPCC, 2007). These projected changes will have broad effects on agriculture (Follett, 2012). Shifts in vegetation zones, increased potential for droughts and floods, elevated rates of soil erosion, and increased photosynthetic rates (from higher CO_2 concentration) represent potential outcomes affecting agriculture, as well as how agriculture affects the broader environment (Climate Change Position Statement Working Group, 2011; Janzen et al., 2011). Moreover, positive feedbacks from climate change—such as accelerated soil organic matter decomposition and release of CH_4 from northern soils—could exacerbate such effects.

Challenges to agriculture associated with climate change are not short term. Momentum in human population growth through the mid-21st century will almost surely result in increased rates of GHG emissions, particularly from the energy sector (IPCC, 2007). Furthermore, even if GHG emissions were to stabilize or decrease, consequences from global climate change would continue well into the next century due to momentum from climate processes and feedbacks (IPCC, 2007; Armour and Roe, 2011). This reality has led to an increased awareness that agriculture has a crucial role to play in responding to climate change, both in mitigating its causes and adapting to its impacts (Climate Change Position Statement Working Group, 2011).

MITIGATING AND ADAPTING TO CLIMATE CHANGE

Recent reviews have provided extensive lists documenting how agricultural practices can mitigate and/or adapt to climate change (CAST, 2011; Eagle et al., 2010; Greenhouse Gas Working Group, 2010; Delgado et al., 2011; Lal et al., 2011; Climate Change Position Statement Working Group, 2011). Broadly, suggested GHG mitigation practices either contribute to soil organic C (SOC) accrual, reduce CH_4 and/or N_2O emissions, or reduce fuel consumption. Adaptation responses to climate change address agroecosystem adjustments to alterations in environmental conditions (Climate Change Position Statement Working Group, 2011). Such responses extend beyond regulating GHG fluxes through management, to address broader themes related to reducing negative impacts on agroecosystems while taking advantage of potential benefits associated with climate change.

Mitigation

The ASA-CSSA-SSSA Greenhouse Gas Working Group provided five broad strategies for mitigating agricultural GHG emissions (Greenhouse Gas Working Group, 2010):

1. Enhance soil C sequestration;
2. Improve N-use efficiency;
3. Increase ruminant digestion efficiency;
4. Capture GHG emissions from manure and other wastes; and
5. Reduce fuel consumption.

These five strategies are well established to either remove GHGs from the atmosphere (1) or reduce GHG emissions from known sources (2, 3, 4, 5). Because each mitigation strategy has been thoroughly addressed in previous reviews, only a synopsis of each is provided here.

ENHANCE SOIL C SEQUESTRATION

Enhancement of soil C sequestration can be achieved by maintaining plant residues on the soil surface, minimizing soil disturbance and erosion, adopting complex cropping systems that provide continuous ground cover, and applying C-rich substrates to soil (Lal and Follett, 2009). The magnitude and rate of soil C sequestration is dependent on various edaphic and climatic factors that directly affect biomass productivity and C retention in soil (Brady and Weil, 1999). In some instances, management practices have had variable effects on soil C dynamics and CH_4 and N_2O flux, resulting in either enhancing (e.g. increased soil C, decreased N_2O emission) or negating (e.g. increased soil C, increased N_2O emission) net GHG emissions. Such variable responses emphasize the importance of inclusive GHG assessments to ascertain GHG tradeoffs associated with management (Eagle et al., 2010).

IMPROVE N-USE EFFICIENCY

Improving N-use efficiency involves the implementation of management practices that make N available in the amount needed at the correct time to meet plant demand (Lal et al., 2011). When successful, such practices result in less reactive N available for potential conversion to N_2O. Numerous management practices are available to improve N-use efficiency, including use of legumes, cover crops, filter strips, and nitrification inhibitors, application of variable-rate technology, and judicious use of soil tests to estimate soil N available for plant uptake (Greenhouse Gas Working Group, 2010).

INCREASE RUMINANT DIGESTION EFFICIENCY

Methane emissions from ruminant livestock depend on many factors, most notably livestock type, diet quality, and feed intake (Westberg et al., 2001). Strategies to reduce CH_4 emissions from livestock include improved feeding practices (e.g. enhancing pasture quality), use of dietary amendments (e.g. edible oils, ionophores, organic acids), and improved genetics (Kebreab et al., 2006). However, the effectiveness of these strategies is often influenced by environmental conditions, soil and plant interactions, animal behavior, and level of management expertise (Murray et al., 2007).

CAPTURE GHG EMISSIONS FROM MANURE AND OTHER WASTES

Livestock manure can be a significant source of CH_4 and N_2O (U.S.-EPA, 2011). Capturing biogas (CH_4, CO_2) from manure through anaerobic digestion increases production efficiencies by utilizing CH_4 as fuel for generating on-site electricity and heat energy (Kebreab et al., 2006). Moreover, residual solid material (sludge) following digestion may be used as fertilizer, thereby supplementing plant nutrient requirements. Additional strategies to reduce GHG fluxes from manure include composting, covering stored manure, altering diet composition, adoption of novel application methods, and using nitrification inhibitors (Külling et al., 2001; Schoenau et al., 2010).

REDUCE FUEL CONSUMPTION

Reduction in fuel consumption directly contributes to lower CO_2 emissions. In this regard, agricultural practices that reduce the number of field passes by farm machinery, such as conservation tillage, lower fuel consumption (West and Marland, 2002). Agricultural practices that reduce applications of synthetic fertilizer and pesticides can reduce upstream CO_2 emissions associated with their manufacture (Hoeppner et al., 2006). Additionally, implementation of efficient irrigation practices (e.g. drip irrigation) and maximizing in-field grain drying prior to harvest serves to increase energy-use efficiency, thereby avoiding CO_2 emissions (Greenhouse Gas Working Group, 2010).

Adaptation

Significant concerns exist regarding the capacity of agroecosystems to provide food, feed, fiber, and fuel, and maintain ecosystem services under anticipated conditions of global climate change. Development and adoption of adaptation strategies will be essential to minimize negative biophysical and socioeconomic consequences, particularly as demand for agricultural products and competition for natural resources increases with a larger human population (Tilman et al., 2011). While research on this topic is in its infancy, select management strategies have been proposed to adapt to global climate change (Climate Change Position Statement Working Group, 2011):

1. Increase crop diversity;
2. Implement efficient irrigation methods;
3. Adopt integrated pest management; and
4. Improve soil management.

Each strategy directly or indirectly addresses adaptation to climate change by responding to changes in long-term temperature and precipitation conditions, annual weather variation, and challenges associated with invasive pests and/or diseases (Follett, 2012). Moreover, the strategies serve to increase production efficiencies while simultaneously improving environmental quality.

INCREASE CROP DIVERSITY

Increasing the number of crops in rotation as well as broadening the tolerance of crops to drought, heat, and nutrient stresses through improved crop varieties can moderate weather-related effects associated with climate change (Climate Change Position Statement Working Group, 2011). Moreover, adoption of annual crop sequencing approaches that optimize production, economic, and resource conservation goals can serve to increase management adaptability in the context of climate-induced change (Hanson et al., 2007). Such cropping systems, which are inherently dynamic in space and time, allow sequencing of crops in a manner to take advantage of available water and nutrients while disrupting weed and disease cycles (Tanaka et al., 2002). Accordingly, dynamic cropping systems can decrease requirements for off-farm inputs (e.g. fertilizer and pesticides) as compared with fixed-sequence and monoculture cropping systems (Tanaka et al., 2005).

IMPLEMENT EFFICIENT IRRIGATION METHODS

Efficient utilization of water for crop growth will be essential in adapting to global climate change. Irrigated agriculture is of particular concern, given its significant production potential and high economic value relative to rainfed production systems, coupled with its vulnerability to depleted water supplies (Hatfield et al., 2011). Adoption of irrigation technology capable of delivering water to crops in space and time in precise doses with minimal loss will increase water- and nutrient-use efficiency (Delgado et al., 2011). Additional strategies for efficient water use include adoption of conservation practices that increase water storage and decrease evaporative demand (Follett, 2012).

ADOPT INTEGRATED PEST MANAGEMENT (IPM)

Climate change has significant potential to increase the complexity of pest and disease management. Anticipated effects of climate change include increased populations, shorter life cycles, range expansion, increased herbivory, and new crop hosts (Chakraborty et al., 2000; Bale et al., 2002). Implementation and/or modification of current IPM strategies will be necessary to address these challenges, and will require the development of new methodologies to adapt IPM to different climatic conditions (Climate Change Position Statement Working Group, 2011).

IMPROVE SOIL MANAGEMENT

Soil management practices that conserve water, minimize erosion, and improve soil function will contribute to increased agroecosystem resilience under anticipated climate change. Generally, management strategies that increase C input to soil, reduce decay rates of soil organic matter, and improve N-use efficiency will contribute positively to these improvements in production efficiency (Eagle et al., 2010; Delgado et al., 2011; Lal et al., 2011; Follett, 2012).

Co-Benefits

Numerous management practices that mitigate GHG emissions or that can be used to adapt to global climate change also enhance agroecosystem function, and accordingly contribute to the achievement of production and environmental goals (Lal and Follett, 2009; Delgado et al., 2011; Lal et al., 2011). Such co-benefits have been strongly associated with practices that enhance soil C sequestration (Janzen, 2005). Accrual of SOC in agricultural lands has been associated with improvements in soil physical, chemical, and biological properties, which affect key soil functions, such as nutrient cycling, filtering and buffering capacity, and regulation of hydrological attributes (Andrews et al., 2004; Franzluebbers, 2010). While precise relationships are difficult to quantify, improvements in soil attributes and related functions have positive effects on agronomic yield and environmental quality (Bauer and Black, 1994; Diaz-Zorita et al., 1999; Wienhold et al., 2006; Lal and Follett, 2009). Such associations have led others to assert that the greatest value from C sequestration may relate more to improvements in soil functions, on-site productivity, and off-site environmental benefits than a reduction in GHG emissions (Duxbury, 1994).

In many respects, mitigation and adaptation strategies focus on conserving C and N within agroecosystems, thereby improving production efficiencies. Carbon and N retained in agroecosystems—and not lost through GHG emissions—increases the likelihood of more efficient use of nutrients, water, energy, and labor, which can result in lower input costs for producers (Delgado et al., 2011). Moreover, management strategies that directly reduce demands for fossil energy and irrigation water translate to critically important economic co-benefits, particularly as these resources become more limiting, and hence more expensive (National Intelligence Council, 2008).

In addition to economic co-benefits associated with improved production efficiencies, select management practices may generate supplemental income for producers through payments from emission trading programs. Such programs, similar to those previously administered by the National Farmers Union (NFU) and Chicago Climate Exchange (CCX), have provided a framework for GHG emitting entities (e.g. power generation companies) to offset their emissions by purchasing credits from entities known to achieve net GHG uptake (Reicosky et al., 2012). When active (2006–2010), the NFU/CCX program provided more than

$7 million in offset payments to U.S. farmers and ranchers employing conservation practices known to sequester atmospheric CO_2 (Dale Enerson, personal communication, 2011). While the future of such programs in the U.S. is unknown at this time, they have the potential to provide valuable economic co-benefits for producers coping with agronomic impacts from climate change.

SUMMARY

Atmospheric concentrations of GHGs have increased significantly since the mid-1700s. The capacity of GHGs to trap outgoing long-wave radiation and emit it back to the earth's surface as heat has contributed to global-scale climate change. Direct effects of climate change are significant and long-lasting, and are projected to affect agriculture through shifts in vegetation zones, increased potential for droughts and floods, elevated rates of soil erosion, and increased photosynthetic rates. Maintaining key agronomic and environmental functions in the future will require deployment of a broad portfolio of management practices that can mitigate GHG emissions and/or adapt to impacts from climate change. Agricultural strategies for mitigating GHG emissions include enhancing soil C sequestration, improving N-use efficiency, increasing ruminant digestion efficiency, capturing GHG emissions from manure and other wastes, and reducing fuel consumption. Though less developed, climate change adaptation strategies specific to agriculture include increasing crop diversity, implementing efficient irrigation methods, adopting integrated pest management, and improving soil management. Significant production, environmental, and economic co-benefits potentially exist through the successful application of mitigation and adaptation practices.

References

Andrews, S.S., Karlen, D.L., Cambardella, C.A., 2004. The soil management assessment framework: a quantitative soil quality evaluation method. Soil Sci. Soc. Am. J. 68, 1945–1962.

Armour, K.C., Roe, G.H., 2011. Climate commitment in an uncertain world. Geophys. Res. Lett. 38 L01707, doi:10.1029/2010GL045850.

Bale, J.S., Masters, G.J., Hodkinson, I.D., Awmack, C., Bezemer, T.M., Brown, V.K., Butterfield, J., Buse, A., Coulson, J.C., Farrar, J., Good, J.E.G., Harrington, R., Hartley, S., Jones, T.H., Lindroth, R.L., Press, M.C., Symrnioudis, I., Watt, A.D., Whittaker, J.B., 2002. Herbivory in global climate change research: direct effects of rising temperatures on insect herbivores. Global Change Biol. 8, 1–16.

Bauer, A., Black, A.L., 1994. Quantification of the effect of soil organic matter content on soil productivity. Soil Sci. Soc. Am. J. 58, 185–193.

Brady, N.C., Weil, R.R., 1999. The Nature and Properties of Soils, twelfth ed. Prentice Hall, Upper Saddle River, NJ, USA.

CAST (Council for Agricultural Science and Technology), 2011. Carbon sequestration and greenhouse gas fluxes in agriculture: challenges and opportunities. Task Force Report No. 142, 106.

Chakraborty, S., Tiedemann, A.V., Teng, P.S., 2000. Climate change: potential impact on plant diseases. Environ. Poll. 108, 317–326.

Climate Change Position Statement Working Group, 2011. Position statement on climate change. Working Group Rep. ASA, CSSA, and SSSA, Madison, WI. Accessed at <https://www.crops.org/files/science-policy/asa-cssa-sssa-climate-change-policy-statement.pdf> (verified December 15, 2011).

Delgado, J.A., Groffman, P.M., Nearing, M.A., Goddard, T., Reicosky, D., Lal, R., Kitchen, N.R., Rice, C.W., Towery, D., Salon, P., 2011. Conservation practices to mitigate and adapt to climate change. J. Soil Water Conserv 66 (4), 118A–129A.

Diaz-Zorita, M., Buschiazzo, D.E., Peinemann, N., 1999. Soil organic matter and wheat productivity in the semiarid Argentine Pampas. Agron. J. 91, 276–279.

Duxbury, J.M., 1994. The significance of agricultural sources of greenhouse gases. Fert. Res. 38, 151–163.

Eagle, A.J., Henry, L.R., Olander, L.P., Haugen-Kozyra, K., Millar, N., Robertson, G.P., 2010. Greenhouse gas mitigation potential of agricultural land management in the United States: a synthesis of the literature. Nicholas Institute for Environmental Policy Solutions. Duke Univ. Durham, NC. Accessed at <http://nicholasinstitute.duke.edu/ecosystem/land/TAGGDLitRev> (verified December 15, 2011).

Enerson D. Personal communication.

Franzluebbers, A.J., 2010. Will we allow soil carbon to feed our needs? Carbon Manage. 1, 237–251.

Follett, R.F., 2012. Beyond mitigation: adaptation of agricultural strategies to overcome projected climate change. In: Liebig, M.A., Franzluebbers, A.J., Follett, R.F. (Eds.), Managing agricultural Greenhouse Gases: Coordinated Agricultural Research through GRACEnet to Address Our Changing Climate. Academic Press, San Diego, CA.

Godfrey, H.C.J., Beddington, J.R., Crute, I.R., Haddad, L., Lawrence, D., Muir, J.F., Pretty, J., Robinson, S., Thomas, S.M., Toulmin, C., 2010. Food security: the challenge of feeding 9 billion people. Science 327, 812–818.

Greenhouse Gas Working Group, 2010. Agriculture's role in greenhouse gas emissions & capture. Greenhouse Gas Working Group Rep. ASA, CSSA, and SSSA, Madison, WI. Accessed at <https://www.crops.org/files/science-policy/ghg-report-august-2010.pdf> (verified December 15, 2011).

Hanson, J.D., Liebig, M.A., Merrill, S.D., Tanaka, D.L., Krupinsky, J.M., Stott, D.E., 2007. Dynamic cropping systems: increasing adaptability amid an uncertain future. Agron. J. 99 (4), 939–943.

Hatfield, J.L., Boote, K.J., Kimball, B.A., Ziska, L.H., Izaurralde, R.C., Ort, D., Thompson, A.M., Wolfe, D., 2011. Climate impacts on agriculture: implications for crop production. Agron. J. 103, 351–370.

Hoeppner, J.W., Entz, M.H., McConkey, B.G., Zentner, R.P., Nagy, C.N., 2006. Energy use and efficiency in two Canadian organic and conventional crop production systems. Renew. Agric. Food Syst. 21, 60–67.

IPCC (Intergovernmental Panel on Climate Change), 2007. Climate change 2007: the physical science basis. 4th Assessment Report Accessed at <http://www.ipcc.ch/publications_and_data/publications_and_data_reports.shtml> (verified December 15, 2011).

Janzen, H.H., 2005. Soil carbon: a measure of ecosystem response in a changing world? Can. J. Soil Sci. 85, 467–480.

Janzen, H.H., Fixen, P.E., Franzluebbers, A.J., Hattey, J., Izaurralde, R.C., Ketterings, Q.M., Lobb, D.A., Schlesinger, W.H., 2011. Global prospects rooted in soil science. Soil Sci. Soc. Am. J. 75, 1–8.

Kebreab, E., Clark, K., Wagner-Riddle, C., France, J., 2006. Methane and nitrous oxide emissions from animal agriculture: a review. Can. J. Anim. Sci. 86, 135–158.

Külling, D.R., Henzi, H.K., Krober, T.F., Neftel, A., Sutter, F., Lischer, P., Kreuzer, M., 2001. Emissions of ammonia, nitrous oxide and methane from different types of dairy manure during storage as affected by dietary protein content. J. Agric. Sci. 137, 235–250.

Lal, R., Delgado, J.A., Groffman, P.M., Millar, N., Dell, C., Rotz, A., 2011. Management to mitigate and adapt to climate change. J. Soil Water Conserv. 66 (4), 276–285.

Lal, R., Follett, R.F., 2009. Soils and climate change. In: Lal, R., Follett, R.F. (Eds.), Soil Carbon Sequestration and the Greenhouse Effect, second ed.). SSSA Spec. Publ. 57. ASA-CSSA-SSSA, Madison, WI, pp. xxi–xxviii.

Murray, P.J., Chadwick, D.C., Newbold, C.J., D.R., Lockyer, 2007. Measurement of methane from grazing animals—the tunnel method. In: Makkar, H.P.S., Vercoe, P.E. (Eds.), Measuring Methane Production from Ruminants. FAO/IAEA. Springer, Netherlands, pp. 105–109.

National Intelligence Council, 2008. Global trends 2025: a transformed world. Office of the Director of National Intelligence. National Intelligence Council, Washington, DC. Accessed at <http://www.dni.gov/nic/PDF_2025/2025_Global_Trends_Final_Report.pdf> (verified December 15, 2011).

NOAA (National Oceanic and Atmospheric Administration), 2011. The NOAA annual greenhouse gas index (AGGI). U.S. Dept. of Commerce, Earth Systems Research Laboratory, Global Monitoring Division. Accessed at <http://www.esrl.noaa.gov/gmd/aggi/index.html> (verified July 26, 2011).

Paustian, K., Antle, J.M., Sheehan, J., Paul, E.A., 2006. Agriculture's role in greenhouse gas mitigation. Accessed at Pew Center on Global Climate Change, Arlington, VA. http://www.pewclimate.org/docUploads/Agriculture's%20Role%20in%20GHG%20Mitigation.pdf (verified December 15, 2011).

Reicosky, D.C., Goddard, T., Enerson, D., Chan, A.S.K., Liebig, M.A., 2012. Greenhouse gas trading activities related to North American agriculture. In: Liebig, M.A., Franzluebbers, A.J., Follett, R.F. (Eds.), Managing Agricultural Greenhouse Gases: Coordinated Agricultural Research through GRACEnet to Address our Changing Climate. Academic Press, San Diego, CA.

Schlesinger, W.H., 1997. Biogeochemistry. Academic Press, San Diego, CA.

Schoenau, J.J., Carley, C., Stumborg, C., Malhi, S.S., 2010. Strategies for maximizing crop recovery of applied manure nitrogen in the Northern Great Plains of North America. In: Malhi, S.S., et al. (Eds.), Recent Trends in Soil Science and Agronomy Research in the Northern Great Plains of North America. Research Signpost, Kerala, India, pp. 95–107.

Tanaka, D.L., Anderson, R.L., Rao, S.C., 2005. Crop sequencing to improve use of precipitation and synergize crop growth. Agron. J. 97, 385–390.

Tanaka, D.L., Krupinsky, J.M., Liebig, M.A., Merrill, S.D., Ries, R.E., Hendrickson, J.R., Johnson, H.A., Hanson, J.D., 2002. Dynamic cropping systems: an adaptable approach to crop production in the Great Plains. Agron. J. 94, 957–961.

Tilman, D., Balzer, C., Hill, J., Befort, B.L., 2011. Global food demand and the sustainable intensification of agriculture. Proc. Natl. Acad. Sci. doi 10.1073/pnas.1116437108.

U.S.-EPA (U.S. Environmental Protection Agency), 2011. Inventory of U.S. greenhouse gas emissions and sinks: 1990–2009. U.S. Environmental Protection Agency, Office of Atmospheric Programs, EPA 430-R-11-005, April 2011. Accessed at <www.epa.gov/climatechange/emissions/usinventoryreport.html> (verified July 26, 2011).

Vitousek, P.M., Aber, J.D., Howarth, R.W., Likens, G.E., Matson, P.A., Schindler, D.W., Schlesinger, W.H., Tilman, D.G., 1997. Human alteration of the global nitrogen cycle: sources and consequences. Ecol. Appl. 7 (3), 737–750.

West, T.O., Marland, G., 2002. A synthesis of carbon sequestration, carbon emissions, and net carbon flux in agriculture: comparing tillage practices in the United States. Agric. Ecosys. Environ. 91, 217–232.

Westberg, H., Lamb, B., Johnson, K.A., Huyler, M., 2001. Inventory of methane emissions from U.S. cattle. J. Geophys. Res. 106 (D12), 12633–12642.

Wienhold, B.J., Pikul Jr., J.L., Liebig, M.A., Mikha, M.M., Varvel, G.E., Doran, J.W., Andrews, S.S., 2006. Cropping system effects on soil quality in the Great Plains: synthesis from a regional project. Renew. Agric. Food Syst. 20 (1), 49–59.

CHAPTER 2

GRACEnet: Addressing Policy Needs through Coordinated Cross-location Research

Charles L. Walthall[1], Steven R. Shafer[1], Michael D. Jawson[2]
[1]Natural Resources and Sustainable Agricultural Systems, Agricultural Research Service, U.S. Department of Agriculture, Beltsville, MD
[2]Upper Midwest Environmental Sciences Center, U.S. Geological Survey, U.S. Department of the Interior, La Crosse, WI

CHAPTER OUTLINE

Abbreviations: CASMGS, Consortium for Agricultural Soils Mitigation of Greenhouse Gases; GWP, global warming potential; GHG, greenhouse gases; NDFU, North Dakota Farmers Union; OSQR, Office of Scientific Quality Review; USGCRP, U.S. Global Change Research Program

WHY GRACEnet?

Agricultural Research Service (ARS) organizes the research conducted by its approximately 2000 scientists into 20 National Programs. National Programs are developed by soliciting input from a broad array of customers and stakeholders, including administrators, congressional appropriators, other scientific agencies and organizations, research partners, agricultural commodity and advocacy groups, producers, processers, and others. One of several venues for

Managing Agricultural Greenhouse Gases. DOI: 10.1016/B978-0-12-386897-8.00002-4

obtaining this input is a workshop for a single National Program at a time, hosted by ARS for customers, stakeholders, and research partners in the research topics addressed by that Program. Perspectives are assembled from attendees about their informational and technological needs around the subject matter to provide the basis for development of an Action Plan that will guide that National Program's research priorities and implementation for 5 years.

ARS and its predecessor agencies have conducted studies of climate and weather in agricultural settings for most of USDA's history. ARS organized a Global Change National Program in the early 1990s, soon after the establishment of the federal government-wide U.S. Global Change Research Program (USGCRP). The ARS Global Change National Program Action Plan prepared before the National Program's first full workshop included research objectives on the C cycle and C storage, reflecting ARS support of and involvement in USGRCP priorities. During 2000, the newly established Soil Resource Management National Program identified ongoing soil C research for continued effort by ARS. ARS National Program Leaders were well aware of the connections between research on soil C, which was long associated with broad environmental and agricultural benefits, and the research occurring in the specific interest of greenhouse gas (GHG) mitigation. As planning began for the first Global Change National Program workshop to be held in 2002, connecting the soil C-related research across the two National Programs was recognized as a need to avoid redundancies in research efforts and to increase assurance that the coordinated work would have real-world impact for stakeholders with interests in soil quality and/or climate change.

Another topic that was a focus of attention at the Global Change workshop was the concept of life-cycle analysis. Presentations and discussions at the workshop on the inputs and outputs to agricultural production systems—especially inputs and management practices that mitigate GHG emissions—were subjects of lively sessions. These discussions brought increasing clarity to the interactions among causes and impacts of climate change in agriculture, and the relationships between actions taken to adapt production systems to climate change and the implications of those actions for GHG mitigation, as well as the challenges to designing applicable research.

Customers and stakeholders of ARS research recognize that sustaining yield is the central purpose of agricultural production, and they are aware of the need for agriculture to protect the environment, whether through water quality enhancement or mitigation of GHG emissions. From a research standpoint then, the challenge was to bring all these concepts together.

The GRACEnet (Greenhouse gas Reduction through Agricultural Carbon Enhancement network) project evolved from previous attempts to organize ARS climate change science around themes that were national in scope, such as the U.S. GHG Inventory. A desire of ARS National Program Leaders was for a research network across the country with relatively stable base funding for long-term research. One lesson learned from earlier attempts establishing such networks was that the intent of the efforts tended to be so ambitious that scientists had a hard time envisioning how they could individually contribute to such a project without substantial new funding and/or a redirection of much of their current research. This pitfall needed to be avoided.

Increasing national and international interest in terrestrial C sequestration was a driving force in the conception of GRACEnet. Other collaborative efforts were also responding to this need. For example, the Consortium for Agricultural Soils Mitigation of Greenhouse Gases (CASMGS, 2011) was developed by federal agency and university scientists to address C sequestration and other GHG issues from agriculture. Numerous journal articles and a series of books (Lal et al., 1998; Follett et al., 2001; Kimble et al., 2002) clearly showed the potential for crop, pasture, and rangeland management to sequester soil C. This interest had progressed to the point that land management, especially agricultural land, became a key element of discussions for C trading markets (Reicosky et al., 2012).

GRACEnet was formed from discussions among ARS National Program Leaders and scientists after the 2002 Global Change National Program workshop. Ideas of accommodating different production systems and creating a scalable design led to consolidation into the following questions: (1) What are the GHG emissions and C sequestration consequences under the production practices that are typical for crops and locations today, i.e. "business as usual" or baseline condition? (2) What kind of management strategies or inputs, within the range of real-world feasibility, can be implemented to increase C sequestration, and how much more C could be sequestered compared to the business-as-usual situation? (3) Similarly, what can be done overall to reduce net GHG emissions (not necessarily the same thing as increasing C sequestration alone)? (4) What package of management decisions and inputs could be implemented to strike a desirable balance among all goals: increasing C sequestration, reducing net GHG emissions, meeting other natural resource-related goals, and sustaining yields? The desire to draw conclusions across agroecosystems required an agreement among participating scientists to conduct a minimum, defined set of measurements and methods, and a willingness to share data to support cross-location activities such as synthesis, integration, modeling, and emissions estimations.

Thus, GRACEnet was conceived to build upon ongoing research to improve soil productivity, while explicitly addressing the challenges and opportunities presented by the increased interest in C sequestration from a climate change perspective. Specifically, the challenges were to develop a program, in which scientists recognized where their activities and expertise were already contributing and would be able to package their information, so that it would be understood and used by both producers and decision makers.

The research vision for GRACEnet was and remains: Knowledge and information used to implement scientifically based agricultural management practices from the field to national policy scales on C sequestration, GHG emissions, and environmental benefits.

Consequently, GRACEnet has had a foundational framework of a national network of soil C sequestration and GHG reduction research projects consisting of USDA-ARS (Agricultural Research Service) laboratories and university and land manager (e.g. farmer and rancher) cooperators. There is a potential for international expansion.

CLEAR OBJECTIVES, COMMUNICATION, ORGANIZATION, LEADERSHIP: ELEMENTS FOR SUCCESSFUL SCIENCE

Organizational issues were deemed important from the outset to ensure that contributing scientists would understand and take ownership of their roles and responsibilities. Continuous, effective communication among contributing scientists was to be maintained via email contact lists, conference calls, and face-to-face meetings. Contributing scientists have taken advantage of professional society meetings for yearly gatherings. Organization was based on the concept of a master umbrella project plan with a principal investigator: scientists seeking to contribute to GRACEnet wrote their own project plans with objectives supporting the umbrella project plan. The umbrella project plan and individual scientist project plans were reviewed for scientific quality using panels of non-ARS scientists convened by the ARS Office of Scientific Quality Review (OSQR), as per all ARS research plans. The umbrella project plan and the individual scientist project plans were required to fit within the ARS organizational structure to provide assurance of benefit to the nation, science, scientists, their laboratory management units, and USDA. Because ARS research is conducted on 5-year cycles, GRACEnet was structured at the outset within this 5-year framework.

Finding trusted scientific leaders at field locations to help implement the project was another key step in developing and maintaining momentum for GRACEnet. ARS National Program Leaders have the responsibility for setting research priorities and direction, but do not directly control field unit budgets. Without new resources to devote to GRACEnet, National Program

Leaders had to truly lead, i.e. make others actually want what s/he wants to achieve success. Leadership from field scientists provided essential technical legitimacy to the success of GRACEnet. The initial leadership team of field scientists consisted of Ron Follett (Fort Collins, CO), Tim Parkin (Ames, IA), Jane Johnson (Morris, MN), and Jeff Smith (Pullman, WA). Other scientists became prominent in the effort as the program progressed. One of the initial assignments of the leadership team was to develop sampling and data management protocols that would be used across the program. Fortuitously, new funding appeared in the FY 2005 budget to hire a scientist to lead the modeling and data management aspects of GRACEnet and to provide a modest amount of financial resources for a few field locations.

The national focus of GRACEnet required a standardized approach to assess C sequestration and GHG emission from different crop and rangeland systems. Plot-level data needed to be scalable across fields, landscapes, watersheds, and regions. Time scales for the C and N processes under investigation also required attention (e.g. diurnally, seasonally, and yearly). Components for successful collaboration among scientists had to address the following:

- Common measurement protocols
- Emphasis on calibration
- Structured experimental designs
- Robust and easily searchable database
- Use and/or development of process models as a goal and as a means of prioritizing data needs for all locations
- Integration and synthesis of results written in forms suitable for delivery to policy makers
- Scientific peer review of manuscripts
- Selection of research sites with relevance to major agroecosystems of the continental U.S.

SIGNIFICANT MILESTONES AND IMPACT

Since establishment of GRACEnet in 2002, participating researchers have significantly expanded GHG mitigation science. With over 250 publications associated with the project, GRACEnet has addressed key information gaps concerning effects of land-based agricultural practices on GHG emissions and soil C dynamics. Improvements in methods and models have also occurred through the work of GRACEnet, thereby enhancing measurement and predictive capacity for estimating GHG emissions across a range of spatial and temporal scales. Many of these accomplishments are summarized in the following chapters of this book. Further, these research-based accomplishments have served as a backbone for important project milestones during the last 10 years. Notable milestones are summarized below.

Cross-location Project Certified

The GRACEnet umbrella project was certified by the ARS OSQR during 2006. The project was among the first formal cross-location projects conducted by the Natural Resources and Sustainable Agriculture Systems programs.

Provided Critical Input for Synthesis Documents

Information generated by GRACEnet has been used to produce synthesis documents for action agencies, such as the Natural Resources Conservation Service, the USDA Global Change Program Office, and other policy makers. Synthesis documents have addressed the feasibility of management practices to sequester C, reduce GHG emissions, and contribute to co-benefits associated with their adoption (e.g. water quality improvement).

Salient synthesis documents produced by GRACEnet researchers have included a series of regional summaries documenting the effects of agricultural management on GHG emission and mitigation potential throughout North America (Franzluebbers and Follett, 2005). The data contained within these regional summaries were a key supporting basis for the rate coefficients used by the Chicago Climate Exchange (CCX) for offset trading, as well as for the development of other C trading protocols. GRACEnet researchers have also contributed to the modeling estimation of N_2O emissions used for the U.S. Agriculture and Forestry Greenhouse Gas Inventory (Global Change Program Office, 2008). This 112 page report and the included appendices inventoried GHG emissions from cropland, livestock, grazed land, and energy use, as well as documented changes in soil carbon stocks. This report was recently updated by extending the analysis to 2008 (Global Change Program Office, 2011).

Led the Development of a Significant White Paper

A white paper led by GRACEnet researchers and co-authored with university colleagues provided a "policy-maker-friendly-language" document suitable to inform decision makers on the state of the science and to use as a reference during the development of C-credit trading policies. This paper drew heavily on GRACEnet results and included an executive summary, which received the journal editor's award for best paper of the year (Morgan et al., 2010). The white paper was subsequently expanded and published as an issue paper (Council for Agricultural Science and Technology, 2011).

Facilitated Development of Voluntary Carbon Credit Program

Information from a regional summary (outlined above) was used in the development of a C credit program for the North Dakota Farmers Union (NDFU) in collaboration with the CCX. Producers participating in the C credit program received payments using estimated C offsets derived from regional summaries. Over the course of the program (2006—2010), approximately $7.4 million in offset payments were accrued to farmers and ranchers throughout the U.S., thereby supporting land management practices that reduced net GHG emissions and improved soil and environmental quality, while concurrently increasing income for agricultural producers.

Served as Foundation for Soil Carbon and Greenhouse Gas Modeling

Data from GRACEnet has been used to expand the Carbon Management Evaluation Tool (COMET; Paustian et al., 2009). The COMET model is widely used in support of research, policy and production decision making affecting C sequestration and GHG emissions. The USDA-Natural Resources Conservation Service (NRCS) is working with a version of COMET to assist landowners with strategies to reduce GHG and increase C sequestration. GRACEnet researchers continue to work collaboratively with the USDA Global Change Program Office and university partners to provide technical input and expertise in the development of science-based methods and technical guidelines for quantifying GHG sources and sinks in the forest and agriculture sectors (U.S. Department of Agriculture, 2011).

Data from GRACEnet have also been used to improve the capacity of DayCent, a biogeochemical model capable of simulating crop yields, soil C changes, N_2O emissions, and NO_3 leaching under irrigated and rainfed cropping systems (Del Grosso et al., 2010). Use of DayCent has expanded considerably in recent years, and is now used to estimate N_2O emissions from cropland soils in the U.S. (Global Change Program Office, 2011).

Animal GRACEnet Established

Scientists in ARS are also building on the GRACEnet concept by conducting collaborative research on GHG emissions from livestock production. An Animal GRACEnet subgroup was formed during 2011 by researchers from 13 ARS locations. Notable efforts by the dairy industry to decrease its GHG emissions and improve its sustainability added further incentive to an already motivated group of scientists.

Members of the Animal GRACEnet group recognized the needs for standardized measurement protocols, accessibility of scientifically vetted GHG data, and development and utilization of models for identifying practical solutions for GHG management that were compatible with livestock air quality issues. Animal GRACEnet researchers have chosen to focus initial efforts on production of fact sheets describing: (1) existing GHG management technologies that are ready to be used by producers, and (2) summaries of livestock system GHG mitigation research showing promise for the future. The fact sheets will serve as brief research summaries, provide summary information to scientists, producers and policy makers, and foster collaboration among scientists. Animal GRACEnet researchers also intend to use the process-based model, DairyGEM (Rotz, 2011), which will provide guidance for the selection of ancillary factors measured coincident with GHGs.

FUTURE: NEW DIRECTIONS AND THE RELEVANCE OF GRACEnet TO SUSTAINABLE AGRICULTURE

Demand by farmers, ranchers, land managers, program managers, policy makers, and scientists continues to be high for scientifically valid (and legally defensible) GHG data. Historical lack of basic GHG and C sequestration data for land-based agricultural practices has limited the ability to manage for C sequestration and GHG emissions. The GRACEnet project has become a key provider of these data. Accordingly, requests from production agriculture continue to be received for a "GRACEnet researcher" focusing on yet another crop or production system. Recent additions to GRACEnet include research focused on orchard systems, wine and table grape production, and horticulture systems.

Simulation models suitable for use as core elements of decision support tools are needed to balance yield expectations, ecosystem services, and stewardship of natural resources. Tools such as Nitrate Leaching and Economic Analysis Package (NLEAP; Shaffer et al., 2010) for improving the efficiency of fertilizer use will contribute to GRACEnet as the project matures to increase its emphasis on the development of low-emission and C-enhancing management strategies for producers. This focus will expand to encompass work with plant scientists—collaboration with the crop genetics community has already begun to couple production systems with new varieties of low GHG emission, climate resilient crops.

Further, no mitigation solution fits all management scenarios and thus, region and crop-specific technologies for reducing GHG intensity are needed. The need to investigate the spatial variability of GHG emissions and C sequestration has emerged as landscape-level management objectives gain wider attention. Applying precision agriculture technologies to manage within-field spatial and temporal variability of GHG emissions and C sequestration is a logical next step.

A critical concern for agriculture pertains to emissions of N_2O, with its global warming potential (GWP) of 298 times that of CO_2 and its prevalent emissions from land-based agricultural management systems (U.S.-EPA, 2011). Emissions of N_2O also merit examination from an economic perspective—applied N not reaching its intended crop destination, whether lost via water transport or air emissions, constitutes an economic loss to the producer. Given the need to increase agricultural production to meet the global challenge of feeding 9 billion people by 2050, gross reduction of GHG emissions from agriculture appears unlikely. However, reduced GHG emission per unit of product (i.e. GHG intensity) seems a more reasonable target. Greater production intensity per unit area of land should be a reasonable goal, thus requiring greater soil productivity—a potential outcome that could be achieved by focusing on enhanced soil C sequestration in agricultural systems. Thus, GRACEnet objectives are consistent with mainline agricultural research to improve the efficiency and profitability of agriculture concurrent with reducing GHG emissions and increasing soil C.

Finally, discussions have taken place on merging the university-based Consortium for Agricultural Soil Mitigation of Greenhouse Gas (CASMGS, 2011) with GRACEnet. Further, the appeal of GRACEnet has already transcended the boundaries of the United States. The publicly available methods of GRACEnet (Follett, 2011) and experimental concepts are providing a sound footing for research developing within the Global Research Alliance on Agricultural Greenhouse Gases (Baker and Follett, 2012; Shafer et al., 2011). International agricultural research and policy organizations frequently inquire about collaborations with ARS from the perspective of seeking to benefit from GRACEnet and offering to share insights outside of U.S. agricultural GHG mitigation science. Thus, it appears that GRACEnet is poised to have international impact.

References

Baker, J.M., Follett, R.F., 2012. Potential GRACEnet linkages with other greenhouse gas and soil carbon research and monitoring programs. In: Liebig, M.A., Franzluebbers, A.J., Follett, R.F. (Eds.), Managing Agricultural Greenhouse Gases: Coordinated Agricultural Research through GRACEnet to Address our Changing Climate. Academic Press, San Diego, CA.

CASMGS, 2011. Consortium for agricultural soils mitigation of greenhouse gases. Available at http://www.casmgs.colostate.edu/about/about.asp (verified October 7, 2011).

Council for Agricultural Science and Technology, 2011. Carbon sequestration and greenhouse gas fluxes in agriculture: challenges and opportunities. Task Force Rep. No. 142. Council for Agricultural Science and Technology, Ames, IA, 106 p.

Del Grosso, S.J., Ogle, S.M., Parton, W.J., Breidt, F.J., 2010. Estimating uncertainty in N_2O emissions from U.S. cropland soils. Global Biogeochem. Cycles. 24, GB1009.

Follett, R.F. (Ed.), 2011. GRACEnet Sampling Protocols. Available at http://www.ars.usda.gov/research/GRACEnet (verified October 7, 2011).

Follett, R.F., Kimble, J.M., Lal, R. (Eds.), 2001. The Potential of U.S. Grazing Lands to Sequester Carbon and Mitigate the Greenhouse Effect. Lewis Publishers, Boca Raton, FL, p. 442.

Franzluebbers, A.J., Follett, R.F., 2005. Greenhouse gas contributions and mitigation potential in agricultural regions of North America: introduction. Soil Tillage Res. 83, 1–8.

Global Change Program Office, 2011. U.S. agriculture and forestry greenhouse gas inventory: 1990–2008, U.S. Department of Agriculture, Office of the Chief Economist. Tech. Bul. No. 1930. GPO, 159.

Global Change Program Office, 2008. U.S. agriculture and forestry greenhouse gas inventory: 1990–2005, U.S. Department of Agriculture, Office of the Chief Economist. Tech. Bul. No. 1921. GPO, 161.

Kimble, J.M., Lal, R., Follett, R.F. (Eds.), 2002. Agricultural Practices and Policies for Carbon Sequestration in Soil. Lewis Publishers, Boca Raton, FL, p. 512.

Lal, R., Kimble, J.M., Follett, R.F., Cole, C.V. (Eds.), 1998. The Potential of U.S. Cropland to Sequester Carbon and Mitigate the Greenhouse Effect. Ann Arbor Press, Chelsea, MI, p. 128.

Morgan, J.A., Follett, R.F., Allen, L.H., Del Grosso, S., Derner, J., Dijkstra, F., Franzluebbers, A., Fry, R., Paustian, K., Schoeneberger, M.M., 2010. Carbon sequestration in agricultural lands of the United States. J. Soil Water Conserv. 65 (1), 6A–13A.

Paustian, K., Brenner, J., Easter, M., Killian, K., Ogle, S., Olson, C., Schuler, J., Vining, R., Williams, S., 2009. Counting carbon on the farm: Reaping the benefits of carbon offset programs. J. Soil Water Conserv 64 (1), 36A–40A.

Reicosky, D.C., Goddard, T., Enerson, D., Chan, A.S.K., Liebig, M.A., 2012. Agricultural greenhouse gas trading markets in North America. In: Liebig, M.A., Franzluebbers, A.J., Follett, R.F. (Eds.), Managing Agricultural Greenhouse Gases: Coordinated Agricultural Research through GRACEnet to Address Our Changing Climate. Academic Press, San Diego, CA.

Rotz, C.A., 2011. DairyGEM: A software tool for assessing emissions and mitigation strategies for dairy production systems. In: Proc. of Western Dairy Air Quality Symposium, Convergency of Science and Regulation. Western States Dairy Producers Trade Assoc., Modesto, CA, pp. 31–35.

Shafer, S.R., Walthall, C.L., Franzluebbers, A.J., Scholten, M., Clark, H., Meijs, J., Reisinger, A., Kazuyuki, Y., Slattery, B., Campbell, I., Mcconkey, B., Angers, D., Soussana, J., Richard, G., 2011. Emergence of the global research alliance on agricultural greenhouse gases. Carbon Manage 66, 209–214.

Shaffer, M.J., Delgado, J.A., Gross, C., Follett, R.F., Gagliardi, P., 2010. Simulation processes for the Nitrogen Loss and Environmental Assessment Package. In: Delgado, J.A., Follett, R.F. (Eds.), Advances in Nitrogen Management for Water Quality. SWCS, Ankeny, IA, pp. 362–373.

U.S. Department of Agriculture, 2011. Development of technical guidelines and scientific methods for quantifying GHG emissions and carbon sequestration for agricultural and forestry activities. U.S. Department of Agriculture, Office of the Chief Economist. Federal Register, vol. 76 (No. 34), 9534–9537. February 18, 2011.

U.S.-EPA (U.S. Environmental Protection Agency), 2011. Inventory of U.S. greenhouse gas emissions and sinks: 1990–2009. U.S. Environmental Protection Agency, Office of Atmospheric Programs, EPA 430-R-11-005, April 2011. Available online at www.epa.gov/climatechange/emissions/usinventoryreport.html (verified July 26, 2011).

SECTION 2

Agricultural Management and Soil Carbon Dynamics

Cropland Management in the Eastern United States for Improved Soil Organic Carbon Sequestration

Curtis J. Dell[1], Jeffrey M. Novak[2]
[1]USDA-ARS-Pasture Systems and Watershed Management Research Unit, University Park, PA
[2]USDA-ARS-Coastal Plains Soil, Water, and Plant Research Center, Florence, SC

CHAPTER OUTLINE

Abbreviations: CT, conventional tillage; NT, no-till; SOC, soil organic carbon

INTRODUCTION

The eastern United States (Figure 3.1) is a very diverse region with respect to geography, soils, and climate. This diversity leads to the production of a wide range of crops using numerous management practices and variable potentials to increase soil organic C (SOC) sequestration (defined here as the accumulation of SOC in response to changing management). The largest extent of cropland lies on Atlantic and Gulf Coastal Plains and Piedmont of the southeastern and mid-Atlantic states; however, portions of the Appalachian Plateau, Ridge and Valley, and New England physiographic provinces are also used for crop production.

The eastern U.S. has a long history of crop production. Prior to the arrival of European settlers in the 17th and 18th centuries, Native Americans produced crops such as corn (*Zea mays*),

Managing Agricultural Greenhouse Gases. DOI: 10.1016/B978-0-12-386897-8.00003-6
2012, Published by Elsevier Inc.

beans (genus *Phaseolus*), and squash (genus *Cucurbita*) in small clearings and established new fields when older fields became depleted. With European colonization, settlers expanded the area under cultivation by clearing forests and established intensive cultivation of cash crops such as corn, cotton (*Gossypium* L.), tobacco (*Nicotina* L.), rice (*Oryzal* L.), indigo (*Indigofera tintoria*), and timber production. Continuous cultivation of fields coupled with poor land stewardship resulted in rapid depletion in nutrients, accelerated soil erosion, and an eventual decline in crop productivity (Busscher et al., 2010). This type of management was commonly practiced because there was always more land available for farming. Unfortunately, it took another century for farmers to be made aware that land availability was limited and that the soil needed good stewardship to replenish nutrients and reduce erosion losses (Bennett and Chapline, 1928). Because of a more favorable climate, the intensification of agriculture and depletion of soil resources was greatest in the southeastern and mid-Atlantic states. However, past soil erosion also degraded productivity of soils throughout the northeastern U.S.

For almost two centuries, fields in the eastern U.S. were prepared for row crop production using some form of conventional tillage (CT, plowing and/or disking). These tillage operations invert topsoil and incorporate crop residue, a practice that is well known to hasten its microbial oxidation and loss from the SOC pool (Reicosky et al., 1995; Hunt et al., 1996). This has been a serious concern because SOC declines in sandy soils of the coastal plains can result in poor soil physical conditions for plant growth (Busscher et al., 1987), low water storage capacities (Peele et al., 1970; Campbell et al., 1974), and reduced capacity of soils to retain nutrients (Pierzynski et al., 2000) and herbicides (Novak et al., 1996). In addition to tilling the soil using conventional practices, SOC declines in the

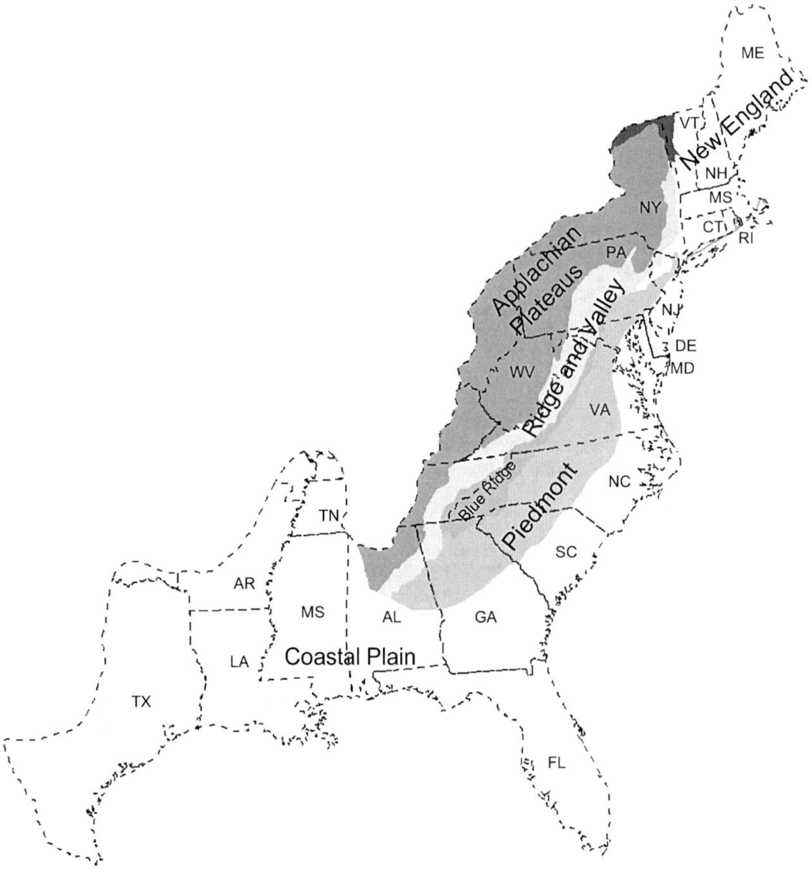

FIGURE 3.1
Physiographic provinces of the Eastern United States.

southeastern U.S. have been made worse in many areas by long-term cotton monoculture because cotton returns small amounts of crop residue to the soil (Causarano et al., 2006; Novak et al., 2009). Other crops such as corn, soybean (*Glycine max* L.), or wheat (*Triticum* L.) can supply more crop residues (>3.1 t/ha) to the soil (Hunt and Matheny, 1993; Karlen et al., 1987); unfortunately, more crop residues may not translate into higher SOC contents especially when soils are tilled using conventional practices (Parton and Schimel, 1987; Wang et al., 2000).

Increased awareness and adoption of conservation measures since the mid-20th century have helped to control erosion and restore productivity of soils throughout the eastern U.S. The adoption of conservation tillage, which involves minimal surface tillage, leaves crop residues to accumulate at the soil surface. The unincorporated crop residues decompose more slowly when compared to soils under traditional tillage operations (Reicosky and Lindstrom, 1993; Lal and Kimble, 1997; Paustian et al., 2000; Bauer et al., 2006). The use of additional management practices such as cover cropping, manure application, and improved crop rotations can also help to reduce soil losses and increase retention of soil organic matter. This chapter will address current crop production management practices and their impact on SOC dynamics in the eastern U.S. and analyze potential for additional SOC sequestration.

GEOGRAPHIC REGIONS OF THE EASTERN U.S

The Coastal Plain is an expansive geophysical province that extends from southern New Jersey along the Atlantic coast through the coast of the Gulf of Mexico to southeastern Texas (Figure 3.1), with cropland comprising about 15% of the total land area in the region (USDA, 2006). The Coastal Plain was formed through a series of sea level rises and recessions and subsequent depressional and erosional forces (Siple, 1967). The landscape is relatively flat and typified by scarps and terraces resulting from changes in ocean level, deposition of sediments, and river dissection over time. The elevation ranges from sea level to about 150 m (Daniels et al., 1999). Ultisols are the dominant Coastal Plain soil order. Stable coastal surfaces developed aged soils that included an eluvial (E) horizon, weathered clays (Daniels et al., 1967a), and a reddened argillic B horizon (Daniels et al., 1967b). Because of the extreme age, abundant rainfall, and humid climate, many of the Ultisols have a high degree of weathering leading to low pH (unless limed), highly weathered clay (Shaw et al., 2004; Novak et al., 2009), low cation exchange capacity (<2 to 4 $cmol_c\,kg^{-1}$; Kleiss, 1994), and low SOC content (0.2 to 0.8 $g\,kg^{-1}$; Hunt et al., 1982; Novak et al., 2009). Coastal Plains sandy soils also commonly have a restrictive subsurface hard layer (Mullins, 2000; Chartres et al., 1990) that can limit root penetration (Busscher et al., 2001). Average precipitation is 1000–1500 $mm\,yr^{-1}$ (increasing north to south), with maximum rainfall in mid-summer in the eastern portion and winter and spring in the west. Average temperatures range from 13 to 20°C (increasing north to south), with the average number of frost-free days ranging from 200 to 305 (USDA, 2006).

The Piedmont extends from the Appalachian Mountains to the Coastal Plain, ranging from Alabama to southeastern Pennsylvania (Figure 3.1). Cropland accounts for 8% of the Southern Piedmont Major Land Resource Area (MLRA) and 28% of Northern Piedmont MLRA (USDA, 2006). The Piedmont can be extensively hilly and contain soils that formed in unstable positions where soil profile expression has been limited (i.e. Inceptisols, Entisols). In more stable positions, such as on gently rolling topography, soils are older and will show more soil profile development (Alfisols and Ultisols). Piedmont soils are often formed from residuum or alluvium along streams and rivers (Daniels et al., 1999) leading to textures that vary from fine-clayey to coarse. Profile horizonation sequences of Piedmont soils are highly variable. Profiles can be composed of kaolinitic, mixed, and smectitic clays and depending on age are low in base saturation due to leaching from the rock parent

TABLE 3.1 Land Area (Hectares) in the Eastern United States Planted to Major Row Crops Reported in the 2007 Census of Agriculture (USDA-National Agricultural Statistic Service). Sugarcane was Produced on 164,200 ha in Louisiana and 153,328 ha in Florida. The Total Land Area in all States Planted to these Crops was 14,620,640 ha

State	Corn (grain)	Corn (silage)	Soybeans	Cotton	Small grains	Rice	All vegetables	Peanuts	Grain sorghum	Tobacco	State total
Alabama	112,048	11,016	72,768	154,939	41,014	NR	7,433	64,162	2,360	NR	465,739
Arkansas	236,775	1,887	1,141,889	346,036	284,760	536,767	5,637	NR	87,655	NR	2,641,406
Connecticut	1,443	24,147	119	NR[†]	3,220	NR	4,167	NR	NR	1,267	34,363
Delaware	75,090	6,353	62,997	NR	35,167	NR	14,179	NR	132	NR	193,919
Florida	13,736	27,005	4,887	32,421	25,486	4,603	91,059	48,048	539	421	248,205
Georgia	181,848	38,657	113,489	403,553	120,432	NR	48,196	210,081	18,101	7,286	1,141,643
Louisiana	292,567	4,087	240,495	135,191	87,774	152,613	7,601	NR	99,381	NR	1,019,708
Maine	1,325	23,516	310	NR	18,205	NR	27,006	NR	NR	NR	70,362
Maryland	186,355	63,979	156,575	NR	83,541	NR	13,024	NR	1,982	171	505,628
Massachusetts	985	13,895	100	NR	NR	NR	6,302	NR	NR	536	21,818
Mississippi	353,815	11,900	579,589	265,701	134,910	74,897	12,438	NR	47,345	NR	1,480,596
New Hampshire	92	12,640	NR	NR	NR	NR	1,366	NR	NR	NR	14,097
New Jersey	33,030	11,528	32,083	NR	134,908	NR	20,510	NR	249	NR	232,307
New York	223,410	507,568	80,909	NR	66,269	NR	64,859	NR	290	NR	943,305
N. Carolina	390,998	56,886	559,221	213,054	224,033	NR	47,398	36,057	3,497	68,884	1,600,028
Pennsylvania	397,205	429,139	174,576	NR	116,122	NR	22,192	NR	1,278	3,194	1,143,706
Rhode Island	17	1,653	NR	NR	NR	NR	964	NR	NR	NR	2,634
S. Carolina	150,886	13,392	179,197	64,110	64,032	NR	10,453	22,814	2,428	8,134	515,446
Tennessee	316,146	52,565	395,284	204,143	104,447	937	11,650	NR	4,093	8,144	1,300,617
Vermont	2,174	87,403	814	NR	279	NR	1,156	NR	NR	NR	91,827
Virginia	162,433	126,295	198,610	23,993	96,571	NR	10,637	8,761	NR	8,457	635,758
Crop total	3,132,377	1,525,511	3,993,913	1,843,142	1,641,171	769,817	428,228	389,923	269,331	106,493	

[†] *NR, None reported*

materials (Daniels et al., 1999). The average precipitation is 940–1525 mm yr^{-1} (increasing north to south). Average temperatures range from 9 to 18°C (also increasing north to south), with the average number of frost-free days ranging from 185 to 275 (USDA, 2006).

The Appalachian Ridge and Valleys extend from northern Alabama through central Pennsylvania (Figure 3.1). Parallel ridges of limestone, shale, and sandstone are separated by narrow to moderately wide valleys that range from nearly level to rolling hills. Soils are typically shallow on ridges, but can be deep and productive in larger valleys. Valley soils are classified as Inceptisols, Alfisols, and Ultisols with loamy or clayey textures and drainage typically ranging from excessively drained to moderately well drained. Croplands occupy approximately 15% of the Ridge and Valley landscape. Average precipitation is 800–1300 mm yr^{-1}, with maximum precipitation from late winter to early summer. Average temperatures are 11–17°C in the southern portion and 7–14°C in the north, with an average of 205 frost-free days in the south and 180 days in the north (USDA, 2006).

Much of northern Pennsylvania and New York lies on the glaciated Appalachian Plateau (Figure 3.1). Soils are primarily formed from glacial till and outwash (April et al., 1986). Soils that formed on semi-stable plateaus are classified as Inceptisols or Alfisols, with loamy texture. These soils range from shallow to moderately deep with drainage ranging from well to very poorly drained. Soils in the outwash areas are classified as Entisols, Inceptisols, or Spodosols, and can be well to excessively well drained especially if the texture is dominated by sands (April et al., 1986). Cropland is typically found on broad plateau tops which are nearly level to moderately sloping and dissected by narrow, steep-walled valleys. Approximately 17% of the land area on the Appalachian Plateau is used for crop production (USDA, 2006). It should be noted that the large amount of rock material on the till surface and in the profile of glacial soils make agricultural production difficult; extremely rocky lands are left for forestry production. Average precipitation is 760–1200 mm y^{-1}, with a large portion as snowfall. The average temperature is 4–10°C, with an average of 165 frost-free days per year (USDA, 2006).

The New England physiographic province (Figure 3.1) comprises the northern-most portion of the eastern United States and is a portion of the Appalachian Highlands. Over 80% of New England is mountainous and forested, with less than 4% of the land used for crop production. The greatest portion of the cultivated land is on gently rolling uplands and coastal lowlands. The dominant cultivated soils are Entisols and Inceptisols formed from glacial till and outwash. The average precipitation is 850–1400 mm y^{-1}. The average temperature is 6–12°C in the southern portion of the region and 4–9°C in the north, with an average of 190 frost-free days in the south decreasing to 160 days in the north portion of the region.

CROPLAND MANAGEMENT IN THE EASTERN U.S.

The 2007 Census of Agriculture (USDA-NASS, 2007) provided the most recent state-by-state listing of land area devoted to the production of major crops (Table 3.1). Corn is the most commonly grown crop throughout the eastern U.S., with approximately 4.6 and 1.5 Mha grown for grain and silage, respectively. A majority of the corn is harvested for grain in the mid-Atlantic and southeastern states, while much of the corn in the northeastern states (generally grown in rotation with multiple years of alfalfa) is harvested for silage. Soybean is the second most widely grown crop in the eastern U.S. (approximately 4 Mha) and accounted for the greatest cropland area in several states (Arkansas, Mississippi, North Carolina, South Carolina, Tennessee, and Virginia). Cotton production in the southeastern states accounted for approximately 1 Mha. Approximately 1.6 Mha were used for small grain production throughout the region. Other major crops included rice (770,000 ha),

TABLE 3.2 Use of Conservation Tillage for Major Crops in Eastern U.S. States where Data are Available[†]

Crop/state	Hectares planted	Conservation tillage[‡]	Reduced tillage[§]	Conventional tillage[¶]	No-till[#]	No-till with >30% residue cover
		percent of hectares			percent of hectares	
Corn (2005)						
Georgia	109,102	59	9	31	32	29
New York	399,941	8	8	84	2	2
North Carolina	303,027	60	11	29	57	54
Pennsylvania	545,393	43	25	32	41	40
Cotton (2007)						
Alabama	161,622	48	21	30	40	35
Arkansas	347,529	12	5	83	9	9
Georgia	416,086	41	16	43	38	27
Louisiana	135,428	7	7	87	5	5
Mississippi	266,673	19	3	78	20	19
North Carolina	202,019	56	15	29	57	46
South Carolina	72,753	47	11	41	48	45
Tennessee	208,024	65	17	17	80	65
Soybeans (2006)						
Arkansas	1,249,220	26	8	66	15	14
Louisiana	351,480	32	12	56	26	25
Mississippi	674,680	41	14	46	35	35
North Carolina	553,480	60	31	10	73	55
Tennessee	463,687	83	5	12	74	72
Virginia	210,080	93	3	4	82	81

[†]Source: Horowitz et al. (2010).
[‡]>30% residue cover on soil surface.
[§]15–30% residue cover on soil surface.
[¶]<15% residue cover on soil surface.
[#]Defined as no-inversion tillage or other mechanical disturbance of the soil.

peanut (*Arachis hypogea* L.) (390,000 ha), sugarcane (*Saccharium officinarum* L.) (318,000 ha), grain sorghum (*Sorghum bicolor* Moench) (270,000 ha), and tobacco (106,000 ha). A wide range of vegetables, including potato (*Solanum tuberosum* L.), are also grown throughout the eastern states, with production of all vegetables accounting for approximately 430,000 ha.

No-till and other conservation tillage (>30% residue cover) and reduced tillage (15–30% residue cover) are used to varying extents within eastern states. While annual data for tillage practices for all major crops in each state are not available, the USDA-Economic Research Service estimates (Horowitz et al., 2010) are available for selected states in years that survey data were obtained for a specific crop (Table 3.2). For corn (last estimated in 2005), conservation tillage usage was as high as 60% in North Carolina and as low as 2% in New York. The low rate of adoption of conservation tillage in New York and the New England states has been attributed to slower warming of soils under residue cover and subsequent impacts on crop establishment in the spring (P. Salon, personal communication). Estimates of tillage usage for cotton production in 2007 (Table 3.2) also indicate large differences among states. Conservation tillage was used on approximately 60% of the acreage in North Carolina and Tennessee and only 10% of the land in Arkansas and Louisiana. Soybean production with conservation tillage (estimated in 2006) ranged from as much as 93% of the acreage in Virginia to 26% in Arkansas.

Poultry, dairy, swine, and beef production are important throughout the eastern U.S., and land application of manure from confined animal operations is a significant C input to soils on many farms throughout the region (Table 3.3). North Carolina, New York, and Pennsylvania produce the greatest quantities of manure in the region (approximately 900,000 Mg C y^{-1}, each, in 1997). With the exception of Connecticut and Vermont, annual statewide manure C production would be $\leq$0.6 Mg C ha^{-1} on available cropland assuming uniform distribution (which does not occur), even in the states with the greatest manure production (Table 3.3). Since only a portion of the manure C will be resistant to degradation and contribute to the formation of stable organic matter in soil, the quantities of manure produced limits the total impact of manure application on C sequestration throughout states within the region. However, ample manure is available for land application in areas with concentrated livestock production and can contribute substantially to SOC sequestration. Application of manure at a rate sufficient to provide the N requirement for a corn crop (145–180 kg N ha^{-1}) typically requires 50,000–80,000 L ha^{-1} of liquid dairy or swine manure, 15–25 Mg of beef manure ha^{-1}, or 2–4 Mg ha^{-1} of chicken litter [fresh weight basis, Penn State University Agronomy Factsheet 55 (online at http://cropsoil.psu.edu/extension/facts/agfact55.pdf)]. Assuming that manure solids and any included bedding materials are 40% C, these application rates represent a C addition to soil of 1.0–1.6 Mg C ha^{-1} for liquid dairy manure (assuming 5% dry matter), 1.8–3.0 Mg C ha^{-1} for beef manure (assuming 30% dry matter), 1.25–2.0 Mg C ha^{-1} for swine slurry (assuming 2.5% dry matter), and 0.3–0.5 Mg C ha^{-1} with poultry litter (assuming 34% dry matter).

Cover crops are used throughout the eastern U.S. to reduce soil erosion and nutrient losses, and they contribute biomass that can potentially lead to greater sequestration of C in soils. The use of cover crops can be especially beneficial following crops that leave limited quantities of residues on the soil surface (i.e. soybean, cotton, and corn harvested for silage). However, comprehensive data on the extent of eastern U.S. cropland planted to cover crops and the type of cover crop used are not available. A recent program in Maryland, as part of that state's efforts to improve water quality in the Chesapeake Bay, has provided financial incentives to farms planting winter cover crops. This resulted in the planting of a state record 161,066 ha of winter cover crops in the fall of 2010 (Maryland Department of Agriculture, 2010).

SYNTHESIS OF PUBLISHED FINDINGS

Results of Previously Reported Research

An earlier review (Franzluebbers, 2005) addressing SOC sequestration potential for cropland in the southeastern U.S. indicated that cropland management had a variable impact on SOC storage. For 96 comparisons at 22 locations with 5 to15 years of NT, an average of 0.41 $\pm$0.46 Mg SOC ha^{-1} yr^{-1} (mean $\pm$ standard deviation) was sequestered in response to the use of NT compared to CT. With an additional 51 comparisons of SOC on conventional and conservation-tillage cropland in the region, SOC sequestration rate was revised to 0.45 $\pm$ 0.04 Mg C ha^{-1} yr^{-1} (mean $\pm$ standard error) at a sampling depth of 20 $\pm$ 1 cm and 11 $\pm$1 y of duration (Franzluebbers, 2010). Soil organic C sequestration rate in studies with cover crops (0.55 $\pm$ 0.06 Mg C ha^{-1} y^{-1}, n = 87), summarized by Franzluebbers (2005), was greater than without cover crops (0.30 $\pm$ 0.05 Mg C ha^{-1} y^{-1}, n = 60). Impacts of manure application on SOC sequestration reported in Franzluebbers (2005) were unclear, with an average increase of 0.26 $\pm$ 2.15 Mg SOC ha^{-1} yr^{-1} with manure (19 comparisons). When only manure application studies longer than 2 years were considered (7 studies), manure additions increased the SOC sequestration rate by 0.72 $\pm$ 0.67 Mg ha^{-1} yr^{-1}. In contrast, an estimated SOC sequestration rate of $-$0.07 $\pm$0.27 Mg ha^{-1} yr^{-1} with conversion from CT to NT was reported for the northeastern U.S. and eastern Canada by Franzluebbers and Follett (2005).

TABLE 3.3 Manure Production from Confined livestock Operations Including Dairy and Beef Cattle, Swine, and Poultry, and Cropland Available for Manure Application in Eastern U.S. States in 1997

	Number of animal units[†]				Manure production (Mg yr^{-1})[‡]						
State	**Dairy**	**Beef**	**Swine**	**Poultry**	**Dairy**	**Beef**	**Swine**	**Poultry**	**Total manure C (Mg yr^{-1})[§]**	**Cropland available for application (ha)[†,¶]**	**Mg C ha^{-1}**
Alabama	14,428	1,283	20,816	379,305	253,933	13,728	258,118	1,327,568	202,645	838,969	0.24
Arkansas	31,795	117,970	125,331	641,716	559,592	1,262,279	1,554,104	2,246,006	533,522	3,331,393	0.16
Connecticut	33,956	7,942	450	16,952	597,626	84,979	5,580	59,332	47,131	11,075	4.26
Delaware	10,077	2,184	4,566	86,875	177,355	23,369	56,618	304,063	54,482	211,256	0.26
Florida	109,078	10,392	3,811	90,306	1,919,773	111,194	47,256	316,071	149,990	270,159	0.56
Georgia	93,134	9,802	71,243	466,868	1,639,158	104,881	883,413	1,634,038	341,764	1,476,675	0.23
Louisiana	43,691	3,650	2,383	57,003	768,962	39,055	29,549	199,511	69,676	1,419,030	0.05
Maine	43,979	9,114	569	23,826	774,030	97,520	7,056	83,391	60,423	149,940	0.40
Maryland	92,809	23,577	9,955	121,458	1,633,438	252,274	123,442	425,103	170,442	611,128	0.28
Massachusetts	28,773	5,395	1,693	3,167	506,405	57,727	20,993	11,085	33,414	53,594	0.62
Mississippi	36,723	75,933	47,036	236,545	646,325	812,483	583,246	827,908	259,781	1,793,075	0.14
N. Hampshire	21,160	3,878	416	1,121	372,416	41,495	5,158	3,924	23,554	38,971	0.60
New Jersey	19,485	4,492	2,549	NR	342,936	48,064	31,608	—	23,240	159,297	0.15
New York	765,239	148,410	9,116	20,686	13,468,206	1,587,987	113,038	72,401	850,496	1,433,114	0.59
N. Carolina	62,491	4,706	1,433,278	66,1202	1,099,842	50,354	17,772,647	2,314,207	942,292	1,869,291	0.50
Pennsylvania	674,726	122,717	141,342	227,680	11,875,178	1,313,072	1,752,641	796,880	892,037	1,659,145	0.54
Rhode Island	2,437	353	323	307	42,891	3,777	4,005	1,075	2,786	4,875	0.57
S. Carolina	19,451	3,324	38,923	174,990	342,338	35,567	482,645	612,465	119,440	730,141	0.16
Tennessee	89,128	127,355	39,635	53,340	1,568,653	1,362,699	491,474	186,690	279,936	1,736,549	0.16
Vermont	200,663	42,050	231	1,217	3,531,669	449,935	2,864	4,260	224,183	199,068	1.13
Virginia	98,503	112,014	50,577	291,627	1,733,653	1,198,550	627,155	1,020,695	385,925	1,088,989	0.35

[†] *Source: Kellogg et al. (2000).*

[‡] *Assuming 17.6, 10.7, 12.4, and 3.5 metric tons manure per animal unit per year for milk cows (as excreted, no bedding), beef heifers (excreted manure plus bedding), hogs for slaughter (excreted manure with in-barn dilution, no-bedding), and broiler chickens (including bedding) (adapted from 2011 Pennsylvania Agronomy Guide).*

[§] *Assuming manure and bedding solids are 40% C, and dry matter content of dairy, beef, swine, and poultry manure are 12, 30, 8, and 34%, respectively (adapted from 2011 Pennsylvania Agronomy Guide).*

[¶] *Includes land used for production of 24 crops including all those widely grown in the eastern U.S.*

However, because data for the northeastern U.S. were not available, that estimate was derived only from research conducted in eastern Canada (Gregorich et al., 2005)

Recent Information from the Southeastern U.S.

Recent studies reporting SOC changes in response to adoption of NT in the southeastern U.S. croplands (Table 3.4) are consistent with the previous estimate for the region (Franzluebbers, 2005). For 37 comparisons of CT and NT with an average duration of 9 years, the SOC sequestration rate was 0.58 ± 0.71 Mg C ha^{-1} y^{-1}(mean $\pm$ standard deviation). Because initial SOC masses were not always reported, sequestration rates were calculated as the difference in SOC mass between NT and CT at the final sampling divided by the duration of the conservation practice. The variation in sequestration among sites may be a reflection of differences in the capacities of individual soils to protect and retain organic matter, but is also likely influenced by crop species and management as well as variability in experimental design including factors such as length of experiment, sampling depth, and numbers of samples obtained.

Soil organic C sequestration rates determined from two extensive field surveys were consistent with rates estimated by Franzluebbers (2005). When sampling 63 sites to 15 cm depths on the Coastal Plain of Virginia, Spargo et al. (2008) showed a positive sequestration of 0.31 ± 0.28 Mg SOC ha^{-1} yr^{-1} (mean $\pm$ standard deviation) with NT. Half of the sites studied had histories of biosolid application. The sequestration rate with NT for fields receiving biosolids was 0.48 ± 0.34 Mg SOC ha^{-1} yr^{-1}, while the rate for fields without biosolid application was 0.11 ± 0.35 Mg SOC ha^{-1} yr^{-1}. An on-farm study with 87 sites across the Piedmont and Coastal Plains of Alabama, Georgia, North Carolina, South Carolina, and Virginia (Causarano et al., 2008) showed an average sequestration rate of 0.52 ± 0.59 Mg C ha^{-1} yr^{-1} with no distinct differences among states or geophysical province. Average SOC mass in the upper 20 cm of soil was 38.9, 27.9, and 22.2 Mg SOC ha^{-1} in pasture, NT, and CT, respectively. Management explained 41.6% of the variation in SOC, while surface horizon clay content explained 5.2%, and mean annual precipitation accounted for only 1% of the variation.

Soil C accumulation with conservation tillage systems that employ some tillage, but retain >30% surface cover, has been studied to a lesser extent than with strict NT. Novak et al. (2009) measured SOC content in a comparison of disk tillage (2 passes, 15 cm deep) with conservation tillage (paratill subsoiling to 40 cm deep and NT planting) across a field in the South Carolina Coastal Plain that was under a cotton/corn rotation. Sampling soils by depth across the field (Figure 3.2) showed that 8 years of conservation tillage led to a 49% increase in SOC in the 0–3 cm depth (1.0 Mg ha^{-1}, significant at $P \leq 0.05$). On the other hand, SOC contents in 3–15 cm depth declined under conservation tillage by 26% (2.8 Mg ha^{-1}, significant at $P \leq 0.05$). While data indicate that minimal residue incorporation into soil with conservation tillage induced SOC stratification, the mixing of crop residues with disk tillage resulted in little change between initial and final SOC contents in either depth increment. A statistically significant reduction in SOC contents in the subsurface (20–25 cm) with NT in Minnesota has also been reported (Dolan et al., 2006).

A limited number of other reports for conservation tillage systems in the southeast have shown net SOC accumulation. A comparison of CT, NT, NT planting with paratill (subsoiling to 40–50 cm) and paratill with disking conducted on the Alabama Coastal Plain (Siri-Prieto et al., 2007) indicated that using both NT and NT with paratill increased SOC concentration in the 0–5 cm depth by approximately 20% over 3 years (no significant change with CT and paratill with disking). However, in the 5–10 cm depth, SOC concentrations for the paratill with NT planting treatment increased slightly over initial SOC concentration while concentration did not change with time when NT was used. When CT, NT, and strip-till (subsoiled to 35 cm in 30 cm-wide strip) were compared in Georgia, Sainju et al. (2006)

TABLE 3.4 Soil C Sequestration from Recent (since 2005) Southeastern US Comparing Tillage or Combinations of Tillage, Cover Crop, and/or Nutrient Source (Adapted from Franzluebbers, 2010)

Location	Soil series or taxonomic group	Duration (years)	Depth (cm)	Crop rotation‡	Organic waste	SOC CT	SOC NT	Reference
						Mg ha^{-1}		
Belle Mina, AL	Decatur SiL†	10	20	CO/RY-CO-CN/RY§	none	37.4	40.1	Sainju et al., 2008
Belle Mina, AL	Decatur SiL	10	20	CO/RY-CO-CN/RY§	poultry litter	43.7	43.7	Sainju et al.,2008
Headland, AL	Dothan LS	3	20	CO/RY-PN/O	none	21.6	23.3	Siri-Prieto et al., 2007
Shorter, AL	Typic and Aquic Paleudults (LS)	2.5	30	CO/BO+RY-CN/ WL+CC	none	23.5	26.2	Terra et al., 2005
Shorter, AL	Typic and Aquic Paleudults (LS)	2.5	30	CO/BO+RY-CN/ WL+CC	dairy manure	29.1	32.6	Terra et al., 2005
AL (Coastal Plain)	Kandiudults/ Haploxeralfs (LS)	13.3	20	CO/RY-PN/RY	none	18.0	20.1	Causarano et al., 2008
AL (Piedmont)	Kanhapludults (SCL)	10	20	CO	none	19.6	25.4	Causarano et al., 2008
Bartow, GA	Dothan SL	2	15	CO/RY§	none	17.6	19.3	Sainju et al., 2007
Tifton, GA	Tifton LS	2	15	CO/RY-CO/RY-PN/ RY§	none	17.7	17.0	Sainju et al., 2007
Fort Valley, GA	Dothan SL	7	30	CO-GS	none	21.4	24.2	Sainju et al., 2006
Fort Valley, GA	Dothan SL	7	30	CO/RY-GS/RY	none	22.6	27.2	Sainju et al., 2006
Fort Valley, GA	Dothan SL	7	30	CO/HV-GS/HV	none	23.1	26.7	Sainju et al., 2006
Fort Valley, GA	Dothan SL	7	30	CO/RY+HV-GS/ RY+HV	none	23.9	27.9	Sainju et al., 2006
Watkinsville, GA	Cecil SL/SCL	7	20	CO/RY-CN/RY-ML/ RY-GS/RY-SB/CC- CN/CC	poultry litter	27.1	32.7	Franzluebbers et al, 2007
Watkinsville, GA	Cecil SL/SCL	3	30	GS/RY (ungrazed)	none	40.6	51.6	Franzluebbers and Stuedemann, 2008
Watkinsville, GA	Cecil SL/SCL	3	30	GS/RY (grazed)	From grazing cattle	46.5	49.5	Franzluebbers and Stuedemann, 2008
Watkinsville, GA	Cecil SL/SCL	3	30	WW/ML (ungrazed)	none	45.6	45.1	Franzluebbers and Stuedemann, 2008
Watkinsville, GA	Cecil SL/SCL	3	30	WW/ML (grazed)	From grazing cattle	42.8	46.6	Franzluebbers and Stuedemann, 2008

GA (Coastal Plain)	Plinthic Kaniudlts (LS)	15.3	20	CO-PN/WW-SB/RY	none	16.7	21.5	Causarano et al., 2008
GA (Piedmont)	Kanhapludults (SL)	8.3	20	GS/WW-SB-CO/O	none	26.7	27.5	Causarano et al., 2008
Harmony, NC	Fairview SCL/ Braddock L	7	20	CN/BL (silage)	Dairy manure	33.5	36.6	Franzluebbers and Brock, 2007
Harmony, NC	Fairview SCL/ Braddock L	7	20	CN/RY (silage)	Dairy manure	33.5	33.1	Franzluebbers and Brock, 2007
Harmony, NC	Fairview SCL/ Braddock L	7	20	CN/BL-SG/RY (silage)	Dairy manure	33.5	39.1	Franzluebbers and Brock, 2007
NC (Coastal Plain)	Quartzsamments/ Kandiudults (LS)	7.5	20	CO-CO/RY-SB/WW	none	18.2	33.6	Causarano et al., 2008
NC (Piedmont)	Kanhapludults (SCL)	12.7	20	PN/WW-CO/WW-SB/ CC	none	25.0	32.6	Causarano et al., 2008
Florence, SC	Norfolk LS	25	7.6	CO-WW/SB-CO	none	10.1	20.3	Bauer et al., 2006
Florence, SC	Norfolk LS	24	15	CO-WW/SB-CO	none	20.6	31.4	Novak et al., 2007
SC (Coastal Plain)	Kaniudults/ Kanhapludults (LS)	16	20	CO-SB	none	24.8	25.3	Causarano et al., 2008
SC (Piedmont)	Typic Kanhapludults (SCL)	14.7	20	CO-SB/WW-ML	none	22.2	26.5	Causarano et al., 2008
VA (Coastal Plain)	Altavista SL	14	15	CO-WW/SB	none	15.9	21.9	Spargo et al., 2008
VA (Coastal Plain)	Altavista SL	9	15	CO-WW/SB	biosolids	20.2	22.9	Spargo et al. , 2008
VA (Coastal Plain)	Bojac LfS	3	15	CO-WW/SB	none	14.1	13.3	Spargo et al. , 2008
VA (Coastal Plain)	Bojac LfS	9	15	CO-WW/SB	biosolids	16.7	19.1	Spargo et al. , 2008
VA (Coastal Plain)	Emporia LfS	11	15	CO-WW/SB	none	17.2	19.3	Spargo et al. , 2008
VA (Coastal Plain)	Emporia LfS	11	15	CO-WW/SB	biosolids	20.2	29.9	Spargo et al. , 2008
VA (Coastal Plain)	Hapludults (SL)	13.3	20	CO-WW/SB	none	20.4	31.4	Causarano et al., 2008
VA (Piedmont)	Kanhapludults/ Kandiudults (L)	9	20	SB-CO/WW	none	27.7	29.9	Causarano et al., 2008
Mean ± std. dev.		8.8 ± 5.6	21 ± 6			25.4± 9.4	29.6± 9.4	

[†]*Soil texture abbreviations: LfS, loamy fine sand; LS, loamy sand; S, sand; SCL, sandy clay loam; SL, sandy loam; SiL, silt loam.*

[‡]*Crop abbreviations: BL, barley; BO, black oat; CC, crimson clover; CN, corn; CO, cotton; FR, forage radish; GS, grain sorghum; HV, hairy vetch; ML, millet; O, oats; PN, peanut; RY, annual rye grass; SB, soybean; SG, small grain (wheat or barley); WL, white lupine; WW winter wheat.*

[§]*Mean includes treatments both with and without rye cover crop, because no significant effect on SOC of using cover crop.*

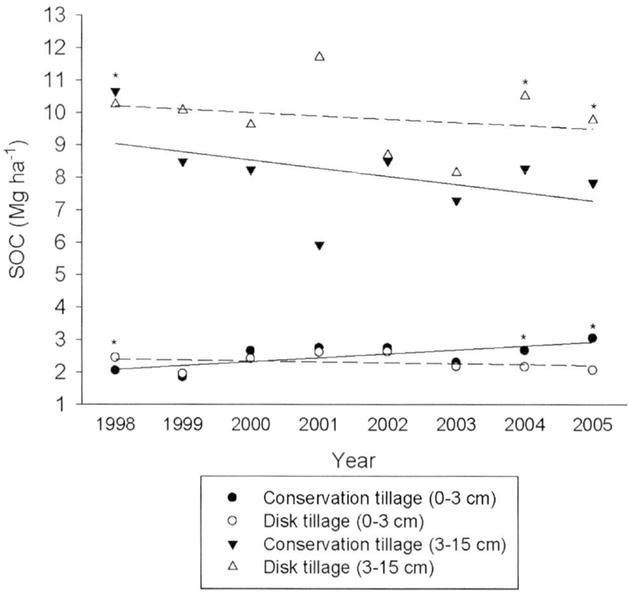

FIGURE 3.2

Median SOC contents (n = 50) in topsoil after 8 years of conservation and disk tillage under a corn/cotton rotation at Florence, South Carolina, indicated SOC gains in the upper 3 cm were offset by declining content in the 3–15 cm depth. Linear regression lines indicate the trend in SOC with time within each depth (solid lines for conservation tillage, dashed lines for disk tillage). The symbol (*) indicates that the tillage treatments are significantly different ($P \leq 0.05$, Mann-Whitney rank sum test) within a depth increment for a given year. (Adapted from Novak et al., 2009.)

observed similar rates of SOC gain with strip-till and NT (0.61 ± 0.47 vs. 0.54 Mg ha^{-1} y^{-1}) compared to CT. In a separate study in Alabama, Sainju et al. (2007) found that, when inorganic fertilizers were used, mulch till (rotary harrowing to 5–7 cm) led to an increase of 4.0 Mg SOC ha^{-1} over 10 years, while NT resulted in an increase of 1.5 Mg SOC ha^{-1} and 1.2 Mg SOC ha^{-1} was lost with CT (moldboard plow/disk/field cultivator). However, when the fertility source was poultry litter the 10-year increase in SOC was similar for the three tillage systems (4–5 Mg ha^{-1}).

While winter cover crops were used in a large number of the studies cited in Table 3.4, only three studies included comparisons to no-cover crop controls. Sainju et al. (2006) reported that rye (*Secale cereal* L.), hairy vetch (*Vicia villosa*) and rye/vetch mixtures as cover crops led to greater SOC in both CT and NT in a study conducted in Georgia. Hairy vetch/rye mixtures resulted in the greatest increase. In CT, SOC increased by about 0.8, 1.6, and 2.4 Mg ha^{-1}, relative to using no cover crop, with rye, hairy vetch, and the mixture during the 7-year study. In NT, SOC was 2.4, 1.9, and 2.6 Mg ha^{-1} greater than the control for rye, vetch, and the mixture. Conversely, cover crop use did not affect SOC mass in studies conducted in Alabama (Sainju et al., 2008) and Georgia (Sainju et al., 2007) despite significant increases in biomass inputs.

The impact of applying manures and other organic amendment on SOC sequestration remains unclear, but data suggest there may be an interaction between NT use and manure addition. Biosolid applications on the Virginia Coastal Plain (Spargo et al., 2008) resulted in greater SOC accumulations in the upper 15 cm compared to unamended fields with both CT and NT (3.3 ± 0.9 and 5.8 ± 4.8 Mg ha^{-1} for CT and NT). Sainju et al. (2008) found that poultry litter application over a 10-year period added approximately 0.3 Mg SOC ha^{-1} yr^{-1} to the upper 20 cm with NT, mulch till, and CT. Watts et al. (2010) reported that the combination of NT and poultry litter application (sufficient to supply 170 kg N ha^{-1} in corn or 45 kg P ha^{-1} in soybean) for 14 years to continuous corn and continuous soybean in northeastern Alabama increased SOC concentration in the 0–5 cm depth by 170 and 104%. The use of NT with inorganic fertilizer increased SOC contents in the 0–5 cm depth by 110% with corn, but did not affect SOC in soybean. With CT, poultry litter application increased SOC in corn plots by 40% but did not affect SOC concentration when applied to soybean plots. Neither NT nor poultry litter impacted SOC concentration in the 5–10 or 10–20 cm depth. Adeli et al. (2007) also observed an apparent interaction of NT and poultry litter application on SOC sequestration in cotton fields on two farms in central Mississippi that had soils with similar texture and initial organic matter concentrations. On the farm using NT, 0.6 Mg SOC ha^{-1} yr^{-1} was sequestered in the upper 15 cm with annual application of 4.5 Mg ha^{-1} of broiler litter and as much as 1.6 Mg SOC ha^{-1} yr^{-1} added with 6.5 Mg ha^{-1} litter and supplemental inorganic N. Soil organic C on the farm using CT was unchanged by broiler litter applications.

A limited number of studies from the southern U.S. have documented soil quality improvement in response to greater accumulation of SOC with the adoption of conservation practices. In their survey of sites across Coastal Plain and Piedmont of several states, Causarano et al. (2008) observed greater aggregate stability and a greater proportion of larger soil aggregates with NT than with CT, suggesting improved soil structure, aeration, and resistance to erosion. They also found a close correlation between total SOC, microbial biomass C, and potentially mineralized C, suggesting enhanced microbial activity and nutrient cycling with increasing SOC. Franzluebbers and Stuedemann (2008) also observed 2–3-fold increases in microbial biomass and potentially mineralized C with NT compared to CT.

Northeastern U.S.

Reports of management impacts on SOC sequestration in northeastern U.S. croplands are very limited, and findings from replicated plot studies that have been specifically designed to follow soil C dynamics are not available. In an on-farm study, Dell et al. (2008) sampled soils from a common series (Hagerstown silt loam) on several farms in the State College, PA, area where CT and NT had been used with silage corn/alfalfa rotations, with and without rye cover crops. They found an average of 51% greater mineral-associated C in the upper 5 cm of fields where NT had been used, but they observed similar accumulations in the 5–10 cm depths and upper Bt horizons of all fields. The difference in mineral-associated and particulate soil C between NT and CT fields suggested a sequestration rate of about 0.5 Mg ha^{-1} yr^{-1} with NT. However, the study used neither repeated measurements nor comparison of paired fields, and a true measure of sequestration was not determined. There was no detectable effect of cover crop usage, possibly because of the reported poor establishment of rye in many years due to unfavorable weather. In a long-term plot study (25 years) conducted near State College, PA, Duiker and Beegle (2006) reported soil organic matter (SOM) concentrations, determined by loss on ignition, that were 72 and 32% greater in the upper 25 cm with NT compared to moldboard plow/disk and chisel/disk systems. Concentration of SOM was very similar in the 5–10 and 10–15 cm depths of NT and moldboard plowed soils, but SOM was 10 to 20% greater in the 5–10 and 10–15 cm layer with chisel tillage compared to either NT or moldboard tillage.

Blanco-Canqui and Lal (2008) and Chatterjee and Lal (2009) reported paired comparisons of CT and NT on several farms throughout Pennsylvania, along with farms in Ohio and Kentucky. They generally observed greater SOC concentration in the upper 10 cm of soil with NT compared to CT and estimated SOC sequestration rates to a 60 cm depth for NT ranging from −2.95 to 4.94 Mg ha^{-1} y^{-1}, but no statistically significant differences ($P \leq 0.05$) could be detected between CT and NT for the entire 60 cm soil profiles on any of the farms. However, Kravchenko and Robertson (2011) showed that the amount of replication reported by Blanco-Canqui and Lal (2008) provided low statistical power and was insufficient to support hypothesis testing. The probability of verifying an SOC change of 10% was ≤10% in all depth increments and the probably of verifying even a 100% change in SOC was only about 50% in the 50–60 cm depth. Kravchenko and Robertson (2011) stressed caution when drawing conclusions about whole-profile SOC changes in cases where high variability at deeper depth masks the identification of significant changes near the soil surface.

Detection of statistically significant impacts of cropland management on SOC sequestration is complicated by the measurement error resulting from spatial variation in SOC concentrations and is dependent on adequate sampling replication (Kravchenko and Robertson, 2011). In the uneven and sloping terrain of the Piedmont, Ridge and Valley, Appalachian Plateau, and New England Uplands, SOC can vary greatly over even short distances. Dell and Sharpley (2006) observed a high degree of spatial variation in SOC concentration in the surface soils (0–5 cm) across a small watershed in the Ridge and Valley Province of central Pennsylvania. Using geostatistical analysis, they determined that soil sampling was required at 10 m or closer intervals to adequately capture the range of spatial variation within fields in that landscape. Observed coefficients of variation for individual fields, sampled at 30 m intervals, indicated that 2- to 5-fold more samples were needed to statically verify changes in SOC that were less than 10% of the original mean.

CARBON SEQUESTRATION POTENTIAL IN THE REGION

Southeastern U.S.

Available data for the southeastern U.S. indicates that, on average, adoption of NT can be expected to sequester approximately 0.5 Mg SOC ha^{-1} yr^{-1} for 10 to 20 years after the elimination of tillage. However, deviation in estimated sequestration rates is sufficiently wide to include some systems where no net accumulation of SOC is achieved, as well as soils with substantially greater sequestration. Data are not sufficient to estimate sequestration rates for conservation tillage systems that utilize some tillage, especially considering the wide range of field operations that are used. Estimates of sequestration potential with the use of cover crops and manure application are also difficult to determine because of high variability among studies and the wide range of practices used.

Northeastern U.S.

Soil carbon sequestration potential in the northeastern U.S. is difficult to estimate because of sparse data availability. Limited on-farm data from Pennsylvania showed greater SOC accumulations near the surface in NT fields compared to adjacent CT fields, but sampling has not been extensive enough to support calculation of SOC sequestration rates. Moreover, data are not available to accurately estimate impacts of manure application or cover crop use on SOC in the region. No-till or other high residue conservation tillage methods have not been widely adopted in New York and the New England states, contributing to the scarcity of data for the northeast.

Research Needs

While conversion to NT or use of cover crops is generally expected to result in increased SOC near the surface, management impacts on SOC below the upper few cm of soil are not as well established. Both sampling depth and sample numbers must be carefully addressed, especially with diverse soil types, in future studies in order to provide sound information about management impacts on SOC below the upper few cm of the soil profile and to correctly estimate whole-profile SOC accumulations. While there has been controversy over appropriate sampling depth (Baker et al., 2007), VandenBygaart et al. (2011) evaluated data from several sites across Canada and determined that sampling to 30 cm was appropriate for capturing SOC changes. Given rooting depths of commonly grown crops, sampling to 30 cm also appears appropriate in the eastern U.S. Adequate replication can become increasingly more important with increasing depth in the soil profile, because SOC concentration decreases and spatial variability tends to increase. Therefore, large sample numbers may be needed to provide adequate statistical power to evaluate SOC stocks (Kravchenko and Robertson, 2011). Lack of statistically significant effects of management practices on SOC accumulations have led to conclusions that the practices have no effect. However, Kravchenko and Robertson (2011) point out the possibility of committing Type II statistical errors (concluding no difference when differences are present) when sample numbers are low and statistical power is limited.

A limited number of studies, such as Novak et al. (2009), have reported that increases in SOC near the surface with the adoption of NT or other conservation tillage practices can be offset by statistically significant reductions in SOC lower in the soil profile. A better knowledge of the extent and cause of these atypical responses is needed to strengthen estimates of SOC sequestration in cropland soils. Longer-term studies are needed to determine if SOC reductions observed in subsurface layers are short-term responses to changing management or if they are sustained over time.

In general, much more information is needed to adequately estimate the potential for increased SOC sequestration in the northeastern U.S. One limitation to increasing SOC stocks in croplands of New York and the New England states is little adoption of NT. Research and education efforts are needed to identify and promote high residue/low disturbance production methods that are appropriate for the region. Strip-till and related practices have been successful in Canada where, like the northeastern U.S., cool spring time soil temperatures are seen as problematic with conventional NT (Vyn and Raimbault, 1992).

Use of cover crops is generally expected to increase SOC, but impacts of cover crop species and management remain poorly defined. Mixtures of cover crops may have a greater impact on SOC than monocultures, but additional research is needed at a greater number of locations with a wider range of cover crop species and management practices.

Application of organic amendments can have a significant impact on SOC levels, but the relationship between manure management practices and sequestration has not been extensively studied in all regions of the eastern U.S. and with a complete range of organic materials. Manure management guidelines have changed in recent years to address water quality impairment, most notably in the Chesapeake Bay watershed, but it is not known if those changes have impacted the sequestration of C added with manures. Manure injection is one technology that offers promise for reducing nutrient transport to surface waters, but research is needed to determine if it has an impact on the sequestration of manure C in soil.

Vegetables are grown on ~0.5 Mha throughout the eastern U.S., but little is known about how management and utilization of conservation practices can impact SOC sequestration in those systems. Sugarcane and rice are additional crops that are grown extensively in portions of the region, with limited information concerning sequestration potential. However, more is known about SOC losses through subsidence in the southern regions where sugarcane is grown than is known about rebuilding SOC levels.

Improvements in soil quality and productivity resulting from practices that sequester SOC merit greater consideration. Additionally, a better understanding of the impact of reducing erosion, and subsequent soil organic matter losses from the field, on regional C budgets is needed.

CONCLUSIONS AND RECOMMENDATIONS

Topography, soil resources, climate, crops, and production methods vary greatly across the eastern U.S., and these factors make estimation of SOC sequestration potential complex. Recent data from the southeastern U.S. has generally been consistent with previously published sequestration rates following adoption of conservation practices in the region (0.41 ± 0.46 Mg SOC ha^{-1} y^{-1}; Franzluebbers, 2005). This range would encompass the limited number of systems where no net gain in SOC occurs. While reductions in SOC at depth with conservation tillage do not appear to be the norm, further research is needed to define the cause, extent, and impact of this phenomenon on regional and national SOC sequestration estimates. To avoid ambiguous results and potentially misleading conclusions, sampling from ongoing and future SOC monitoring should be obtained to a depth of at least 30 cm and replication from all sampling depths should be sufficient to provide adequate power for conclusive statistical testing. In general, much more information is needed to predict SOC sequestration potential with conservation practices in the northeastern U.S.

Reducing soil erosion remains the primary benefit of conservation tillage and NT, regardless of the potential for SOC sequestration with these practices. Controlling erosion and subsequent redistribution of SOC is essential to maintain productivity of soil resources. Sequestration of SOC and mitigation of climate change are valuable "side" benefits of NT and other conservation tillage systems, but even in situations where conservation practices do not result in a net gain in SOC, agronomic and environmental benefits of controlling soil erosion remain compelling reasons for the use of the practices.

The combined use of NT and cover crops and the applications of manures or other C-rich organic amendments to a broader land base represent the best potential for increased SOC sequestration in eastern U.S. cropland soils. The greatest sequestration is likely to be achieved by conversion of marginally productive croplands to perennial vegetation (see Chapter 5) or, based on recent reports, by the use of recalcitrant biochars to increase SOC sequestration (Spokas, 2010; Novak and Busscher, 2011).

References

Adeli, A., Sistani, K.R., Rowe, D.E., Tewolde, H., 2007. Effects of broiler litter application to no-till cotton on selected soil properties. Soil Sci. Soc. Am. J. 71, 974–983.

April, H.A., Hluchy, M.E., Newton, R.M., 1986. The nature of vermiculite in Adirondack soils and till. Clays Clay Miner 34, 549–556.

Bauer, P.J., Frederick, J.R., Novak, J.M., Hunt, P.G., 2006. Soil CO_2 flux from a Norfolk loamy sand after 25 years of conventional and conservation tillage. Soil Tillage Res. 90, 205–211.

Baker, J.M., Ochsner, T.E., Venterea, R.T., Griffis, T.J., 2007. Tillage and soil carbon sequestration—what do we really know? Agric. Ecosystem. Environ. 118, 1–5.

Bennett, H.H., Chapline, W.R., 1928. Soil Erosion: A National Menace. United States Department of Agriculture. Circular 33.

Blanco-Canqui, H., Lal, R., 2008. No-tillage and soil profile carbon sequestration: an on-farm assessment. Soil Sci. Soc. Am. J. 72, 693–701.

Busscher, W.J., Spivey, L.D., Campbell, R.B., 1987. Estimation of soil strength properties for critical rooting conditions. Soil Tillage Res. 9, 377–386.

Busscher, W.J., Frederick, J.R., Bauer, P.J., 2001. Effect of penetration resistance and timing of rain on grain yield of narrow-row corn in a coastal plain loamy sand. Soil Tillage Res. 63, 15–24.

Busscher, W.J., Schomberg, H., Raper, R.L., 2010. Soil and water conservation in the southeastern United States: a look at conservation practices past, present and future. In: Zobeck, T.M., Schllinger, W.F. (Eds.), Soil and water conservation in the United States. Soil Sci. Soc. Am. Spec. Pub. 60, Madison, WI, pp. 183–200.

Campbell, R.B., Reicosky, D.C., Doty, C.W., 1974. Physical properties and tillage of Paleudults in the southeastern Coastal Plains. J. Soil Water Conserv. 29 (5), 220–224.

Causarano, H.J., Franzluebbers, A.J., Reeves, D.W., Shaw, J.N., 2006. Soil organic carbon sequestration in cotton production systems of the Southeastern United States: a review. J. Environ. Qual. 35, 1374–1383.

Causarano, H.J., Franzluebbers, A.J., Shaw, J.N., Reeves, D.W., Raper, R.L., Wood, C.W., 2008. Soil organic carbon fractions and aggregation in the Southern Piedmont and Coast Plain. Soil Sci. Soc. Am. J. 72, 221–230.

Chartres, C.J., Kirby, J.M., Raupach, M., 1990. Poorly-ordered silica and aluminosilicates as temporary cementing agents in hard-setting soils. Soil Sci. Soc. Am. J. 54, 1060–1067.

Chatterjee, A., Lal, R., 2009. On farm assessment of tillage impact on soil carbon and associated soil quality parameters. Soil Tillage Res. 104, 270–277.

Daniels, R.B., Gamble, E.E., Nelson, L.A., 1967a. Relationship between A2 horizon characteristics and drainage in some fine loamy Ultisols. Soil Sci. 104, 365–369.

Daniels, R.B., Gamble, E.E., 1967b. The edge effect in some Ultisols in the North Carolina Coastal Plain. Geoderma. 1, 117–124.

Daniels, R.B., Buol, S.W., Kleiss, H.J., Ditzler, C.A., 1999. Soil systems in North Carolina. North Carolina State University, Technical Bullettin 314. NCSU, Raleigh, NC.

Dell, C.J., Sharpely, A.N., 2006. Spatial variation of soil organic carbon in a northeastern U.S. watershed. J. Soil Water Conserv. 61, 129–136.

Dell, C.J., Salon, P.R., Franks, C.D., Benham, E.C., Plowden, Y., 2008. No-till and cover crop impacts on soil carbon and associated properties on Pennsylvania dairy farms. J. Soil Water Conserv. 63, 137–142.

Duiker, S.W., Beegle, D.B., 2006. Soil fertility distribution in long-term no-till, chisel/disk and moldboard plow/disk systems. Soil Tillage Res. 88, 30–41.

Franzluebbers, A.J., 2005. Soil organic carbon sequestration and agricultural greenhouse gas emissions in the southeastern USA. Soil Tillage Res. 83, 120–147.

Franzluebbers, A.J., 2010. Achieving soil organic carbon sequestration with conservation agricultural systems in the southeastern United States. Soil Sci. Soc. Am. J. 347–357.

Franzluebbers, A.J., Brock, B.G., 2007. Surface soil responses to silage cropping intensity on a Typic Kanhapludult in the Piedmont of North Carolina. Soil Tillage Res. 93, 126–137.

Franzluebbers, A.J., Follett, R.F., 2005. Greenhouse gas contribution and mitigation potential in agricultural regions of North America: Introduction. Soil Tillage Res. 83, 1–8.

Franzluebbers, A.J., Stuedemann, J.A., 2008. Early response of soil organic fractions to tillage and integrated crop-livestock production. Soil Sci. Soc. Am. J. 72, 613–625.

Gregorich, E.G., Rochette, P., VandenBygaart, A.J., Angers, D.A., 2005. Greenhouse gas contributions of agricultural soils and potential mitigation practices in Eastern Canada. Soil Tillage Res. 83, 53–72.

Horowitz, J., Ebel, R., Ueda, K., 2010. "No-till" farming is a growing practice. USDA-ERS, Economic Information Bulletin, Number 70.

Hunt, P.G., Matheny, T.A., Campbell, R.B., Parsons, J.E., 1982. Ethylene accumulation in southeastern coastal plain soils: soil characteristics and oxidative-reductive involvement. Comm. Soil Sci. Plant Anal. 13, 267–278.

Hunt, P.G., Matheny, T.A., 1993. Dry matter and nitrogen accumulation in determinant soybean grown on low-nitrogen soils of the southeastern United States. Comm. Soil Sci. Plant. Anal. 24, 1271–1280.

Hunt, P.G., Karlen, D.L., Matheny, T.A., Quisenberry, V.L., 1996. Changes in carbon content of a Norfolk loamy sand after 14 years of conservation or conventional tillage. J. Soil Water Conserv. 51, 255–258.

Karlen, D.L., Sadler, E.J., Camp, C.R., 1987. Dry matter, nitrogen, phosphorus, and potassium accumulation rates by corn on Norfolk loamy sand. Agron. J. 79, 649–656.

Kellogg, R.L., Lander, C.L., Moffitt, D.C., Gollehon, N., 2000. Manure nutrients relative to the capacity of cropland and pastureland to assimilate nutrients (2000). USDA-NRCS-ERS nps00–0579 at <http://www.nhq.nrcs.usda.gov/land/index/publication.html>

Kleiss, H.J., 1994. Relationship between geomorphic surfaces and low activity clay on the North Carolina coastal plain. Soil Sci. 157, 373–378.

Kravchenko, A.N., Robertson, G.P., 2011. Whole-profile soil carbon stocks: the danger of assuming too much from analyses of too little. Soil Sci. Soc. Am. J. 75, 235–240.

Lal, R., Kimble, J.M., 1997. Conservation tillage for carbon sequestration. Nutr. Cycling Agroecosyst. 49, 243–253.

Maryland Department of Agriculture, 2010. Governor O'Malley Announces Record Cover Crop Acreage Planted by Maryland Farmers. At www.mda.state.md.us/article.php?i=29978.

Mullins, C.E., 2000. Hardsetting soils. G-65-g-87. In: Sumner, M.E. (Ed.), Handbook of Soil Science. CRC, Boca Raton, FL.

Novak, J.M., Watts, D.W., Hunt, P.G., 1996. Long-term tillage effects on atrazine and fluometuron sorption in Coastal Plain soils. Agric. Ecosyst. Environ. 60, 165–173.

Novak, J.M., Frederick, J.R., Bauer, P.J., Watts, D.W., 2009. Rebuilding organic carbon contents in Coastal Plain soils using conservation tillage systems. Soil Sci. Soc. Am. J. 73, 622–629.

Novak, J.M., Busscher, W.J., 2011. Selection and use of designer biochars to improve characteristics of Southeastern USA Coastal Plain degraded soils. In: Lee, J.W. (Ed.), Advanced Biofuels and Bioproducts. Springer Science. In press.

Parton, W.J., Schimel, D., 1987. Analysis of factors controlling soil organic matter levels in Great Plains grasslands. Soil Sci. Soc. Am. J. 51, 1173–1179.

Paustian, K., Six, J., Elliott, E.T., Hunt, H.W., 2000. Management options for reducing CO_2 emissions from agricultural soils. Biogeochemistry 48, 147–163.

Peele, T.C., Beale, O.W., Lesesne, F.F., 1970. The physical properties of some South Carolina soils. South Carolina Exp. Station Tech. Bull. 1037.

Pierzynski, G.M., Sims, J.T., Vance, G.F., 2000. Soils and Environmental Quality, second ed. CRC Press, Boca Raton, FL.

Reicosky, D.C., Kemper, W.D., Langdale, G.W., Douglas Jr., C.L., Rasmussen, P.E., 1995. Soil organic matter changes resulting from tillage and biomass production. J. Soil Water Conserv. 50, 253–261.

Reicosky, D.C., Lindstrom, M.J., 1993. Effects of fall tillage method on short-term carbon dioxide flux from soil. Agron. J. 85, 1237–1243.

Sainju, U.M., Singh, B.P., Whitehead, W.F., Wang, S., 2006. Carbon supply and storage in tilled and nontilled soils as influenced by cover crops and nitrogen fertilization. J. Environ. Qual. 35, 1507–1517.

Sainju, U.M., Schomberg, H.H., Singh, B.P., Whitehead, W.F., Tillman, P.G., Lachnicht-Weyers, S.L., 2007. Cover crop effect on soil carbon fractions under conservation tillage cotton. Soil Tillage Res. 96, 205–218.

Sainju, U.M., Senwo, Z.N., Nyakatawa, E.Z., Tazisong, I.A., Reddy, K.C., 2008. Tillage cropping systems, and nitrogen fertilizer effects on soil carbon sequestration and fractions. J. Environ. Qual. 37, 880–888.

Shaw, J.N., West, L.T., Bosch, D.D., Truman, C.C., Leigh, D.S., 2004. Parent material influence on soil distribution and genesis in a Paleudult and Kandiudult complex, southeastern USA. 2004. Catena 57, 157–174.

Siple, G.E., 1967. Geology and groundwater of the Savannah River plant and vicinity South Carolina. U.S. Geol. Surv. Water Supply Papers 1941. USGS, Washington, DC.

Siri-Prieto, G., Reeves, D.W., Raper, R.L., 2007. Tillage systems for a cotton-peanut rotation with winter-annual grazing: impacts on soil carbon, nitrogen and physical properties. Soil Tillage Res. 96, 260–268.

Spargo, J.T., Alley, M.A., Follett, R.F., Wallace, J.V., 2008. Soil carbon sequestration with continuous no-till management of grain cropping systems in the Virginia coastal plain. Soil Tillage Res. 100, 133–140.

Spokas, K., 2010. Review of the stability of biochars in soils: predictability of O: C molar ratios. Carbon Manage 1 (2), 289–303.

Terra, J.A., Reeves, D.W., Shaw, J.N., Raper, R.L., 2005. Impacts of landscape attributes on carbon sequestration during the transition from conventional to conservation management practices on a Coastal Plain field. J. Soil Water Conserv. 60, 438–446.

USDA., 2006. USDA Agricultural Handbook 29: Online at <http://soils.usda.gov/MLRAexplorer>

USDA-NASS, 2007. 2007 Census of Agriculture: Online at <http://www.agcensus.usda.gov>

VandenBygaart, A.J., Bremer, E., McConkey, B.G., Ellert, B.H., Janzen, H.H., Angers, D.A., Carter, M.R., Drury, C.F., Lafond, G.P., McKenzie, R.H., 2011. Impact of sampling depth on differences in soil carbon stocks in long-term agroecosystem experiments. Soil Sci. Soc. Am. J. 226–234.

Vyn, T.J., Raimbault, B.A., 1992. Evaluation of strip tillage systems for corn production in Ontario. Soil Tillage Res. 23, 163–167.

Wang, Y., Admunson, R., 2000. Seasonal and altitudinal variation in decomposition and soil organic matter inferred from radiocarbon measurements of CO_2 flux. Global Biogeochem. Cycles 14, 199–211.

Watts, D.B., Torbert, H.A., Prior, S.A., Huluka, G., 2010. Long-term tillage and poultry litter impacts soil carbon and nitrogen mineralization and fertility. Soil Sci. Soc. Am. J. 74, 1239–1247.

CHAPTER 4

Soil Carbon Sequestration in Central U.S. Agroecosystems

Cynthia A. Cambardella[1], Jane M. F. Johnson[2], Gary E. Varvel[3]
[1]USDA-Agricultural Research Service, National Laboratory for Agriculture and the Environment, Ames, IA
[2]USDA-Agricultural Research Service, North Central Soil Conservation Res. Lab., Morris, MN
[3]USDA- Agricultural Research Service, Agroecosystem Management Research Unit, Lincoln, NE

CHAPTER OUTLINE

Abbreviations: C, carbon; CO_2, carbon dioxide; SOC, soil organic carbon; CT, conventional tillage; MLRA, Major Land Resource Area; NRCS, Natural Resource Conservation Service; NT, no-tillage; Pg, petagram; Tg, teragram; U.S., United States; USDA, United States Department of Agriculture

INTRODUCTION

Land-management strategies that foster C sequestration in agroecosystems will continue to be an important strategy for the reduction of atmospheric carbon dioxide (CO_2) in the coming decades (Smith, 2004). Atmospheric CO_2 content is currently increasing at the rate of ~4.1 Pg C yr^{-1} (IPCC, 2007) with a small fraction estimated to come directly from agricultural activity (Lal, 2007; Morgan et al., 2010). Increasing the amount of protected, stabilized SOC is an implicit goal for land managers seeking to sequester atmospheric CO_2 in agricultural soils.

The central U.S., commonly referred to as the "Corn Belt", contains some of the most productive agricultural soils in the world, predominantly Mollisols and Alfisols derived from

Managing Agricultural Greenhouse Gases. DOI: 10.1016/B978-0-12-386897-8.00004-8

glacial till or loess deposits developed on native prairie and native hardwood forest vegetation (Johnson et al., 2005). The north central part of this region is dominated by poorly drained landscapes where the installation of artificial subsurface tile drainage has been used since the beginning of the 20th century as a management practice to improve crop production. Cropland and pasture are the primary agricultural land uses in the central U.S., where more than two-thirds of the landscape is dominated by corn (*Zea mays* L.)—soybean (*Glycine max* [L.] Merr.) cropping systems (USDA, 1994). Relatively small changes in cropland SOC in this region have the potential to contribute relatively large reductions in greenhouse gas emissions on a national scale (Johnson et al., 2005).

In the past decade, many literature reviews and synthesis papers have been published on the potential of various land management strategies to sequester SOC in agricultural systems (Lal et al., 1998; Post and Kwan, 2000; Reicosky et al., 2000; Eve et al., 2002; West and Post, 2002; West and Marland, 2002; Sperow et al., 2003; Smith, 2004; Franzluebbers, 2005, 2010; Liebig et al., 2005; Ogle et al., 2005; Causarano et al., 2006; McLauchlan, 2006; VandenBygaart and Angers, 2006; Baker et al., 2007; Bolinder et al., 2007; Christopher and Lal, 2007; West and Six, 2007; West et al., 2008; Nelson et al., 2009; Snyder et al., 2009; Johnson et al., 2011), several of which focus specifically on the central region of the U.S. (Johnson et al., 2005; Grace et al., 2006; Baker et al., 2007; Johnson et al., 2007a; Fissore et al., 2010). The objective of this chapter is to review and synthesize recent literature related to C storage in cropland soils of the central U.S. in order to define our current state of knowledge on the impact of agricultural management on soil C sequestration. A separate discussion of the impacts of harvesting crop biomass for bioenergy on SOC content in central U.S. cropland soils can be found in Johnson and Novak (2012).

CLIMATE, SOIL, AND LAND-USE CHARACTERISTICS

The central U.S. is defined in this review according to boundaries developed by the U.S. Forest Service (Bailey, 1995), which divided the U.S. into ecosystem provinces based primarily on native vegetation. Eastern Broadleaf Forest provinces dominate the eastern half of the central region and Prairie Parkland provinces the western half. The USDA-NRCS divides the central region into major land resource areas (MLRA) that provide detailed information on soil and climate interaction related to crop production (USDA-NRCS, 1997). Ecologically defined similarities of the central U.S. comprise all or most of Illinois, Indiana, Iowa, Kentucky, Michigan, Missouri, Minnesota, Ohio, Tennessee, West Virginia, and Wisconsin; the eastern edge of Kansas, Nebraska, Oklahoma, and Texas; and the western edge of Pennsylvania (Figure 4.1). Johnson et al. (2005) and Franzluebbers (2005) provided a detailed summary of the characteristics of the ecoregion provinces, MLRAs, and dominant soils within the region.

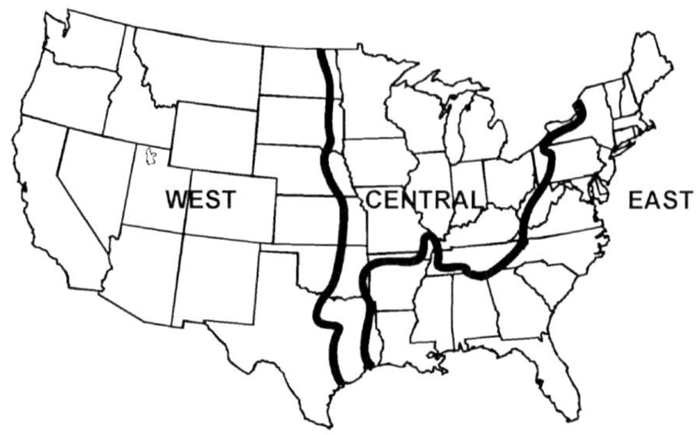

FIGURE 4.1
Approximate regional delineations for syntheses used in this volume.

Climate and parent material determine the dominant soil type and native vegetation that occur and can be managed in an ecoregion (Jenny, 1941; Franzluebbers and Follett, 2005). Mean annual temperature in the central U.S. increases from <5°C in the cool humid north to >20°C in the warm humid southern part of the region The frost-free period ranges from <90 days in the extreme north to >150 days in the southern parts of the region (Johnson et al., 2005; Franzluebbers, 2005). Mean annual precipitation increases from 50 to 130 cm moving west to east across the region with the majority of the precipitation falling in the summer months (Owensby et al., 2001). Soils in the Prairie Parkland ecoregion provinces are primarily Mollisols and Alfisols, with a few Entisols occurring in the Iowa and Missouri Deep Loess Hills (MLRA 107). The Eastern Broadleaf Forest provinces are dominated by Alfisols, although Mollisols occur in the Northern Illinois and Indiana Till plain (MLRA 110) and the Illinois and Iowa Deep Loess (MLRA 108). Ultisols and Inceptisols are found along the far eastern and southwestern edges of the region, primarily in the Alleghany Mountain and Plateau regions (MLRA 125–128), the Texas Prairie (MLRA 85 and 86), and the Coastal Plain (MLRA 133 and 155) areas. A few Vertisols exist in the Texas Claypan (MLRA 87) and Coastal Prairie (MLRA 150) areas (Bailey, 1995; Johnson et al., 2005; Franzluebbers, 2005). More than 75% of the cropland in the central U.S. is planted to corn and soybean, with Iowa, Illinois, Minnesota, and eastern Nebraska being the largest corn producers and Iowa, Illinois, Minnesota, and Indiana the largest soybean producers in the U.S. Hay in the region is produced mostly in Missouri and Wisconsin and a small amount of wheat (*Triticum aestivum* L.) is produced primarily in Minnesota (USDA, 1994; Johnson et al., 2005). The majority of the land-use along the southwestern edge (eastern Texas and Oklahoma) of the region is pasture (Franzluebbers, 2005; Franzluebbers et al., 2012). A range of agricultural practices exist within the central region of the U.S. that provide options for C sequestration and the predominant ones are described in the following.

MANAGEMENT PRACTICES AFFECTING CARBON STORAGE

Agricultural land management options recommended to foster C sequestration include increasing crop rotation complexity, reducing or eliminating tillage, optimizing fertility management, and conversion of marginally productive land to perennial vegetation. Changes in SOC content vary as a result of complex interactions among different factors including climate, baseline soil C levels and agricultural management practices. Land-use change from native ecosystems to intensive agriculture has led to large decreases in SOC content worldwide in the last two centuries (Alvarez, 2005) and contributed to increases in global atmospheric CO_2 concentrations (Houghton et al., 1983). Intensive cultivation of native grassland and forest soils in the Midwestern U.S. resulted in 40–60% loss of surface SOC during the 20th century (Jenny, 1941; Haas et al., 1957). For a wet prairie site in Wisconsin recently converted to agriculture in the middle of the 20th century, Jelinski and Kucharik (2009) reported a 35% loss of SOC at a depth of 0–10 cm. David et al. (2009) confirmed historic estimates of SOC loss using data from tile-drained fields at 19 locations in central Illinois cropped to corn and soybean. Most of the SOC loss occurred in the top 50 cm and prior to the 1950s, with no change in SOC as a result of synthetic fertilizer application from 1957 to 2002.

Loss of SOC related to land-use change can be attributed to several phenomena. First, native perennial biomass is removed and/or buried and frequently replaced with annual species. Annual crop species have fewer photosynthetically active days compared to perennial species, which can take advantage of early and late season growing conditions in temperate climates (Baker and Griffis, 2009). Second, annual species in general tend to put less biomass below ground compared to perennials (Zan et al., 2001; Bolinder et al., 2002; Johnson et al., 2006), resulting in a reduction in photosynthate entering the soil to build SOC. Third, inversion tillage (i.e. moldboard plowing) buries residue which can accelerate decomposition (Guérif et al., 2001; Burgess et al., 2002) in addition to stimulating an ephemeral CO_2 flux from soil (Reicosky and Lindstrom, 1993; Rochette et al., 1999; Grandy and Robertson, 2006; Reicosky and Archer, 2007).

Crop Rotation Complexity

Mono- and bi-crop rotations are common throughout the central U.S., a situation that presents numerous opportunities to enhance crop rotation complexity as a means to sequester C in agricultural soils (Table 4.1). Enhancing rotation complexity to three or more crops sequestered an average of 0.20 Mg C ha^{-1} yr^{-1}, but changing from continuous corn to corn–soybean rotation resulted in no net gain in SOC (West and Post, 2002). Crop type dramatically influences the amount and decomposability of shoot and root residues returned to the soil. Compared to other annual agricultural crops grown in the central U.S., corn returns a high level of C to the soil, exceeding C inputs from soybean by >50% (Allmaras et al., 2004; Huggins et al., 2007). Lower quantity of soybean residue C and more rapid decomposition rate of soybean residue due to high N concentration limit the retention of soybean residue as SOC (Huggins et al., 2007). Varvel and Wilhelm (2008) reported that continuous corn had greater SOC (0–30 cm) than corn–soybean and continuous soybean when irrigated for 14 years. In a companion study without irrigation at a different location in eastern Nebraska, Varvel and Wilhelm (2010) reported SOC content (0–30 cm) was greatest under continuous corn, followed by corn–soybean, and lowest under continuous soybean. A subsequent study conducted at the same location in eastern Nebraska revealed that total profile SOC content (0–100 cm) was significantly greatest under continuous corn, followed by corn–soybean, and lowest under continuous soybean (Varvel and Wilhelm, 2011). Similar to surface SOC content (0–30 cm) in Nebraska, continuous corn rotations in northwestern Illinois had significantly greater SOC than corn–soybean after 23 years (Jagadamma et al., 2007). However, there was no significant difference in SOC between rotations at deeper soil depths (30–90 cm), which resulted in no difference in total soil profile (0–90 cm) SOC content between continuous corn and corn–soybean rotation. Summarizing results from

TABLE 4.1 Summary of Selected Studies (2004–2011) Evaluating Effect of Enhanced Crop Rotation Complexity on C Sequestration in Central U.S. Cropland

State	Duration (years)	Annual crop[a]	Cover crop[a]	Perennial crop[a]	Depth (cm)	Study
IA	4	Corn, soybean, wheat	Clover		0–18	Singer et al. (2004)
IA	7	Corn, soybean		Alfalfa	0–30	Al-Kaisi et al. (2005)
NE	18	Corn, sorghum, soybean	Clover, oat		0–30	Varvel (2006)
IA	3	Corn, soybean	Oat, rye		0–50	Kaspar et al. (2006)
IL	5	Corn, soybean	Hairy vetch, rye		0–30	Villamil et al. (2006)
SD	16	Corn, soybean, wheat		Alfalfa	0–15	Pikul et al. (2008)
NE	14	Corn, soybean			0–30	Varvel and Wilhelm (2008)
NE	24	Corn, soybean			0–30	Varvel and Wilhelm (2010)
IA, NE	35	Corn, soybean, wheat		Pine,red cedar, cottonwood	0–30	Hernandez-Ramirez et al. (2010)
WI	12	Corn, soybean, wheat		Alfalfa, poplar	0–100	Syswerda et al. (2011)
NE	19	Corn, soybean			0–100	Varvel and Wilhelm (2011)

[a]*Crop taxonomy: alfalfa* (M. sativa); *bromegrass* (R. mermis *Leyss); clover* (T. pretense *L. and* T. alexandrinum *L.); corn* (Z. mays *L.); cottonwood* (P. deltoides); *hairy vetch* (V. villesa *Roth.); oat* (A. sativa *L.); pine* (P. sylvestris *L. and* P. strobus *L.); poplar* (Populus x euramericana *c.v. Eugenei); rye* (S. cereal *L.); soybean* (G. max *L. Merr.); switchgrass* (P. virgatum *L.); wheat* (T. aestivum).

three field sites in Illinois, Coulter et al. (2009) reported greater surface SOC content (0–30 cm) for continuous corn than for corn–soybean at the Dekalb research site, but not at the Urbana or Dixon Springs sites.

In many studies, changes in SOC with time have not been presented. Only a few studies have reported explicit temporal changes. Varvel and Wilhelm (2008) reported a non-significant increase in SOC content (0–30 cm) during 14 years of continuous corn and no trend for corn–soybean near Shelton, NE. In an 18-year study near Mead, NE, the effects of crop rotation and N fertilizer management on SOC levels in corn-based cropping rotations at several points in time relative to a baseline SOC level were evaluated (Varvel, 2006). Many of the SOC gains measured during the first 8 years of the study were lost during the next 10 years in all but the 4-year crop rotation that included small grains and clover.

Cover crops also increase crop rotation complexity and extend the duration of photosynthetic capacity in annual crop rotations, thereby increasing organic C inputs to soil and subsequently increasing the potential for soil C sequestration. Cover crops can also provide other important ecosystem services when planted within corn–soybean and extended cropping rotations. Environmental benefits such as decreased soil erosion (Kaspar et al., 2001) and decreased nitrate leaching (Kladivko et al., 2004; Strock et al., 2004; Kaspar et al., 2007) have been consistently demonstrated. Winter rye (*Secale cereale* L.) has been reported to increase SOC in areas with relatively mild winters, such as Georgia (Sainju et al., 2002). Greater soil C storage with cover crops has been more difficult to demonstrate in Midwest cropland soils, especially in the cool, moist regions of the upper Midwest. Kaspar et al. (2006) evaluated the benefits to soil C of a rye, oat (*Avena sativa* L.) and mixed oat/rye cover crop interseeded into soybean in a corn–soybean rotation in Iowa. The rye cover crop increased SOC relative to the control in the soybean phase, but decreased SOC relative to the control in the corn phase. Averaged over both rotation phases, winter cover crops did not increase SOC relative to the no-cover-crop control. In south-central Illinois, winter rye and hairy-vetch (*Vicia villosa* Roth) cover crops were planted in a corn–soybean rotation under NT management. Treatments included rye planted after corn and soybean, rye planted after corn and hairy-vetch after soybean, and rye planted after corn and a rye/hairy-vetch combination planted after soybean. Treatments supplying additional N as hairy-vetch residues increased soil organic matter (SOM) by 9% to a depth of 30 cm relative to a corn–soybean control without cover crop (Villamil et al., 2006). Despite the fact that rye produced a greater amount of above- and belowground residue with a wider C/N ratio than hairy vetch (Ranells and Wagger, 1997), the biomass could not be converted to SOM without N from the hairy-vetch at this field site. The balance of increased C inputs from cover crop residue and increased soil respiration due to enhanced microbial decomposition of cover crop residue will determine whether a cropping system will function as a source or sink for C. Bavin et al. (2009) compared soil respiration for a corn–soybean rotation managed with CT and strip tillage plus a rye cover crop for a field site in south-central Minnesota. Strip tillage plus cover crop had been managed for 2 years prior to the experiment. Soil respiration was greater in the cover cropped system due to increased decomposition rates from the cover crop residue. The authors cautioned that C dynamics in the strip till plus cover crop treatment may not have been at steady state, suggesting higher respiration rate could be a transient response to the change in tillage and cover crop C inputs.

Crop rotations that include forage legumes and small grains are more complex and biologically diverse than continuous monocultures of corn or simple 2-year combinations of corn and soybean. Integrated, extended crop rotations have been shown to increase SOC compared to mono- or bi-crop rotations (Singer et al., 2004; Al-Kaisi et al., 2005; Marriott and Wander, 2006a; Teasdale, 2007) with impacts being especially evident in the biologically active fractions of SOM (Marriott and Wander, 2006b; Tu et al., 2006). Pikul et al. (2008) determined SOC content (0–15 cm) for chisel-plowed continuous corn, corn–soybean, and corn–soybean–winter wheat (*Triticum aestivum*)–alfalfa (*Medicago sativa*) cropping systems in eastern South Dakota with high, mid, and zero N fertilizer additions. After 16 years, a significant effect of N fertilizer and crop rotation on SOC was found. Surface soil under continuous corn with high fertilizer N lost 2.3 Mg SOC ha^{-1} and the extended rotation that included a small grain and forage legume gained 0.3 Mg SOC ha^{-1}. In Michigan, Syswerda et al. (2011) assessed SOC concentration and content to a depth of 100 cm for corn–soybean–winter wheat rotations managed with different tillage, nutrient and herbicide management protocols. Annual cropping systems included rotations managed with (1) conventional inputs and CT, (2) conventional inputs and NT, (3) reduced chemical inputs and CT, and (4) organic inputs only and CT. The latter two treatments include a leguminous winter cover crop grown following corn and wheat. The reduced input system received one-third of the N and pesticide applied to the conventional systems. After 12 years, total profile SOC content (0–100 cm) did not differ among the annual cropping systems, although SOC concentration in the A/Ap horizon (~20 cm depth) was greater under organic management than conventional and reduced input management.

Including or expanding perennial species in agricultural landscapes is a strategy for increasing SOC content, reducing soil erosion, and improving off-site water quality (Brown et al., 2000; Nelson et al., 2006; Johnson et al., 2007b; Evans and Cohen, 2009; Wilhelm et al., 2010). Perennial plants can potentially sequester more C than annual plants, because they produce more (1) belowground biomass that increases C input (Zan et al., 2001; Bolinder et al., 2002) and (2) surface cover that can reduce off-site sediment and nutrient loads (Dabney et al., 1999; Lee et al., 2003; Self-Davis et al., 2003). Conversion from row crops to perennial grass has been shown to increase SOC (Lal et al., 1998; Zan et al., 2001; Heaton et al., 2004). Benefits derived from planting former cropland to perennial grass can be retained if conservation tillage practices are used during the conversion to annual row crops. Follett et al. (2009) reported that total SOC content (0–30 cm) did not change during a 6-year period following conversion of a smooth bromegrass (*Bromus inermis* Leyss) grassland to rainfed corn production using NT management.

Planting trees into degraded cropland is also a viable strategy to enhance C sequestration (Garten, 2002; Vesterdal et al., 2002). Sauer et al. (2007) reported that SOC storage was 57% greater in afforested soils of a mature shelterbelt compared to adjacent CT cropped soils. In a companion study, Hernandez-Ramirez et al. (2011) assessed SOC for a 35-year-old coniferous shelterbelt in eastern Nebraska and a 35-year-old coniferous forest plantation in northwestern Iowa. Afforested surface SOC had a turnover time of 45–55 years and a large percentage (21%) was associated with particulate organic matter, most of which (79%) was derived from tree C inputs. In Michigan, SOC content (0–100 cm) was measured after 12 years of perennial poplar (*Populus x euramericana*) and alfalfa systems and corn-based annual cropping systems (Syswerda et al., 2011). Total profile SOC content was 28% greater in the perennial systems than the annual systems, although the difference was not statistically significant. Concentration of SOC in the A/Ap horizon (~20 cm depth) did not differ between perennial and annual systems. A previous study at this site reported higher near-surface SOC content under poplar and alfalfa systems, although this study was sampled prior to cutting of poplar trees in 1999 (Robertson et al., 2000). Syswerda et al. (2011) attributed the difference in SOC storage under the poplar trees to enhanced decomposition due to microclimate changes after tree harvest. Whether perennials are added to the landscape in buffer strips, wind breaks, conservation reserve program (CRP), or as dedicated bioenergy crops, opportunities exist to mitigate global climate change by sequestering SOC.

Tillage, Residue, and Manure Management

Conservation tillage (Table 4.2) to increase SOC content of C-depleted agricultural soils has been accepted as a key strategy for stabilizing global atmospheric CO_2 concentrations over the

TABLE 4.2 Summary of Selected Studies (2004–2011) Evaluating Effect of Conservation Tillage on C Sequestration in Central U.S. Cropland

State	Duration (years)	Crop rotation[a]	Conventional tillage[b]	Conservation tillage[b]	Depth (cm)	Study
MN	13	CC	CP, MP	NT	0–30	Allmaras et al. (2004)
IA	7	C–S	CP	NT	0–30	Al-Kaisi et al. (2005)
MN	23	C–S	CP, MP	NT	0–45	Dolan et al. (2006)
IN	24	CC, C–S	CP, STI–CP	NT, ST–NT	0–100	Omononde et al. (2006)
MN	10–15	C–S	CP, MP	BT, NT	0–60	Venterea et al. (2006)
IN	28	CC, C–S	MP	NT	0–100	Gal et al. (2007)
MN	14	CC, C–S, SS	CP, MP	NT	0–45	Huggins et al. (2007)
IL	11	C–S	MP	NT	0–50	Yang et al. (2008)
WI	10		AT	BT	0–25	Jelinski and Kucharik (2009)
OH	6	CC	CP, MP	NT	0–30	Ussiri and Lal (2009)
OH	42–44	CC, C–S	CP	NT	0–40	Mishra et al. (2010)
NE	18	CC, C–S, SS	CP, DP, MP	NT, RT, ST	0–30	Varvel and Wilhelm (2010)
WI	12	C–S–W	MP	NT	0–100	Syswerda et al. (2011)
NE	18	CC, C–S, SS	CP, DP, MP	NT, RT, ST	0–100	Varvel and Wilhelm (2011)

[a]*Crop rotation: CC = monoculture corn (*Z. mays *L.); C–S = corn–soybean (*Glycine max *L. Merr.); SS = monoculture soybean; corn–soybean–wheat (*Triticum aestivum*).*
[b]*Tillage: AT = annual tillage; BT = biennial tillage; CP = chisel plow; DP = disk plow; MP = moldboard plow; NT = no tillage; RT = ridge tillage; ST = subtill; STI–CP = intermittent chisel plow; ST–NT = short-term no tillage.*

next 50 years (Baker et al., 2007; Lal, 2007; Morgan et al., 2010). Adoption of reduced tillage practices in the U.S. from 1990 to 2004 resulted in net fossil emissions reduction of 2.4 Tg C (Nelson et al., 2009). Type and intensity of tillage directly controls substrate availability to soil microorganisms and rate of decomposition of substrates by affecting the quantity and distribution of plant residues and roots (Huggins et al., 2007). Tillage factors can also exert indirect control on microbial decomposition processes by influencing soil aeration, water content, soil temperature, and especially soil aggregate properties. Reducing intensity of soil disturbance with adoption of conservation tillage practices promotes accumulation of labile forms of C by rendering these compounds less available to microbial decomposition through physical protection within stable soil aggregates (Elliott, 1986; Van Veen and Kuikman, 1990).

The United States Department of Agriculture (USDA) originally promoted conservation tillage to minimize erosion through residue retention. They defined conservation tillage as any tillage method that leaves at least 30% crop residue cover on the soil surface after planting (Franzluebbers, 2004). The most extreme form of conservation tillage is NT, in which soil is not disturbed throughout the year and all crop residues are left on the soil surface (Baker et al., 2007). Surface SOC sequestration rates for conversion of CT to NT in the central U.S. Corn Belt averaged 0.40 ± 0.61 Mg C ha^{-1} yr^{-1} (Johnson et al., 2005). For the eastern Corn Belt, SOC gains after converting from plow tillage to NT were 0.2–0.6 Mg ha^{-1} yr^{-1} (Dick et al., 1998). In Texas, conservation tillage increased SOC by 0.28 Mg C ha^{-1} yr^{-1} after converting from more intensive tillage (Martens et al., 2005). Al-Kaisi et al. (2005) observed that reducing tillage intensity enhanced C and N sequestration in corn–soybean cropping systems in Iowa. These changes were measured in the upper 30 cm of the soil profile across five different soil associations in Iowa and incorporated a number of different tillage systems. No detectable difference in SOC occurred among soil associations caused by different tillage operations. Varvel and Wilhelm (2010, 2011) assessed SOC to a depth of 30 cm and 100 cm for six tillage systems in rainfed corn cropping systems in eastern Nebraska. After 16 years of tillage management, they reported that NT had the greatest SOC and plowed soil the lowest SOC content at both depths.

Tillage can impact the distribution of SOC within the soil profile more than its net accumulation (Yang and Wander, 1999; Gal et al., 2007). With more than a decade of plowing,

^{13}C natural abundance studies have shown that the crop residues are distributed relatively uniformly throughout the plow layer (Clapp et al., 2000). Yang et al. (2008) studied the influence of soil depth on SOC sequestration in a corn–soybean rotation managed with NT or moldboard plowing in south-central Illinois. Statistically greater SOC concentration (g C kg^{-1}) occurred with NT than with moldboard plowing in the 0–5 cm depth, but the tillage effect was reversed in the subsurface (20–30 cm). Near-surface SOC content (Mg C ha^{-1}) was also greater for NT than for moldboard plowing, but soil profile SOC storage (0–30 cm) was not significantly different between the two tillage treatments due to greater SOC concentration at depth in the moldboard plowed treatment. Syswerda et al. (2011) evaluated SOC content for a corn–soybean–winter wheat rotation managed with NT and CT. Soil organic C content was significantly greater under NT than CT in the A/Ap horizon (~20 cm depth), but there was no statistically significant difference between NT and CT systems when the entire soil profile (0–100 cm) was considered. The authors suggested that high spatial variability and lower concentration of SOC in deeper horizons (Bt2/C) limited the detection of statistically significant differences between tillage systems.

Tillage impacts on SOC have also been demonstrated to extend beyond the depth of the plow layer. In some cases, the effect of greater SOC in surface layers under NT has been balanced by greater SOC in deeper layers with plowing, leading to equivalent SOC stocks for NT and CT (Yang and Wander, 1999; Blanco-Canqui and Lal, 2008). Huggins et al. (2007) assessed the interactive effects of tillage and crop rotation on SOC in southern Minnesota cropping systems—SOC content was measured after 14 years to a depth of 45 cm for continuous corn (CC), continuous soybean (SS), and corn–soybean (CS) under moldboard plow, chisel plow, and NT. Tillage effects on SOC were greatest in CC, in which chisel plow had 26% and NT 20% greater SOC than moldboard plow, whereas SOC in SS was similar across tillage treatments. As much as one-third of the greater SOC under CC for chisel plow and NT compared with moldboard plow occurred below tillage operating depths. Substantial loss of SOC was estimated (1.6 Mg SOC ha^{-1} yr^{-1}) despite reduced SOC decay rate with reduced tillage and a high level of C input with CC. Gal et al. (2007) reported that SOC below the plow layer (30–50 cm) was 23% greater with CT than with NT for a long-term (28 years) corn–soybean rotation in Indiana. Surface SOC content (0–15 cm) was greater under NT than under CT and there was no difference in SOC at 15–30 cm between the two tillage treatments. Accrual of SOC below the plow layer in CT partially offset the greater surface SOC in NT, however, still resulting in a statistically greater SOC content under NT than under CT at a depth of 0–100 cm, whether calculated by equivalent soil volume or soil mass. Dolan et al. (2006) measured SOC differences after 23 years of corn–soybean rotation in southern Minnesota in response to tillage, residue removal, and N fertilization. Surface soil (0–20 cm) of NT had 30% greater SOC than moldboard plow and chisel plow treatments. The tillage trend was reversed at 20–25 cm depth, in which moldboard plow had 50% greater SOC than NT. No significant difference in SOC occurred among tillage treatments for the 0–50 cm depth. Nitrogen fertilizer did not increase SOC, and plots where the residue was returned rather than harvested generally had greater SOC. Greater accrual of SOC below the plow layer may have been partly due to differences in corn root distribution between CT and NT soils (Gal et al., 2007; Huggins et al., 2007) and subsequently greater root density and biomass in NT surface soil (0–25 cm) (Holanda et al., 1998). Allmaras et al. (2004) reported that rhizodeposition increased when there was tillage in the system. Evaluations of tillage practices on C sequestration must consider that limiting soil sampling to surface soils can bias results in favor of reduced tillage due to tillage-induced rooting distribution differences (Baker et al., 2007) and redistribution of C to deeper depths (Staricka et al., 1991).

Adoption of NT management is not uniform across the central U.S. In the cooler, moist regions of the Midwest, continuous NT for corn production can be hampered by reduced grain yield caused by delayed seedling emergence (Swan et al., 1996). Rotational tillage may help overcome these problems while minimizing unnecessary tillage (Venterea et al., 2006).

In the more temperate eastern Corn Belt, NT has been more widely adopted, although farmers will often use intermittent tillage, such as plowing after soybean in a corn–soybean rotation (Omonode et al., 2006). Venterea et al. (2006) quantified SOC content in a corn–soybean rotation after 10–15 years of CT (moldboard plow and chisel plow in alternate years), continuous NT, and biennial tillage (chisel plow before soybean only) in Minnesota. The CT treatment had lower SOC content (0–30 cm) than NT and biennial tillage treatments. However, there was no significant difference in SOC between CT and NT to a depth of 60 cm. Omonode et al. (2006) studied the effects of several tillage treatments, including long-term (24 years) NT, short-term (7 years) NT after 14 years of moldboard plow, long-term (24 years) chisel plow, and intermittent chisel plow after 14 years of moldboard plow, in continuous corn and corn–soybean rotations for a prairie-derived soil in Indiana. After 24 years, there was no significant difference in SOC content between the two crop rotations averaged across all tillage treatments. Significant tillage effects on SOC were observed in the upper 50 cm, with continuous NT having greater SOC (151 Mg ha^{-1}) than continuous chisel plow (108 Mg ha^{-1}). Results suggested that combinations of moldboard plowing and intermittent chisel plowing or short-term NT fostered greater SOC storage than 24 years of continuous chisel plowing. Jelinski and Kucharik (2009) sampled corn and soybean fields on a floodplain soil in southern Wisconsin to estimate changes in SOC due to conversion from annual tillage to biennial tillage. No significant difference in SOC (0–25 cm) occurred with 10 years of biennial tillage compared with annual tillage.

Although SOC storage does not always occur in response to eliminating tillage, there is evidence that changes in tillage management alters C cycling processes. For example, results from native forest-derived soils in northern Appalachia using ^{13}C natural abundance indicated that most of the C in NT soils was corn derived, whereas the majority of the C in CT soils was old, forest-derived C (Mishra et al., 2010). Similarly in Ohio, Ussiri and Lal (2009) reported that SOC increased after 43 years of continuous corn under NT compared to CT and most (68–74%) of the SOC in the NT soil was corn derived. In Minnesota, Allmaras et al. (2004) found greater corn-derived C under NT compared to moldboard plow or chisel plow, primarily because the humification rate of corn-derived C was about twice as high in the NT system. Huggins et al. (2007) used natural ^{13}C abundance at a southern Minnesota field site to quantify corn-derived SOC (0–45 cm) after 14 years of cropping to continuous corn or corn–soybean. Significantly greater corn-derived C was found under NT and chisel plow tillage than under moldboard plow tillage for continuous corn and corn–soybean cropping systems. Benefits of reduced tillage management for SOC storage were derived through retention of newer, corn-derived C.

The impact of tillage on SOC storage can also interact with other management practices, such as application of manure. Pendell et al. (2006) measured SOC change (0–30 cm) after 10 years for six continuous corn cropping systems under NT or CT management and amended with beef cattle (*Bos taurus*) manure or chemical fertilizer N in northeastern Kansas. Greater SOC content was reported (1) under NT than under CT for manure and chemical fertilizer treatments and (2) with beef cattle manure than with chemical fertilizer across NT and CT soils. Hernandez-Ramirez et al. (2009) estimated SOC response in continuous corn and corn–soybean cropping systems amended with liquid swine (*Sus scrofa* L.) manure or urea ammonium nitrate fertilizer (UAN) prior to corn planting in northeastern Indiana. Surface SOC concentration (0–15 cm) was greatest in the spring prior to planting of continuous corn after fall application of liquid swine manure and lowest in the spring after soybean in the corn–soybean rotation. There were no significant differences in total profile SOC storage (0–100 cm) among any treatments. Lopez-Bellido et al. (2010) evaluated SOC storage after 38 years of NT and CT continuous and rotated corn with and without manure amendment. For these native forest-derived soils in northern Appalachia, continuous corn managed with NT or moldboard plow maintained SOC, but SOC sequestration was only observed for NT continuous corn plus manure.

Nitrogen Fertilizer Application

Application of synthetic N fertilizer increases crop residue production, and is therefore expected to result in greater organic C input to the soil and subsequent enhancement of SOC sequestration (Christopher and Lal, 2007). However, studies evaluating the effect of N application on SOC have produced dichotomous results. In some agroecosystems, N application stimulated soil C sequestration (Gregorich et al., 1996, Halvorson et al., 1999; Liebig et al., 2002; Varvel, 2006; Pikul et al., 2008; Varvel and Wilhelm, 2008), but other evaluations found no effect (Huggins and Fuchs, 1997; Halvorson et al., 2002; Dolan et al., 2006; Khan et al., 2007; Coulter et al., 2009; Russell et al., 2009). Conflicting results can be attributed to whether enhanced decomposition occurs following application of N fertilizer. For example, greater decomposition of enhanced C input with application of N fertilizer was demonstrated for Iowa cropping systems, in which no significant change in SOC storage occurred over a 12-year period (Russell et al., 2005). Decomposition rates of crop residue and SOC were also greater with N fertilizer application for other Midwest cropping systems (Huggins et al., 1998, 2007). In corn-based cropping systems in Iowa, Russell et al. (2009) reported that corn residue C:N declined and decomposition rate significantly increased with N fertilization. These studies demonstrated that N fertilization can impact decomposition rate of organic inputs and SOC, as well as the magnitude of residue C input.

Nitrogen fertilization can also impact N_2O loss from agricultural ecosystems (Cavigelli and Parkin, 2012). Bavin et al. (2009) compared N_2O flux for corn–soybean cropping systems in south-central Minnesota. Emission of N_2O was affected more by N fertilization management and fertilizer type than tillage. Johnson et al. (2010) observed that the spring thaw accounted for 65% of the annual N_2O emissions for three different management systems (plowed, chemically fertilized corn–soybean rotations; strip-tilled corn–soybean–winter wheat/alfalfa rotations with or without chemical fertilization) in Minnesota with <6% due to the application of N fertilizer. Global warming potential of N_2O is ~300 times greater than CO_2 (Houghton et al., 2001), highlighting the importance of enhancing our understanding of agricultural management effects on N_2O emissions (Johnson et al., 2010).

CLIMATE CHANGE AND MANAGEMENT INTERACTIONS

Carbon accumulation in soil depends on soil C inputs exceeding soil C losses (e.g. CO_2 and CH_4 efflux, erosion, leaching). Elevated atmospheric CO_2 concentration is expected to increase plant growth of important agricultural crops (Hatfield et al., 2008) and subsequently increase soil C inputs via transfer of plant C into soil (Jastrow et al., 2005; Prior et al., 2005). An increase in global temperature is likely to negatively affect SOC (Bellamy et al., 2005) by influencing soil organic matter decomposition and mineralization rates (Davidson et al., 2000; Conant et al., 2008), as well as soil and root respiration (Kirschbaum, 2004; Jones et al., 2005). Agricultural management impacts on these processes can determine whether a soil functions as a net source or sink for atmospheric CO_2. In central Illinois, Peralta and Wander (2008) used free-air concentration enrichment (FACE) technology to investigate the effects of elevated CO_2 on SOC in a corn–soybean rotation managed with conservation tillage. They observed no differences in total or labile SOC after 3 years of CO_2 enrichment despite increased aboveground and root biomass under elevated CO_2. Soil respiration rate was greater under elevated CO_2, suggesting more rapid mineralization of new C inputs in the CO_2-enriched system. Total and labile SOC declined over the course of the experiment in both the CO_2-enriched and ambient-CO_2 treatments (Peralta and Wander, 2008). Moran and Jastrow (2010) sampled the soil in the same FACE experiment 3 years later and confirmed the initial findings of Peralta and Wander (2008). Further, they reported no increases in labile SOC, aggregate-protected SOC, or silt- and clay-associated SOC under elevated CO_2. These results suggest that despite increased plant biomass, there was limited potential for long-term stabilization of plant C as SOC under elevated CO_2 in this corn—soybean system, because of rapid mineralization of new C inputs.

Information about maximum soil C storage capacity is important for development of improved agricultural management strategies to enhance C sequestration. One strategy used to obtain this information is to compare current stocks of SOC in cropland soil to native, undisturbed soil. Huggins et al. (2007) assessed tillage effects and crop sequence effects on SOC dynamics using natural ^{13}C abundance in corn and soybean cropping systems. Soil samples were collected after 14 years in each treatment and in fallowed alleyways. Alleyway soils were assumed to be at steady state with respect to SOC and were used as a surrogate for native, undisturbed baseline samples for initial SOC. All of the tillage and cropping treatments lost SOC compared to initial SOC, but conservation tillage and NT lost the least. A study conducted on two long-term experiments at the Kellogg Biological Station's Long-Term Ecological Research site in southwest Michigan evaluated changes in SOC using different agricultural management practices and in virgin grassland (Senthilkumar et al., 2009). Soil organic C declined under conventional management practices relative to the baseline, but SOC did not change under conservation practices of NT and planting of cover crops. Conservation practices appeared to have only prevented SOC losses compared with conventional management practices, rather than facilitating SOC sequestration. Greater losses were often associated with higher baseline C values, while SOC gains were more likely to be observed if baseline C were low. Greatest SOC loss relative to baseline was observed at native grassland sites, which had greater than twice the SOC at the start of experiments than agricultural sites. Other studies have also observed that greater SOC losses with time relative to baseline SOC levels have been associated with greater initial SOC contents (VandenBygaart et al., 2002; Bellamy et al., 2005; Tan et al., 2006). Interpretation of data on SOC sequestration comparing cropland to native soils must consider the complex interaction of climate change variables, such as increased atmospheric CO_2 and higher temperature, with agricultural management.

Since the turn of the 21st century, agriculture in the central U.S. has been asked to meet the increasing and competing demands for food, animal feed, and bioenergy. Concurrently, changes in land use or agricultural management practices to enhance C sequestration in C-depleted agricultural soils and C storage in terrestrial biomass have been considered important tools to reduce atmospheric CO_2 over the next several decades (Niu and Duiker, 2006; Fissore et al., 2010). As more CRP land converts to annual crop production and C trading becomes intensified, estimates of soil C stocks at local, landscape, and regional scales need to be made, despite great uncertainty with inadequate measurement frequency of soil profile C in the region. A number of alternative approaches to obtain accurate and reliable information on C sequestration potential for Midwest agroecosystems have been published in recent years, including application of land-use data from publically available databases, compilation of metadata from published studies (Eve et al., 2002; Sperow et al., 2003; West et al., 2008; Nelson et al., 2009; Fissore et al., 2010), and process-based simulation modeling (Grace et al., 2006; Pan et al., 2010; Lu and Zhuang, 2010; Liu et al., 2011).

CONCLUSIONS

This chapter reviewed recent literature related to C storage in cropland soils of the central U.S. to define the current state of knowledge of agricultural management impacts on SOC sequestration. The most widely adopted management practice to enhance SOC sequestration in this region is conservation tillage. Recent research has demonstrated that NT management increases surface SOC compared to plow tillage, but some studies indicate subsoil C content is higher under plowing. There is evidence that changes in tillage management alters C cycling processes, resulting in greater retention of corn-derived C in NT compared to plowed systems. However, higher subsoil C content in plowed soils can result in similar soil total profile SOC contents for NT and plowed soils in some studies. For the few studies that measured SOC content at the beginning of a field experiment, SOC either remained unchanged or declined in both NT and plowed soils. These results suggest that NT does not necessarily facilitate C sequestration, but rather simply prevents additional SOC loss often associated with CT. Enhancing crop rotation complexity is another successful strategy to foster SOC sequestration in corn-based cropping systems of the Midwest, especially by including small grains and forage legumes in extended rotations and through planting of perennial grasses and trees on marginally productive cropland. Planting cover crops is a viable option for SOC sequestration in this region, although it has been difficult to measure significant changes in SOC due to cover cropping in corn-based cropping systems. Some studies reported that N fertilizer application increased SOC, others reported decreased SOC, and still others reported no change in SOC. Carbon sequestration in response to N fertilizer application depends on interactive processes that simultaneously increase crop C inputs and enhance decomposition rate of crop residues and SOM. Although we have made some progress in the past decade towards defining the effects of crop rotation complexity, tillage, N fertilizer application, and planting perennial grasses on surface and subsurface SOC, we still have a relatively poor understanding of the mechanisms that determine SOC changes in response to land management practices. The following areas deserve increased research attention:

1. Roots are important sources of organic matter to soil, yet the rates of C flux through root systems to soils remains largely unknown. These belowground fluxes, including roots and rhizodeposits, are important for stimulating biological activity, maintaining soil C pools and enhancing aggregate structure of soils. Specific needs are:
    a. Improved methods for quantifying roots and rhizodeposits in the field.
    b. Detailed information on the quantity and quality of aboveground-, root-, and rhizodeposit-derived C inputs to soil.
    c. Better information on the microbial processes that mediate transformation of soil C inputs into SOC and the effect of land-use practices and climate change on these processes.
2. Long-term (>10-year) field studies that include adequate sampling of whole profile SOC at the beginning and throughout the study to enable accurate calculation of SOC changes across time. Some specific needs are:
    a. More rigorous evaluations of individual and integrative effects of conservation tillage, N application, and higher crop rotation complexity on soil C storage in typical and alternative productions systems.
    b. Evaluations of how planting perennial grasses and trees on marginal cropland or within extended cropping rotations might offset C losses within intensive annual cropping of landscapes.
3. Coordinated efforts to implement and maintain long-term field studies in multiple ecoregions that encompass gradients in temperature, moisture, and other abiotic drivers to estimate SOC sequestration potential and can be used to parameterize, validate, and verify simulation models.
4. Studies that improve understanding of the interactive effects of agricultural management practices, increased atmospheric CO_2, and changing climate variables on C sequestration in central U.S. agroecosystems.

References

Al-Kaisi, M.M., Yin, X., Licht, M.A., 2005. Soil carbon and nitrogen changes as influenced by tillage and cropping systems in some Iowa soils. Agric. Ecosys. Environ. 105, 635–647.

Allmaras, R.R., Linden, D.R., Clapp, C.E., 2004. Corn residue transformations into root and soil carbon as related to nitrogen, tillage, and stover management. Soil Sci. Soc. Am. J. 68, 1366–1375.

Alvarez, R., 2005. A review of nitrogen fertilizer and conservation tillage effects on soil organic carbon storage. Soil Use Manage. 21, 38–52.

Bailey, R.G., 1995. Description of the Ecoregions of the United States, 2nd edn. USDA Forest Service Misc. Publ. 1391, 108.

Baker, J.M., Ochsner, T.E., Venterea, R.T., Griffis, T.J., 2007. Tillage and soil carbon sequestration—what do we really know? Agric. Ecosyst. Environ. 118, 1–5.

Baker, J.M., Griffis, T.J., 2009. Evaluating the potential use of winter cover crops in corn–soybean systems for sustainable co-production of food and fuel. Agric. For. Meteor. 149, 2120–2132.

Bavin, T.K., Griffis, T.J., Baker, J.M., Venterea, R.T., 2009. Impact of reduced tillage and cover cropping on the greenhouse gas budget of a maize/soybean rotation ecosystem. Agric. Ecosyst. Environ. 134, 234–242.

Bellamy, P.H., Loveland, P.J., Bradley, R.I., Lark, R.M., Kirk, G.J.D., 2005. Carbon losses from all soils across England and Wales 1978–2003. Nature 437, 245–248.

Blanco-Canqui, H., Lal, R., 2008. No-tillage and soil-profile carbon sequestration: an on-farm assessment. Soil Sci. Soc. Am. J. 72, 693–701.

Bolinder, M.A., Janzen, H.H., Gregorich, E.G., Angers, D.A., VandenBygaart, A.J., 2007. An approach for estimating net primary productivity and annual carbon inputs to soil for common agricultural crops in Canada. Agric. Ecosyst. Environ. 118, 29–42.

Bolinder, M.A., Angers, D.A., Belanger, G., Michaud, R., Laverdiere, M.R., 2002. Root biomass and shoot to root ratios of perennial forage crops in eastern Canada. Can. J. Plant Sci. 82, 731–737.

Brown, R.A., Rosenberg, N.J., Hays, C.J., Easterling, W.E., Mearns, L.O., 2000. Potential production and environmental effects of switchgrass and traditional crops under current and greenhouse-altered climate in the central United States: a simulation study. Agric. Ecosyst. Environ. 78, 31–47.

Burgess, M.S., Mehuys, G.R., Madramootoo, C.A., 2002. Decomposition of grain–corn residues (*Zea mays* L.): a litterbag study under three tillage systems. Can. J. Soil Sci. 82, 127–138.

Causarano, H.J., Franzleubbers, A.J., Reeves, D.W., Shaw, J.M., 2006. Soil organic carbon sequestration in cotton production systems of the southeastern United States: a review. J. Environ. Qual. 35, 1374–1383.

Cavigelli, M.A., Parkin, T.B., 2012. Cropland management contributions to GHG flux: central and eastern U.S. In: Liebig, M.A., Franzluebbers, A.J., Follett, R.F. (Eds.), Managing Agricultural Greenhouse Gases: Coordinated Agricultural Research through GRACEnet to Address Our Changing Climate. Academic Press, San Diego, CA, pp. 177–234.

Christopher, S.F., Lal, R., 2007. Nitrogen management affects carbon sequestration in North American cropland soils. Crit. Rev. Plant Sci. 26, 45–64.

Clapp, C.E., Allmaras, R.R., Layese, M.F., Linden, D.R., Dowdy, R.H., 2000. Soil organic carbon and 13C abundance as related to tillage, crop residue, and nitrogen fertilization under continuous corn management in Minnesota. Soil Tillage Res. 55, 127–142.

Conant, R.T., Drijber, R.A., Haddix, M.L., Parton, W.J., Paul, E.A., Plante, A.F., Six, J., Steinweg, J.M., 2008. Sensitivity of organic matter decomposition to warming varies with its quality. Global Change Biol. 14, 868–877.

Coulter, J.A., Nafziger, E.D., Wander, M.M., 2009. Soil organic matter response to cropping system and nitrogen fertilization. Agron. J. 101, 592–599.

Dabney, S.M., Liu, Z., Lane, M., Douglas, J., Zhu, J., Flanagan, D.C., 1999. Landscape benching from tillage erosion between grass hedges. Soil Tillage Res. 51, 219–231.

David, M.B., McIsaac, G.F., Darmody, R.G., Omonode, R.A., 2009. Long-term changes in Mollisol organic carbon and nitrogen. J. Environ. Qual. 38, 200–211.

Davidson, E.A., Trumbore, S.E., Amundson, R., 2000. Soil warming and organic carbon content. Nature 408, 858–861.

Dick, W.A., Blevins, R.L., Frye, W.W., Peters, S.E., Christenson, D.R., Pierce, F.J., Vitosh, M.L., 1998. Impacts of agricultural management practices on C sequestration in forest-derived soils of the eastern Corn Belt. Soil Tillage Res. 47, 235–244.

Dolan, M.S., Clapp, C.E., Allmaras, R.R., Baker, J.M., Molina, J.A.E., 2006. Soil organic carbon and nitrogen in a Minnesota soil as related to tillage, residue and nitrogen management. Soil Tillage Res. 89, 221–231.

Elliott, E.T., 1986. Aggregate structure and carbon, nitrogen and phosphorous in native and cultivated soils. Soil Sci. Soc. Am. J. 50, 627–633.

Evans, J.M., Cohen, M.J., 2009. Regional water resource implications of bioethanol production in the southeastern United States. Global Change Biol. 15, 2261–2273.

Eve, M.D., Sperow, M., Paustian, K., Follett, R.F., 2002. National-scale estimation of changes in soil carbon stocks on agricultural lands. Environ. Pollut. 116, 431–438.

Fissore, C., Espeleta, J., Nater, E.A., Hobbie, S.E., Reich, P.B., 2010. Limited potential for terrestrial carbon sequestration to offset fossil-fuel emissions in the upper Midwestern US. Front. Ecol. Environ. 8, 409–413.

Follett, R.F., Varvel, G.E., Kimble, J.M., Vogel, K.P., 2009. No-till corn after bromegrass: effect on soil carbon and soil aggregates. Agron. J. 101, 261–268.

Franzluebbers, A.J., 2004. Tillage and residue management effects on soil organic matter. In: Magdoff, F., Weil, R.R. (Eds.), Soil Organic Matter in Sustainable Agriculture. CRC Press, Boca Raton, FL, pp. 227–268.

Franzluebbers, A.J., 2005. Soil organic carbon sequestration and agricultural greenhouse gas emissions in the southeastern USA. Soil Tillage Res. 83, 120–147.

Franzluebbers, A.J., 2010. Achieving soil organic carbon sequestration with conservation agricultural systems in the southeastern United States. Soil Sci. Soc. Am. J. 74, 347–357.

Franzluebbers, A.J., Follett, R.F., 2005. Greenhouse gas contributions and mitigation potential in agricultural regions of North America: introduction. Soil Tillage Res. 83, 1–8.

Franzluebbers, A.J., Owens, L.B., Sigua, G.C., Cambardella, C.A., Haney, R.L., 2012. Soil organic carbon under pasture management. In: Liebig, M.A., Franzluebbers, A.J., Follett, R.F. (Eds.), Managing Agricultural Greenhouse Gases: Coordinated Agricultural Research through GRACEnet to Address our Changing Climate. Academic Press, San Diego, CA, pp. 127–152.

Gal, A., Vyn, T.J., Micheli, E., Kladivko, E.J., McFee, W.W., 2007. Soil carbon and nitrogen accumulation with long-term no-till versus moldboard plowing overestimated with tilled-zone sampling depths. Soil Tillage Res. 96, 42–51.

Garten, C.T., 2002. Soil carbon storage beneath recently established tree plantations in Tennessee and South Carolina. USA. Biomass Bioenergy 23, 93–102.

Grace, P.R., Colunga-Garcia, M., Gage, S.H., Robertson, G.P., Safir, G.R., 2006. The potential impact of agricultural management and climate change on soil organic carbon of the north central region of the United States. Ecosystems 9, 816–827.

Grandy, A.S., Robertson, G.P., 2006. Initial cultivation of a temperate-region soil immediately accelerates aggregate turnover and CO_2 and N_2O fluxes. Global Change Biol. 12, 1507–1520.

Gregorich, E.G., Ellert, B.H., Drury, C.F., Liang, B.C., 1996. Fertilization effects on soil organic matter turnover and corn residue C storage. Soil Sci. Soc. Am. J. 60, 472–476.

Guérif, J., Richard, G., Durr, C., Machet, J.M., Recous, S., Roger-Estrade, J., 2001. A review of tillage effects on crop residue management, seedbed conditions and seedling establishment. Soil Tillage Res. 61, 13–32.

Haas, H.J., Evans, C.E., Miles, E.F., 1957. Nitrogen and carbon changes in the Great Plains soils as influenced by cropping and soil treatments. USDA Tech. Bull. No. 1164, U.S. Gov. Print. Office, Washington, DC.

Halvorson, A.D., Reule, C.A., Follett, R.F., 1999. Nitrogen fertilization effects on soil carbon and nitrogen in a dryland cropping system. Soil Sci. Soc. Am. J. 63, 91–917.

Halvorson, A.D., Wienhold, B.J., Black, A.L., 2002. Tillage, nitrogen, and cropping system effects on soil carbon sequestration. Soil Sci. Soc. Am. J. 66, 906–912.

Hatfield, J.L., Boote, K.J., Fay, P., Hahn, L., Izaurralde, C., Kimball, B.A., Mader, T., Morgan, J., Ort, D., Polley, W., Thomson, A., Wolfe, D., 2008, Agriculture. 362 p. In: The Effects of Climate Change on Agriculture, Land Resources, Water Resources, and Biodiversity in the United States. A report by the U.S. Climate Change Science Program and the Subcommittee on Global Change Research, Washington, DC.

Heaton, E., Voigt, T., Long, S.P., 2004. A quantitative review comparing the yields of two candidate C4 perennial biomass crops in relation to nitrogen, temperature and water. Biomass Bioenergy 27, 21–30.

Hernandez-Ramirez, G., Brouder, S.M., Smith, D.R., Van Scoyoc, G.E., 2009. Carbon and nitrogen dynamics in an eastern Corn Belt soil: N source and rotation. Soil Sci. Soc. Am. J. 73, 128–137.

Hernandez-Ramirez, G., Sauer, T.J., Cambardella, C.A., Brandle, J.R., James, D.E., 2011. Carbon sources and dynamics in afforested and cultivated Corn Belt soils. Soil Sci. Soc. Am. J. 75, 216–225.

Holanda, F.S.R., Mengel, D.B., Paula, M.B., Carvaho, J.G., Bertoni, J.C., 1998. Influence of crop rotations and tillage systems on phosphorus and potassium stratification and root distribution in the soil profile. Commun. Soil Sci. Plant Anal. 29, 2383–2394.

Houghton, R.A., Hobbie, J.E., Melillo, J.M., Moore, B., Peterson, B.J., Shaver, G.R., Woodwell, G.M., 1983. Changes in the carbon content of terrestrial biota and soils between 1860 and 1980: a net release of CO_2 to the atmosphere. Ecological Monogr. 53, 235–262.

Houghton, Ding, J.T.,Y., Griggs, D.J., Noguer, M., van der Linden, P.J., Dai, X., Maskell, K., Johnson, C.A. (Eds.), 2001. Climate Change 2001: The Scientific Basis. Cambridge Univ. Press, Cambridge, UK.

Huggins, D.A., Clapp, C.E., Allmaras, R.R., Lamb, J.A., Layese, M.F., 1998. Carbon dynamics in corn–soybean sequences as estimated from natural carbon-13 abundance. Soil Sci. Soc. Am. J. 62, 195–203.

Huggins, D.R., Allmaras, R.R., Clapp, C.E., Lamb, J.A., Randall, G.W., 2007. Corn–soybean sequence and tillage effects on soil carbon dynamics and storage. Soil Sci. Soc. Am. J. 71, 145–154.

Huggins, D.R., Fuchs, D.J., 1997. Long-term N management effects on corn yield and soil C of an Aquic Haplustoll in Minnesota. In: Paul, E.A., Paustian, K., Elliott, E.T., Cole, C.V. (Eds.), Soil Organic Matter in Temperate Agroecosystems. CRC Press, Boca Raton, Florida, USA.

IPCC, 2007. Climate Change 2007: Mitigation of Climate Change. Working Group III. Cambridge University Press, Cambridge, UK.

Jagadamma, S.R., Lal, R., Hoeft, R.G., Nafziger, E.D., Adec, E.A., 2007. Nitrogen fertilization and cropping systems effects on soil organic carbon and total nitrogen pools under chisel-plow tillage in Illinois. Soil Tillage Res. 95, 348–356.

Jastrow, J.D., Miller, R.M., Matamala, R., Norby, R.J., Boutton, T.W., Rice, C.W., Owensby, C.E., 2005. Elevated atmospheric carbon dioxide increases soil carbon. Global Change Biol. 11, 2057–2064.

Jelinski, N.A., Kucharik, C.J., 2009. Land-use effects on soil carbon and nitrogen on a U.S. midwestern floodplain. Soil Sci. Soc. Am. J. 73, 217–225.

Jenny, H., 1941. Factors of Soil Formation. McGraw-Hill, New York.

Johnson, J.M.F., Reicosky, D.C., Allmaras, R.R., Sauer, T.J., Venterea, R.T., Dell, C.J., 2005. Greenhouse gas contributions and mitigation potential of agriculture in the central USA. Soil Tillage Res. 83, 73–94.

Johnson, J.M.F., Allmaras, R.R., Reicosky, D.C., 2006. Estimating source carbon from crop residues, roots and rhizodeposits using the national grain-yield database. Agron. J. 98, 622–636.

Johnson, J.M.F., Franzluebbers, A.J., Weyers, S.L., Reicosky, D.C., 2007a. Agricultural opportunities to mitigate greenhouse gas emissions. Environ. Pollut.150, 107–142.

Johnson, J.M.F., Coleman, M.D., Gesch, R.W., Jaradat, A.A., Mitchell, R., Reicosky, D.C., Wilhelm, W.W., 2007b. Biomass-bioenergy crops in the United States: a changing paradigm. Amer. J. Plant Sci. Biotechnol. 1, 1–28.

Johnson, J.M.F., Archer, D., Barbour, N., 2010. Greenhouse gas emission from contrasting management scenarios in the Northern Corn Belt. Soil Sci. Soc. Am. J. 74, 396–406.

Johnson, J.M.F., Archer, D.W., Karlen, D.L., Weyers, S.L., Wilhelm, W.W., 2011. Soil management implications of producing biofuel feedstock. In: Hatfield, J.L., Sauer, T.J. (Eds.), Soil Management: Building a Stable Base for Agriculture. ASA Series. ASA and SSSA, Madison, WI, pp. 371–390.

Johnson, J.M.F., Novak, J.M., 2012. Sustainable bioenergy feedstock production systems: integrating C dynamics, erosion, water quality and greenhouse gas production. In: Liebig, M.A., Franzluebbers, A.J., Follett, R.F. (Eds.), Managing Agricultural Greenhouse Gases: Coordinated Agricultural Research through GRACEnet to Address our Changing Climate. Academic Press, San Diego, CA, pp. 153–174.

Jones, C., McConnell, C., Coleman, K., Cox, P., Falloon, P., Jenkinson, D., 2005. Global climate change and soil carbon stocks: predictions from two contrasting models for the turnover of organic carbon in soil. Global Change Biol. 11, 154–166.

Kaspar, T.C., Radke, J.K., Laflen, J.M., 2001. Small grain cover crops and wheel traffic effects on infiltration, runoff, and erosion. J. Soil Water Conserv. 56, 160–164.

Kaspar, T.C., Parkin, T.B., Jaynes, D.B., Cambardella, C.A., Meek, D.W., Jung, Y.S., 2006. Examining changes in soil organic carbon with oat and rye cover crops using terrain covariates. Soil Sci. Soc. Am. J. 70, 1168–1177.

Kaspar, T.C., Jaynes, D.B., Parkin, T.B., Moorman, T.B., 2007. Rye cover crop and gamagrass strip effects on NO_3 concentration and load in tile drainage. J. Environ. Qual. 36, 1503–1511.

Khan, S.A., Mulvaney, R.L., Ellsworth, T.R., Boast, C.W., 2007. The myth of nitrogen fertilization for soil carbon sequestration. J. Environ. Qual. 36, 1821–1832.

Kirschbaum, M.U.F., 2004. Soil respiration under prolonged soil warming: are rate reductions caused by acclimation or substrate loss? Global Change Biol. 10, 1870–1877.

Kladivko, E.J., Frankenberger, J.R., Jaynes, D.B., Meek, D.W., Jenkinson, B.J., Fausey, N.R., 2004. Nitrate leaching to subsurface drains as affected by drain spacing and changes in crop production system. J. Environ. Qual. 33, 1803–1813.

Lal, R., 2007. Soil science and the carbon civilization. Soil Sci. Soc. Am. J. 71, 1425–1437.

Lal, R., Henderlong, P., Flowers, M., 1998. Forages and row cropping effects on soil organic carbon and nitrogen contents. In: Lal, R., et al. (Eds.), Management of C Sequestration in Soil. CRC Press, Boca Raton, FL, pp. 365–379.

Lee, K.H., Isenhart, T.M., Schultz, R.C., 2003. Sediment and nutrient in an established multi-species riparian buffer. J. Soil Water Conserv. 58, 1–8.

Liebig, M.A., Varvel, G.E., Doran, J.W., Wienhold, B.J., 2002. Crop sequence and nitrogen fertilization effects on soil properties in the western Corn Belt. Soil Sci. Soc. Am. J. 66, 596–601.

Liebig, M.A., Morgan, J.A., Reeder, J.D., Ellert, B.H., Gollany, H.T., Schuman, G.E., 2005. Greenhouse gas contributions and mitigation potential of agricultural practices in northwestern USA and western Canada. Soil Tillage Res. 83, 25–52.

Liu, S., Tan, Z., Li, Z., Zhao, S., Yuan, W., 2011. Are soils of Iowa USA currently a carbon sink or source? Simulated changes in SOC stock from 1972 to 2007. Agricult. Ecosys. Environ. 140, 106–112.

López-Bellido, R.J., Lal, R., Owens, L.B., López-Bellido, L., 2010. Does North Appalachian agriculture contribute to soil carbon sequestration? Agricult. Ecosys. Environ. 137, 373–376.

Lu, X., Zhuang, Q., 2010. Evaluating climate impacts on carbon balance of the terrestrial ecosystems in the Midwest of the United States with a process-based ecosystem model. Mitig. Adapt. Strateg. Global Change. 15, 467–487.

Marriot, E.M., Wander, M.M., 2006a. Total and labile soil organic matter in organic and conventional farming systems. Soil Sci. Soc. Am. J. 70, 950–959.

Marriot, E.M., Wander, M.M., 2006b. Qualitative and quantitative differences in particulate organic matter fractions in organic and conventional farming systems. Soil Biol. Biochem. 38, 1527–1536.

Martens, D.A., Emmerich, W., McLain, J.E.T., Johnsen, T.N., 2005. Atmospheric carbon mitigation potential of agricultural management in the southwestern USA. Soil Tillage Res. 83, 95–119.

McLauchlan, K., 2006. The nature and longevity of agricultural impacts on soil carbon and nutrients: a review. Ecosystems 9, 1364–1382.

Mishra, U., Lal, R., Slater, B., Calhoun, F., Liu, D.S., Van Meirvenne, M., 2010. Predicting soil organic carbon stock using profile depth distribution functions and ordinary kriging. Soil Sci. Soc. Am. J. 73, 614–621.

Moran, K.Y., Jastrow, J.D., 2010. Elevated carbon dioxide does not offset loss of soil carbon from a corn–soybean agroecosystem. Environ. Pollut. 158, 1088–1094.

Morgan, J.A., Follett, R.F., Allen Jr., L.H., Del Grosso, S., Derner, J.D., Dijkstra, F., Franzluebbers, A., Fry, R., Paustian, K., Schoeneberger, M.M., 2010. Carbon sequestration in agricultural lands of the United States. J. Soil Water Conserv. 65, 6A–13A.

Nelson, R.G., Ascough II, J.C., Langemeier, M.R., 2006. Environmental and economic analysis of switchgrass production for water quality improvement in northeast Kansas. J. Environ. Manage. 79, 336–347.

Nelson, R.G., Hellwinckel, C.M., Brandt, C.C., West, T.O., De La Torre Ugarte, D.G., Marland, G., 2009. Energy use and carbon dioxide emissions from cropland production in the United States, 1990–2004. J. Environ. Qual. 38, 418–425.

Niu, X.Z., Duiker, S.W., 2006. Carbon sequestration potential by afforestation of marginal agricultural land in the Midwestern US. Forest Ecol. Manage. 223, 415–427.

Ogle, S.M., Breidt, F.J., Paustian, K., 2005. Agricultural management impacts on soil organic carbon storage under moist and dry climatic conditions of temperate and tropical regions. Biogeochemistry 72, 87–121.

Omonode, R.A., Gal, A., Stott, D.E., Abney, T.S., Vyn, T.J., 2006. Short-term versus continuous chisel and no-till effects on soil carbon and nitrogen. Soil Sci. Soc. Am. J. 70, 419–425.

Owensby, J., Heim, R.J., Burgin, M., Ezell, D., 2001. Climatography of the U.S. No. 81—Supplement #3 Maps of Annual 1961–1990 Normal Temperature, Precipitation and Degree Days NOAA (posted May 29, 2001; verified July 13, 2004). http://www.ncdc.noaa.gov/oa/documentlibrary/clim81supp3/clim81.html.

Pan, Z., Andrade, D., Segal, M., Wimberley, J., McKinney, N., Takle, E., 2010. Uncertainty in future soil carbon trends at a central U.S. site under an ensemble of GCM scenario climates. Ecological Modeling 221, 876–881.

Pendell, D.L., Williams, J.R., Rice, C.W., Nelson, R.G., Boyles, S.B., 2006. Economic feasibility of no-tillage and manure for soil carbon sequestration in corn production in northeastern Kansas. J. Environ. Qual. 35, 1364–1373.

Peralta, A.L., Wander, M.M., 2008. Soil organic matter dynamics under soybean exposed to elevated CO_2. Plant Soil. 303, 69–81.

Pikul, J.L., Johnson, J.M.F., Schumacher, T.E., Vigil, M., Riedell, W.E., 2008. Change in surface soil carbon under rotated corn in Eastern South Dakota. Soil Sci. Soc. Am. J. 72, 1738–1744.

Post, W.M., Kwon, K.C., 2000. Soil carbon sequestration and land-use change: processes and potential. Global Change Biol. 6, 317–327.

Prior, S.A., Runion, G.B., Rogers, H.H., Torbert, H.A., Reeves, D.W., 2005. Elevated atmospheric CO_2 effects on biomass production and soil carbon in conventional and conservation cropping systems. Global Change Biol. 11, 657–665.

Ranells, N.N., Wagger, M.G., 1997. Grass-legume bicultures as winter annual cover crops. Agron. J. 89, 659–665.

Reicosky, D.C., Lindstrom, M.J., 1993. Fall tillage method: effect on short-term carbon dioxide flux from soil. Agron. J. 85, 1237–1243.

Reicosky, D.C., Hatfield, J.L., Sass, R.L., 2000. Agricultural contributions to greenhouse gas emissions. In: Reddy, K.R., Hodges, H.F. (Eds.), Climate Change and Global Crop Production. ACBI International, London, pp. 37–55.

Reicosky, D.C., Archer, D.W., 2007. Moldboard plow tillage depth and short-term carbon dioxide release. Soil Tillage Res. 94, 109–121.

Robertson, G.P., Paul, E.A., Harwood, R.R., 2000. Greenhouse gases in intensive agriculture: contributions of individual gases to the radiative forcing of the atmosphere. Science 289, 1922–1925.

Rochette, P., Flanagan, L.B., Gregorich, E.G., 1999. Separating soil respiration into plant and soil components using analyses of the natural abundance of carbon-13. Soil Sci. Soc. Am. J. 63, 1207–1213.

Russell, A.E., Laird, D.A., Parkin, T.B., Mallarino, A.P., 2005. Impact of nitrogen fertilization and cropping system on carbon sequestration in Midwestern Mollisols. Soil Sci. Soc. Am. J. 69, 413–422.

Russell, A.E., Cambardella, C.A., Laird, D.A., Jaynes, D.B., Meek, D.W., 2009. Nitrogen fertilizer effects on soil carbon balances in Midwestern U.S. agricultural systems. Ecological Applic. 19, 1102–1113.

Sainju, U.M., Singh, B.P., Whitehead, W.F., 2002. Long-term effects of tillage, cover crops, and nitrogen fertilization on organic carbon and nitrogen concentrations in sandy loam soils in Georgia. USA. Soil Tillage Res. 63, 167–179.

Sauer, T.J., Cambardella, C.A., Brandle, J.R., 2007. Soil carbon and tree litter dynamics in a red cedar–Scotch pine shelterbelt. Agrofor. Syst. 71, 163–174.

Self-Davis, M.L., Moore, P.A., Daniels, T.C., Nichols, D.J., Sauer, T.J., West, C.P., Aiken, G.E., Edwards, D.R., 2003. Forage species and canopy cover effects on runoff from small plots. J. Soil Water Conserv. 58, 349–359.

Senthilkumar, S., Basso, B., Kravchenko, A.N., Robertson, G.P., 2009. Contemporary evidence of soil carbon loss in the U.S. Corn Belt. Soil Sci. Soc. Am. J. 73, 2078–2086.

Singer, J.W., Kohler, K.A., Liebman, M., Richard, T.L., Cambardella, C.A., Buhler, D.D., 2004. Tillage and compost affect yield of corn, soybean, and wheat and soil concentrations of organic matter, phosphorus, and potassium. Agron. J. 96, 531–537.

Smith, P., 2004. Carbon sequestration in croplands: the potential in Europe and the global context. Europ. J. Agron. 20, 229–236.

Snyder, C.S., Bruulsema, R.W., Jensen, T.L., Fixen, P.E., 2009. Review of greenhouse gas emissions from crop production systems and fertilizer management effects. Agricult. Ecosyst. Environ. 133, 247–266.

Sperow, M., Eve, M., Paustian, K., 2003. Potential soil C sequestration on US agricultural soils. Clim. Change. 57, 319–339.

Staricka, J., Allmaras, R.R., Nelson, W.W., 1991. Spatial variation of crop residue incorporation by tillage. Soil Sci. Soc. Am. J. 55, 1668–1674.

Strock, J.S., Porter, P.M., Russelle, M.P., 2004. Cover cropping to reduce nitrate loss through subsurface drainage in the northern U.S. Corn Belt. J. Environ. Qual. 33, 1010–1016.

Swan, J.B., Kaspar, T.C., Erbach, D.C., 1996. Seed-row residue management for corn establishment in the northern US Corn Belt. Soil Tillage Res. 40, 55–72.

Syswerda, S.P., Corbin, A.T., Mokma, D., Kravchenko, A.N., Robertson, G.P., 2011. Agricultural management and soil carbon storage in surface vs. deep layers. Soil Sci. Soc. Am. J. 75, 92–101.

Tan, Z., Lal, R., Liu, S., 2006. Using experimental and geospatial data estimate regional carbon sequestration potential under no-till management. Soil Sci. 171, 950–959.

Teasdale, J.R., 2007. Strategies for soil conservation in no-tillage and organic farming systems. J. Soil Water Conserv. 62, 144A–147A.

Tu, C., Louws, F.J., Creamer, N.G., Mueller, J.P., Brownie, C., Fager, K., Bell, M., Hu, S., 2006. Responses of soil microbial biomass and N availability to transition strategies from conventional to organic farming systems. Agric. Ecosys. Environ. 113, 206–215.

Ussiri, D.A.N., Lal, R., 2009. Long-term tillage effects on soil carbon storage and carbon dioxide emissions in continuous corn cropping system from an alfisol in Ohio. Soil Tillage Res. 104, 39–47.

USDA, 1994. Major World Crop Areas and Climatic Profiles: Agricultural Handbook 664. [Online]. United States Department of Agriculture, World Agricultural Outlook Board, Joint Agricultural Weather Facility. Available at. http://www.usda.gov/oce/weather/pubs/Other/MWCACP/MajorWorldCropAreas.pdf (posted 2006; verified May 25, 2011).

USDA-NRCS, 1997. Broad land cover/use, by state. United States Department of Agriculture Natural Resources Conservation Service, Resource Assessment Division (posted June 2001; verified July 13, 2004). http://www.nrcs.usda.gov/technical/land/cover_use.html.

VandenBygaart, A.J., Yang, X.M., Kay, B.D., Aspinall, J.D., 2002. Variability in carbon sequestration potential in no-till soil landscape of southern Ontario. Soil Tillage Res. 65, 231–241.

VandenBygaart, A.J., Angers, D.A., 2006. Towards accurate measurements of soil organic carbon stock change in agroecosystems. Can. J. Soil Sci. 86, 465–471.

Van Veen, J.A., Kuikman, P.J., 1990. Soil structural aspects of decomposition of organic matter by microorganisms. Biogeochemistry 11, 213–223.

Varvel, G.E., 2006. Soil organic carbon changes in diversified rotations of the western Corn Belt. Soil Sci. Soc. Am. J. 70, 426–433.

Varvel, G.E., Wilhelm, W.W., 2008. Soil carbon levels in irrigated western Corn Belt rotations. Agron. J. 100, 1180–1184.

Varvel, G.E., Wilhelm, W.W., 2010. Long-term soil organic carbon as affected by tillage and cropping system. Soil Sci. Soc. Am. J. 74, 915–921.

Varvel, G.E., Wilhelm, W.W., 2011. No-tillage increases soil profile carbon and nitrogen under long-term rainfed cropping systems. Soil Tillage Res. 114, 28–36.

Venterea, R.T., Baker, J.M., Dolan, M.S., Spokas, K.A., 2006. Carbon and nitrogen storage are greater under biennial tillage in a Minnesota corn–soybean rotation. Soil Sci. Soc. Am. J. 70, 1752–1762.

Vesterdal, L., Ritter, E., Gundersen, P., 2002. Change in soil organic carbon following afforestation of former arable land. For. Ecol. Manage. 169, 137–147.

Villamil, M.B., Bollero, G.A., Darmody, R.G., Simmons, F.W., Bullock, D.G., 2006. No-till corn/soybean systems including winter cover crops: effects on soil properties. Soil Sci. Soc. Am. J. 70, 1936–1944.

West, T.O., Post, W.M., 2002. Soil organic carbon sequestration rates by tillage and crop rotation: a global data analysis. Soil Sci. Soc. Am. J. 66, 1930–1946.

West, T.O., Marland, G., 2002. A synthesis of carbon sequestration, carbon emissions, and net carbon flux in agriculture: comparing tillage practices in the United States. Agric. Ecosyst. Environ. 91, 217–232.

West, T.O., Six, J., 2007. Considering the influence of sequestration duration and carbon saturation on estimates of soil carbon capacity. Clim. Change. 80, 25–41.

West, T.O., Brandt, C.C., Wilson, B.S., Hellwinckel, C.M., Tyler, D.D., Marland, G., De La Torre Ugarte, D.G., Larson, J.A., Nelson, R., 2008. Estimating regional changes in soil carbon with high spatial resolution. Soil Sci. Soc. Am. J. 72, 285–294.

Wilhelm, W.W., Hess, J.R., Karlen, D.L., Johnson, J.M.F., Muth, D.J., Baker, J.M., Gollany, H.T., Novak, J.M., Stott, D.E., Varvel, G.E., 2010. Review: balancing limiting factors & economic drivers for sustainable Midwestern US agricultural residue feedstock supplies. Indust. Biotechnol. 6, 271–287.

Yang, X.M., Wander, M.M., 1999. Tillage effects on soil organic carbon distribution and storage in a silt loam soil in Illinois. Soil Tillage Res. 52, 1–9.

Yang, X.M., Drury, C.R., Wander, M.M., Kay, B.D., 2008. Evaluating the effect of tillage on carbon sequestration using the minimum detectable difference concept. Pedosphere 18, 421–430.

Zan, C.S., Fyles, J.W., Girouard, P., Samson, R.A., 2001. Carbon sequestration in perennial bioenergy, annual corn and uncultivated systems in southern Quebec. Agric. Ecosyst. Environ 86, 135–144.

CHAPTER 5

Agricultural Management and Soil Carbon Dynamics: Western U.S. Croplands

Harold P. Collins[1], Maysoon M. Mikha[2], Tabitha T. Brown[3], Jeffrey L. Smith[4], David Huggins[4], Upendra M. Sainju[5]
[1]USDA-ARS, Vegetable and Forage Crops Research Unit, Prosser, WA
[2]USDA-ARS, Central Great Plains Research Station, Akron, CO 80720
[3]Dept. Crop & Soil Sciences, Washington State University, Pullman, WA
[4]USDA-ARS, Washington State University, Pullman, WA
[5]USDA-ARS, Northern Plains Agricultural Research Laboratory, Sidney, MT

CHAPTER OUTLINE

INTRODUCTION

Soils are the largest pool of carbon (C) in the terrestrial environment (Jobbagy and Jackson, 2000; Schlesinger, 1995). The amount of C stored in soils is twice the amount of C in the atmosphere and three times the amount of C stored in living plants (Kimble and Stewart, 1995). Therefore, a change in the size of the soil C pool could significantly alter current atmospheric CO_2 concentrations (Wang et al., 1999). Carbon stored in soils is derived from litter, root inputs, sediment deposits, and exogenous applications of manures/mulches, while losses result from microbial degradation of soil organic matter, eluviation, and erosion (Entry and Emmingham, 1998). As an ecosystem approaches maturity, C sequestration potential is controlled by climate, topography, soil type, and vegetation (Harmon et al., 1990;

Managing Agricultural Greenhouse Gases. DOI: 10.1016/B978-0-12-386897-8.00005-X
2012, Published by Elsevier Inc.

Dewar, 1991; Van Cleve et al., 1993). At equilibrium, the rate and amount of C added to the soil from plant residues and roots, organic amendments as well as erosion deposits, are equal to the rate and amount of C lost through organic matter decomposition and soil erosion processes (Henderson, 1995; Paustian et al., 1997).

Within limits, C in soil increases with increasing soil water and decreases with temperature (Wang et al., 1999). The effect of soil water is much greater than the effect of soil temperature (Liski et al., 1999). Increasing water within a temperature zone can increase plant production and, thus, C input to soils via increased plant litter and root production (Liski et al., 1999), but increasing water also can reduce soil C through enhanced decomposition.

Land-use changes can impact the amount of C stored in the soil by altering C inputs and losses. Conversion of native vegetation to agricultural cropping has resulted in both substantial C transfer to the atmosphere and loss of soil C (Lal et al., 1999; Wang et al., 1999).

Agricultural practices that can partially restore depleted soil organic carbon (SOC) include: (1) adoption of conservation tillage including no-tillage; (2) intensification of cropping by eliminating fallow, increasing cover crops and including more perennial vegetation (Sperow et al., 2003); and (3) improving biomass production through the use of soil amendments (manures), fertilizers and high yielding crop varieties (Lal et al., 1998; Follett, 2001).

CLIMATE, CROP, AND CROPPING PRACTICE CHARACTERISTICS

Climate

The climate of the western U.S. is characterized mainly as semiarid, with the southwest portion of the region classified as arid; each of these areas exhibits a diverse range in annual precipitation and temperature that significantly influences crop production and C storage. The following information was obtained from the Western Regional Climate Center (http://www.wrcc.dri.edu/). In general, precipitation decreases from east to west across the region. The average midwinter temperature ranges in the extreme north (ND) are −8 to −13°C, while summer temperatures range from 18 to 23°C. In the southern prairie (OK, TX) winter temperatures average 10 to 13°C, with little variation in summer temperatures from 27 to 29°C. Precipitation along the western border of the prairie region ranges from 46 cm yr^{-1} in the north to 64 cm yr^{-1} in the south. Most precipitation occurs in the early summer. Precipitation averages about 50 cm yr^{-1} in the southern plains and 25 cm yr^{-1} in the northern plains, with extreme year-to-year variations.

The intermountain arid region of the west shows considerable climatic variation between its northern and southern boundaries. In NM, AZ, and southeastern CA the greatest precipitation occurs in the summer months, with an average annual precipitation from 8 cm in the valleys to 76 cm yr^{-1} in the mountains of AZ, NM, NV, and UT. North of the UT–AZ line, the summer months usually are very dry; maximum precipitation occurs in the winter and early spring. In the desert valleys west of Great Salt Lake, mean annual precipitation averages 10 cm.

The Pacific Northwest states of WA, OR, and ID are bounded on the east by the Rocky Mountains and on the west by the Pacific Ocean. The climate is influenced by the region's mountain ranges. In WA and OR, the north to south Cascade Range delineates a wetter coastal climate from a drier inland continental climate. Precipitation across the region ranges from 15 to 500 cm yr^{-1} with most occurring between October and April. The valleys west of the Cascade Range receive $\geq$75 cm yr^{-1} of precipitation, whereas east of the Cascades precipitation is $\leq$50 cm yr^{-1}. Temperature in the coastal zone is mild, but east of the Cascades and in southern Idaho, temperature is warmer in the summer and has larger annual ranges. Many areas of the Pacific Northwest could be classified as Mediterranean if it were not for the cold winter temperatures. Although the northern plateaus are generally arid, some of the mountainous areas of central Washington and Idaho receive ≥150 cm of precipitation annually. Throughout the western U.S., the irregular occurrence of precipitation is the principal factor for the development of irrigation to support agriculture.

Crop Production

Agriculture in the western U.S. is as diverse as the climate and topography. The type of crops produced varies depending on rainfall, irrigation, soil, elevation, and temperature extremes. The total harvested cropland in the western U.S. was estimated at 57 Mha in 2007 (NASS, 2007). In the most arid regions (AZ, NM, NV, TX, and UT) without irrigation, production agriculture generally supports only livestock grazing. Texas is a major cattle and sheep raising area, as well as the nation's largest producer of cotton (*Gossypium hirsutum* L.). Climate is a major factor influencing the distribution of cropping systems within the dryland (i.e. rainfed) cropping regions of the PNW (Papendick, 1996; Paustian et al., 1997). Areas that receive <375 mm of water annually are typically managed using a 2-year winter wheat (*Triticum aestivium* L.)—summer fallow rotation (Rasmussen et al., 1998; McCool and Roe, 2005). Areas with rainfall of 375 to 450 mm are typically managed using a 2- to 5-year rotation that often includes winter wheat—spring barley (*Hordeum vulgare* L.) or spring wheat—summer fallow (Papendick, 1996; Rasmussen et al., 1998; McCool and Roe, 2005). Fallow is discontinued and annual cropping is practiced in areas that receive >450 mm of precipitation. Approximately 80—90% of cropland in the Northern and Central Plains region is farmed under dryland conditions and <20% has irrigation. The traditional dryland rotation is winter wheat—fallow under conventional tillage (CT). In higher rainfall areas, crops grown in rotation with winter wheat are corn (*Zea maize* L.), spring wheat, proso millet (*Panicum miliaceum* L.), sorghum (*Sorghum bicolor* L.), sunflower (*Helianthus annuus* L.), barley, oat (*Avena sativa* L.), and pea (*Pisum sativum* L.). In eastern Montana and North Dakota the dominant dryland cropping system is spring wheat—fallow under conventional tillage. This rotation is being slowly replaced by continuous cropping of spring wheat, or in 1- to 4-year crop rotations with winter wheat, barley, pea, lentil (*Lens culinaris* L.), canola (*Brassica napus* L.), and yellow mustard (*Brassica juncea* L.). Annual forages, such as barley hay, Austrian winter pea hay, foxtail millet hay (*Setaria italic* L.), and perennial forages, such as alfalfa (*Medicago sativa* L.), are also included in rotation with cereals.

Irrigation dominates (>90%) crop production in AZ, CA, and NV, with ~64% in ID, NM, UT, and WY, ~36% in CO, NE, OR, and WA, and <25% in KS, MT, ND, OK, SD, and TX (NASS, 2007). CA has the largest area under irrigated production with >3.5 M ha (NASS, 2007). Irrigation in the areas of the PNW, CA, and the southwest (AZ, NM, NV, UT) allows for the production of perennial fruits, nuts, and annual vegetables as well as grain, alfalfa, and grass hay. California and Arizona are also major producers of citrus crops. Irrigated cropping systems in the Great Plains region incorporate sugar beet (*Beta vulgaris* L.), malt barley, potato (*Solanum tuberosum* L.), corn, soybean (*Glycine max* L.), and dry bean (*Phaseolus vulgaris* L.) in rotation.

Tillage

Tillage in the western U.S. is highly variable due to the diversity of crops. Grain crops are significant in all states and are mostly grown under conventional tillage by plowing. Area under no-tilled grains is increasing in the Pacific Northwest, although it is <10%. Area under full-width inversion tillage is similar to that under winter grain crop production but area under reduced tillage (e.g. chisel, sweeps, field cultivators) tends to follow that under spring crops (grain or legume) in the higher rainfall areas where annual dryland cropping is practiced. Overall, conservation tillage regime has been widely used for most crop rotations (Kok et al., 2009). Conservation tillage practices leave at least 30% of the soil surface covered by crop residues and reduce soil disturbance and SOC decomposition (Jian et al., 2005). In the drier regions, grain production with wheat—fallow systems is common and tillage includes several rod-weeding operations and disking in the fallow year. In the production of hay and grass seed, tillage is a minor component. However, most of the other crops produced in the Pacific Northwest use some form of tillage to prepare seed beds and control weeds.

MANAGEMENT EFFECTS ON SOIL C STORAGE: DRYLAND SYSTEMS

Conversion of Native Ecosystems to Agriculture

Soil disturbances resulting from land-use changes have been shown to modify the turnover of C and the formation of soil organic matter (SOM) (Huggins et al., 1998Collins; et al., 2000; Dinesh et al., 2003). Conversion of native ecosystems to agricultural production usually results in a loss of SOM. Such losses are well documented for the Great Plains and Corn Belt (Paustian et al., 1997), with few examples in the Pacific Northwest (Rasmussen et al., 1980; Rasmussen and Collins, 1991).

Conversion of native lands to agricultural production in the Pacific Northwest resulted in SOC loss of 35% during the first 40 years of cultivation (Sievers and Holtz, 1926). Purakayastha et al. (2008) reported a 56% reduction in SOC during 100 years of conventional tillage. The rate of profile (0 to 125 cm depth) SOC decline following conversion of native vegetation to cropland ranged from 0.53 to 1.07 Mg C ha^{-1} yr^{-1} in the higher and 0.33 to 0.76 Mg C ha^{-1} yr^{-1} in the intermediate precipitation zones of the PNW (Brown and Huggins, 2011). In the low precipitation region of the Pacific Northwest, profile SOC declined by 0.16 to 1.15 Mg C ha^{-1} yr^{-1}, but these data represent a more recent history of cultivation (3 to 9 years) compared to the high and intermediate precipitation regions (35 to 100 years). The 0.24 Mg C ha^{-1} yr^{-1} SOC gain observed by Rodman (1988) in a footslope position profile is indicative of the role erosion processes play in SOC content and distribution. The large range in SOC changes reflect not only the differences in biomass production and agricultural management, but also various soil depths sampled, time under cultivation, erosion history, and climatic conditions that exist within the Pacific Northwest. This emphasizes the need for a more standardized protocol as most of the data ranges reported here were not sampled specifically for evaluating SOC stocks.

Conversion of Conventional Tillage (CT) to No-Tillage (NT) or Reduced Tillage (RT)

Increasing concern about the sustainability of crop production systems and environmental quality has emphasized the need to develop and implement management strategies that maintain and protect soil, water, and air resources. Tillage affects the amount of SOM buildup in two fundamental ways: (1) through the physical disturbance and mixing of soil and the exposure of soil aggregates to disruptive forces and (2) through controlling the incorporation and distribution of plant residues in the soil profile. Adoption of NT following CT generally results in SOC increases, but in some instances little or no increases are observed. In the Pacific Northwest, increases in profile SOC (20 cm) ranged from 0.03 to 1.95 Mg C ha^{-1} yr^{-1} and averages for different agroclimatic zones ranged from 0.21 to 0.71 Mg C ha^{-1} yr^{-1} (Brown and Huggins, 2011). These values are within the range of 0.3 to 0.8 Mg C ha^{-1} yr^{-1} reported previously (Follett, 2001; West and Post, 2002; Liebig et al., 2005) from comparison of NT and CT cropland. Purakayastha et al. (2008) found that SOC (0 to 20 cm) from an unreplicated field survey was significantly higher under NT compared to CT, but no significant differences between 4 and 28 years of NT were observed. However, particulate organic C (POC) in the surface 5 cm was significantly higher in the 28-year NT site (8.1 Mg C ha^{-1}) compared to the 4-year NT site (6.3 Mg C ha^{-1}). In the surface 15 to 20 cm, changes in SOC following change from CT to RT ranged from gains of 0.02 to 0.07 Mg C ha^{-1} yr^{-1} to losses of 0.08 Mg C ha^{-1} yr^{-1}. Increased SOC storage in the fine organic matter (FOM) fraction of RT soils (i.e. sweep) compared to moldboard plowed soils ranged from 0.16 to

0.18 Mg C ha^{-1} yr^{-1} at N fertilizer rates of 45 and 180 kg N ha^{-1}, respectively, in a 60 cm profile under long-term (44 years) wheat–fallow system (Gollany et al., 2005). However, tillage had a larger effect on FOM than N fertilization (Gollany et al., 2005). On a profile SOC basis, the 0.05 Mg C ha^{-1} yr^{-1} rate of SOC increase under RT estimated by Liebig et al. (2005) for the northwestern U.S. appears reasonable for the dryland Pacific Northwest. It is also important to note that in a few instances NT was reported to have less (Fuentes et al., 2004; Kennedy and Schillinger, 2006) and RT more SOC than the CT counterpart (Rasmussen and Rhode, 1988; Machado et al., 2006). This can be explained by residue burial to a greater depth due to tillage and shows that limited depth of soil sampling could result in over- or underestimation of SOC sequestration potential from changes in tillage management. Machado et al. (2006) in eastern Oregon found that tillage influenced the amount and distribution of SOC in the soil profile. Compared to historical records (1931), CT reduced SOC in all layers except the 10–20 cm depth increment, where plowing deposited the majority of crop residue. Under NT, SOC concentration closely mirrored a grass pasture system, with the highest SOC in the 0–10 cm layer and SOC decreasing with soil depth. Kennedy and Schillinger (2006) reported depth and landscape differences in SOC response to NT compared to traditional tillage (TT). The SOC content of NT soils was similar to TT at summit positions, but greater than TT in the surface 0 to 5 cm at toe and side slope landscape positions. This trend was not always observed in the 5 to 10 cm depth increment (Kennedy and Schillinger, 2006).

In central Montana, Sainju et al. (2006a) reported that NT increased SOC at the 0–5 cm depth by 0.2 Mg ha^{-1} yr^{-1} compared with CT after 6 years. They found that SOC at 0 to 20 cm increased by 0.6 Mg ha^{-1} in NT, but was reduced by 0.8 Mg ha^{-1} in CT from 1998 to 2003. In eastern Montana, Aase and Pikul (1995) observed that SOC declined at a rate of 0.4 Mg ha^{-1} yr^{-1} in CT spring wheat–fallow compared with a decline of 0.1 Mg ha^{-1} yr^{-1} in NT continuous spring wheat from 1983 to 1993. Sainju et al. (2007b) found that SOC at 0–20 cm was 8.4 Mg ha^{-1} greater in NT continuous spring wheat than in CT spring wheat–fallow after 21 years. They reported that SOC declined more in CT spring wheat–fallow than in NT continuous spring wheat from 1983 to 2004. When total (organic + inorganic) C was considered, C storage increased by 0.19 Mg ha^{-1} yr^{-1} in NT continuous spring wheat, but decreased by 0.27 Mg ha^{-1} yr^{-1} in CT spring wheat–fallow during the same period. They observed greater increases in POC, potential C mineralization, and microbial biomass C relative to SOC in NT continuous spring wheat than in CT spring wheat–fallow. However, SOC was not altered by tillage and cropping sequence after 3 years of study (Sainju et al., 2008, 2010).

In the central Great Plains (Akron, CO), on Weld silt loam soil (Aridic Paleustolls), NT increased SOC at the 0–5 cm depth by 0.2 Mg ha^{-1} compared with CT after 15 years of winter wheat–fallow rotation (Mikha et al., 2010). In NT, fallow elimination significantly increased SOC by 0.3 Mg ha^{-1} at 0–5 cm and by 0.24 Mg ha^{-1} at 0–20 cm, with wheat–corn–millet rotation compared with wheat–fallow rotation. The retention of SOC declined as the fallow frequency increased in rotation. Fabrizzi et al. (2007) evaluated four long-term sites (Table 5.1) in Kansas (Tribune, Hays, Manhattan, and Parsons) managed under variable fertilizer rates and tillage regimes including CT, RT and NT. They found the greatest increase in SOC at 0 to 5 cm for N-fertilized NT treatment. At 0 to 15 and 0 to 30 cm, no significant differences in SOC were

TABLE 5.1 Description of the Experimental Sites Reported by Fabrizzi et al. (2007)

Site	Location	Soil series/type	Year after initiation
Tribune	Southwest	Richfield silt loam (Aridic Argiustolls)	16
Hays	North central	Harney silt loam soil (Typic Argiustoll)	37
Manhattan	Northeast	Muir silt loam (Cumulic Haplustoll)	29
Parsons	Southeast	Parsons silt loam (Mollic Albaqualfs)	20

observed between tillage systems, except at the Manhattan site, although in most cases NT tended to result in greater SOC content. The SOC levels in CT and RT at Hays indicated a loss of C from the system and a gain in NT (0.02 Mg C ha^{-1} yr^{-1}). Previous land use was important when evaluating the effect of management on soil C. For example at Tribune, all tillage systems had a negative effect on C sequestration, but C loss was lower in NT (4%) than in either RT (7%) or CT (9%). This experiment was initiated on a site broken from native prairie sod, and had higher initial soil C concentrations than the other sites, which had been under cultivation before the establishment of the tillage systems. At the Parsons site, 20 years of N addition at 140 kg N ha^{-1} increased C sequestration by 0.11 Mg C ha^{-1} yr^{-1} relative to 0 N rate (Fabrizzi et al., 2007).

Previous research reported that NT practices improve soil aggregation, C sequestration, and aggregate stability (Mikha and Rice, 2004; McVay et al., 2006; Zhang et al., 2007) which results in increased water infiltration and resistance to wind and water erosion (Zhang et al., 2007). Macroaggregate stability (>250 μm diam.) is particularly responsive to management practices (Jiao et al., 2006; Zibilske and Bradford, 2007). The loss of macroaggregate-occluded organic matter is a primary source of carbon lost as a result of changes in management practices (Six et al., 2002; Mikha and Rice 2004; Jiao et al., 2006; Zibilske and Bradford, 2007). Continuous cropping with reduced fallow frequency and NT exhibits a positive effect on macroaggregate formation and stabilization as well as particulate organic matter (POM) and SOC (Mikha et al., 2010).

Fallow and Crop Rotations

Historical SOC losses from fallow cropping every other year have ranged from 0.09 to 0.65 Mg C ha^{-1} yr^{-1} within the surface 30 cm (Horner et al., 1960) but more recent estimates exhibit a narrower range of SOC loss (0.09 to 0.12 Mg C ha^{-1} yr^{-1}) (Rasmussen and Albrecht, 1998). Schillinger et al. (2007) observed that conversion from CT wheat–fallow to annual NT continuous wheat increased SOC to near native values in the top 5 cm but was not different in the surface 10 cm layer among NT cropping systems with various annual crop rotations in wheat–fallow regions of Washington state. Horner et al. (1960) reported surface SOC changes under annual wheat cropping with tillage to range from a loss of 0.25 Mg C ha^{-1} yr^{-1} to a gain of 0.04 Mg C ha^{-1} yr^{-1} and wheat in rotation with spring pea or oats to exhibit a relatively larger potential for SOC losses (0.07 to 0.47 Mg C ha^{-1} yr^{-1}). Greater SOC loss has been observed in unfertilized compared to N fertilized conditions but the rate and direction of change were site specific depending also on duration of residue and crop rotation management (Horner et al., 1960; Gollany et al., 2006). Gollany et al. (2006) and Wuest et al. (2005) reported that long-term (75-year) application of organic amendments (manure management and pea vine residue) on a Walla Walla silt loam (Typic Haploxeroll) was more effective in increasing the surface 60 cm SOC stocks than N fertilization or residue management for wheat–fallow plots under moldboard plowing. This was mostly due to the significant increases in FOM in the surface Ap horizon (0 to 26 cm).

Recent profile SOC data addressing the impact of current Pacific Northwest crop rotations, alternative cropping system practices and fertilization are lacking for many of the dryland cropping systems. In central Montana, Sainju et al. (2006b) found that cropping sequence did not influence SOC but POC at 0 to 5 cm was 0.6 to 0.8 Mg ha^{-1} greater in spring wheat–spring wheat–fallow and spring wheat–pea–fallow than in spring wheat–fallow after 6 years. Similarly, microbial biomass C (MBC) at 0 to 5 cm was 26% greater in spring wheat–spring wheat–fallow than in spring wheat–fallow (Sainju et al., 2007a). In eastern Montana, Sainju et al. (2011) reported that SOC at 0–5 cm was greater in continuous spring wheat than in spring wheat–barley hay–pea in both NT and CT after 4 years, probably a result of lower biomass of pea than spring wheat and removal of aboveground biomass of barley for hay.

In the central Great Plains, Akron, CO, on Weld silt loam soil (Aridic Paleustolls), Mikha et al. (2010) found that continuous cropping with reduced fallow significantly increased SOC compared with rotations that included fallow every second and third year. Including a legume in rotation (wheat–corn–millet–pea) did not increase SOC relative to summer fallow rotations in NT (Mikha et al., 2010). The low SOC accumulation with pea in rotation could be due to fast decomposition rate of pea residue, relative to other crops in rotation. After 15 years, MBC at 0 to 5 cm was 42% greater in continuous cropping (wheat–corn–millet) than in wheat–corn–fallow and 72% greater than wheat–fallow (Acosta-Martínez et al., 2007). Continuous cropping (wheat–corn–millet) significantly increased soil particulate organic matter (POM) at 0 to 5 cm depth by 29% compared to wheat–corn–fallow and by 21% compared with wheat–corn–millet–fallow. This indicates that the differences in soil POM among cropping sequences increased as fallow frequency increased in rotations (Mikha et al., 2010). This observation is consistent with the regional study, eight long-term sites throughout the Great Plains and the western Corn Belt (Table 5.2), reported by Mikha et al. (2006), who observed an increase in soil POM level with the reduction in tillage intensity and fallow frequency.

At the study site in Mandan, ND, on a Temvik-Wilton silt loam soil (Typic and Pachic Haplustolls), Halvorson et al. (2002a) reported that 12 years of fallow occurrence in cropping systems, even with NT management, could result in losses of SOC. Similarly, Mikha et al. (2006) also reported that decreased fallow frequency causes an increase in SOC, within the 0 to 7.5 cm depth. Overall, SOC loss is likely to occur during the fallow period (Cihacek and Ulmer, 1995) because fallow period signifies a phase of continued microbial activity and residue decomposition with no crop residue input. In addition, soil may be susceptible to SOC loss by wind erosion during the fallow period (Haas et al., 1975). The relative difference in SOC between the continuous cropping (wheat–corn–millet) and wheat–fallow cropping systems, Weld silt loam soil at Akron, CO, increased during the years between 1999 and 2005. Bowman et al. (1999) reported approximately 13% greater SOC, in the 0–15 cm depth, was associated with continuous rotation (wheat–corn–millet) compared with the wheat–fallow cropping systems and no differences in SOC among different cropping intensities below the 15 cm soil depth. However, Benjamin et al. (2008) reported a 30% increase of SOC with continuous cropping system (wheat–corn–millet) compared with the wheat–fallow rotation averaged over the 37 cm soil depth. They also found differences in SOC between wheat-–corn–millet and wheat–fallow rotation at the 20 to 28 cm depth. The changes in SOC over time indicate that a fallow elimination continues to add SOC to the soil system and that the SOC is translocated deeper in the soil (Benjamin et al., 2008).

Compared to annual cropping systems, the inclusion of a perennial crop into an otherwise annual crop rotation (mixed perennial–annual rotation) was estimated to increase profile SOC stocks (0 to 92 cm layer) by 0.66 to 1.97 Mg C ha^{-1} yr^{-1} in higher precipitation regions of

TABLE 5.2 **Site location and soil series from the regional study, throughout the Great Plains and the western Corn Belt, reported by Mikha et al. (2006)**

Site location	Soil series
Akron, CO	Weld silt loam
Brookings, SD	Barnes sandy clay loam
Bushland, TX	Pullman silty clay loam
Fargo, ND	Fargo silty clay
Mandan, ND	Wilton silt loam
Mead, NE	Sharpsburg silty clay loam
Sidney, MT	Willams loam
Swift Current, SK	Swinton silt loam

the dryland Pacific Northwest (Brown and Huggins, 2011). Liebig et al. (2005) previously estimated a 0.94 ± 0.86 Mg C ha^{-1} yr^{-1} regional rate of SOC gain for conversion of cropland or reclaimed mining land to grass. Sainju and Lenssen (2011) reported that perennial legume forages, such as alfalfa, increased SOC by 2.2 to 5.2 Mg ha^{-1} and POC by 1.1 to 1.2 Mg ha^{-1} at 0–15 cm compared with durum–annual forage sequences after 4 years in eastern Montana. The corresponding increase in potential C mineralization at 0 to 120 cm was 46 to 84 kg ha^{-1} and MBC at 0 to 15 cm was 6 to 80 kg ha^{-1}. They attributed these to be a result of increased belowground biomass (root + rhizodeposit) C and a relatively undisturbed soil condition in alfalfa compared with durum–annual forages.

Potter (2006) compared soils collected from several perennial and annual cropping systems in central Texas on Houston Black Series (Udic Haplusterts) to archived (1949) soil samples. Tilled soils had lost nearly 60% of SOC in the surface 30 cm due to long-term agricultural practices compared to the adjacent native grasslands. Restoring the sites to a perennial coastal Bermuda grass (*Cynondon dactylon* (L.) Pers.) sequestered 13 to 20 Mg ha^{-1} after 39 and 55 years of cropping, respectively, averaging 0.36 Mg C ha^{-1} yr^{-1}. Two other sites evaluated had been in continuous row cropping of small grains, sorghum, corn and cotton since 1949. After 55 years of continuous cropping C sequestered in the surface 30 cm averaged 0.14 Mg C ha^{-1} yr^{-1}, roughly 40% of the C sequestered in the perennial Bermuda grass. In a related study, Potter and Derner (2006) showed that SOC in agricultural soils had a 40 to 60% reduction compared to native grassland soils. The native grassland soils contained 13.7 to 18.4 Mg C ha^{-1} in the surface 5 cm, while the agricultural soils ranged from 7.3 to 8.6 Mg C ha^{-1}. Particulate organic carbon comprised 15 to 18% of SOC in the 10 to 40 cm depth increment in the native grassland soils and 7 to 16% of SOC in the agricultural soils.

Conservation Reserve Program

The Conservation Reserve Program (CRP) is a voluntary USDA program of the U.S. Farm Bill that encourages farmers to convert highly erodible cropland or other environmentally sensitive areas to conservation vegetation, such as introduced or native grasses, trees, filter strips, or riparian buffers. Follett et al. (2001) reported C sequestration under CRP in areas of summer rainfall to range between 0.6 and 1.0 Mg C ha^{-1} yr^{-1}. In NE, KS, and TX, Gebhart et al. (1994) found that SOC accumulated at a rate of 1.1 Mg C ha^{-1} yr^{-1} in 2 m profile. Purakayastha et al. (2008) noted SOC gains of 0.35 and 0.03 Mg C ha^{-1} yr^{-1} at 0–5 cm and 5–10 cm, respectively, after 11 years in CRP management. In addition, Sanchez de-Leon (2007) observed that after 23 years in CRP, the SOC content of CRP ground remained 2.4, 4.1, and 3.4 Mg C ha^{-1} below those of native Palouse prairie in the 0 to 10, 10 to 20, and 20 to 30 cm depth increments, respectively. Lee et al. (2007) found that the amount of C sequestered was dependent upon the type of N fertilizer applied. They reported a C sequestration potential of 2.4 Mg C ha^{-1} yr^{-1} with NH_4NO_3 fertilizers and 4.0 Mg C ha^{-1} yr^{-1} for manure N in a 0.9 m profile under CRP lands in South Dakota.

Machado et al. (2006) in eastern Oregon found that after 73 years, grassland pasture in NT with large annual additions of grass residue had higher (82 Mg ha^{-1}) SOC concentrations than CT winter wheat fallow (50 Mg ha^{-1}) or for fertilized CT continuous crop winter wheat systems (75 Mg ha^{-1}). Bronson et al. (2004) in soils of the Southern High Plains found that SOC and total N content in CRP land was greater than that of nearby cropland soils in the 0 to 5 cm layer. They reported SOC in CRP soils averaged 5.7 Mg ha^{-1} compared to 3.4 Mg ha^{-1} for irrigated cotton soil. Below 5 cm, they found no differences in SOC storage.

In Wyoming, Reeder et al. (1998) estimated a 60–75% decrease in SOC of the 0 to 20 cm depth after 6 years following cultivation of a native grass land due to tillage. They also observed a significant increase in SOC (~0.4 Mg C ha^{-1} yr^{-1}) after 4 years of reestablishing a fertilized grassland (34 kg N ha^{-1}) after 60 years cropping. In central Kansas, Huang et al. (2002) reported an increase in SOC (~0.14 Mg C ha^{-1} yr^{-1}) at the 0–5 cm depth after 10 years of grass on CRP lands compared with adjacent cropland. However, as the CRP lands were converted to CT croplands, the surface 0–5 cm depth lost about 1.5 Mg C ha^{-1} where the 5–10 cm depth gained about 1.1 Mg C ha^{-1}. In eastern Washington, after 4 to 7 years of CRP establishment, Staben et al. (1997) observed no differences in SOC between CRP land and a wheat–fallow rotation at the 0 to 7.5 cm depth. However, the potential mineralizable C pool was 55% greater with CRP than wheat–fallow soils, which was probably due to buildup of higher quality SOM that increased the potential for microbial activity.

MANAGEMENT EFFECTS ON SOIL C STORAGE: IRRIGATED SYSTEMS

In arid and semiarid environments of the western U.S., crop production requires irrigation to increase plant production to the point where cropping becomes economically viable (Lal et al., 1998). Irrigated crops produce twice as much plant biomass as rain-fed crops (Bucks et al., 1990; Howell, 2000). Irrigation increases C input to soils by producing higher crop yields that contribute to stable SOC through greater inputs from plant residues and root systems and increased aggregate formation (De Gryze et al., 2005; Kong et al., 2005; Gillabel et al., 2007). Estimates of SOC accumulation resulting from irrigation in the western U.S. range between 0.25 and 0.52 Mg C ha^{-1} yr^{-1} (Eve et al., 2002). The increase in SOC in irrigated agricultural soils over native soils is contrary to responses in rain-fed agricultural soils (Paustian et al., 1997; Entry et al., 2002; Cochran et al., 2007). Semiarid shrub–steppe ecosystems differ from other native systems (forests, permanent grasslands or native prairies) because of the relatively small amount of annual precipitation and historically lower levels of C inputs and storage in soil (Entry et al., 2002). The effect of irrigation on soil C and N dynamics has been quantified for only a few agricultural systems (Leuking and Schepers, 1985; Entry et al., 2002; Gillabel et al., 2007).

Intensively managed irrigated crop or pasture lands have the potential for C gain through the use of improved grazing regimes, improved fertilization practices and irrigation management (Follett, 2001; Bruce et al., 1999). Entry et al. (2002) found that irrigated soils in southern Idaho under different agricultural practices (pasture and conservation tillage) exhibited significant increases in SOC over the native sagebrush ecosystem. They estimated a gain of 9.5 Mg C ha^{-1} over 30 years if the land currently managed with conventional tillage adopted conservation approaches. Carbon sequestration is expected to increase if efficient water use allows the expansion of irrigated agriculture. Land-use shifts from arid native vegetation could sequester 8.0 Mg C ha^{-1} and assuming 10% expansion of irrigated agriculture, 7.2×10^6 Mg C (0.01% of the total C emitted in the next 30 years) could potentially be sequestered in Pacific Northwestern soils.

Cochran et al. (2007) reported that converting the native shrub–steppe to irrigated organic vegetable production (sweet corn–pea rotation) in the Columbia Basin of eastern Washington increased SOC 0.7 g C kg^{-1} soil above the native soil on an Adkins silt loam (Xerollic Camborthid) after the first year of cropping. After 3 years of cropping and the incorporation of crop residues (10 Mg ha^{-1}) and stable compost (34 Mg ha^{-1}) SOC increased to 6.4 g g^{-1}, 2.1 g C kg^{-1} soil above the native soil in the surface 20 cm layer. The increase in SOC was attributed to the resistant nature of the added compost. Acid hydrolysis of the compost showed that 73% of compost C comprised the resistant fraction. This fraction was composed of aromatic humics and lignin which are slow to decompose. Laboratory incubation of the compost showed that 4.3% of the total C was mineralized, suggesting that 0.4 Mg C ha^{-1} could be lost during the growing season though decomposition. Collins et al., (2010) found that 18 years of cultivation of an irrigated Quincy sand (Xeric Torripsamment) in eastern WA consisting of potato-based rotations that included corn and wheat increased soil C 30% above the native SOC (3.0 g kg^{-1}).

Soils developed under native vegetation in the Sandhills of Nebraska with low values of SOC showed increased C concentrations after 15 years of irrigation while soils with higher initial C and N concentrations decreased only slightly (Lueking and Schepers, 1985). Gillabel et al. (2007) found that irrigation increased C inputs compared with no irrigation in semiarid southwestern Nebraska. They found that SOC content in the top 20 cm was almost 25% higher under irrigation (33.0 Mg C ha^{-1}) than under dryland production (26.6 Mg C ha^{-1}). Further, they reported an average C accumulation rate of 0.19 Mg ha^{-1} yr^{-1} over 33 years for the 0 to 20 cm layer under irrigation compared with no irrigation. This was at the lower end of estimates by Eve et al. (2002) but 20% higher than the values estimated by Lal et al. (1998).

King et al. (2009) reported that furrow irrigation in a California field had a positive effect on SOC storage, showing a net increase of 0.03 Mg C ha^{-1} yr^{-1}. This increase was in line with current estimates of C sequestration in global agricultural systems that range between 0.03 and 0.65 Mg C ha^{-1} yr^{-1} (Christopher and Lal, 2007; Hutchinson et al., 2007; Izaurralde et al., 2007). Poch et al. (2006) found similar results on SOC storage (0.02 Mg C ha^{-1} yr^{-1}) for a furrow-irrigated corn field in the central valley of CA. These C gains were 20% lower than the estimated yearly C sequestration rates for other long-term experiments (Eve et al., 2002). Wu et al. (2008) found that SOC concentrations were nearly 40% greater in an irrigated San Joaquin Valley soil above the C stored in native soil after 55 years of irrigated farming. After 90 years of irrigation, SOC increased 70% above the native SOC. The soil inorganic C (SIC) concentrations to a depth of 1 m showed an opposite trend: a decrease after 55 years of irrigation in the San Joaquin Valley but an increase in Imperial Valley after 85 years.

Major shifts in crop production will occur as farmers gear up to supply the demand for biomass feedstocks supporting biofuel production. These shifts will change agroecosystem services related to water use, carbon use and storage, nutrient cycling and greenhouse gas (GHG) emissions that have direct consequences on air, water, and soil quality. Perennial herbaceous crops such as switchgrass (*Panicum virgatum* L.) are important sources of cellulosic biomass for the bioenergy industry. Removal of this biomass may adversely affect C dynamics in soil. Collins et al. (2010) used the natural ^{13}C abundance of soils to calculate the quantity and turnover of C inputs in irrigated fields cropped to switchgrass. Three years of irrigated switchgrass on a Quincy sand soil (Xerollic Torripsamment) showed a 20% increase in SOC in the 0 to 15 cm depth increment with no significant change below 15 cm. The average accrual rate of switchgrass-derived SOC was estimated at 0.5 Mg ha^{-1} yr^{-1}. Mean residence time (turnover) for the active (C_a) and slow (C_s) C pools averaged 27 days and 2.4 years, respectively. Estimates of the mean residence time of the native C_3—C under the irrigated C_4 monocultures of switchgrass were greater than 60 years in the 0—15 cm and 30—55 years in the 15 to 30 cm depth increments. Liebig et al. (2008) also found accrual rates of 1.1 Mg C ha^{-1} yr^{-1} with much of that increase occurring in the surface 30 cm. Zan et al. (2001) found that switchgrass increased SOC by 3 Mg ha^{-1} yr^{-1} when compared to a corn field after 4 years of production. They further reported that corn biomass C contributed 0.8 Mg ha^{-1} yr^{-1} to SOC.

GAPS IN KNOWLEDGE

Long-term research designed to quantify soil organic C content under different management practices and agroecological conditions is limited. Rates of SOC sequestration may peak within 5 to 10 years and approach a new steady state 20 to 50 years following a management change (West and Post, 2002) or until the soil storage capacity is reached (Lal, 2004). Establishment of long-term sites with major agricultural systems and alternative management to business as usual for the diverse cropping systems of the western U.S. is needed. Overall it appears that (1) SOC databases are lacking for many areas of western U.S. dryland cropping systems; (2) baseline sampling of SOC prior to management practice is largely nonexistent; (3) soil erosion processes having large impact on SOC have not been quantified; and (4) inconsistent sampling methodologies and analyses have been used, thereby contributing to large variability in potential SOC sequestration rates (Brown and Huggins, 2011).

Sampling Methodology

Sampling biases can arise if disparate soil sampling occurs, such as sampling after a recent addition of biomass from residues after harvest. Crop residues can mix with soil sample that might overestimate SOC concentration. We suggest that soil sampling should take place prior to C inputs from crop residues. In the northern Great Plains, spring soil sampling, however, may not be feasible due to limited time available for planting crops. If a fall sampling is used for C analysis, it is suggested that soil sampling be done consistently after crop harvest in the fall every year to reduce any seasonal variation.

Depth of Sampling

Changes in SOC following a shift in agricultural management practice are often more pronounced at the surface than the subsurface layer (Follett, 2001; West and Post, 2002). However, changes occurring in the subsurface layer and the whole soil profile (minimum of 150 cm) need to be accounted for. The depth of soil sampling can have a major effect on changes in soil C stock among treatments (Franzluebbers, 2010). VandenBygaart et al. (2011) reported that soils should be sampled at different depths for evaluating the effects of management practices on SOC in agroecosystem experiments. They recommended that for evaluating the effect of fallow on SOC, soils should be sampled to a depth of 15 cm, for tillage to a depth of 30 cm, and for perennial crops to a depth of 45 cm. This is because of the differences in statistical significance observed among various management practices at various depths. For example, incorporation of crop residue to a depth below 15 cm due to tillage results in greater C storage in the subsoil layer. Similarly, greater root biomass of perennial than annual crops results in greater SOC below 30 cm. Kravchenko and Robertson (2011) and Syswerda et al. (2011) found that most of the significant differences in SOC among management practices occurred at the surface compared with subsurface soil layers or the whole soil profile due to high variability in the subsurface layers. As a result, they suggested that treatment effects on SOC should be evaluated at the individual soil depth rather than the whole soil profile.

Labile Carbon

Carbon stored in undecomposed and partially decomposed crop residues has largely been ignored when assessing soil C sequestration. In fact, attempts have been usually made to remove labile C, as it is assumed that these residues will quickly decompose and not contribute to stored SOC. However, observations of long-term NT fields suggest a different reality. Annual additions of surface residue and roots in NT management can result in the accumulation of a mulch layer that might influence SOC. The CropSyst model was used to simulate differences in C storage between tillage and cropping practices (C. Stockle, Washington State University, Personal Communication). Model results showed that SOC increased in 5 to 6 years in NT due to addition of crop residue and roots. For example, the model predicted an increase of 1.0 Mg C ha^{-1} in NT compared to CT in St. John and Pullman, WA. The size of this residue C was influenced by tillage crop rotations. For example, near Pullman, increases in residue C were predicted to be 1.2 Mg C ha^{-1} under continuous cereals compared to 0.7 Mg C ha^{-1} under peas every 3 years. Although labile C pools are sensitive to annual changes in C inputs, they represent a new steady state of C stocks that were absent under tillage.

Some of the parameters used to measure changes in SOC and biological soil quality are particulate organic C (POC), microbial biomass C (MBC), and potential C mineralization (PCM). Since SOC has a large pool size and inherent spatial variability, it changes slowly with management practices (Franzluebbers et al., 1995). As a result, measurement of SOC alone does not adequately reflect changes in soil quality and nutrient status (Franzluebbers et al., 1995; Bezdicek et al., 1996). Measurement of biologically active fractions of SOC, such as MBC and PCM that change rapidly with time, could better reflect changes in soil quality and productivity, because they alter nutrient dynamics due to immobilization–mineralization (Saffigna et al., 1989; Bremner and Van Kissel, 1992). These fractions can provide an assessment of soil organic matter changes induced by management practices, such as tillage and cropping systems (Campbell et al., 1989). Similarly, POC has been considered as an intermediate fraction of SOC between active and slow fractions that changes due to management practices (Cambardella and Elliott, 1992). The POC also provides substrates for microorganisms and influences soil aggregation (Six et al., 1999).

Inorganic Carbon

Studies on soil C sequestration have been focused mostly on changes in SOC. However, in dryland cropping systems, soil inorganic C (SIC) can dominate C storage more than SOC at the subsoil layers. In eastern Montana, observations have been noted about increasing SIC and decreasing SOC with depth (to 120 cm) under dryland cropping systems (unpublished data). Little is known about the effects of management practices on SIC storage.

Sainju et al. (2007a) reported greater SIC at 0 to 20 cm depth in NT continuous spring wheat than in CT spring wheat–fallow in eastern Montana. They speculated that surface placement of fertilizer and lower soil water content probably increased SIC in NT continuous spring wheat compared with CT spring wheat–fallow. Cropping can increase the formation of $CaCO_3$ compared with fallow by increasing plant and microbial activities, thereby increasing SIC (Cerling, 1984; Monger, 2002). Addition of soil amendments, such as fertilizer containing Ca that could lead to increased formation of $CaCO_3$, can also increase SIC (Amundson and Lund, 1987; Mikhailova and Post, 2006). In contrast, Cihacek and Ulmer (2002) found that SIC at the surface soil was greater in CT than in NT soils in the northern Great Plains.

Because of the higher concentration of SIC, especially at lower depths, under dryland cropping systems, soil total (organic + inorganic) C instead of SOC alone may be needed to evaluate the effects of management practices on soil C storage. Further studies are needed to examine if soil total C instead of SOC could be used for measuring changes in soil C storage as affected by management practices, especially under dryland cropping systems.

Erosion

Most comparative studies of management impacts (e.g. tillage, crop rotation) on SOC budgets emphasize differences in biological processes (e.g. primary production, decomposition) that operate at the pedon scale. In many situations, however, historic and current soil erosion processes have likely increased variability in profile and landscape-scale SOC distribution (Pennock et al., 1994). These are increasingly recognized as major determinants of field-scale SOC budgets (De Jong and Kachanoski, 1988; Starr et al., 1999; Fang et al., 2006; Chaplot et al., 2009). Erosion processes may be as important as other ecosystem processes controlling C flux, such as tillage and microbial decomposition (Christopher and Lal 2007; Izaurralde et al., 2007). Van Oost et al. (2007) also suggested that erosion may also lead to the stabilization of C in depositional sites through burial. Erosion-induced deposition and buried C may amount from 0.4 to 0.6 $Gt\,C\,yr^{-1}$ (Lal, 2004).

Conversion of native vegetation to cropland in the Palouse region of the Pacific Northwest has resulted in historical soil erosion rates of over 25 $Mg\,ha^{-1}\,yr^{-1}$. This may have removed 100% of the topsoil from 10% of the landscape, with another 25 to 75% loss from 60% of agricultural land (USDA, 1978). Purakayastha et al. (2008) estimated that 50 to 70% of SOC had been lost from upland soils. Not often recognized or measured, however, is the effect of soil erosion processes on within-field soil deposition and likely site-specific SOC levels.

Soil erosion is a selective process and transported sediments are enriched in clay-sized particles (Ongley et al., 1981) and particulate and dissolved organic C (Lal, 1995). Erosion-related processes laterally redistribute SOC, thereby removing and/or depositing SOC from different field locations. This can be as important as biological processes for determining SOC stocks at a specific site. Huggins et al. (2011) concluded that historical soil erosion processes in dryland farming regions of the inland Pacific Northwest challenge the common expectation that SOC will be linearly related to C inputs (Larson et al., 1972; Rasmussen and Collins, 1991; Huggins et al., 1998). In this study, the close proximity of areas with low and high surface to subsoil SOC ratios indicated a coupling of soil detachment, transport, and deposition within the field that contributed to the large within-field variability of SOC stocks.

Redistribution of SOC due to erosion processes (historical or contemporary) has important implications for methods used to assess changes in SOC due to management practice. Baseline sampling at the initiation of the experiment is required to evaluate treatment-induced changes in SOC over time at a given location. Soil sampling must occur throughout the soil profile at targeted depth increments to capture all SOC changes, along with associated bulk density to express SOC on a mass per unit volume basis. Furthermore, the importance of sampling at various depths that extend beyond the surface is important for evaluating changes in SOC when shifting tillage from CT to NT in eroded soils (Huggins et al., 2007; Baker et al., 2007; Huggins et al., 2011). Coupling baseline and depth-increment sampling representing landscapes over time will enable field-scale assessment of treatment effects on SOC that include biological as well as physical processes (VandenBygaart et al., 2002, 2006; Huggins et al., 2011). The sampling methodologies described will lead to greater understanding of field-scale variations of SOC that arise from the interaction of biophysical processes (e.g. C inputs from crop residues and roots; decomposition; soil erosion driven by water, wind and tillage).This will be important to quantifying SOC sequestration, for developing sophisticated SOC models, and for promoting improved land use and management decisions for precision conservation practices.

SYNTHESIS

For decades, predominant dryland cropping systems in the Great Plains of western U.S. have been winter wheat–summer fallow management in the CT system. The system promoted SOC decomposition and soil losses through erosion. It has been well documented that tillage (1) enhances residue decomposition by incorporating the residue into the soil, (2) exposes previously protected SOC to soil fauna by destroying soil aggregates, and (3) increases soil losses due to wind and water erosion. Adopting NT increases residue accumulation and surface SOC due to less soil disturbance, less residue incorporation and oxidation, and decreased risks of soil erosion, in addition to improving soil water content during the fallow period compared with the CT system. Although only partly adopted in the western U.S., NT has shown improved soil water conservation and increased SOC storage. Increased cropping intensity and crop rotation that includes perennial crops combined with reduced tillage and fallow periods can increase the amount of crop residue returned to the soil, increase SOC content, and reduce the potential for soil erosion.

Irrigation can increase crop yields and economic viability of agriculture in arid and semiarid environments where plant growth is limited by available water. Irrigation also increases C input to soils via increased litter and root production. However, the potential of irrigation to cause a net increase of C storage is tempered by C loss as CO_2 emitted to the atmosphere as a result of (1) fertilizer manufacture, storage, transport, and application, (2) pumping irrigation water, (3) farm operations such as tillage and planting, (4) dissolved carbonate in irrigation water, and (5) increased C mineralization from soil organic matter and crop residue due to increased microbial activity as a result of greater soil water content. The SOC levels, however, will be determined by a balance between C inputs from crop residues and soil amendments and rate of C mineralization. Intensively managed irrigated crop or pasture lands have the potential for C gain through the use of improved grazing regimes, improved fertilization practices, and irrigation management. Irrigated lands produce approximately twice as much plant biomass as rain-fed agricultural production systems. A substantial reduction of atmospheric CO_2 could be attained if policy makers and agricultural experts recognize the potential benefit of land and water management strategies. Lands could be more purposely used for their greatest good, be that food production, carbon storage, native habitat, or other uses.

We recommend that future research directions to improve C storage in western agricultural systems over the next 5–10 years include:

- Increased precision farming of crops
- Greater use of buffer zones
- Increased use of no-till
- Increase in fertilizer use efficiency
- Precision irrigation and fertilizer application
- Development of organic amendments and fertilizers
- Better organic farming practices (BMPs)
- Increase in tree fruit fertilizer use efficiency

References

Aase, J.K., Pikul Jr., J.L., 1995. Crop and soil response to long-term tillage practices in the northern Great Plains. Agron. J. 87, 652–656.

Acosta-Martínez, V., Mikha, M.M., Vigil, M.V., 2007. Microbial communities and enzyme activities in soils under alternative crop rotations compared to wheat–fallow for the Central Great Plains. Applied Soil Ecol. 37, 41–52.

Amundson, R.G., Lund, L.J., 1987. The stable isotope chemistry of a native and irrigated Typic Natrargid in the San Joaquin Valley of California. Soil Sci. Soc. Am. J. 51, 761–767.

Baker, J.M., Ochsner, T.E., Venterea, R.T., Griffis, T.J., 2007. Tillage and soil carbon sequestration: what do we really know? Agric. Ecosyst. Environ 118, 1–5.

Benjamin, J.G., Mikha, M.M., Vigil, M.F., 2008. Organic carbon effects on soil physical and hydraulic properties in semiarid climate. Soil Sci. Soc. Am. J. 72, 1357–1362.

Bezdicek, D.F., Papendick, R.I., Lal, R., 1996. Introduction: importance of soil quality to health and sustainable land management. In: Doran, J.W., Jones, A.J. (Eds.), Methods of Assessing Soil Quality. SSSA Spec. Publ. 49. Soil Science Society of America, Madison, WI, pp. 1–18.

Bowman, R.A., Vigil, M.F., Nielsen, D.C., Anderson, R.L., 1999. Soil organic matter changes in intensively cropped dryland systems. Soil Sci. Soc. Am. J. 63, 186–191.

Bremner, E., Van Kessel, C., 1992. Plant-available nitrogen from lentil and wheat residues during a subsequent growing season. Soil Sci. Soc. Am. J. 56, 1155–1160.

Bronson, K.F., Zobeck, T.M., Chua, T.T., Acosta-Martinez, V., van Pelt, R.S., Booker, J.D., 2004. Carbon and nitrogen pools of southern high plains cropland and grassland soils. Soil Sci. Soc. Am. J. 68, 1695–1704.

Brown, T.T., Huggins, D.R. 2011. Accepted. Soil carbon sequestration in the dryland cropping region of the Pacific Northwest. J. Soil Water Conserv.

Bruce, J.P., Frome, M., Haites, E., Janzen, H., Lal, R., Paustian, K., 1999. Carbon sequestration in soils. J. Soil Water Conserv 59, 382–389.

Bucks, D.A., Sammis, T.W., Dickey, G.L., 1990. Irrigation for arid areas. In: Hoffman, G.J., et al. (Eds.), Management of Farm Irrigation Systems. ASAE, St. Joseph, MI, pp. 449–548.

Cambardella, C.A., Elliott, E.T., 1992. Particulate soil organic matter changes across a grassland cultivation sequence. Soil Sci. Soc. Am. J. 56, 777–783.

Campbell, C.A., Biederbeck, V.O., Schnitzer, M., Selles, F., Zentner, R.P., 1989. Effects of 6 years of zero tillage and N fertilizer management on changes in soil quality of an orthic brown chernozem in southeastern Saskatchewan. Soil Tillage Res. 14, 39–52.

Cerling, T.E., 1984. The stable isotopic composition of modern soil carbonate and its relationship to climate. Earth Planet. Sci. Lett. 71, 229–240.

Chaplot, V., Podwojewski, P., Phachomphon, K., Valentin, C., 2009. Soil erosion impact on soil organic carbon spatial variability on steep tropical slopes. Soil Sci. Soc. Am. J. 73, 769–779.

Christopher, S.F., Lal, R., 2007. Nitrogen management affects carbon sequestration in North American cropland soils. Crit. Rev. Plant Sci. 26, 45–64.

Cihacek, L.J., Ulmer, M.G., 1995. Estimated soil organic carbon losses from long-term crop-fallow in the northern Great Plains of the USA. In: Lal, R., et al. (Eds.), Advances in Ssoil Sciences: Soil Management and Greenhouse Effect. Lewis Publishers, CRC Press, Boca Raton, FL, pp. 85–92.

Cihacek, L.J., Ulmer, M.G., 2002. Effect of tillage on inorganic carbon storage in soils of the Northern Great Plains of the U.S. In: Lal, R., et al. (Eds.), Agricultural Practices and Policies for Carbon Sequestration in Soil. Lewis Publ., CRC Press, Boca Raton, FL, pp. 63–69.

Cochran, R.L., Collins, H.P., Kennedy, A.C., Bezdicek, D.F., 2007. Soil carbon pools and fluxes following land conversion in a semi-arid shrub-steppe ecosystem. Biol. Fertil. Soils 43, 479–489.

Collins, H.P., Elliott, E.T., Paustian, K., Bundy, L.G., Dick, W.A., Huggins, D.R., Smucker, A.J.M., Paul, E.A., 2000. Soil carbon pools and fluxes in long-term Corn-Belt agroecosystems. Soil Biol. Biochem. 32, 157–168.

Collins, H.P., Fransen, S., Smith, J., Alva, A.K., Kruger, C., Granatstein, D., 2010. Carbon sequestration under irrigated switchgrass (*Panicum virgatum*) production. Soil Sci. Soc. Am. J. 74, 2049–2058.

De Gryze, S., Six, J., Brits, C., Merckx, R., 2005. A quantification of short-term macroaggregate dynamics: influences of wheat residue input and texture. Soil Biol. Biochem. 37, 55–66.

De Jong, E., Kachanoski, R.G., 1988. The importance of erosion in the carbon balance of prairie soils. Can. J. Soil Sci. 68, 111–119.

Dewar, R.C., 1991. Analytical model of carbon storage in the trees, soils, and wood products of managed forests. Tree Physiol. 8, 239–258.

Entry, J.A., Emmingham, W.H., 1998. Influence of forest age on forms of carbon in Douglas-fir soils in the Oregon Coast Range. Can. J. For. Res. 28, 390–395.

Entry, J.A., Sojka, R.E., Shewmaker, G.E., 2002. Management of irrigated agriculture to increase organic carbon storage in soils. Soil Sci. Soc. Am. J. 66, 1957–1964.

Eve, M.D., Sperow, M., Howerton, K., Paustian, K., Follett, R.F., 2002. Predicted impact of management on soil carbon storage for each cropland region of the conterminous US. J. Soil Water Cons. 57, 196–204.

Fabrizzi, K.P., Rice, C.W., Schlegel, A., Sweeney, D., Peterson, D., Thompson, C., 2007. Soil carbon sequestration in Kansas: long-term effect of tillage, N fertilization, and crop rotation. <http://www.agcarbonmarkets.com/documents/KSUCarbonSoils_January2007.pdf>

Fang, H.J., Cheng, S.L., Zhang, X.P., Liang, A.Z., Yang, X.M., Drury, C.F., 2006. Impact of soil redistribution in a sloping landscape on carbon sequestration in northeast China. Land Degrad. Develop. 17, 89–96.

Follett, R.F., 2001. Soil management concepts and carbon sequestration in cropland soils. Soil Tillage Res. 61, 77–92.

Follett, R.F., Samson-Liebig., E., Kimble, J.M., Pruessner, E.G., Waltman, S., 2001. Carbon sequestration under CRP in the historic grassland soils of the U.S. In: Lal, R. (Ed.), Soil Carbon Sequestration and the Greenhouse Effect. Soil Sci. Soc. Amer. Special Publ. #57, Madison, WI, pp. 27–40.

Franzluebbers, A.J., 2010. Achieving soil organic carbon sequestration with conservation agricultural systems in the southeastern United States. Soil Sci. Soc. Am. J. 74, 347–357.

Franzluebbers, A.J., Hons, F.M., Zuberer, D.A., 1995. Soil organic carbon, microbial biomass, and mineralizable carbon and nitrogen in sorghum. Soil Sci. Soc. Am. J. 59, 460–466.

Fuentes, J.P., Flury, M., Bezdicek, D.F., 2004. Hydraulic properties in a silt loam soil under natural prairie, conventional till, and no-till. Soil Sci. Soc. Am. J. 68, 1679–1688.

Gebhart, D.L., Johnson, H.B., Mayeux, H.S., Polley, H.W., 1994. The CRP increases in soil organic carbon. J. Soil Water Conserv. 49, 488–492.

Gillabel, J., Denef, K., Brenner, J., Merckx, R., Paustian, K., 2007. Carbon sequestration and soil aggregation in center-pivot irrigated and dryland cultivated farming systems. Soil Sci. Soc. Am. J. 71, 1020–1028.

Gollany, H.T., Allmaras, R.R., Copeland, S.M., Albrecht, S.L., Douglas Jr., C.L., 2005. Tillage and nitrogen fertilizer influence on carbon and soluble silica relations in a Pacific Northwest Mollisol. Soil Sci. Soc. Am. J. 69, 1102–1109.

Gollany, H.T., Allmaras, R.R., Copeland, S.M., Albrecht, S.L., Douglas Jr., C.L., 2006. Incorporated source carbon and nitrogen fertilization effects on carbon storage and soluble silica in a Haploxeroll. Soil Sci. 171 (8), 585–597.

Haas, H.J., Evans, C.E., Miles, E.F., 1975. Nitrogen and carbon changes in Great Plains soils as influenced by cropping and soil treatments. USDA Tech. Bull. 1164. U.S. Govt. Print. Office, Washington, DC.

Halvorson, A.D., Wienhold, B.J., Black, A.L., 2002a. Tillage, nitrogen, and cropping systems effects on soil carbon sequestration. Soil Sci. Soc. Am. J. 66, 906–912.

Harmon, M.E., Ferrell, W., Franklin, J.F., 1990. Effects on carbon storage of conversion of old growth forests to young growth forests. Science 247, 699–702.

Henderson, G.S., 1995. Soil organic matter: a link between forest management and productivity. In: McFee, W.F., Kelley, J.M. (Eds.), Carbon Forms and Functions in Forest Soils. SSSA, Madison, WI, pp. 419–436.

Horner, G.M., Oveson, M.M., Baker, G.O., Pawson, W.W., 1960. Effect of cropping practices on yield, soil organic matter and erosion in the Pacific Northwest Wheat Region. Washington, Idaho, and Oregon Agric. Exp. Station and ARS-USDA Coop. Bull. No. 1, Pullman, WA.

Howell, T.A., 2000. Irrigations role in enhancing water use efficiency. In: Evans, R.G. (Ed.), National Irrigation Symposium. Am. Soc. Engin, St. Joseph, MI, pp. 67–80.

Huang, X., Skidmore, E.L., Tibke, G.L., 2002. Soil quality of two Kansas soils as influenced by the conservation reserve program. J. Soil Water Cons. 57, 344–350.

Huggins, D.R., Kemanian, A.R., Brown, T.T., D.P. Uberuaga. 2011 Spatial heterogeneity of soil organic carbon: relationships to soil properties and terrain attributes. Unpublished data. In review.

Huggins, D.R., Buyanovsky, G.A., Wagner, G.H., Brown, J.R., Darmody, R.G., Peck, T.R., Lesoing, G.W., Vanotti, M.B., Bundy, L.G., 1998. Soil organic C in the tallgrass prairie-derived region of the corn belt: effects of long-term crop management. Soil Tillage Res. 47, 219–234.

Huggins, D.R., Allmaras, R.R., Clapp, C.E., Lamb, J.A., Randall, G.W., 2007. Corn-soybean sequence and tillage effects on soil carbon dynamics and storage. Soil Sci. Soc. Am. J. 71, 145–154.

Hutchinson, J.J., Campbell, C.A., Desjardins, R.L., 2007. Some perspectives on carbon sequestration potential in agriculture. Agric. Forest Meteor. 142, 288–302.

Izaurralde, R.C., Williams, J.R., Post, W.M., Thomson, A.M., McGill, W.B., Owens, L.B., Lal, R., 2007. Long-term modeling of soil C erosion and sequestration at the small watershed scale. Climatic Change 80, 73–90.

Jian, A.K., West, T.O., Yang, X., Post, W.M., 2005. Assessing the impact of changes in climate and CO2 on potential carbon sequestration in agricultural soils. Geophys. Res. Let. 32, 1–4. L19711, doi:10.1029/2005GL023922.

Jiao, Y., Whalen, J.K., Hendershot, W.H., 2006. No-tillage and manure application increase aggregation and improve nutrient retention in a sandy-loam soil. Geoderma 134, 24–33.

Jobbagy, E.G., Jackson, R.B., 2000. The vertical distribution of organic carbon and its relation to climate and vegetation. Ecol. Appl. 10, 423–436.

Kennedy, A.C., Schillinger, W.F., 2006. Soil quality and water intake in traditional-till vs. no-till paired farms in Washington's Palouse Region. Soil Sci. Soc. Am. J. 70, 940–949.

Kimble, L., Stewart, B.A., 1995. World soils as a source or sink for radiatively active gases. In: Lal, R. (Ed.), Soils and Global Change. CRC Press, Boca Raton, FL, pp. 1–7.

King, A.P., Evatt, K.J., Six, J., Poch, R.M., Rolston, D.E., Hopmans, J.W., 2009. Annual carbon and nitrogen loadings for a furrow-irrigated field. Agric. Water Manage 96, 925–930.

Kok, H., Papendick, R.I., Saxton, K.E., 2009. STEEP: impact of long-term conservation farming research and education in Pacific Northwest wheatlands. J. Soil Water Conserv. 64, 253–264.

Kong, A.Y.Y., Six, J., Bryant, D.C., Denison, R.F., van Kessel, C., 2005. The relationship between carbon input, aggregation, and soil organic carbon stabilization in sustainable cropping systems. Soil Sci. Soc. Am. J. 69, 1078–1085.

Kravchenko, A.N., Robertson, G.P., 2011. Whole-profile soil carbon stocks: the danger of assuming too much from analyses of too little. Soil Sci. Soc. Am. J. 75, 235–2040.

Lal, R., 1995. Global soil erosion by water and carbon dynamics. In: Lal, R., Kimble, J.M., Follett, R.F., Stewart, B.A. (Eds.), Management of Carbon Sequestration in Soil. CRC Press, Boca Raton, FL, pp. 131–142.

Lal, R., 2004. Soil carbon sequestration impacts on global climate change and food security. Science 304, 1623–1627.

Lal, R., Kimble, J.M., Follett, R.F., Cole, C.V., 1998. The potential of U.S. cropland to sequester carbon and mitigate the greenhouse effect. Sleeping Bear Press, Inc., Chelsea.

Lal, R., Follett, R.F., Kimble, J., Cole, C.V., 1999. Managing U.S. cropland to sequester carbon in soil. J. Soil Water Conserv. 59, 374–381.

Larson, W.E., Clapp, C.E., Pierre, W.H., Morachan, Y.B., 1972. Effect of increasing amounts of organic residues on continuous corn: II. Organic carbon, nitrogen, phosphorus, and sulfur. Agron. J. 64, 204–208.

Lee, D.K., Owens, V.N., Doolittle, J.J., 2007. Switchgrass and soil carbon sequestration response to ammonium nitrate, manure, and harvest frequency on conservation reserve program land. Agron. J. 99, 462–468.

Liebig, M.A., Morgan, J.A., Reeder, J.D., Ellert, B.H., Gollany, H.T., Schuman, G.E., 2005. Greenhouse gas contributions and mitigation potential of agricultural practices in northwestern USA and western Canada. Soil Till. Res. 83, 25–52.

Liebig, M.A., Schmer, M.R., Vogel, K.P., Mitchell, R.B., 2008. Soil carbon storage by switchgrass grown for bioenergy. Bioenergy Res. 1, 215–222.

Liski, J., Iivesniemi, H., Makela, A., Westman, C.J., 1999. CO_2 emissions from soil in response to climatic warming are overestimated—the decomposition of old soil organic matter is tolerant of temperature. Ambio 28, 171–174.

Lueking, M.A., Schepers, J.S., 1985. Changes in soil carbon and nitrogen due to irrigation development in Nebraska's sandhill soils. Soil Sci. Soc. Am. J. 49, 626–630.

Machado, S., Rhinhart, K., Petrie, S., 2006. Long-term cropping system effects on carbon sequestration in eastern Oregon. J. Environ. Qual 35, 1548–1553.

McCool, D.K., Roe, R.D., 2005. Long-term erosion trends on cropland in the Pacific Northwest. ASAE Section Meeting Paper No. PNW05–1002. 2005 Pacific Northwest Section Meeting, Alberta, Canada. 22–24 September 2005. ASAE, St. Joseph, MI.

McCool, D.K., Huggins, D.R., Saxton, K.E., Kennedy, A.C., 2001. Factors affecting agricultural sustainability in the Pacific Northwest, USA. In: Stott, D.E., Mohtar, R.H., Steinhardt, G.C. (Eds.), Sustaining the Global Farm. The 10th International Soil Conservation Organization Meeting, 1999, West Lafayette, IL 24–29 May 1999. Purdue University and the USDA-ARS National Soil Erosion Research Laboratory.

McVay, K.A., Budde, J.A., Fabrizzi, K., Mikha, M.M., Rice, C.W., Schlegel, A.J., Peterson, D.E., Sweeney, D.W., Thompson, C., 2006. Management effects on soil physical properties in long-term tillage studies in Kansas. Soil Sci. Soc. Am. J. 70, 434–438.

Mikha, M.M., Rice, C.W., 2004. Tillage and manure effects on soil and aggregate-associated carbon and nitrogen. Soil Sci. Soc. Am. J. 68, 809–816.

Mikha, M.M., Benjamin, J.B., Vigil, M.F., Nielsen, D.C., 2010. Cropping intensity impacts on soil aggregation and carbon sequestration in the central Great Plains. Soil Sci. Soc. Am. J. 74, 1712–1719.

Mikha, M.M., Vigil, M.F., Liebig, M.A., Bowman, R.A., McConkey, B., Deibert, E.J., Pikul Jr., J.L., 2006. Cropping system influences on soil chemical properties and soil quality in the Great Plains. Renew. Agri. Food Sys. 21, 26–35.

Mikhailova, E.A., Post, C.J., 2006. Effects of land use on soil inorganic carbon stocks in the Russian Chernozem. J. Environ. Qual 35, 1384–1388.

Monger, C.H., 2002. Pedogenic carbonate: linked between biotic and abiotic CaCO3. In Proc. 17th World Congress of Soil Science, 14–21, August 2002. Bangkok, Thailand. pp. 897–1 to 897-9.

NASS., 2007. Census of agriculture-State data. National Agricultural Statistics Service, Washington, DC.

Ongley, E.D., Bynoe, M.C., Percival, J.B., 1981. Physical and geochemical characteristics of suspended solids, Wilton Creek, Ontario. Can. J. Earth Sci. 18, 1365–1379.

Papendick, R.I., 1996. Farming systems and conservation needs in the Northwest wheat region. Am. J. Alt. Agric. 11 (2&3), 52–57.

Paustian, K., Collins, H.P., Paul, E.A., 1997. Management controls on soil carbon. In: Paul, E.A. (Ed.), Soil Organic Matter in Temperate Agroecosystems: Long-term Experiments in North America. Lewis Publishers, CRC Press, Boca Raton, FL, pp. 15–49.

Paustian, K., Andren, O., Janzen, H.H., Lal, R., Smith, P., Tian, G., Tiessen, H., Van Noordwijk, M., Woomer, P.L., 1997. Agricultural soils as a sink to mitigate CO2 emissions. Soil Use Manage 13, 230–244.

Pennock, D.J., Anderson, D.W., de Jong, E., 1994. Landscape scale changes in indicators of soil quality due to cultivation in Saskatchewan, Canada. Geoderma 64, 1–19.

Poch, R.M., Hopmans, J.W., Six, J.W., Rolston, D.E., McIntyre, J.L., 2006. Considerations of a field-scale soil carbon budget for furrow irrigation. Agric. Ecosys. Environ. 113, 391–398.

Potter, K.N., 2006. Soil carbon content after 55 years of management of a Vertisol in central Texas. J. Soil Water Conserv. 61, 338–343.

Potter, K.N., Derner, J.D., 2006. Soil carbon pools in central Texas: prairies, restored grasslands, and croplands. Soil Water Conserv. 61, 124–128.

Purakayastha, T.J., Huggins, D.R., Smith, J.L., 2008. Carbon sequestration in native prairie, perennial grass, no-till, and cultivated Palouse silt loam. Soil Sci. Soc. Am. J. 72 (2), 534–540.

Rasmussen, P.E., Rohde, C.R., 1988. Long-term tillage and nitrogen fertilization effects on organic nitrogen and carbon in a semiarid soil. Soil Sci. Soc. Am. J. 52, 1114–1117.

Rasmussen, P.E., Collins, H.P., 1991. Long-term impacts of tillage, fertilizer, and crop residue on soil organic matter in temperate semi-arid regions. Adv. Agron 45, 93–134.

Rasmussen, P.E., Albrecht, S.L., 1998. Crop management effects on organic carbon in semi-arid Pacific Northwest soils. In: Lal, R. (Ed.), Management of Carbon Sequestration in Soil. Advances in Soil Science. CRC Press, Boca Raton, FL, pp. 209–219.

Rasmussen, P.E., Albrecht, S.L., Smiley, R.W., 1998. Soil C and N changes under tillage and cropping systems in semi-arid Pacific Northwest agriculture. Soil Tillage Res. 47, 197–205.

Reeder, J.D., Schuman, G.E., Bowman, R.A., 1998. Soil C and N changes on conservation reserve program lands in the Central Great Plains. Soil Tillage Res. 47, 339–349.

Rodman, A.W., 1988. The effect of slope position, aspect, and cultivation on organic carbon distribution in the Palouse. M.S. Thesis. Washington State University, Pullman, WA, p 164.

Saffigna, P.G., Powlson, D.S., Brookes, P.C., Thomas, G.A., 1989. Influence of sorghum residues and tillage on soil organic matter and soil microbial biomass in an Australian Vertisol. Soil Biol. Biochem. 21, 759–765.

Sainju, U.M., Lenssen, A.W., Caesar-TonThat, T., Waddell, J., 2006a. Tillage and crop rotation effects on dryland soil and residue carbon and nitrogen. Soil Sci. Soc. Am. J. 70, 668–678.

Sainju, U.M., Lenssen, A.W., Caesar-TonThat, T., Waddell, J., 2006b. Carbon sequestration in dryland soils and plant residue as influenced by tillage and crop rotation. J. Environ. Qual. 35, 1341–1347.

Sainju, U.M., Lenssen, A.W., Caesar-TonThat, T., Waddell, J., 2007b. Dryland plant biomass and soil carbon and nitrogen fractions on transient land as influenced by tillage and crop rotation. Soil Tillage Res. 93, 452–461.

Sainju, U.M., Lenssen, A.W., Caesar-TonThat, T., Jabro, J.D., Lartey, R.T., Evans, R.G., Allen, B.L., 2011. Dryland residue and soil organic matter as influenced by tillage, crop rotation, and cultural practice. Plant Soil 338, 27–41.

Sainju, U.M., Lenssen, A.W., 2011. Dryland soil carbon dynamics under alfalfa and durum-forage cropping sequences. Soil and Tillage Research 113, 30–37.

Sainju, U.M., Jabro, J.D., Stevens, W.B., 2008. Soil carbon dioxide emission and carbon sequestration as influenced by irrigation, tillage, cropping system, and nitrogen fertilization. J. Environ. Qual. 37, 98–106.

Sainju, U.M., Caesar-Thonthat, T., Lenssen, A.W., Evans, R.G., Kohlberg, R., 2007a. Long-term tillage and cropping sequence effects on dryland residue and soil carbon fractions. Soil Sci. Soc. Am. J 71, 1730–1739.

Sainju, U.M., Stevens, W.B., Caesar-ThonThat, T., Jabro, J.D., 2010. Land use and management practices impact on plant biomass carbon and soil carbon dioxide emission. Soil Sci. Soc. Am. J. 74, 1613–1622.

Sanchez de-Leon, Y., 2007. Carbon dynamics and earthworm populations in grassland ecosystems of the Palouse region. Ph.D. Dissertation. University of Idaho, Moscow, ID, p 178.

Schillinger, W.F., Kennedy, A.C., Young, D.L., 2007. Eight years of annual no-till cropping in Washington's winter wheat –summer fallow region. Agric. Ecosys. Env 120, 345–358.

Schlesinger, W.M., et al., 1995. An overview of the C cycle. In: Lal, R. (Ed.), Soils and Global Change. CRC Press, Boca Raton, FL, pp. 9–26.

Sievers, F.J., Holtz, H.F., 1926. The significance of nitrogen in soil organic matter relationships. Washington Agric. Exp. Station Bull. No. 206, Pullman, WA, p 43.

Six, J., Feller, C., Denef, K., Ogle, S.M., Moraes Sa, J.C., Alerecht, A., 2002. Soil organic matter, biota and aggregation in temperate and tropical soil-effect of no-tillage. Agronomie 22, 755–775.

Six, J., Elliott, E.T., Paustian, K., 1999. Aggregate and soil organic matter dynamics under conventional and no-tillage systems. Soil Sci. Soc. Am. J. 63, 1350–1358.

Sperow, M., Eve, M., Paustian, K., 2003. Potential soil C sequestration on US agricultural soils. Climate Change 57, 319–339.

Staben, M.L., Bezdicek, D.F., Smith, J.L., Fauci, M.F., 1997. Assessment of soil quality in conservation reserve program and wheat–fallow soils. Soil Sci. Soc. Am. J. 61, 124–130.

Starr, G.C., Lal, R., Malone, R., Hotheem, D., Owens, L., Kimble, J., 2000. Modeling soil carbon transported by water erosion processes. Land Degrad. Develop. 11, 83–91.

Syswerda, S.P., Corbin, A.T., Mokma, D.J., Kravchenko, A.N., Robertson, G.P., 2011. Agricultural management and soil carbon storage in surface vs. deep layers. Soil Sci. Soc. Am. J. 75, 92–101.

United States Department of Agriculture, Natural Resources Conservation Service, 2006. Land Resource Regions and Major Land Resource Areas of the United States, the Caribbean, and the Pacific Basin. U.S. Department of Agriculture Handbook, p. 296.

USDA., 1978. Palouse Cooperative River Basin Study, Soil Conservation Service, Forest Service, and Economics, Statistics and Cooperative Service. U.S. Government Printing Office, Washington, DC, p. 182.

Van Cleve, K., Dryness, C.T., Marion, G.M., Erickson, R., 1993. Control of soil development on the Tanana River floodplain, interior, Alaska. Can. J. For. Res. 23, 941–955.

Van Oost, K., Quine, T.A., Govers, G., De Gryze, S., Six, J., Harden, J.W., Ritchie, J.C., McCarty, G.W., Heckrath, G., Kosmas, C., Giraldez, J.V., Marques da Silva, J.R., Merckx, R., 2007. The impact of agricultural soil erosion on the global carbon cycle. Science 318, 626–629.

Vanden Bygaart, A.J., 2006. Monitoring soil organic carbon stock changes in agricultural landscapes: issues and a proposed approach. Can. J. Soil Sci. 86, 451–463.

Vanden Bygaart, A.J., Yang, X.M., Kay, B.D., Aspinall, J.D., 2002. Variability in carbon sequestration potential in no-till soil landscapes of southern Ontario. Soil Tillage Res. 65, 231–241.

VandenBygaart, A.J., Bremer, E., McConkey, B.G., Ellert, B.H., Angers, D.A., Carter, M.R., Drury, C.F., Lafond, G.P., McKenzie, R.H., 2011. Impact of sampling depth on differences in soil carbon stocks in long-term agro-ecosystem experiments. Soil Sci. Soc. Am. J. 75, 226–234.

Wang, Y., Amundson, R., Trumbore, S., 1999. The impact of land use change on C turnover in soils. Gobal Biogeochem. Cycles 13, 47–57.

West, T.O., Post, W.M., 2002. Soil organic carbon sequestration rates by tillage and crop rotation: a global data analysis. Soil Sci. Soc. Am. J. 66, 1930–1946.

Wu, L.S., Wood, Y., Jiang, P.P., Li, L.Q., Pan, G.X., Lu, J.H., Chang, A.C., Enlone, H.A., 2008. Carbon Sequestration and Dynamics of Two Irrigated Agricultural Soils in California. Soil Sci. Soc. Am. J. 72, 808–814.

Wuest, S.B., Caesar-TonThat, T.C., Wroght, S.F., Williams, J.D., 2005. Organic matter addition, N, and residue burning effects on infiltration, biological, and physical properties of an intensively tilled silt-loam soil. Soil Tillage Res. 84, 154–167.

Zan, C.S., Fyles, J.W., Girouard, P., Samson, R.A., 2001. Carbon sequestration in perennial bioenergy, annual corn and uncultivated systems in southern Quebec. Agric. Ecosyst. Environ. 86, 135–144.

Zhang, G.S., Chan, K.Y., Oates, A., Heenan, D.P., Huang, G.B., 2007. Relationship between soil structure and runoff/soil loss after 24 years of conventional tillage. Soil Tillage Res. 92, 122–128.

Zibilske, L.M., Bradford, J.M., 2007. Soil aggregation, aggregate carbon and nitrogen, and moisture retention induced by conservation tillage. Soil Sci. Soc. Am. J. 71, 793–802.

CHAPTER 6

Soil Carbon Dynamics and Rangeland Management

Justin D. Derner[1], Virginia L. Jin[2]
[1]USDA-Agricultural Research Service, Rangeland Resources Research Unit, Cheyenne, WY and Fort Collins, CO
[2]USDA-Agricultural Research Service, Agroecosystem Management Research Unit, Lincoln, NE

CHAPTER OUTLINE

Abbreviations: CH_4, methane; CO_2, carbon dioxide; EPA, Environmental Protection Agency; GHGs, greenhouse gases; IPCC, International Panel on Climate Change; N_2O, nitrous oxide; SOC, soil organic carbon; SIC, soil inorganic carbon; U.S., United States

DEFINITION AND EXTENT OF U.S. RANGELANDS

Grazing lands occupy one-third of the total land base in the United States (U.S.), covering an estimated 316 to 336 Mha (Lal et al., 2003; USDA-ERS, 2006). Of this area, 80% is privately owned (214 to 257 Mha) and 20% is publicly owned (27 to 59 Mha) (Sobecki et al., 2001; USDA-NRCS, 2003) (Figure 6.1). Although a small proportion of grazing lands occur as grazed forest or hay land, most grazing lands in the conterminous U.S. occur west of the 95th meridian as rangelands (~80%), with the remainder occurring eastward as more intensively managed improved pastures (Mitchell, 2000; Schnabel et al., 2001). Over half of the total land base in the 17 western states of the conterminous U.S. is classified as rangeland. This chapter focuses on rangelands, which are defined as uncultivated lands managed with minimal inputs and consisting of extensively grazed native or naturalized plant species representative of historic climax vegetation (Figure 6.2; USDA-NRCS, 2003; Follett and Reed, 2010).

Rangelands represent one of the largest and most diverse land resources in the U.S., and encompass broad environmental gradients in temperature and precipitation (Liebig et al., 2012). Rangelands significantly impact both rural and national economies through the domestic livestock industry, with contributions assessed at $32 billion to the national economy in 2009 (USDA-ERS, 2010). A broad array of ecosystem goods and services, including

Managing Agricultural Greenhouse Gases. DOI: 10.1016/B978-0-12-386897-8.00006-1
2012, Published by Elsevier Inc.

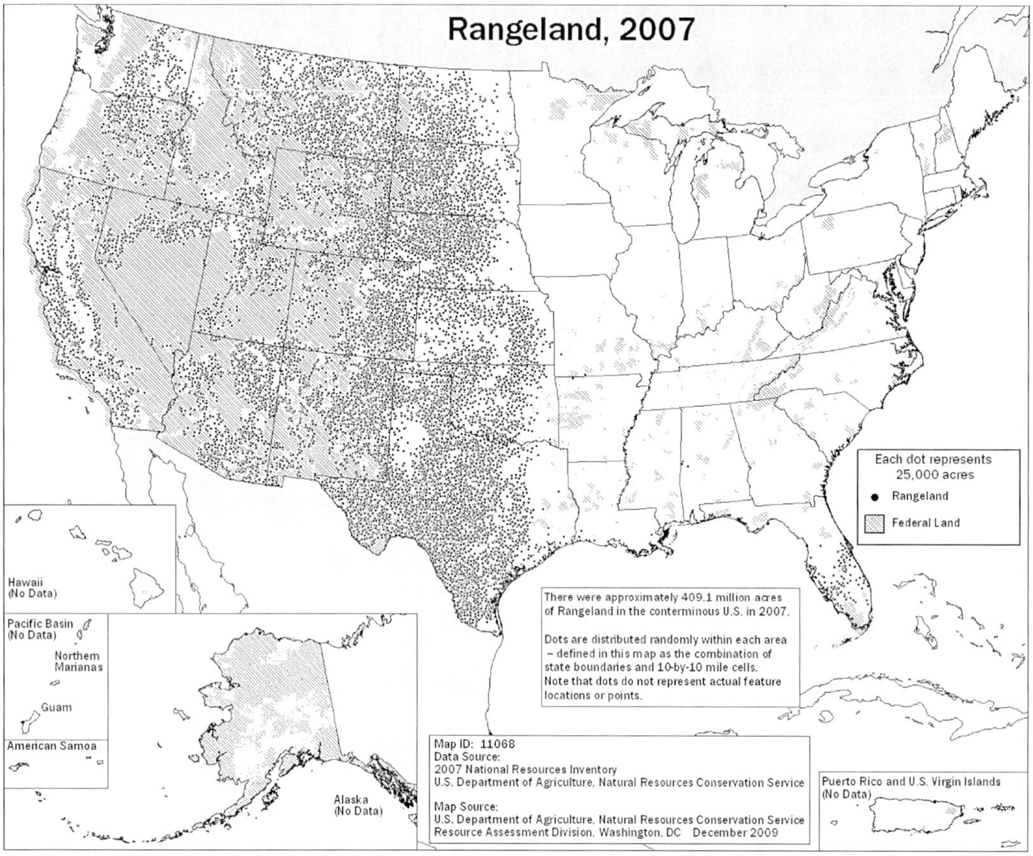

FIGURE 6.1
Extent of privately and publicly owned rangeland in the U.S. in 2007 (USDA-NRCS, 2009).

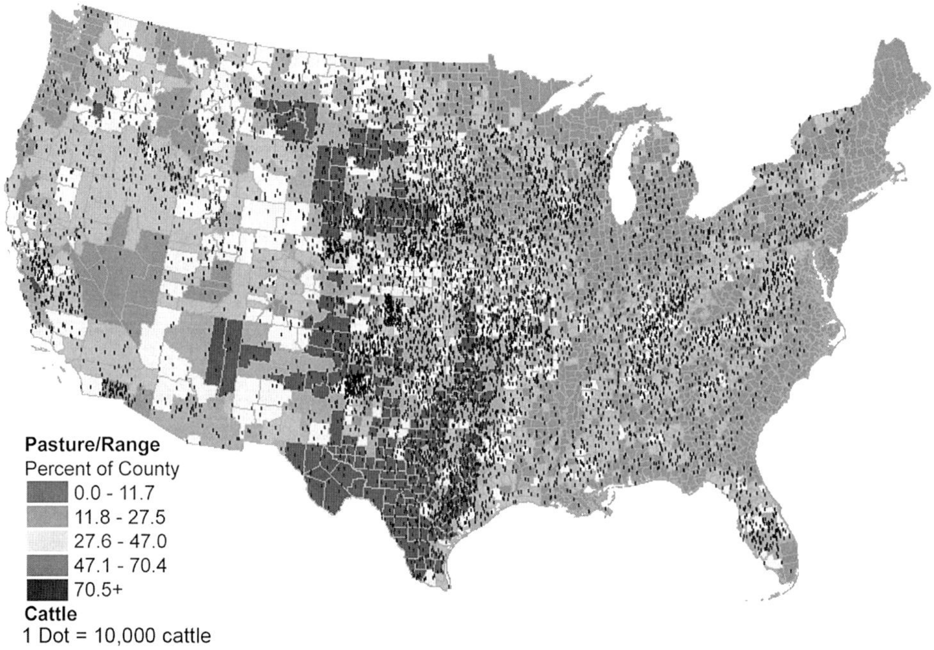

FIGURE 6.2
Extent of grazing land as a proportion of total county area and livestock production cattle inventory in the U.S. (one dot = 10,000 cattle; USGCRP, 2009). Please see color plate section at the back of the book.

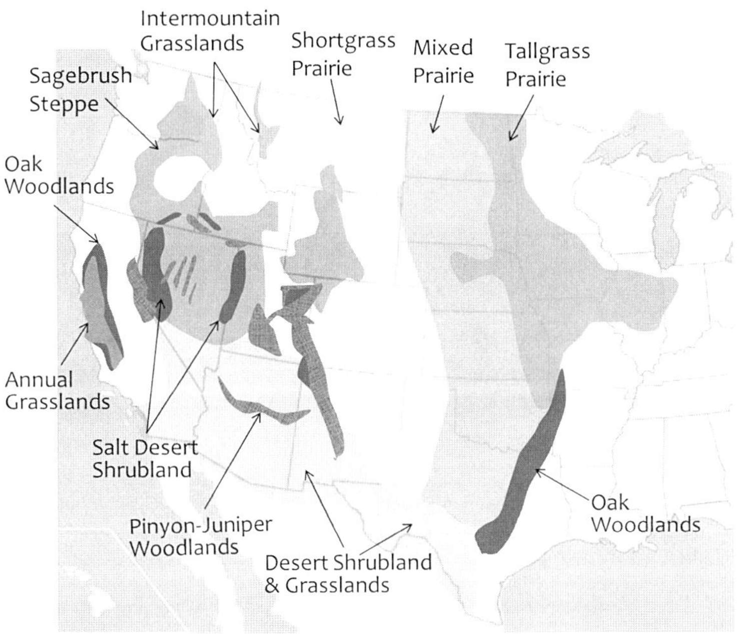

FIGURE 6.3
Major rangelands ecosystems of the U.S. (from Kuchler, 1964; courtesy of K. Launchbaugh). Please see color plate section at the back of the book.

livestock production, nutrient cycling, climate mediation, watershed function (flood control, storage, filtering), water quality and quantity, biodiversity, recreation, wildlife habitat, viewsheds, airsheds, and C sequestration, are provided from rangelands to society (Millennium Ecosystem Assessment, 2005; Havstad et al., 2007). Rangelands are characterized by a high diversity of climatic conditions and plant communities, including natural grasslands, savannas, most deserts, tundra, alpine plant communities, coastal and freshwater marshes, and wet meadows (USDA-NRCS, 2003). For classification purposes, rangelands are often separated into major ecosystems (Figure 6.3). Within any given rangeland ecosystem, landscapes can exhibit high diversity in plant communities, depending on topography. Further, plant community dynamics and associated ecological processes within these distinct ecological sites can be depicted by state transitions that occur between several different alternative plant community states in response to disturbance or management (Figure 6.4).

RANGELAND GHG MITIGATION POTENTIAL

Rangelands harbor considerable potential to mitigate climate impacts resulting from rising atmospheric levels of various greenhouse gases (GHGs), such as carbon dioxide (CO_2), methane (CH_4), and nitrous oxide (N_2O) due to an extensive land area (Schimel et al., 1990; Ojima et al., 1993; Conant et al., 2001), low-input management practices, and adaptive management facilitating active responses to highly variable within-year and between-year precipitation. Globally, rangelands could capture up to 2 to 4% of annual anthropogenic GHG emissions and up to 20% of the CO_2 released annually from global deforestation and land-use change (Derner and Schuman, 2007; Follett and Reed, 2010). Whether rangelands function as GHG sinks or sources will be determined by complex interactions between climate, vegetation, and grazing management and their effects on soils (both organic and inorganic C pools) and livestock (i.e. CH_4 emissions from enteric fermentation, N_2O emissions from manure) (Derner et al., 2006; Derner and Schuman, 2007; Ingram et al., 2008; Follett and Reed, 2010; Liebig et al., 2010a; Morgan et al., 2010). Rangelands are typically characterized by short periods of high C uptake during the growing season (2–3 months) and long periods of C balance or small losses during the remainder of the year (Svejcar et al., 2008), resulting in substantial interannual variability in net ecosystem exchange (Svejcar et al., 2008; Zhang et al., 2010).

The U.S. GHG emission inventory by the Environmental Protection Agency (EPA) groups emission estimates for six sectors defined by the International Panel on Climate Change

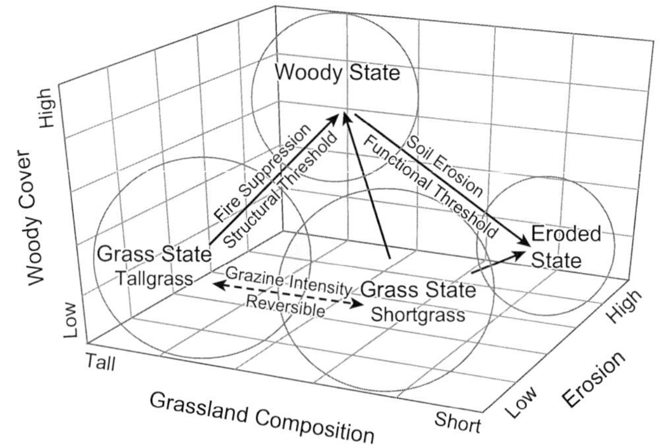

FIGURE 6.4
Example of a conceptual state-and-transition model showing selected alternative stable states that may occur on a given ecological site. Dashed and solid arrows show reversible and nonreversible transitions, respectively (from Briske et al., 2005).

(IPCC): Energy; Industrial Processes; Solvent Use; Agriculture; Land Use, Land-Use Change, and Forestry; and Waste. Emissions from livestock and rangelands are embedded within the Agriculture and Land Use, Land-Use Change, and Forestry sectors, respectively. Here, we report emissions estimates from the 2008 National GHG Inventory (EPA, 2011) from unconfined grazing systems for the following categories: enteric fermentation, manure management, and grasslands (inclusive of rangeland and improved pasture) (Table 6.1). In this inventory, grasslands on federal lands are included in the managed land base, but C stock changes are not estimated on these lands because federal grassland areas are assumed to have negligible changes in C due to limited land use and management change (EPA, 2011). Unconfined livestock values for enteric fermentation and manure management are for emissions from cattle, horses, sheep, and goats (Annex 3; EPA, 2011).

The dominant mechanisms for GHG mitigation in rangelands will be through C storage in soils and by minimization of livestock-related non-CO_2 emissions. For rangelands specifically, N_2O emissions from manure deposition is the primary GHG source (Liebig et al., 2012). Rangeland CH_4 fluxes from soil and areas of deposited manure are relatively small (~3% of total grazing land GHG emissions) due to predominantly aerobic conditions on most pastures

TABLE 6.1 Estimated 2008 Greenhouse Gases in Grasslands and Unconfined Livestock in the Conterminous US (EPA, 2011)[a]

	Livestock category (Tg CO_2 equivalents)				
Sources	*Cattle*	*Sheep*	*Goats*	*Horses*	*Total*
Enteric fermentation $[CH_4]$[b]	89.1	1.0	0.3	3.6	94.0
Manure management $[CH_4 + N_2O]$[b]	1.95	0.43	0.05	1.22	3.7
Grassland soil [direct + indirect N_2O]	–	–	–	–	61.7
Total greenhouse gas emissions	Unconfined livestock + grassland				159.4
	Total from agriculture[c]				427.5
	Total U.S.				6,957
Sinks	Grasslands remaining grasslands $[CO_2]$				(8.7)
	Lands converted to grasslands $[CO_2]$				(24.2)
Emissions mitigated (%)	Unconfined livestock + grassland emissions				21%
	Total agricultural emissions				8%
	Total U.S. emissions				0.5%

[a]*Totals may not sum due to independent rounding.*
[b]*Excluding contributions from confined livestock (beef cattle, dairy cattle, swine, poultry).*
[c]*Crop lands, grazing lands (including rangeland), and all livestock (confined + unconfined) emissions included; forest lands excluded.*

and ranges. In the conterminous U.S., modeled estimates of emissions from grasslands (including rangelands and improved pastures) and unconfined livestock accounted for approximately 37% of total agricultural GHG emissions in 2008 (Table 6.1) (EPA, 2011). Grasslands soils and vegetation combined, however, offset 21% of the total emissions estimated from grasslands and unconfined livestock.

Empirical studies measuring the net effects of GHG emissions from rangelands, however, are limited (Liebig et al., 2012). Direct measurements suggest that grazing management has minimal effects on rangeland N_2O emissions (Liebig et al., 2010a; Wolf et al., 2010). Similar to IPCC results, measured CH_4 emissions from the soil are minimal in grazing systems (Liebig et al., 2010a), with the dominant source being CH_4 emitted from livestock enteric fermentation. Although animal density on the land will control livestock-related CH_4 emissions per unit area, simply transferring those animals to another part of the rangeland landscape is not likely to result in a general reduction of CH_4 emissions from this ecosystem.

Rangeland C sequestration rates on a per unit area basis are relatively low relative to croplands, improved pastures, or forested lands, but can still have a substantial impact on overall GHG emissions because of their large land base. For specific categories within rangelands (i.e. soils, vegetation, livestock), potential mitigation or emission pathways are not well defined (Figure 6.5). While much research has focused on rangeland soil organic carbon (SOC) pools, little is known about alternative pathways for soil C storage or emission, such as soil accumulation of pyrogenic C derived from burning (i.e. charcoal, or black C) or CO_2 fluxes from the soil inorganic C (SIC) pool. For rangeland vegetation, management and local climate condition affect plant community dynamics and subsequent plant C inputs into soils (Derner et al., 2006; Ingram et al., 2008; Zhang et al., 2010). Two factors, however, have emerged as the primary controls on the fate of SOC in rangelands: (1) plant productivity (i.e. above- and belowground biomass quantity, plant nutrient quality) (Derner et al., 2006; Derner and Hart, 2007; Piñeiro et al., 2010); and (2) direct and indirect effects of grazing on vegetation composition (Derner and Schuman, 2007; Bagchi and Ritchie, 2010) (Figure 6.5). Management of rangeland plant communities will also be impacted by state changes after disturbances such as fire, plant invasions (i.e. non-native annuals, woody plants) (Figure 6.4), or managed introductions of desirable species (i.e. N-fixing legumes). Finally, non-CO_2 GHG emissions from unconfined livestock will be affected most strongly by stocking rate. Stocking rate also controls C and nutrient inputs from manure into soils as well as the level of physical disturbance related to hoof action or overgrazing, which can lead to increased wind and water

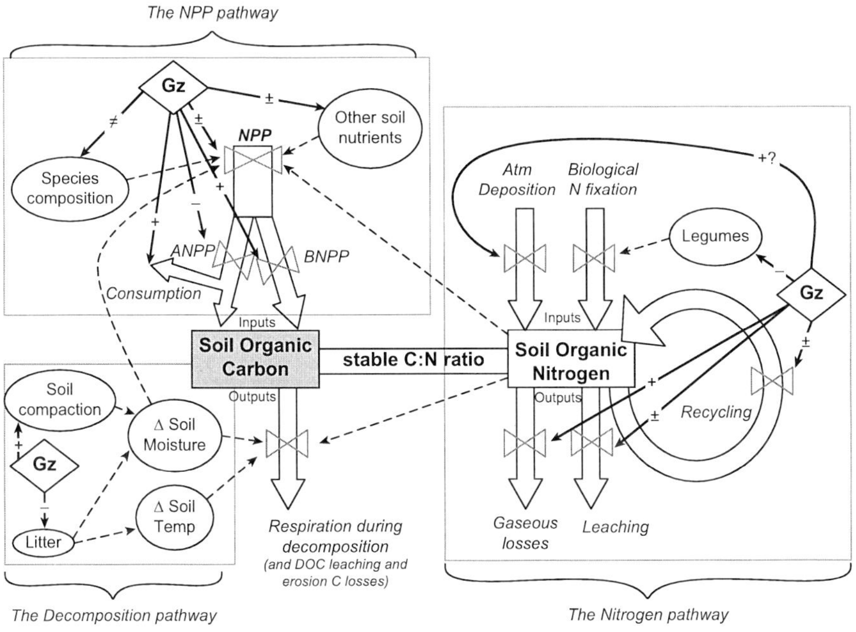

FIGURE 6.5
Effects of grazing on soil organic carbon pools via changes in net primary production (NPP), nitrogen stocks, nitrogen or cycling, and/or decomposition. ANPP, aboveground NPP; BNPP, belowground NPP; C:N, carbon-to-nitrogen ratio; DOC, dissolved organic carbon; Gz, grazing (from Piñeiro et al., 2010).

erosion of rangeland soils. The GHG impacts of changes in vegetation species composition and supplemental feeding practices on livestock emissions have focused primarily on C dynamics, while the full complement of GHG emissions have rarely been addressed (Liebig et al., 2005, 2010a). Long-term management for the sustainable use of rangelands will depend ultimately on elucidating the mechanisms driving rangeland processes to improve our ability to predict ecosystem-level responses to projected changes in temperature and precipitation.

RANGELAND C DISTRIBUTION: VEGETATION AND SOILS

The distribution of C between vegetation and various soil pools varies greatly across U.S. rangelands, reflecting the tremendous variability in climate and plant community types (Figure 6.3). Plant aboveground biomass can store <5% of total ecosystem C in non-woody plant dominated ecosystems (Derner et al., 2006) to >25% in pinyon-juniper dominated ecosystems (Rau et al., 2010). Soils contain the majority of the C in rangelands. Total soil C (organic plus inorganic) in rangeland soils to 1 m depth range from 90 to 266 $Mg\,C\,ha^{-1}$ (Guo et al., 2006; Figure 6.6A, B). Soil organic C concentration of rangeland soils varies from <1 to >10% (Janzen, 2001). Rangelands tend to be C sinks (Schlesinger, 1997; USDA, 2008), with SOC sequestration rates of up to 0.5 $Mg\,C\,ha^{-1}\,yr^{-1}$ (Schuman et al., 1999; Derner and Schuman, 2007; Morgan et al., 2010; Liebig et al., 2010a).

Greater than 90% of stored C will be sequestered in the soil as SOC derived from decomposing plant matter and microbial biomass, but two other potential pathways for C sequestration are: (1) the formation of charcoal (i.e. black C) which is stored in the SOC pool, and (2) the formation of secondary carbonates which are stored in the soil inorganic C (SIC) pool. Vegetation composition in U.S. rangelands is commonly controlled using prescribed fires, particularly in preventing or slowing the invasion of woody plant encroachment (Scholes and Archer, 1997). In the southern Great Plains, two to three prescribed fire events can significantly increase soil black C content (Dai et al., 2005). Black C produced from burning aboveground biomass results in extremely stable forms of organic C (i.e. highly aromatic, few functional groups), with mean residence times of >10,000 years (Kuhlbusch, 1998; Swift, 2001). In rangeland soils worldwide, black C can make up 5 to 35% of the total SOC pool (Skjemstad et al., 1999; Glaser and Amelung, 2003; Dai et al., 2005). Its small particle size (<53 μm) also makes black C relatively mobile, resulting in long-term storage of this highly resistant C deeper in the soil profile (Skjemstad, 1999; Dai et al., 2005).

The uptake or release of CO_2 from SIC pools could also contribute to the role of rangeland soil as a C source or sink. In contrast to the rest of the U.S., a vast pool of SIC occurs as carbonates in semiarid and arid rangeland soils (Monger and Martinez-Rios, 2001; Emmerich, 2003; Guo et al., 2006) (Figure 6.6A, B). The prevalence of SIC in rangelands is promoted by the combination of more xeric conditions and the predominance of calcareous soil parent materials, contributing to calcification and secondary carbonate formation (Batjes and Sombroek, 1997). In various Land Resource Regions (LRR; USDA-NRCS) encompassing rangeland and/or grazing land (Figure 6.7), SIC can equal or exceed SOC stores in the top 2 meters of the soil profile (Guo et al., 2006) (Table 6.2). Inorganic C storage or loss pathways, however, are poorly understood (Follett et al., 2001; Liebig et al., 2006; Svejcar et al., 2008), which is reflected in the wide range of predicted SIC sequestration values (0.0012–0.120 $Mg\,C\,ha^{-1}\,yr^{-1}$; Gile et al., 1981; Reheis et al., 1992, 1995; Schlesinger, 1997).

RANGELAND MANAGEMENT IMPACTS

Management practices that maximize soil C storage while minimizing both direct and indirect emissions of non-CO_2 GHGs from grazing livestock will have the greatest impacts on rangeland GHG mitigation potential. Range management has long recognized vegetation as the pivotal management component affecting soil quality (i.e. quantity and quality of above- and

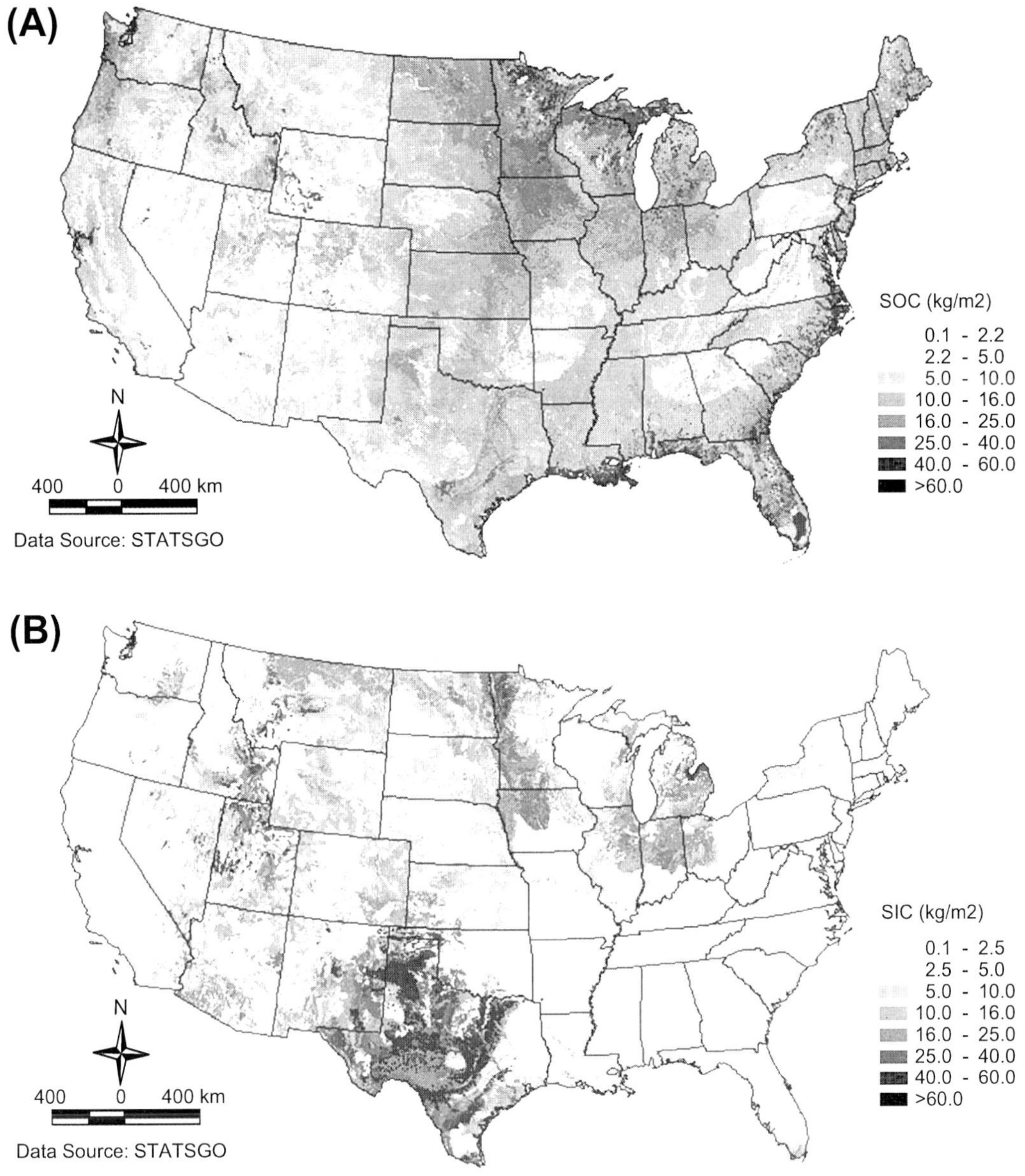

FIGURE 6.6
Distribution of (A) soil organic C (SOC) and (B) soil inorganic C (SIC) to 2 m soil depth in the conterminous U.S. (from Guo et al., 2006). Please see color plate section at the back of the book.

belowground C inputs), livestock productivity (i.e. quantity, quality, and timing of available forage), and overall ecosystem sustainability (i.e. productivity of plant and livestock, soil erosion control, resistance to invasive plant species and subsequent impacts on altered fire regime). Range management approaches have evolved over time from linear Clementsian succession-based concepts towards more complex state-and-transition models that better incorporate disturbance, climate, and multi-directional changes in plant community dynamics (Westoby, 1980; Young and Clements, 2001; Briske et al., 2005). Rates of SOC sequestration decrease with longevity of a management practice (Derner and Schuman, 2007), indicating that ecosystems reach a "steady state" and that changes in management or inputs may be required to sequester additional C (Conant et al., 2001, 2003; Swift, 2001). Key considerations for SOC sequestration in rangelands are: (1) the aboveground C pool is a minor component of total ecosystem C storage and mean residence time of this C pool is only a few years, so yearly variations in aboveground biomass minimally affect C storage; (2) most SOC is recalcitrant

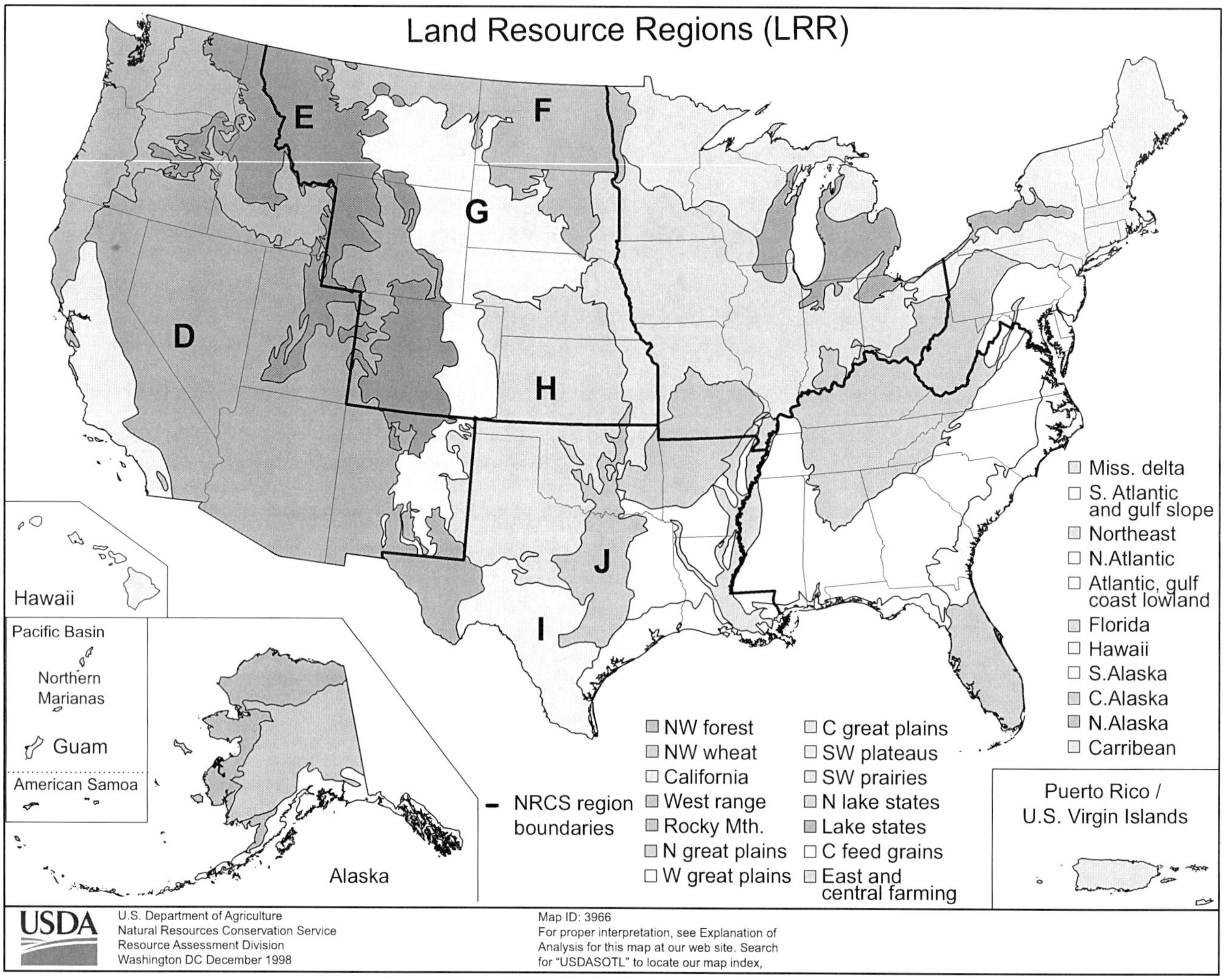

FIGURE 6.7
Land Resource Regions (LRRs) in the U.S. (USDA-NRCS). LRRs in Table 6.2 are labeled. Please see color plate section at the back of the book.

TABLE 6.2 Mean (±1 Standard Deviation) Soil Organic C (SOC) and Soil Inorganic C (SIC) Storage in Western Land Resources Regions (LRRs) that Include Grazing Lands and/or Rangelands (adapted from Guo et al., 2006)

LRR	Region description	Area	SOC (Mg C ha^{-1})		SIC (Mg C ha^{-1})	
		Mha	*0–20 cm*	*0–100 cm*	*0–20 cm*	*0–100 cm*
D	Western range and irrigated	127	14 ± 16	35 ± 52	11 ± 24	57 ± 112
E	Rocky Mountain range and irrigated	52	28 ± 27	66 ± 88	3 ± 11	24 ± 78
F	Northern Great Plains spring wheat	35	50 ± 29	117 ± 72	5 ± 12	73 ± 100
G	Western Great Plains range and irrigated	52	20 ± 12	50 ± 30	7 ± 17	46 ± 83
H	Central Great Plains winter wheat and range	58	27 ± 12	80 ± 42	8 ± 20	65 ± 134
I	Southwest plateaus and plains range and cotton	17	24 ± 10	71 ± 62	45 ± 48	195 ± 219
J	Southwestern prairies cotton and forage	14	24 ± 12	81 ± 60	28 ± 46	155 ± 242

and well protected from minor natural disturbances; microbial biomass and particulate or light-fraction organic C are most sensitive to management or land-use change, whereas chemically resistant organic C and soil carbonates are least sensitive (Allen et al., 2010); (3) major pathways of SOC accumulation include rhizodeposition belowground, surface deposition of animal feces, decaying litter from aboveground vegetation, and the activity of soil biota (bacteria, fungi, protists, and fauna) promoting humic substance synthesis, aggregate formation, and subsequent long-term SOC stabilization (Six et al., 2002; Jones and Donnelly, 2004; Ingram et al., 2008, Piñerio et al., 2010); and (4) large perturbations in the SOC pool can occur with major soil disturbances (i.e. tillage, wind and water erosion, surface denudation with overgrazing). These impacts can occur naturally with extreme weather conditions (i.e. drought) or through poor management decisions that reduce the vigor of plant communities (Follett et al., 2001).

Grazing Management

Improved grazing management is central to rangeland health (Derner and Schuman, 2007; Follett and Reed, 2010; Morgan et al., 2010). Improved management strategies include using appropriate stocking rate and forage utilization, timing grazing to avoid the months of high C uptake and adjusting the frequency of grazing (i.e. destocking during drought conditions), and implementing adaptive management practices to promote active responses to highly variable within-year and between-year precipitation. Shifts in plant community composition due to grazing can influence SOC. For example, SOC was greater in grazed compared to ungrazed areas in northern mixed grass prairie (Frank et al., 1995; Manley et al., 1995; Schuman et al., 1999; Reeder and Schuman, 2002; Ganjegunte et al., 2005, Liebig et al., 2006, 2010a), partially due to the increased dominance of the grazing-resistant species blue grama (*Bouteloua gracilis*) which produces greater root biomass in the upper soil profile compared to the midgrass species it replaces under grazing (Derner et al., 2006).

Rangeland SOC response to stocking rate and grazing intensity is variable (Smoliak et al., 1972; Wood and Blackburn, 1984; Warren et al., 1986; Biondini et al., 1998; Schuman et al., 1999; Liebig et al., 2006, 2010a). Observed changes in SOC due to grazing have been attributed to differential grazing impacts across environmental gradients (Derner et al., 2006) as well as to management-by-climate interactions (e.g. drought) (Ingram et al., 2008). For example, heavy grazing alone on restored rangelands can slow down the rate of soil C accumulation (Fuhlendorf et al., 2002), but heavy grazing during drought conditions can result in overall losses in SOC (Schnabel et al., 2001; Ingram et al., 2008).

The response of SOC to a specific grazing method has been investigated sparsely, at best. Two studies suggest an increase in SOC with rotational grazing compared with continuous season-long grazing (Conant et al., 2003; Teague et al., 2010), but another study found no differences in SOC between these grazing systems (Manley et al., 1995). The majority of available scientific evidence suggests that rotational grazing has no direct impact on broad-scale vegetation production or composition (Briske et al., 2008). Any change in SOC under rotational grazing, therefore, is expected to reflect vegetation changes that are independent of stocking rate.

N-fixing Legumes

Many rangelands are N-limited, so increasing N inputs by interseeding N-fixing legumes can increase forage production and quality as well as C sequestration (Mortenson et al., 2004, 2005; Liebig et al. 2010b). The introduction of N-fixing legumes as an alternative to N-fertilization has been the subject of research for decades (Tesar and Jakobs 1972; Heinrichs 1975; Kruger and Vigil 1979; Berdahl et al., 1989), providing critical insights to both shorter- and longer-term management impacts of interseeding legumes in rangelands. At one long-term study site in northern mixed-grass prairie, interseeding yellow-flowered alfalfa (*Medicago sativa* ssp. *falcata*) increased soil C by 4–17% across three interseeding dates (Mortenson et al.,

2004), corresponding to C sequestration rates of 1.56, 0.65, and 0.33 Mg C ha^{-1} yr^{-1}, for 3-, 14-, and 36-year post-interseeding, respectively. In addition, interseeding significantly increased soil total N, aboveground production, and N forage quality (Mortenson et al., 2004, 2005) did not impact soil N_2O emissions (Liebig et al., 2012).

Prescribed Fire

Burning can indirectly contribute to rangeland C sequestration by enhancing nutrient availability for the following season's plant growth, which can exceed C combustion losses (Swift, 2001). Burning, therefore, indirectly affects photosynthesis (Knapp, 1985; Svejcar and Browning, 1988; Bremmer and Ham, 2010), soil and canopy respiration (Knapp et al., 1998; Bremmer and Ham, 2010), and can alter species composition (Pacala et al., 2007; Boutton et al., 2009). Carbon losses from burning herbaceous plant-dominated grazing lands is a minor component of the total annual ecosystem C emitted (Owensby et al., 2006; Bremmer and Ham, 2010). Burning rangelands with a significant above ground woody component, however, can result in immediate substantial ecosystem C losses (Rau et al., 2010). Further, C losses in western U.S. rangelands via burning of woody biomass could be compounded by additional losses of surface soil SOC derived from historical (i.e. >100 years) woody plant encroachment (Boutton et al., 2009; Neff et al., 2009).

INTERACTIONS BETWEEN MANAGEMENT AND CLIMATE

Climate and management can influence soil C and GHG emissions on rangelands (Schuman et al., 1999; Follett et al., 2001; Jones and Donnelly, 2004; Derner et al., 2006; Derner and Schuman, 2007; Ingram et al., 2008; Svejcar et al., 2008; Bremmer and Ham, 2010; Liebig et al., 2010a; Piñeiro et al., 2010; Rau et al., 2010; Zhang et al., 2010). Specifically, both short-term weather conditions (e.g. droughts) (Ciais et al., 2005; Soussana and Lüschert, 2007; Ingram et al., 2008; Svejcar et al., 2008; Zhang et al., 2010) and long-term changes in the global environment (e.g. rising temperature, altered precipitation patterns, rising atmospheric CO_2 concentrations) can reduce soil quality by altering plant and microbial community compositions (Jin and Evans, 2010; Jin et al., 2011) and changing forage quality (Milchunas et al., 2005; Soussana and Lüschert, 2007; Hatfield et al., 2008; Morgan et al., 2008, 2010). Adaptive management practices that optimize rangeland vegetation and soil responses to changing environmental conditions, therefore, play a significant role in determining whether rangelands will lose or sequester C (Ingram et al., 2008; Svejcar et al., 2008).

Projected increases in temperature due to climate change may constrain rangeland C storage or cause rangelands to become C sources, particularly in drought-prone regions. For example, in the Great Plains region under non-drought conditions, rangelands to the west are generally C sources and eastern rangelands are C sinks. In years when drought affects >65% of the region, however, the Great Plains becomes an overall C source (Zhang et al., 2010). In contrast, the interaction of warmer temperatures with rising atmospheric CO_2 over the past century has enhanced net primary production (Hatfield et al., 2008) which could increase rangeland C sequestration. Such positive effects of climate change on productivity are expected at northern latitudes and at high altitudes where temperature is an important factor limiting production and annual precipitation is not expected to decline. Increased C inputs due to CO_2-enhanced plant growth, however, can also stimulate microbial decomposition of older SOC pools in the short term (i.e. priming effect; Kuzyakov 2002) causing soils to become C sources (Carney et al., 2007). If warming accelerates, the CO_2 benefit to plant productivity and water use efficiency could be offset further by increased desiccation stresses that constrain or reduce plant productivity, especially in the southwestern quadrant of North America (Seager et al., 2007). Other phenomena with important implications for GHG mitigation, such as woody plant encroachment (e.g. Boutton et al., 2009; Neff et al., 2009), likely have been derived from several of these factors (Morgan et al., 2008, 2010; Van Auken, 2009).

KNOWLEDGE GAPS AND FUTURE RESEARCH NEEDS

Knowledge gaps, both research and informational, exist for mitigation of GHGs in rangelands (Derner and Schuman, 2007; Morgan et al., 2010). First, from the research perspective, almost all the prior experimental efforts have focused on individual GHGs, particularly CO_2. Emissions of CH_4 and N_2O, however, might counteract SOC sequestration and lead to positive or negative GHG balances (Liebig et al., 2010a), but such studies of total GHG budgets are lacking. Second, very limited research efforts have addressed mitigation of GHGs in arid rangelands, particularly shrublands. Third, interactions between management and environment on GHGs in rangelands have revealed interesting preliminary findings (Ingram et al., 2008), but further quantification is needed for more robust interpretations and guidelines for recommendations. Fourth, adaptive management practices that more closely match forage demand with forage availability in highly variable environments (i.e. high intra- and inter-annual precipitation variability) are based largely on experiential knowledge rather than empirical evidence. Finally, simulation models used in broad national inventory estimates may not be appropriate for smaller-scale, enterprise-level (i.e. individual ranch) GHG accounting, complicating how conservation practices are selected for targeting land areas that provide the largest potential returns to the general public.

References

Allen, D.E., Pringle, M.J., Page, K.L., Dalal, R.C., 2010. A review of sampling designs for the measurement of soil organic carbon in Australian grazing lands. Rangeland Journal 32, 227–246.

Bagchi, S., Ritchie, M.E., 2010. Introduced grazers can restrict potential soil carbon sequestration through impacts on plant community composition. Ecol. Lett. 13, 959–968.

Batjes, N.H., Sombroek, W.G., 1997. Possibilities for carbon sequestration in tropical and subtropical soils. Global Change Biol. 3, 161–173.

Berdahl, J.D., Wilton, A.C., Frank, A.B., 1989. Survival and agronomic performance of 25 alfalfa cultivars and strains interseeded into rangelands. J. Range. Manage. 42, 312–316.

Biondini, M.E., Patton, B.D., Nyren, P.E., 1998. Grazing intensity and ecosystem processes in a northern mixed-grass prairie, USA. Ecol. Applic. 8, 469–479.

Boutton, T.W., Liao, J.D., Filley, T.R., Archer, S.R., 2009. Belowground carbon storage and dynamics accompanying woody plant encroachment in a subtropical savanna. In: Lal, R., Follett, R.F. (Eds.), Soil Carbon Sequestration and the Greenhouse Effect, second ed. SSSA Spec. Publ. 57. ASA-CSSA-SSSA, Madison, WI, pp. 181–205.

Bremmer, D.J., Ham, J.M., 2010. Net carbon fluxes over burned and unburned native tallgrass prairie. Range. Ecol. Manage. 63, 72–81.

Briske, D.D., Derner, J.D., Brown, J.R., Fuhlendorf, S.D., Teague, W.R., Havstad, K.M., Gillen, R.L., Ash, A.J., Willms, W.D., 2008. Rotational grazing on rangelands: reconciliation of perception and experimental evidence. Range. Ecol. Manage. 61, 3–17.

Briske, D.D., Fuhlendorf, S.D., Smeins, F.E., 2005. State-and-transition models, thresholds, and rangeland health: a synthesis of ecological concepts and perspectives. Range. Ecol. Manage. 58, 1–10.

Carney, K.M., Hungate, B.A., Drake, B.G., Megonigal, J.P., 2007. Altered soil microbial community at elevated CO_2 leads to loss of soil carbon. Proc. Nat. Acad. Sci. 104, 4990–4995.

Ciais, P., Reichstein, M., Viovy, N., Granier, A., Ogee, J., Allard, V., Aubinet, M., Buchmann, N., Bernhofer, C., Carrara, A., Chevallier, F., De Noblet, N., Friend, A.D., Friedlingstein, P., Grunwald, T., Heinesch, B., Keronen, P., Knohl, A., Krinner, G., Lousatu, D., Manaca, G., Matteucci, G., Miglietta, F., Ourcival, J.M., Papale, D., Pilegaard, K., Rambal, S., Seufert, G., J-Soussana, F., Sanz, M.J., Schulze, D.E., Vesala, T., Valentini, R., 2005. An unprecedented reduction in the primary productivity of Europe during 2003 caused by heat and drought. Nature 437, 529–532.

Conant, R.T., Six, J., Paustian, K., 2003. Land use effects on soil carbon fractions in the southeastern United States. I. Management-intensive versus extensive grazing. Biol. Fertil. Soils 38, 386–392.

Conant, R.T., Paustian, K., Elliot, E.T., 2001. Grassland management and conversion into grassland: effects on soil carbon. Ecol. Applic. 11, 343–355.

Dai, X., Boutton, T.W., Glaser, B., Ansley, R.J., Zech, W., 2005. Black carbon in a temperate mixed-grass savanna. Soil Biol. Biochem. 37, 1879–1881.

Derner, J.D., Schuman, G.E., 2007. Carbon sequestration and rangelands: a synthesis of land management and precipitation effects. J. Soil Water Conserv. 62, 77–85.

Derner, J.D., Hart, R.H., 2007. Grazing-induced modifications to peak standing crop in northern mixed-grass prairie. Range. Ecol. Manage. 60, 270–276.

Derner, J.D., Boutton, T.W., Briske, D.D., 2006. Grazing and ecosystem carbon storage in the North American Great Plains. Plant Soil 280, 77–90.

Emmerich, W.E., 2003. Carbon dioxide fluxes in a semiarid environment with high carbonate soils. Agric. For. Meteorol. 116, 91–102.

EPA (Environmental Protection Agency), 2011. Draft inventory of U.S. greenhouse gas emissions and sinks: 1990–2009. http://www.epa.gov/climatechange/emissions/usinventoryreport.html. accessed March 7, 2011.

Follett, R.F., Reed, D., 2010. Soil carbon sequestration in grazing lands: societal benefits and policy implications. Range. Ecol. Manage. 63, 4–15.

Follett, R.F., Kimble, J.M., Lal, R., 2001. The Potential of U.S. Grazing Lands to Sequester Carbon and Mitigate the Greenhouse Effect. CRC Press, Boca Raton, FL.

Frank, A.B., Tanaka, D.L., Hofmann, L., Follett, R.F., 1995. Soil carbon and nitrogen of Northern Great Plains grasslands as influenced by long-term grazing. J. Range. Manage. 48, 470–474.

Fuhlendorf, S.D., Zhang, H., Tunnell, T.R., Engle, D.M., Cross, A.F., 2002. Effects of grazing on restoration of southern mixed prairie soils. Restor. Ecol. 10, 401–407.

Ganjegunte, G.K., Vance, G.F., Preston, C.M., Schuman, G.E., Ingram, L.J., Stahl, P.D., Welker, J.M., 2005. Soil organic carbon composition in a northern mixed-grass prairie: effects of grazing. Soil Sci. Soc. Am. J. 69, 1746–1756.

Gile, L.H., Hawley, J.W., Grossman, R.B., 1981. Soils and Geomorphology in the Basin and Range Area of Southern New Mexico—Guidebook to the Desert Project. New Mexico Bureau of Mines and Mineral Resources. Memoir. 39, Socorro, NM.

Glaser, B., Amelung, W., 2003. Pyrogenic carbon in native grassland soils along a climosequence in North America. Global Biogeochem. Cycles 17, 331–338.

Guo, Y., Amundson, R., Gong, P., Yu, Q., 2006. Quantity and spatial variability of soil carbon in the conterminous United States. Soil Sci. Soc. Am. J. 70, 590–600.

Hatfield, J.L., Boote, K.J., Kimball, B.A., Wolfe, D.W., Ort, D.R., Izaurralde, C.R., Thomson, A.M., Morgan, J.A., Polley, H.W., Fay, P.A., Mader, T.L., Hahn, G.L., 2008. Agriculture. In: The Effects of Climate Change on Agriculture, Land Resources, Water Resources, and Biodiversity. U.S. Climate Change Science Program and the Subcommittee on Global Change Research, Washington, DC, p. 362.

Havstad, K.M., Peters, D.P.C., Skaggs, R., Brown, J., Betelmeyer, B., Fredrickson, E., Herrick, J., Wright, J., 2007. Ecological services to and from rangelands of the United States. Ecol. Econ 64, 261–268.

Heinrichs, D.H., 1975. Potential of legumes for rangelands. In: Campbell, R.S., Herbel, C.H. (Eds.), Improved Range Plants. Range Symposium Series 1. Society for Range Management, Denver, Colorado, pp. 50–61.

Ingram, L.J., Stahl, P.D., Schuman, G.E., Buyer, J.S., Vance, G.F., Ganjegunte, G.K., Welker, J.M., Derner, J.D., 2008. Grazing impacts on soil carbon and microbial communities in a mixed-grass ecosystem. Soil Sci. Soc. Am. J. 72, 939–948.

Janzen, H.H., 2001. Soil science on the Canadian prairies—peering into the future from a century ago. Can. J. Soil Sci. 81, 489–503.

Jin, V.L., Evans, R.D., 2010. Microbial ^{13}C utilization patterns via stable isotope probing of phospholipid biomarkers in Mojave Desert soils exposed to ambient and elevated atmospheric CO_2. Glob. Change Biol. 16, 2334–2344.

Jin, V.L., Schaeffer, S.M., Ziegler, S.E., Evans, R.D., 2011. Soil water availability and microsite mediate fungal and bacterial phospholipid fatty acid biomarker abundances in Mojave Desert soils exposed to elevated atmospheric CO_2. J. Geophys. Res. 116 G02001, doi:10.1029/2010JG001564.

Jones, M.B., Donnelly, A., 2004. Carbon sequestration in temperate grassland ecosystems and the influence of management, climate and elevated CO_2. New Phytol. 164, 423–439.

Knapp, A.K., 1985. Effect of fire and drought on the ecophysiology of *Andropogon gerardii* and *Panicum virgatum* in a tallgrass prairie. J. Ecol. 66, 1309–1320.

Knapp, A.K., Conrad, S.L., Blair, J.M., 1998. Determinants of soil CO_2 flux from a sub-humid grassland: effect of fire and fire history. Ecol. Applic. 8, 760–770.

Kruger, C.F., Vigil, F.R., 1979. Interseeding alfalfa for grassland improvement in the Northern Great Plains. Report of the Twenty-Sixth Alfalfa Improvement Conference, Brookings, South Dakota.

Kuchler, A.S., 1964. Potential natural vegetation of the conterminous United States. American Geographical Society, Special Publication No. 36.

Kuhlbusch, T.A.J., 1998. Black carbon and the carbon cycle. Science 280, 1903–1904.

Kuzyakov, Y., 2002. Review: factors affecting rhizosphere priming effects. J. Plant Nutr. Soil Sci. 165, 382–396.

Lal, R., Follett, R.F., Kimble, J.M., 2003. Achieving soil carbon sequestration in the United States: a challenge to the policy makers. J. Soil Sci. 168, 827–845.

Liebig, M.A., Dong, X., McLain, J.E.T., Dell, C.J., 2012. Greenhouse gas flux from managed grasslands in the U.S. In: Liebig, M.A., Franzluebbers, A.J., Follett, R.F. (Eds.), Managing Agricultural Greenhouse Gases: Coordinated Agricultural Research through GRACEnet to Address our Changing Climate, Academic Press.

Liebig, M.A., Morgan, J.A., Reeder, J.D., Ellert, B.H., Gollany, H.T., Schuman, G.E., 2005. Greenhouse gas contributions and mitigation potential of agricultural practices in northwestern USA and western Canada. Soil Tillage Res. 83, 25–52.

Liebig, M.A., Gross, J.R., Kronberg, S.L., Hanson, J.D., Frank, A.B., Phillips, R.L., 2006. Soil response to long-term grazing in the northern Great Plains of North America. Agric. Ecosyst. Environ. 115, 270–276.

Liebig, M.A., Gross, J.R., Kronberg, S.L., Phillips, R.L., Hanson, J.D., 2010a. Grazing management contributions to net global warming potential: a long-term evaluation in the northern Great Plains. J. Environ. Qual. 39, 799–809.

Liebig, M.A., Hendrickson, J.R., Berdahl, J.D., 2010b. Response of soil carbon and nitrogen to transplanted alfalfa in North Dakota rangeland. Can. J. Soil Sci. 90, 523–526.

Manley, J.T., Schuman, G.E., Reeder, J.D., Hart, R.H., 1995. Rangeland soil carbon and nitrogen response to grazing. J. Soil Water Conserv. 50, 294–298.

Milchunas, D.G., Mosier, A.R., Morgan, J.A., LeCain, D.R., King, J.Y., Nelson, J.A., 2005. Elevated CO_2 and defoliation effects on a shortgrass steppe: forage quality versus quantity for ruminants. Agric. Ecosyst. Environ. 111, 166–184.

Millennium Ecosystem Assessment, 2005. Ecosystems and Human Well-being Synthesis. http://www.millenniumassessment.org/documents/document.356.aspx.pdf. accessed Feb. 22, 2011.

Mitchell, J.E., 2000. Rangeland resource trends in the United States: a technical document supporting the 2000 USDA Forest Service RPA Assessment. Source: Gen. Tech. Rep. RMRS-GTR-68. U.S. Department of Agriculture, Forest Service, Rocky Mountain Research Station, Fort Collins, CO.

Monger, H.C., Martinez-Rios, J.J., 2001. Inorganic carbon sequestration in grazing lands. In: The Potential of U.S. Grazing Lands to Sequester Carbon and Mitigate the Greenhouse Effect. CRC Press, Boca Raton, FL, pp. 87–118.

Morgan, J.A., Derner, J.D., Milchunas, D.G., Pendall, E., 2008. Management implications of global change for Great Plains rangelands. Rangelands 30 (3), 18–22.

Morgan, J.A., Follett, R.F., Allen Jr., L.H., Grosso, S.D., Derner, J.D., Dijkstra, F., Franzluebbers, A., Fry, R., Paustian, K., Schoeneberger, M.M., 2010. Carbon sequestration in agricultural lands of the United States. J. Soil Water Conserv. 65, 6A–13A.

Mortenson, M.C., Schuman, G.E., Ingram, L.J., 2004. Carbon sequestration in rangelands interseeded with yellow-flowering alfalfa (*Medicago sativa* ssp. *falcata*). Environ. Manage. 33, S475–S481.

Mortenson, M.C., Schuman, G.E., Ingram, L.J., Nayigihugu, V., Hess, B.W., 2005. Forage production and quality of a mixed-grass rangeland interseeded with *Medicago sativa* ssp. *falcata*. Range. Ecol. Manage. 58, 505–513.

Neff, J.C., Barger, N.N., Baisden, W.T., Fernandez, D.P., Asner, G.P., 2009. Soil carbon responses to expanding pinyon-juniper populations in southern Utah. Ecol. Applic. 19, 1405–1416.

Ojima, D.S., Dirks, B.O., Glenn, E.P., Owensby, C.E., Scurlock, J.O., 1993. Assessment of C budget for grasslands and drylands of the world. Water Air Soil Pollut. 70, 95–109.

Owensby, C.E., Ham, J.M., Auen, L.M., 2006. Fluxes of CO_2 from grazed and ungrazed tallgrass prairie. Range. Ecol. Manage. 59, 111–127.

Pacala, S., Birdsey, R.A., Bridgham, S.D., Conant, R.T., Davis, K., Hales, B., Houghton, R.A., Jenkins, J.C., Johnston, M., Marland, G., Paustian, K., Caspersen, J., Socolow, R., Tol, R.S.J., 2007. The North American carbon budget past and present. In: King, A.W., Dilling, L., Zimmerman, G.P., Fairman, D.M., Houghton, R.A., Marland, G., Rose, A.Z., T.J. (Eds.), Synthesis and Assessment Product 2.2, The First State of the Carbon Cycle Report: The North American Carbon Budget and Implications for the Global Carbon Cycle. National Oceanic and Atmospheric Administration, National Climatic Data Center, Asheville, NC, USA, pp. 29–36.

Piñeiro, G., Paruelo, J.M., Oesterheld, M., Jobbágy, E.G., 2010. Pathways of grazing effects on soil organic carbon and nitrogen. Range. Ecol. Manage. 63, 109–119.

Rau, B.M., Tausch, R., Reiner, A., Johnson, D.W., Chambers, J.C., Blank, R.R., Lucchesi, A., 2010. Influence of prescribed fire on ecosystem biomass, carbon, and nitrogen in a pinyon juniper woodland. Range. Ecol. Manage. 63, 197–202.

Reeder, J.D., Schuman, G.E., 2002. Influence of livestock grazing on C sequestration in semi-arid mixed-grass and short-grass rangelands. Environ. Pollut. 116, 457–463.

Reheis, M.C., Goodmacher, J.C., Harden, J.W., McFadden, L.D., Rockwell, T.K., Shroba, R.R., Sowers, J.M., Taylor, E.M., 1995. Quaternary soils and dust deposition in southern Nevada and California. Geol. Soc. Am. Bull. 107, 1003–1022.

Reheis, M.C., Sowers, J.M., Taylor, E.M., McFadden, L.D., Harden, J.W., 1992. Morphology and genesis of carbonate soils on the Kyle Canyon fan, Nevada, USA. Geoderma. 52, 303–342.

Schimel, D.S., Parton, W.J., Kittel, T.G., Ojima, D.S., Cole, C.V., 1990. Grassland biogeochemistry: links to atmospheric processes. Clim. Change 17, 13–25.

Schlesinger, W.H., 1997. Biogeochemistry: Analysis of Global Change, second ed. Academic Press, San Diego, CA.

Schnabel, R.R., Franzluebbers, A.J., Stout, W.L., Sanderson, M.A., Stuedemann, J.A., 2001. The effects of pasture management practices. In: Follett, R.F., Kimble, J.M., Lal, R. (Eds.)., The Potential of U.S. Grazing Lands to Sequester Carbon and Mitigate the Greenhouse Effect. CRC Press, Boca Raton, FL, pp. 291–322.

Scholes, R.J., Archer, S.R., 1997. Tree–grass interactions in savannas. Ann. Rev. Ecol. System 28, 517–544.

Schuman, G.E., Reeder, J.D., Manley, J.T., Hart, R.H., Manley, W.A., 1999. Impact of grazing management on the carbon and nitrogen balance of a mixed-grass rangeland. Ecol. Applic. 9, 65–71.

Seager, R., Ting, M., Held, I., Kushnir, Y., Lu, J., Vecchi, G., Huang, H., Harnik, N., Leetmaa, A., Lau, N., Li, C., Velez, J., Naik, N., 2007. Model projections of an imminent transition to a more arid climate in southwestern North America. Science (Washington, DC) 316, 1181–1184.

Six, J., Feller, C., Denef, K., Ogle, S.M., de Moraes, J.C., Albrecht, A., 2002. Soil organic matter, biota and aggregation in temperate and tropical soils—effects of no-tillage. Agronomy 22, 755–775.

Skjemstad, J.O., Taylor, J.A., Smernik, R.J., 1999. Estimation of charcoal (char) in soils. Comm. Soil Sci. Plant Anal. 30, 2283–2298.

Smoliak, S., Dormaar, J.F., Johnston, A., 1972. Long-term grazing effects on *Stipa-Bouteloua* prairie soils. J. Range. Manage. 25, 246–250.

Sobecki, T.M., Moffitt, D.L., Stone, J., Franks, C.D., Mendenhall, A.G., 2001. A broad-scale perspective on the extent, distribution, and characteristics of U.S. grazing lands. In: Follett, R.F., Kimble, J.M., Lal, R. (Eds.). The Potential of U.S. Grazing Lands to Sequester Carbon and Mitigate the Greenhouse Effect. CRC Press, Boca Raton, FL, pp. 21–63.

Soussana, J.-F., Lüschert, A., 2007. Temperate grasslands and global atmospheric change: a review. Grass Forage Sci. 62, 127–134.

Svejcar, T., Angell, R., Bradford, J.A., Dugas, W., Emmerich, W., Frank, A.B., Gilmanov, T., Haferkamp, M., Johnson, D.A., Mayeux, H., Mielnick, P., Morgan, J., Saliendra, N.Z., Schuman, G.E., Sims, P.L., Snyder, K., 2008. Carbon fluxes on North American rangelands. Range. Ecol. Manage. 61, 465–474.

Svejcar, T.J., Browning, J.A., 1988. Growth and gas exchange of *Andropogon gerardii* as influenced by burning. J. Range. Manage. 41, 239–244.

Swift, R.S., 2001. Sequestration of carbon by soil. J. Soil Sci. 166, 858–871.

Teague, W.R., Dowhower, S.L., Baker, S.A., Ansley, R.J., Kreuter, U.P., Conover, D.M., Waggoner, J.A., 2010. Soil and herbaceous plant responses to summer patch burns under continuous and rotational grazing. Agric. Ecosyst. Environ. 137, 113–123.

Tesar, M.B., Jakobs, J.A., 1972. Establishing the stand. In: Hanson, C.H. (Ed.), Alfalfa Science and Technology, Agronomy Monograph No. 15. American Society of Agronomy, Madison, Wisconsin, pp. 415–435.

USDA (United States Department of Agriculture), 2008. Agriculture and Forestry Greenhouse Gas Inventory: 1990–2005. Global Change Program Office, Office of the Chief Economist, U.S. Department of Agriculture. U.S. Tech. Bull. No. 1921.

USDA (US Department of Agriculture), 2009. Summary report: 2007 National Resources Inventory. Washington, DC, Natural Resources Conservation Service. Ames, IA: Center for Survey Statistics and Methodology, Iowa State University.

USDA-ERS (Economic Research Service), 2006. Major uses of land in the United States, 2002. Econ. Info. Bull. No. 14.

USDA-ERS, 2010. U.S. Beef and Cattle Industry: Background Statistics and Information. http://www.ers.usda.gov/news/BSECoverage.htm. accessed Jan. 12, 2011.

USDA-NRCS (Natural Resources Conservation Service), 2003. National Range and Pasture Handbook. 190-V1-NRPH, rev. 1.

USGCRP (United States Global Change Research Program), 2009. Global Climate Change Impacts in the United States. In: Karl, T.R., Melillo, J.M., Peterson, T.C. (Eds.). Cambridge University Press.

Van Auken, O.W., 2009. Causes of consequences of woody plant encroachment into western North American grasslands. J. Environ. Manage. 90, 2931–2942.

Warren, S.D., Blackburn, W.H., Taylor Jr., C.A., 1986. Soil hydrologic response to number of pastures and stocking density under intensive rotation grazing. J. Range. Manage. 39, 500–504.

Westoby, M., 1980. Elements of a theory of vegetation dynamics in arid rangelands. Isr. J. Bot. 28, 169–194.

Wolf, B., Zheng, X., Bruggemann, N., Dannenmann, M., Han, X., Sutton, M.A., Wu, H., Yao, Z., Butterbach-Bahl, K., 2010. Grazing-induced reduction of natural nitrous oxide release from continental steppe. Nature 464, 881–884.

Wood, M.K., Blackburn, W.H., 1984. Vegetation and soil responses to cattle grazing systems in the Texas rolling plains. J. Range. Manage. 37, 303–308.

Young, J.A., Clements, C.D., 2001. Range research: the second generation. J. Range Manage. 54, 115–121.

Zhang, L., Wylie, B.K., Ji, L., Gilmanov, T.G., Tieszen, L.L., 2010. Climate driven interannual variability in net ecosystem exchange in the northern Great Plains grasslands. Range. Ecol. Manage. 63, 40–50.

CHAPTER 7

Soil Organic Carbon under Pasture Management

Alan J. Franzluebbers[1], Lloyd B. Owens[2], Gilbert C. Sigua[3], Cynthia A. Cambardella[4], Richard L. Haney[5]

[1]USDA—Agricultural Research Service, Watkinsville GA
[2]USDA—Agricultural Research Service, Coshocton OH
[3]USDA—Agricultural Research Service, Brooksville FL
[4]USDA—Agricultural Research Service, Ames IA, USA
[5]USDA—Agricultural Research Service, Temple TX

CHAPTER OUTLINE

CLIMATE AND SOILS OF THE EASTERN U.S.

The eastern U.S. is classified broadly in the moist, subtropical, mid-latitude climate zone of the Köppen climate classification system. The region can be broadly subdivided into a cool-humid region in the north and a warm-humid region in the south (Figure 7.1). It is characterized by moderate to high precipitation, i.e. generally >750 mm. Precipitation is greatest in the Southern Appalachian region of western North Carolina and north Georgia (1750−2000 mm) and in the Gulf Coast region of Louisiana, Mississippi, Alabama, and Florida (1500−1750 mm). Mean annual temperature varies from 5°C in the north to 25°C in the south. Precipitation-to-potential evapotranspiration (P/PET) is generally ≥1, although there can be significant periods in summer in which precipitation is less than potential evapotranspiration (Figure 7.1). Excess precipitation often occurs during winter.

Soils in the eastern U.S. were predominantly developed under deciduous forest ecosystems, but portions of the western and northern fringes of the region were developed under prairie. Soils vary widely in the region, including Mollisols in the western portion of the region, Alfisols in the central portion of the region, Spodosols and Inceptisols in the northeastern portion of the region, Ultisols in the southeastern portion of the region, Vertisols sporadically

Managing Agricultural Greenhouse Gases. DOI: 10.1016/B978-0-12-386897-8.00007-3

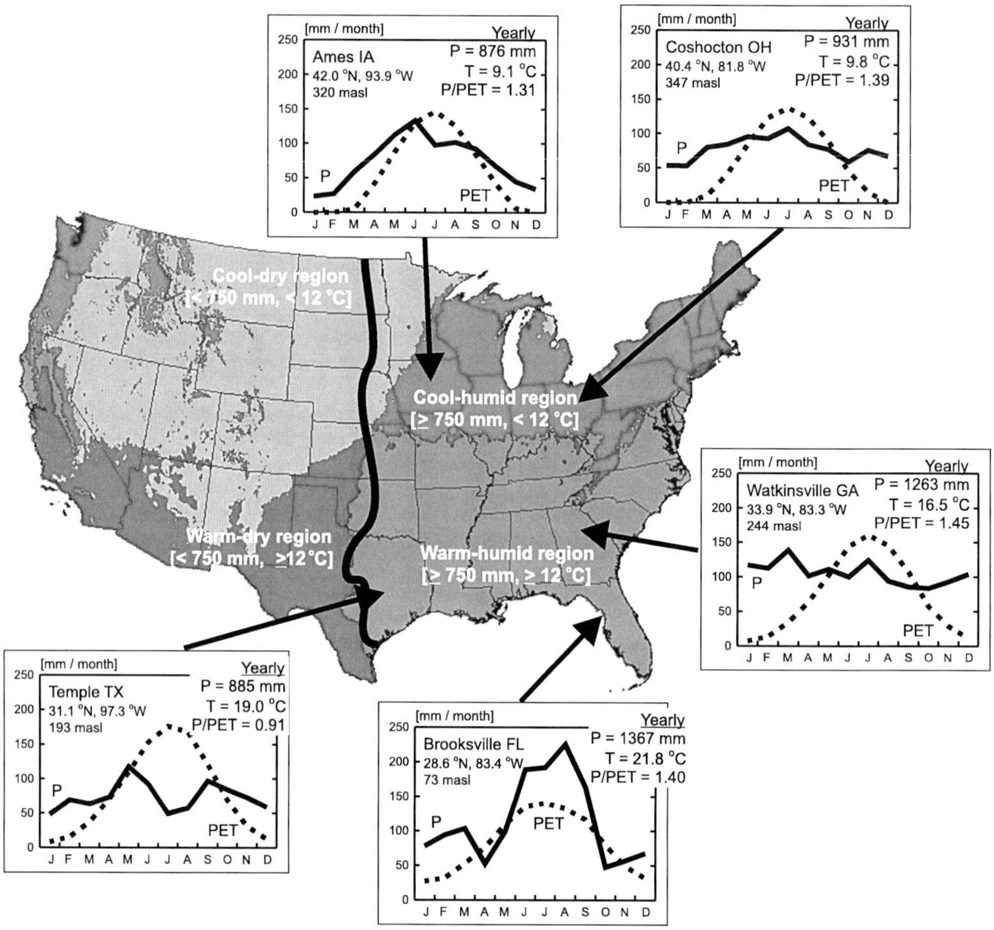

FIGURE 7.1
Climatic characteristics with focus on specific locations in the cool-humid and warm-humid regions of the eastern U.S. P is precipitation (mm) , T is temperature (°C), and PET is potential evapotranspiration (mm).

throughout, Entisols in major floodplains, and Histosols in Florida, coastal shorelines, and Great Lakes regions.

DESCRIPTION OF PASTURE TYPES, USES, AND EXTENT

Grasslands in the eastern U.S. are diverse, depending on climatic conditions, previous land use, landscape position, dominant soil types, and management inputs. Grassland pastures are often relegated to landscape positions, in which more profitable horticultural and field crops cannot be grown. Unfortunately, a large portion of pastures in the eastern U.S. are relatively unmanaged or managed with minimal attention to achieve full production potential. Management could be improved in general, but some specific strategies will be noted in this chapter with regards to SOC sequestration.

Pastures occupy a significant portion of farmland throughout the eastern U.S. (Table 7.1). Across 31 states in the eastern U.S. (Figure 7.1), farmland occupies 122.2 Mha (or 41% of the total land area) (USDA-NASS, 2011). Of this farmland, crops are grown on 65% of the land (79.6 Mha), woodlands occupy 17% of the land (20.5 Mha), permanent pastures occupy 13% of the land (15.7 Mha), and conservation reserve lands occupy 4% of the land (5.3 Mha). Total grazing land composed of perennial pastures, conservation reserve, and grazed cropland and

TABLE 7.1 Percentage of Land in Farming Activities among 31 States in the Eastern U.S. Data from USDA-NASS (2011)

Distribution (%)	Total land as farmland	Farmland as cropland	Farmland as perennial pasture	Farmland as woodland
0–10	ME, NH		DE, IN, IL, MI, ME, NH, NJ, CT, RI, MA	IA, IL, IN, MN, DE
10–20	RI, MA, CT, NJ		OH, WI, NY, VT, MD, PA, MN, IA, NC, GA, SC	OH, MI, LA, MO, AR, NJ, MD, TN, WI
20–30	VT, NY, WV, SC, FL, NC, PA, GA, MI, AL, LA	WV, NH	AR, KY, MS, LA, TN, VA, AL, MO	NY, PA, KY, FL, NC, VA
30–40	VA, MD, MS	FL, AL, RI, MA, ME	WV, FL	CT, MS, GA, AL, SC, WV
40–50	TN, AR, WI	CT, VA, VT, SC, GA, MS		VT, MA, RI, ME
50–60	MN, OH, KY	KY, TN, MO, NC, LA		NH
60–70	IN, MO	NY, AR, PA, WI, NJ, MD		
70–80	IL	OH, MI		
80–90	IA	MN, DE, IA, IN, IL		

woodland accounts for 31.2 Mha in the region. Large differences in these land-use percentages occur among states, as noted in Table 7.1.

Compared with the U.S. in total, these 31 eastern states occupy only 33% of the 916.2 Mha of total land area, 33% of the 373.3 Mha of farmland, 9% of the 165.5 Mha of perennial pasture or rangeland, and 34% of the 15.6 Mha of conservation reserve land (USDA-NASS, 2011). However, they occupy 48% of the 164.5 Mha of cropland and 67% of the 34.4 Mha of woodland on farms. The eastern U.S. region also accounts for 65% of the 307 million people in the U.S., 62% of the 2.2 million farms, 58% of the nearly 1 million farms with cattle (39% of the 96.3 million head of cattle and calves), and 84% of the 70,000 farms in the country with dairy cows (54% of the 9.3 million dairy cows) (U.S. Census Bureau, 2011).

In Iowa and most of the Midwest, cool-season grass pastures are the base forage for beef cow herds. Cool-season pastures produce growth in spring and early summer, and therefore continuous stocking leads to underutilization in spring and early summer and overutilization in late summer and autumn. A problem with continuous grazing is that as ungrazed forage matures its nutritive value declines. Forage species of importance in Iowa are alfalfa (*Medicago sativa*), red clover (*Trifolium pratense*), birdsfoot trefoil (*Lotus corniculatus*), orchardgrass (*Dactylis glomerata*), reed canarygrass (*Phalaris arundinacea*), and smooth brome (*Bromus inermis*). Some additional forage species of interest are kura clover (*Trifolium ambiguum*), annual medics (*Medicago* sp.), tall fescue (*Lolium arundinaceum*), perennial ryegrass (*Lolium perenne*), timothy (*Phleum pratense*), and Eastern gamagrass (*Tripsacum dactyloides*).

In Ohio, grazing of pastures is typically with continuously stocked cow-calf herds. Rotational grazing by dairy herds does occur, especially among Amish producers. Grazing by horses, sheep, and other species (e.g. alpacas and llamas) also occurs. Common pasture species are orchardgrass, tall fescue (endophyte infected), Kentucky bluegrass (*Poa pratensis*), perennial ryegrass, and a variety of non-desirable species. Most improved pastures also contain alfalfa, white clover (*Trifolium repens*), and/or red clover. Ohio has approximately 0.7 Mha of pasture (~12% of the agricultural land), although these lands are more common in the southeast part of the state. The Appalachian Foothills have rolling to steep unglaciated topography that is poorly suited to cropping, but suitable for pasture.

In Texas, there is an increasing trend for conversion of native rangeland and croplands to non-native pastures (Wilkins et al., 2009). Much of this conversion is to "Coastal" bermudagrass (*Cynodon dactylon*) used either for grazing or for hay harvest. Non-native pastures now account for ~4.4 Mha; the third largest land-use category in Texas. Pastures are typically continuously grazed with small herds of cows and calves, as well as with an increasing population of goats.

The Piedmont region of the southeastern U.S. (17 Mha in AL, GA, SC, NC, and VA) has pastures (primarily of tall fescue, common bermudagrass, and a variety of non-desirable species) that are typically continuously grazed with small herds of cows and calves. Pastures are often on highly eroded land following decades of continuous cropping; many contain terraces resulting from the rehabilitation efforts in the mid-20th century prior to pasture conversion. Obviously there are also some pastures in the area with improved species composition (e.g. hybrid bermudagrass, novel-endophyte-infected tall fescue), improved soil fertility, improved grazing methods to balance consumption and conservation, and optimum stocking rate and limited hay harvest. Many forage species are able to grow in the mild climate of the southeastern U.S., including cool-season perennial grasses [tall fescue, orchardgrass, bluegrass, perennial ryegrass, timothy, and rescuegrass (*Bromus catharticus*)], cool-season perennial legumes (alfalfa, red clover, and white clover), warm-season perennial grasses [bahiagrass (*Paspalum notatum*), bermudagrass, dallisgrass (*Paspalum dilatatum*), johnsongrass (*Sorghum halepense*), big bluestem (*Andropogon gerardii*), eastern gamagrass, indiangrass (*Sorghastrum nutans*), little bluestem (*Schizachyrium scoparium*), and switchgrass (*Panicum virgatum*)], and warm-season perennial legumes [perennial peanut (*Arachis glabrata*) and sericea lespedeza (*Lespedeza cuneata*)]. Cool-season annual forages include annual ryegrass (*Lolium annuum*), barley (*Hordeum vulgare*), oat (*Avena sativa*), rye (*Secale cereale*), triticale (×*Triticosecale rimpaui*), wheat (*Triticum aestivum*), arrowleaf clover (*Trifolium vesiculosum*), ball clover (*Trifolium nigrescens*), berseem clover (*Trifolium alexandrinum*), crimson clover (*Trifolium incarnatum*), rose clover (*Trifolium hirtum*), subterranean clover (*Trifolium subterraneum*), vetch (*Vicia villosa*), and winter pea (*Pisum sativum*). Warm-season annual forages include browntop millet (*Urochloa ramosa*), crabgrass (*Digitaria sanguinalis*), corn (*Zea mays*), foxtail millet (*Setaria italica*), pearl millet (*Pennisetum glaucum*), sorghum (*Sorghum bicolor*), teff (*Eragrostis tef*), annual lespedeza (*Kummerowia stipulacea*), and soybean (*Glycine max*).

In the subtropical region of Florida, even greater choices are available on the species of plants used for forage. Warm-season perennials of bahiagrass, bermudagrass, stargrass (*Cynodon plectostachyus*), and St. Augustine grass (*Stenotaphrum secundatum*) are the most dominant pasture types. Other warm-season grasses, in addition to those described earlier for other regions, include atra paspalum (*Paspalum atratum*), hybrid brachiaria (*Brachiaria brizantha x B. ruziziensis*), digitgrass (*Digitaria eriantha*), elephantgrass (*Pennisetum purpureum*), guineagrass (*Urochloa maxima*), limpograss (*Hemarthria altissima*), paragrass (*Urochloa mutica*), rhodesgrass (*Chloris gayana*), broomsedge bluestem (*Andropogon virginicus*), chalky bluestem (*Andropogon capillipes*), Florida paspalum (*Paspalum floridanum*), hairy panicum (*Panicum anceps*), blue maindencane (*Amphicarpum purshii*), lopsided indiangrass (*Sorghastrum secundum*), pineland threeawn (*Aristida stricta*), and Florida bluestem (*Andropogon floridanus*). Other cool-season legumes not described earlier include black medic (*Medicago lupulina*), lupine (*Lupinus* sp.), Persian clover (*Trifolium resupinatum*), and sweetclover (*Melilotus officinalis*). Other warm-season legumes not described earlier include deer vetch (*Lotus crassifolius*), alyce clover (*Alysicarpus ovalifolius*), carpon desmodium (*Desmodium heterocarpon*), cowpea (*Vigna unguiculata*), creeping beggarweed (*Desmodium incanum*), hairy indigo (*Indigofera hirsuta*), lablab (*Lablab purpureus*), phasey bean (*Macroptilium lathyroides*), and savanna stylo (*Stylosanthes guianensis*).

LONG-TERM LAND-USE EFFECTS ON SOIL ORGANIC C

Pasture land use can store nearly as much C in soils as forested land, at least in the southeastern U.S. (Franzluebbers, 2005). Additional recent literature supports this high storage potential of pastures. From a compilation of 35 observations across multiple studies in the southeastern U.S., SOC sequestration with pasture establishment (typically compared with cultivated cropland) was 0.84 ± 0.11 Mg C ha^{-1} yr^{-1} (Franzluebbers, 2010a). In a survey of 29 farms in Alabama, Georgia, South Carolina, North Carolina, and Virginia, SOC under pasture was significantly greater than under conventional-tillage cropping at depths of 0–5, 5–12.5, and 0–20 cm depths (Figure 7.2). Surface accumulation of SOC was most dominant due to long-term residue cover and lack of soil disturbance (Causarano et al., 2008). In Ohio, an orchardgrass pasture had SOC concentration at a depth of 0–6 cm of 26.4 g kg^{-1}, while four cropped fields contained 20.0 ± 3.6 g kg^{-1} and an oak–hickory–beech–maple forest had 43.0 g kg^{-1} (Wilcox et al., 2002). Soil from *Lumbricus terrestris* middens in this study contained an average of 180% greater coarse litter, 39% greater particulate organic matter, 35% greater soil microbial biomass N, and 19% greater SOC than bulk soil across all these land-use systems. This suggests that more biologically diverse and active surface soils, irrespective of management, can be sources of fertility with high SOC. In Texas on a Udic Haplustert, SOC concentration was 55.7 ± 16.4 g kg^{-1} in the surface 10 cm among four perennial grass species and only 18.5 g kg^{-1} in a cultivated field of corn–wheat rotation (Haney et al., 2010). Biologically active fractions of mineralizable and microbial biomass C and N followed a similar pattern of response.

A total of 267 soil profiles across Georgia were examined for differences in SOC by land use, soil order, and Major Land Resource Region (MLRA) (Franzluebbers, 2010b). These data were originally reported in Perkins (1987). At 5 and 10 cm depths, SOC concentration was greater under pasture and forest than under cropping. At 20, 30, and 40 cm depths, SOC concentration was greater under pasture than under forest and cropping. At 50 and 100 cm depths, SOC concentration was greater under pasture than under cropping; forest was intermediate. When concentration of SOC was converted to content in 205 Ultisol soil profiles, land-use differences appeared in some, but not all, MLRAs (Table 7.2). Pasture land use contained significantly greater SOC stock than cropping land use in Blue Ridge, Piedmont, and Coastal Plain MLRAs, but not in Ridge/Valley and Flatwoods MLRAs. Across all soil orders (205 Ultisols, 35 Alfisols, 11 Entisols, 8 Inceptisols, 6 Spodosols, and 2 Mollisols) and MLRAs, cropping land use had lower SOC content than other land uses at 0–10, 0–30, and 0–100 cm depths. Soil organic C content was not different between pasture and forested land uses. Although this evaluation provided some insight into the potential for land use to alter SOC, there are many management strategies within a land use that could alter SOC content, but these management differences could not be identified from this survey source.

In a natural wetland soil in Florida, drainage and conversion to bahiagrass pasture resulted in a dramatic decline in SOC with 60 years of paired land-use differences (Sigua et al., 2009). Soil

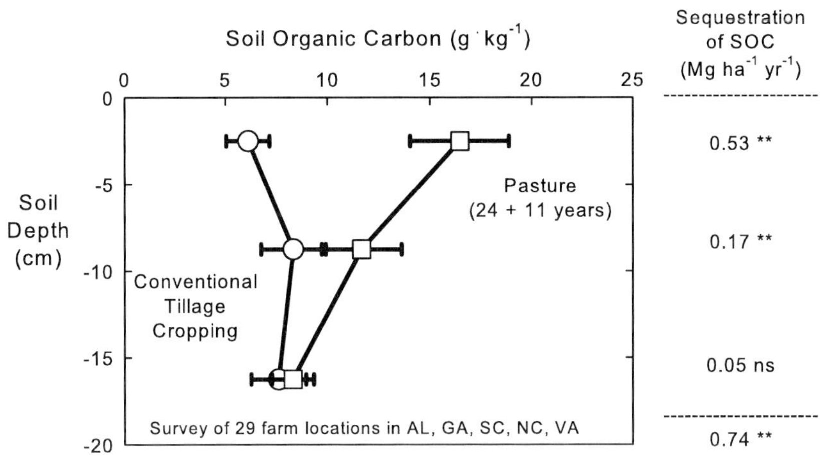

FIGURE 7.2
Soil organic C concentration (graphic) by depth and land use and sequestration rate of soil organic C as a result of pasture management (data column at right; calculated from the difference between land use) by depth. ** indicates significance at $p < 0.01$. Data from Causarano et al. (2008).

TABLE 7.2 Stock of Soil Organic C at Various Depths among Major Land Resource Areas (MLRA; among Ultisols) and General Land-Use Category. All is for all Soil Orders and MLRAs. Data from Perkins (1987) and analyzed in Franzluebbers (2010a).

			Stock of soil organic C (Mg ha^{-1})		
MLRA	*Land use*	*No. obs.*	*0–10 cm*	*0–30 cm*	*0–100 cm*
Blue Ridge	Crop	7	21.5	46.7	80.7
	Pasture	2	34.8	76.7	120.4
	Forest	5	26.4	48.9	76.7
	Alternate	3	37.3	72.1	108.6
	$LSD_{(p\,=\,0.05)}$		13.4*	24.7*	33.7*
Ridge and Valley	Crop	13	23.6	49.0	76.7
	Pasture	4	29.0	48.2	69.2
	Forest	4	27.1	52.9	79.6
	Alternate	4	26.5	59.2	99.1
	$LSD_{(p\,=\,0.05)}$		13.1	24.9	29.7*
Piedmont	Crop	27	19.5	40.0	64.6
	Pasture	8	29.6	53.8	78.6
	Forest	12	28.0	55.9	86.4
	$LSD_{(p\,=\,0.05)}$		5.1*	9.6*	13.9*
Coastal Plain	Crop	67	16.8	34.7	57.1
	Pasture	13	25.0	47.2	70.1
	Forest	12	25.6	48.9	83.6
	Alternate	10	20.9	46.2	78.1
	$LSD_{(p\,=\,0.05)}$		7.0*	13.6*	23.8*
Flatwoods	Crop	2	15.1	25.8	33.4
	Pasture	2	9.2	20.9	34.0
	Forest	10	17.7	28.5	42.6
	$LSD_{(p\,=\,0.05)}$		10.4	14.9	18.3
All	Crop	130	19.1	39.9	66.1
	Pasture	49	25.9	52.5	86.2
	Forest	68	25.0	48.3	77.7
	Alternate	20	25.2	53.2	85.5
	$LSD_{(p\,=\,0.05)}$		3.7*	7.4*	12.9*

**indicates a difference between at least two pairs of means.*
Alternate land use was a mixture of peach and apple orchards, idle land, native weeds, and rushes.

organic C at a depth of 0–20 cm was 180 g kg^{-1} under natural wetland and 7.8 g kg^{-1} under grazed pasture. Another sampling event comparing natural wetland with the initial return to wetland condition following 60 years of pasture management resulted in differences in SOC concentration at a depth of 0–20 cm (149 ± 69 vs. 21 ± 5 g kg^{-1}, respectively), but no differences between land uses at various depth increments to 100 cm (average of 6 ± 3 g kg^{-1}). Clearly this wetland-area study emphasizes the unique site-specific nature of land-use conversion on SOC compared with more upland conditions.

Other land-use comparisons of SOC have been conducted in countries with similar environmental conditions as the eastern U.S. In the Netherlands, a Fluventic Eutrodept under long-term (>70 years) pasture had greater SOC concentration than under cultivated cropland at 3–10 cm depth (34 ± 2 vs. 11 ±1 g kg^{-1}) and at 10–20 cm depth (20 ± 1 vs. 10 ± 1 g kg^{-1}) (Pulleman and Marinissen, 2004). Concentrations converged eventually at 30–50 cm depth. On a Vertic Hyploxeroll in Oregon (1024 mm annual precipitation), organic C at the end of 11 years of replicated and randomized treatment arrangement was significantly greater in the surface 15 cm of soil under perennial ryegrass/subclover (41.0 Mg ha^{-1}) than under Douglas fir plantation (34.1 Mg ha^{-1}) and intermediate under agroforestry combining pasture and woods (37.2 Mg ha^{-1}) (Sharrow and Ismail, 2004). At a depth of 15–45 cm, no differences among treatments occurred, but statistical significance among treatments was maintained when calculated to 0–45 cm depth. However, total system C accumulation (above- and belowground components) followed the order: agroforestry (109.3 Mg C ha^{-1}) > ryegrass/subclover pasture (103.5 Mg C ha^{-1}) = Douglas fir plantation (101.1 Mg C ha^{-1}).

A study of various agricultural land-use conditions on selected farms in Brazil resulted in estimates of both SOC sequestration and loss compared with the condition of native vegetation (Carvalho et al., 2010). Non-degraded pasture under fertile soil condition had greater SOC to 30 cm depth than native vegetation at one site and was not different at another site. At a site with degraded pasture, SOC was lower than with native vegetation. Converting pasture to annual cropping resulted in lower SOC than in non-degraded pasture, but SOC was similar with cropping as with degraded pasture. Implementing an integrated crop–livestock system on these sites following a few years of sole cropping resulted in an increase in SOC of 7.4 ± 4.6 Mg C ha^{-1}. In Chile, cropping systems rotated with 2 to 5 years of pasture had greater total and macroaggregate-associated organic C than continuous cropping systems (Sandoval et al., 2007).

PASTURE MANAGEMENT EFFECTS ON SOIL ORGANIC C

Type of pasture management can have a significant influence on SOC sequestration (Franzluebbers, 2010c). For example, if soil fertility is limited, then fertilization can increase primary forage production and subsequent SOC sequestration. In tall fescue pastures in Georgia, SOC was 45.0 Mg ha^{-1} with low fertilization (134-15-56 N-P-K ha^{-1} yr^{-1}) and 47.6 Mg ha^{-1} with high fertilization (336-37-139 N-P-K ha^{-1} yr^{-1}); a large portion of this difference ($P < 0.05$) was a result of accumulation in the particulate organic C fraction. A low to moderate level of fertilization may be more sustainable from an environmental standpoint, since there would be fewer opportunities for leaching, runoff, and denitrification of N, as well as avoiding the high CO_2 cost embedded in production, application, and liming components (0.98 kg CO_2-C kg N^{-1}; West and Marland, 2002). The optimum level of pasture fertilization should also consider the long-term accumulation of organic N (Franzluebbers and Stuedemann, 2010); pasture fertilization in early years feeds not only the plant, but also the organic N pool, while in later years nutrients can be supplied to pasture plants primarily through steady-state mineralization of organic N.

Fertilization of pasture with an organic amendment should increase SOC, given the high C content of the amendment relative to its N-fertilizer value. However, the evidence available to support this effect is not overwhelmingly strong. Two on-farm surveys of pastures in Alabama and Oklahoma found that SOC was 5.7 Mg ha^{-1} greater after one to three decades with broiler litter application than without (Sharpley et al., 1993; Kingery et al., 1994). In a 12-year pasture study in Georgia, SOC was statistically greater with broiler litter than without in only one of four management scenarios (Franzluebbers and Stuedemann, 2010). The calculated rate of SOC sequestration with broiler litter was 0.21 ±0.43 Mg C ha^{-1} yr^{-1} among the four regimes, which was an average retention of 9% of C applied as broiler litter. In a review of literature, retention of C from manure application was estimated as 7 ± 5% in thermic regions and 23 ± 15% in temperate or frigid regions (Franzluebbers and Doraiswamy, 2007), suggesting that manure application could have a more positive impact on SOC accumulation in the northern half of the eastern U.S. due to temperature limitation on decomposition.

Stocking rate of pastures is a key management choice that affects short-term harvest of forage, pasture productivity in the long-term, residual herbage mass, and the consequence of these aboveground responses on soil quality (Monaghan et al., 2005). Unfortunately there have been few studies looking at the impact of cattle stocking rate on long-term changes in SOC in the eastern U.S. At the western fringe of the region in southcentral Oklahoma, SOC declined with increasing stocking rate on a Durant loam (Udertic Argiustoll), but increased with increasing stocking rate on a Teller silt loam (Potter et al., 2001). This effect occurred

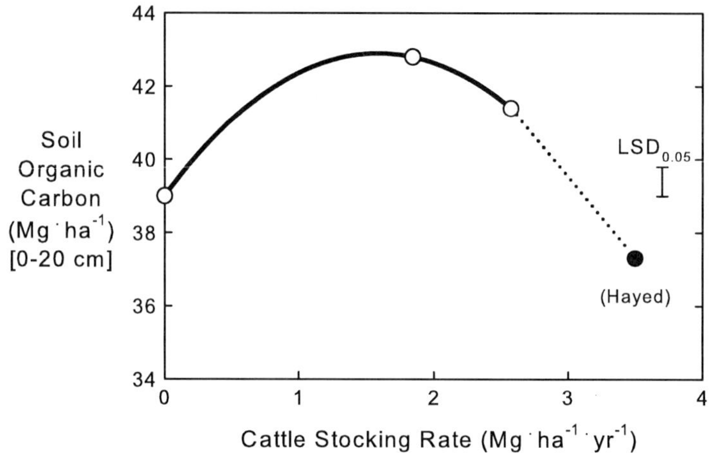

FIGURE 7.3

Soil organic C as affected by 5 years of grazing management (i.e. cattle stocking rate differences) on a Typic Kanhapludult in Georgia. Data from Franzluebbers et al. (2001). Note: A hayed treatment was arbitrarily placed at the highest stocking rate simply to illustrate the effect of increasing pressure on pasture utilization, but the treatment did not have cattle.

throughout the soil profile and was measured at the end of 10 years of grazing management. In Georgia on Typic Kanhapludults, SOC increased to an optimum level with increasing stocking rate, but declined with further pressure (Figure 7.3). Consistent stocking rate effects on SOC were present at the soil surface (Franzluebbers et al., 2001) and throughout the soil profile at the end of 5 years (Franzluebbers and Stuedemann, 2005a) and at the end of 12 years (Franzluebbers and Stuedemann, 2009).

Rotational stocking (or management-intensive grazing) has increasingly received popular-press attention as a potential strategy to rejuvenate degraded pastures, increase productivity, increase profit, and possibly sequester SOC (Henning et al., 2000; Beetz and Rhinehart, 2010). Unfortunately though, there are only a few research data to support the claim of greater SOC. From a field survey of eight pastures in Virginia, Conant et al. (2003) reported 8.2 ± 4.5 Mg ha^{-1} greater SOC under pastures with management-intensive grazing than with extensive grazing. This equated to an SOC sequestration rate of 0.61 Mg C ha^{-1} yr^{-1} during an evaluation period of 14 ± 11 years. There are no other research data in the eastern U.S. on this topic, but some observations have been made in other parts of the world. In a survey of pastures in three counties in northern Texas, SOC was greater with rotational than with continuous grazing (W.R. Teague, personal communication, 2011). At the end of 4 years of differential grazing method in New South Wales, Australia, microarthropod abundance was greater with high intensity–short duration grazing than with set stocking (Tom et al., 2006). However, no differences were found in bulk density, earthworm abundance, soil microbial biomass, and respiration.

Choice of botanical composition of pastures could likely also affect the potential to sequester SOC. Warm-season perennial species are able to fix large quantities of CO_2 during the summer and cool-season perennial species do so in the spring and autumn. Optimizing the botanical composition of pastures with a diversity of growth habits could lead to greater forage productivity, as well as enhance SOC. However, few data exist to compare the change in SOC with one forage species versus another. In Georgia, SOC was greater under grazed tall fescue than under hayed or grazed bermudagrass (Franzluebbers et al., 2000; Franzluebbers, 2010c). Timing of forage dormancy likely has a controlling influence on how soil water is utilized by forages, a consequence of which might limit or allow microbial decomposition of SOC. In cool-season pastures of Pennsylvania, emission of CO_2 from the soil during winter months was 0.9–1.0 Mg C ha^{-1} (Skinner, 2007). Over the course of 4 years of evaluation, net ecosystem exchange of CO_2 with the atmosphere in mixed pastures (dominated by cool-season grasses) of Pennsylvania led to removal of 0.27 Mg C ha^{-1} yr^{-1} (Skinner, 2008). As seen

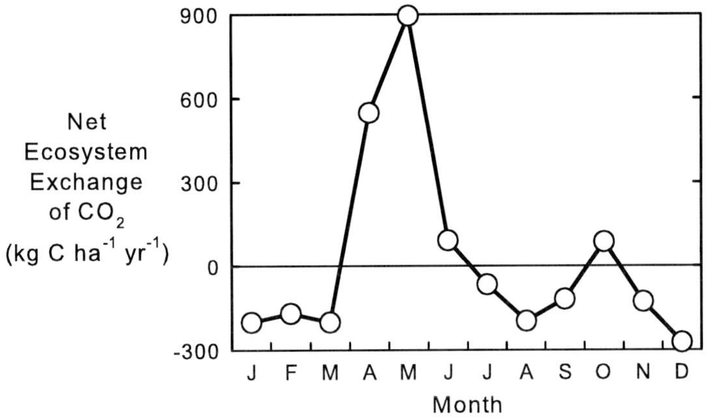

FIGURE 7.4
Net ecosystem exchange of CO_2 with the atmosphere in mixed pastures (dominated by cool-season grasses) on Typic Hapludults in Pennsylvania. Positive values indicate CO_2 removal from the atmosphere and negative values indicate release to the atmosphere. Data from Skinner (2008).

in Figure 7.4, peak net removal from the atmosphere occurred in April—May at the time of peak forage production.

Endophyte association of tall fescue has also been shown to affect SOC. Pastures with high frequency of wild-type endophyte have been observed with greater SOC and/or C fractions than pastures with low frequency of endophyte in Georgia (Franzluebbers et al., 1999; Franzluebbers and Stuedemann, 2005b) and in Kentucky (Siegrist and McCulley, 2008; Handayani et al., 2011). With the development of novel-endophyte-infected tall fescue having low toxicity to grazing animals, there is a need to determine its role in SOC sequestration, considering the widespread recommendation to destroy old stands of wild-type-infected tall fescue pastures and sow novel-endophyte-infected tall fescue pastures.

Winter-annual pastures are sometimes planted solely for grazing or as a grazed cover crop in the southeastern U.S. At the end of 3 years of planting a mixture of wheat and rye for winter grazing in Arkansas, Anders et al. (2010) reported that surface soil contained a greater concentration of water-stable aggregates and a greater content of aggregate-associated SOC when planted with no-tillage and disk tillage than with traditional tillage (chisel and disk plowing). The effect of soil disturbance with tillage on SOC stock is similar to that reported in cropping systems (Puget and Lal, 2004; Franzluebbers, 2010a). During 3 years of evaluation in Georgia, SOC was generally not affected whether winter or summer cover crops were grazed by cattle or not (Franzluebbers and Stuedemann, 2008). This study suggested that grazing of cover crops could be integrated into a cropping system without detriment to soil resources, especially if managed with conservation tillage to preserve high surface soil organic matter following long-term pasture termination.

SPATIAL DISTRIBUTION OF SOIL ORGANIC C

Vertical distribution of SOC is important in conservation agricultural systems (Franzluebbers, 2010d). When agricultural soils are not disturbed with tillage implements, such as with no-tillage cropping or pasture management, high surface SOC occurs and stratification ratios of SOC are often >2 (Franzluebbers, 2009). Accumulation of surface SOC can occur rapidly with conversion of long-term conventionally tilled cropland to grassland (Figure 7.5). The example in Figure 7.5 illustrates that tilled cropland has uniform distribution of SOC in the surface 25 cm, but increasingly greater stratification of SOC with prairie restoration, leading up to highly stratified SOC under remnant prairie.

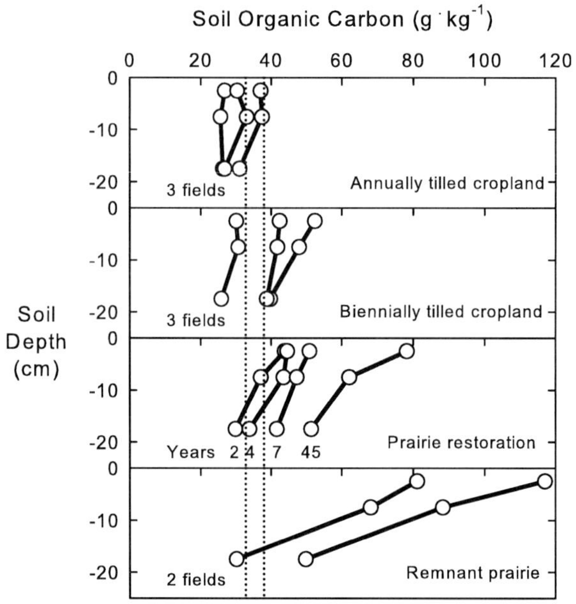

FIGURE 7.5

Soil organic C concentration in the surface 25 cm of a Wacousta silty clay loam (fine-silty, mixed, superactive, mesic Typic Endoaquoll) in Wisconsin as affected by land use. The two dotted lines represent ±1 standard error from all observations at 10–25 cm depth. Data from Jelinski and Kucharik (2009).

Grazing-induced spatial (horizontal) variation in SOC can also be highly significant due to traffic patterns, dung deposition, and forage avoidance. During 12 years of grazing in Georgia, SOC in the surface 12 cm was nearly 4 Mg ha^{-1} greater near shade and water sources than farther away in paddocks (Figure 7.6), which was statistically and practically significant considering the stock of C was ~43 Mg ha^{-1} throughout the pasture. Total soil N was about 600 kg N ha^{-1} greater near shade and water than farther away (Franzluebbers and Stuedemann, 2010), suggesting that redistribution of organic C and N could potentially affect leaching and denitrification processes. In North Carolina, cattle camping areas had greater SOC than non-camping areas at a depth of 0–5 cm, but not at a depth of 5–15 cm (Iyyemperumal et al., 2007). In Florida, shades, water troughs, and mineral feeders had accumulating effects on inorganic N and extractable P (and presumably organic C) (Sigua and Coleman, 2006).

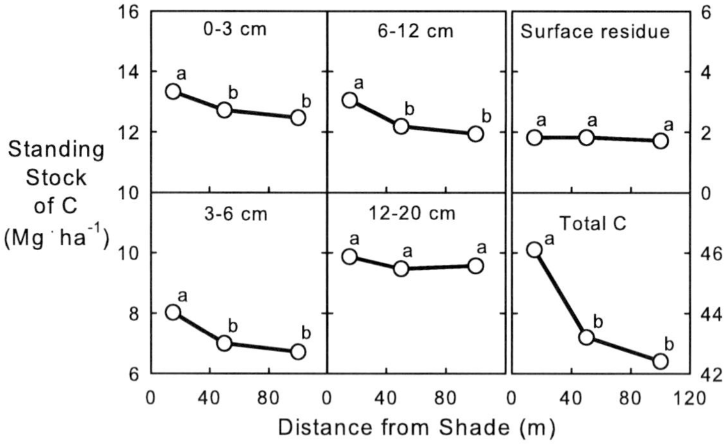

FIGURE 7.6

Surface residue and soil organic C distribution (as indicated by distance from shade) within grazed pastures at different soil depth increments at the end of 5 years of bermudagrass management on a Typic Kanhapludult in Georgia. Data from Franzluebbers and Stuedemann (2010).

Spatial variance of SOC in grazed pastures was equal to neighboring wooded land uses on Davidson loam and clay loam (fine, kaolinitic, thermic Rhodic Kandiudults) in Georgia (Worsham et al., 2010). Soil organic C content of the surface 7.5 cm was 22.2 ± 0.6 Mg C ha^{-1} in one pasture and 16.6 ± 0.4 Mg C ha^{-1} in another pasture (mean ± standard error). The coefficient of variation of SOC ($n = 64$) in pastures was 2.6 to 3.0% compared with values of 2.6 to 4.4% in hardwood and pine forests. Spatial structure evaluated from samples collected at distances of 10 to 100 m of separation was weak, with large nugget-to-sill ratios, suggesting little autocorrelation in samples collected at this moderate spatial scale. In Florida, bahiagrass pasture exhibited significant spatial variation due to slope aspect and slope position, but much of this variation was associated with clay concentration in soil (i.e. increasing clay concentration above a threshold of 8% increased SOC) (Sigua and Coleman, 2010).

TEMPORAL CHANGES IN SOIL ORGANIC C

Soil organic C is expected to accumulate with time with restoration of surface cover and plant input from grassland, especially under initially degraded soil conditions. Indeed, SOC sequestration was estimated at 0.45 Mg C ha^{-1} yr^{-1} in a tall-grass prairie restoration chronosequence in Texas (Potter et al., 1999). Also, accumulation of SOC following mine reclamation in Ohio was highly significant in the surface 15 cm of soil under chronosequences of both pasture and forested land uses (Shrestha and Lal, 2010). The rate of SOC sequestration appeared to be linear at all depths, with the vast majority of C sequestration in the surface 5 cm (68%) and an additional 15% occurring in the 5–15 cm depth (Figure 7.7). These results indicate the strong temporal trend that occurs near the soil surface under pasture management and the much greater random variation that occurs with increasing depth in the soil profile. In addition, these results do not suggest saturation of SOC within the first 25 years of reclamation

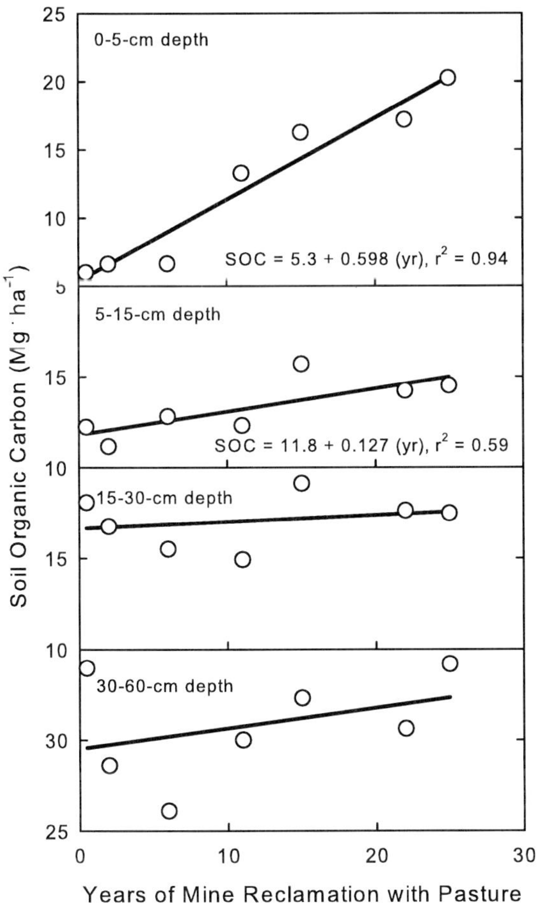

FIGURE 7.7
Soil organic C content as affected by years of mine reclamation with pasture on sites with previous soils classified as Morristown silty clay loam (loamy-skeletal, mixed, active, calcareous, mesic Typic Udorthent) in Ohio. Regressions were significant only for soil organic C at 0–5 cm and 5–15 cm depths, but intercept and slope were 5.3 Mg C ha^{-1} and 0.598 Mg C ha^{-1} yr^{-1} at a depth of 0–5 cm, 11.8 and 0.127 at 5–15 cm, 16.6 and 0.037 at 15–30 cm, 29.5 and 0.112 at 30–60 cm, and 63.3 and 0.875 at 0–60 cm. Data from Shrestha and Lal (2010).

from an initially low status. The concept of soil C saturation has been described for sites wherein little or no increase in steady-state SOC content occurs with increasing C input (Stewart et al., 2007).

A chronosequence study of naturalized grassland following abandonment of tilled cropland in Minnesota also showed accumulation of SOC (Knops and Bradley, 2009). This study indicated a linear increase in SOC up to 74 years of age, and therefore saturation of SOC may require at least a century to occur (Figure 7.8). The only statistically significant change in SOC with time occurred in the surface 20 cm, although the insignificant changes in SOC at depths from 20 to 100 cm did not overwhelm the strong surface effect, resulting in a statistically significant increase of 0.375 Mg C ha^{-1} yr^{-1} at a depth of 0–100 cm. Accumulation of surface SOC at 0–20 cm depth (0.168 Mg C ha^{-1} yr^{-1}) accounted for 45% of the observed change to a depth of 1 m. Lack of significant change in SOC with depth must be viewed cautiously, since random variation is often greater with depth (Kravchenko and Robertson, 2011). However, this chronosequence evaluation in Minnesota provided a firm basis to suggest that, under grassland restoration conditions, surface soil changes are most pronounced, and that the yearly rate of change per unit of initial SOC is relatively stable at all depths (i.e. $1.0 \pm 0.3\%$ yr^{-1}). It is curious that saturation of SOC did not occur within the first 74 years of this evaluation, despite the soil being of coarse texture. It is possible that with more intensive management, resulting in much greater rates of SOC sequestration during this time period, saturation may have eventually occurred. This is an area in need of further research.

In a bermudagrass/tall fescue pasture in Georgia, surface SOC accumulation was dependent on weather extremes that altered plant production and decomposition (Franzluebbers and

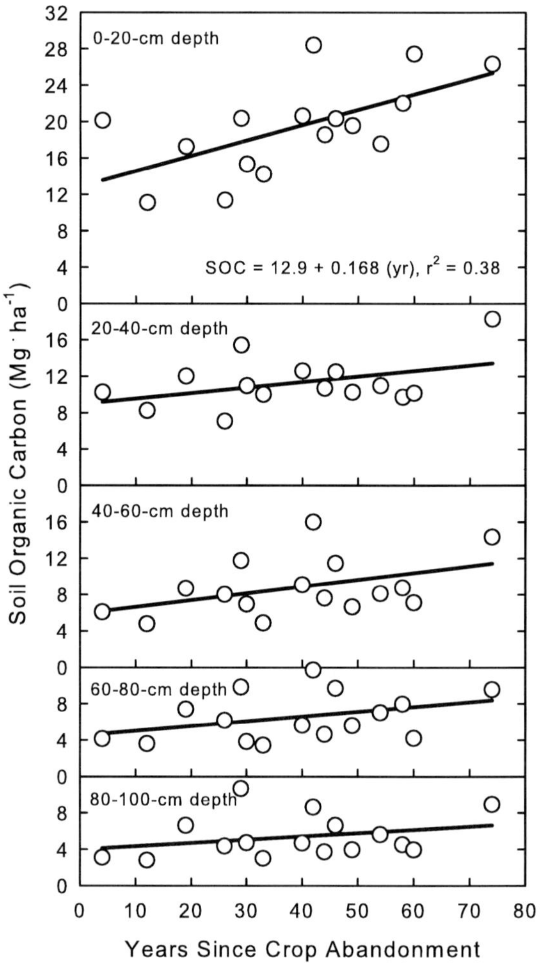

FIGURE 7.8

Soil organic C content as affected by years since crop abandonment (or alternatively as years since naturalized grassland establishment) on a Sartell and Nymore sand (mixed, frigid Typic Udipsamments) and Zimmerman sand (mixed, frigid Lamellic Udipsamment) in Minnesota. Regressions were significant only for soil organic C at 0–20 cm and 0–100 cm depths, but intercept and slope were 12.9 Mg C ha^{-1} and 0.168 Mg C ha^{-1} yr^{-1} at a depth of 0–20 cm, 9.0 and 0.060 at 20–40 cm, 5.9 and 0.075 at 40–60 cm, 4.5 and 0.052 at 60–80 cm, 4.0 and 0.035 at 80–100 cm, and 35.3 and 0.375 at 0–100 cm. Data from Knops and Bradley (2009).

Stuedemann, 2010). Soil organic C increased by an average of 0.94 Mg C ha^{-1} yr^{-1} during years 0 to 5 of pasture management (1409 mm rainfall yr^{-1}), declined by 0.69 Mg C ha^{-1} yr^{-1} during years 5 to 8 (1047 mm rainfall yr^{-1}), and increased by 0.70 Mg C ha^{-1} yr^{-1} during years 8 to 12 (1288 mm rainfall yr^{-1}). Unfortunately, 12 years of evaluation does not appear to be adequate to conclude whether a steady-state condition with soil C saturation might have occurred. However, this study did provide a controlled experimental condition with random replication of treatments that could be measured with time without having to assume equal starting conditions. Chronosequence evaluations have to make rather large assumptions of equal soil and environmental conditions, when in fact those conditions may not have been present. Therefore, chronosequence evaluations should be made more robust with numerous field-site samplings within a time period and numerous samplings of a continuous gradient of management ages.

Other long-term evaluations of SOC change have been conducted in countries with similar environmental conditions as the eastern U.S. In New Zealand, re-sampling of 31 soil profiles after 21 ± 3 years under long-term pasture management systems (predominantly intensive dairy on flat land) resulted in a mean SOC loss rate of 1.06 ± 0.39 Mg C ha^{-1} yr^{-1} (Schipper et al., 2007). In an extension of this repeated sampling survey approach (27 ± 8 years) in New Zealand, a significantly positive change in SOC to a depth of 60 cm was detected for sloped, dry-stocked pastures (0.70 ± 0.28 Mg C ha^{-1} yr^{-1}; n = 15), no significant change was detected for flat, dry-stocked pastures (−0.33 ± 0.19 Mg C ha^{-1} yr^{-1}; n = 27) and dry-stocked, tussock pastures (−0.03 ± 0.22 Mg C ha^{-1} yr^{-1}; n = 8), and a significantly negative change was detected for flat dairy pastures (−1.00 ± 0.20 Mg C ha^{-1} yr^{-1}; n = 27) (Schipper et al., 2010). The reason for the divergence in response was attributed to the recently greater N fertilizer input that may have stimulated decomposition of the initially high soil organic matter on flat dairy sites compared with the inclusion of legumes and high animal-manure C input in sloped pastures. Initial SOC (0–60 cm) among sites was also much greater in flat dairy sites (191 ±12 Mg C ha^{-1}) than in sloped, dry-stocked sites (132 ± 14 Mg C ha^{-1}).

IMPACT OF SEQUESTERED SOIL ORGANIC C IN PASTURES ON OTHER ENVIRONMENTAL ISSUES

Pastures, especially those with improved fertility and grazing management, have greater surface SOC than fields with row crops on similar soil types (Franzluebbers, 2005; Figure 7.2). In fact, stratification ratio of SOC with depth is often greater under pastures than under conventional- or even conservation-tillage cropland (Causarano et al., 2008). High stratification of SOC buffers soil and water quality against typical perturbations in agricultural systems, e.g. moderate fertilizer, organic amendment, and pesticide applications (Franzluebbers, 2008). Perturbations of concern still remain with excessively high nutrient applications from fertilizer and manure inputs that can cause leaching of nitrate to groundwater, runoff of dissolved P to surface water bodies, and emission of N_2O to the atmosphere.

High stratification of SOC with depth under pasture and conservation-tillage cropland has been shown to reduce water runoff volume and loss of nutrients in several studies across small plots, fields, and water catchments (Franzluebbers, 2008). In water catchments (4.8 ± 2.7 ha each) in eastern Texas, total N loss in surface runoff among 8 years was 15.8 ±13.9 kg N ha^{-1} yr^{-1} in conventionally cropped fields and 0.9 ±1.8 kg N ha^{-1} yr^{-1} in pastures (Harmel et al., 2009). Total P loss was 1.9 ±1.5 kg P ha^{-1} yr^{-1} in cropped fields and 1.5 ± 2.6 kg P ha^{-1} yr^{-1} in pastures. Dissolved N loss was also greater in cropped fields than in pastures (12.6 ±12.9 vs. 0.3 ± 0.5 kg N ha^{-1} yr^{-1}), but dissolved P loss was similar between cropped fields and pastures (0.9 ± 0.9 vs. 1.4 ± 2.5 kg P ha^{-1} yr^{-1}). Elevated dissolved P loss in pastures with surface amendment of fertilizer or animal manures can be a water quality concern; hence, mitigation has been attempted with periodic inversion tillage (Sharpley, 2003), as well as with injection of fertilizer or manure below the soil surface (Pote et al., 2003, 2006; Warren et al., 2008). When P fertilizer is applied to pastures a few days prior to runoff-producing precipitation events, dissolved P in runoff can be elevated (Owens and Shipitalo, 2006). Mixing animal manure with aluminum sulfate can control NH_3 loss and P availability, but does not appear to affect C loss from manure (DeLaune et al., 2004). Further work is needed to understand how various SOC fractions interact with nutrients and their dynamics.

High surface SOC concentration under pastures can result in greater sediment-borne C concentration during runoff events from pastures compared with cropland (Owens and Shipitalo, 2011). However, concentration of sediment-borne C from an individual runoff event was not well correlated with the total amount of sediment-borne C lost during a runoff event. Rather, loss of sediment-borne C was well correlated with total sediment mass loss ($r = 0.98$, $p < 0.001$) and dissolved organic C loss ($r = 0.94$, $p < 0.001$) on an event basis. These relationships indicate that the way to reduce runoff C loss from pastures, both sediment-borne and dissolved, is to reduce sediment mass loss (i.e. soil erosion). Maintaining robust vegetative cover, both in winter and in summer, is a key factor in avoiding sediment loss from pastures.

Hay feeding stations can be a spatially isolated source of nutrients and denudation of vegetation. On silt loam soils in Ohio, winter feeding of hay on pastures led to greater inputs of P and K than summer grazing only, which resulted in greater water runoff volume (99 vs. 16 mm yr^{-1}) and greater surface losses (kg ha^{-1} yr^{-1}) of P (0.7 vs. 0.1), Ca (7.1 vs. 0.5), Mg (4.2 vs. 0.4), Na (1.4 vs. 0.1), K (32.1 vs. 2.2), Cl (10.9 vs. 0.8), and C (118.7 vs. 3.9) (Owens et al., 2003).

In Florida, land-use changes from wetland to pasture not only affected SOC, but also organic P in soil (Figure 7.9). The implication of this land-use-dependent relationship is that pasture soils with lower SOC content than native wetland soils will have experienced a significant loss of P to the environment during aerobic decomposition of surface organic matter (Sigua et al., 2009). Loss of P can have significant eutrophication effects on nearby water systems (Gathumbi et al., 2005).

Contamination of surface waters with fecal-borne pathogens derived from pastures can be of great concern when waters are shared with municipal and recreational uses (Chesters and Schierow, 1985; Myers et al., 1985). Survival of enteric bacteria from cattle feces on pasture was determined in New Zealand, and results summarized as follows (Sinton et al., 2007). Environmental conditions following deposition were of great importance in how organisms survived or proliferated. Ranking of times necessary for 90% inactivation were: *Campylobacter jejuni* (6.2 d from deposition) < fecal streptococci (35 d) < *Salmonella enterica* (38 d) < *Escherichia coli* (48 d) < enterococci (56 d). Fecal-pat temperature was important, but desiccation was the dominant factor in controlling bacterial survival. Of 20 monthly leaching

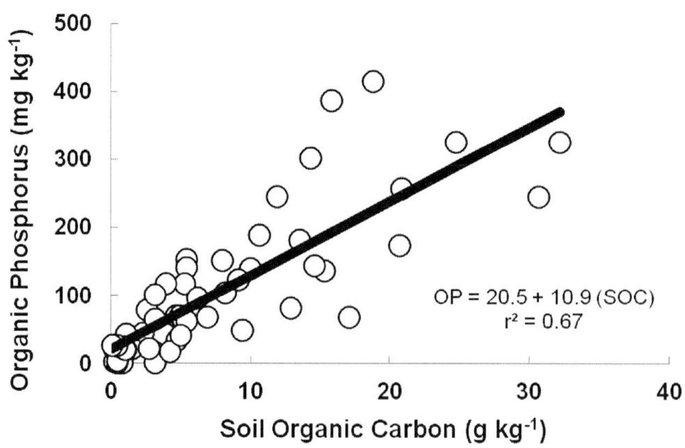

FIGURE 7.9

Relationship of organically bound P with soil organic C in pasture and wetlands soils (Terric Medisaprists) in Florida. Data from Sigua et al. (2009).

losses of *E. coli*, 16 were <10% of total counts in the fecal pat, and 12 were <1%. Drainage losses of *C. jejuni* (generally <1%) were detected for only 1 to 2 months. In tall fescue pasture on a Typic Hapludult in Alabama, water runoff during rainfall simulations (110 mm h^{-1}) contained significantly greater *E. coli* counts when broiler litter was broadcast than when subsurface banded (Sistani et al., 2009). *E. coli* counts when rainfall simulation occurred 8–22 d after broiler litter was applied were about one-half of those when rainfall simulation occurred 1 d after broiler litter was applied. Further work is needed to understand how pasture structure (botanical composition, residual forage mass, surface residue content, etc.) might affect runoff relationships and survival and transport of fecal-borne pathogens.

Trace-gas emission (N_2O and CH_4) from pastures depends on a diversity of factors, such as soil texture, previous history, soil pH, SOC, forage species, fertilization, forage harvest management, intensity of grazing, grazing method, etc. Many of these issues will be covered in Chapter 11 of this book.

KEY RESEARCH ISSUES

Improved pasture management has the potential to sequester a large quantity of SOC, not only directly due to better fertilization, grazing method, and forage type and utilization practices, but also indirectly when viewed as a more valuable land use on par with other agricultural land uses. There are opportunities to create resilient, profitable, productive, and environmentally sustainable agricultural systems with perennial forages as part of the foundation (Steiner and Franzluebbers, 2009; Izaurralde et al., 2011). Improved, contemporary, integrated crop–livestock systems can be designed to take advantage of the conservation benefits of perennial forages, the digestion capabilities of ruminants, the unique landscape characteristics of a farm, the resource-efficient nutrient cycling dynamics of mixed crop–forage systems with legumes and high nutrient-demand cereal and vegetable crops, the diversity of farm income, the resiliency of agricultural systems to climate change and other perturbations, and the foundational support for local food systems (Russelle et al., 2007; Sulc and Tracy, 2007; Lasley et al., 2009; Singer et al., 2009).

Many research opportunities remain to be explored for pasture management systems in the eastern U.S. Some of these include:

- Comprehensive grazing intensity studies are needed at several different locations to understand the multiple interactive effects of long-term stocking rate differences on soil, water, air, plant, and animal responses.
- Comprehensive grazing method studies are needed at strategically selected sites throughout the eastern U.S. to better represent continuous and rotational stocking treatments relevant to an area. Response variables need to be robust in generating widely applicable data, which can then be transferable, developed into process-based models, and support strong recommendations and policy decisions.
- Analyses of potential changes in SOC and other environmental quality responses are needed in studies of season of grazing, deferment, species composition of pastures, and how riparian areas might be utilized.
- Expanded research on regional changes in SOC stocks, including paired-land-use sampling surveys, repeated sampling surveys of the same land use, and developing a network of long-term, high-intensity, mechanistic-level research sites on stock changes in SOC and greenhouse gas emissions.
- Better understanding of the role of soil order, soil texture, landscape position, history of land use, soil pH, and climatic conditions on the potential of pasture management systems to sequester SOC.
- Quantifying the change in SOC that can be expected across a diversity of edaphic conditions from the implementation of a wider range of improved pasture management

approaches (e.g. grazing method, stocking rate, optimum residual mass, grass/legume mixtures, etc.).

We propose that an expanded network of comprehensive pasture-management research be conducted in a coordinated manner so that knowledge gaps and data deficiencies in how pasture management affects SOC can be rapidly filled. The potential benefits of an expanded research approach to producers are numerous (greater production, profit, and sustainability), but the benefits to the environment and society (cleaner water, reduced greenhouse gas emissions, enhanced soil biological diversity) should be considered wise investments for the future sustainability of American agriculture.

Acknowledgments

We appreciate the excellent technical support from Steve Knapp. This report is part of GRACEnet cross-location research (Greenhouse gas Reduction through Agricultural Carbon Enhancement network).

References

Anders, M.M., Beck, P.A., Watkins, B.K., Gunter, S.A., Lusby, K.S., Hubbell, D.S., 2010. Soil aggregates and their associated carbon and nitrogen content in winter annual pastures. Soil Sci. Soc. Am. J 74, 1339–1347.

Beetz, A.E., Rinehart, L., 2010. Rotational grazing. Nat. Center Appropr. Techn., Fayetteville, AR. p. 12. Accessed at www.attra.ncat.org/attra-pub/rotgraze.html. Verified: 23 February 23, 2011.

Carvalho, J.L.N., Raucci, G.S., Cerri, C.E.P., Bernoux, M., Feigl, B.J., Wruck, F.J., Cerri, C.C., 2010. Impact of pasture, agriculture and crop-livestock systems on soil C stocks in Brazil. Soil Till. Res. 110, 175–186.

Causarano, H.J., Franzluebbers, A.J., Shaw, J.N., Reeves, D.W., Raper, R.L., Wood, C.W., 2008. Soil organic carbon fractions and aggregation in the Southern Piedmont and Coastal Plain. Soil Sci. Soc. Am. J 72, 221–230.

Chesters, G., Schierow, L.-J., 1985. A primer on nonpoint pollution. J. Soil Water Conserv 40, 9–13.

Conant, R.T., Six, J., Paustian, K., 2003. Land use effects on soil carbon fractions in the southeastern United States: I. Management-intensive versus extensive grazing. Biol. Fertil. Soils 38, 386–392.

DeLaune, P.B., Moore Jr., P.A., Daniel, T.C., Lemunyon, J.L., 2004. Effect of chemical and microbial amendments on ammonia volatilization from composting poultry litter. J. Environ. Qual 33, 728–734.

Franzluebbers, A.J., 2005. Soil organic carbon sequestration and agricultural greenhouse gas emissions in the southeastern USA. Soil Till. Res. 83, 120–147.

Franzluebbers, A.J., 2008. Linking soil and water quality in conservation agricultural systems. J. Integr. Biosci 6, 15–29.

Franzluebbers, A.J., 2009. Linking soil organic carbon and environmental quality through conservation tillage and residue management. In: Lal, R., Follett, R.F. (Eds.), Soil Carbon Sequestration and the Greenhouse Effect, second ed. Soil Sci. Soc. Am. Spec. Publ. No. 57, Madison, WI, pp. 263–289.

Franzluebbers, A.J., 2010a. Achieving soil organic carbon sequestration with conservation agricultural systems in the southeastern United States. Soil Sci. Soc. Am. J. 74, 347–357.

Franzluebbers, A.J., 2010b. Depth distribution of soil organic carbon as a signature of soil quality. In Proc. 19th World Congr. Soil Sci., 1–6 August 2010, Brisbane, Australia. DVD, p. 4.

Franzluebbers, A.J., 2010c. Soil organic carbon in managed pastures of the southeastern United States of America. In: Abberton, M., Conant, R., Batello, C. (Eds.), Grassland Carbon Sequestration: Management, Policy and Economics. Integrated Crop Manage, vol. 11. United Nations Food Agric, Org., Rome, Italy, pp. 163–175.

Franzluebbers, A.J., 2010d. Will we allow soil carbon to feed our needs? Carbon Manage 1, 237–251.

Franzluebbers, A.J., Doraiswamy, P.C., 2007. Carbon sequestration and land degradation. In: Sivakumar, M.V.K., Ndiangu'ui, N. (Eds.), Climate and Land Degradation. Springer-Verlag, New York, pp. 337–354.

Franzluebbers, A.J., Nazih, N., Stuedemann, J.A., Fuhrmann, J.J., Schomberg, H.H., Hartel, P.G., 1999. Soil carbon and nitrogen pools under low- and high-endophyte-infected tall fescue. Soil Sci. Soc. Am. J 63, 1687–1694.

Franzluebbers, A.J., Stuedemann, J.A., 2005a. Bermudagrass management in the Southern Piedmont USA: VII. Soil-profile organic carbon and total nitrogen. Soil Sci. Soc. Am. J 69, 1455–1462.

Franzluebbers, A.J., Stuedemann, J.A., 2005b. Soil carbon and nitrogen pools in response to tall fescue endophyte infection, fertilization, and cultivar. Soil Sci. Soc. Am. J 69, 396–403.

Franzluebbers, A.J., Stuedemann, J.A., 2008. Early response of soil organic fractions to tillage and integrated crop-livestock production. Soil Sci. Soc. Am. J 72, 613–625.

Franzluebbers, A.J., Stuedemann, J.A., 2009. Soil-profile organic carbon and total nitrogen during 12 years of pasture management in the Southern Piedmont USA. Agric. Ecosyst. Environ 129, 28–36.

Franzluebbers, A.J., Stuedemann, J.A., 2010. Surface soil changes during twelve years of pasture management in the Southern Piedmont USA. Soil Sci. Soc. Am. J 74, 2131–2141.

Franzluebbers, A.J., Stuedemann, J.A., Schomberg, H.H., Wilkinson, S.R., 2000. Soil organic C and N pools under long-term pasture management in the Southern Piedmont USA. Soil Biol. Biochem 32, 469–478.

Franzluebbers, A.J., Stuedemann, J.A., Wilkinson, S.R., 2001. Bermudagrass management in the Southern Piedmont USA. I. Soil and surface residue carbon and sulfur. Soil Sci. Soc. Am. J 65, 834–841.

Gathumbi, S.M., Bohlen, P.J., Graetz, D.A., 2005. Nutrient enrichment of wetland vegetation and sediments in subtropical pastures. Soil Sci. Soc. Am. J 69, 539–548.

Handayani, I.P., Coyne, M.S., Phillips, T.D., 2011. Soil organic carbon fractions differ in two contrasting tall fescue systems. Plant Soil 338, 43–50.

Haney, R.L., Kiniry, J.R., Johnson, M.V.V., 2010. Soil microbial activity under different grass species: underground impacts of biofuel cropping. Agric. Ecosyst. Environ 139, 754–758.

Harmel, R.D., Smith, D.R., Haney, R.L., Dozier, M., 2009. Nitrogen and phosphorus runoff from cropland and pasture fields fertilized with poultry litter. J. Soil Water Conserv 64, 400–412.

Henning, J., Lacefield, G., Rasnake, M., Burris, R., Johns, J., Johnson, K., Turner, L., 2000. Rotational grazing. Coop. Ext. Serv. Univ. Kentucky, Lexington, KY, 16 p.

Iyyemperumal, K., Israel, D.W., Shi, W., 2007. Soil microbial biomass, activity and potential nitrogen mineralization in a pasture: impact of stock camping activity. Soil Biol. Biochem 39, 149–157.

Izaurralde, R.C., Thomson, A.M., Morgan, J., Fay, P., Polley, W., Hatfield, J.L., 2011. Climate impacts on agriculture: Implications for forage and rangeland production. Agron. J. 103, 371–381.

Jelinski, N.A., Kucharik, C.J., 2009. Land-use effects on soil carbon and nitrogen on a U.S. midwestern floodplain. Soil Sci. Soc. Am. J. 73, 217–225.

Kingery, W.L., Wood, C.W., Delaney, D.P., Williams, J.C., Mullins, G.L., 1994. Impact of long-term land application of broiler litter on environmentally related soil properties. J. Environ. Qual. 23, 139–147.

Knops, J.M., Bradley, K.L., 2009. Soil carbon and nitrogen accumulation and vertical distribution across a 74-year chronosequence. Soil Sci. Soc. Am. J. 73, 2096–2104.

Kravchenko, A.N., Robertson, G.P., 2011. Whole-profile soil carbon stocks: the danger of assuming too much from analyses of too little. Soil Sci. Soc. Am. J. 75, 235–240.

Lasley, P., Hogberg, M., Helsel, Z., Larson, A., 2009. People, grassland, and livestock in revitalized rural communities. In: Wedin, W.F., Fales, S.L. (Eds.), Grassland: Quietness and Strength for a New American Agriculture. ASA-CSSA-SSSA, Madison, WI, pp. 173–188.

Monaghan, R.M., Paton, R.J., Smith, L.C., Drewry, J.J., Littlejohn, R.P., 2005. The impacts of nitrogen fertilisation and increased stocking rate on pasture yield, soil physical condition and nutrient losses in drainage from a cattle-grazed pasture. N.Z. J. Agric. Res. 48, 227–240

Myers, C.F., Meek, J., Tuller, S., Weinberg, A., 1985. Nonpoint sources of water pollution. J. Soil Water Conserv 40, 14–18.

Owens, L.B., Shipitalo, M.J., 2006. Surface and subsurface phosphorus losses from fertilized pasture systems in Ohio. J. Environ. Qual. 35, 1101–1109.

Owens, L.B., Shipitalo, M.J., 2011. Sediment-bound and dissolved carbon concentration and transport from a small pastured watershed. Agric. Ecosyst. Environ. 141, 162–166.

Owens, L.B., van Keuren, R.W., Edwards, W.M., 2003. Non-nitrogen nutrient inputs and outputs for fertilized pastures in silt loam soils in four small Ohio watersheds. Agric. Ecosyst. Environ. 97, 117–130.

Perkins, H.F., 1987. Characterization data for selected Georgia soils. Spec. Publ. 43, Georgia Agric. Exp. Sta. Univ. Georgia, Athens GA, p. 545.

Pote, D.H., Kingery, W.L., Aiken, G.E., Han, F.X., Moore Jr., P.A., 2006. Incorporating granular inorganic fertilizer into perennial grassland soils to improve water quality. J. Soil Water Conserv 61, 1–7.

Pote, D.H., Kingery, W.L., Aiken, G.E., Han, F.X., Moore Jr., P.A., Buddington, K., 2003. Water-quality effects of incorporating poultry litter into perennial grassland soils. J. Environ. Qual 32, 2392–2398.

Potter, K.N., Daniel, J.A., Altom, W., Torbert, H.A., 2001. Stocking rate effect on soil carbon and nitrogen in degraded soils. J. Soil Water Conserv. 56, 233–236.

Potter, K.N., Torbert, H.H., Johnson, H.B., Tischler, C.R., 1999. Carbon storage after long-term grass establishment on degraded soils. Soil Sci. 164, 718–725.

Puget, P., Lal, R., 2004. Soil organic carbon and nitrogen in a Mollisol in central Ohio as affected by tillage and land use. Soil Till. Res. 80, 201–213.

Pulleman, M.M., Marinissen, J.C.Y., 2004. Physical protection of mineralizable C in aggregates from long-term pasture and arable soil. Geoderma 120, 273–282.

Russelle, M.P., Entz, M.H., Franzluebbers, A.J., 2007. Reconsidering integrated crop-livestock systems in North America. Agron. J. 99, 325–334.

Sandoval, M.A., Stolpe, N.B., Zagal, E.M., Mardones, M., 2007. The effect of crop-pasture rotations on the C, N and S contents of soil aggregates and structural stability in a volcanic soil of south-central Chile. Acta Agric. Scand. 57, 255–262.

Schipper, L.A., Baisden, W.T., Parfitt, R.L., Ross, C., Claydon, J.J., Arnold, G., 2007. Large losses of soil C and N from soil profiles under pasture in New Zealand during the past 20 years. Global Change Biol. 13, 1138–1144.

Schipper, L.A., Parfitt, R.L., Ross, C., Baisden, W.T., Claydon, J.J., Fraser, S., 2010. Gains and losses in C and N stocks of New Zealand pasture soils depend on land use. Agric. Ecosyst. Environ. 139, 611–617.

Sharpley, A.N., Smith, S.J., Bain, W.R., 1993. Nitrogen and phosphorus fate from long-term poultry litter applications to Oklahoma soils. Soil Sci. Soc. Am. J. 57, 1131–1137.

Sharrow, S.H., Ismail, S., 2004. Carbon and nitrogen storage in agroforests, tree plantations, and pastures in western Oregon, USA. Agroforestry Syst. 60, 123–130.

Shrestha, R.K., Lal, R., 2010. Carbon and nitrogen pools in reclaimed land under forest and pasture ecosystems in Ohio, USA. Geoderma 157, 196–205.

Siegrist, J.A., McCulley, R.L., 2008. Endophyte effects on litter decomposition rates and soil carbon pools in southeastern U.S. tall fescue pasture. Paper COS 78–10 at 93rd Ecol. Soc. Am. Ann. Mtg. 3–8 Aug. 2008, Milwaukee, WI.

Sigua, G.C., Coleman, S.W., 2006. Sustainable management of nutrients in forage-based pasture soils: effect of animal congregation sites. J. Soils Sediments 6, 249–253.

Sigua, G.C., Coleman, S.W., 2010. Spatial distribution of soil carbon in pastures with cow-calf operation: effects of slope aspect and slope position. J. Soils Sediments 10, 240–247.

Sigua, G.C., Coleman, S.W., Albano, J., 2009. Beef cattle pasture to wetland reconversion: impact on soil organic carbon and phosphorus dynamics. Ecol. Eng. 35, 1231–1236.

Singer, J.W., Franzluebbers, A.J., Karlen, D.L., 2009. Grass-based farming systems: soil conservation and environmental quality. In: Wedin, W.F., Fales, S.L. (Eds.), Grassland: Quietness and Strength for a New American Agriculture. ASA-CSSA-SSSA, Madison, WI, pp. 121–136.

Sinton, L.W., Braithwaite, R.R., Hall, C.H., Mackenzie, M.L., 2007. Survival of indicator and pathogenic bacteria in bovine feces on pasture. Appl. Environ. Microbiol. 73, 7917–7925.

Sistani, K.R., Torbert, H.A., Way, T.R., Bolster, C.H., Pote, D.H., Warren, J.G., 2009. Broiler litter application method and runoff timing effects on nutrient and *Escherichia coli* losses from tall fescue pasture. J. Environ. Qual. 38, 1216–1223.

Skinner, R.H., 2007. Winter carbon dioxide fluxes in humid-temperate pastures. Agric. For. Meteor. 144, 32–43.

Skinner, R.H., 2008. High biomass removal limits carbon sequestration potential of mature temperate pastures. J. Environ. Qual. 37, 1319–1326.

Steiner, J.L., Franzluebbers, A.J., 2009. Farming with grass—for people, for profit, for production, for protection. J. Soil Water Conserv 64, 75A–80A.

Stewart, C.E., Paustian, K., Conant, R.T., Plante, A.F., Six, J., 2007. Soil carbon saturation: concept, evidence and evaluation. Biogeochemistry 86, 19–31.

Sulc, R.M., Tracy, B.F., 2007. Integrated crop–livestock systems in the U.S. Corn Belt. Agron. J. 99, 335–345.

Tom, N., Raman, A., Hodgkins, D.S., Nicol, H., 2006. Populations of soil organisms under continuous set stocked and high intensity-short duration rotational grazing practices in the central tablelands of New South Wales (Australia). N.Z.J. Agric. Res 49, 261–272.

U.S. Census Bureau, 2011. State and county quick facts. Accessed at. http://quickfacts.census.gov/qfd/index.html Verified: February 16, 2011.

USDA-NASS (National Agricultural Statistics Service), 2011. 2007 Census of Agriculture Report. United States Department of Agriculture, Washington, DC. Accessed at http://www.agcensus.usda.gov/ (Verified: February 15, 2011).

Warren, J.G., Sistani, K.R., Way, T.R., Mays, D.A., Pote, D.H., 2008. A new method of poultry litter application to perennial pasture: subsurface banding. Soil Sci. Soc. Am. J. 72, 1831–1837.

West, T.O., Marland, G., 2002. A synthesis of carbon sequestration, carbon emissions, and net carbon flux in agriculture: comparing tillage practices in the United States. Agric. Ecosyst. Environ. 91, 217–232.

Wilcox, C.S., Dominguez, J., Parmelee, R.W., McCartney, D.A., 2002. Soil carbon and nitrogen dynamics in *Lumbricus terrestris* L. middens in four arable, a pasture, and a forest ecosystem. Biol. Fertil. Soils 36, 26–34.

Wilkins, R.N., Snelgrove, A.G., Fitzsimons, B.C., Stevener, B.M., Skow, K.L., Anderson, R.E., Dube, A.M., 2009. Texas Land Trends. Texas A&M Inst. Renew. Nat. Res. Texas A&M Univ. Accessed at http://texaslandtrends.org (Verified: March 23, 2011).

Worsham, L., Markewitz, D., Nibbelink, N., 2010. Incorporating spatial dependence into estimates of soil carbon contents under different land covers. Soil Sci. Soc. Am. J. 74, 635–646.

CHAPTER 8

Sustainable Bioenergy Feedstock Production Systems: Integrating Carbon Dynamics, Erosion, Water Quality, and Greenhouse Gas Production

Jane M. F. Johnson[1], Jeffrey M. Novak[2]
[1]USDA-Agricultural Research Service, North Central Soil Conservation Research Lab, Morris, MN
[2]USDA-Agricultural Research Service, Coastal Plains Research Center, Florence, SC

CHAPTER OUTLINE

NB: The U.S. Department of Agriculture offers its programs to all eligible persons regardless of race, color, age, sex, or national origin, and is an equal opportunity employer.

Abbreviations: ARS, Agricultural Research Service; CO_2, carbon dioxide; CRP, Conservation Reserve Program; GRACEnet, Greenhouse gas Reduction through Agricultural Carbon Enhancement network; GHG, greenhouse gas; HPEC, herbaceous perennial energy crops; CH_4, methane; N_2O, nitrous oxide; REAP, Renewable Energy Assessment Project; SOC, soil organic carbon; SOM, soil organic matter; WPEC, woody perennial energy crops

Managing Agricultural Greenhouse Gases. DOI: 10.1016/B978-0-12-386897-8.00008-5
2012, Published by Elsevier Inc.

INTRODUCTION

There are multiple forces driving the paradigm shift toward using more renewable energy, although justifications are shrouded in debate and controversy. First, energy, national security, and rural development are toted frequently as drivers for expanding the use of biofuels; extensive socio-political and economic implication have been discussed (Charles et al., 2007; Steenhof and McInnis, 2008; Demirbas, 2009; Hoekman, 2009; Londo et al., 2010). Second, fossil fuels are a finite resource, yet fuel consumption and prices are increasing (Duncan and Youngquist, 1999; Pimentel et al., 2004; Charles et al., 2007; Nehring, 2009; OCHA, 2010; U.S. EIA, 2010). Third, burning fossil fuel is currently a major contributor to GHG emission, which is linked to global warming and climate change (IPCC, 2007a). In response to these inter-related drivers, multiple renewable energy sources are being developed and expanded to help meet the growing energy demands.

Hydropower, wind, solar, geothermal, biomass, and hydrogen are renewable energy sources available for power and/or liquid fuel production. Unlike some of the renewable energy forms, biomass can be used for production of multiple fuel sources, including heat, power, electricity, and liquid fuels. From 2009 to 2025, the demand for power and liquid fuels is predicted to increase by 25% and 10%, respectively (U.S. EIA, 2011a), with renewable energy sources expected to provide a larger percentage of the overall energy consumed (BRDB, 2008; Hoekman, 2009). Renewable biomass feedstocks include crop biomass (grain and crop residues), dedicated woody and herbaceous energy crops, and organic waste (e.g. manure) products (Wilhelm et al., 2004; Perlack et al., 2005; Hill et al., 2006; Johnson et al., 2007c; Lal and Pimentel, 2007; BRDB, 2008; Lal, 2009). Although biomass fuel currently represents only a small percentage of the total U.S. energy consumption, future demand might represent a substantial ($>500\,Tg\,yr^{-1}$) amount of biomass (Perlack et al., 2005; BRDB, 2008). Thus, a serious effort to expand dedicated bioenergy feedstock production and/or integrate with traditional food crops is expected to meet the growing feedstock demand. Potentially, this could place stress on the environment due to land-use changes, increased tillage, expanded fertilizer use, and increased (direct or indirect) GHG production.

Ideally, renewable energy sources provide C-neutral or C-negative energy; thereby, mitigating GHG emission (IPCC, 2007b; Johnson et al., 2007c; BRDB, 2008; Hattori and Morita, 2010). However, the extent that bioenergy exacerbates or mitigates GHG emission is controversial and highly dependent upon feedstock (e.g. corn grain or perennial grass), production systems (e.g. tillage and nutrient management), energy conversion platforms (e.g. fermentation or thermo-chemical), and assumptions used in life-cycle analysis (Farrell et al., 2006; Fargione et al., 2008, 2010; Steenhof and McInnis, 2008; Borjesson, 2009; Malça and Freire, 2011). For example, life-cycle analyses vary in how they treat co-products such as dry distiller's grain, conversion efficiency of feedstock to ethanol and assumptions concerning potential indirect land-use changes.

Across the wide range of the U.S.'s climatic regions, there is a vast array of native and agronomic ecosystems for production of biomass feedstocks. Recently, a regional biomass feedstock roadmap for the U.S. was developed, based on feedstocks, land use and environmental concerns (Braun et al., 2010).The eastern region is dominated by forest and special uses (urban), the central region has the largest percentage of cropland, while the western region is mixed but has the largest acreage of grassland, as well as substantial forest and special-use acreages (http://www.ers.usda.gov/Data/MajorLandUses/map.htm) (Figure 8.1). A combination of bioenergy feedstocks may be available in any of the three regions. Based on existing land-use patterns, woody perennial energy crops (WPEC) would be the expected major feedstock in eastern and portions of the western regions; whereas, herbaceous perennial energy crops (HPEC) would be a major feedstock in other areas of the western and portions of the central regions. Due to the high acreage of cropland, crop residues are expected to be a major feedstock in the central region. Local specialty crops such as hay from grass-seed production in the Pacific Northwest or rice bagasse could provide local or regional scale feedstocks for biofuel

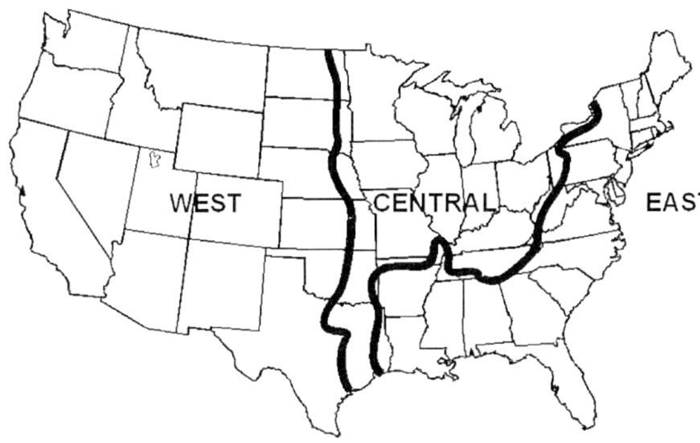

FIGURE 8.1
Delineation of three broad regions in the U.S.

production (Johnson et al., 2010c). However, across all regions and feedstock types there is a need to safeguard against soil organic carbon (SOC) loss and soil movement via wind and water erosion (Johnson et al., 2010c).

If biomass harvest decreases SOC content (Cowie et al., 2006) or increases GHG emission, then the benefit of displacing fossil fuel would be reduced or lost (Adler et al., 2007). Unfortunately, there is a dearth of data on the direct response to harvesting crop residues on GHG emission compared to returning residues to the soil. Jacinthe and Lal (2003) compared N_2O flux from no-till plots amended with 0, 8, and 16 Mg ha^{-1} wheat (*Triticum aestivum* L.) straw, and with or without N fertilization. The presence of straw and N fertilizer on bare silt loam soil had highest N_2O emission. In contrast, in a multi-year field study on a sandy clay loam, the addition of wheat, barley (*Hordeum vulgare* L.), canola (*Brassica napus* L.) or pea (*Pisum sativum* L.) residue had negligible impact on N_2O flux, but there was an interaction with N application (Malhi and Lemke, 2007). Likewise, two cycles of returning or harvesting corn (*Zea mays* L.) stover residue on a loam soil had negligible impact on N_2O flux (Johnson and Barbour, 2010). Clearly, the diversity of responses points to a need for research to examine the impact of harvesting biomass feedstocks on GHG emission. Fortunately, we are aware of several recently initiated studies across the U.S. that will help assess biomass harvest on GHG emission and SOC content. For example, experiments in Minnesota, Iowa, and South Dakota are assessing the impact of corn stover removal on direct GHG emission in conjunction with the USDA-Agricultural Research Service (ARS) Renewable Energy Assessment Project (REAP) (http://www.ars.usda.gov/research/programs/programs.htm?np_code=212&docid=21224) and a broad integrated study funded through USDA and DOE (Karlen, 2010; Wilhelm et al., 2010).

Historically, maintaining SOC was not an important consideration since typical agricultural practices (e.g. land clearing, tillage, monoculture, replacing perennials with annuals, etc.) have contributed to its decline (Reicosky et al., 1995; Lal et al., 1998b; Balesdent et al., 2000; West and Post, 2002; Franzluebbers and Follett, 2005; Novak et al., 2009a). As a result of land-use change, SOC content declined between 20 and 75% compared to pre-agriculture levels (Bruce et al., 1999; Slobodian et al., 2002; Johnson et al., 2005; Liebig et al., 2005b). The magnitude of SOC decline varied greatly in these studies due to differences in regional climate, years under cultivation, and slope position. Substantial SOC decline has raised concerns that harvesting bioenergy feedstocks, especially over-harvesting crop residues, could further exacerbate SOC losses (Wilhelm et al., 2004; Lal, 2008b) and cause other agronomic or environmental damage (Wilhelm et al., 2004, 2010; Blanco-Canqui et al., 2006; Blanco-Canqui and Lal, 2007a, b, 2009a, b; Johnson et al., 2007c, 2010a, b; Blanco-Canqui, 2010). Only a small portion of C from plant biomass is converted into stable SOC, none-the-less reducing biomass inputs can reduce SOC (Wilhelm et al., 2004; Johnson et al., 2006). A loss of SOC undermines the agricultural sector's ability to mitigate climate change through C sequestration (IPCC, 2007b).

Carbon stored in SOM pools is not released quickly to the atmosphere as CO_2; therefore, increasing SOM content is a viable GHG mitigation strategy (Paustian et al., 1997). Thus, limiting harvest rates of annual or perennial feedstocks and/or adding compensatory management are needed to maintain SOC.

Soil organic C is explicitly linked to C inputs (e.g. plant biomass); thus a prerequisite for sustainable bioenergy is establishing harvest rates that avoid SOC reduction. Soil organic C content is also a function of decomposition. Reducing biomass inputs or increasing decomposition can shift the equilibrium toward declining SOC contents (Lal, 2009; Johnson et al., 2010b). Based on empirical literature from several regions, crops, and soil types in the U.S., Johnson et al. (2010b) estimated that 2.5 ± 1.7 Mg C ha^{-1} (6.2 ± 4.35 Mg dry biomass ha^{-1}) needs to be returned annually to maintain SOC content. Conservation practices that protect SOC also reduce the risk of soil erosion and avoid related water quality degradation (Johnson et al., 2010a). Soil organic C influences many soil biological, chemical and physical properties and processes that are important to many soil functions (Johnson et al., 2007c, 2010b; Lal et al., 2007a, b; Lal, 2008a, 2009; Cruse et al., 2009). Therefore, sustainable biofuel crop production systems must also be managed to protect the soil resource and its functions to serve as effective media for crop production.

CHALLENGES OF HISTORICAL BIOENERGY FEEDSTOCKS

In the U.S., corn-grain ethanol has been a long-term reliable feedstock for industrial ethanol production (Farrell et al., 2006; Hattori and Morita, 2010). However, there has been extensive and unpleasant societal discourse on producing corn-grain ethanol, primarily due to the economic cost of its production (Pimentel and Patzek, 2005; Pimentel and Pimentel, 2008) and the highly publicized and controversial competition with human food (Hattori and Morita, 2010; Meyer, 2010). Other concerns about corn-grain ethanol have been raised about water quality and its consumption for ethanol industrial production (Committee on Water Implications of Biofuels Production in the United States National Research Council, 2008; Singh and Kumar, 2011), land-use change (CAST, 2007; Fargione et al., 2008; Searchinger et al., 2008) and GHG emission (Borjesson, 2009; Fargione et al., 2010). Much of the controversy on water use has focused on industrial production, even though the amount of water used during conversion is minor compared to the water needed to grow corn (Wilhelm et al., 2010). Many of the mentioned environmental risks from annual cropping systems are related to management (e.g. tillage, fertilizer management) that would still exist if corn were grown for food. Furthermore, it is noteworthy that corn-grain ethanol production only represents a small fraction of the total U.S.'s current and projected liquid fuel consumption (Perlack et al., 2005; Committee on Water Implications of Biofuels Production in the United States National Research Council, 2008; Hoekman, 2009; U.S. EIA, 2011b, a). The Renewable Fuels Standard mandate in the Energy Independence and Security Act of 2007 capped corn-grain ethanol at 56.8 billion L of the desired annual 136.3 billion L of biofuel by 2022 and 227 billion L by 2030 (U.S. Congress, 2007). Recently, this mandate was reduced to 24.6 billion L, which is more aligned with current projected production capacity (Coyle, 2010). Substantial amounts of non-grain biomass will be needed to meet feedstock demand, not just for cellulosic ethanol (BRDB, 2008), but also for renewable heat and power demands.

In the U.S. by 2030, potentially >500 Tg of dry biomass could be harvested annually from agriculture for cellulosic ethanol production (Perlack et al., 2005). Corn stover represents a majority of the identified annual non-grain biomass supply (Nelson, 2002; Perlack et al., 2005). Additional biomass sources are agricultural waste (e.g. hulls, shells), forestry waste products, and dedicated lignocellulosic biofuel crops (Perlack et al., 2005; Graham et al., 2007; Lal and Pimentel, 2007; BRDB, 2008). Dedicated biofuel crops or so-named second-generation feedstocks include sweet sorghum (*Sorghum vulgare* Pers.), HPEC [i.e. switchgrass (*Panicum Virgatum* L.), and giant miscanthus (*Miscanthus giganteus*)] and WPEC [i.e. hybrid poplar (*Populas* spp.), eucalyptus (*Eucalyptus* spp.) and willow (*Salix* spp.)] (Perlack et al., 2005; Graham et al., 2007; BRDB, 2008). As will be discussed, there are limited land and water resources available to produce these feedstocks, due to competition from societal and environmental needs.

Arable land is a limited resource; therefore, efficient land use and management are needed to avoid deleterious impacts on our natural resources or undue competition among demands for agricultural products (food, feed, fiber, and fuel) (Karlen et al., 2009; Wilhelm et al., 2010). By definition, a sustainable bioenergy system is managed to minimize or avoid potential risks (e.g. loss of SOC, water quality, etc.) and optimize benefits such as mitigating GHG emission (CAST, 2007; Johnson et al., 2007c, 2010a; Cruse and Herndl, 2009; Hattori and Morita, 2010; Wilhelm et al., 2010). In the U.S. there is about 374 Mha of total farm land, of which about 30 Mha could be shifted for dedicated biofuel crop production (Dicks et al., 2009). Indeed, availability of this additional land for biofuel production is a key determinant for biofuel sustainability and its growth (Gopalakrishnan et al., 2009). Therefore, current production lands must be efficiently utilized. The most efficient use of land area varies among and within geographic regions (Braun et al., 2010). For example, it has been proposed in the Midwest to utilize double or relay cropping to grow both a food [e.g. soybean (*Glycine max* L.) Merr.] and energy crop [e.g. camelina (*Camelina sativa* L.)] on the same parcel of land in a single growing season (Russ Gesch, personal communication, 2010).

Over the past two centuries, cropland in the U.S. has experienced severe soil erosion (Bennett, 1939; USDA, 1989). While the rate of soil loss has declined in the past few decades, cropland still is losing 4.7 to 7.4 Mg ha^{-1} yr^{-1} from wind or water erosion (USDA-NRCS, 2009), which in many cases exceeds the soil tolerance level "T" for U.S. soils (Hudson, 1982; USDA-NRCS, 1997, 2009). As noted, the overzealous harvest of annual crop residues will likely cause increased soil erosion and sediment and nutrient loading resulting in decreased quality of recipient surface water systems (Pimentel and Kounang, 1998; Blanco-Canqui et al., 2009; Cruse et al., 2009; Cruse and Herndl, 2009). Erosion not only displaces valuable topsoil, but the off-site environmental impacts and losses in crop productivity (CAST, 1982; Cruse et al., 2009) are estimated to cost $17 to 27 billion yr^{-1} (Pimentel et al., 1995).

MANAGEMENT OF ROW CROPS FOR SUSTAINABILITY—AVOIDING/MITIGATING RISKS

Soil cover and residue management have long been recognized as means to prevent or reduce erosion risk (Oschwald et al., 1978). No-tillage management alone is insufficient to minimize run-off and soil erosion, especially if sufficient residue is not available (Blanco-Canqui et al., 2009; Karlen et al., 2009). The risk of increased soil erosion due to residue harvest is widely recognized. Therefore, predictions of harvestable crop residues typically exclude highly erodible lands and are constrained to maintain erosion at or below tolerable soil loss (T) (Nelson, 2002; Nelson et al., 2004; Perlack et al., 2005; Graham et al., 2007; BRDB, 2008). Erosion control is a critical factor when considering how much residue can be harvested from row crops. A limited amount of crop residue might be available for removal without risking erosional soil losses (Blanco-Canqui et al., 2009; Johnson et al., 2010a). However, basing harvest limits solely on avoiding erosion may not provide sufficient biomass input to maintain SOC (Wilhelm et al., 2007, 2010). Therefore, applying conservation management approaches that coincidentally build SOM while avoiding erosion and provide other environmental services may be needed to offset risks associated with crop residue harvest (Johnson et al., 2010a; Wilhelm et al., 2010).

A case study based on Iowa soils and yields (10.2 Mg ha^{-1} corn grain yield) using RUSLE2 and CQESTR simulation models demonstrated that converting conventionally tilled fields to no tillage allowed about 50% of the corn stover to be harvested without increasing erosion or losing SOC (Wilhelm et al., 2010). This case study also modeled using cover crops in conjunction with no tillage; the results suggested that the maximum harvest rate could be sustained while concomitantly protecting soil resource from erosion and SOC loss. Since corn cobs represent about 20% of the stover mass, harvesting grain and cobs instead of total stover is another strategy to harvest both a food and an energy crop from the same parcel of land (Varvel and Wilhelm, 2008; Halvorson and Johnson, 2009; Wilhelm et al., 2011). Cob and grain harvest did not increase erosion or nutrient run-off, since sufficient stover remained on the soil surface (Wienhold and Gilley, 2010). Although the type and capacity of the conversion facility, storage, transportation, and other economic issues are relevant, they are beyond the scope of this review to recommend best row-crop harvest strategies in a given locale.

Strategies to reduce erosion in row crops can also maintain or build SOC. For example, reducing or eliminating tillage can reduce erosion losses while also rebuilding SOC (Reicosky et al., 1995; Hunt et al., 1996; Wander et al., 1998; Deen and Kataki, 2003; Novak et al., 2009a). Unincorporated crop residues decompose slower than when buried by inversion tillage, such as disking or moldboard plowing (Reicosky and Lindstrom, 1993; Lal and Kimble, 1997; Paustian et al., 2000; Burgess et al., 2002). Surface residue placement commonly results in accumulation of SOC at the immediate soil surface (Hunt et al., 1996; Dolan et al., 2006; Novak et al., 2007; Blanco-Canqui and Lal, 2008) with some studies reporting SOC declines at deeper topsoil depths due to reduced residue incorporation (Wander et al., 1998; Deen and Kataki, 2003; Novak et al., 2009a). Other strategies for rebuilding SOC contents include adding mulches, compost, and manures to soils (Tiessen et al., 1994; Gregorich et al., 1996; Johnson et al., 2007b; Viaud et al., 2010). The combination of animal manure and no-till management increased SOC, especially in the soil surface compared to using conventional tillage practices (Bissonnette et al., 2001; Jiao et al., 2006). Unfortunately, studies have also reported that the SOC increases using reduced tillage and mulches are not long-lasting, but must be continually resupplied with fresh residue to overcome decomposition (Parton et al., 1987; Wang et al., 2000). Some have questioned the ability of reduced or no tillage to sequester SOC (Baker et al., 2007) and studies that measured C in horizons below the depth of tillage in row crops have not necessarily shown an increase in SOC through the profile (Venterea et al., 2006; Blanco-Canqui and Lal, 2008). In contrast, others have reported increases in SOC in a 1 m profile comparing no tillage to moldboard plowing (Gal et al., 2007). Furthermore, Kravchenko and Robertson (2011) delineated that differences in SOC may be erroneously missed due to inadequate replication to compensate for the inherent variability in SOC. These authors went on to provide a strong argument for using power analysis and making comparison by horizon to improve sensitivity of statistical analysis. A rigorous assessment of SOC throughout the profile is important for understanding the ability of an agroecosystem to mitigate GHG emissions through soil C sequestration.

MANAGING FOR SUSTAINABLE BIOENERGY WITH PERENNIALS

Row crops are essential for food production, but opportunities exist to diversify the landscape by integrating HPEC and WPEC into the agroecosystem. Compared to annual row crops, HPEC and WPEC have a greater potential to increase SOC, reduce soil erosion and improve off-site water quality. These beneficial outcomes strengthen the rationale for considering them as bioenergy feedstock (Brown et al., 2000; Nelson et al., 2006; Johnson et al., 2007c; Evans and Cohen, 2009). Perennial grasses and trees provide year-round cover, extensive rooting systems and an increased level of raindrop interception, which collectively contributes to reduced erosion and run-off losses (Thompson and Luckman, 1993; Meyer et al., 1995; Kort et al., 1998; Pimentel and Kounang, 1998; Dabney et al., 1999; Self-Davis et al., 2003). Strips of trees and switchgrass planted in various landscape positions (field edges, wetlands, riparian buffers) can effectively reduce 80 to 97% of sediment, N and P loads (Lee et al., 2003). Taking advantage of the environmental benefits with perennial species is consistent with a "sustainable landscape vision" proposed by Karlen et al. (2009) and further described by Wilhelm et al. (2010). This vision integrates economic, environmental, and social aspects of agriculture and selectively allows harvests only from those areas where crop residue exceeds that needed to maintain soil resource condition. Integrating food and fuel production systems, therefore, could mitigate surface and groundwater quality issues through nutrient capture and reduced run-off, and provide wildlife habitat for pollinators (Karlen et al., 2009; Wilhelm et al., 2010). For example, Mitchell et al. (2010) estimated that harvesting switchgrass grown only in the corners of center pivots without irrigation could provide enough feedstock within a 40 km radius to operate a 190 million L cellulosic ethanol plant. Such a strategic approach converts underutilized areas into viable parcels for HPEC production without major land-use shifts, which is consistent with the landscape vision for providing food, feed, fiber, and fuel.

Perennials for erosion control have been utilized by the Conservation Reserve Program (CRP). Conversion of marginal croplands to CRP played a dramatic role in reducing soil erosion from >45 $Mg\,ha^{-1}\,yr^{-1}$ to about 2.5 $Mg\,ha^{-1}\,yr^{-1}$ (Taylor and Lacewell, 2009). In the U.S. approximately 15 Mha of land are enrolled under the CRP program. Land under CRP protection is considered marginal land that is unfit for prime agricultural use because of poor quality characteristics such as high slopes, shallow soil, or poor physical and chemical characteristics that result in low crop yields (USDA, 1989). Taylor and Lacewell (2009) estimated that if only 20% of CRP lands were brought back into row crop production, annual soil erosion could increase by nearly 190 Tg. In contrast, converting CRP or abandoned agricultural land into large-scale production of WPEC or HPEC (Campbell et al., 2008) would improve nutrient depleted soils (Frank et al., 2004), and reduce GHG emission due to higher amounts of C sequestration (Coleman et al., 2004; Johnson et al., 2007b, c). Perennials have a greater potential to sequester soil C for several reasons, even if a portion is harvested as bioenergy feedstock. Typically, perennials compared to annual crop species have a greater root biomass (Zan et al., 2001; Bolinder et al., 2002). Biochemically, roots tend to be more recalcitrant than shoots (Johnson et al., 2007a). Perennials can also utilize more photosynthetically available days resulting in more atmospheric CO_2 converted into plant C (Baker and Griffis, 2009). Thus, more C enters the soil under perennials compared to annual crops. As an example, switchgrass is capable of adding significant amounts of root C to about 90 cm depth (Ma et al., 2000; Garten et al., 2010). In fact, the top 30 cm of soil below a >20-year-old switchgrass stand had about 15 $Mg\,ha^{-1}$ root biomass (Al-Kaisi and Grote, 2007). Belowground deposition from a switchgrass root system can add up to 1.2 $Mg\,C\,ha^{-1}\,yr^{-1}$ to the top 30 cm of soil (Liebig et al., 2005a, 2008). An unpublished study on a Norfolk sandy loam in the Coastal Plain region of SC showed dramatic increases in profile SOC content in as few as 2 years of switchgrass growth (Figure 8.2). Likewise, Hansen et al. (2004) attributed the observed change in SOC of 0.71 to 1.03 $Mg\,C\,ha^{-1}\,yr^{-1}$ in the top 100 cm of soil to deep miscanthus roots. Changing from annual crops to perennial species such as miscanthus, switchgrass, and tall fescue (*Festuca arundinacea*) has been reported to increase SOC content by 0.49 to 0.75 $Mg\,C\,ha^{-1}\,yr^{-1}$ (Lal et al., 1998a; Zan et al., 2001; Heaton et al., 2004; King et al., 2004).

Research monitoring changes in SOC contents under WPEC production systems has shown mixed results. Both Grigal and Berguson (1998) and Hansen (1993) reported that SOC content initially declined after establishment of a WPEC, but as the trees matured (10 to 18 years), SOC increased 1 to 1.6 $Mg\,C\,ha^{-1}\,yr^{-1}$. In contrast, Jug et al. (1999) reported an SOC increase at one of four sites in Germany under willow, poplar, and aspen (*P. tremula* × *P. tremuloides* cv. *Astria*—AS) production. Makeschin (1994) also reported mixed SOC responses under WPEC. The review by Johnson et al. (2007c) discussed additional environmental considerations related to establishment and management of HPEC and WPEC. While crop biomass sources and activities can rebuild SOC levels, researchers are also examining direct and novel approaches to increase SOC levels.

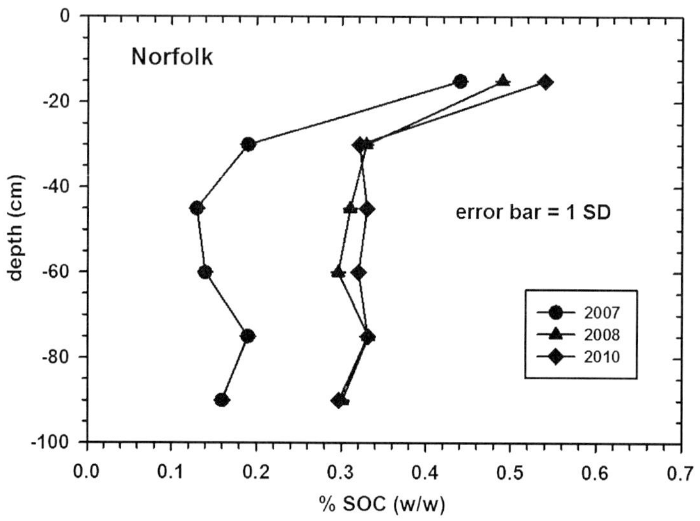

FIGURE 8.2
Soil profile SOC concentration of a Norfolk loamy sand measured before planting (2007) and after 1 (2008) and 2 years (2009) of switchgrass production (Novak and Frederick, unpublished data, 2011). Error bars represent one standard deviation of the mean (SD). In some cases the error bars are obscured by the symbols.

MANAGING FOR SUSTAINABLE BIOENERGY WITH A NOVEL AMENDMENT

Rebuilding SOC levels using reduced tillage, mulches, or by higher residue input and even perennials may take many years to have a measurable impact. Only a fraction of the organic C inputs are converted into stable SOC (Wilhelm et al., 2004; Johnson et al., 2006; Novak et al., 2009a). In contrast to fresh organic C inputs that cycle relatively quickly through the soil, biochars may provide a more stable soil amendment (Lehmann, 2007; Laird, 2008; Busscher et al., 2010). Thus, biochars are a novel approach for rebuilding SOC.

Biochar is a charcoal-like product produced through thermochemical conversion under low oxygen conditions (Antal and Gronli, 2003; Laird, 2008; Brown, 2009; Spokas, 2010). Biochars have a highly recalcitrant structure (Cheng et al., 2008; Kuzyakov et al., 2009), which stems from its highly aromatic composition (Glaser et al., 2002; Novak et al., 2009b) and a low O:C molar ratio (0.25 to 0.6) (Spokas, 2010). Although, pyrogenic C sources like biochar should not be considered humic substances, they do contribute to the organic material within the soil phase defined as SOC (Laird et al., 2008). Because biochars can have a very long residence time in soils (Laird, 2008), they may be used to reduce atmospheric CO_2 concentrations by sequestering C in soils (Sombroek et al., 2003; Lehmann et al., 2006; Liang et al., 2006; Fowles, 2007; Gaunt and Lehmann, 2008; Spokas, 2010; Woolf et al., 2010). A best-case model scenario predicted a maximum sustainable potential for C mitigation from biochar systems at 1.6 Pg yr^{-1}, which is equivalent to 12% of global CO_2 emissions (Woolf et al., 2010). This model assumed that it took about 40 years to reach maximize biomass pyrolysis and biochar production. This abatement strategy also has the potential to improve soil quality (see below), making biochar application a provocative consideration among future candidates of mitigation strategies.

There are reports that in addition to sequestering C, biochars can improve soil fertility (Glaser et al., 2002; Lehmann et al., 2003; Steiner et al., 2007; Novak et al., 2009b), increase soil moisture storage (Glaser et al., 2002; Novak et al., 2009b), and boost crop yields (Day et al., 2005; Steiner et al., 2007; Chan et al., 2008). Biochar properties are related to feedstock and pyrolysis conditions (Chan and Xu, 2009; Sohi et al., 2009), which can influence their quality as a soil amendment (Novak et al., 2009c; Spokas, 2010). For example, several biochars produced from different feedstocks and at different pyrolysis temperatures were laboratory incubated for 127 days in the Norfolk loamy sand, which is an extremely weathered Ultisol

TABLE 8.1 Mean Soil Organic Carbon (SOC) Concentration in a Norfolk Loamy Sand (Ap Horizon) After a 127-day Incubation at Ambient Laboratory Conditions Following the Addition of 2% biochar (w w^{-1}) (Novak et al., Unpublished data)[†]

		SOC	
		g kg^{-1}	
Feedstock	**Pyrolysis (°C)**	**Mean**	**SD**
Control	—	2.81e[‡]	0.08
Switchgrass	250	12.91c	0.34
	500	19.64a	0.07
Poultry litter	350	11.07d	0.75
	700	10.28d	0.41
Hardwood	450–600	17.18b	0.63

[†]*Biochar was added at 20 g kg^{-1} into Norfolk Ap; incubated at 10% (w w^{-1}) soil water content, and then every 30 d leached (4× total) with 1.2–1.3 pore volumes of deionized H_2O.*
[‡]*Means followed by a different letter differed based on Fisher LSD pair-wise multiple comparison procedures at $P = 0.05$.*

from the SC middle Coastal Plain region (Table 8.1). Adding 2% biochars (w w^{-1}) increased SOC compared to the control by as much as six-fold (Table 8.1) (Novak et al., 2009c). Similarly, other laboratory incubation experiments found that amending with biochars increased SOC content (Kimetu and Lehmann, 2010; Laird et al., 2010).

While biochars may increase SOC content, they should be applied discriminately to soil. Biochars produced at high pyrolysis temperatures (500 to 700°C) can be alkaline. Although alkaline biochars may be suitable for use in buffered acidic soils, their application to a poorly buffered loamy sand dramatically raised soil pH levels from 5.9 in the control to as high as 10 (Novak et al., 2009c). In addition, there are few studies that assayed soil microbial and macro-invertebrate communities' response to biochar application. A high temperate poultry-litter biochar impeded earthworm survival and growth (Liesch et al., 2010). Furthermore, biochars are expensive to apply, especially at high application rates ($300 at 112 Mg ha^{-1}) (Williams and Arnott, 2010). Therefore, it may be more financially prudent to invest in a biochar with definite chemical and physical properties that can target specific soil problems (Steinbeiss et al., 2009; Atkinson et al., 2010; Novak and Busscher, 2011). Applying appropriate biochars to soil would avoid pernicious biological, chemical and physical legacies.

SUMMARY AND RESEARCH NEEDS

Corn grain is the historical ethanol feedstock. However, to meet existing production mandates, next generation biofuel dedicated feedstocks and/or crop residue are under strong consideration. A regionally specific, balanced, and integrated landscape approach is critical for sustainable biofuel production that protects soil resources and crop productivity, and reduces GHG emissions. However, there are many questions that need to be addressed if this lofty goal is to be achieved. For example, harvest rates for crop residue and perennials alike are needed that avoid exacerbating erosion and related risks, such as loss of SOM or direct or indirect acceleration of GHG emission. A broad integrated study funded through USDA and DOE is under way to answer some of these questions and to provide data for simulation and predictive models (Karlen, 2010; Wilhelm et al., 2010). Empirical data on direct impacts of harvesting non-grain biomass (including WPEC and HPEC) on GHG emission is sorely needed. Fortunately, research is under way through projects such as the USDA-ARS-REAP and GRACEnet (Greenhouse gas Reduction through Agricultural Carbon Enhancement net; http://www.ars.usda.gov/research/programs/programs.htm?np_code=212&docid=21223) as well as university projects. Viable and sustainable bioenergy systems will be highly diverse across the country, fitted to the feedstocks available and the size and scope of the conversion platform. On a national scale, large-scale cellulosic ethanol production is of interest; however, there is also interest on a local scale for using feedstocks in an institutional or medium-size industrial facility (http://renewables.morris.umn.edu/biomass/). Research is also needed to assess potential benefits and risks of thermochemical co-products and their use as soil amendments. The recalcitrance of biochar makes land application a strong strategy for long-term C sequestration and reducing atmospheric CO_2 concentrations. However, additional information is needed so biochars can be designed with properties that will safely infuse fertility into depleted soils.

A simple one-size-fits-all sustainable bioenergy system does not exist. However, conservation-based management and forethought will promote biofuel development and utilization, all of which can mitigate GHG emission, provide environmental services beneficial to natural resources, and provide domestic and renewable energy. Efficient land use will be needed to provide sufficient food, feed, fiber, and fuel.

Acknowledgments

The authors express sincere gratitude to all participants who prepared biochars and the incubation studies. They also wish to express their appreciation to B. Burmeister for careful proof-reading. Biochar production and characterization was supported through the United States Department of Agriculture, ARS, GRACEnet program. This chapter contributes to the United States Department of Agriculture, ARS, GRACEnet, and REAP projects.

References

Adler, P.R., Del Grosso, S.J., Parton, W.J., 2007. Life-cycle assessment of net greenhouse-gas flux for bioenergy cropping systems. Ecol. Applic. 17, 675–691.

Al-Kaisi, M.M., Grote, J.B., 2007. Cropping systems effects on improving soil carbon stocks of exposed subsoil. Soil Sci. Soc. Am. J. 71, 1381–1388.

Antal, M.J., Gronli, M., 2003. The art, science, and technology of charcoal production. Ind. Eng. Chem. Res. 42, 1619–1640.

Atkinson, C., Fitzgerald, J., Hipps, N., 2010. Potential mechanisms for achieving agricultural benefits from biochar application to temperate soils: a review. Plant Soil 337, 1–18.

Baker, J.M., Griffis, T.J., 2009. Evaluating the potential use of winter cover crops in corn-soybean systems for sustainable co-production of food and fuel. Agric. For. Meteor. 149, 2120–2132.

Baker, J.M., Ochsner, T.E., Venterea, R.T., Griffis, T.J., 2007. Tillage and soil carbon sequestration—what do we really know? Agric. Ecosys. Environ. 118, 1–5.

Balesdent, J., Chenu, C., Balabane, M., 2000. Relationship of soil organic matter dynamics to physical protection and tillage. Soil Tillage Res. 53, 215–230.

Bennett, H.H., 1939. Soil Conservation. McGraw-Hill, New York, NY.

Bissonnette, N., Angers, D.A., Simard, R.R., Lafond, J., 2001. Interactive effects of management practices on water-stable aggregation and organic matter of a Humic Gleysol. Can. J. Soil Sci. 81, 545–551.

Blanco-Canqui, H., 2010. Energy crops and their implications on soil and environment. Agron. J. 102, 403–419.

Blanco-Canqui, H., Lal, R., 2007a. Impacts of long-term wheat straw management on soil hydraulic properties under no-tillage. Soil Sci. Soc. Am. J. 71, 1166–1173.

Blanco-Canqui, H., Lal, R., 2007b. Soil and crop response to harvesting corn residues for biofuel production. Geoderma 141, 355–362.

Blanco-Canqui, H., Lal, R., 2008. No-tillage and soil-profile carbon sequestration: an on-farm assessment. Soil Sci. Soc. Am. J. 72, 693–701.

Blanco-Canqui, H., Lal, R., 2009a. Corn stover removal for expanded uses reduces soil fertility and structural stability. Soil Sci. Soc. Am. J. 73, 418–426.

Blanco-Canqui, H., Lal, R., 2009b. Crop residue removal impacts on soil productivity and environmental quality. Crit. Rev. Plant Sci. 28, 139–163.

Blanco-Canqui, H., Stephenson, R.J., Nelson, N.O., Presley, D.R., 2009. Wheat and sorghum residue removal for expanded uses increases sediment and nutrient loss in runoff. J. Environ. Qual. 38, 2365–2372.

Blanco-Canqui, H., Lal, R., Post, W.M., Izaurralde, R.C., Owens, L.B., 2006. Changes in soil carbon and structural properties due to stover removal from no-till corn plots. Soil Sci. 171, 468–482.

Bolinder, M.A., Angers, D.A., Belanger, G., Michaud, R., Laverdiere, M.R., 2002. Root biomass and shoot to root ratios of perennial forage crops in eastern Canada. Can. J. Plant Sci. 82, 731–737.

Borjesson, P., 2009. Good or bad bioethanol from a greenhouse gas perspective—what determines this? Appl. Energy 86, 589–594.

Braun, R., Karlen, D.L., Johnson, D. (Eds.), 2010. Sustainable alternative fuel feedstock opportunities, challenges and roadmaps for six U.S. regions: Proc. of the Sustainable Feedstocks for Advanced Biofuel Workshop. Soil Water Conserv. Soc., Ankeny, IA.

BRDB, 2008. Increasing feedstock production for biofuels: economic drivers, environmental implications, and the role of research. [Online]. Biomass Research and Development Board (BRBD). Available at: <http://www.usbiomassboard.gov/pdfs/increasing_feedstock_revised.pdf> (verified May 25, 2011).

Brown, R., 2009. Biochar production technology. In: Lehmann, J., Joseph, S. (Eds.), Biochar for Environmental Management, Science and Technology. Earthscan Publ., London, UK, pp. 127–146.

Brown, R.A., Rosenberg, N.J., Hays, C.J., Easterling, W.E., Mearns, L.O., 2000. Potential production and environmental effects of switchgrass and traditional crops under current and greenhouse-altered climate in the central United States: a simulation study. Agric. Ecosyst. Environ. 78, 31–47.

Bruce, J.P., Frome, M., Haites, M., Janzen, H.H., Lal, R., Paustian, K., 1999. Carbon sequestration in soils. J. Soil Water Conserv. 54, 382–389.

Burgess, M.S., Mehuys, G.R., Madramootoo, C.A., 2002. Decomposition of grain-corn residues (*Zea mays* L.): a litterbag study under three tillage systems. Can. J. Soil Sci. 82, 127–138.

Busscher, W.J., Novak, J.M., Evans, D.E., Watts, D.W., Niandou, M.A.S., Ahmedna, M., 2010. Influence of pecan biochar on physical properties of a Norfolk loamy sand. Soil Sci. 175, 10–14.

Campbell, J.E., Lobell, D.B., Genova, R.C., Field, C.B., 2008. The global potential of bioenergy on abandoned agriculture lands. Environ. Sci. Technol. 42, 5791–5794.

CAST., 1982. Soil erosion: it's agricultural, environmental, and socio-economic implications. Council for Agricultural Science and Technology (CAST), Ames, IA.

CAST., 2007. Convergence of agriculture and energy. II. Producing cellulosic biomass for biofuels. [Online]. Council for Agricultural Science and Technology. Available at: <http://www.cast-science.org/websiteUploads/publicationPDFs/CAST%20Commentary%202007-2%207o12145.pdf> (verified January 19, 2011).

Chan, K.Y., Xu, Z., 2009. Biochar: nutrient properties and their enhancement. In: Lehmann, J., Joseph, S. (Eds.), Biochar for Environmental Management, Science and Technology. Earthscan Publ., London, UK, pp. 67–84.

Chan, K.Y., Van Zwieten, L., Meszaros, I., Downie, A., Joseph, S., 2008. Using poultry litter biochars as soil amendments. Aust. J. Soil Res. 46, 437–444.

Charles, M.B., Ryan, R., Ryan, N., Oloruntoba, R., 2007. Public policy and biofuels: the way forward? Energy Policy 35, 5737–5746.

Cheng, C.-H., Lehmann, J., Thies, J.E., Burton, S.D., 2008. Stability of black carbon in soils across a climatic gradient. J. Geophys. Res. 113: G02027, doi:02010.01029/02007jg000642.

Coleman, M.D., Isebrands, J.G., Tolsted, D.N., Tolbert, V.R., 2004. Comparing soil carbon of short rotation poplar plantations with agricultural crops and woodlots in North Central United States. Environ. Manage. 33 (Suppl. 1), S299–S308.

Committee on Water Implications of Biofuels Production in the United States National Research Council (Ed.), 2008. Water implications of biofuels production in the United States. National Academies Press, Washington, DC.

Cowie, A.L., Smith, P., Johnson, D., 2006. Does soil carbon loss in biomass production systems negate the greenhouse benefits of bioenergy? Mitigation Adapt. Strat. Global Change 11, 979–1002.

Coyle, W.T., 2010. Next-generation biofuels: near-term challenges and implications for agriculture. Bio-01–01. [Online]. USDA-Economic Research Service. Available at: <http://wwwers.usda.gov/publications/BIO0101/BIO0101.pdf> (posted May 2010; verified January 24, 2011).

Cruse, R.M., Herndl, C.G., 2009. Balancing corn stover harvest for biofuels with soil and water conservation. J. Soil Water Conserv. 64, 286–291.

Cruse, R.M., Cruse, M.J., Reicosky, D.C., 2009. Soil quality impacts of residue removal for biofuel feedstock, In: Lal, R., Stewart, B.A. (Eds.), Advances in Soil Science. Soil Quality and Biofuel Production. CRC Press Taylor and Francis Group, Boca Raton, FL, pp. 45–62.

Dabney, S.M., Liu, Z., Lane, M., Douglas, J., Zhu, J., Flanagan, D.C., 1999. Landscape benching from tillage erosion between grass hedges. Soil Tillage Res. 51, 219–231.

Day, D., Evans, R.J., Lee, J.W., Reicosky, D., 2005. Economical CO_2, SOx, and NOx capture from fossil-fuel utilization with combined renewable hydrogen production and large-scale carbon sequestration. Energy 30, 2558–2579.

Deen, W., Kataki, P.K., 2003. Carbon sequestration in a long-term conventional versus conservation tillage experiment. Soil Tillage Res. 74, 143–150.

Demirbas, A., 2009. Political, economic and environmental impacts of biofuels: a review. Appl. Energy 86, S108–S117.

Dicks, M.R., Campiche, J.L., Ugarte, D.D.L.T., Hellwinckel, C.M., Bryant, H.L., Richardson, J.W., 2009. Land use implications of expanding biofuel demand. J. Agric. Appl. Econ. 41, 435–453.

Dolan, M.S., Clapp, C.E., Allmaras, R.R., Baker, J.M., Molina, J.A.E., 2006. Soil organic carbon and nitrogen in a Minnesota soil as related to tillage, residue and nitrogen management. Soil Tillage Res. 89, 221–231.

Duncan, R.C., Youngquist, W., 1999. Encircling the peak of world oil production. Natural Resources Res. 8, 219–232.

Evans, J.M., Cohen, M.J., 2009. Regional water resource implications of bioethanol production in the Southeastern United States. Global Change Biol. 15, 2261–2273.

Fargione, J., Hill, J., Tilman, D., Polasky, S., Hawthorne, P., 2008. Land clearing and the biofuel carbon debt. Science 319, 1235–1238.

Fargione, J.E., Plevin, R.J., Hill, J.D., 2010. The ecological impact of biofuels. Annu. Rev. Ecol. Evol. Syst. 41, 351–377.

Farrell, A.E., Plevin, R.J., Turner, B.T., Jones, A.D., O'Hare, M., Kammen, D.M., 2006. Ethanol can contribute to energy and environmental goals. Science 311, 506–508.

Fowles, M., 2007. Black carbon sequestration as an alternative to bioenergy. Biom. Bioenergy 31, 426–432.

Frank, A.B., Berdahl, J.D., Hanson, J.D., Liebig, M.A., Johnson, H.A., 2004. Biomass and carbon partitioning in switchgrass. Crop Sci. 44, 1391–1396.

Franzluebbers, A.J., Follett, R.F., 2005. Greenhouse gas contributions and mitigation potential in agricultural regions of North America: Introduction. Soil Tillage Res. 83, 1–8.

Gal, A., Vyn, T.J., Micheli, E., Kladivko, E.J., McFee, W.W., 2007. Soil carbon and nitrogen accumulation with long-term no-till versus moldboard plowing overestimated with tilled-zone sampling depths. Soil Tillage Res. 96, 42–51.

Garten Jr., C.T., Smith, J.L., Tyler, D.D., Amonette, J.E., Bailey, V.L., Brice, D.J., Castro, H.F., Graham, R.L., Gunderson, C.A., Izaurralde, R.C., Jardine, P.M., Jastrow, J.D., Kerley, M.K., Matamala, R., Mayes, M.A., Metting, F.B., Miller, R.M., Moran, K.K., Post III, W.M., Sands, R.D., Schadt, C.W., Phillips, J.R., Thomson, A.M., T. Vugteveen, West, T.O., Wullschleger, S.D., 2010. Intra-annual changes in biomass, carbon, and nitrogen dynamics at 4-year old switchgrass field trials in west Tennessee, USA. Agric. Ecosys. Environ. 136, 177–184.

Gaunt, J.L., Lehmann, J., 2008. Energy balance and emissions associated with biochar sequestration and pyrolysis bioenergy production. Environ. Sci. Tech. 42, 4152–4158.

Glaser, B., Lehmann, J., Zech, W., 2002. Ameliorating physical and chemical properties of highly weathered soils in the tropics with charcoal—a review. Bio. Fert. Soils 35, 219–230.

Gopalakrishnan, G., Negri, M.C., Wang, M., Wu, M., Snyder, S.W., LaFreniere, L., 2009. Biofuels, land, and water: a systems approach to sustainability. Environ. Sci. Technol. 43, 6094–6100.

Graham, R.L., Nelson, R., Sheehan, J., Perlack, R.D., Wright, L.L., 2007. Current and potential U.S. corn stover supplies. Agron. J. 99, 1–11.

Gregorich, E.G., Ellert, B.H., Drury, C.F., Liang, B.C., 1996. Fertilization effects on soil organic matter turnover and corn residue C storage. Soil Sci. Soc. Am. J. 60, 472–476.

Grigal, D.F., Berguson, W.E., 1998. Soil carbon changes associated with short-rotation systems. Biom. Bioenergy 14, 371–377.

Halvorson, A.D., Johnson, J.M.F., 2009. Corn cob characteristics in irrigated central Great Plains studies. Agron. J. 101, 390–399.

Hansen, E.A., 1993. Soil carbon sequestration beneath hybrid popular plantations in the North Central United States. Biom. Bioenergy 5, 431–436.

Hansen, E.M., Christensen, B.T., Jensen, L.S., Kristensen, K., 2004. Carbon sequestration in soil beneath long-term Miscanthus plantations as determined by 13C abundance. Biom. Bioenergy 26, 97–105.

Hattori, T., Morita, S., 2010. Energy crops for sustainable bioethanol production; which, where and how? Plant Prod. Sci. 13, 221–234.

Heaton, E., Voigt, T., Long, S.P., 2004. A quantitative review comparing the yields of two candidate C4 perennial biomass crops in relation to nitrogen, temperature and water. Biom. Bioenergy 27, 21–30.

Hill, J., Nelson, E., Tilman, D., Polasky, S., Tiffany, D., 2006. Environmental, economic, and energetic costs and benefits of biodiesel and ethanol biofuels. PNAS 103, 11206–11210.

Hoekman, S.K., 2009. Biofuels in the U.S.—challenges and opportunities. Renewable Energy 34, 14–22.

Hudson, N.W., 1982. Soil erosion and conservation in the tropics. ASA Spec. Publ. No. 43. ASA, Madison, WI.

Hunt, P.G., Karlen, D.L., Matheny, T.A., Quisenberry, V.L., 1996. Changes in carbon content of a Norfolk loamy sand after 14 years of conservation or conservation tillage. J. Soil Water Conserv. 51, 255–258.

IPCC., 2007a. Intergovernmental Panel on Climate Change WGI, Fourth assessment report, Climate Change 2007: The physical science basis, Summary for policymakers [Online] Available at: <http://www.ipcc.ch/pdf/assessment-report/ar4/syr/ar4_syr_spm.pdf> (posted 5 February, 2007; verified February 11, 2011).

IPCC., 2007b. Summary for policymakers. In: Climate Change 2007: Mitigation. Contribution of Working Group III to the Fourth assessment report of the Intergovernmental Panel on Climate Change [Online]. Cambridge University Press Available at: <http://www.ipcc.ch/publications_and_data/ar4/wg3/en/spm.html> (verified January 18, 2011).

Jacinthe, P.A., Lal, R., 2003. Nitrogen fertilization of wheat residue affecting nitrous oxide and methane emission from a central. Ohio Luvisol. Biol. Fertil. Soils 37, 338–347.

Jiao, Y., Whalen, J.K., Hendershot, W.H., 2006. No-tillage and manure applications increase aggregation and improve nutrient retention in a sandy-loam soil. Geoderma 134, 24–33.

Johnson, J.M.F., Barbour N.W., 2010. Crop yield and greenhouse gas responses to stover harvest on glacial till Mollisol. 19th World Congr. Soil Sci.: soil solutions for a changing world. 1–6 August 2010, Brisbane, Australia, pp. 36–39.

Johnson, J.M.F., Allmaras, R.R., Reicosky, D.C., 2006. Estimating source carbon from crop residues, roots and rhizodeposits using the national grain-yield database. Agron. J. 98, 622–636.

Johnson, J.M.F., Barbour, N.W., Weyers, S.L., 2007a. Chemical composition of crop biomass impacts its decomposition. Soil Sci. Soc. Am. J. 71, 155–162.

Johnson, J.M.F., Karlen, D.L., Andrews, S.S., 2010a. Conservation considerations for sustainable bioenergy feedstock production: if, what, where, and how much? J. Soil Water Conserv. 65 (4), 88A–91A.

Johnson, J.M.F., Franzluebbers, A.J., Weyers, S.L., Reicosky, D.C., 2007b. Agricultural opportunities to mitigate greenhouse gas emissions. Environ. Pollut. 150, 107–124.

Johnson, J.M.F., Papiernik, S.K., Mikha, M.M., Spokas, K., Tomer, M.D., Weyers, S.L., 2010b. Soil processes and residue harvest management. In: Lal, R., Stewart, B.A. (Eds.), Carbon Management, Fuels, and Soil Quality. Taylor and Francis, LLC, New York, NY, pp. 1–44.

Johnson, J.M.F., Reicosky, D.C., Allmaras, R.R., Sauer, T.J., Venterea, R.T., Dell, C.J., 2005. Greenhouse gas contributions and mitigation potential of agriculture in the central USA. Soil Tillage Res. 83, 73–94.

Johnson, J.M.F., Arriaga, F.J., Banowetz, G.M., Huggins, D.R., Laird, D.A., Ottman, M., Wienhold, B.J., 2010c. Crop residues of the contiguous United States: balancing feedstock and soil needs with conservation tillage, cover crops, and biochar. In: Braun, R. (Ed.), Sustainable Alternative Fuel Feedstock Opportunities, Challenges and Roadmaps for Six U.S. Regions: Proc. of the Sustainable Feedstocks for Advanced Biofuel Workshop, Soil Water Conserv. Soc, Ankenny, IA, pp. 82–104.

Johnson, J.M.F., Coleman, M.D., Gesch, R.W., Jaradat, A.A., Mitchell, R., Reicosky, D.C., Wilhelm, W.W., 2007c. Biomass-bioenergy crops in the United States: a changing paradigm. The Amer. J. Plant Sci. Biotechnol. 1, 1–28.

Jug, A., Makeschin, F., Rehfuess, K.E., Hofmann-Schielle, C., 1999. Short-rotation plantations of balsam poplars, aspen and willows on former arable land in the Federal Republic of Germany. III. Soil ecological effects. For. Ecol. Manage. 121, 85–99.

Karlen, D.L., 2010. Corn stover feedstock trials to support predictive modeling. Glob. Change Biol. Bioenergy 2, 235–247.

Karlen, D.L., Lal, R., Follett, R.F., Kimble, J.M., Hatfield, J.L., Miranowski, J.M., Cambardella, C.A., Manale, A., Anex, R.P., Rice, C.W., 2009. Crop residues: the rest of the story. Environ. Sci. Technol. 43, 8011–8015.

Kimetu, J.M., Lehmann, J., 2010. Stability and stabilisation of biochar and green manure in soil with different organic carbon contents. Aust. J. Soil Res. 48, 577–585.

King, J.A., Bradley, R.I., Harrison, R., Carter, A.D., 2004. Carbon sequestration and saving potential associated with changes to the management of agricultural soils in England. Soil Use Manage. 20, 394–402.

Kort, J., Collins, M., Ditsch, D., 1998. A review of soil erosion potential associated with biomass crops. Biom. Bioenergy 14, 351–359.

Kravchenko, A.N., Robertson, G.P., 2011. Whole-profile soil carbon stocks: the danger of assuming too much from analyses of too little. Soil Sci. Soc. Am. J. 75, 235–240.

Kuzyakov, Y., Subbotina, I., Chen, H., Bogomolova, I., Xu, X., 2009. Black carbon decomposition and incorporation into soil microbial biomass estimated by 14C labeling. Soil Biol. Biochem. 41, 210–219.

Laird, D.A., 2008. The charcoal vision: a win win win scenario for simultaneously producing bioenergy, permanently sequestering carbon, while improving soil and water quality. Agron. J. 100, 178–181.

Laird, D.A., Chappell, M.A., Martens, D.A., Wernshaw, R.L., Thompson, M.L., 2008. Distinguishing black carbon from biogenic humic substances in soil clay fractions. Geoderma 143, 115–122.

Laird, D.A., Fleming, P., Davis, D.D., Horton, R., Wang, B., Karlen, D.L., 2010. Impact of biochar amendments on the quality of a typical Midwestern agricultural soil. Geoderma 158, 443–449.

Lal, R., 2008a. Carbon sequestration. Phil. Trans. Royal Soc. B: Biol. Sci. 363, 815–830.

Lal, R., 2008b. Crop residues as soil amendments and feedstock for bioethanol production. Waste Manage. 28, 747–758.

Lal, R., 2009. Soil quality impacts of residue removal for bioethanol production. Soil Tillage Res. 102, 233–241.

Lal, R., Kimble, J.M., 1997. Conservation tillage for carbon sequestration. Nutr. Cycl. Agroecosys. 49, 243–253.

Lal, R., Pimentel, D., 2007. Biofuels from crop residues. Soil Tillage Res. 93, 237–238.

Lal, R., Henderlong, P., Flowers, M., et al., 1998a. Forages and row cropping effects on soil organic carbon and nitrogen contents. In: Lal, R.S. (Ed.), Management of C Sequestration in Soil. CRC Press, Boca Roca, FL, pp. 365–379.

Lal, R., Reicosky, D.C., Hanson, J.D., 2007a. Evolution of the plow over 10,000 years and the rationale for no-till farming. Soil Tillage Res. 93, 1–12.

Lal, R., Follett, R.F., Stewart, B.A., Kimble, J.M., 2007b. Soil carbon sequestration to mitigate climate change and advance food security. Soil Sci. 172, 943–956.

Lal, R., Kimble, J.M., Follett, R.F., Cole, C.V., 1998b. Potential of U.S. cropland to sequester carbon and mitigate the greenhouse effect. Ann Arbor Press, an imprint of Sleeping Bear Press, Chelsea, MI.

Lee, K.H., Isenhart, T.M., Schultz, R.C., 2003. Sediment and nutrient in an established multi-species riparian buffer. J. Soil Water Conserv. 58, 1–8.

Lehmann, J., 2007. Bio-energy in the black. Frontiers Ecol. Environ. 5, 381–387.

Lehmann, J., Gaunt, J., Rondon, M., 2006. Bio-char sequestration in terrestrial ecosystems—a review. Mitigation Adapt. Strat. Glob. Clim. Change 11, 403–427.

Lehmann, J., da Silva Jr., J.P., Steiner, C., Nehls, T., Zech, W., Glaser, B., 2003. Nutrient availability and leaching in an archaeological Anthrosol and a Ferralsol of the Central Amazon Basin: fertilizer, manure and charcoal amendments. Plant Soil 249, 343–357.

Liang, B., Lehmann, J., Solomon, D., Kinyangi, J., Grossman, J., O'Neill, B., Skjemstad, J.O., Thies, J., Luizao, F.J., Petersen, J., Neves, E.G., 2006. Black carbon increases cation exchange capacity in soils. Soil Sci. Soc. Am. J. 70, 1719–1730.

Liebig, M.A., Johnson, H.A., Hanson, J.D., Frank, A.B., 2005a. Soil carbon under switchgrass stands and cultivated cropland. Biom. Bioenergy 28, 347–354.

Liebig, M., Schmer, M., Vogel, K., Mitchell, R., 2008. Soil carbon storage by switchgrass grown for bioenergy. BioEnergy Res. 1, 215–222.

Liebig, M.A., Morgan, J.A., Reeder, J.D., Ellert, B.H., Gollany, H.T., Schuman, G.E., 2005b. Greenhouse gas contributions and mitigation potential of agricultural practices in northwestern USA and western Canada. Soil Tillage Res. 83, 25–52.

Liesch, A.M., Weyers, S.L., Gaskins, J.W., Das, K.C., 2010. Impact of two different biochars on earthworm growth and survival. Ann. Environ. Sci. 4, 1–9.

Londo, M., Lensink, S., Wakker, A., Fischer, G., Prieler, S., van Velthuizen, H., de Wit, M., Faaij, A., Junginger, M., Berndes, G., Hansson, J., Egeskog, A., Duer, H., Lundbaek, J., Wisniewski, G., Kupczyk, A., Könighofer, K., 2010. The REFUEL EU road map for biofuels in transport: application of the project's tools to some short-term policy issues. Biom. Bioenergy 34, 244–250.

Ma, Z., Wood, C.W., Bransby, D.I., 2000. Impacts of soil management on root characteristics of switchgrass. Biom. Bioenergy 18, 105–112.

Makeschin, F., 1994. Effects of energy forestry on soils. Biom. Bioenergy 6, 63–79.

Malça, J., Freire, F., 2011. Life-cycle studies of biodiesel in Europe: a review addressing the variability of results and modeling issues. Renewable Sustainable Energy Rev. 15, 338–351.

Malhi, S.S., Lemke, R., 2007. Tillage, crop residue and N fertilizer effects on crop yield, nutrient uptake, soil quality and nitrous oxide gas emissions in a second 4-yr rotation cycle. Soil Tillage Res. 96, 269–283.

Meyer, L.D., Dabney, S.M., Harmon, W.C., 1995. Sediment-trapping effectiveness of stiff-grass hedges. Trans. ASAE 38, 809–815.

Meyer, P.W., 2010. Biofuel Review: Part 4—Food vs. fuel and profit vs. hunger. [Online]. IEEE-USA Today's Engineer. Available at: <http://www.todaysengineer.org/2010/Jun/biofuels-pt4.asp> (verified January 18, 2011).

Mitchell, R.B., Owens, V., Gutterson, N., Richard, E.P., Barney, J.N., 2010. Herbaceous perennials: placement, benefits and incorporation challenges in diversified landscapes. In: Braun, R. (Ed.), Sustainable Alternative Fuel Feedstock Opportunities, Challenges and Roadmaps for Six U.S. Regions: Proc. of the Sustainable Feedstocks for Advanced Biofuel Workshop. Soil Water Conserv. Soc, Ankenny, IA.

Nehring, R., 2009. Traversing the mountaintop: world fossil fuel production to 2050. Philos. Trans. R. Soc. Lond. B. Biol. Sci. 364, 3067–3079.

Nelson, R.G., 2002. Resource assessment and removal analysis for corn stover and wheat straw in the Eastern and Midwestern United States—rainfall and wind-induced soil erosion methodology. Biom. Bioenergy 22, 349–363.

Nelson, R.G., Ascough Ii, J.C., Langemeier, M.R., 2006. Environmental and economic analysis of switchgrass production for water quality improvement in northeast Kansas. J. Environ. Manage. 79, 336–347.

Nelson, R.G., Walsh, M., Sheehan, J.J., Graham, R., 2004. Methodology for estimating removable quantities of agricultural residues for bioenergy and bioproduct use. Appl. Biochem. Biotech 113, 13–26.

Novak, J.M., Busscher, W.J., 2011. Selection and use of designer biochars to improve characteristics of southeastern USA Coastal Plain degraded soils. (In press.). In: Lee, J.W. (Ed.), Advanced Biofuels and Bioproducts. Springer, New York, NY.

Novak, J.M., Bauer, P.J., Hunt, P.G., 2007. Carbon dynamics under long-term conservation and disk tillage management in a Norfolk loamy sand. Soil Sci. Soc. Am. J. 71, 453–456.

Novak, J.M., Frederick, J.R., Bauer, P.J., Watts, D.W., 2009a. Rebuilding organic carbon contents in Coastal Plain soils using conservation tillage systems. Soil Sci. Soc. Am. J. 73, 622–629.

Novak, J.M., Busscher, W.J., Laird, D.L., Ahmedna, M., Watts, D.W., Niandou, M.A.S., 2009b. Impact of biochar amendment on fertility of a Southeastern Coastal Plain soil. Soil Sci. 174, 105–112.

Novak, J.M., Lima, I.M., Xing, B., Gaskin, J.W., Steiner, C., Das, K.C., Ahmedna, M., Rehrah, D., Watts, D.W., Busscher, W.J., Schomberg, H., 2009c. Characterization of designer biochar produced at different temperatures and their effects on a loamy sand. Ann. Environ. Sci. 3, 195–206.

OCHA., 2010. OCHA Occasional policy briefing series. Brief No. 3: Energy security and humanitarian action: key emerging trends and challenges. [Online]. Policy Development and Studies Branch, UN Office for the Coordination of Humanitarian Affairs (OCHA). Available at: <http://ochanet.unocha.org/p/Documents/OCHA%20OPB%20Energy%2011Nov10%20fnl-2.pdf> (posted September 2010; verified June 30, 2011).

ASA Spec. Publ. no. 31Oschwald, W.R., Stelly, M., Kral, D.M., Nauseef, J.H. (Eds.), Crop residue management systems. ASA, CSSA, SSSA, Madison, WI.

Parton, W.J., Schimel, D., Cole, C.V., Ojima, D.S., 1987. Analysis of factors controlling soil organic matter levels in Great Plains grassland. Soil Sci. Soc. Am. J. 51, 1173–1179.

Paustian, K., Collins, H.P., Paul, E.A., Paul, E.A., et al., 1997. Management controls on soil carbon. Soil Organic Matter in Temperate Agroecosystems: Long-term Experiments in North America. CRC Press, Boca Raton, FL, pp. 15–49.

Paustian, K., Six, J., Elliott, E.T., Hunt, H.W., 2000. Management options for reducing CO_2 emissions from agricultural soils. Biogeochemistry 48, 147–163.

Perlack, R.D., Wright, L.L., Turhollow, A., Graham, R.L., Stokes, B., Erbach, D.C., 2005. Biomass as feedstock for a bioenergy and bioproducts industry: the technical feasibility of a billion-ton annual supply. [Online]. U.S. Department of Energy and U.S. Department of Agriculture. Available at: <http://www.eere.energy.gov/biomass/pdfs/final_billionton_vision_report2.pdf> (posted 15 July, 2005; verified May 16, 2011).

Pimentel, D., Kounang, N., 1998. Ecology of soil erosion in ecosystems. Ecosystems 1, 416–426.

Pimentel, D., Patzek, T.W., 2005. Ethanol production using corn, switchgrass, and wood; biodiesel production using soybean and sunflower. Natural Resources Res. 14, 65–76.

Pimentel, D., Pimentel, M., 2008. Corn and cellulosic ethanol cause major problems. Energies 1, 35–37.

Pimentel, D., Harvey, C., Resosudarmo, P., Sinclair, K., Kurz, D., McNair, M., Crist, S., Shpritz, L., Fitton, L., Saffouri, R., Blair, R., 1995. Environmental and economic costs of soil erosion and conservation benefits. Science 267, 1117–1123.

Pimentel, D., Pleasant, A., Barron, J., Gaudioso, J., Pollock, N., Chae, E., Kim, Y., Lassiter, A., Schiavoni, C., Jackson, A., Lee, M., Eaton, A., 2004. US energy conservation and efficiency: benefits and costs. Environ. Develop. Sustain. 6, 279–305.

Reicosky, D.C., Lindstrom, M.J., 1993. Fall tillage method: effect on short-term carbon dioxide flux from soil. Agron. J. 85, 1237–1243.

Reicosky, D.C., Kemper, W.D., Langdale, G.W., Douglas Jr., C.W., Rasmussen, P.E., 1995. Soil organic matter changes resulting from tillage and biomass production. J. Soil Water Conserv. 50, 253–261.

Searchinger, T., Heimlich, R., Houghton, R.A., Dong, F., Elobeid, A., Fabiosa, J., Tokgoz, S., Hayes, D., T- Yu, H., 2008. Use of U.S. croplands for biofuels increases greenhouse gases through emissions from land-use change. Science 319, 1238–1240.

Self-Davis, M.L., Moore, P.A., Daniels, T.C., Nichols, D.J., Sauer, T.J., West, C.P., Aiken, G.E., Edwards, D.R., 2003. Forage species and canopy cover effects on runoff from small plots. J. Soil Water Conserv. 58, 349–359.

Singh, S., Kumar, A., 2011. Development of water requirement factors for biomass conversion pathway. Bioresource Technol. 102, 1316–1328.

Slobodian, N., Van Rees, K., Pennock, D., 2002. Cultivation-induced effects on belowground biomass and organic carbon. Soil Sci. Soc. Am. J. 66, 924–930.

Sohi, S., Lopez-Capel, E., Krull, E., Bol, R., 2009. Biochar, climate change and soil: a review to guide future research. CSIRO Land and Water Science Report 05/09. [Online]. CSIRO. Available at: <http://www.csiro.au/files/files/poei.pdf> (posted February 2009; verified January 30, 2011).

Sombroek, W.M., Ruivo, M.L., Fearnside, P.M., Glaser, B., Lehmann, J., 2003. Amazonian dark earths as carbon stores and sinks. In: Lehmann, J. (Ed.), Amazonian Dark Earths: Origins, Properties, Management. Kluwer Acad. Publ., Dordrecht, The Netherlands, pp. 125–139.

Spokas, K.A., 2010. Review of the stability of biochar in soils: predictability of O: C molar ratios. Carbon Manage. 1, 289–303.

Steenhof, P.A., McInnis, B.C., 2008. A comparison of alternative technologies to de-carbonize Canada's passenger transportation sector. Technol. Forecasting Social Change 75, 1260–1278.

Steinbeiss, S., Gleixner, G., Antonietti, M., 2009. Effect of biochar amendment on soil carbon balance and soil microbial activity. Soil Biol. Biochem 41, 1301–1310.

Steiner, C., Teixeira, W., Lehmann, J., Nehls, T., de Macêdo, J., Blum, W., Zech, W., 2007. Long term effects of manure, charcoal and mineral fertilization on crop production and fertility on a highly weathered Central Amazonian upland soil. Plant Soil 291, 275–290.

Taylor, C.R., Lacewell, R.D., 2009. Effects of corn ethanol and soy-based biodiesel production on soil erosion and return of conservation reserve program land to crop production. Bioenergy policy briefs BPB 070409. [Online]. Auburn University and Texas A&M. Available at: <https://sites.auburn.edu/academic/ag/group/bioenergy/CRP%20%20Erosion%20Effects/CRP%20and%20Erosion%20Policy%20Brief.pdf> (posted July 2009; verified January 24, 2011).

Thompson, R.C., Luckman, P.G., 1993. Performance of biological erosion control in New Zealand soft rock hill terrain. Agrofor. Syst. 21, 191–211.

Tiessen, H., Cuevas, E., Chacon, P., 1994. The role of soil organic matter in sustaining soil fertility. Nature 371, 783–785.

U.S. Congress, 2007. Energy Independence and Security Act of 2007. [Online]. U.S. Congress. Available at: <http://frwebgate.access.gpo.gov/cgi-bin/getdoc.cgi?dbname=110_cong_bills&docid=f:h6enr.txt.pdf> (verified January 18, 2011).

U.S. EIA, 2010. World marketed energy consumption. [Online]. United States Energy Information Administration, Department of Energy. Available at: <www.eia.doe.gov/oiaf/ieo/index.html> (verified December 13, 2010).

U.S. EIA, 2011a. Annual energy outlook 2011 early release overview. [Online]. United States Energy Information Administration. Department of Energy. Available at: http://www.eia.gov/forecasts/aeo/pdf/0383er(2011).pdf (verified January 14, 2011).

U.S. EIA, 2011b. Short term energy outlook. [Online]. United States Energy Information Administration, Department of Energy. Available at: <http://www.eia.doe.gov/emeu/steo/pub/steo_full.pdf> (posted January, 2011; verified January 14, 2011).

USDA-NRCS., 1997. Tolerable erosion levels "T". [Online]. USDA-Natural Resource Conservation Service. Available at: <http://www.mn.nrcs.usda.gov/technical/nri/findings/erosion_t.htm> (verified May 12, 2011).

USDA-NRCS., 2009. National Resources Inventory—2007 NRI. [Online]. United States Department of Agriculture, Natural Resources Conservation Service. Available at: <http://www.nrcs.usda.gov/technical/nri/2007/nri07erosion.html> (posted December 17, 2009; verified June 30, 2011).

USDA., 1989. The second RCA appraisal: soil, water, and related resources on nonfederal land in the United States: analysis of conditions and trends. United States Department of Agriculture, Washington, DC.

Varvel, G.E., Wilhelm, W.W., 2008. Cob biomass production in the western Corn Belt. Bioenergy Res. 1, 223–228.

Venterea, R.T., Baker, J.M., Dolan, M.S., Spokas, K.A., 2006. Carbon and nitrogen storage are greater under biennial tillage in a Minnesota corn-soybean rotation. Soil Sci. Soc. Am. J. 70, 1752–1762.

Viaud, V., Angers, D.A., Parnadeau, V., Morvan, T., Menasseri-Aubry, S., 2010. Response of organic matter to reduced tillage and animal manure in a temperate loamy soil. Soil Use Manage. doi:10.111/j.1475–2743.2010.00314.x.

Wander, M.M., Bidart, M.G., Aref, S., 1998. Tillage impacts on the depth distribution of total and particulate organic matter in three Illinois soils. Soil Sci. Am. J. 62, 1704–1711.

Wang, Y., Amundson, R., X- Niu, F., 2000. Seasonal and altitudinal variation in decomposition of soil organic matter inferred from radiocarbon measurements of soil CO2 flux. Global Biogeochem. Cycles 14, 199–211.

West, T.O., Post, W.M., 2002. Soil organic carbon sequestration rates by tillage and crop rotation: a global data analysis. Soil Sci. Soc. Am. J. 66, 1930–1946.

Wienhold, B.J., Gilley, J.E., 2010. Cob removal effect on sediment and runoff nutrient loss from a silt loam soil. Agron. J. 102, 1448–1452.

Wilhelm, W.W., Johnson, J.M.F., Karlen, D.L., Lightle, D.T., 2007. Corn stover to sustain soil organic carbon further constrains biomass supply. Agron. J. 99, 1665–1667.

Wilhelm, W.W., Johnson, J.M.F., Hatfield, J.L., Voorhees, W.B., Linden, D.R., 2004. Crop and soil productivity response to corn residue removal: a literature review. Agron. J. 96, 1–17.

Wilhelm, W.W., Hess, J.R., Karlen, D.L., Johnson, J.M.F., Muth, D.J., Baker, J.M., Gollany, H.T., Novak, J.M., Stott, D.E., Varvel, G.E., 2010. Review: balancing limiting factors & economic drivers for sustainable Midwestern US agricultural residue feedstock supplies. Ind. Biotechnol. 6, 271–287.

Wilhelm, W., Johnson, J., Lightle, D., Karlen, D., Novak, J., Barbour, N., Laird, D., Baker, J., Ochsner, T., Halvorson, A., Archer, D., Arriaga, F., 2011. Vertical distribution of corn stover dry mass grown at several US locations. BioEnergy Res. 4, 11–21.

Williams, M.M., Arnott, J.C., 2010. A comparison of variable economic costs associated with two proposed biochar application methods. Ann. Environ. Sci. 4, 23–30.

Woolf, D., Amonette, J.E., Street-Perrott, F.A., Lehmann, J., Joseph, S., 2010. Sustainable biochar to mitigate global climate change. Nat. Commun. 1, 56. doi.org/10.1038/ncomms1053.

Zan, C.S., Fyles, J.W., Girouard, P., Samson, R.A., 2001. Carbon sequestration in perennial bioenergy, annual corn and uncultivated systems in southern Quebec. Agric. Ecosyst. Environ. 86, 135–144.

SECTION 3

Agricultural Management and Greenhouse Gas Flux

CHAPTER 9

Cropland Management Contributions to Greenhouse Gas Flux: Central and Eastern U.S.

Michel A. Cavigelli[1], **Timothy B. Parkin**[2]
[1]Sustainable Agricultural Systems Laboratory, USDA-ARS, Henry A. Wallace Beltsville Agricultural Research Center, Beltsville, MD
[2]National Laboratory for Agriculture and the Environment, USDA-ARS, Ames, IA

CHAPTER OUTLINE

Abbreviations: AA, anhydrous ammonia; AN, ammonium nitrate; APP, ammonium polyphosphate; BMP, best management practice; BAU, business as usual; CO_2, carbon dioxide; CP, chisel plow; CT, conventional tillage; DAP, diammonium phosphate; N_2, dinitrogen; GWP, global warming potential; GHG, greenhouse gas; GRACEnet, Greenhouse gas Reduction through Agricultural Carbon Enhancement network; LSM, liquid swine manure; LDM, liquid dairy manure; MAXC, maximum C sequestration CH_4, methane; MP, moldboard plow; NEE, net ecosystem exchange; NO, nitric oxide; N, nitrogen; N_2O, nitrous oxide; NFT-NSS, non-flow

Managing Agricultural Greenhouse Gases. DOI: 10.1016/B978-0-12-386897-8.00009-7
2012, Published by Elsevier Inc.

through, non-steady state; NT, no tillage; OGGB, optimum greenhouse gas benefits; P, phosphorus; PCU, polymer coated urea; PET, potential evapotranspiration; PL, poultry litter; PT, precision tillage; ST, strip tillage; UAN, urea ammonium nitrate; WFPS, water-filled pore space

INTRODUCTION

Atmospheric concentrations of the three major biogenic greenhouse gases (GHGs) carbon dioxide (CO_2), methane (CH_4) and nitrous oxide (N_2O) have been increasing at unprecedented rates since the beginning of the industrial revolution, contributing to disruptive climate change around the globe (IPCC, 2007). Nitrous oxide is also a natural catalyst of stratospheric ozone decay (Rasmussen and Khalil, 1986; Cicerone, 1987). While the atmospheric concentrations of CH_4 (1.74–1.87 ppm) and N_2O (321–322 ppb) are substantially lower than that of CO_2 (385 ppm), the impacts of CH_4 and N_2O on global warming on a mass basis are 25 and 298 times greater, respectively, than that of CO_2 over a 100-year time horizon (USEPA, 2011).

Agriculture is the dominant source of CH_4 and N_2O to the atmosphere, contributing about 50 and 60%, respectively, of these two gases when both direct and indirect emissions are accounted for (IPCC, 2007). The major agricultural CH_4 sources are enteric fermentation in livestock (accounting for about 20% of total U.S. CH_4 emissions), manure management (prior to being applied to soils; 7% of total U.S. emissions), and rice cultivation (about 1% of total U.S. emissions). Non-flooded agricultural soils are not considered an important source of CH_4 in the U.S. and the world (USEPA, 2011). By contrast, agricultural soils are the most important source of anthropogenic N_2O, accounting for 69% of total emissions in the U.S. in 2009 (USEPA, 2011). In addition, N_2O emissions from agriculture on a global basis are expected to increase by 35 to 60% by 2030 (IPCC, 2007) as agricultural production is challenged to meet the needs of an increasing global human population.

Most agricultural soils are not an important source of atmospheric CO_2. While there are large fluxes of CO_2 into and out of the biosphere in all agricultural systems, the net impact of agriculture on biogenic production or consumption of CO_2 is negligible (IPCC, 2007). Nonetheless, individual cropping systems may have positive or negative net ecosystem exchange (NEE).

While agriculture is considered to contribute about 6% of total U.S. (USEPA, 2011) and 10–12% of global GHG emissions (IPCC, 2007), these assessments do not include GHGs released during the production, transport, distribution, and application of agricultural inputs such as fertilizers and pesticides, nor do they account for CO_2 emissions from equipment operation. Additionally, GHGs released when native systems are replaced with agricultural systems are not accounted for in the agricultural sector (IPCC, 2007). Such emissions are allocated to the industrial sector and land-use change categories, respectively, in global and national GHG assessments. A complete assessment of the impact of agriculture on climate change needs to consider these "non-agricultural" GHG emissions.

In this chapter, we summarize recent research results on the effectiveness of GHG mitigation strategies in croplands in the eastern and central U.S. We focus on soil N_2O and CH_4 emissions, and address NEE briefly; we do not address non-soil emissions of GHGs. While we recognize the importance of the latter, these are beyond the scope of this chapter. Moreover, very few studies in the U.S. have adequately assessed global warming potential (GWP) of various GHG mitigation options.

FACTORS CONTROLLING N_2O AND CH_4 FLUX FROM SOILS

Nitrification and denitrification are generally considered to be the primary sources of N_2O in soils. While other processes, including nitrifier denitrification and chemodenitrification, are also recognized as sources of N_2O, their contributions in agricultural soils are uncertain (Bateman and Baggs, 2005; Kool et al., 2010).

Nitrification is the conversion of ammonium to nitrite and then nitrate by obligate aerobic bacteria. Nitrification rates in soil are highest when ammonium is readily available and soil moisture is sufficient to provide rapid diffusion of ammonium but not so high that oxygen diffusion is limited, i.e. ~60% water-filled pore space (WFPS) (Linn and Doran, 1984). Nitrous oxide production during nitrification occurs when nitrite is used as an electron acceptor, possibly as a detoxification mechanism (Poth and Focht, 1985). The proportion of N_2O to NO_3 produced during nitrification increases as oxygen partial pressure decreases (Goreau et al., 1980). Thus, N_2O production during nitrification tends to increase as oxygen concentration in soils limits the overall rate of nitrification.

Denitrification is an anaerobic respiration process in which nitrogen oxides (primarily nitrate) are used as terminal electron acceptors and thereby reduced to gaseous forms (NO, N_2O, N_2) when oxygen is limiting (Tiedje, 1988). Denitrification rate in soil increases exponentially as soil oxygen level decreases (Parkin and Tiedje, 1984), with the highest rates often occurring under anaerobic conditions as long as nitrate and carbon are present in sufficient quantities (Linn and Doran, 1984). Since nitrate production (via nitrification) stops under anaerobic conditions, denitrification may eventually be limited under anaerobic conditions by lack of nitrate.

While dinitrogen (N_2) is usually the primary end product of denitrification, N_2O, an intermediate in the denitrification process, is released when conditions favor the induction and/or activity of denitrification enzymes leading to its production more than the induction and/or activity of nitrous oxide reductase, the enzyme that reduces N_2O to N_2. Such conditions may exist during the onset of denitrifying conditions following a precipitation or irrigation event (Firestone and Tiedje, 1979; Dendooven and Anderson, 1994; Sexstone et al., 1985), under microaerobic or low pH conditions, or when the ratio of available C:NO_3 is low (Firestone and Davidson, 1989; Betlach and Tiedje, 1981). Thus, the ratio of $N_2O/(N_2O + N_2)$ produced during denitrification may vary from 0 to 1 and this ratio changes with time during any particular denitrification event, although these dynamics are rarely well characterized in the field (Dendooven and Anderson, 1994). In addition, the impact of environmental regulators of the denitrifier enzymes might differ with the community composition of the denitrifying bacteria (Cavigelli and Robertson, 2000, 2001; Philippot et al., 2007), but the impacts of agricultural management on the functional characteristics of denitrifier (and nitrifier) communities is poorly understood (Kramer et al., 2006; Philippot et al., 2007; Avrahami and Bohannan, 2009).

Nitrification and denitrification may occur simultaneously and in close proximity in soil (Kuenen and Robertson, 1994) despite substantial differences in factors regulating the activity of the nitrifying and denitrifying bacteria. The contribution of each process to N_2O emissions depends on soil type, WFPS, N inputs and other factors (Bateman and Baggs, 2005) and thus can vary with time. Measuring the contribution of each process to N_2O emissions in the field is difficult, so there are few studies where N_2O source is well characterized (Bateman and Baggs, 2005). Nonetheless, in the relatively moist conditions predominating in the eastern and central U.S., denitrification is generally considered to be the dominant source of N_2O emissions from most soils, although nitrification may be the dominant source in soils receiving ammoniacal fertilizers (Venterea, 2007).

Methane flux dynamics in soil are complex. While non-flooded soils are typically considered to be net consumers of atmospheric CH_4 (Steudler et al., 1989; Mosier et al., 1991; Prieme et al., 1997), non-flooded systems can exhibit net methane production, especially following rainfall and temporary flooding and/or manure additions (Chan and Parkin, 2001). Methane oxidation is an aerobic process that is carried out in soils by the methanotrophs and ammonia-oxidizing bacteria, the latter being capable of using CH_4 as an alternative substrate for ammonia monooxygenase (Suzuki et al., 1976; Hyman and Wood, 1983; Jones and

Morita, 1983; Ward, 1987; Willison et al., 1995). Methane production is an anaerobic process resulting from CO_2 reduction and/or acetate fermentation (Topp and Pattey, 1997). Because it is an anaerobic process, methanogenesis is sometimes assumed to be negligible in non-flooded soils. However, there is growing evidence to the contrary showing CH_4 produced by methanogenesis serves as the substrate for methanotrophs (Megraw and Knowles, 1987; Yavitt et al., 1995; Wang and Bettany, 1997; Conrad, 1995; Sey et al., 2008). Thus, the net CH_4 flux from soils reflects the net balance between CH_4 production and consumption in soils.

While processes controlling CH_4 and N_2O production in soils are well studied, field scale fluxes show enormous temporal and spatial variability due to complex interactions among factors regulating production and consumption of CH_4 and N_2O. This regulatory complexity and the delay between production of gases in soil and their emission from soil (Wagner-Riddle et al., 2008) often results in poor or no correlation between individual regulating factors (e.g. soil mineral N, moisture, WFPS, and temperature) and soil GHG fluxes (Rochette et al., 2004; Kaspar and Parkin, 2011). Accordingly, spatial and temporal variability of soil GHG fluxes are recognized as important challenges to collecting robust soil GHG flux data (Parkin and Kaspar, 2003; Parkin, 2008).

While this complexity can make it difficult to generalize about the impacts of management on GHG emissions (Snyder et al., 2009), the highest rates of N_2O emission from soils generally occur when soil mineral N (esp. nitrate) is high and moisture is above 60% WFPS but below saturation (Granli and Bockman, 1994). In croplands, these conditions are most likely to exist when rainfall (or irrigation) events occur soon after N fertilizer or manure application, or legume cover crop termination, or during freeze–thaw events (Christensen and Tiedje, 1990; Singurindy et al., 2007; Parkin and Kaspar, 2006).

PROPOSED GWP MITIGATION OPTIONS FOR CROPLANDS

A number of GHG mitigation options have been proposed for agricultural systems (Snyder et al., 2009; Eagle et al., 2011). Strategies designed to reduce soil N_2O emissions are generally based on creating environmental conditions that limit the activities of nitrifying and denitrifying bacteria and by minimizing soil mineral N levels, especially at times that soils are likely to be wet. These are, in general, the same strategies proposed for increasing nitrogen-use efficiency—applying the right N fertilizers (or other sources of N) at the right rate, time, and place (Roberts, 2007)—without sacrificing crop production. Additional strategies include reducing the frequency of high N demanding crops in a crop rotation, incorporating non-legume cover crops into crop rotations to take up residual soil mineral N, and manipulating soil moisture via irrigation or drainage management. One strategy that has received limited attention but might have potential to mitigate N_2O emissions is limiting the ratio of N_2O produced per unit of mineral N nitrified or denitrified (Kramer et al., 2006), either by increasing the C:NO_3 ratio in soils (using organic N fertilizers such as legume cover crops and/or animal manures) to favor the conversion of nitrate to N_2 rather than N_2O, or by selecting for soil nitrifier or denitrifier communities with lower N_2O production potentials (Kramer et al., 2006; Philippot et al., 2007; Avrahami and Bohannan, 2009). However, it is possible that higher C:NO_3 ratios could also increase N_2O emissions since organic matter can also promote nitrite-driven reactions that produce N_2O (Venterea, 2007).

Greenhouse gas mitigation strategies designed to increase soil C sequestration (i.e. reduce net CO_2 emissions) include reducing tillage and increasing the use of perennial and/or cover crops in a rotation (Sperow et al., 2003). These strategies, however, may sometimes increase soil N_2O emissions compared to the systems they replace (Robertson and Grace, 2004; Rochette, 2008) and thereby negate at least a portion of soil C sequestration benefits. Strategies to reduce GHG emissions from non-flooded cropland soils generally do not consider CH_4 emissions given their minor contribution to global emissions (USEPA, 2011).

SOIL, CLIMATE, AND AGRICULTURAL CHARACTERISTICS OF EASTERN AND CENTRAL U.S.

Eastern and central U.S. (Figure 9.1) is a large and diverse region encompassing 54% of the U.S.'s 128 million hectares of cropland (USDA-NASS, 2011). In 2010, the area encompassed 70% of U.S. land area planted to corn, 76% planted to soybean, 53% planted to alfalfa (*Medicago sativa* L.), 41% planted to cotton (*Gossypium hirsutum* L.), and 85% planted to peanuts (*Arachis hypogaea* L.). In 2000, cropland accounted for about 27% of all land area in eastern and central U.S. There is great diversity among states in the region in the proportion of land area devoted to cropland. More than 50% of total cropland in the region is in just five central states (Illinois, Indiana, Iowa, Minnesota, and Missouri) while 15 states, mostly in the northeast and mid-Atlantic, together account for only about 10% of total cropland in the region (USDA-NASS, 2011; statistics presented here do not include cropland in portions of states not fully included in eastern and central U.S. as delineated in Figure 9.1).

The northern portions of eastern and central U.S. have a humid continental climate, being cold in the winter and mild to warm in the summer. Annual potential evapotranspiration (PET) is less than annual precipitation (P). Southern portions of the region have a humid subtropical climate, being mild in the winter and hot in the summer, with PET >>P. The southern tip of Florida has a tropical wet/dry season climate. The northeast portion of the U.S. has rocky soils with considerable slopes. Topography in the southeast is diverse, including flat lowlands in the Mississippi Valley and Coastal Plains, mountainous regions in the Appalachians, and uplands in the Piedmont regions between the coasts and mountains. Soils are accordingly diverse but tend to have relatively low soil organic matter. Constraints to crop production in northern regions include short growing seasons and wet soils while southern regions are most often limited by heat and low soil moisture despite relatively high precipitation. The northwest portion of the region, known as the U.S. Corn Belt, has intermediate weather conditions with some of the best soils and highest agricultural productivity in the world (Franzluebbers and Follett, 2005). According to Del Grosso et al. (2006), the Corn Belt also contributes substantial N_2O emissions to the U.S. budget because of the large proportion of land used for agriculture, relatively high N application rates in this area, and humid climate conditions.

PERFORMANCE OF MITIGATION OPTIONS IN EASTERN AND CENTRAL U.S.

Earlier literature reviews noted that there was insufficient data to make generalizations about management impacts on N_2O and CH_4 emissions in eastern and central U.S. (Franzluebbers, 2005; Gregorich et al., 2005; Johnson et al., 2005). Since 2005, a large amount of data on N_2O

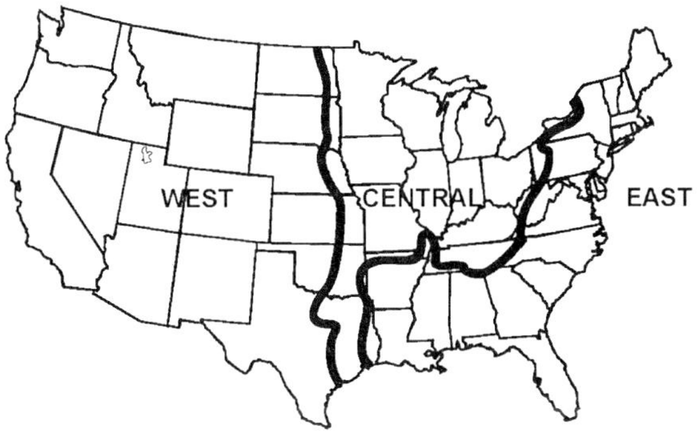

Region	State	Reported Field Studies N_2O	CH_4	NEE
East	Alabama	1	1	-
	Maryland	1	-	-
	New York	2	-	-
	Pennsylvania	1	-	-
Central	Illinois	-	-	1
	Indiana	4	3	-
	Iowa	3	1	1
	Kentucky	1	1	-
	Michigan	3	-	-
	Minnesota	8	2	1
	Missouri	1	-	-
	Nebraska	1	1	1
	Ohio	1	1	-
	Total	27	10	4

FIGURE 9.1
Number of cropland studies published since 2005 addressing impacts of management practices on soil N_2O and CH_4 emissions and net ecosystem exchange (NEE) in states within the eastern and central U.S.

and CH_4 fluxes has been collected in the eastern and central U.S., in part due to the coordinated efforts of the USDA-ARS Greenhouse gas Reduction through Agricultural Carbon Enhancement network (GRACEnet) (Follett et al., 2005; Jawson et al., 2005). We found 27 field studies comparing N_2O emissions and 10 studies comparing CH_4 emissions under various management practices in cropland soils in eastern and central U.S. during this time period.

In this chapter we summarize results from these studies to assess the effectiveness of GHG mitigation strategies in croplands in the eastern and central U.S. Almost all studies use non-flow through, non-steady state (NFT-NSS) chambers. Since the design and implementation of NFT-NSS chambers have been shown to impact measured emission rates (Venterea et al., 2009) and limit the ability to compare emissions among studies (Rochette and Eriksen-Hamel, 2007), we do not attempt to provide a formal meta-analysis of these data. Instead, we summarize the findings from individual studies, arranging our discussion by potential GHG mitigation strategies. The results of field studies published since the 2005 summaries were compiled and are summarized in Tables 9.1–9.8, along with unpublished results from Maryland. We do not include studies in which N_2O or CH_4 emissions were measured fewer than 10 times per year as those studies are not likely to include sufficient temporal resolution to provide robust data.

N RATE EFFECTS: SYNTHETIC N FERTILIZERS

Nitrogen fertilizer application rate is perhaps the best single predictor of soil N_2O emission from croplands in eastern and central US (Snyder et al., 2009; Millar et al., 2010a; Eagle et al., 2011). The IPCC Tier 1 approach to predicting soil N_2O emissions, which is based on previous data syntheses, assumes a linear increase in N_2O emissions (0.01 kg N_2O-N per kg fertilizer N added) with increasing N application rate (IPCC, 2007). A review of studies conducted in eastern Canada and northeastern U.S. also found a linear relationship between N application rate and N_2O emissions (0.0119 kg N_2O-N per kg fertilizer N added) (Gregorich et al., 2005). However, an earlier review article described an exponential relationship between N application rate and N_2O emissions (Bouwman et al., 2001).

An exponential relationship between N application rate and N_2O emissions has also been observed in three recent studies conducted in Michigan using high resolution N fertilizer gradients (6 to 9 N application rates, from 0 to 291 kg N ha^{-1} yr^{-1}). Researchers observed an exponential increase in seasonal N_2O emission with increasing N fertilizer application rate in corn (McSwiney and Robertson, 2005; Hoben et al., 2011; Table 9.1) and wheat (*Triticum aestivum* L.) (Millar et al., personal communication). In eight on-farm trials in corn, Hoben et al. (2011) found maximum and economically optimum corn grain yields were reached at N fertilizer rates of 167 and 154 kg N ha^{-1}, respectively, and seasonal N_2O emissions increased exponentially at rates >180 kg N ha^{-1}. Results were summarized using the equation, $y = 4.46e^{0.0062x}$, where y is average daily N_2O emissions (g N ha^{-1} d^{-1}) and x is N application rate (kg N ha^{-1}). Millar et al. (2010a) note that many earlier studies used too few N fertilizer rates to adequately distinguish between linear and exponential response curves. Recent studies conducted in Ontario and New Brunswick, Canada, also support the exponential model (Zebarth et al., 2008; Ma et al., 2010). Interestingly, Zebarth et al. (2008) propose the exponential N_2O response curves may be due to an increase in both denitrification rate and the ratio of $N_2O/(N_2O + N_2)$ produced during denitrification as soil NO_3-N level increases, an observation that supports the concept that the $N_2O/(N_2O + N_2)$ ratio increases as the C:NO_3 ratio decreases.

The results from the Michigan studies have been used to develop an N fertilizer reduction protocol that its authors claim could reduce N_2O emissions from fertilized croplands in the U.S. Midwest by more than 50% while maintaining crop yields (Millar et al., 2010b). This

TABLE 9.1 Impact of Nitrogen Fertilizer and Manure Application Rate on Nitrous Oxide Emissions in Croplands of Eastern and Central U.S. (Publications Since 2005)

Nitrogen													
Location	Previous crop	Crop	Rate (kg N ha^{-1})	Source[a]	Placement[b]	Timing[c]	Primary tillage[d]	Time in treatment (yr)[e]	Soil type	Study period[f]	N_2O flux (kg N_2O-N ha^{-1} yr^{-1})[g]	Sig.[h]	N_2O loss (% N applied)[g]	Reference
Michigan	Corn	Corn	0	—	—	—	CP	0	Loam, sandy loam	May–Jul/ Oct 2001–2003	~2.6	NR	—	McSwiney and Robertson (2005)
"	"	"	34	UAN	Injected	Planting	"	"	"	"	~3.7	"	~3.2	"
"	"	"	67	"	Injected or b'cast + incorp'd	Planting (34), sidedress (remainder)	"	"	"	"	~6.6	"	~6.0	"
"	"	"	101	"	"	"	"	"	"	"	~7.7	"	~5.0	"
"	"	"	134	"	"	"	"	"	"	"	~18.6	"	~11.9	"
"	"	"	168	"	"	"	"	"	"	"	~9.5	"	~4.1	"
"	"	"	202	"	"	"	"	"	"	"	~9.9	"	~3.6	"
"	"	"	246	"	"	"	"	"	"	"	~18.3	"	~6.4	"
"	"	"	291	"	"	"	"	"	"	"	~19.3	"	~5.7	"
Michigan	Soybean	Corn	0	—	—	—	CP	0	Loam, fine sandy loam	Apr/May–Sep 2007–2008	~1.8	NR	—	Hoben et al. (2011)
"	"	"	45	Granular urea	B'cast + incorp'd	Preplant	"	"	"	"	~2.2	"	~0.8	"
"	"	"	90	"	"	"	"	"	"	"	~2.9	"	~1.2	"
"	"	"	135	"	"	"	"	"	"	"	~4.4	"	~1.9	"
"	"	"	180	"	"	"	"	"	"	"	~6.6	"	~2.6	"
"	"	"	225	"	"	"	"	"	"	"	~9.5	"	~3.4	"
Iowa	Soybean	Rye + oat cover–corn	202	LSM (101) + UAN (101)	Injected	Fall: LSM; sidedress: UAN	NT	8	Silty clay loam, loam	Oct 2005–Oct 2006	~5.1	c	~2.5	Jarecki et al. (2009)
"	"	"	303	LSM (202) + UAN (101)	"	"	"	"	"	"	~8.4	b	~2.8	"
"	"	"	370	LSM (303) + UAN (67)	"	"	"	"	"	"	~12.1	a	~3.3	"

[a]LSM = liquid swine manure; UAN = urea ammonium nitrate; numbers in parentheses are kg N ha^{-1} of material applied.
[b]B'cast = broadcast; incorp'd = incorporated.
[c]Numbers in parentheses are kg N ha^{-1} of material applied at given time.
[d]CP = chisel plow; NT = no-tillage.
[e]Number of years in NT or other treatment at the beginning of study period.
[f]Seasonal sampling indicated by month–month year–year; annual sampling indicated by month–year month–year.
[g]Values preceded by ~ are estimated from graphs in publications.
[h]Statistical significance as reported by authors. Numbers followed by a different letter are significantly different (usually $P < 0.05$) for a given reference; NR = not reported.

approach has the added likely benefits of reducing releases of reactive N other than N_2O, such as leached nitrate and N in runoff and erosion. Adviento-Borbe et al. (2007), however, caution against relying solely on this kind of simple approach to mitigating N_2O emissions in developing cropping systems that mitigate GHG emissions. As discussed in more detail later, they show that combining increased N fertilizer application rate along with other agronomic practices designed to increase system performance (e.g. crop varieties, planting dates and densities, fertilizer rates and timing, tillage, and irrigation scheduling) can result in greater crop yield without increasing N_2O emissions per unit of production.

All recent studies showing an exponential relationship between N fertilizer application rate and N_2O emission have been conducted in tilled soils in northern regions (Michigan, Ontario, New Brunswick). While it is theoretically compelling to expect soil N_2O emissions would increase exponentially when N application rates exceed plant and soil N uptake capacity (Grant et al., 2006), areas further south or systems under no-till (NT) management lack necessary data to support such conjecture. Given the warmer conditions and lower soil C levels in many regions further south, it is possible that an exponential N_2O response to N rate may be even stronger in the southern U.S. than in the north. By contrast, NT systems often have higher soil C levels than conventional till (CT) systems, which might impact the shape of the response curve. Additional studies conducted in diverse soils, climates, and management scenarios are needed to more fully quantify the impact of N fertilizer application rate on soil N_2O emission.

N RATE EFFECTS: ANIMAL MANURES

There is only one recently published study in eastern and central U.S. evaluating the impact of animal manure application rate on soil N_2O emissions (Table 9.1). Consistent with results discussed above for N fertilizers, Jarecki et al. (2009) observed an exponential increase in N_2O emissions when N fertilizer was augmented with increasing rates of liquid swine manure (LSM). The proportion of total N applied emitted as N_2O (not corrected for a zero control) increased from 2.52 to 3.28% as N input increased from 202 to 370 kg N ha^{-1}.

N SOURCE EFFECTS: N FERTILIZERS AND INHIBITORS

Globally, about 40% of total fertilizer N is applied as urea, 26% as anhydrous ammonia, 21% as urea ammonium nitrate (UAN), and 4% as ammonium nitrate (Bouwman et al., 2002). Different fertilizer N sources might impact soil N_2O emissions by providing nitrification and/or denitrification substrates in different proportions and/or with different release rates. While various studies have shown differences in N_2O emissions with N fertilizer source, review articles have shown results vary by soil type, climate, and application method, rate and timing (Eichner, 1990; Granli and Bockman, 1994; Stehfest and Bouwman, 2006). When these factors are accounted for, there are few generalizable differences in fertilizer source impacts on N_2O emission (Stehfest and Bouwman, 2006).

Nonetheless, there may be N source impacts on N_2O emissions that are regionally important. Studies conducted in Minnesota, for example, show that urea can reduce N_2O emissions in NT, conservation till (CP in Table 9.2) and moldboard plowed (MP) corn–soybean rotations (Venterea et al., 2005) and conventionally tilled continuous corn (Venterea et al., 2010) by 50% or more compared to anhydrous ammonia (Table 9.2). While fertilizer type was confounded with application method in these two studies, Venterea et al. (2010) found soil nitrite concentration and dissolved organic carbon were higher following anhydrous ammonia injection than when urea was incorporated. They speculate that elevated N_2O emission following anhydrous ammonia application is due to reactions mediated by nitrification involving nitrite favored at lower pH.

Another study conducted in Minnesota showed delayed release fertilizers might provide N_2O mitigation. In a loamy sand soil cropped to potato, Hyatt et al. (2010) showed one single

TABLE 9.2 Impact of Nitrogen Fertilizer, Manure Source, and Nitrification Inhibitor on Nitrous Oxide Emissions in Croplands of Eastern and Central U.S. (Publications Since 2005)

Nitrogen

Location	Previous crop	Crop	Rate (kg N ha^{-1})	Source[a]	Placement[b]	Timing	Primary tillage[c]	Time in treatment (yr)[d]	Soil type	Study period[e]	N_2O flux (kg N_2O-N ha^{-1} yr^{-1})[f]	Sig.[g]	N_2O loss (% N applied)[h]	Reference
Minnesota	Soybean	Corn	120	Urea	B'cast	Sidedress	MP	13	Silt loam	Apr–Nov 2004	1.4	ab	1.2	Venterea et al. (2005)
"	"	"	"	"	"	"	CP or disk	"	"	"	1.4	a	1.2	"
"	"	"	"	"	"	"	NT	"	"	"	2.2	c	1.8	"
"	"	"	"	UAN	"	Preplant	MP	"	"	"	2.4	bc	2.0	"
"	"	"	"	"	"	"	CP or disk	"	"	"	2.5	bc	2.1	"
"	"	"	"	"	"	"	NT	"	"	"	2.5	c	2.1	"
"	"	"	"	AA	Injected	"	MP	"	"	"	10.5	e	8.8	"
"	"	"	"	"	"	"	CP or disk	"	"	"	11.4	e	9.5	"
"	"	"	"	"	"	"	NT	"	"	"	5.9	d	4.9	"
Minnesota	Corn–corn	Corn–corn	146	AA	Injected	Preplant	CP	15	Silt loam	Apr–Oct/Nov 2005–2007	2.15	a	1.5	Venterea et al. (2010)
"	"	"	"	Urea	B'cast + incorp'd	"	"	"	"	"	1.01	b	0.7	"
"	Corn–soybean	Corn–soybean	146	AA	Injected	Preplant	CP	15	Silt loam	Apr–Oct/Nov 2005–2006	1.37	a	0.9	"
"	"	"	"	Urea	B'cast + incorp'd	"	"	"	"	or 2005–2007	0.8	b	0.5	"
Minnesota	Rye or rye and mustard	Potato	0	—	—	—	MP	0	Loamy sand	Apr/May–Sep 2008–2009	0.6	"	—	Hyatt et al. (2010)
"		"	270	Urea + AN	Banded + fertigation	Planting, emergence + 4–5 dates	"	"	"	"	1.36	a	0.37	"
"	"	"	"	PCU 1	B'cast + incorp'd	Preplant	"	"	"	"	0.83	b	0.13	"
"	"	"	"	PCU 2	"	"	"	"	"	"	1.13	ab	0.27	"
Missouri	Soybean	Corn	0	—	—	—	ST and NT	NR	Silt loam	Apr–Sep 2009–2010	1.32	b	—	Nash (2010)
"	"	"	140	Urea	B'cast	Preplant	"	"	"	"	5.21	a	2.8	"
"	"	"	"	PCU 2	"	"	"	"	"	"	5.48	a	3.0	"
Iowa	"	Corn	125–168	AA, DAP, nitrapyrin	Injected	November	ST	0	Silty clay loam	Nov 2005 – Oct 2007	6.15	ns	4.2	Parkin and Hatfield (2010)
"	"	"	"	AA, DAP	"	"	"	"	"	"	5.88	ns	4.0	"

Continued

TABLE 9.2 Impact of Nitrogen Fertilizer, Manure Source, and Nitrification Inhibitor on Nitrous Oxide Emissions in Croplands of Eastern and Central U.S. (Publications Since 2005)—continued

Nitrogen														
Location	*Previous crop*	*Crop*	*Rate (kg N ha^{-1})*	*Source*[a]	*Placement*[b]	*Timing*	*Primary tillage*[c]	*Time in treatment (yr)*[d]	*Soil type*	*Study period*[e]	*N_2O flux (kg N_2O-N ha^{-1} yr^{-1})*[f]	*Sig.*[g]	*N_2O loss (% N applied)*[h]	*Reference*
Indiana	Corn	Corn	157	UAN	Injected	Side-dress	CP	17	Silty clay loam, silt loam	Mar–Nov/ Dec 2005–2006	6.40	ns	4.1	Hernandez-Ramirez et al. (2009)
"	"	"	255	LSM	"	Preplant	"	"	"	"	8.17	ns	3.2	"
Iowa	Soybean	Rye+ oat cover – corn	175	UAN	Injected	Side-dress	NT	8	Silty clay loam, loam	Oct 2005–May 2006	~5.1	ns	2.9	Jarecki et al. (2009)
"	"	"	202	LSM (101) + UAN (101)	"	Fall: LSM; sidedress: UAN	"	"	"	"	~5.1	ns	2.5	"
Kentucky	Soybean	Corn	0	–	B'cast	Preplant	NT	~10	Silt loam	Apr–Sep 2007–2008	1.6	b	–	Sistani et al. (2010)
"	"	"	179	UAN	"	"	"	"	"	"	5.3	a	2.1	"
"	"	"	200–204	LSM	"	"	"	"	"	"	5.1	a	1.8	"
Pennsylvania	Corn	Corn	224	AN	B'cast + incorp'd	Preplant	CT	17	Silt loam	Apr–Oct/ Dec 2006–2007	3.6	b	1.6	Adviento-Borbe et al. (2010)
"	"	"	225	LDM	"	"	"	"	"	"	5.5	a	2.4	
"	Alfalfa	Corn	90	AN	"	"	"	"	"	"	5.8	a	6.4	"
"	"	"	120	LDM	"	"	"	"	"	"	5.7	a	4.8	"
Maryland	Wheat–corn–rye	Soy-bean	–	UAN (corn, wheat)	–	–	CP	14	Silt loam	Apr–Dec 2010	0.48	b	–	Cavigelli et al. (unpub'd)
"	Wheat–vetch–corn–rye	"	"	Poultry litter (corn, wheat)	"	"	MP	"	"	"	1.96	a	–	

[a]AA = anhydrous ammonia; AN = ammonium nitrate; DAP = diammonium phosphate; LDM = liquid dairy manure; LSM = liquid swine manure; PCU 1 = polymer-coated urea, 42% N w/w; PCU 2 = polymer-coated urea, 44% N w/w; UAN = urea ammonium nitrate; numbers in parentheses are kg N ha^{-1} of material applied.

[b]B'cast = broadcast; incorp'd = incorporated.

[c]CP = chisel plow; CT = conventional tillage, primary implement not specified; MP = moldboard plow; NT = no-tillage; ST = strip tillage.

[d]Number of years in NT or other treatment at the beginning of study period; NR = not reported.

[e]Seasonal sampling indicated by month–month year–year; annual sampling indicated by month–year month–year.

[f]Values preceded by ~ are estimated from graphs in publications.

[g]Statistical significance as reported by authors or inferred from reports. Numbers followed by a different letter are significantly different (usually $P < 0.05$) for a given reference or portions of references separated by horizontal lines.

[h]Calculated without a 0 N control except where a 0 N control is indicated in the N rate column.

application of a 42% N w/w polymer coated urea (PCU) reduced seasonal N_2O emissions by 39% compared to urea plus ammonium nitrate applied in seven annual split applications during a 3-year trial. A second PCU, with 44% N w/w, had no impact on N_2O emissions compared to the urea plus ammonium nitrate treatment. The authors attributed the difference in performance to a lower N release rate in the 42% N than the 44% N PCU. Nash (2010), in a study conducted in Missouri using the 44% N PCU, found no effect of PCU compared to urea on soil seasonal N_2O emissions in corn during 2 years with relatively high precipitation. Given that N_2O emissions were decreased by 49% compared to urea using the 44% N PCU in Colorado (Halvorson et al., 2010), and that N release rates from PCU depend on soil conditions, additional studies are warranted.

Nitrification inhibitors have also been proposed as tools to reduce soil N_2O emissions (Snyder et al., 2009; Eagle et al., 2011). Early studies showed nitrapyrin can reduce N_2O emissions from spring-applied ammoniacal fertilizers (Aulakh et al., 1984; Bremner et al., 1981; Bronson et al., 1992; McTaggart et al., 1997). These studies, however, measured impacts of nitrification inhibitors on soil N_2O emissions for less than 1 year and sometimes for as few as 50 days after application. In the only recent study testing the impact of nitrification inhibitors on soil N_2O emissions in the eastern and central U.S., Parkin and Hatfield (2010) found nitrapyrin reduced soil N_2O emissions by 50% during the late fall and early spring following fall application of anhydrous ammonia in one of two years in Iowa (from 1.24 to 0.62 kg N_2O-N ha^{-1}). However, when results were compared for a full year following application of N fertilizer, there was no impact of nitrapyrin on N_2O emissions (Table 9.2). While nitrapyrin seems to have blocked nitrification for part of the year, its persistence and effectiveness is strongly controlled by soil moisture and temperature conditions, such that using nitrification inhibitors to mitigate N_2O emissions is likely to give variable results by year and season. This study illustrates the value of measuring N_2O emissions throughout the year and the potential limitations of short-term studies.

N SOURCE EFFECTS: SYNTHETIC N FERTILIZERS VS. ANIMAL MANURES

Animal manures are an important source of agricultural nutrients, including N, and contribute to soil N_2O emissions. On a global basis, about 38% of N is applied as animal manures (Bouwman et al., 2002). In the U.S. in 2006, animal manures were applied on 12% of cropland area planted to corn, 9% planted to oats, 7% planted to hay, 4% planted to peanuts, 2.6% planted to cotton, and less than 2% planted to soybean and wheat (MacDonald et al., 2009). Based on historical records, Davidson (2009) estimated between 1860 and 2005, 2.0% and 2.5% of animal manure and fertilizer nitrogen, respectively, has been released to the atmosphere as N_2O. Managing manure to reduce N_2O emissions will become more important as livestock numbers increase to meet human demand for animal products.

Animal manures are often thought to increase soil N_2O emissions compared to N fertilizers since they provide a substantial amount of C, which stimulates soil heterotrophic activity, thereby depleting soil oxygen levels and augmenting denitrification. The expected impact of manure on N_2O production, however, may be mitigated because the amount of N_2O released as a proportion of denitrification end products decreases with increasing C:NO_3 ratio (Firestone and Davidson, 1989; Kramer et al., 2006).

There are five recent reports comparing N_2O emissions following the application of liquid manure and mineral fertilizer in eastern and central U.S. These studies show few differences in N_2O emissions based on N source (Table 9.2). Adviento-Borbe et al. (2010) found seasonal N_2O emissions were 35% lower when continuous corn was fertilized with ammonium nitrate rather than liquid dairy manure (LDM), although 2-year means were driven by significant differences in 1 year. There were no differences in seasonal N_2O emissions due to N source in the same experiment when corn followed alfalfa even though N fertilizer rate was reduced to account for N provided by alfalfa. Hernandez-Ramirez et al. (2009) found N_2O emission in Indiana was similar in tilled continuous corn whether UAN was applied at sidedress or LSM was applied in the spring. Jarecki et al. (2009) found no differences in annual N_2O emissions following application of UAN in the spring or LSM in the fall to a rye (*Secale cereale* L.) and oat (*Avena sativa* L.) cover followed by corn. Furthermore, Sistani et al. (2010) found no differences in seasonal N_2O emission when UAN or LSM was surface applied in NT corn fields in Kentucky over 2 years.

Since there is a residual impact on soil C and N pools when manures are applied to soil (Mallory and Griffin, 2007; Cavigelli and Dao, 2008), it is important to consider long-term impacts in manure application studies. Chang et al. (1998), for example, found very high rates of N_2O emissions—up to 56 kg N ha^{-1} yr^{-1}—in plots that had received annual manure applications of 180 kg N ha^{-1} yr^{-1} for 21 years. They attributed the high rates to elevated soil C and nitrate due to manure applications. In a more recent study, seasonal N_2O emissions during the soybean phase of a corn–soybean–wheat rotation, when no N was applied, were about four times greater from a soil with a 14-year history of moderate poultry litter application than from the same soil with no history of manure application in Maryland (Table 9.2; Cavigelli et al., unpublished data). While there were system-level differences among treatments in this study, differences in N_2O emissions were likely due to greater C and N release in soil compared to soil without a history of poultry litter application (Spargo et al., 2011). Variable results from studies comparing N_2O emissions following manure and N fertilizer applications suggest additional studies are needed, including studies designed to better understand factors controlling these N_2O emissions.

N TIMING: FERTILIZERS AND MANURES

Improving timing of N application relative to crop growth might reduce N_2O emissions by providing mineral N at high levels primarily when crop demand is high. The only recent study on the impact of N fertilizer application timing on N_2O emissions in eastern and central U.S. found seasonal N_2O emissions in Indiana over 3 years were similar regardless of whether N fertilizer was applied prior to corn planting, or in a split application (Table 9.3; Smith et al., 2011). The only study from the region comparing manure application timing found injecting LSM to soils in the fall in Indiana resulted in 60% lower seasonal N_2O emission than injecting the same amount of manure in the spring (Hernandez-Ramirez et al., 2009). However, these results might simply reflect greater leaching or other losses of N following fall rather than spring manure application. Such N losses, of course, could contribute to environmental degradation, including indirect N_2O emissions. Given that N fertilizers are applied in the fall to about 30% of land planted to corn in the U.S. (CAST, 2004), and that nitrogen-use efficiency can be improved by side dressing N, additional studies comparing N timing impacts on N_2O emissions are warranted.

N PLACEMENT: FERTILIZERS AND MANURES

The impact of N source placement depth on soil N_2O emissions is unclear. In NT systems, especially, deep placement of N sources might decrease N_2O emissions by placing N at depths where microbial activity is relatively low due to low soil C and temperature (Nash, 2010). Conversely, shallow placement of N sources might reduce N_2O emissions by placing N where oxygen concentration is greater than at depth (Drury et al., 2006). By the same reasoning, broadcasting N sources on the surface might reduce N_2O emissions compared to injecting N, although losses of N via NH_3 volatilization need to be considered. Depth or method of incorporating legume cover crop residues might also impact soil N_2O emissions (Loecke and Robertson, 2009).

TABLE 9.3 Impact of Nitrogen Fertilizer and Animal Manure Application Timing and Placement on Nitrous Oxide Fluxes in Croplands of Eastern and Central U.S. (Publications Since 2005)

Nitrogen														
Location	Previous crop	Crop	Rate (kg N ha^{-1})	Source[a]	Placement[b]	Timing[c]	Primary tillage[d]	Time in treatment (yr)[e]	Soil type	Study period[f]	N_2O flux (kg N_2O-N ha^{-1} yr^{-1})	Sig.[g]	N_2O loss (% N applied)[h]	Reference
Indiana	Soybean	Corn	193	UAN (168), DAP (25)	Knifed (168), banded (25)	Preplant (168), planting (25)	NT	>10	Silt loam	Apr 2004–Oct 2007	3.79	ns	2.0	Smith et al. (2011)
"	"	"	168	UAN (168),	Knifed (84), dribbled (84)	Preplant (84), V3–V4 (84)	"	"	"	"	3.87	ns	2.3	"
Indiana	Corn	Corn	255	LSM	Injector	Fall	CP	17	Silty clay loam, silt loam	Mar–Nov/Dec 2005–2006	3.29	b	1.3	Hernandez-Ramirez et al. (2009)
"	"	"	"	"	"	Spring	"	"	"	"	8.17	a	3.2	"
Missouri	Soybean	Corn	140	Urea, PCU 2	Deep banded	Preplant	ST	NR	Silt loam	Apr–Aug/Sep 2009–2010	3.66	ns	2.6	Nash (2010)
"	"	"	"	"	B'cast	"	NT	"	"	"	4.34	ns	3.1	"
Kentucky	Soybean	Corn	200–204	LSM	B'cast	Preplant	NT	~10	Silt loam	Apr–Sep 2007–2008	5.1	ns	1.8	Sistani et al. (2010)
"	"	"	"	"	Injected	"	"	"	"	"	6.5	ns	2.5	"
"	"	"	"	"	Aeration	"	"	"	"	"	5.9	ns	2.2	"

[a]DAP = diammonium phosphate; LSM = liquid swine manure; PCU 2 = polymer-coated urea, 44% N w/w; UAN = urea ammonium nitrate; numbers in parentheses are kg N ha^{-1} of material applied.
[b]B'cast = broadcast.
[c]Numbers in parentheses are kg N ha^{-1} of material applied at given time; V3, V4 are corn growth stages.
[d]CP = chisel plow; NT = no-tillage; ST = strip tillage.
[e]Number of years in NT or other treatment at the beginning of study period; NR = not reported.
[f]Seasonal sampling indicated by month–month year–year; annual sampling indicated by month–year month–year.
[g]Statistical significance as reported by authors or estimated. Numbers followed by a different letter are significantly different (usually $P < 0.05$) for a given reference.
[h]Calculated without a 0 N control.

The only recent field study comparing N fertilizer placement impacts on N_2O emissions in eastern and central U.S. compared broadcast and deep banded (15 cm) UAN and PCU in NT and strip tilled systems in Missouri (Nash, 2010), where no differences in N_2O emissions were observed among N placement methods regardless of N fertilizer source (Table 9.3). Nash speculated high plant residue levels reduced mineral N availability when N sources were surface applied, such that N_2O emissions were similar whether deep banded or surface applied.

Soil incorporation of animal manures has long been considered a best management practice (BMP) to reduce NH_3 losses by volatilization (Thompson and Meisinger, 2002) and N and P losses via surface runoff (Eghball and Gilley, 1999). Recent advances in manure injection technology (Coelho et al., 2006; Warren et al., 2008) are providing an option to improve manure management in diverse systems, including NT cropping systems. However, manure injection may increase soil N_2O emissions due to increased soil–manure contact compared to surface application (Flessa and Beese, 2000; Wulf et al., 2002). The only recent study comparing manure placement conducted in eastern and central U.S., however, found no consistent difference in seasonal N_2O emission whether LSM was broadcast, injected to a depth of 20 cm, or applied with an aerator to a depth of 20 cm in NT corn in Kentucky (Table 9.3; Sistani et al., 2010). In 1 year, N_2O emissions were greater when injected (8.2 kg N ha^{-1}) than when surface applied (2.9 g kg N ha^{-1}), but the opposite was true the previous year. This result highlights the interaction between annual weather events and management practices on soil N_2O emissions.

COVER CROPS

142 Cover crops are crops grown to protect the soil from erosion, mitigate losses of nutrients via leaching and runoff, and/or to provide biologically fixed nitrogen; they are generally not harvested (Clark, 2007). Cover crops are most often grown during the late fall, winter and early spring in the eastern and central U.S., a time when precipitation is usually much greater than PET, and erosion, leaching, and runoff losses can be high. Non-legume cover crops planted after high N demanding crops such as corn have proven to be effective at reducing soil mineral N and moisture levels during the winter and early spring (Shipley et al., 1992), both factors that should aid in reducing soil N_2O emissions.

Grass Cover Crops

Results from four recent studies conducted in eastern and central U.S. show few impacts of grass cover crops on soil N_2O emissions (Table 9.4). Jarecki et al. (2009) found that a rye + oat cover crop planted in the fall in Iowa had no impact on annual soil N_2O emissions during the growth of the cover crop and a succeeding corn crop regardless of whether manure had been applied in the fall or UAN had been applied in the spring. However, injecting the manure in this 1-year experiment damaged the growing cover crop, thereby likely reducing its effectiveness at assimilating soil nitrate and water. Three studies designed to test the impact of a rye cover crop following corn showed no impact on N_2O emissions in Iowa over 2 full years (Parkin and Kaspar, 2006), in Indiana over 3 full years (Smith et al., 2011), or during freeze–thaw periods in New York in 1 of 2 years (Dietzel et al., 2011). Parkin and Kaspar (2006) note that, while rye reduced soil nitrate concentration, it did so during a time period when N_2O emissions were not high in plots with no rye. In the New York study, the rye cover crop resulted in a 65% decrease in N_2O emissions during freeze–thaw cycles compared to bare ground in 1 of 2 years. This occurred during a relatively mild winter when N_2O emissions were, on average, 30% lower than in a colder winter. The authors suggest milder winters will be more common with global climate change such that cover crops may become more effective tools in limiting soil N_2O emissions (Dietzel et al., 2011). The impact of grass cover crops on N_2O emissions in the warmer portions of the eastern and central U.S., however, is unknown.

TABLE 9.4 Impact of Cover Crops On Nitrous Oxide Emissions in Croplands of Eastern and Central U.S. (Publications Since 2005)

Nitrogen

Location	Previous crop	Crop	Rate (kg N ha^{-1})	Source[a]	Placement[b]	Timing[c]	Primary tillage[d]	Time in treatment (yr)[e]	Soil type	Study period[f]	N_2O flux (kg N_2O-N ha^{-1} yr^{-1})[g]	Sig.[h]	N_2O loss (% N applied)[i]	Reference
Iowa	Soybean	Corn	175, 303 (mean)	UAN (175 or 101), LSM (202)	Injected	Fall: LSM; spring: UAN	NT	8	Silty clay loam, loam	Oct 2005–Oct 2006	~7.6	ns	~3.2	Jarecki et al. (2009)
"	"	Rye + oat cover–corn	"	"	"	"	"	"	"	"	~6.8	ns	~2.8	"
Iowa	Corn–soybean	Corn–soybean	215 (corn)	APP (13), UAN (202)	Banded	Planting (13), sidedress (202)	NT	8	Silty clay loam, loam, clay loam	Apr 2003–Feb 2005	7.08	ns	6.6	Parkin and Kaspar (2006)
"	Corn–rye–soybean or corn–soybean–rye	Corn–rye–soybean or corn–soybean–rye	"	"	"	"	"	"	"	"	8.38	ns	7.8	"
Indiana	Corn–soybean	Corn–soybean	193 (corn)	UAN (168), DAP (25)	Knifed (168), banded (25)	Preplant (168), planting (25)	NT	>10	Silt loam	Apr 2004–Oct 2007	2.66	ns	1.4	Smith et al. (2011)
"	"	Corn–soybean–annual ryegrass	"	"	"	"	"	"	"	"	3.24	ns	1.7	"

[a]APP = ammonium polyphosphate; DAP = diammonium phosphate; LSM = liquid swine manure; UAN = urea ammonium nitrate; numbers in parentheses are kg N ha^{-1} of material applied.
[b]NR = not reported; numbers in parentheses are kg N ha^{-1} of material applied.
[c]Numbers in parentheses are kg N ha^{-1} of material applied at given time.
[d]CP = chisel plow; NT = no-tillage.
[e]Number of years in NT or other treatment at the beginning of study period.
[f]Seasonal sampling indicated by month–month year–year; annual sampling indicated by month–year month–year.
[g]Values preceded by ~ are estimated from graphs in publications.
[h]Statistical significance as reported by authors or inferred from reports. Numbers followed by a different letter are significantly different (usually $P < 0.05$) for a given reference or portions of references separated by horizontal lines.
[i]Calculated without a 0 N control.

Legume Cover Crops

Since about 30% of energy use in U.S. agriculture is used for production and transport of fertilizers (Miranowski, 2005), substituting legume cover crops for at least a portion of N fertilizer applications might provide global warming mitigation by reducing energy use. However, legume cover crops can contribute to soil N_2O emissions by increasing soil C and nitrate levels after incorporation. While there is evidence from laboratory studies that soil denitrification rates increase when annual legume cover crops are incorporated into soil (Aulakh et al., 1991; Rosecrance et al., 2000), there are no data from the region on the impact of annual legume cover crops on soil N_2O emissions. One study conducted in Pennsylvania found mixed impacts of a perennial legume cover crop on soil N_2O emissions (Adviento-Borbe et al., 2010). Seasonal N_2O emission during the corn growing season was greater in corn following an alfalfa crop than in continuous corn in 1 of 2 years, even though the ammonium nitrate application rate was adjusted to account for the N provided by alfalfa (Table 9.2). When LDM was used instead of ammonium nitrate, there was no difference in N_2O emission in continuous corn vs. corn following alfalfa. Long-term impacts of alfalfa and manure on soil C and N pools were speculated to contribute to (1) greater N_2O emission in corn following alfalfa when N fertilizer was applied, and (2) the lack of difference in N_2O emission among treatments when LDM was applied.

BIOCHAR

Biochar, a co-product of pyrolysis, has the potential to provide long-term C sequestration (Laird, 2008; Steinbeiss et al., 2009) and may also decrease soil N_2O emissions by absorbing soil water, thereby increasing the oxygen concentration of surrounding soil (Yanai et al., 2007), or by increasing the soil C:N ratio and thereby decreasing net N mineralization and nitrification (Lehmann, 2007). To date, there are no published studies on the impact of field applications of biochar on N_2O or CH_4 emissions in the eastern and central U.S.

TILLAGE REGIME

While soil C can be sequestered at rates of 0.40 ± 0.61 Mg ha^{-1} yr^{-1} and 0.42 ± 0.46 Mg ha^{-1} yr^{-1} under NT compared to CT in central and southeastern U.S., respectively (Johnson et al., 2005; Franzluebbers, 2005), NT can also result in increased N_2O emissions in some cases. In a review paper, Rochette (2008) found N_2O emissions to be 50% greater under NT than CT in poorly aerated soils; a result not found in well- and moderately drained soils, where differences between NT and CT were negligible. The authors suggested greater N_2O emissions in poorly aerated NT soils resulted from greater denitrification. Six et al. (2004) found N_2O emissions were greater under NT than CT management during the first 5 to 10 years in NT and that N_2O emissions accounted for the bulk of GWP in NT systems. However, when NT had been in place for greater than 20 years in humid climates, soils under NT management had lower N_2O emissions than those under CT management. Reduced N_2O emissions under long-term NT may result from improved soil quality, i.e. greater aggregate stability, which increases oxygen diffusion into soil. In addition, NT soils are cooler and have less extreme wet/dry cycles than tilled soils, factors that tend to reduce N_2O emissions. Also, crop residues left on the soil surface in NT can provide insulation that decreases soil freezing depth and intensity, two factors that are positively correlated to N_2O production during thawing events (Wagner-Riddle et al., 2007).

Nine reports describing the impact of tillage on soil N_2O emissions have been published in the last 6 years from eastern and central U.S. (Table 9.5 and 9.2). Five studies from Michigan, Iowa, Alabama, Indiana, and Maryland found no difference in seasonal or annual N_2O emission between NT and CT on moderately to well-drained soils after 8 to 14 years in NT (Grandy et al., 2006; Parkin and Kaspar, 2006; Smith et al., 2010, 2011; Cavigelli et al.,

TABLE 9.5 Impact of Tillage Practices On Nitrous Oxide Emissions in Croplands of Eastern and Central U.S. (Publications Since 2005)

Location	Previous crop	Crop	Rate (kg N ha^{-1})	Source[a]	Placement[b]	Timing[c]	Primary tillage[d]	Time in treatment (yr)[e]	Soil type	Study period[f]	N_2O flux (kg N_2O–N ha^{-1} yr^{-1})[g]	Sig.[h]	N_2O loss (% N applied)	Reference
							Nitrogen							
Michigan	Varied	Corn–soybean or corn–soybean–wheat	Corn: 84–163; wheat: 56–71	Corn: AN, UAN	E'cast	Planting, sidedress	MP, CP	2	Sandy loam	1991–1994, 1996–2002	1.19	ns	1.43	Grandy et al. (2006)
"	"	"	"	"	"	"	NT	1	"	"	1.32	ns	1.59	"
Iowa	Corn–soybean	Corn–soybean	215 (corn)	APP(13), UAN (202)	Banded	Planting (13), sidedress (202)	CP	1	Silty clay loam, loam, clay loam	Apr 2003–Feb 2005	7.58	ns	7.1	Parkin and Kaspar (2006)
"	"	"	"	"	"	"	NT	8	"	"	7.08	ns	6.6	"
Alabama	Sorghum	Soybean	0	—	—	—	CP	10	Silt loam	Apr–Oct 2007	0.46	ns	—	Smith et al. (2010)
"	"	Soybean–cover crop	0	"	"	"	NT	"	"	"	0.73	ns	—	"
Indiana	Corn–soybean	Corn–soybean	193 (corn)	UAN (168), DAP (25)	Knifed (168), banded (25)	Preplant (168), planting (25)	CP	0	Silt loam	Apr 2004–Oct 2007	3.10	a	1.6	Smith et al. (2011)
"	"	"	107 (corn)	UAN	Knifed	Planting	PT	0	"	"	1.92	b	1.8	"
"	"	"	193 (corn)	UAN (168), DAP (25)	Knifed (168), banded (25)	Preplant (168), planting (25)	NT	>10	"	"	2.66	ab	1.4	"
Maryland	Corn–soybean–wheat	Corn–rye–soybean–oats or wheat	168 (corn), 100 (oats, wheat)	AN (25, corn); UAN (remainder)	Knifed (corn); b'cast (oat, wheat)	Planting (25), sidedress (143), b'cast (100)	CP	13	Silt loam	Apr–Dec 2008, 2010	0.92	ns	1.0	Cavigelli et al. (unpub'd)
"	"	"	"	"	"	"	NT	"	"	"	0.80	ns	0.9	"

Continued

TABLE 9.5 Impact of Tillage Practices On Nitrous Oxide Emissions in Croplands of Eastern and Central U.S. (Publications Since 2005)—continued

Location	Previous crop	Crop	Rate (kg N ha^{-1})	Source[a]	Placement[b]	Timing[c]	Primary tillage[d]	Time in treatment (yr)[e]	Soil type	Study period[f]	N_2O flux (kg N_2O–N ha^{-1} yr^{-1})[g]	Sig.[h]	N_2O loss (% N applied)	Reference
							Nitrogen							
New York	Corn	Corn residue	NR	LDM	Injected	Dec 17, 2004	MP, Nov 4, 2004	0	Clay loam	Nov 2004–Apr 2005	0.86	a	—	Singurindy et al. (2009)
"	"	"	"	"	"	"	MP, Dec 16, 2004	"	"	"	0.64	b	—	"
Indiana	Corn or soybean	Corn	258–296	AN (36), UAN (222–260)	Banded (36), injected (222–260)	Planting (36), sidedress (222–260)	MP	29	Silty clay loam	Apr/Jun–Sep 2004–2006	5.61	a	2.1	Omonode et al. (2011)
"	"	"	"	"	"	"	CP	"	"	"	7.25	a	2.7	"
"	"	"	"	"	"	"	NT	"	"	"	3.37	b	1.2	"
Ohio	Corn	Corn	200	Starter (16), AA (184)	Banded (16), injected (184)	Planting (16), sidedress (184)	MP	43	Silt loam	Nov 2004–Nov 2005	1.82	a	0.9	Ussiri et al. (2009)
"	"	"	"	"	"	"	CP	"	"	"	1.96	a	1.0	"
"	"	"	"	"	"	"	NT	"	"	"	0.94	b	0.5	"

[a]AA = anhydrous ammonia; AN = ammonium nitrate; APP = ammonium polyphosphate; DAP = diammonium phosphate; LDM = liquid dairy manure; UAN = urea ammonium nitrate; numbers in parentheses are kg N ha^{-1} of material applied.
[b]B'cast = broadcast.
[c]Numbers in parentheses are kg N ha^{-1} of material applied at given time.
[d]CP = chisel plow; MP = moldboard plow; NT = no-tillage; PT = precision tillage.
[e]Number of years in NT or other treatment at the beginning of study period.
[f]Seasonal sampling indicated by month–month year–year; annual sampling indicated by month–year month–year.
[g]Statistical significance as reported by authors or inferred from reports. Numbers followed by a different letter are significantly different (usually $P < 0.05$) for a given reference.
[h]Calculated without a 0 N control.

unpublished data). These results are consistent with Rochette (2008) who found few differences in N_2O emissions between NT and CT in moderately and well-drained soils. Interestingly, Smith et al. (2011) found precision tillage, which is similar to zone tillage, reduced N_2O emission by 38% compared to CT soils in Indiana; the authors did not comment on possible reasons for this difference. Also of interest, Singurindy et al. (2009) showed time of fall moldboard plowing impacted winter N_2O emissions in New York. Nitrous oxide emissions were 20 to 30% lower for 35 to 50 days in a soil plowed in November than in an unplowed soil, due to lowered bulk density and greater aeration in the short term. However, when the untilled soil was plowed in December and LDM applied to both soils, N_2O emissions during freeze–thaw cycles were greater in the soil tilled in November than December due to lower aeration in the soil plowed in November.

Two recent studies from the eastern and central U.S. are consistent with Six et al. (2004), showing that after more than 20 years in NT management N_2O emissions are lower in NT than CT. Omonode et al. (2011) found N_2O emissions were about 40 and 57% lower in NT than in MP and chisel plow (CP) systems, respectively, after 29 years on a silty clay loam in Indiana. The authors attributed lower N_2O emissions to lower soil organic C decomposition due to less mixing of crop residues with soil, and cooler soil temperatures in NT than the tilled systems. Ussiri et al. (2009) also found N_2O emissions were about 50% lower in NT than in MP and CT plots after 43 years of consistent management in a silt loam soil in Ohio. They attributed lower N_2O emissions in NT to greater oxygen diffusion into NT soils, which had lower bulk density and greater water-stable aggregates than tilled soils. Interestingly, these studies were conducted on a poorly drained (albeit tile-drained) and a somewhat poorly drained soil, which, based on earlier results, might be expected to have greater N_2O emissions under NT than CT (Rochette, 2008). These reports, while illustrating the value of long-term (>20-year) studies, also raise the question of whether the long-term impacts of NT identified by Six et al. (2004) might be independent of soil drainage class in contrast to shorter-term impacts.

Venterea et al. (2005) showed the impact of tillage on seasonal N_2O emissions can vary with N fertilizer type in Minnesota (Table 9.2). When urea was broadcast in a corn field, N_2O emission was greater in NT than in MP and CP by an average of 68% over two seasons. The opposite was true for injected anhydrous ammonia; seasonal N_2O emission was on average 85% greater in MP and CP than in NT. There was no difference in N_2O emission between NT, MP, and CP systems when the N source was UAN.

CROP SELECTION

Crop species differ in their N uptake patterns, N fertilizer needs, water relations, rhizodeposition, and other factors that impact N_2O production in soils (Sey et al., 2010). The impact of crop selection can be substantial; in cropping systems to which mineral N fertilizers are applied, annual or seasonal N_2O emissions were 2.3 to 5.0 times greater for corn than soybean (mean, weighted by number of years: 2.7) in five recently published studies from eastern and central U.S. (Table 9.6). This is not surprising since N fertilizers are not typically applied for soybean while N applications for corn are generally substantial. Seasonal N_2O emissions were similar regardless of whether the corn was in continuous monoculture, part of a corn–soybean rotation (Adviento-Borbe et al., 2007; Venterea et al., 2010; Omonode et al., 2011), or where N fertilizer was reduced to account for N supplied by a preceding soybean crop (Hernandez-Ramirez et al., 2009; Table 9.6).

As a result of differences in N_2O emission between corn and soybean, corn–soybean rotations typically have substantially lower N_2O emissions than continuous corn (Table 9.6; Hernandez-Ramirez et al., 2009; Venterea et al., 2010). Venterea et al. (2010) noted that there was an interaction between N fertilizer source applied to corn and crop rotation impacts on N_2O emissions, with the difference between a corn–soybean and a continuous corn rotation

TABLE 9.6 Impact of Crop Selection On Nitrous Oxide Emissions in Croplands of Eastern and Central U.S. (Publications Since 2005)

Nitrogen														
Location	*Previous crop*	*Crop*[a]	*Rate (kg N ha^{-1})*	*Source*[b]	*Placement*[c]	*Timing*[d]	*Primary tillage*[e]	*Time in treatment (yr)*[f]	*Soil type*	*Study period*[g]	*N_2O flux (kg N_2O-N ha^{-1} yr^{-1})*[h]	*Sig.*[i]	*N_2O loss (% N applied)*	*Reference*
Iowa	Soybean	Corn	215	APP (13), UAN (202)	Banded	Planting (13), sidedress (202)	CP, NT	8	Silty clay loam, loam, clay loam	April 2004 – Feb 2005	10.2	a	4.7	Parkin and Kaspar (2006)
"	Corn	Soybean	–	–	–	–	"	"	"	"	4.49	b	–	"
Nebraska	Corn	Corn	240	AN	B'cast + incorp'd	Preplant (~120), V6 (~120)	CP	6	Silty clay loam	Apr–Dec 2005	1.38	a	0.6	Adviento-Borbe et al. (2007)
"	Corn	Soybean	0	–	–	–	"	"	"	"	0.4	b	–	"
"	Corn	Corn	310	AN, UAN	B'cast + incorp'd or irrigated	Preplant, V6 , V10, before tasseling, on residue (50 UAN)	"	"	"	"	1.8	a	0.6	"
"	Corn	Soybean	130	AN	B'cast + incorp'd	R3.5	"	"	"	"	1.49	a	1.1	"
Indiana	Corn	Corn	157	UAN	Injected	June	CP	7	Silty clay loam, silt loam	Mar–Nov/ Dec 2005 –2006	6.40	a	4.1	Hernandez-Ramirez et al. (2009)
"	Soybean	Corn	135	UAN	"	"	"	"	"	"	4.89	a	3.6	"
"	Corn	Soybean	0	–	–	–	"	"	"	"	0.98	c	–	"
"	Corn–soybean	Corn–soybean	135 (corn)	UAN	Injected	June	"	"	"	"	2.94	b	2.1	"
Indiana	Soybean	Corn	193	UAN (168), DAP (25)	Injected (168), banded (25)	Preplant (168), planting (25)	NT	>10	Silt loam	Apr 2004–Oct 2007	3.79	a	1.8	Smith et al. (2011)
"	Corn	Soybean	0	–	–	–	"	"	"	"	1.52	b	–	"
"	Native grass	Native grass	0	–	–	–	"	0	"	"	0.57	b	–	"
"	Sorg.-Sudan.	Sorghum-Sudan.	0–168	Urea	B'cast	June	"	"	"	"	2.61	a	1.8	"
"	Corn–soybean	Corn–soybean	See above	See above	See above	See above	"	"	"	"	2.66	a	1.8	"

Maryland	Wheat–soybean	Corn	168	UAN	Knifed	Planting (25), sidedress (143)	CP	15	Silt loam	Apr–Dec 2010	1.57	a	0.9	Cavigelli et al. (unpub'd)
"	Corn–rye	Soybean	–	–	–	–	"	"	"	"	0.48	b	–	"
"	Wheat–vetch	Corn	185	PL (85), vetch (100)	B'cast + incorp'd	Preplant	MP	"	"	"	1.65	a	0.9	"
"	Corn–rye	Soybean	–	–	–	–	"	"	"	"	1.96	a	–	"
"	Corn–rye–soybean	Corn–rye–soybean	85	PL	B'cast + incorp'd	Preplant	MP	"	"	"	1.59	a	3.7	"
"	Corn–rye–soybean–wheat/vetch	Corn–rye–soybean–oat/vetch	185 (corn), 85 (oat)	PL (85, corn, oat) vetch (100)	"	"	"	"	"	"	1.57	a	1.7	"
"	Corn–rye–soybean–wheat/alfalfa–alfalfa–alfalfa	Corn–rye–soybean–oat/alfalfa–alfalfa–alfalfa	225 (corn), 85 (oat)	PL (85, corn, oat) alfalfa (140)	"	"	"	"	"	"	0.76	b	1.5	"
Indiana	Corn	Corn	258–296	AN (36), UAN (222–260)	Banded (36), injected (222–260)	Planting (36), sidedress (222–260)	MP, CP, NT (mean)	29	Silty clay loam	Apr/Jun–Sep 2004–2006	6.01	ns	2.2	Omonode et al. (2011)
"	Soybean	Corn	"	"	"	"	"	"	"	"	4.81	ns	1.8	"

[a]Only studies reporting results of statistical analyses comparing crop types are included.
[b]AN = ammonium nitrate; APP = ammonium polyphosphate; DAP = diammonium phosphate; PL = poultry litter; UAN = urea ammonium nitrate; numbers in parentheses are kg N ha^{-1} of material applied.
[c]B'cast = broadcast; incorp'd = incorporated.
[d]Numbers in parentheses are kg N ha^{-1} of material applied at given time; V6, V10 are corn growth stages; R3.5 is soybean growth stage.
[e]CP = chisel plow; MP = moldboard plow; NT = no-tillage.
[f]Number of years in NT or other treatment at the beginning of study period.
[g]Seasonal sampling indicated by month–month year–year; annual sampling indicated by month–year month–year.
[h]Statistical significance as reported by authors or estimated. Numbers followed by a different letter are significantly different (usually $P < 0.05$) for a given reference or portions of references separated by horizontal lines.
[i]Calculated without a 0 N control except where a 0 N control is indicated in the N rate column.

being attenuated when the N source was urea compared to anhydrous ammonia. The 26% increase in corn acreage in the U.S. between 2002 and 2007 (USDA NASS, 2011)—driven by increased biofuel demand—likely resulted in an increase in N_2O emissions.

The impact of fertilizers rather than crop selection per se on N_2O emissions in these studies is illustrated by Adviento-Borbe et al. (2007). When corn and soybean were fertilized according to recommended practices (i.e. no N fertilizer applied to soybean), N_2O emissions under corn were 3.5 times as great as under soybean. When soybean was managed intensively (i.e. 130 kg N ha^{-1} applied), N_2O emissions under soybean were no different than under corn (Table 9.6). The impact of crop selection on N_2O emission may be attenuated when animal manures are applied repeatedly over time. In corn–soybean–wheat rotations in Maryland, seasonal N_2O emission in a system to which N fertilizer was added was 3.3 times greater in corn than soybean, while seasonal N_2O emission was similar in corn and soybean when poultry litter (PL) was applied, even though PL was only applied during the corn and wheat phases of the rotation (Table 9.6; Cavigelli et al., unpublished data). Relatively high N_2O emission during the soybean phase of the rotation was likely due to greater residual C and N pools in the system receiving PL (Spargo et al., 2011).

In this same study, N_2O emissions decreased with length and complexity of crop rotation during the 1 year for which data are currently available. Seasonal N_2O emission in a 6-year corn–rye–soybean–wheat–alfalfa–alfalfa–alfalfa rotation was 52% lower than in a 3-year corn–rye–soybean–wheat–vetch (*Vicia villosa* Roth.) rotation and a 2-year corn–rye–soybean–vetch rotation in 2010 (Table 9.6). These three systems were managed organically, relying on PL inputs to corn (along with legume cover crops) and wheat. Lower N_2O emissions in the longer rotation were likely due to the high N-demanding crops and associated N inputs being present for only 2 of 6 years in the longer rotation, compared to 1 of every 2 or 2 of every 3 years in the shorter rotations. In Minnesota, Johnson et al. (2010), however, found length and complexity of crop rotation did not impact N_2O emissions. However, treatment effects were overshadowed by N_2O emissions at spring thaw, which accounted for 65% of annual emissions in this study.

CROPPING SYSTEMS

Since N_2O emissions in croplands are influenced by complex interactions among diverse management practices, soil type, and annual weather patterns, one approach to identifying effective GHG mitigation strategies is to measure GHG emissions on cropping systems that combine a number of GHG mitigation options or BMPs. Three recent publications report on such studies from the eastern and central U.S. (Table 9.7). Adviento-Borbe et al. (2007) measured soil N_2O emissions in continuous corn and a corn–soybean rotation managed using recommended or intensive practices (i.e. crop varieties, planting dates and densities, fertilizer rates and timing, tillage, and irrigation scheduling). They found N_2O emissions were similar in continuous corn in the recommended and intensive systems in 2 years. During the first year of the study, however, the intensive system had more than double the N_2O emission as the recommended system, likely due to residual impacts of management the previous fall (Table 9.7). In the corn–soybean rotation, there was no difference in N_2O emissions in the 2 corn years but emissions in the soybean year were almost four times as great in the intensive as in the recommended system, as noted earlier. Despite differences in N_2O emissions, soil C sequestration was greater in the intensive than the recommended systems such that GWP was similar in the two systems. In addition, crop yields were greater in the intensive system, suggesting that the intensive system might provide greater crop production per unit of GWP. Johnson et al. (2010) evaluated GHG emissions from three contrasting cropping systems in southwest Minnesota using GRACEnet-defined scenarios: business as usual (BAU), maximum C sequestration (MAXC), and optimum greenhouse gas benefits (OGGB). They found no differences in soil N_2O emissions despite fairly substantial differences in management among

TABLE 9.7 Impact of Cropping System On Nitrous Oxide Emissions in Croplands of Eastern and Central U.S. (Publications Since 2005)

			Nitrogen											
Location	*System*[a]	*Crop*	*Rate (kg N ha^{-1})*	*Source*[b]	*Placement*[c]	*Timing*[d]	*Primary tillage*[e]	*Time in treatment (yr)*[f]	*Soil type*	*Study period*[g]	*N_2O flux (kg N_2O-N ha^{-1} yr^{-1})*	*Sig.*[h]	*N_2O loss (% N applied)*[i]	*Reference*
Nebraska	Rec	Corn–corn	200–240	AN	B'cast + incorp'd	Preplant (~120), V6 (~120)	CP	6	Silty clay loam	Apr–Oct 2004–2005	1.87	a	0.8–0.9	Adviento-Borbe et al. (2007)
"	Int	"	280–310	AN	B'cast + incorp'd or irrigated	Preplant, V6, V10, tasseling, on residue	"	"	"	"	2.16	a	0.7–0.8	
"	Rec	Corn–soybean	140 (corn)	AN	B'cast + incorp'd	Preplant (~120), V6 (~120)	"	"	"	"	1.42	b	2.0	"
"	Int	"	230 (corn), 130 (soybean)	AN	B'cast + incorp'd or irrigated	Preplant, V6, V10, tasseling, on residue	"	"	"	"	2.03	a	0.6	"
"	Rec	Corn–corn	180	AN	B'cast + incorp'd	Preplant (~120), V6 (~120)	"	"	"	Apr–Oct 2003	3.92	b	2.2	"
"	Int	"	250	AN	B'cast + incorp'd or irrigated	Preplant, V6, V10, tasseling, on residue	"	"	"	"	9.24	a	3.7	"
Minnesota	Cont	Soybean	–	–	–	–	CP	NR	Silt loam	June–Oct 2004	1.32	ns	–	Bavin et al. (2009)
"	Expt	"	"	"	"	"	ST	"	"	Jul–Aug 2004	2.17	ns	–	"
"	Cont	Corn	112	AA	Knifed	Preplant	CP	"	"	May–Aug 2005	11.6	a	10.4	"
"	Expt	"	"	Urea	B'cast	"	ST	"	"	"	7.85	b	7.0	"

Continued

TABLE 9.7 Impact of Cropping System On Nitrous Oxide Emissions in Croplands of Eastern and Central U.S. (Publications Since 2005)—continued

Nitrogen														
Location	System[a]	Crop	Rate (kg N ha^{-1})	Source[b]	Placement[c]	Timing[d]	Primary tillage[e]	Time in treatment (yr)[f]	Soil type	Study period[g]	N_2O flux (kg N_2O-N ha^{-1} yr^{-1})	Sig.[h]	N_2O loss (% N applied)[i]	Reference
Minnesota	BAU	Corn–soybean	140–150 (corn), 0–11 (soybean)	AN or AA (corn); AN (soybean)	Banded (11), b'cast (AN), injected (AA)	Planting (11); sidedress (129–139; corn)	MP (corn), CP (soy)	2	Glacial till: loam, silty clay	Apr 2004 – Apr 2007	4.87	ns	6.1–6.5	Johnson et al. (2010)
"	MAXC	Corn–soybean–wheat/alfalfa–alfalfa	11 (corn, soybean); 81–86 (wheat)	AN	Banded (11) b'cast (81–86)	Planting	ST (corn, alfalfa)	"	loam, clay loam	"	4.81	ns	18.7	"
"	OGGB	"	0	–	–	–	"	"	"	"	5.04	ns	–	"

[a]Rec = recommended; Int = integrated; Cont = control; Expt = experimental; BAU = business as usual; MAXC = maximum carbon sequestration; OGGB = optimum greenhouse gas benefits. See text for management details for each system.
[b]AA = anhydrous ammonia; AN = ammonium nitrate; numbers in parentheses are kg N ha^{-1} of material applied.
[c]B'cast = broadcast; incorp'd = incorporated.
[d]V6, V10 refer to corn growth stages; numbers in parentheses are kg N ha^{-1} of material applied at given time.
[e]CP = chisel plow; MP = moldboard plow; ST = strip tillage.
[f]Number of years in treatment at the beginning of study period; NR = not reported.
[g]Seasonal sampling indicated by month–month year–year; annual sampling indicated by month–year month–year.
[h]Statistical significance as reported by authors or inferred from reports. Numbers followed by a different letter are significantly different (usually $P < 0.05$) for a given reference or portions of references separated by a horizontal line.
[i]Calculated without a 0 N control.

the systems, including no fertilizer application in OGGB (Table 9.7). The authors attributed the lack of treatment differences to high N_2O emissions (65% of annual emissions) during spring thaw in all systems. This study, in particular, raises the specter that not measuring N_2O emissions year round can provide misleading treatment effects, at least in more northern climates.

Bavin et al. (2009), using a BMP approach, found different results in each of 2 years in a corn–soybean rotation in Minnesota. Seasonal N_2O emission in 2004 during the soybean phase of the rotation was similar in an experimental system employing a rye cover crop, strip tillage, and urea fertilizer and in a control treatment including no cover crop, chisel plowing, and anhydrous ammonia fertilizer. As in most studies, no N fertilizer was applied to soybean. In 2005, when the crop was corn, the experimental treatment had seasonal N_2O emissions 32% lower than the control treatment, suggesting the use of urea rather than anhydrous ammonia contributed to lower N_2O emissions, as found by Venterea et al. (2005, 2010). Additional systems-level experiments are needed to identify combinations of farming practices to reduce N_2O emissions from cropland soils.

METHANE

There have been relatively few studies measuring CH_4 fluxes from croplands in the eastern and central U.S. (Table 9.8). Venterea et al. (2005) found an interaction between N source and tillage such that, with CT, CH_4 uptake was greater with anhydrous ammonium than urea and was greater with UAN than anhydrous ammonium; and with NT, CH_4 uptake was greater with urea and UAN than anhydrous ammonia. While Sistani et al. (2010) found no difference in CH_4 flux when UAN or LSM were applied in the spring, Hernandez-Ramirez et al. (2009) found a positive CH_4 flux from soils when LSM was applied and a negative flux when UAN was the source of N. However, the N application rate was 98 kg N ha^{-1} greater for the LSM than UAN. Sistani et al. (2010) found injecting LSM increased seasonal CH_4 emission 1.6- to 15-fold, depending on year, compared to aerator or broadcast application. No application timing impacts on CH_4 flux were found when LSM was spring or fall applied (Hernandez-Ramirez et al., 2009) or when UAN was applied at planting or as a split application (Smith et al., 2011).

Three studies found no differences in CH_4 flux between NT and CT (Omonode et al., 2007; Smith et al., 2010, 2011), and one study found no tillage impact when urea was the N source (Venterea et al., 2005). However, when the N source was anhydrous ammonia, CH_4 uptake was greater with MP than NT and when the N source was UAN, CH_4 uptake was greater with CP than MP (Venterea et al., 2005). In addition, Ussiri et al. (2009) found NT soil was a sink for CH_4 while MP and CP soils were sources of CH_4 after 29 years. Differences were attributed to lower soil bulk density, which likely increased gas exchange in NT compared to the tilled systems. Smith et al. (2011) found a cover crop did not impact CH_4 flux while Adviento-Borbe et al. (2007) found CH_4 uptake was 11 times greater in continuous corn than in corn following soybean, but only in 1 of 3 years. These authors also found no impact of recommended or intensive management on CH_4 flux. Similarly, Johnson et al. (2010) found no impact of three diverse management systems on CH_4 flux.

CARBON DIOXIDE

Net ecosystem exchange (NEE) can be measured directly using micrometeorological techniques (Hollinger et al., 2005; Hernandez-Ramirez et al., 2011) or can be inferred by measuring changes in soil carbon. Changes in soil C in croplands are reviewed in companion chapters in this volume. There are only a handful of recent studies using meteorological techniques to evaluate the impact of management on NEE in croplands in the central U.S. (we found no recent studies from the eastern U.S.). These studies found conservation tillage

TABLE 9.8 Impact of Various Management Practices on Methane Fluxes in Croplands of Eastern and Central U.S. (Publications Since 2005)

			Nitrogen										
Location	Previous crop or system[a]	Crop	Rate (kg N ha^{-1})	Source[b]	Placement[c]	Timing[d]	Primary tillage[e]	Time in treatment (yr)[f]	Soil type	Study period[g]	CH_4 flux (g C ha^{-1} yr^{-1})[h]	Sig[i]	Reference
Minnesota	Soybean	Corn	120	Urea	B'cast	Sidedress	MP	12	Silt loam	Apr–Nov 2004	−93	ab	Venterea et al. (2005)
"	"	"	"	"	"	"	CP or disk	"	"	"	−212	bcd	"
"	"	"	"	"	"	"	NT	"	"	"	−206	bcd	"
"	"	"	"	UAN	"	Preplant	MP	"	"	"	−103	abc	"
"	"	"	"	"	"	"	CP or disk	"	"	"	−263	d	"
"	"	"	"	"	"	"	NT	"	"	"	−206	bcd	"
"	"	"	"	AA	Injected	"	MP	"	"	"	−227	cd	"
"	"	"	"	"	"	"	CP or disk	"	"	"	−124	abc	"
"	"	"	"	"	"	"	NT	"	"	"	−62	a	"
Kentucky	Soybean	Corn	0	–	"	Preplant	NT	~10	Silt loam	Apr–Sep 2007–2008	350	b	Sistani et al. (2010)
"	"	"	179	UAN	B'cast	"	"	"	"	"	450	b	"
"	"	"	200–204	LSM	"	"	"	"	"	"	700	b	"
"	"	"	"	"	Aeration	"	"	"	"	"	1500	b	"
"	"	"	"	"	injected	"	"	"	"	"	10,600	a	"
Indiana	Corn	Corn	255	LSM	Injected	Fall	CP	17	Silty clay loam, silt loam	Mar–Nov/ Dec 2005–2006	159	ab	Hernandez-Ramirez et al. (2009)
"	"	"	"	"	"	Spring	"	"	"	"	329	a	"
"	"	"	157	UAN	"	Sidedress	"	"	"	"	−177	b	"
"	Soybean	Corn	"	"	"	"	"	"	"	"	−128	ab	"
"	Corn	Soybean	–	–	–	–	"	"	"	"	−27	ab	"
Indiana	Corn–soybean	Corn–soybean	193 (corn)	UAN (168), DAP (25)	Knifed (168), banded (25)	Preplant (168), planting (25)	CP	0	Silt loam	Apr 2004–Oct 2007	−270	ns	Smith et al. (2011)
"	"	"	107 (corn)	UAN	Knifed	Planting	PT	0	"	"	−210	ns	"
"	"	"	193 (corn)	UAN (168), DAP (25)	Knifed (168), banded (25)	Preplant (168), planting (25)	NT	>10	"	"	−230	ns	"

”	”	”	168 (corn)	UAN (168)	Knifed (84), dribbled (84)	Preplant (84), V3–V4 (84)	”	”	”	”	−210	ns	”
”	”	Corn–ann. ryegrass–soybean	193 (corn)	UAN (168), DAP (25)	Knifed (168), banded (25)	Preplant (168), planting (25)	”	”	”	”	−330	ns	”
”	”	Native grass	Native grass	0	–	–	–	”	0	”	−170	ns	”
”	”	Sorg.-Sudan.	Sorghum-Sudan.	Urea (0–168)	B’cast	June	”	”	”	”	−120	ns	”
Indiana	Corn or soybean	Corn	258–296	AN (36), UAN (222–260)	Banded (36), injected (222–260)	Planting (36), sidedress (222–260)	MP	29	Silty clay loam	Apr/Jun–Sep 2004–2006	−110	ns	Omonode et al. (2007)
”	”	”	”	”	”	”	CP	”	”	”	−422	ns	”
”	”	”	”	”	”	”	NT	”	”	”	253	ns	”
Alabama	Sorghum	Soybean	0	–	–	–	CP	10	Silt loam	Apr–Oct 2007	193	ns	Smith et al. (2010)
”	”	Soybean–cover crop	0	”	”	”	NT	”	”	”	−9.5	ns	”
Ohio	Corn	Corn	200	AN (16), AA (184)	Banded (16), injected (184)	Planting (16), sidedress (184)	MP	43	Silt loam	Nov 2004–Nov 2005	2.76	b	Ussiri et al. (2009)
”	”	”	”	”	”	”	CP	”	”	”	2.27	b	”
”	”	”	”	”	”	”	NT	”	”	”	−0.32	a	”
Nebraska	Rec.	Corn–corn	200–240	AN	B’cast + incorp’d	Preplant (~120), V6 (~120)	CP	6	Silty clay loam	Apr–Oct 2004–2005	−3880	ns	Adviento-Borbe et al. (2007)
”	Int	”	280–310	”	B’cast + incorp’d or irrigated	Preplant, V6, V10, tasseling, on residue	”	”	”	”	−3760	ns	”
”	Rec	Corn–soybean	140 (corn)	”	B’cast + incorp’d	Preplant (~120), V6 (~120)	”	”	”	”	−4340	ns	”
”	Int	”	230 (corn), 130 (soybean)	”	B’cast + incorp’d or irrigated	Preplant, V6, V10, tasseling, on residue	”	”	”	”	−2480	ns	”
”	Rec + Int	Corn–corn	See above	AN	See above	See above	”	”	”	Apr–Oct 2003	−4825	a	”
”	Rec + Int	Corn after soybean	”	”	See above	See above	”	”	”	”	−425	b	”

Continued

TABLE 9.8 Impact of Various Management Practices on Methane Fluxes in Croplands of Eastern and Central U.S. (Publications Since 2005)—continued

Location	Previous crop or system[a]	Crop	Rate (kg N ha^{-1})	Source[b]	Placement[c]	Timing[d]	Primary tillage[e]	Time in treatment (yr)[f]	Soil type	Study period[g]	CH_4 flux (g C ha^{-1} yr^{-1})[h]	Sig[i]	Reference
					Nitrogen								
Minnesota	BAU	Corn–soybean	140–150 (corn), 0–11 (soybean)	AN or AA (corn); AN (soybean)	Banded (11), b'cast (AN), injected (AA)	Planting (11); sidedress (129–139; corn)	MP (corn), CP (soy)	2	Glacial till: loam, silty clay loam, clay loam	Apr 2004–Apr 2007	−560	ns	Johnson et al. (2010)
"	MAXC	Corn–soybean–wheat/alfalfa–alfalfa	11 (corn, soybean); 81–86 (wheat)	AN	Banded (11) b'cast (81–86)	Planting	ST (corn, alfalfa)	"	"	"	0	ns	"
"	OGGB	"	0	–	–	–	"	"	"	"	−190	ns	"

[a]Rec = recommended; Int = integrated; BAU = business as usual; MAXC = maximum carbon sequestration; OGGB = optimum greenhouse gas benefits. See text for management details for each system.
[b]AA = anhydrous ammonia; AN = ammonium nitrate; DAP = diammonium phosphate; LSM = liquid swine manure; UAN = urea ammonium nitrate; numbers in parentheses are kg N ha^{-1} of material applied.
[c]B'cast = broadcast; incorp'd = incorporated.
[d]V3, V4, V6, V10 refer to corn growth stages; numbers in parentheses are kg N ha^{-1} of material applied at given time.
[e]CP = chisel plow; MP = moldboard plow; NT = no-tillage; PT = precision tillage; ST = strip tillage.
[f]Number of years in NT or other treatment at the beginning of study period.
[g]Seasonal sampling indicated by month–month year–year; annual sampling indicated by month–year month–year.
[h]Positive values indicate CH_4 emission; negative values indicate CH_4 uptake.
[i]Statistical significance as reported by authors or inferred from reports. Numbers followed by a different letter are significantly different (usually $P < 0.05$) for a given reference or portions of references separated by horizontal lines.

practices under corn–soybean rotations apparently have no significant effect on annual NEE whether measured during tillage management transition (Baker and Griffis, 2005; Verma et al., 2005; Glenn et al., 2010) or after long-term management (Hollinger et al., 2005). Similarly, there were few differences in annual NEE of corn–soybean rotations under rain-fed versus irrigated systems (Verma et al., 2005). These multi-year studies typically evaluated corn and soybean cultivation only in alternative years as part of a single site biennial rotation system. Thus, results are confounded by effects of inter-annual climate variation. One study in which NEE was measured concomitantly in corn and soybean fields found annual NEE ($g\,C\,m^{-2}\,yr^{-1}$) was -466 ± 38 for corn and -13 ± 39 for soybean over 3 years (Hernandez-Ramirez et al., 2011). Based on these results and estimated C balance, soybean appears to induce depletion of soil C, while corn seems to be C neutral. Thus, crop selection seems to have an important effect on NEE.

Ecosystem C balances reported in existing studies show fluctuations from net source to neutral outcomes, demonstrating NEE over cropland is highly variable and dynamic. In investigating seasonal CO_2 flux from a variety of ecosystems, Falge et al. (2002) demonstrate substantial differences in the temporal patterns of assimilatory and respiratory processes among ecosystems. They state temperature, moisture, and light affect NEE differently through impacts on gross primary production and ecosystem respiration. Consequently, there is a need to understand factors influencing seasonality of gross primary production, ecosystem respiration and NEE. Since ecosystem respiration is the sum of autotrophic respiration and heterotrophic respiration, similar logic dictates that driving factors may differentially control the temporal dynamics of these two processes. This argues for detailed characterizations of these heterotrophic and autotrophic respiratory processes in order to provide a better understanding of feedbacks on NEE.

It is worth noting that NEE cannot be measured using the flux chamber techniques commonly used to measure soil gas fluxes. This is because flux chambers only measure CO_2 outputs and do not account for CO_2 inputs into the system via fixation during photosynthesis. In addition, soil CO_2 emission measured using chambers is the sum of autotrophic (largely roots) and heterotrophic respiration, which cannot easily be differentiated. Thus, soil CO_2 fluxes cannot be used to calculate the GWP of agricultural systems, despite a number of recent studies that have summed soil CO_2, CH_4 and N_2O emissions on a CO_2 equivalent basis and referred to that value as GWP.

INFORMATION GAPS AND FUTURE RESEARCH NEEDS

While the number of publications reporting results of seasonal or annual N_2O and CH_4 fluxes in eastern and central U.S. has increased considerably since 2005, the bulk of publications are from northern portions of the region (Figure 9.1). We found only one publication addressing N_2O and CH_4 emissions from states further south than Missouri, Kentucky, and Maryland. Also, four northern states (Indiana, Iowa, Michigan, and Minnesota) account for two-thirds of studies on N_2O emissions published since 2005. In addition, there are a limited number of studies comparing particular GHG mitigation options. Results among studies are sometimes inconsistent, reflecting differences in soil types, climate, weather patterns during years of study, management history, and management details that could impact N_2O and CH_4 fluxes. Additional studies are needed from the region to account for regional variability in climate, soils, and cropping systems.

Even within many studies, inconsistent results among years reflect the strong impact of weather on GHG fluxes, especially timing of rainfall relative to implementation of management practices (Adviento-Borbe et al., 2007, 2010; Sistani et al., 2010; Dietzel et al., 2011). Since most published studies report results from 1 to 3 years, inter-annual variability, which can be greater than treatment effects in the short term (Adviento-Borbe et al., 2007; Bavin et al.,

2009), is not adequately accounted for in much of the literature. As noted above, there are differences between short-term and long-term impacts of tillage management on N_2O emissions (Six et al., 2004; Table 9.5). Chang et al. (1998) and our review suggest the same may be true for manure applications (Tables 9.2, 9.6; Cavigelli et al., unpublished data). Altogether, these issues indicate that more long-term studies are needed to understand the impacts of management on GHG emissions.

Many studies report results from data collected only during the main crop growth period despite clear evidence that significant N_2O emissions occur after crops have matured (Zebarth et al., 2008) or during freeze–thaw cycles, at least in northern regions (Wagner-Riddle et al., 2007; Johnson et al., 2010). Since there can be a strong interaction between management, weather, and N_2O emissions during freeze–thaw cycles (Wagner-Riddle et al., 2007; Dietzel et al., 2011), not including measurements during this time period could both underestimate annual emissions and skew treatment rankings (Hyatt et al., 2010; Jarecki et al., 2009). This is especially true of soils to which manure has been applied, since N_2O emissions can be more than double those receiving no manure during freeze–thaw cycles (Phillips, 2007). Accordingly, year-round sampling seems warranted in northern regions.

It remains unclear, however, what regions experience substantial N_2O emissions during freeze–thaw cycles. Some regions, including eastern Maryland and central Indiana, do not seem to experience substantial N_2O emissions during spring thaw, likely due to shallow or non-existent soil freezing (Djurickovic, 2010; Smith et al., 2011). In these regions, targeting gas sampling to crop growth periods may be adequate, though additional data are needed for confirmation.

Almost all studies included in this summary were conducted using NFT-NSS chambers, likely because they are technically simple and because a large number of treatments can be compared over relatively short distances. The design of NFT-NSS chambers is known to impact gas flux measurements (Rochette and Erikson-Hamel, 2008; Venterea et al., 2009), and despite standardization efforts in GRACEnet and other programs (Parkin and Venterea, 2010; Rochette, 2011), the selection of chamber design is often constrained by other considerations such as resource availability. Among studies reviewed here, chamber deployment methods, headspace sampling interval, and sampling frequency relative to factors known to impact N_2O and CH_4 emissions (timing of N applications, rainfall, freeze–thaw) vary considerably. Moreover, only a few studies specify frame heights above the soil surface (Rochette et al., 2008), which is necessary to account for differences in frame volume resulting from soil subsidence or the depth to which an individual frame is inserted into the ground. Differences in soil surface microtopography and depth of frame insertion impact chamber volume, which can substantially influence gas flux values (Cavigelli, unpublished data). Finally, the method of calculating GHG fluxes can impact estimates by up to 35% (Venterea et al., 2010). These methodological influences on the data impact the confidence of quantitative generalizations about current mitigation options and indicate the need for further standardization of methods (Rochette, 2011).

Improved characterization of temporal and spatial variability of GHG fluxes is also needed. Automated sampling systems can help improve the temporal resolution of GHG flux data (Parkin and Kaspar, 2003, 2004) but these systems require greater resources and technical expertise than manual methods. Therefore, manual sampling from NFT-NSS chambers is likely to continue at many locations. However, periodically sampling with high temporal resolution and/or use of automated systems can help identify manual sampling schemes that provide more accurate data with a minimum increase in sampling intensity (Parkin and Kaspar, 2003, 2004; Parkin, 2008). Accounting for temporal resolution is important because, among other things, the timing of peak N_2O or CH_4 fluxes following various events, such as N source application and rainfall events, can vary by treatment (Sistani et al., 2010). Therefore, a sampling scheme that does not provide sufficient temporal resolution might impact results independent of true treatment effects. Automated sampling methods can also be used to better characterize or capture spatial variability of GHG fluxes. While more sophisticated micrometeorological methods can integrate information across large areas that contain high spatial variability and are well suited to characterizing temporal variability for a limited number of locations, they are not well suited to characterizing spatial variability. Additional methodological issues are addressed in more detail in separate chapters (Venterea and Parkin, 2012; Skinner and Wagner-Riddle, 2012).

CONCLUSIONS

Since agricultural soils account for 69% of global N_2O emissions, and are predicted to increase in the near future, it is important to identify farming practices that reduce N_2O emissions. One of the more promising N_2O mitigation options for croplands in the eastern and central U.S. is applying N fertilizer at rates that do not exceed crop N needs. This strategy may have a larger impact on atmospheric N_2O concentrations than the most recent IPCC report suggests, as a number of studies indicate N_2O emissions increase exponentially with increasing N fertilizer application rate rather than linearly as the IPCC (2007) report suggests. While there is only one study showing a similar pattern for animal manure application rate, applying animal manures at rates that do not exceed crop N needs should also help reduce N_2O emissions, although the long-term impacts of manure application on N_2O emissions need to be more fully evaluated. Crop selection is another management practice for which there is sufficient evidence to suggest a clear impact on N_2O emissions in eastern and central U.S. Reducing the frequency of high N-demanding crops in a crop rotation seems to be a robust way to reduce N_2O emissions. The only other management practice for which there is evidence of reduced N_2O emissions is long-term (>20-year) NT management.

There is little and/or inconsistent evidence that other means of increasing nitrogen-use efficiency—N fertilizer and manure timing and placement, nitrification inhibitors, delayed release N fertilizers, and cover crops—have a significant impact on soil N_2O emissions in eastern and central U.S. In addition, there was no evidence that N_2O emissions are different when animal manures vs. fertilizer are used as an N source. However, UAN and/or urea seem to decrease N_2O emissions compared to anhydrous ammonia in a few studies conducted in Minnesota.

While there is some evidence that CH_4 emissions are impacted by N source—emissions were greater when animal manure was injected than when N fertilizer was broadcast applied, and there was a complex interaction between N fertilizer and tillage regime—we found few treatment differences, in part because there are only a limited number of studies investigating cropland management impacts on CH_4 emissions.

There is a clear need for additional studies evaluating cropland management on GHG fluxes in all regions of eastern and central U.S., especially in southern regions. Long-term studies are essential for identifying long-term impacts of management on GHG fluxes, as these may differ substantially from short-term impacts, as indicated by a limited number of studies on long-term NT and manure application. A detailed meta-analysis of management, climate, and soil factors impacting soil N_2O emissions would be beneficial, provided methodological variations among studies were taken into account.

Acknowledgments

We thank the following people who have helped collect the N_2O emissions data from the Maryland research site: Anne Conklin, Gary Cooke, Milutin Djurickovic, Erwin Fuentes, Grace Garst, Ariela Haber, Nate Jackson, Jessica Kelly, Jaimie Kemper, Mikhail Kozlov, Victoria Lake, Jill Lau, Chris Rasmann, John Spargo, Rachel Steele, Matthew Stocker, and Nate Zuk.

References

Adviento-Borbe, M.A.A., Haddix, M.L., Binder, D.L., Walters, D.T., Dobermann, A., 2007. Soil greenhouse gas fluxes and global warming potential in four high-yielding maize systems. Global Change Biol. 13, 1972–1988.

Adviento-Borbe, M.A.A., Kaye, J.P., Bruns, M.A., McDanial, M.D., McCoy, M., Harkcom, S., 2010. Soil greenhouse gas and ammonia emissions in long-term maize-based cropping system. Soil Sci. Soc. Am. J. 74, 1623–1634.

Aulakh, M.S., Doran, J.W., Walters, D.T., Power, J.F., 1991. Legume residue and soil water effects on denitrification in soils of different textures. Soil Biol. Biochem. 23, 1161–1167.

Aulakh, M.S., Rennie, D.A., Paul, E.A., 1984. Acetylene and N-serve effects upon N_2O emissions from NH_4 and NO_3 treated soils under aerobic and anaerobic conditions. Soil Biol. Biochem. 16, 351–356.

Avrahami, S., Bohannan, B.J.M., 2009. N_2O emission rates in a California meadow soil are influenced by fertilizer level, soil moisture and the community structure of ammonia-oxidizing bacteria. Global Change Biol. 15, 643–655.

Baker, J.M., Griffis, T.J., 2005. Examining strategies to improve the carbon balance of corn/soybean agriculture using eddy covariance and mass balance techniques. Agric. Forest Meteorol. 128, 163–177.

Bateman, E.J., Baggs, E.M., 2005. Contributions of nitrification and denitrification to N_2O emissions from soils at different water-filled pore space. Bio. Fertil. Soils 41, 379–388.

Bavin, T.K., Griffis, T.J., Baker, J.M., Venterea, R.T., 2009. Impact of reduced tillage and cover cropping on the greenhouse gas budget of a maize/soybean rotation ecosystem. Agric. Ecosyst. Environ. 134, 234–242.

Betlach, M.R., Tiedje, J.M., 1981. Kinetic explanation for accumulation of nitrite, nitric oxide, and nitrous oxide during bacterial denitrification. Appl. Environ. Microbiol. 42, 1074–1084.

Bouwman, A.F., Boumans, L.J.M., Batjes, N.H., 2002. Emissions of N_2O and NO from fertilized fields: summary of available measurement data. Global Biogeochem. Cycles 16, 1058–1072.

Bremner, J.M., Breitenbeck, G.A., Blackmer, A.M., 1981. Effect of nitrapyrin on emission of nitrous oxide from soils fertilized with anhydrous ammonia. Geophys. Res. Lett. 100, 2785–2800.

Bronson, K.F., Mosier, A.R., Bishnoi, S.R., 1992. Nitrous oxide emissions in irrigated corn as affected by nitrification inhibitors. Soil Sci. Soc. Am. J. 56, 161–165.

Cavigelli, M.A., Dao, T.H., 2008. Residual impact of raw and composted poultry litter on soil carbon pools. J. Integr. Biosci. 6, 30–40.

Cavigelli, M.A., Robertson, G.P., 2000. The functional significance of denitrifier community composition in a terrestrial ecosystem. Ecology 81, 1402–1414.

Cavigelli, M.A., Robertson, G.P., 2001. Role of denitrifier diversity in rates of nitrous oxide consumption in a terrestrial ecosystem. Soil Biol. Biochem. 33, 297–310.

Chan, A.S.K., Parkin, T.B., 2001. Methane oxidation and production activity in soils from natural and agricultural ecosystems. J. Environ. Qual. 30, 1896–1903.

Chang, C., Cho, C.M., Janzen, H.H., 1998. Nitrous oxide emission from long-term manured soils. Soil Sci. Soc. Am. J. 62, 677–682.

Christensen, S., Simkins, S., Tiedje, J.M., 1990. Temporal patterns of soil denitrification: their stability and causes. Soil Sci. Soc. Am. J. 54, 1614–1618.

Cicerone, R.J., 1987. Changes in stratospheric ozone. Science 237, 35–41.

Clark, A., 2007. Managing Cover Crops Profitably, second ed. Sustainable Agriculture Network Series, Beltsville, MD.

Coelho, B.R.B., Roy, R.C., Bruin, A.J., 2006. Nitrogen recovery and partitioning with different rates and methods of sidedressed manure. Soil Sci. Soc. Am. J. 70, 464–473.

Council for Agricultural Science and Technology, 2004. Climate change and greenhouse gas mitigation: challenges and opportunities for agriculture. Task Force Report 141.

Davidson, E.A., 2009. The contribution of manure and fertilizer nitrogen to atmospheric nitrous oxide since 1860. Nature Geosci. 2, 659–662.

Del Grosso, S.J., Parton, W.J., Mosier, A.R., Walsh, M.K., Ojima, D.S., Thornton, P.E., 2006. DAYCENT national-scale simulations of nitrous oxide emissions from cropped soils in the United States. J. Environ. Qual. 35, 1451–1460.

Dendooven, L., Anderson, J.M., 1994. Dynamics of reduction enzymes involved in the denitrification process in pasture soil. Soil Biol. Biochem. 26, 1501–1506.

Dietzel, R., Wolfe, D., Thies, J.E., 2011. The influence of winter soil cover on spring nitrous oxide emissions from an agricultural soil. Soil Biol. Biochem. 43, 1989–1991.

Djurickovic, M.S., 2010. Global warming potential of corn, soybean, and wheat production in organic, chisel till, and no-till farming systems. M.S. thesis, Hood College.

Drury, C.F., Reynolds, W.D., Tan, C.S., Welacky, T.W., Calder, W., McLaughlin, N.B., 2006. Emissions of nitrous oxide and carbon dioxide: influence of tillage type and nitrogen placement depth. Soil Sci. Soc. Am. J. 70, 570–581.

Eagle, A.J., Henry, L.R., Olander, L.P., Haugen-Kozyra, K., Millar, N., Robertson, G.P., 2011. Greenhouse gas mitigation potential of agricultural land management in the United States: a synthesis of the literature. Nicholas Institute for Environmental Policy Solutions, Durham, NC.

Eghball, B., Gilley, J.E., 1999. Phosphorus and nitrogen in runoff following beef cattle manure or compost application. J. Environ. Qual. 28, 1201–1210.

Eichner, M.J., 1990. Nitrous oxide emissions from fertilized soils: summary of available data. J. Environ. Qual. 19, 272–280.

Falge, E., Baldocchi, D., Tenhunen, J., Aubinet, M., Bakwin, P., Berbigier, P., Bernhofer, C., Burba, G., Clement, R., Davis, K., Elbers, J., Goldstein, A., Grelle, A., Granier, A., Guomundsson, J., Hollinger, D., Kowalski, A., Katul, G., Law, B., Malhi, Y., Meyers, T., Monson, R., Munger, J., Oechel, W., Kyaw Tha Paw, U., Pilegaard, K., Rannik, U., Rebmann, C., Suyker, A., Valentini, R., Wilson, K., Wofsy, S., 2002. Seasonality of ecosystem respiration and gross primary production as derived from FLUXNED measurements. Ag. Forest Meteorol. 113, 53–74.

Firestone, M.K., Davidson, E.A., 1989. Microbiological basis of NO and N_2O production and consumption in soil. In: Andreae, M.O., Schimel, D.S. (Eds.), Exchange of Trace Gases between Terrestrial Ecosystems and the Atmosphere. John Wiley and Sons, Berlin, Germany, pp. 7–21.

Firestone, M.K., Tiedje, J.M., 1979. Temporal change in N_2O and N_2 from denitrification following onset of anaerobiosis. Appl. Environ. Microbiol. 38, 673–679.

Flessa, H., Beese, F., 2000. Laboratory estimates of trace gas emissions following surface application and injection of cattle slurry. J. Environ. Qual. 29, 262–268.

Follett, R.F., Shafer, S.R., Jawson, M.D., Franzluebbers, A.J., 2005. Research and implementation needs to mitigate greenhouse gas emissions from agriculture in the USA. Soil Tillage Res. 83, 159–166.

Franzluebbers, A.J., 2005. Soil organic carbon sequestration and agricultural greenhouse gas emissions in the southeastern USA. Soil Tillage Res. 83, 120–147.

Franzluebbers, A.J., Follett, R.F., 2005. Greenhouse gas contributions and mitigation potential in agricultural regions of North America: introduction. Soil Tillage Res. 83, 1–8.

Glenn, A.J., Amiro, B.D., Tenuta, M., Stewart, S.E., Wagner-Riddle, C., 2010. Carbon dioxide exchange in a northern Prairie cropping system over three years. Agric. Forest Meteorol. 150, 908–918.

Goreau, T.J., Kaplan, W.A., Wofsy, S.C., McElroy, M.B., Valois, F.W., Watson, S.W., 1980. Production of NO_2- and N_2O by nitrifying bacteria at reduced concentrations of oxygen. Appl. Environ. Microbiol. 40, 526–532.

Grandy, A.S., Loecke, T.D., Parr, S., Robertson, G.P., 2006. Long-term trends in nitrous oxide emissions, soil nitrogen, and crop yields of till and no-till cropping systems. J. Environ. Qual. 35, 1487–1495.

Granli, T., Bockman, O.C., 1994. Nitrous oxide from agriculture. Norwegian J. Agric. Sci. 12, 7–127.

Grant, R.F., Pattey, E., Goddard, T.W., Kryzanowski, L.M., Puurveen, H., 2006. Modeling the effects of fertilizer application rate on nitrous oxide emissions. Soil Sci. Soc. Am. J. 70, 235–248.

Gregorich, E.G., Rochette, P., VandenBygaart, A.J., Angers, D.A., 2005. Greenhouse gas contributions of agricultural soils and potential mitigation practices in Eastern Canada. Soil Tillage Res. 83, 53–72.

Halvorson, A.D., Del Grosso, S.J., Alluvione, F., 2010. Tillage and inorganic nitrogen source effects on nitrous oxide emissions from irrigated cropping systems. Soil Sci. Soc. Am. J. 74, 436–445.

Hernandez-Ramirez, G., Brouder, S.M., Smith, D.R., Van Scoyoc, G.E., 2009. Nitrous oxide production in an eastern Corn Belt soil: weather, nitrogen source, and rotation. J. Environ. Qual. 38, 841–854.

Hernandez-Ramirez, B., Hatfield, J.L., Parkin, T.B., Sauer, T.J., Prueger, J.H., 2011. Carbon dioxide fluxes in corn-soybean rotations in the Midwestern U.S.: Inter- and intra-annual variations, and biophysical controls. Ag. Forest Meteorol. 151, 1831–1842.

Hoben, J.P., Gehl, R.J., Millar, N., Grace, P.R., Robertson, G.P., 2011. Nonlinear nitrous oxide (N_2O) response to nitrogen fertilizer in on-farm corn crops of the US Midwest. Glob. Change Biol. 17, 1140–1152.

Hollinger, S.E., Bernacchi, C.J., Meyers, T.P., 2005. Carbon budget of mature no-till ecosystem in North Central Region of the United States. Ag. Forest Meteorol. 130, 59–69.

Hyatt, C.R., Venterea, R.T., Rosen, C.J., McNearney, M., Wilson, M.L., Dolan, M.S., 2010. Polymer-coated urea maintains potato yields and reduces nitrous oxide emissions in a Minnesota loamy sand. Soil Sci. Soc. Am. J. 74, 419–428.

Hyman, M.R., Wood, P.M., 1983. Methane oxidation by *Nitrosomonas europaea*. Biochem. J. 212, 31–37.

Intergovernmental Panel on Climate Change, 2007. Contribution of Working Group I to the Fourth Assessment Report of the Intergovernmental Panel on Climate Change, 2007. In: Solomon, S., Qin, D., Manning, M., Chen, Z., Marquis, M., Averyt, K.B., Tignor, M., Miller, H.L. (Eds.), Contribution of Working Group I to the

Fourth Assessment Report of the Intergovernmental Panel on Climate Change, 2007. Cambridge University Press, Cambridge, United Kingdom and New York, NY, USA.

Jarecki, M.K., Parkin, T.B., Chan, A.S.K., Kaspar, T.C., Moorman, T.B., Singer, J.W., Kerr, B.J., Hatfield, J.L., Jones, R., 2009. Cover crop effects on nitrous oxide emission from a manure-treated Mollisol. Agric. Ecosyst. Environ. 134, 29–35.

Jawson, M.D., Shafer, S.R., Franzluebbers, A.J., Parkin, T.B., Follett, R.F., 2005. GRACEnet: greenhouse gas reduction through agricultural carbon enhancement network. Soil Tillage Res. 83, 167–172.

Johnson, J.M.F., Reicosky, D.C., Allmaras, R.R., Sauer, T.J., Venterea, R.T., Dell, C.J., 2005. Greenhouse gas contributions and mitigation potential of agriculture in the central USA. Soil Tillage Res. 83, 73–94.

Johnson, J.M.F., Archer, D., Barbour, N., 2010. Greenhouse gas emission from contrasting management scenarios in the northern Corn Belt. Soil Sci. Soc. Am. J., 396–406.

Jones, R.D., Morita, R.Y., 1983. Methane oxidation by *Nitrococcus oceanus* and *Nitrosomonas europaea*. Appl. Environ. Microbiol. 45, 401–410.

Kaspar, T.C., Parkin, T.B., 2011. Soil CO_2 flux in response to wheel traffic in a no-till system. Soil Sci. Soc. Am.. J. In press.

Kool, D.M., Dolfing, J., Wrage, N., Van Groenigen, J.W., 2010. Nitrifier denitrification as a distinct and significant source of nitrous oxide from soil. Soil Biol. Biochem. 43, 174–178.

Kramer, S.B., Reganold, J.P., Glover, J.D., Bohannan, B.J.M., Mooney, H.A., 2006. Reduced nitrate leaching and enhanced denitrifier activity and efficiency in organically fertilized soils. Proc. Nat. Acad. Sci. 103, 4522–4527.

Laird, D.A., 2008. The charcoal vision: a win-win-win scenario for simultaneously producing bioenergy, permanently sequestering carbon, while improving soil and water quality. Agron. J. 100, 178–181.

Lehmann, J., 2007. Bio-energy in the black. Front. Ecol. Environ. 5, 381–387.

Linn, D.M., Doran, J.W., 1984. Effect of water-filled pores pace on carbon dioxide and nitrous oxide production in tilled and nontilled soils. Soil Sci. Soc. Am. J. 48, 1267–1272.

Loecke, T.D., Robertson, G.P., 2009. Soil resource heterogeneity in terms of litter aggregation promotes nitrous oxide fluxes and slows decomposition. Soil Biol. Biochem. 41, 228–235.

Ma, B.L., Wu, T.Y., Tremblay, N., Deen, W., Morrison, M.J., McLaughlin, N.B., Gregorich, E.G., Stewart, G., 2010. Nitrous oxide fluxes from corn fields: on-farm assessment of the amount and timing of nitrogen fertilizer. Global Change Biol. 16, 156–170.

MacDonald, J.M., Ribaudo, M.O., Livingston, M.J., Beckman, J., Huang, W., 2009. Manure use for fertilizer and for energy. Report to Congress. USDA-ERS.

Mallory, E.B., Griffin, T.S., 2007. Impacts of soil amendment history on nitrogen availability from manure and fertilizer. Soil Sci. Soc. Am. J. 71, 964–973.

McTaggart, I.P., Clayton, H., Parker, J., Swan, L., Smith, K.A., 1997. Nitrous oxide emissions from grassland and spring barley, following N fertilizer application with and without nitrification inhibitors. Biol. Fert. Soils. 25, 261–268.

McSwiney, C.P., Robertson, G.P., 2005. Nonlinear response of N_2O flux to incremental fertilizer addition in a continuous maize (*Zea mays* L.) cropping system. Global Change Biol. 11, 1712–1719.

Megraw, S.R., Knowles, R., 1987. Methane production and consumption in a cultivated humisol. Biol. Fertil. Soils. 5, 56–60.

Millar, N., Robertson, G.P., Grace, P.R., Gehl, R.J., Hoben, J.P., 2010a. Nitrogen fertilizer management for nitrous oxide (N_2O) mitigation in intensive corn (maize) production: an emissions reduction protocol for US Midwest agriculture. Mitig. Adapt. Strateg. Glob. Change. 15, 185–204.

Millar, N., Robertson, G.P., Grace, P.R., Gehl, R.J., Hoben, J.P., 2010b. Nitrogen fertilizer rate management as a nitrous oxide mitigation strategy: development of a nitrous oxide emission reduction protocol (NERP). 19th World Congress of Soil Science, , Brisbane, Australia. August 1–6.

Miranowski, J.A., 2005. Energy consumption in US agriculture. In: Outlaw, J.L., Collins, K.J., Duffield, J.A. (Eds.), Agriculture as a Producer and Consumer of Energy. CAB International, Cambridge, MA.

Mosier, A., Schimel, D., Valentine, D., Bronson, K., Parton, W., 1991. Methane and nitrous oxide fluxes in native, fertilized and cultivated grasslands. Nature 350, 330–332.

Nash, P.R., 2010. Alternative tillage and nitrogen management options to increase crop production and reduce nitrous oxide emissions from claypan soils. M.S. thesis. University of Missouri-Columbia.

Omonode, R.A., Smith, D.R., Gal, A., Vyn, T.J., 2011. Soil nitrous oxide emissions in corn following three decades of tillage and rotation treatments. Soil Sci. Soc. Am. J. 75, 152–163.

Omonode, R.A., Vyn, T.J., Smith, D.R., Hegymegi, P., Gal, A., 2007. Soil carbon dioxide and methane fluxes from long-term tillage systems in continuous corn and corn-soybean rotations. Soil Tillage Res. 95, 182–195.

Parkin, T.B., 2008. Effect of sampling frequency on estimates of cumulative nitrous oxide emissions. J. Environ. Qual. 37, 1390–1395.

Parkin, T.B., Hatfield, J.L., 2010. Influence of nitrapyrin on N_2O losses from soil receiving fall-applied anhydrous ammonia. Agric. Ecosyst. Environ. 136, 81–86.

Parkin, T.B., Kaspar, T.C., 2003. Temperature controls on diurnal carbon dioxide flux: implications for estimating soil carbon loss. Soil Sci. Soc. Am. J. 67, 1763–1772.

Parkin, T.B., Kaspar, T.C., 2004. Temporal variability of soil carbon dioxide flux: effect of sampling frequency on cumulative carbon loss estimation. Soil Sci. Soc. Am. J. 68, 1234–1241.

Parkin, T.B., Kaspar, T.C., 2006. Nitrous oxide emissions from corn-soybean systems in the Midwest. J. Environ. Qual. 35, 1496–1506.

Parkin, T.B., Tiedje, J.M., 1984. Application of a soil core method to investigate the effect of oxygen concentration on denitrification. Soil Biol. Biochem. 16, 331–334.

Parkin, T.B., Venterea, R.T., 2010. Chamber-based trace gas flux measurements. Beltsville, MD. Available at. In: Follett, R.F. (Ed.), USDA-ARS GRACEnet Project Protocols, pp. 1–39. www.ars.usda.gov/research/GRACEnet (verified September 27, 2011).

Philippot, L., Hallin, S., Schloter, M., 2007. Ecology of denitrifying prokaryotes in agricultural soil. Adv. Agron. 96, 249–305.

Phillips, R.L., 2007. Organic agriculture and nitrous oxide emissions at sub-zero soil temperatures. J. Environ. Qual. 36, 23–30.

Poth, M., Focht, D.D., 1985. ^{15}N kinetic analysis of N_2O production by *Nitrosomonas europaea*: an examination of nitrifier denitrification. Appl. Environ. Microbiol. 49, 1134–1141.

Prieme, A., Christensen, S., Dobbie, K.E., Smith, K.A., 1997. Slow increase in rate of methane oxidation in soils with time following land use change from arable agriculture to woodland. Soil Biol. Biochem. 29, 1269–1273.

Rasmussen, R.A., Khalil, M.A.K., 1986. Atmospheric trace gases: trends and distributions over the last decade. Science 232, 1623–1624.

Roberts, T.L., 2007. Right product, right rate, right time and right place...the foundation of best management practices for fertilizer. In: Fertilizer Best Management Practices. General Principles, Strategy for their Adoption, and Voluntary Initiatives vs. Regulations. Proc.. of IFA International Workshop. Brussels, Belgium. International Fertilizer Industry Association, Paris, France, pp. 7–9. March 2007.

Robertson, G.P., Grace, P.R., 2004. Greenhouse gas fluxes in tropical and temperate agriculture: the need for a full-cost accounting of global warming potentials. Environ. Develop. Sustain. 6, 51–63.

Rochette, P., 2008. No-till only increases N_2O emissions in poorly-aerated soils. Soil Tillage Res. 101, 97–100.

Rochette, P., 2011. Towards a standard non-steady-state chamber methodology for measuring soil N_2O emissions. Animal Feed Sci. Technol. 166–167, 141–146.

Rochette, P., Angers, D.A., Chantigny, M.H., Gagnon, B., Bertrand, N., 2008. N_2O fluxes in soils of contrasting textures fertilized with liquid and solid dairy cattle manures. Can. J. Soil Sci. 88, 175–187.

Rochette, P., Angers, D.A., Belanger, G., Cantigny, M.H., Prevost, D., Levesque, G., 2004. Emissions of N_2O from alfalfa and soybean crops in eastern Canada. Soil Sci. Soc. Am. J. 68, 493–506.

Rochette, P., Eriksen-Hamel, N.S., 2008. Chamber measurements of soil nitrous oxide flux: are absolute values reliable? Soil Sci. Soc. Am. J. 72, 331–342.

Rosecrance, R.C., McCarty, G.W., Shelton, D.R., Teasdale, J.R., 2000. Denitrification and N mineralization from hairy vetch (*Vicia villosa* Roth) and rye (*Secale cereal* L.) cover crop monocultures and bicultures. Plant Soil. 227, 283–290.

Sexstone, A.J., Parkin, T.B., Tiedje, J.M., 1985. Temporal response of soil denitrification rates to rainfall and irrigation. Soil Sci. Soc. Am. J. 49, 99–103.

Sey, B.K., Manceur, A.M., Whalen, J.K., Gregorich, E.G., Rochette, P., 2008. Small-scale heterogeneity in carbon dioxide, nitrous oxide and methane production from aggregates of a cultivated sandy-loam soil. Soil Biol. Biochem. 40, 2468–2473.

Sey, B.K., Manceur, A.M., Whalen, J.K., Gregorich, E.G., Rochette, P., 2010. Root-derived respiration and nitrous oxide production as affected by crop phenology and nitrogen fertilization. Plant Soil. 326, 369–379.

Shipley, P.R., Meisinger, J.J., Decker, A.M., 1992. Conserving residual corn fertilizer nitrogen with winter cover crops. Agron. J. 84, 869–876.

Singurindy, O., Molodovskaya, M., Richards, B.K., Steenhuis, T.S., 2009. Nitrous oxide emission at low temperatures from manure-amended soils under corn (*Zea mays* L.). Agric. Ecosyst. Environ. 132, 74–81.

Sistani, K.R., Warren, J.G., Lovanh, N., Higgins, S., Shearer, S., 2010. Greenhouse gas emissions from swine effluent applied to soil by different methods. Soil Sci. Soc. Am. J. 74, 429–435.

Six, J., Ogle, S.M., Breidt, F.J., Conant, R.T., Mosier, A.R., Paustian, K., 2004. The potential to mitigate global warming with no-tillage management is only realized when practiced in the long term. Global Change Biol. 10, 155–160.

Skinner, R.H., Wagner-Riddle, C., 2012. Micrometeorological methods for assessing greenhouse gas flux. In: Liebig, M.A., Franzluebbers, A.J., Follett, R.F. (Eds.), Managing Agricultural Greenhouse Gases: Coordinated Agricultural Research through GRACEnet to Address our Changing Climate. Academic Press, San Diego, CA.

Smith, D.R., Hernandez-Ramirez, G., Armstrong, S.D., Bucholtz, D.L., Stott, D.E., 2011. Fertilizer and tillage management impacts of non-carbon-dioxide greenhouse gas emissions. Soil Sci. Soc. Am. J. 75, 1070–1082.

Smith, K.E., Runion, G.B., Prior, S.A., Rogers, H.H., Torbert, H.A., 2010. Effects of elevated CO_2 and agricultural management on flux of greenhouse gases from soil. Soil Sci. 175, 349–356.

Snyder, C.S., Bruulsema, T.W., Jensen, T.L., Fixen, P.E., 2009. Review of greenhouse gas emissions from crop production systems and fertilizer management effects. Agric. Ecosyst. Environ. 133, 247–266.

Spargo, J.T., Cavigelli, M.A., Mirsky, S.B., Maul, J.E., Meisinger, J.J., 2011. Mineralizable soil nitrogen and labile soil organic matter in diverse long-term cropping systems. Nutr. Cycl. Agroecosyst. 90, 253–266.

Sperow, M., Eve, M., Paustian, K., 2003. Potential soil C sequestration on U.S. agricultural soils. Climatic Change. 57, 319–339.

Stehfest, E., Bouwman, L., 2006. N_2O and NO emission from agricultural fields and soils under natural vegetation: summarizing available measurement data and modeling of global annual emissions. Nutr. Cycl. Agroecosyst. 74, 207–228.

Steinbeiss, S., Gleixner, G., Antonietti, M., 2009. Effect of biochar amendment on soil carbon balance and soil microbial activity. Soil Biol. Biochem. 6, 1301–1310.

Steudler, P.A., Bowden, R.D., Melillo, J.M., Aber, J.D., 1989. Influence of nitrogen fertilisation on methane uptake in temperate forest soils. Nature 341, 314–315.

Suzuki, I.S., Kwok, S.C., Dular, U., 1976. Competitive inhibition of ammonia oxidation in *Nitrosomonas europaea* by methane, carbon monoxide of methanol. FEBS Lett. 72, 117–120.

Thompson, R.B., Meisinger, J.J., 2002. Management factors affecting ammonia volatilization from land-applied cattle slurry in the Mid-Atlantic USA. J. Environ. Qual. 31, 1329–1338.

Tiedje, J.M., 1988. Ecology of denitrification and dissimilatory nitrate reduction to ammonium. In: Zehnder, J.B. (Ed.), Biology of Anaerobic Microorganisms. John Wiley & Sons, Inc., Hoboken, NJ, pp. 179–244.

Topp, E., Pattey, E., 1997. Soils as sources and sinks for atmospheric methane. Can. J. Soil Sci. 77, 167–178.

U.SDepartment of Agriculture. National Agricultural Statistics Service. 2011. Quick Stats. http://quickstats.nass.usda.gov/ (verified July 5, 2011).

Ussiri, D.A.N., Lal, R., Jarecki, M.K., 2009. Nitrous oxide and methane emissions from long-term tillage under a continuous corn cropping system in Ohio. Soil Tillage Res. 104, 247–255.

U.S. Environmental Protection Agency, 2011. Inventory of U.S. greenhouse gas emissions and sinks: 1990–2009.

Venterea, R.T., 2007. Nitrite-driven nitrous oxide production under aerobic soil conditions: kinetics and biochemical controls. Global Change Biol. 13, 1789–1809.

Venterea, R.T., Burger, M., Spokas, K.A., 2005. Nitrogen oxide and methane emissions under varying tillage and fertilizer management. J. Environ. Qual. 34, 1467–1477.

Venterea, R.T., Dolan, M.S., Ochsner, T.E., 2010. Urea decreases nitrous oxide emissions compared with anhydrous ammonia in a Minnesota corn cropping system. Soil Sci. Soc. Am. J. 74, 407–418.

Venterea, R.T., Parkin, T.B., 2012. Quantifying biases in non-steady state chamber measurements of soil-atmosphere gas exchange. In: Liebig, M.A., Franzluebbers, A.J., Follett, R.F. (Eds.), Managing Agricultural Greenhouse Gases: Coordinated Agricultural Research through GRACEnet to Address our Changing Climate. Academic Press, San Diego, CA.

Venterea, R.T., Spokas, K.A., Baker, J.M., 2009. Accuracy and precision analysis of chamber-based nitrous oxide gas flux estimates. Soil Sci. Soc. Am. J. 73, 1087–1093.

Verma, S.B., Dobermann, A., Cassman, K.G., Walters, D.T., Knops, J.M., Arkebauer, T.J., Suyker, A.E., Burba, G.G., Amos, B., Yang, H., Ginting, D., Hubbard, K.G., Gitelson, A.A., Walter-Shea, E.A., 2005. Annual carbon dioxide exchange in irrigated and rainfed maize-based agroecosystems. Agric. Forest Meteorol. 131, 77–96.

Wagner-Riddle, C., Furon, A., McLaughlin, N.L., Lee, I., Barbeau, J., Jayasundara, S., Parkin, G., von Bertoldi, P., Warland, J., 2007. Intensive measurement of nitrous oxide emissions from a corn-soybean-wheat rotation under two contrasting management systems over 5 years. Global Change Biol. 13, 1722–1736.

Wagner-Riddle, C., Hu, Q.C., van Bochove, E., Jayasundara, S., 2008. Linking nitrous oxide flux during spring thaw to nitrate denitrification in the soil profile. Soil Sci. Soc. Am. J. 72, 908–916.

Wang, F.L., Bettany, J.R., 1997. Methane emission from Canadian prairie and forest soils under short term flooding conditions. Nutr. Cycl. Agroecosyst. 49, 197–202.

Ward, B.B., 1987. Kinetic studies on ammonia and methane oxidation by *Nitrosococcus oceanus*. Arch. Microbiol. 147, 126–133.

Warren, J.G., Sistani, K.R., Way, T.R., Mays, D.A., Pote, D.H., 2008. A new method of poultry litter application to perennial pasture: subsurface banding. Soil Sci. Soc. Am. J. 72, 1831–1837.

Willison, T.W., Webster, C.P., Goulding, K.W.T., Powlson, D.S., 1995. Methane oxidation in temperate soils: effects of land use and the chemical form of nitrogen fertilizer. Chemosphere 30, 539–546.

Wulf, S., Maeting, M., Clemens, J., 2002. Application technique and slurry co-fermentation effects on ammonia, nitrous oxide, and methane emissions after spreading: II. Greenhouse gas emissions. J. Environ. Qual. 31, 1795–1801.

Yanai, Y., Toyota, K., Okazaki, M., 2007. Effects of charcoal addition on N_2O emissions from soil resulting from rewetting air-dried soil in short-term laboratory experiments. Soil Sci. Plant Nutr. 53, 181–188.

Yavitt, J.B., Fahey, T.J., Simmons, J.A., 1995. Methane and carbon dioxide dynamics in a northern hardwood ecosystem. Soil Sci. Soc. Am. J. 59, 796–804.

Zebarth, B.J., Rochette, P., Burton, D.L., 2008. N_2O emissions from spring barley production as influenced by fertilizer nitrogen rate. Can. J. Soil Sci. 88, 197–205.

CHAPTER 10

Management to Reduce Greenhouse Gas Emissions in Western U.S. Croplands

Ardell D. Halvorson[1], Kerri L. Steenwerth[2], Emma C. Suddick[3], Mark A. Liebig[4], Jeffery L. Smith[5], Kevin F. Bronson[6], Harold P. Collins[7]
[1]USDA-ARS, Fort Collins, CO
[2]USDA-ARS, RMI North, Davis, CA
[3]University of California, Dept. of Plant Sciences, Davis, CA
[4]USDA-ARS, Northern Great Plains Research Laboratory, Mandan, ND
[5]USDA-ARS, Washington State University, Pullman, WA
[6]USDA-ARS, U.S. Arid-Land Agricultural Research Center, Maricopa, AZ
[7]USDA-ARS, Vegetable and Forage Crops Research Unit, Prosser, WA

CHAPTER OUTLINE

NB: Mention of trade names or commercial products in this publication is solely for the purpose of providing specific information and does not imply recommendation or endorsement by the U.S. Department of Agriculture. USDA is an equal opportunity provider and employer.

Abbreviations: CC, continuous corn; CSb, corn–soybean rotation; CT, conventional tillage practices; CO_2, carbon dioxide; CH_4, methane; FI, furrow irrigation; MT, minimum-tillage; NT,

Managing Agricultural Greenhouse Gases. DOI: 10.1016/B978-0-12-386897-8.00010-3
2012, Published by Elsevier Inc.

no-tillage; N_2O, nitrous oxide; SD, surface drip irrigation; SSD, subsurface drip irrigation; SOC, soil organic carbon; WFPS, water-filled pore space.

INTRODUCTION

Rationale and Approach

Agriculture contributes about 6% of the total U.S. greenhouse gas (GHG) emissions in the form of carbon dioxide (CO_2), nitrous oxide (N_2O) and methane (CH_4), and approximately 67% of the total U.S. N_2O emissions (USEPA, 2010). Since N_2O has a global warming potential (GWP) approximately 298 times greater than CO_2 (Solomon et al., 2007), development of management practices to reduce N_2O emissions from cropland plays an important role in mitigating GHG emissions. Methane production in agricultural croplands is relatively low with most of the CH_4 emissions coming from animal agriculture, manure management, and rice production (Greenhouse Gas Working Group, 2010). The GWP of CH_4 is approximately 25 times greater than that of CO_2 (Solomon et al., 2007).

Among the western states, California is the second largest emitter of GHG emissions in the U.S., and tenth largest in the world, with ~500 million metric tons (MMT) of $CO_{2equiv.}$ emitted annually (Bemis and Allen, 2005), 8% of which is associated with agricultural activities. In the western U.S., crop production is highly vulnerable to climate change as it impacts water supply (Barnett et al., 2004; Sheppard, et al., 2002; Hatfield et al., 2011).

The geographic area identified as the western U.S. (Figure 10.1) includes all croplands west of the eastern edge of the U.S. Great Plains to the Pacific Ocean as delineated by Franzluebbers et al. (2005). Research conducted in the eastern portions of North and South Dakota, Nebraska, Kansas, central-eastern Oklahoma, and central-eastern Texas with precipitation levels greater than about 500 mm are excluded. Liebig et al. (2005) and Martens et al. (2005) previously presented a comprehensive summary of soil organic carbon (SOC) dynamics and CO_2, CH_4, and N_2O flux from croplands in the western U.S. prior to 2005. Additionally, the Greenhouse Gas Working Group (2010) conducted an assessment of agriculture's role in GHG emissions and capture for the U.S. The report concluded that there was a major need for additional research to develop sound management strategies to mitigate GHG emissions from agricultural cropping systems. Outcomes from such research were necessary to extrapolate site-specific GHG data to regional and national scales using processed-based models. The purpose of this chapter is to provide an update of management practices that could be used in western U.S. croplands to mitigate GHG emissions, with emphasis on GHG studies reported since 2005.

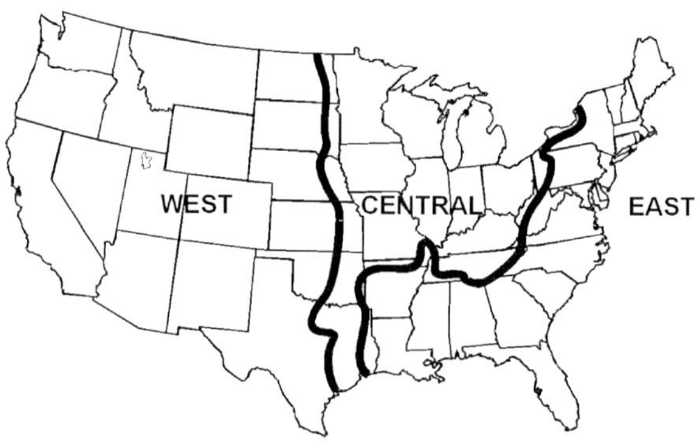

FIGURE 10.1
Map showing the approximate regional delineation for the western U.S. used in this chapter.

CHARACTERIZATION OF CLIMATE, CROPS, AND MANAGEMENT PRACTICES

Climate variation in the western U.S. is significant, with mean annual temperature ranging from 4 to 16°C in the Great Plains states and 6 to 49°C in the Pacific area, and mean annual precipitation ranging from <150 to about 500 mm (Greenhouse Gas Working Group, 2010). Crops in the western U.S. are produced primarily in areas with arid and semiarid climates; however, California includes a Mediterranean climate region with extensive crop production including vegetables, orchards, and vineyards. Cropland areas to the east and west of the Rocky Mountains are located in a semiarid steppe climate zone; a mid-latitude desert climate throughout most of Nevada, southern Utah, southern California, and west-central Arizona; a Mediterranean (dry in summer, wet in winter) climate in California and the Palouse region of eastern Oregon and Washington, and a marine west-coast climate in western Oregon and Washington (Bailey, 1995). Crops grown in the western U.S. are diverse, and include small grains, corn (*Zea mays* L.), cotton (*Gossypium hirsutum* L.), rice (*Oryza sativa* L.), oilseeds, legumes, vegetables, nuts, and fruits. Accordingly, cropping systems within this geographic area reflect this diversity. Dryland (rainfed) crop production systems are dominant in the semiarid Great Plains states, with considerable irrigation in the central and southern Great Plains, southwest states, California, and Pacific Northwest (PNW) states. Estimated total harvested cropland in 2007 in the western U.S. was 57 million ha (NASS, 2007). Harvested cropland in irrigated agriculture was small (15 million ha, 27%) compared to the dryland cropping area (42 million ha, 73%), but crops produced by irrigated agriculture account for a high proportion of the economic returns from crop production (about 4.5 times more than non-irrigated crops nationally) because of their high economic value (Hoffman and Evans, 2007). More than 95% of the harvested cropland in AZ, CA, and NV are irrigated, >64% in ID, NM, UT, and WY, >36% in CO, NE, OR, and WA, and <25% in KS, MT, ND, OK, SD, and TX (NASS, 2007). In the western U.S., about 58% of the irrigation systems are sprinkler systems, 35% gravity/furrow systems, and 6.8% drip, trickle, or low-flow micro-sprinkler systems (NASS, 2007).

Tillage systems used in the western U.S. for crop production are highly variable due to the diversity of crops. Plow tillage and tillage to form seedbeds are common practices in furrow-irrigated areas, with adoption of no-till (NT) production systems increasing slowly. The Great Plains states have had a significant increase in reduced and NT production practices with an increase in cropping intensity from the traditional crop–fallow system of farming. Fertilizers are essential inputs for economic crop production in the western U.S. Organic fertilizer inputs, such as animal manures, are used on croplands located close to dairies and animal feeding operations. The use of green manure crops and cover crops to improve nitrogen (N) use efficiency is becoming a more common practice. California agricultural management practices are intensive with the majority of cropping practices using conventional tillage (CT) operations, high inputs of inorganic N fertilizers, and furrow irrigation (Suddick et al., 2010). Additional information on crop production systems in the western U.S. may be found in Liebig et al. (2005) and Martens et al. (2005).

MANAGEMENT EFFECTS ON GREENHOUSE GAS (CO_2, CH_4, N_2O) FLUX

CO_2 Flux

DRYLAND

Carbon dioxide flux from soil (efflux) integrates root respiration and microbial decomposition of SOC. In dryland cropping systems, management interventions of tillage, cropping intensity, and crop type significantly influence CO_2 efflux (Martens et al., 2005; Liebig et al., 2005). Subsequent investigations have found CT increases CO_2 efflux compared to NT by 8 to 61%

within paired cropping systems due to greater physical release of CO_2 under CT and slower residue decomposition under NT (Jabro et al., 2008; Sainju et al., 2008, 2010). Within NT cropping systems, CO_2 efflux was found not to be affected by cropping intensity (Frank et al., 2006), fallow type (Liebig et al., 2010), or inclusion of pea (*Pisum sativum* L.) in rotation (Sainju et al., 2010). In a different study, however, Sainju et al. (2008) found growing season CO_2 efflux greater under rye (*Secale cereal* L.) (80 kg C ha^{-1} d^{-1}) compared to pea (55 kg C ha^{-1} d^{-1}), and greater under cropping (68 kg C ha^{-1} d^{-1}) compared to fallow (38 kg C ha^{-1} d^{-1}) when evaluated for a single growing season under NT management without N fertilizer. A single growing season contrast of irrigated and dryland barley (*Hordeum vulgare* L. and *Hordeum distichon* L.) treatments under NT showed CO_2 efflux to be 10% greater under irrigation (55 kg C ha^{-1} d^{-1}) than dryland (50 kg C ha^{-1} d^{-1}); a response caused by increased soil water status under irrigated conditions (Jabro et al., 2008). Evaluations in North Dakota and Montana over five site-years found N fertilization to have a negligible effect on CO_2 efflux under dryland cropping ($r = -0.26$; $P = 0.54$) (Phillips et al., 2009; Sainju et al., 2010, 2008).

IRRIGATED

Mosier et al. (2006) reported a slight reduction in growing season CO_2 flux on an irrigated Fort Collins clay loam soil in northern Colorado by converting from CT to NT, continuous corn (CC) production system. Averaged over N rates (4) and years (3), growing season (183–186 d) CO_2 flux for CT averaged 16.0 kg C ha^{-1} d^{-1} and NT averaged 16.2 kg C ha^{-1} d^{-1}. During the corn phase of a NT corn–soybean (*Glycine max* L.) (CSb) rotation, growing season CO_2 flux averaged 16.6 kg C ha^{-1} d^{-1}. Rate of N application had no effect on measured growing season CO_2 flux. During two non-crop periods (176–208 d), CO_2 flux averaged 5.4, 3.3, and 3.2 kg C ha^{-1} d^{-1} for the CT–CC, NT–CC, and NT–CSb rotations, respectively. Alluvione et al. (2009) reported 14% lower growing season CO_2 flux from an irrigated NT–CC system (13.5 kg C ha^{-1} d^{-1}) than from a CT–CC system (15.6 kg C ha^{-1} d^{-1}) when averaged over N rates and 2 years. They reported no effect of N rate on CO_2 flux in any of the cropping systems evaluated. Halvorson et al. (2010a) also reported higher growing season CO_2 flux from CT–CC (14.1 kg C ha^{-1} d^{-1}) than from NT–CC (12.7 kg C ha^{-1} d^{-1}) when averaged over N rates and 2 years. Nitrogen fertilization had no effect on CO_2 flux in this study. During the corn phase of two NT rotations in 2008, fertilizer N source treatments had no effect on CO_2 flux in a NT corn–dry bean (*Phaseolus vulgaris* L.) rotation (average 11.3 kg C ha^{-1} d^{-1}) and in a NT corn–barley rotation (11.7 kg C ha^{-1} d^{-1}). Halvorson et al. (2010b) found no significant differences in growing season CO_2 flux among six N sources in an irrigated, NT continuous corn system, with an average 2-year CO_2 flux of 11.5 kg C ha^{-1} d^{-1}.

Steenwerth and Belina (2008a) observed increased CO_2 flux following a tillage event in CT cover cropped vineyard soils (annual fluxes of 4.2 kg C ha^{-1} d^{-1} for CT fallow; 7.3 and 8.0 kg C ha^{-1} d^{-1} for minimum till (MT) rye and Trios 102 (*Triticale x triosecale*) cover crops, respectively, in Greenfield, CA. Burger et al. (2005) observed increased CO_2 flux following furrow irrigation of a Yolo silt loam soil in both a conventional [tomato (*Lycopersicon esculentum* Mill.)–wheat (*Triticum aestivum* L.) with fertilizer] and organic (tomato–corn–winter legume with compost) systems, with growing season emissions ranging from 19.5 to 25.7 kg C ha^{-1} d^{-1} and 14.3 to 21.4 kg C ha^{-1} d^{-1} for sprinkler irrigated organic and conventional systems, respectively. At the same site, post-irrigation growing season CO_2 flux for organic tomato systems ranged from 12 to 36 kg C ha^{-1} d^{-1} for tomato crops with a winter fallow rotation and 18–54 kg C ha^{-1} d^{-1} with a winter cover crop (Barrios-Masias et al., 2010). Although CO_2 flux and other GHGs are responsive to soil rewetting, few studies have focused on mitigating CO_2 flux through modified irrigation practices. Kallenbach et al. (2010) compared CO_2 emission from furrow (FI) vs. subsurface drip (SSD) irrigation in a tomato cropping system over 1 year. In the tomato growing season, mean CO_2 flux from subsurface drip (14 kg C ha^{-1} d^{-1}) and furrow (16 kg C ha^{-1} d^{-1}) irrigation were similar. In the furrow irrigated treatment over this same period, CO_2 flux was greater from soils that had 3 preceding years of a leguminous cover crop (23 and 20 kg C ha^{-1} d^{-1} for FI and SSD irrigation, respectively) than fallowed soils (i.e. no cover crop), suggesting that the cover crop derived C and N inputs and the cover crop's low C:N ratio contributed to these greater emissions. During the rainy season, FI and SSD had similar CO_2 flux without cover crop (6 kg C ha^{-1} d^{-1}), and with a cover crop, FI had a greater CO_2 flux than SSD (10 and 6 kg C ha^{-1} d^{-1}, respectively).

Carbon dioxide flux can vary spatially, with scale, and with tillage type. Lee et al. (2009) compared temporal and spatial variation of mean seasonal CO_2 flux across a tomato seedbed. Seasonal CO_2 flux ranged from 4.6 to 46.9 and 4.8 to 52.4 kg C ha^{-1} d^{-1} for CT and MT treatments near Davis, CA, respectively, with significant variation in CO_2 flux across furrow, crop row, and side dress seedbed positions (CT: 19.6 bed; 15.3 crop row; 33.4 kg C ha^{-1} d^{-1} side dress, and MT: 18.6 bed, 16.6 crop row; 21.5 kg C ha^{-1} d^{-1} side dress). Large pulses of CO_2 following irrigation and planting suggested seasonal variation in CO_2 coincided with patterns of soil rewetting and crop growth through autotrophic root respiration and heterotrophic respiration from rhizodeposition and root turnover. Although CO_2 efflux differed between tillage type and specific crop, seasonal patterns in CO_2 emission at the field scale were not affected by tillage type.

Steenwerth et al. (2010) observed changes to annual CO_2 emission over 2 years from a wine grape (*Vitis vinifera* L.) vineyard (Napa, CA) where a cover crop was mowed (CC + mow), tilled (CC + till), or resident vegetation was tilled (RV + till). Year 1 CO_2 flux for the CC + till (24.7 kg C ha^{-1} d^{-1}) and RV + till (23.0 kg C ha^{-1} d^{-1}) were lower compared to the second year (27.7 and 30.1 kg C ha^{-1} d^{-1} for CC + till and RV + till, respectively). However, the CC + mow (26.2 kg C ha^{-1} d^{-1}) treatment decreased slightly in the second year (23.5 kg C ha^{-1} d^{-1}). Differences between years were attributed to greater soil moisture content following changes in precipitation. Carlisle et al. (2006) investigated the effect of the conversion of oak woodland to a vineyard on soil CO_2 respiration. Higher seasonal and annual rates of soil CO_2 respiration from oak woodland (43.2 kg C ha^{-1} d^{-1}; 15.67 ± 1.44 Mg C ha^{-1} yr^{-1}) were observed compared to vineyard (19.2 kg C ha^{-1} d^{-1}; 7.02 ± 0.58 Mg C ha^{-1} yr^{-1}).

High CO_2 fluxes in the Pacific Northwest (PNW) occur under irrigation where the soil is moist and warm and fertilization ranges from 200 to 300 kg N ha^{-1}. In Washington, Halie-Mariam et al. (2008) found total CO_2 flux from sweet corn and potato (*Solanum tuberosum* L.) integrated over two cropping seasons averaged 23.5 kg C ha^{-1} d^{-1} and 17.9 kg C ha^{-1} d^{-1}, respectively. Cumulative CO_2 effluxes from sweet corn and potato fields were 17 and 13 times higher, respectively, than adjacent non-irrigated, native shrub steppe vegetation.

A summary of CO_2 emissions during extended periods of measurement are presented in Table 10.1 for many of the studies discussed above. Table 10.1 serves as a quick reference to daily CO_2 fluxes associated with many of the diverse cropping systems in the western U.S.

CH_4 Flux

DRYLAND

Dryland cropping systems in arid and semiarid regions of the U.S. have been identified as net sinks for atmospheric CH_4 (Franzluebbers et al., 2006). Though CH_4 uptake (i.e. negative CH_4 flux) rates are generally small under dryland cropping systems (<10 g C ha^{-1} d^{-1}), previous research has found CH_4 flux to be affected by cropping patterns, tillage intensity, and N fertilization (Martens et al., 2005; Liebig et al., 2005). Mid-May application of N fertilizer to corn was found to increase CH_4 uptake by 29% relative to N fertilizer applied in early April (Phillips et al., 2009). An investigation of CH_4 uptake under chemical and green fallow found no difference in CH_4 flux between fallow treatments over a 19-month fallow period (Liebig

TABLE 10.1 Summary of Carbon Dioxide (CO_2) Daily Flux from Many Crop Management Practices in Western U.S. Crop Production Systems

Location	Management practice	Measurement period	CO_2 flux (kg C ha^{-1} d^{-1})	Reference
Dryland (non-irrigated) Cropland				
Sidney, MT	Rye vs. pea cropping systems Crop period vs. fallow period	GS[†]	80 vs. 55 68 vs. 38	Sainju et al., 2010
Sidney, MT	Dryland vs. irrigated barley	GS	55 vs. 50	Jabro et al., 2008
Irrigated Cropland				
Fort Collins, CO	CT–CC vs. NT–CC NT corn–soybean (CSb)	GS GS	16.0 vs. 16.2 16.6	Mosier et al., 2006
Fort Collins, CO	CT–CC, NT–CC, NT–CSb	Non-crop period	5.4, 3.3, 3.2	Mosier et al., 2006
Fort Collins, CO	NT–CC vs. CT–CC	GS	13.5 vs. 15.6	Alluvione et al., 2009
Fort Collins, CO	NT–CC vs. CT–CC NT corn–barley (corn phase) NT corn–dry bean (corn phase)	GS GS GS	12.7 vs. 14.1 11.3 11.7	Halvorson et al., 2010a
Greenfield, CA	CT fallow vineyard MT rye cover crop vineyard MT Trios cover crop	Annual	4.2 7.3 8.0	Steenwerth and Belina, 2008a
Davis, CA	Conventional tomato–wheat Organic tomato–wheat	GS	19.5–25.7 14.3–21.4	Burger et al., 2005
Davis, CA	Organic tomatoes with fallow Org. tomatoes with cover crop	GS	12–36 18–54	Barrios-Masias et al., 2010
Davis, CA	Furrow irrigated tomato subsurface drip tomato	GS	16 14	Kallenbach et al., 2010
Davis, CA	Furrow and drip w/o cover crop Furrow irrig. with cover crop Subsurface drip irrig. with cover crop	Rainy season	6 10 6	Kallenbach et al., 2010
Davis, CA	Tomato seedbed location: CT–bed CT–crop row CT–side dress MT–bed MT–crop row MT–side dress	GS	19.6 15.3 33.4 18.6 16.6 21.5	Lee et al., 2009
Napa Valley, CA	Vineyard management: Cover crop mowed Cover crop tilled Resident vegetation tilled	Annual	23.5–26.2 24.7–27.7 23.0–30.1	Steenwerth et al., 2010
Napa Valley, CA	Conversion of oak woodland to vineyard	Annual	43.2 vs. 19.2	Carlisle et al., 2006
Patterson, WA	Sweet corn Potato	GS	23.5 17.9	Halie-Mariam et al., 2008

[†]*GS, growing season.*

et al., 2010). Among the three GHGs summarized in this review, CH_4 flux within dryland cropping systems is the least studied.

IRRIGATED

Mosier et al. (2006) found no significant differences between NT–CC and CT–CC irrigated production systems and N rate treatments in growing season (185 d) CH_4 fluxes with a range

TABLE 10.2 Summary of Methane (CH_4) Daily Flux from Several Crop Management Practices in Western U.S. Crop Production Systems

Location	Management practice	Measurement period	CH_4 flux ($g\ C\ ha^{-1}\ d^{-1}$)	Reference
Irrigated cropland				
Fort Collins, CO	Range: NT–CC and CT–CC Avg. tillage and cropping systems	GS[†] Non-crop	−0.5 to +0.5 −0.3	Mosier et al., 2006
Fort Collins, CO	NT–CC CT–CC	GS	0.12 0.01	Alluvione et al., 2009
Fort Collins, CO	Avg. NT and CT cropping systems	GS	0.10 to 0.13	Halvorson et al., 2010a, b
Patterson, WA	Sweet corn production Potato production	GS	−1.7 −2.3	Halie-Mariam et al., 2008

[†]*GS, growing season.*

of 0.48 to −0.48 $g\,C\,ha^{-1}\,d^{-1}$ from an irrigated Fort Collins clay loam soil in northern Colorado. During the non-crop period (192 d), average CH_4 flux was −0.33 $g\,C\,ha^{-1}\,d^{-1}$ for all tillage and cropping systems. Alluvione et al. (2009) reported slightly higher CH_4 flux from NT–CC (0.12 $g\,C\,ha^{-1}\,d^{-1}$) than from CT–CC (0.01 $g\,C\,ha^{-1}\,d^{-1}$) on a clay loam soil. Nitrogen fertilizer rate had no effect on CH_4 flux. Halvorson et al. (2010a, b) found no influence of tillage system or N fertility treatment on growing season (160 d) CH_4 flux from CT and NT cropping systems, with average fluxes of 0.10 to 0.13 $g\,C\,ha^{-1}\,d^{-1}$. In Washington under irrigated cropping, Halie-Mariam et al. (2008) reported CH_4 uptake from sweet corn and potato plots at rates of −1.7 and −2.3 $g\,CH_4\text{-}C\text{-}ha^{-1}\,d^{-1}$, respectively.

Methane emissions derived from the anaerobic decomposition of organic material by methanogenic bacteria in agricultural systems and cultural practices such as the cultivation of rice fields are relatively small, representing less than 2% of the total CH_4 emissions from California (Bemis and Allen, 2005). Management of rice in California generally includes the combination of both flooding and straw amendments which enhance annual CH_4 emissions (Fitzgerald et al., 2000; Martens et al., 2005). Using a combination of micrometeorological and chamber methods, McMillan et al. (2007) found CH_4 emissions to increase during vegetative rice growth and decrease over ripening and reproductive phases, which have been attributed to ebullition and allocation of C to rooting structures and rhizodeposition during the initial growth phases. Over the course of the study, emitted CH_4 was 65.7 to 77.7 $kg\,CH_4\text{-}C\,ha^{-1}$ (219–259 $g\,CO_{2equiv.}\,m^{-2}$). Concurrent assessments of CO_2 flux indicated 26 to 31% of the radiative impact of emitted CH_4 was offset by C sequestration in the rice paddy, highlighting the importance of measuring net system GHG flux for accurate assessment of mitigation potential.

A summary of CH_4 flux during extended periods of measurement is presented in Table 10.2 for several of the studies discussed above. Table 10.2 serves as a quick reference to daily CH_4 flux associated with some of the diverse cropping systems in the western U.S.

N_2O Flux

DRYLAND

Previous research has found N_2O flux from dryland cropping systems to rarely exceed 5 $g\,N\,ha^{-1}\,d^{-1}$ (Liebig et al., 2005). Recent research confirms this finding, underscoring that dryland cropping systems are a minor source of N_2O. Emission of N_2O from dryland corn was 21% lower when fertilized in mid-May as compared to early April, indicating the importance of fertilization timing on soil GHG emissions (Phillips et al., 2009). Method of N application also affects N_2O emission, with the lowest flux for broadcasted N (2.8 g N ha^{-1} d^{-1}), intermediate for banded N (5.0 g N ha^{-1} d^{-1}), and greatest for nested N (6.1 g N ha^{-1} d^{-1}) (Engel et al., 2010). In a 2-year tillage, cropping, and N fertilization study, N_2O emissions were concentrated following N fertilization and during freeze–thaw cycles in Montana and did not differ between CT and NT within a winter wheat–fallow cropping sequence (Dusenbury et al., 2008). Further, N_2O flux increased by 0.2 g N ha^{-1} d^{-1} with every 50 kg N ha^{-1} applied ($r = 0.82$), with the fraction of N fertilizer lost as N_2O never exceeding 0.45% in any treatment. No difference in N_2O emission was found between chemical- and green-fallow management systems in central North Dakota (Liebig et al., 2010).

IRRIGATED

In irrigated corn studies in semiarid northern Colorado (Liu et al., 2005; Mosier et al., 2006; Halvorson et al., 2008), linear increases in growing season N_2O emissions were found with increasing N fertilizer rates ranging from 0 to 224 kg N ha^{-1} over 5 years. Inclusion of soybean or dry bean in a corn-based NT rotation resulted in increased N_2O emissions during the corn phase of the rotation compared to the continuous corn system. When the growing season N_2O flux data (Mosier et al., 2006; Halvorson et al., 2008) were averaged over the 5 years by Halvorson et al. (2009), growing season emissions per kg N applied were greater from the CT–CC production system (6.6 g N ha^{-1}; intercept = 214 g N ha^{-1}) than from the NT–CC production system (5.7 g N ha^{-1}; intercept = 151 g N ha^{-1}). During the corn phase of the NT corn–bean rotations, emissions were 5.4 g N ha^{-1} (intercept = 517 g N ha^{-1}). Mosier et al. (2006) found N_2O emissions varied annually among three growing seasons, with no differences in growing season N_2O emissions between CT and NT systems in 2 years and greater emissions from CT than NT in 1 year. Averaged over the three growing seasons (185 d) and N rates (0–224 kg N/ha), N_2O fluxes were 5.97 g N ha^{-1} d^{-1} for CT–CC, 5.03 g N ha^{-1} d^{-1} for the NT–CC, and 5.01 g N ha^{-1} d^{-1} for the NT corn–soybean. Mean N_2O flux during the two non-crop periods (192 d) of the rotations were 0.67 g N ha^{-1} d^{-1} for CT–CC, 0.43 g N ha^{-1} d^{-1} for the NT–CC, and 0.36 g N ha^{-1} d^{-1} for the NT corn–soybean (Mosier et al., 2006). Halvorson et al. (2008) found a significant difference between growing season (172 d) N_2O fluxes when averaged over N rates and years between CT–CC (4.12 g N ha^{-1} d^{-1}) and NT–CC (3.56 g N ha^{-1} d^{-1}). They also reported the 2-year average growing season N_2O flux from NT corn–barley (156 d) was 2.66 g N ha^{-1} d^{-1}, and 4.79 g N ha^{-1} d^{-1} from NT corn–dry bean (159 d) rotations. The daily flux value for NT corn–dry bean reflects a higher N_2O flux from corn following dry bean, similar to the increased emissions reported by Mosier et al. (2006) for the NT corn–soybean rotation.

Liu et al. (2006) found that deep placement (10–15 cm soil depth) of urea ammonium-nitrate (UAN) fertilizer lowered N_2O emissions (50%) than placement of UAN on the soil surface or at a 5 cm soil depth in both CT and NT systems. Venterea and Stanenas (2008) also suggest placing N fertilizer below a 15 cm soil depth or biologically active zone in reduced tillage systems may reduce N_2O emissions.

The influence of fertilizer N source on N_2O emissions from cropping systems has received little attention. Polymer-coated, controlled release, stabilized, and slow-release N sources are available as alternatives to the commonly used N sources such as dry granular urea and ammonium nitrate, liquid UAN, and anhydrous ammonia, but generally with a higher cost. Halvorson et al. (2010a) reported higher growing season (160 d) N_2O flux from CT–CC (14.3 g N ha^{-1} d^{-1}) than from NT–CC (4.57 g N ha^{-1} d^{-1}), with no difference between dry granular urea (14.4 g N ha^{-1} d^{-1}) and polymer-coated urea (PCU) (14.1 g N ha^{-1} d^{-1})

fertilizers applied at 246 kg N ha^{-1} in CT–CC, but a significant reduction in growing season N_2O flux in NT–CC with PCU (3.08 g N ha^{-1} d^{-1}) compared with dry granular urea (6.05 g N ha^{-1} d^{-1}). They also reported a stabilized urea N source containing urease and nitrification inhibitors applied at 246 kg N ha^{-1} reduced growing season (161 d) N_2O flux (2.54 and 2.27 g N ha^{-1} d^{-1}) compared to dry granular urea (5.14 and 4.92 g N ha^{-1} d^{-1}) during the corn phase of irrigated NT corn–barley and corn–dry bean rotations, respectively.

Halvorson et al. (2010b) found significant reductions in growing season (160 d) N_2O emissions with the use of alternative N fertilizer sources compared to dry granular urea in irrigated NT–CC, including lower emissions with UAN than with urea. Daily N_2O fluxes averaged over two growing seasons with the application of 246 and 202 kg N/ha in 2007 and 2008, respectively, were 5.03, 2.09, 3.24, 2.09, 3.15, 1.61, and 0.91 g N ha^{-1} d^{-1} for urea, ESN®, Duration III®, SuperU®, UAN, UAN + AgrotainPlus®, and a check treatment (no N added), respectively. Duration III and ESN are polymer-coated urea products and SuperU and AgrotainPlus contain urease and nitrification inhibitors. Compared to urea, SuperU reduced growing season N_2O emissions 48%, ESN 34%, Duration III 31%, UAN 27%, and UAN + AgrotainPlus 53% when averaged over 2 years. Compared to UAN, UAN + AgrotainPlus reduced growing season N_2O emissions 35% and SuperU 29% (Halvorson et al., 2010b). Halie-Mariam et al. (2008) found N_2O losses accounted for 0.5% (0.55 kg N ha^{-1}) of the applied fertilizer (112 kg N ha^{-1}) in corn and 0.3% (0.59 kg N ha^{-1}) of the 224 kg N ha^{-1} fertilizer applied to potato in the PNW.

Yabaji et al. (2009) measured $N_2O + N_2$ emissions from drip-irrigated cotton with UAN fertigation on an Acuff sandy clay loam in Lubbock, TX, in 2005 and 2006. Emissions of $N_2O + N_2$ ranged from 491 g N_2O-N ha^{-1} with no N applied to 1745 g N_2O-N with 73 kg N ha^{-1} of UAN fertilizer injections, with no statistical differences between treatments ($P > 0.05$). Bronson et al. (2011) measured N_2O emissions from center-pivot-irrigated cotton on a Pullman clay loam in west Texas with one-time UAN applications at first-square of 0, 45, 90, and 134 kg N ha^{-1}. During a 34 d measurement period, N_2O fluxes were 139 and 913 g N_2O-N ha^{-1} for the 0 and 134 kg N ha^{-1} treatments, respectively (significantly different at $P < 0.05$). Nitrous oxide loss in this study was only 0.5% of applied N, which was below the IPCC (2006) default value of 1.0%.

Increased annual N_2O emissions from a cover cropped vineyard (rye, 694 g N ha^{-1} yr^{-1} and Trios, 565 g N ha^{-1} yr^{-1}) as compared to a non cover cropped vineyard (467 g N ha^{-1} yr^{-1}) have been documented (Steenwerth and Belina, 2008b). Garland et al. (2011) assessed direct N_2O emissions from the transition of CT to conservational tillage (MT) in a cover cropped vineyard. Annual emissions of N_2O were greater, but not significantly so in the MT system (0.19 kg N ha^{-1} in vine row and 0.11 kg N ha^{-1} in tractor row) compared to the CT system (0.13 kg N ha^{-1} in vine row and 0.07 kg N ha^{-1} in tractor row), where cover crop mowing and incorporation (1.6 g N ha^{-1} d^{-1} in CT and 7.7 g N ha^{-1} d^{-1} in MT, peak events) had the largest influence on emissions.

Kong et al. (2007) addressed the impact of conventional, low input, and organic management systems in a maize–tomato rotation on cumulative growing season N_2O flux and soil aggregate dynamics (labeled with ^{15}N). The form of fertilized N varied by cropping system [conventional: 60 kg N ha^{-1} as NPK, 220 kg N ha^{-1} as urea; organic: 373 kg N ha^{-1} composted manure; low input and organic: 100 kg N ha^{-1} of vetch (*Vicia villosa* Roth) and pea]. Peak flux following the first and second fertilization events was greater in the conventional system (1st: 8.3 g N ha^{-1} d^{-1}; 2nd: 14.5 g N ha^{-1} d^{-1}) compared to the organic (1st: 0 g N ha^{-1} d^{-1}; 2nd: 0.6 g N ha^{-1} d^{-1}) and low input systems (1st: 2.0 g N ha^{-1} d^{-1}; 2nd: 0 g N ha^{1} d^{-1}). Cumulative N_2O flux corresponded to the average mean residence time of the soil aggregate fractions in each cropping rotation, such that the shorter the mean residence time, the higher the N_2O emissions over the growing season.

Tillage and irrigation practices influence annual N_2O emissions, interacting with other cropping practices such as fertilization and cover crops to either enhance or diminish overall emission rates in annual crops (e.g. Kong et al., 2007, 2009; Lee et al. 2006, 2009; Kallenbach et al. 2010). Kong et al. (2009) studied the effect of tillage type (i.e. MT vs. CT) and cropping systems on soil stabilization of newly added N and N_2O flux. Minimum tillage was implemented during the year of study, but all treatments had been in CT for the previous 11 years. Recovery of fertilizer- or cover crop-derived N in the macro- and micro-aggregates was greater under MT than CT, such that under MT, macro- and micro-aggregates retained 56.0 and 56.6% more newly added N than under CT, respectively. In general, MT soils had 43% greater N_2O flux (28.9 g N_2O-N ha^{-1} d^{-1}) than the CT soils in May, whereas N_2O flux was highest from the MT-conventional system in June (39.5 g^{-1} N_2O-N ha^{-1} d^{-1}). In the low input and organic systems, MT had greater mean residence times than CT. In contrast, the conventional system had greater emissions from MT than CT, which was attributed to incorporation of new N into the unstable sand and silt fractions, and the lowest mean N residence time among all treatments.

Spatial and seasonal variation imposed by agricultural management also can affect N emission budgets while masking the effects of management. Lee et al. (2009) found tillage influenced N_2O emissions only in the side dress position in a tomato crop where fertilizer was added, such that MT had greater N_2O emissions than CT. However, this effect was not apparent at the field scale. In general, the mean N_2O flux from MT decreased from 8.53 kg N ha^{-1} yr^{-1} in maize (2004) to 3.84 kg N ha^{-1} yr^{-1} in sunflower (*Helianthus annuus* L.) (2005) to 2.03 kg N ha^{-1} yr^{-1} in chickpea (*Cicer arietinum* L.) (2006), whereas in CT, N_2O fluxes were 3.83, 2.91, and 1.28 kg N ha^{-1} yr^{-1} in these same crop varieties and years, respectively. This decrease was associated with differences in fertilizer N requirements for each crop. Season was also important, where 4 to 45% and 3 to 18% of total annual N_2O emission from CT and MT, respectively, were documented in the winter/fallow season, and greatest flux rates were observed after fertilization during the growing season (i.e. 161.2 g N_2O-N ha^{-1} d^{-1} in maize, 2004; 66.8 g N_2O-N ha^{-1} d^{-1} in sunflower).

Little research on mitigation options for irrigation practices has been performed, despite the clear link between N_2O flux and soil rewetting. In a 1-year tomato crop study, Kallenbach et al. (2010) compared the effect of furrow (FI) vs. subsurface drip (SSD) irrigation with or without cover crops (CC) on event-based N_2O emissions. Highest peak flux (763.6 g N ha^{-1} d^{-1}) was observed following the first precipitation event of the season within the FI + CC and lowest flux (0.07 g N ha^{-1} d^{-1}) during the summer from the SSD + CC. Subsurface drip decreased emissions by 50% compared to FI with or without a cover crop. The continuous, relatively constant soil moisture achieved with subsurface drip irrigation and the improved matching of applied N fertilizer with crop N demands led to its comparatively lower N_2O emissions. Suddick et al. (2011) conducted a similar short-term study in an almond (Terminalia c atappa L.) orchard investigating the impact of three irrigation systems on N_2O emissions [SSD, microjet sprinkler (MS) and surface drip (SD)], where SSD exhibited lower flux (11.6 g N ha^{-1} d^{-1}) compared to the other irrigation systems (18.7 g N ha^{-1} d^{-1} MS and 170.7 g N ha^{-1} d^{-1} SD). In a similar study in an almond orchard, Smart et al. (2011) found a reduction in N_2O emissions during a fertigation event with a microjet sprinkler irrigation system compared to a surface drip irrigation system. During a fall fertigation event (application of 34 kg N ha^{-1}), N_2O fluxes (25 d) were 3.32 g N ha^{-1} d^{-1} for SD and 1.96 g N ha^{-1} d^{-1} for MS irrigation. During a spring fertigation event, N_2O fluxes (31 d) were 6.16 g N ha^{-1} d^{-1} for SD and 3.94 g N ha^{-1} d^{-1} for MS irrigation. These data suggest converting from surface drip irrigation to microjet sprinkler or subsurface drip irrigation has potential to reduce N_2O emissions considerably. Furthermore, cumulative emissions following a 4-day fertilization event (50 kg ha^{-1} UAN, 32% N) in an almond orchard, also exhibited lower N_2O emissions within the SSD irrigation system compared to SD or MS systems (Suddick et al., 2011), thus corroborating Kallenbach et al. (2010). Steenwerth and Belina (2010) observed an increase in N_2O flux following fertigation within the vine row of a California vineyard over a 3-day event.

Fluxes were greater, however, in soils treated with herbicide to remove weeds compared to cultivated soils, where the cultivated soils exhibited greater SOM content and the presence of weeds may have provided a labile C source to facilitate a greater conversion of N_2O to N_2.

A summary of N_2O flux during extended periods of measurement are presented in Table 10.3 for many of the studies discussed above. Table 10.3 serves as a quick reference to daily N_2O flux associated with many of the diverse cropping systems in the western U.S.

MANURE APPLICATIONS

Few field studies have been conducted to characterize GHG gas emissions from soils amended with dairy manure in the western U.S. Collins et al. (2011) reported seasonal CO_2 flux patterns were similar among treatments incorporating manure compared to synthetic fertilizers showing a steady increase during the growing season as a silage crop matured. Mean seasonal uptake of CH_4 was -0.6 g CH_4-C ha^{-1} d^{-1}, except when manure was applied. Methane flux for 2 days after manure applications was ~58-fold higher than the average CH_4 uptake of synthetic fertilizer treatments. Methane flux for 2 days following application of digested effluent and liquid dairy manure averaged 33 g CH_4-C ha^{-1} d^{-1}, and was attributed to the release of dissolved CH_4-C in liquid manure source material and not from soil. They also reported that the N_2O emitted represented 0.03 to 0.12% of the total N applied.

SYNTHESIS

Published literature from GHG emissions studies in the western U.S. since 2005 is still quite limited. Data available to develop management strategies to mitigate GHG emissions from croplands are limited to a few local or regional areas for a given production system. Development of management strategies to mitigate GHGs across the western U.S. as a whole would be difficult due to inherent variability of climate (temperature and precipitation), variability in annual GHG emissions, diverse cropping systems and management practices, and soil types as shown by the studies reported here (Tables 10.1–10.3).

Management practices to mitigate CO_2 emissions from croplands include conversion from CT to NT to enhance SOC sequestration and reduce CO_2 emissions as shown by research from the Great Plains region, but may not be effective in other regions. Management practices that sequester CO_2 in the soil could offset emissions of N_2O from N application. Managing manure applications to croplands at the right rate, form, and time would help reduce GHG emissions and increase SOC sequestration to offset N_2O and CH_4 emissions. Growing season CO_2 fluxes are generally not affected by N fertilization rate or source of inorganic N applied, nor are CH_4 fluxes. Non-growing season CO_2 fluxes tend to be greater in tilled cropping systems than NT systems. Based on the studies examined, there appears to be fewer CO_2 emissions from dryland cropping systems than from irrigated systems, with a reduction in CO_2 emissions by reducing water application in irrigated systems. In particular, furrow irrigation tends to have greater CO_2 emissions than sprinkler irrigation. Cover cropping tends to have greater CO_2 emissions than fallow systems without cover crops, but sequesters more SOC; therefore, there is a need to use a net C balance approach when evaluating these systems.

Methane fluxes from western U.S. croplands tend to be small compared to CO_2 and N_2O fluxes. Many studies show a slight soil uptake of CH_4 during the growing season, but net annual flux values are typically near zero. Managing irrigation water and straw management in rice cropping systems to sequester SOC can offset CH_4 emissions by 26 to 30%. Manure application can result in greater CH_4 flux than with inorganic N fertilizer use.

Using crop management practices that maintain a low level of NO_3-N in the surface 15 cm of soil appears to reduce N_2O emissions throughout the western U.S. Irrigation water applications should be managed in irrigated systems to avoid high levels of WFPS under irrigated crop production to reduce loss of N_2O through denitrification. Selecting alternative N sources, such

TABLE 10.3 Summary of Nitrous Oxide (N_2O) Daily Flux from Many Crop Management Practices in Western U.S. Crop Production Systems

Location	Management practice	Measurement period	N_2O flux (g N ha^{-1} d^{-1})	Reference
Dryland cropland				
Bozeman, MT	N placement: broadcast	GS[†]	2.8	Engel et al., 2010
	Band		5.0	
	Nested		6.1	
Bozeman, MT	Increase in emissions/50 kg N/ha applied in wheat–fallow	GS	0.2 with a max. of 0.48	Dusenbury et al., 2008
Irrigated cropland				
Fort Collins, CO	Cropping season: CT–CC	GS	5.97	Mosier et al., 2006
	NT–CC	Non-crop	5.03	
	NT corn–soybean	Non-crop	5.01	
	Non-crop period: CT–CC	Non-crop	0.67	
	NT–CC		0.43	
	NT corn–soybean		0.36	
Fort Collins, CO	Crop rotation: CT–CC	GS	4.12	Halvorson et al., 2008
	NT–CC		3.56	
	NT corn–barley		2.66	
	NT corn–dry bean		4.79	
Fort Collins, CO	Crop system vs. N source:	GS	14.4	Halvorson et al., 2010a
	CT–CC with granular urea		14.1	
	CT–CC polymer-coated urea		6.05	
	NT–CC with granular urea		3.08	
	NT–CC polymer-coated urea			
Fort Collins, CO	N source in NT–CC: Urea, ESN, Duration III, SuperU, UAN, UAN + AgrotainPlus, and control (no-N), respectively	GS	5.03, 2.89, 3.24, 2.09, 3.15, 1.64, 0.94	Halvorson et al., 2010b
Fort Collins, CO	CT–CC	GS	6.6[‡]	Halvorson et al., 2009; Mosier et al., 2006; Halvorson et al., 2008
	NT–CC		5.7[‡]	
	NT corn–bean rotation		5.4[‡]	
Patterson, WA	Sweet corn: CT with N fertilization	GS	15.4	Halie-Mariam et al., 2008
			5.4	
	CT without N fertilization		8.9	
	MT with N fertilization		4.2	
	MT without N fertilization		9.7	
	Potatoes: CT with N fertilization		4.8	
	CT without N fertilization		12.2	
	MT with N fertilization		5.0	
	MT without N fertilization			
Lubbock, TX	Cotton without N applied	34 days	4.1	Bronson et al., 2010
	Cotton with 134 kg N ha^{-1}		26.9	
Davis, CA	Non-cover crop vineyard	Annual	1.59	Steenwerth and Belina, 2008b
	Rye cover crop vineyard		2.31	
	Trios cover crop vineyard		1.94	
Davis, CA	CT cover crop vineyard	Annual	0.36	Garland et al., 2011
	Vine row		0.19	
	Tractor row		0.52	
	MT cover crop vineyard		0.30	
	Vine row			
	Tractor row			

TABLE 10.3 Summary of Nitrous Oxide (N_2O) Daily Flux from Many Crop Management Practices in Western U.S. Crop Production Systems—continued

Location	Management practice	Measurement period	N_2O flux (g N ha^{-1} d^{-1})	Reference
Davis, CA	N fertilized corn, MT vs. CT	Annual	23.4 vs. 10.5	Lee et al., 2009
	N fertilized sunflower, MT vs.		10.5 vs. 8.0	
	CT		5.6 vs. 3.5	
	N fertilized chickpea, MT vs. CT			
Davis, CA	N fertilized almond orchard,	GS	11.6	Suddick et al.,
	Subsurface drip irrigation		18.7	2011
	Microjet sprinkler irrigation		170.7	
	Surface drip irrigation			
Davis, CA	Fall fertigation almonds,	GS (25 d)	3.32	Smart et al.,
	Surface drip irrigation	GS (34 d)	1.96	2011
	Microjet sprinkler irrigation		6.16	
	Spring fertigation almonds		3.94	
	Surface drip irrigation			
	Microjet irrigation			

†*GS, growing season.*
‡*Five-year average increase in growing season N_2O-N emission per kg N applied (g N_2O-N ha^{-1} kg^{-1} N applied).*

as enhanced-efficiency N fertilizers, has potential to reduce N_2O emissions in western cropping systems. Managing N fertilization (source, rate, time, and placement) to reduce residual soil NO_3-N in the surface 15 cm of soil and improving NUE by reducing N fertilization rates should reduce N_2O emissions. Converting from surface drip irrigation system to microjet sprinkler or subsurface drip irrigation systems has potential to reduce N_2O emissions during fertigation of vineyards and orchards.

Modeling of GHG emissions resulting from crop management practices may be one way to develop and test potential mitigation management practices. For example, the DAYCENT model was calibrated and validated to evaluate mitigation options for annual cropping systems [beans, corn, cotton, safflower (*Carthamus tinctorius* L.), sunflower, tomato and wheat] in California's Central Valley (De Gryze et al., 2010). The potential for farming systems to reduce net soil GHG emissions relative to conventional practices was greatest under manure application (4577 kg CO_2-eq ha^{-1} yr^{-1}), followed by cover cropping (three cover cropped systems: 1072 ± 272 , 2201 ± 82, 2499 ± 47 kg CO_2-eq ha^{-1} yr^{-1}), and conservation tillage (two conservation tillage systems: 336 ± 47, 550 ± 123 kg CO_2-eq ha^{-1} yr^{-1}). Model analysis showed that these net reductions in CO_2-eq ha^{-1} yr^{-1} were due largely to long-term increases in SOC and less so due to reductions in N_2O emissions, underscoring the importance of using farming practices that stabilize SOC and evaluating net changes in soil C and GHG emissions to calculate GWP of farming systems.

GAPS IN GREENHOUSE GAS MITIGATION KNOWLEDGE

The western U.S. has a large amount of land in agricultural production. The area is geologically, climatically, and agriculturally diverse. Such diversity in growing conditions contributes to the need for GHG information by region and cropping system. As pointed out by the Greenhouse Gas Working Group (2010), GHG flux data from cropping systems is very limited and needs to be monitored before GHG mitigation strategies can be developed. This review of the western U.S. confirms the lack of GHG data for developing mitigation practices and improving GHG models. The vastly different topographies, climates, and cropping systems and paucity of published GHG studies make it extremely difficult to estimate regional GHG fluxes. There is a need for more extensive GHG research under all types of crop production systems, soils, and climates. More studies are needed to further understand and predict the effects of tillage, crop rotations, N management, irrigation management, and dryland practices on GHG flux. New research is under way in many western states, but the data are not yet available or published.

Nitrous oxide flux from agricultural land is mainly a function of fertilizer inputs and environmental conditions. Fertigation is probably the most efficient way to apply N fertilizer for center pivots and drip irrigation systems in the western U.S. (Yabaji et al., 2009); however, little N_2O emission data exist for this practice. Flood/furrow irrigation is common for crops like corn, wheat, and alfalfa (*Medicago sativa* L.), but CH_4 and N_2O emission data are lacking for these systems as well.

Not only is GHG information limited from the major cropping systems, but a significant gap in knowledge is evident for the expanding organic agriculture sector which uses many organic products to fertilize crops. Analyzing organic farming systems in relation to GHG emissions is needed, especially in light of some evidence showing that organic systems may reduce GHG emissions (De Gryze, et al., 2010). El-Hage Scialabba and Muller-Lindenlauf (2010) estimated a 20% reduction in N_2O emissions and a 40–72% increase in SOC sequestration by converting to organic systems.

Other aspects that would improve our understanding of GHG emissions and mitigation options include pairing life-cycle analysis (LCA) for cropping systems with soil GHG emissions or net ecosystem fluxes to better characterize the agricultural GHG footprint. An LCA conducted by Archer and Halvorson (2010) on the studies of Mosier et al. (2006) and Halvorson et al. (2008) showed that the NT–CC system with adequate N fertilization reduced GWP more than the NT corn–bean rotations, but had a lower economic return than the NT corn–bean rotations, with the CT–CC system having the highest GWP at optimum levels of N fertilization. This analysis showed the potential of the NT systems to reduce GWP compared to CT systems with similar economic returns. Life-cycle analysis as shown above takes into account the sustainability of a proposed GWP mitigation practice, and provides guidance to help decision and policymakers in establishing sustainable protocols to mitigate GWP.

Acknowledgments

We thank Robert Dungan, Jane Johnson, and Richard Farrell for their helpful comments to improve a previous version of the manuscript. This publication is based on work supported by the USDA Agricultural Research Service GRACEnet Project.

References

Alluvione, F., Halvorson, A.D., Del Grosso, S.J., 2009. Nitrogen, tillage, and crop rotation effects on carbon dioxide and methane fluxes from irrigated cropping systems. J. Environ. Qual. 38, 2023–2033.

Archer, D.W., Halvorson, A.D., 2010. Greenhouse gas mitigation economics for irrigated cropping systems in northeastern Colorado. Soil Sci. Soc. Am. J. 74 (2), 446–452.

Bailey, R.G., 1995. Description of the ecoregions of the United States, second ed.). Forest Service. Misc. Publ. 1391, USDA, p. 108.

Barnett, R., Malone, W., Pennell, D., Stammer, B., Semtner Washington, W., 2004. The effects of climate change on water resources in the West: introduction and overview. Climatic Change 62, 1–11.

Barrios-Masias, F.H., Cantwell, M.I., Jackson, L.E., 2010. Cultivar mixtures of processing tomato in an organic agroecosystem. Organic Agriculture 1 (1), 17–30.

Bemis, G., Allen, J., 2005. Inventory of California Greenhouse Gas Emissions and Sinks: 1990 to 2002 Update. California Energy Commission (CEC) Publication CEC-600-2005-025.

Bronson, K.F., 2011. Nitrous oxide emissions from irrigated cotton in Texas and Arizona. 2011 Annual Meetings. [CD-ROM computer file] Amer. Soc. Agron, Madison, WI.

Burger, M., Jackson, L.E., Lundquist, E.J., Louie, D.T., Miller, R.L., Rolston, D.E., Scow, K.M., 2005. Microbial responses and nitrous oxide emissions during wetting and drying of organically and conventionally managed soil under tomatoes. Biol. Fert. Soils 42, 109–118.

Carlisle, E.A., Steenwerth, K.L., Smart, D.R., 2006. Effects of land use on soil respiration: conversion of oak woodlands to vineyards. J. Environ. Qual. 35, 1396–1404.

Collins, H.P., Streubel, J.D., Alva, A.K., Frear, C., Chen, S., Fransen, S.F., Kruger, C., Granatstein, D., 2011. Greenhouse gas emissions from an irrigated silt loam soil amended with anaerobic digested dairy manure. Soil Sci. Soc. Amer. J. 75, 2206–2216.

De Gryze, S., Wolf, A., Kaffka, S.R., Mitchell, J., Rolston, D.E., Temple, S.R., Lee, J., Six, J., 2010. Simulating greenhouse gas budgets of four California cropping systems under conventional and alternative management. Ecolog. Applic. 20, 1805–1819.

Dusenbury, M.P., Engel, R.E., Miller, P.R., Lemke, R.L., Wallander, R., 2008. Nitrous oxide emission from a northern Great Plains soil as influenced by nitrogen management and cropping systems. J. Environ. Qual. 37, 542–550.

El-Hage Scialabba, N., Muller-Lindenlauf, M., 2010. Organic agriculture and climate change. Renew. Agric. Food Syst. 25, 158–169.

Engel, R., Liang, D.L., Wallander, R., Bembenek, B., 2010. Influence of urea placement on nitrous oxide production from a silt loam soil. J. Environ. Qual. 39, 115–125.

Fitzgerald, G.J., Scow, K.M., Hill, J.E., 2000. Fallow season straw and water management effects on methane emissions in California rice. Global Biogeochem. Cycl. 14, 767–776.

Frank, A.B., Liebig, M.A., Tanaka, D.L., 2006. Management effects on soil CO_2 efflux in northern semiarid grassland and cropland. Soil Tillage Res. 89, 78–85.

Franzluebbers, A.J., Campbell, J.P., Follett, R.F., 2005. Greenhouse gas contributions and mitigation potential in agricultural regions of North America: introduction. Soil Tillage Res. 83, 1–8.

Franzluebbers, A.J., Follett, R.F., Johnson, J.M.F., Liebig, M.A., Gregorich, E.G., Parkin, T.B., Smith, J.L., Del Grosso, S.J., Jawson, M.D., Martens, D.A., 2006. Agricultural exhaust: a reason to invest in soil. J. Soil Water Conserv. 61 (3), 98–101.

Garland, G.M., Suddick, E.C., Burger, M., Horwath, W.R., Six, J., 2011. Direct N_2O emissions following the transition from conventional till to no-till in a cover cropped Mediterranean vineyard. Agric. Ecosyst. Environ. 141, 234–239.

Greenhouse Gas Working Group, 2010. Agriculture's role in greenhouse gas emissions & capture. Greenhouse Gas Working Group Rep. ASA, CSSA, and SSSA, Madison, WI.

Haile-Mariam, S., Collins, H.P., Higgins, S.S., 2008. Greenhouse gas fluxes from an irrigated sweet corn (Zea mays L.)-potato (Solanum tuberosum L.) rotation. J. Environ. Qual. 37, 759–771.

Halvorson, A.D., Del Grosso, S.J., Reule, C.A., 2008. Nitrogen, tillage, and crop rotation effects on nitrous oxide emissions from irrigated cropping systems. J. Environ. Qual. 37, 1337–1344.

Halvorson, A.D., Del Grosso, S.J., Alluvione, F., 2009. Nitrogen rate and source effects on nitrous oxide emissions from irrigated cropping systems in Colorado. Better Crops with Plant Food 93 (1), 16–18.

Halvorson, A.D., Del Grosso, S.J., Alluvione, F., 2010a. Tillage and inorganic nitrogen source effects on nitrous oxide emissions from irrigated cropping systems. Soil Sci. Soc. Am. J. 74 (2), 436–445.

Halvorson, A.D., Del Grosso, S.J., Alluvione, F., 2010b. Nitrogen source effects on nitrous oxide emissions from irrigated no-till corn. J. Environ. Qual. 39, 1554–1562.

Hatfield, J.L., Boote, K.J., Kimball, B.A., Ziska, L.H., Izaurralde, R.C., Ort, D., Thompson, A.M., Wolfe, D., 2011. Climate impacts on agriculture: implications for crop production. Agron. J. 103, 351–370.

Hoffman, G.J., Evans, R.G., 2007. Introduction. Chapter 1. In: Hoffman, G.J., Evans, R.G., Jensen, M.E., Martin, D.L., Elliot, R.L. (Eds.), Design and Operation of Farm Irrigation Systems. Amer. Soc. Agric. Biologic. Engineers, St. Joseph, MI, pp. 1–32.

IPCC., 2006. Guidelines for National Greenhouse Gas Inventories. Volume 4, Agriculture, Forestry and Other Land Use. In: Eggleston, H.S., Buendia, L., Miwa, K., Ngara, T., Tanabe, K. (Eds.), Guidelines for National Greenhouse Gas Inventories. Volume 4, Agriculture, Forestry and Other Land Use. IGES, Japan.

Jabro, J.D., Sainju, U., Stevens, W.B., Evans, R.G., 2008. Carbon dioxide flux as affected by tillage and irrigation in soil converted from perennial forages to annual crops. J. Environ. Manage. 88, 1478–1484.

Kallenbach, C.M., Rolston, D.E., Horwath, W.R., 2010. Cover cropping affects soil N_2O and CO_2 emissions differently depending on type of irrigation. Agric. Ecosyst. Environ. 137, 251–260.

Kong, A.Y.Y., Fonte, S.J., van Kessel, C., Six, J., 2007. Soil aggregates control N cycling efficiency in long-term conventional and alternative cropping systems. Nutr. Cycl. Agroecosyst. 79, 45–58.

Kong, A.Y.Y., Fonte, S.J., van Kessel, C., Six, J., 2009. Transitioning from standard to minimum tillage: trade-offs between soil organic matter stabilization, nitrous oxide emissions, and N availability in irrigated cropping systems. Soil Tillage Res. 104, 256–262.

Lee, J., Hopmans, J.W., van Kessel, C., King, A.P., Evatt, K.J., Louie, D., Rolston, D.E., Six, J., 2009. Tillage and seasonal emissions of CO_2, N_2O and NO across a seed bed and at the field scale in a Mediterranean climate. Agric. Ecosyst. Environ. 129, 378–390.

Lee, J., Six, J., King, A.P., van Kessel, C., Rolston, D.E., 2006. Tillage and field scale controls on greenhouse gas emissions. J. Environ. Qual. 35, 714–725.

Liebig, M.A., Tanaka, D.L., Gross, J.R., 2010. Fallow effects on soil carbon and greenhouse gas flux in central North Dakota. Soil Sci. Soc. Am. J. 74 (2), 358–365.

Liebig, M.A., Morgan, J.A., Reeder, J.D., Ellert, B.H., Gollany, H.T., Schuman, G.E., 2005. Greenhouse gas contributions and mitigation potential of agricultural practices in northwestern USA and western Canada. Soil Tillage Res. 83, 25–52.

Liu, L.J., Mosier, A.R., Halvorson, A.D., Zhang, F.S., 2005. Tillage and nitrogen application effects on nitrous and nitric oxide emissions from irrigated corn fields. Plant and Soil 276, 235–249.

Liu, L.J., Mosier, A.R., Halvorson, A.D., Zhang, F.S., 2006. The impact of nitrogen placement and tillage on NO, N_2O, CH_4 and CO_2 fluxes from a clay loam soil. Plant and Soil 280, 177–188.

Martens, D.A., Emmerich, W., McLain, J.E.T., Johnsen, T.N., 2005. Atmospheric carbon mitigation potential of agricultural management in the southwestern USA. Soil Tillage Res. 83, 95–119.

McMillan, A.M.S., Goulden, M.L., Tyler, S.C., 2007. Stoichiometry of CH_4 and CO_2 flux in a California rice paddy. J. Geophys. Res.-Biogeosci. 112.

Mosier, A.R., Halvorson, A.D., Reule, C.A., Liu, X.J., 2006. Net global warming potential and greenhouse gas intensity in irrigated cropping systems in northeastern Colorado. J. Environ. Qual. 35, 1584–1598.

NASS., 2007. Census of agriculture-State data. National Agricultural Statistics Service, Washington, DC.

Phillips, R.L., Tanaka, D.L., Archer, D.W., Hanson, J.D., 2009. Fertilizer application timing influences greenhouse gas fluxes over a growing season. J. Environ. Qual. 38, 1569–1579.

Sainju, U.M., Jabro, J.D., Stevens, W.B., 2008. Soil carbon dioxide emission and carbon content as affected by irrigation, tillage, cropping system, and nitrogen fertilization. J. Environ. Qual. 37, 98–106.

Sainju, U.M., Jabro, J.D., Caesar-TonThat, T., 2010. Tillage, cropping sequence, and nitrogen fertilization effects on dryland soil carbon dioxide emission and carbon content. J. Environ. Qual. 39, 935–945.

Sheppard, P.R., Comrie, A.C., Packin, G.D., Angersbach, K., Hughes, M.K., 2002. The climate of the US Southwest. Clim. Res. 21, 219–238.

Smart, D.R., Alsina, M.M., Wolff, M.W., Matiasek, M.G., Schellenberg, D.S., Edstrom, J.P., Brown, P.H., Scow, K.M., 2011. N_2O emissions and water management in California perennial crops. In: Guo, L., Gunasekara, A., McConnell, L. (Eds.), Understanding Greenhouse Gas Emissions for Agriculture Management. American Chemical Society Symposium Series. Chapter, 13. American Chemical Society, Washington, DC, pp. 227–255.

Solomon, S., Qin, D., Manning, M., Chen, Z., Marquis, M.M., Averyt, K.B., Tignor, M., Miller, H.L. (Eds.), 2007. Contribution of Working Group I to the Fourth Assessment Report of the Intergovernmental Panel on Climate Change. 2007 Cambridge University Press New York, NY. http://www.ipcc.ch/publications_and_data/ar4/wg1/en/contents.html (verified May 11, 2011).

Steenwerth, K., Belina, K.M., 2008a. Cover crops enhance soil organic matter, carbon dynamics and microbiological function in a vineyard agroecosystem. Applied Soil Ecol. 40, 359–369.

Steenwerth, K., Belina, K.M., 2008b. Cover crops and cultivation: impacts on soil N dynamics and microbiological function in a Mediterranean vineyard agroecosystem. Applied Soil Ecol. 40, 370–380.

Steenwerth, K.L., Belina, K.M., 2010. Vineyard weed management practices influence nitrate leaching and nitrous oxide emissions. Agric. Ecosyst. Environ. 138, 127–131.

Steenwerth, K.L., Pierce, D.L., Carlisle, E.A., Spencer, R.G., Smart, D.R., 2010. A vineyard agroecosystem: disturbance and precipitation affect soil respiration under Mediterranean conditions. Soil Sci. Soc. Amer. J. 74, 231–239.

Suddick, E., Scow, K.M., Horwath, W.R., Jackson, L.E., Smart, D.R., Mitchell, J.P., Six, J., 2010. The potential for California soils to sequester carbon and reduce greenhouse gas emissions: a holistic approach. Burlington: Academic Press, Advan. Agron 107, 123–162.

Suddick, E.C., Steenwerth, K., Garland, G.M., Smart, D.R., Six, J., 2011. Discerning agricultural management effects on nitrous oxide emissions from conventional and alternative cropping systems: a California case study. Chapter 12, 203–226. In L. Guo, A. Gunasekara, and L. McConnell (eds.), Understanding Greenhouse Gas Emissions for Agriculture Management. American Chemical Society Symposium Series; American Chemical Society, Washington, DC.

USEPA, 2010. Inventory of U.S. greenhouse gas emissions and sinks: 1990–2008. U.S. Environmental Protection Agency, 1200 Pennsylvania Ave. N.W. Washington, DC 20460. EPA 430-R-10-006. http://epa.gov/climatechange/emissions/usinventoryreport.html (verified May 11, 2011).

Venterea, R.T., Stanenas, A.J., 2008. Profile analysis and modeling of reduced tillage effects on soil nitrous oxide flux. J. Environ. Qual. 37, 1360–1367.

Yabaji, R., Nusz, J.W., Bronson, K.F., Malapati, A., Booker, J.D., Nichols, R.L., Thompson, T.L., 2009. Nitrogen management for subsurface drip irrigated cotton: ammonium thiosufalte, timing, and canopy reflectance. Soil Sci. Soc. Am. J. 73, 589–597.

CHAPTER 11

Greenhouse Gas Flux from Managed Grasslands in the U.S.

Mark A. Liebig[1], Xuejun Dong[2], Jean E. T. McLain[3], Curtis J. Dell[4]
[1]USDA-ARS, Northern Great Plains Research Laboratory, Mandan, ND
[2]North Dakota State University Central Grasslands Research Extension Center, Streeter, ND
[3]USDA-ARS U.S. Arid-Land Agricultural Research Center, Maricopa, AZ
[4]USDA-ARS Pasture Systems and Watershed Management Research Unit, University Park, PA

CHAPTER OUTLINE

NB: Mention of trade names or commercial products in this publication is solely for the purpose of providing specific information and does not imply recommendation or endorsement by the U.S. Department of Agriculture. USDA is an equal opportunity provider and employer.

Abbreviations: CO_2, carbon dioxide; CO_{2e}, carbon dioxide equivalent; GWP, global warming potential; GHG, greenhouse gas; MAP, mean annual precipitation; CH_4, methane; NEE, net ecosystem exchange; N_2O, nitrous oxide; PPR, prairie pothole region

INTRODUCTION

Grasslands provide a multitude of services with long-term societal benefits (Wedin and Fales, 2009). Grasslands serve as a source of forage for livestock, contributing over $40 billion to the U.S. economy in 2009 (USDA, 2011), and also provide significant traditional environmental services, such as soil and water conservation (Singer et al., 2009) and maintenance of wildlife

Managing Agricultural Greenhouse Gases. DOI: 10.1016/B978-0-12-386897-8.00011-5
2012, Published by Elsevier Inc.

habitat (Niemuth et al., 2007). Emerging services provided by grasslands include use for recreation and aesthetic purposes, energy generation, and mitigation of and/or adaptation to climate change (Sanderson et al., 2009). The multifunctionality of grasslands underscores their central role to affect agricultural sustainability and environmental quality, while highlighting the importance of proper management to balance the many services they provide.

Developing robust estimates of net global warming potential (GWP) for managed grasslands in the U.S. has become increasingly important as terrestrial ecosystems are looked upon to serve as cost-effective sinks for mitigating climate change (Fynn et al., 2010). Grasslands are particularly attractive in this role given their large spatial domain and documented capacity to sequester atmospheric carbon dioxide (CO_2) as soil organic carbon (SOC) (Derner and Schuman, 2007; Follett and Reed, 2010). However, assurance of effective greenhouse gas (GHG) mitigation strategies requires more than accurate estimates of C sequestration. Estimates of methane (CH_4) and nitrous oxide (N_2O) flux from production systems are essential to accurately ascertain management effects on net GWP. Management activities have known variable effects on SOC dynamics and CH_4 and N_2O flux, with activities often interacting to either enhance (e.g. increased SOC, decreased N_2O emission) or negate (e.g. increased SOC, increased N_2O emission) GHG mitigation potential (Eagle et al., 2010). Accordingly, inclusive GHG assessments of grasslands are needed to identify potential GHG trade-offs associated with management.

Understanding GHG mitigation potential of managed grasslands in the U.S. is essential to develop sound policy recommendations associated with this emerging ecosystem service. To date, recent published syntheses addressing management effects on GHG flux from U.S. grasslands are lacking. To address this need, we summarized grassland management effects on CO_2, CH_4, and N_2O flux using published data from rangelands and pasturelands throughout the U.S. In addition to the summary, off-site and/or indirect GHG emissions associated with livestock production are discussed, along with current limitations of the subject matter and pressing research needs. This chapter is intended to serve as an update to previous syntheses (Franzluebbers and Follett, 2005).

MANAGED GRASSLANDS IN THE U.S.: DESCRIPTION, EXTENT, AND USE

For the purposes of this chapter, discussion of managed grasslands will focus on rangelands and pasturelands, which are the dominant grassland types in the U.S. Rangelands are defined as land on which the climax or potential plant cover is composed principally of native grasses, grasslike plants, forbs or shrubs suitable for grazing and browsing, and introduced forage species that are managed as rangeland. Conversely, pasturelands represent land managed primarily for the production of introduced forage plants for livestock grazing, with management consisting of fertilization, weed control, irrigation, reseeding or renovation, and control of grazing (USDA, 2009).

Rangelands and pasturelands in the U.S. encompass a vast area, comprise many different ecosystems, and are exposed to a wide range of environmental conditions (Wedin and Fales, 2009; Fynn et al., 2010). Based on the 2007 National Resources Inventory (USDA, 2009), non-federal rangelands and pasturelands in the continental U.S. occupy 166 and 48 Mha, respectively, with publicly owned grasslands accounting for an additional 27 Mha (Follett and Reed, 2010). Rangeland and pastureland in Alaska, Hawaii, Puerto Rico, Guam, Virgin Islands, and the Northern Mariana Islands occupy approximately 0.8 Mha (U.S. Census Bureau, 2011). Collectively, rangelands and pasturelands account for 31% of total U.S. land area (USDA, 2009). Rangelands are found almost exclusively in arid and semiarid regions west of the 98th meridian, whereas pasturelands are located primarily in the humid eastern half of the U.S. (Figure 11.1). Major rangeland types include mixed-grass prairie and shortgrass steppe of the Great Plains, shrub steppe of the Great Basin, tropical/subtropical desert of Texas, New Mexico,

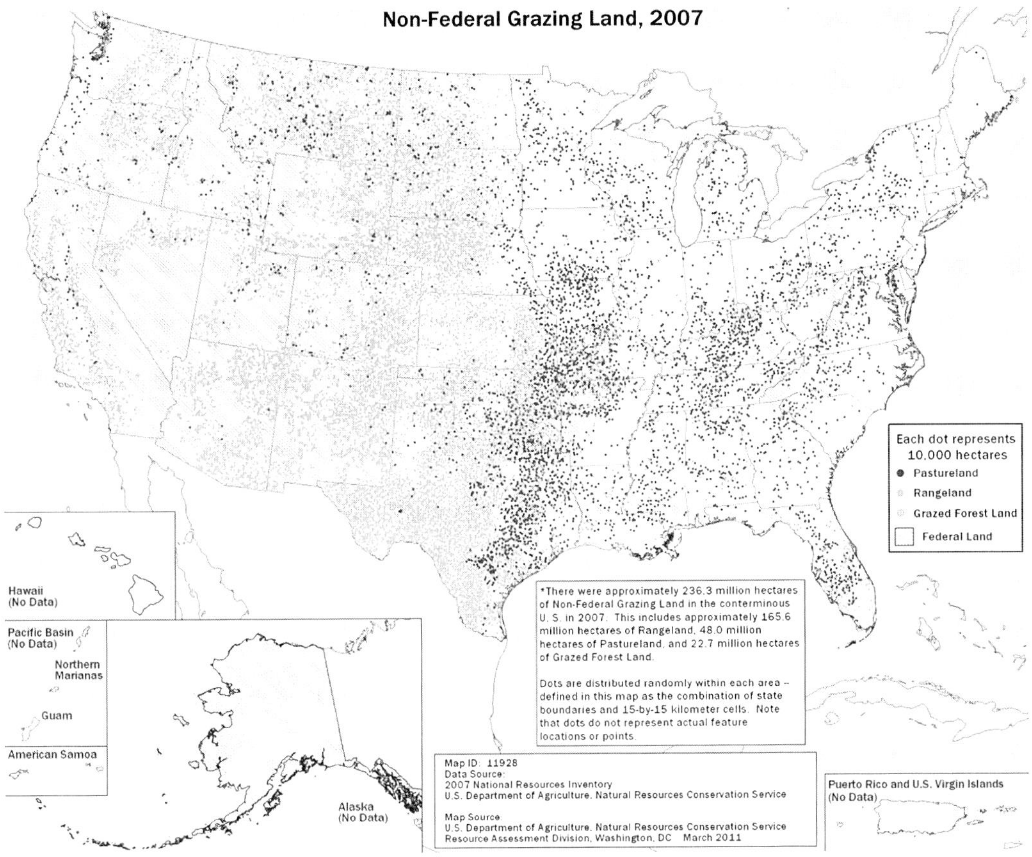

FIGURE 11.1
Distribution of private pastureland, rangeland, and grazed forest land in the 48 contiguous United States, 2007.
[adapted from USDA (2009)]. Please see color plate section at the back of the book.

Arizona, Nevada, and California, and the California annual grasslands (Martens et al., 2005; Liebig et al., 2005; Havstad et al., 2009). Composition of pasturelands throughout the U.S. varies considerably based on production system (beef or dairy production) and forage utilization method (permanent pasture, cropland pasture, or hay) (Sanderson et al., 2009). Given their more intensive management and presence of favorable environmental conditions, pasturelands typically have much higher biomass production per unit area than rangelands (Follett and Reed, 2010). For detailed descriptions of U.S. rangelands and pasturelands, readers are encouraged to review Wedin and Fales (2009).

Cattle and sheep are the predominant livestock that graze non-federal rangeland in the U.S. (U.S. Census Bureau, 2011). Total U.S. cattle inventory for 2010 was approximately 100 million head, while total U.S. inventory of sheep and lambs in 2010 was approximately six million head (USDA, 2011). Unfortunately, reliable estimates of cattle and sheep on U.S. rangeland and pastureland in any given year are difficult to ascertain, as are the number of livestock grazing cropland. In 2009, forage production (representing the sum of all dry hay and haylage/greenchop) in the top 18 forage producing states in the U.S. was approximately 90.5 Tg, with a mean yield of 1.0 ± 0.4 Mg ha^{-1} (USDA, 2011).

GREENHOUSE GAS FLUX FROM U.S. GRASSLANDS

Rangelands

Frequent drought stress, daily and seasonal temperature extremes, low organic matter content, and low nutrient reserves found in arid and semiarid rangelands traditionally led researchers to believe these ecosystems were not significant sinks or sources of trace gases. As a result, arid and semiarid regions were, until recently, overlooked in GHG inventories (Bowden, 1986; Potter et al., 1996). Recent work has revealed 5 to 40% of global soil atmospheric exchange of CO_2, CH_4, N_2O, and NO_x may occur in semiarid and arid regions, but for each of these gases there are fewer than a dozen studies from a limited number of locations to support the estimates (Galbally et al., 2008). Therefore, there is a poorly quantified understanding of the contribution of these regions to global trace gas cycles.

CO_2 EFFLUX AND NET ECOSYSTEM EXCHANGE

Evaluations of CO_2 flux from agroecosystems have generally occurred at two distinct spatial domains: point-scale and ecosystem level. Point-scale assessments quantify CO_2 efflux, typically using chamber-based methods, and thereby consider only microbial and plant root emissions. Conversely, ecosystem level assessments, which utilize micrometeorological methods, integrate both soil and plant CO_2 emissions as well as assimilation of CO_2 into plant biomass, resulting in estimates of net ecosystem exchange (NEE). Given the different processes involved, point-scale and ecosystem level assessments will be addressed separately.

Previous summaries of CO_2 efflux from U.S. rangelands have shown flux events to be strongly influenced by soil moisture and temperature, and positively affected by annual burning, lack of grazing, and elevated CO_2 (Liebig et al., 2005; Martens et al., 2005). Recently published assessments of CO_2 efflux from U.S. rangelands have evaluated responses to soil moisture, N addition, plant species, and woody encroachment (Table 11.1). In a single growing season under mixed-grass prairie, Chimner and Welker (2005) found CO_2 efflux lowest under imposed restrictions to moisture (0.5 × MAP) (18.5 kg CO_2-C ha^{-1} d^{-1}) and greatest under added snow (29.4 kg CO_2-C ha^{-1} d^{-1}). Positive responses to added moisture were also observed under subtropical savanna in Texas, where McCulley et al. (2007) found irrigation of grass and woody shrubs to increase CO_2 efflux from 48 to 131% compared to rainfed conditions. Added N in the form of cattle urine was found to increase CO_2 efflux by 46% relative to an unamended control in mixed-grass prairie (Liebig et al., 2008), whereas moderate additions of synthetic N (40 kg N ha^{-1} yr^{-1}) under tallgrass prairie in Minnesota had a negligible effect on CO_2 efflux (<1% increase) (Adair et al., 2009). In the same study by Adair et al. (2009), CO_2 efflux was greater under a 16 species treatment (22.3 kg CO_2-C ha^{-1} d^{-1}) compared to a single species planting (19.1 kg CO_2-C ha^{-1} d^{-1}), suggesting a positive relationship between plant diversity and microbial and root respiration. Hydropedologic zones in the prairie pothole region (PPR) of North Dakota, each differing in plant species, was found to strongly affect CO_2 efflux, with summertime emissions greatest from mixed-grass prairie (14.3 kg CO_2-C ha^{-1} d^{-1}), intermediate within a wheatgrass—needlegrass (*Thinopyrum* A. Löve—*Achnatherum* P. Beauv.) mixture (9.9 kg CO_2-C ha^{-1} d^{-1}), and least under hydrophytes (4.5 kg CO_2—C ha^{-1} d^{-1}) (Phillips and Beeri, 2008). Within a honey mesquite (*Prosopis glandulosa* Torr.) rangeland in Arizona, CO_2 efflux under N-fixing shrubs was similar to grass (12.8 vs. 12.4 kg CO_2-C ha^{-1} d^{-1}, respectively) (McLain et al., 2008). In contrast, CO_2 efflux under shrubs within subtropical savanna was 52% greater (40.6 kg CO_2-C ha^{-1} d^{-1}) than grass (26.7 kg CO_2-C ha^{-1} d^{-1}). Elevated GHG flux under shrubs have been attributed to their positive influence on soil moisture and plant nutrients, which contribute to islands of fertility relative to areas outside the shrub canopy (Archer et al., 2001).

Published summaries of NEE have found U.S. rangelands to serve as small net sinks for atmospheric CO_2, but with considerable annual variation depending on aboveground biomass production, dormant season CO_2 efflux, and presence/absence of calcite soils (Liebig et al., 2005; Martens et al., 2005). Recently published estimates of NEE for U.S. rangelands generally concur with previous summaries, with net CO_2 assimilation observed across studies for sagebrush and shortgrass steppe, and mixed-grass and tallgrass prairie (Figure 11.2). Carbon dioxide assimilation rates in these rangeland types vary from 0.2 to 1.1 Mg C ha^{-1} yr^{-1}, with considerable variation observed within mixed-grass and tallgrass

TABLE 11.1 Carbon Dioxide Efflux from U.S. Rangeland and Pasture (post-2004 publications)

Location	Ecosystem/ Pasture type	Study period	Precipitation during study (mm)	Management/Treatment	CO_2 efflux (kg CO_2-C ha^{-1} d^{-1})	Reference[†]
Rangeland						
Arizona	Mesquite	2003–2004 (Jul–Sep)	168 (2003); 104 (2004)	Shrub, grazed	12.8	1
				Grass/Forb mixture, grazed	12.4	
Wyoming	Mixed-grass	2003 (April–Sep)	213	Control	27.8	2
				0.5 × MAP[‡]	18.5	
				1.5 × MAP	25.2	
				Added snow	29.4	
North Dakota	Mixed-grass	2001–2003	822	Grazed (0.39 steer ha^{-1})	28.0	3
North Dakota	Mixed-grass	2005 (Jul–Aug)	117	Control	55.0	4
				Cattle urine (w/o tannin)	80.4	
				Cattle urine (w/ tannin)	69.6	
North Dakota	Mixed-grass	2003 (12 Jul, 3 Aug)		Monitoring	14.3	5
	Wheatgrass–needlegrass mix				9.9	
	Hydrophytes				4.5	
Minnesota	Tallgrass	1999–2006	NR[§]	1 species, no added N	19.0	6
				1 species, 4 g N m^{-2} yr^{-1}	19.1	
				16 species, no added N	22.2	
				16 species, 4 g N m^{-2} yr^{-1}	22.4	
Texas	Subtropical savanna	1996–1997	645	Grass, rainfed	21.5	7
				Grass, irrigated	31.9	
				Woody encroach., rainfed	24.5	
				Woody encroach., irrigated	56.7	
California	Coastal prairie	2004–2005	1250	Monitoring	28.2	8
Pasture						
Ohio	Reclaimed mineland, Grass/ legume pasture	2005–2006	1084	Grazed	30.7	9
				Hayed	16.7	

[†] 1, McLain et al. (2008); 2, Chimner and Welker (2005); 3, Frank et al. (2006); 4, Liebig et al. (2008); 5, Phillips and Beeri (2008); 6, Adair et al. (2009); 7. McCulley et al. (2007); 8, Sanderman and Amundson (2010); 9, Shrestha et al. (2009).
[‡] MAP, Mean annual precipitation.
[§] NR, Not reported.

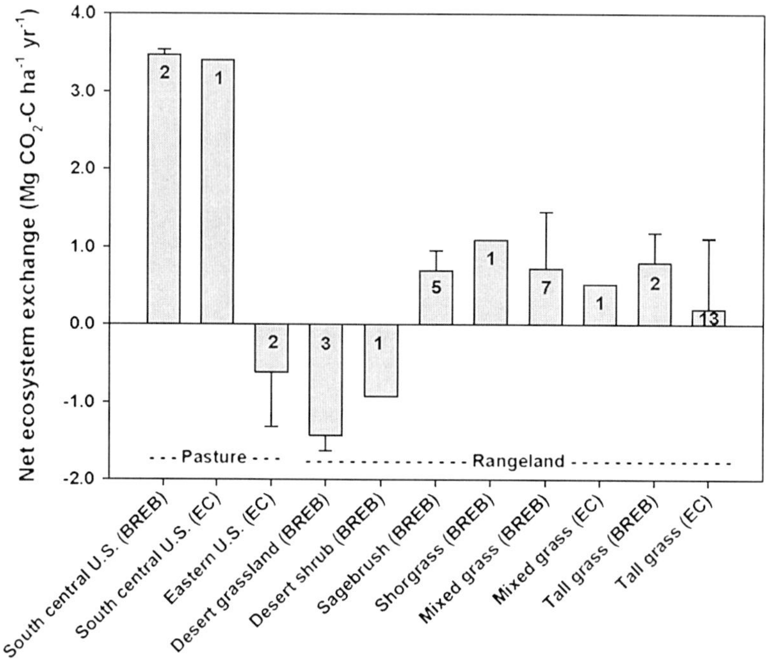

FIGURE 11.2
Net ecosystem exchange (NEE) from pastureland and rangeland in the U.S. by region or rangeland type using either Bowen Ratio Energy Balance (BREB) or Eddy Covariance (EC) micrometeorological methods. Values within each bar represent number of observations used to calculate mean. Where applicable, associated bars reflect ±1 standard error. Positive NEE implies net CO_2 uptake. Data summarized from Bremer and Ham (2010), Dugas et al. (1999), Emmerich (2003), Falge et al. (2003), Frank (2002), Frank and Karn (2005), Gilmanov et al. (2006), Gilmanov et al. (2003a), Gilmanov et al. (2003b), Kjelgaard et al. (2008), Novick et al. (2004), Owensby et al. (2006), Skinner (2008), Suyker and Verma (2001), Suyker et al. (2003), Svejcar et al. (2008), and Xu and Baldocchi (2004).

prairie. Rangeland types found to possess negative NEE included desert grassland and shrub, with net emissions ranging from -0.9 to -1.4 Mg C ha^{-1} yr^{-1} (Figure 11.2). Regardless of rangeland type, sensitivity to drought can have an overriding effect on NEE, with rangelands serving as net CO_2 sinks in non-drought years and sources in drought years (Gilmanov et al., 2006; Svejcar et al., 2008; Polley et al., 2010; Derner and Jin, 2012). Zhang et al. (2010) found Northern Great Plains rangelands to become net CO_2 sources when 65% of the area was drought affected. Accordingly, given the prevalence of drought in U.S. rangelands, caution should be exercised when evaluating outcomes from short-term, non-drought affected studies.

Despite the growing number of multi-year NEE estimates for major rangeland types in the U.S., a lack of evaluations exists for specific rangeland treatments. Grazing was found to have a negligible effect on CO_2 assimilation under tallgrass prairie (-0.1 to 0.1 Mg C ha^{-1} yr^{-1}) (Owensby et al., 2006). Also under tallgrass prairie, Bremer and Ham (2010) found annual and biennial burning to result in net CO_2 loss at rates of -1.95 and -0.98 Mg C ha^{-1} yr^{-1}, respectively. Woody encroachment of mixed-grass prairie in North Dakota was found to result in net assimilation of CO_2 (0.24 Mg C ha^{-1} yr^{-1}) (Frank and Karn, 2005), while Scott et al. (2006) observed net CO_2 loss with woody encroachment in Arizona shrubland. In a novel study by Leffler et al. (2007), presence of Mormon crickets (*Anabrus simplex*) decreased NEE up to 60% in Great Basin rangeland ecosystems. Desert grassland in Arizona invaded by Lehmann lovegrass (*Eragrostis lehmanniana* Nees) was found to assimilate CO_2 at a rate greater than native bush muhly (*Muhlenbergia porter* Scribn. ex Beal) (3.0 vs. 1.6 kg C ha^{-1} d^{-1}, respectively), an outcome attributed to lower ecosystem respiration from lovegrass (Hamerlynck et al., 2010).

CH_4 FLUX

Previous reviews of CH_4 flux in U.S. rangelands have noted a paucity of field research data (Liebig et al., 2005; Martens et al., 2005), resulting in only gross approximations of large-scale CH_4 cycling within rangeland ecosystems. Though upland soils in rangelands are generally regarded as net CH_4 sinks, uncertainty in estimating terrestrial CH_4 cycling is exacerbated through the competing activities of CH_4-oxidizing (methanotrophic) and CH_4-producing (methanogenic) microorganisms (McLain and Ahmann, 2008), which are highly sensitive to agricultural practices, including fertilization and trampling by rangeland animals (Smith et al., 2000). Moreover, activities of these microorganisms are strongly influenced by soil moisture, with reduced moisture generally increasing net CH_4 uptake due to increased gas diffusion rates and microbial access to oxygen and atmospheric CH_4 (Singh et al., 2010).

Within mixed-grass prairie of North Dakota, decreased cattle stocking rate had a negligible effect on CH_4 uptake, with rates ranging from -6.3 to -5.7 g CH_4-C ha^{-1} d^{-1} for stocking rates of 0.39 to 2.3 steer ha^{-1} yr^{-1}, respectively (Liebig et al., 2010) (Table 11.2). At the same site, Liebig et al. (2008) found CH_4 uptake within urine spots to be 26% lower compared to a control site without urine, an outcome likely caused by competitive binding of NH_4^+ to the CH_4-oxidizing enzyme produced by methanotrophs (Dunfield and Knowles, 1995). In grazed southern Arizona rangelands, McLain et al. (2008) reported CH_4 uptake under mesquite, but net CH_4 production under grassland and dead mesquite stumps. Methane uptake under mesquite was associated with reduced soil bulk density and increased soil moisture (McLain and Martens, 2006), while CH_4 production from grassland and woody detritus was likely caused by termite (*Reticulitermes tibialis* Banks) activity. Arizona rangeland under grass not affected by termite activity, however, can serve as a net consumer of atmospheric CH_4 at rates of -3.6 to -5.9 g CH_4-C ha^{-1} d^{-1} (McLain and Martens, 2006). Evaluations in Arizona identified strong seasonal trends in net CH_4 flux, with fluxes near zero during the hot, dry summer months, followed by consumption induced by monsoon summer rains from July through September, and continued consumption throughout the winter.

Within the PPR of the northern Plains, it is not uncommon for grasslands and wetlands to be juxtaposed within a landscape but utilized for similar purposes, such as grazing (Knutsen and Euliss, 2001). Differences in biogeochemical attributes between upland grasslands and wetlands in the PPR was apparent through evaluations by Phillips and Beeri (2008), where upland sites were net CH_4 sinks (-0.8 to -10.6 g CH_4-C ha^{-1} d^{-1}) but depositional sites dominated by hydrophytic vegetation were large CH_4 sources (1.96 kg CH_4-C ha^{-1} d^{-1}) (Table 11.2). Additional studies on trace gas exchange confirm the role of PPR wetlands to serve as seasonal sources of CH_4 (Gleason et al., 2009). Moreover, evaluations of growing season CH_4 flux from wetlands in California found grazing to increase CH_4 emission by over three-fold compared to wetlands not grazed (1.78 vs. 0.5 kg CH_4-C ha^{-1} d^{-1}, respectively) (Oates et al., 2008), an outcome possibly caused by C deposition from livestock manure coupled with a greater potential for anaerobic soil conditions due to trampling.

N_2O EMISSION

Grasslands play a major role at both national and continental scales for the N_2O total emission into the atmosphere (Jarvis, 1997). In 2008, N_2O emission from U.S. grasslands was estimated at 62 Tg CO_{2e} yr^{-1}, accounting for 14% of total N_2O emission from agricultural activities (U.S.-EPA, 2010). Nitrification and denitrification, generally believed to be the most important microbiological sources for N_2O production under aerobic and anaerobic soil conditions, respectively, are both strongly influenced by temperature, moisture, pH, and organic matter availability (Bouwman, 1990). The variability of these edaphic factors within grassland soils results in net N_2O fluxes that are highly variable in space and time and thus difficult to quantify at a large scale.

TABLE 11.2 Methane Flux From U.S. Rangeland and Pasture (post-2004 publications)

Location	Ecosystem/Pasture type	Study period	Precipitation during study (mm)	Management/Treatment	CH_4 flux (g CH_4-C ha^{-1} d^{-1})	Reference[†]
Rangeland						
Arizona	Mixed grasses, forbs	Jul 2002–Sep 2003	373	Monitoring	−5.9 (1.6)[‡]	1
	Mesquite		338		−4.6 (1.2)	
	Perennial sacaton grass		278		−3.6 (1.2)	
Arizona	Mesquite	2003–2004 (Jul–Sep)	168 (2003); 104 (2004)	Honey mesquite	−2.0 (0.6)	2
				Grasses, forbs	0.6 (0.6)	
				Dead mesquite stump	0.9 (0.9)	
North Dakota	Mixed-grass	2003–2006	443	Grazed (0.39 steer ha^{-1})	−6.3 (0.3)	3
	Mixed-grass			Grazed (1.1 steer ha^{-1})	−6.0 (0.3)	
	Crested wheatgrass			Grazed (1.1–2.3 steer ha^{-1}); 45 kg N ha^{-1} yr^{-1}	−5.7 (0.2)	
North Dakota	Mixed-grass	2005 (Jul –Aug)	117	Control	−7.2 (0.5)	4
				Cattle urine (w/o tannin)	−5.3 (0.2)	
				Cattle urine (w/ tannin)	−2.9 (1.0)	
North Dakota	Mixed-grass	2003 (12 Jul, 3 Aug)	0	Monitoring	−0.8	5
	Wheatgrass–needlegrass mix				−10.6	
	Hydrophytes				1960	
North Dakota	Hydrophytes, perennial grasses	2003 (May–Oct)	217	Monitoring	435 (290)	6
California	Hydrophytes	2000 (May–Nov), 2002 (May–Aug)	NR[§]	Grazed wetland	1776	7
				Ungrazed wetland	504	
Pasture						
Michigan	Grass lichen mix	1995–1996	837	Control, no added N	−3.7 (0.5)	8
				Added N (1 g N m^{-2} yr^{-1})	−3.9 (0.7)	
Ohio	Reclaimed mineland, Grass/ legume pasture	2005–2006	1084	Grazed	−3.3	9
				Hayed	−4.2	
Missouri	Bromegrass	2004–2006	NR	Ungrazed	−3.6	10

[†]*1, McLain and Martens (2006); 2, McLain et al. (2008); 3, Liebig et al. (2010); 4, Liebig et al. (2008); 5, Phillips and Beeri (2008); 6, Gleason et al. (2009); 7, Oates et al. (2008); 8, Ambus and Robertson (2006); 9, Shrestha et al. (2009); 10, Nkongolo et al. (2010).*

[‡]*Negative values imply net CH_4 uptake. Values in parentheses represent mean standard error.*

[§]*NR, Not reported.*

TABLE 11.3 Nitrous Oxide Flux from U.S. Rangeland and Pasture (post-2004 publications)

Location	Ecosystem/Pasture type	Study period	Precipitation during study (mm)	Management/Treatment	N_2O flux (g N_2O-N ha^{-1} d^{-1})	Reference[†]
Rangeland						
Arizona	Mixed grasses, forbs	2002 monsoon (Jul–Sep)	238	Monitoring	0.3 (0.7)[‡]	1
	Mesquite		238		3.2 (2.1)	
	Perennial sacaton grass		238		0.6 (0.8)	
	Mixed grasses, forbs	2003 monsoon (Jul–Sep)	95		0.9 (0.8)	
	Mesquite		61		1.3 (0.8)	
	Perennial sacaton grass		95		0.6 (0.5)	
Arizona	Mesquite	2003–2004 (Jul–Sep)	168 (2003); 104 (2004)	Honey mesquite	3.8 (09)	2
				Grasses, forbs	0.8 (0.3)	
				Dead mesquite stump	0.9 (0.2)	
Wyoming	Downy brome	2003 (9 d in Aug)	25.4	Monitoring	1.0 (0.4)	3
	Western wheatgrass				0.4 (0.7)	
North Dakota	Mixed-grass	2003–2006	443	Grazed (0.39 steer ha^{-1})	3.5 (0.3)	4
	Mixed-grass			Grazed (1.1 steer ha^{-1})	3.3 (0.3)	
	Crested wheatgrass			Grazed (1.1–2.3 steer ha^{-1}); 45 kg N ha^{-1} yr^{-1}	8.2 (0.7)	
North Dakota	Mixed-grass	2005 (Jul–Aug)	117	Control	4.6 (1.2)	5
				Cattle urine (w/o tannin)	4.6 (1.9)	
				Cattle urine (w/ tannin)	4.3 (2.6)	
North Dakota	Mixed-grass	2003 (12 Jul, 3 Aug)	0	Monitoring	1.0	6
	Wheatgrass–needlegrass mix				1.4	
	Hydrophytes				6.5	

Continued

TABLE 11.3 Nitrous Oxide Flux from U.S. Rangeland and Pasture (post-2004 publications)—continued

Location	Ecosystem/Pasture type	Study period	Precipitation during study (mm)	Management/Treatment	N_2O flux (g N_2O-N ha^{-1} d^{-1})	Reference[†]
North Dakota	Hydrophytes, perennial grasses	2003 (May–Oct)	217	Monitoring	4.4 (1.4)	7
California	Hydrophytes	2000 (May–Nov), 2002 (May–Aug)	NR[§]	Grazed wetland	4.9	8
				Ungrazed wetland	4.4	
Pasture						
Michigan	Grass lichen mix	1995–1996	837	Control, no added N	0.9 (0.2)	9
				Added N (1 g N m^{-2} yr^{-1})	1.0 (<0.1)	
Ohio	Reclaimed mineland, Grass/legume pasture	2005–2006	1084	Grazed	10.6	10
				Hayed	7.0	
Missouri	Bromegrass	2004–2006	NR	Ungrazed	2.3–7.7	11
Arkansas	Bermudagrass	2000–2001	1420	Control, no poultry litter	3.3 (0.8)	12
				Added poultry litter (215 and 73 kg N ha^{-1} in 2000 and 2001, respectively)	4.8 (1.8)	

[†] *1, McLain and Martens (2006); 2, McLain et al. (2008); 3, Norton et al. (2008); 4, Liebig et al. (2010); 5, Liebig et al. (2008); 6, Phillips and Beeri (2008); 7, Gleason et al. (2009); 8, Oates et al. (2008); 9, Ambus and Robertson (2006); 10, Shrestha et al. (2009); 11, Nkongolo et al. (2010); Sauer et al. (2009).*

[‡] *Values in parentheses represent mean standard error.*

[§] *NR, Not reported.*

Previous literature summaries addressing N_2O fluxes from grassland have reported several general trends: (1) uncertainty in annual flux estimates derived from chamber measurements may be as high as 50% due to temporal and spatial variability (Flechard et al., 2007); (2) among agroecosystems, fluxes are generally lowest from rangelands (Liebig et al., 2005); (3) differences in fluxes among rangeland ecosystems appear to be driven by soil moisture, N availability, and temperature (Martens et al., 2005; McLain and Martens, 2006); and (4) grassland emissions are affected by grazing, but net flux can be increased or decreased, depending on stocking rate, grazing system, and season (Allard et al., 2007).

Recent research has found N_2O emission from U.S. rangelands to be affected by precipitation dynamics, plant species, N fertilization, and hydropedologic zones (Table 11.3). McLain and Martens (2006) observed a 2.5-fold decline in N_2O emission under honey mesquite between consecutive monsoon seasons in southern Arizona (3.2 to 1.3 g N_2O ha^{-1} d^{-1}); a reduction corresponding with a nearly four-fold decrease in precipitation between seasons (238 vs. 61 mm). Concurrent evaluations under grass-dominant ecosystems did not result in a similar decline in N_2O emission between seasons, suggesting the importance of precipitation in mediating N dynamics under mesquite. The role of mesquite to fix N, thereby altering N dynamics, was underscored in findings by McLain et al. (2008), where N_2O emission under mesquite canopy was over four-fold greater (3.8 g N_2O ha^{-1} d^{-1}) than under grasses or woody detritus (0.8–0.9 g N_2O ha^{-1} d^{-1}) during seasonal monsoons. In addition to confirming mesquite-induced enhancements in N_2O emission, these findings suggest semiarid grazing lands represent significant N_2O sources during certain seasons and conditions.

Previously reported aridland N_2O production was thought to result from respiratory denitrification, a process whereby nitrate (NO_3^-) is used as a terminal electron acceptor by anoxic denitrifying bacteria in C- and N-rich microsites dominated by mesquite (Virginia et al., 1982). However, recent studies have reported N_2O production to be significantly reduced by additions of an inhibitor of eukaryotic activity, indicating that much of the N_2O production in semiarid soils may result from denitrification activity of aerobic heterotrophic fungi (McLain and Martens, 2005; Stursova et al., 2006; Kool et al., 2011). Fungi have the ability to metabolize at higher temperatures and lower water potentials than either plants or bacteria (Allen, 2007). The ability to tolerate conditions of low soil moisture may introduce an obvious advantage for fungal populations in semiarid soils (Clark et al., 2009), and may account in part for the higher N_2O production in southeastern Arizona grassland sites during periods of reduced monsoon precipitation (Table 11.3).

Encroachment of non native vegetation on U.S. rangelands has the potential to alter nutrient dynamics in soil from very conservative cycles characterized by high plant and microbial uptake to more open, mineralizing cycles with nutrients in excess of plant and microbial acquisition (Haynes and Williams, 1993). In Wyoming, N_2O flux from cheatgrass (Bromus tectorum L.) was greater than that observed under western wheatgrass [Pascopyrum smithii (Rydb.) A. Löve], illustrating the potential for N losses following invasion of non-native grasses (Norton et al., 2008).

Stocking rate was found not to affect N_2O emission from mixed grass prairie in North Dakota (Liebig et al., 2010), nor was N_2O production increased within simulated urine patches in a short-term study (Liebig et al., 2008) (Table 11.3). While elevated N_2O emissions may be expected under increased stocking rate, findings by Wolf et al. (2010) suggest grazing can counteract potential N-induced emissions on rangelands by reducing surface biomass, resulting in more extreme soil temperatures, lower soil moisture, and corresponding inhibition of microbial activity responsible for N_2O emissions. Such findings indicate grazing effects on rangeland N_2O flux can be, under certain conditions, independent of livestock N application. Application of fertilizer N to rangelands, however, has been found to consistently stimulate N_2O emissions (Flechard et al., 2007). In support of previous findings, Liebig et al.

(2010) observed over two-fold greater N_2O emission from fertilized crested wheatgrass [*Agropyron desertorum* (Fisch. ex. Link) Schult] compared to unfertilized mixed grass prairie (8.2 and 3.4 g N_2O ha^{-1} d^{-1}, respectively) (Table 11.3).

Biogeochemical differences among hydropedologic zones in the PPR of North Dakota contributed to five-fold greater N_2O emission from wetland areas dominated by hydrophytic vegetation (6.5 g N_2O ha^{-1} d^{-1}) compared to upland sites characterized by mixed-grass species (1.2 g N_2O ha^{-1} d^{-1}) (Phillips and Beeri, 2008) (Table 11.3). Gleason et al. (2009) observed similar N_2O emission rates for wetlands in the PPR during the growing season (4.4 ± 1.4 g N_2O ha^{-1} d^{-1}), as did Oates et al. (2008) for grazed and ungrazed wetlands in California (4.9 and 4.4 g N_2O ha^{-1} d^{-1}, respectively). Despite limited data, it appears that C sequestration benefits associated with wetlands may be offset by amplified emissions of N_2O and CH_4 (Bridgham et al., 2006).

Pasturelands

Earlier syntheses indicated a lack of GHG flux measurements from U.S. pastures (Franzluebbers and Follett, 2005). A review of literature published since 2004 indicated data were still limited.

CO_2 EFFLUX AND NET ECOSYSTEM EXCHANGE

Carbon dioxide efflux from grass/legume pasture in Ohio was found to be nearly double with grazing (30.7 kg CO_2-C ha^{-1} d^{-1}) compared to haying (16.7 kg CO_2-C ha^{-1} d^{-1}) (Shrestha et al., 2009) (Table 11.1). Limited assessments of NEE for pastures suggest a large CO_2 sink capacity in the south central U.S. (≥3.4 Mg C ha^{-1} yr^{-1}) (Dugas et al., 1999; Gilmanov et al., 2003b), whereas assessments in Pennsylvania and North Carolina have found cool-season grass pastures to contribute to net CO_2 loss (−0.27 to −0.97 Mg C ha^{-1} yr^{-1}) (Novick et al., 2004; Skinner, 2008) (Figure 11.2).

CH_4 FLUX AND N_2O EMISSION

Recently published assessments of CH_4 flux from U.S. pasturelands are consistent in direction (net uptake) and strikingly similar in magnitude (−3.3 to −4.2 g CH_4-C ha^{-1} d^{-1}; $n = 5$) (Table 11.2). Management effects on CH_4 uptake under pasture, however, are modest. Added N to pasture in Michigan had no effect on CH_4 flux (Ambus and Robertson, 2006). Grazing contributed to slightly less CH_4 uptake (−3.3 g CH_4-C ha^{-1} d^{-1}) compared to haying (−4.2 g CH_4-C ha^{-1} d^{-1}) in a grass/legume pasture in Ohio (Shrestha et al., 2009). While both pasture treatments in the Ohio study were net CH_4 sinks, the grazed treatment was a CH_4 source in autumn while the hayed treatment was a source in spring. An ungrazed smooth bromegrass (*Bromus inermis* Leyss.) pasture in central Missouri was found to be a net CH_4 sink in 2 of 3 years, with a mean annual uptake rate of −3.6 g CH_4-C ha^{-1} d^{-1} (Nkongolo et al., 2010).

Compared to rangeland, N_2O emissions from pasture have the potential to be substantial in magnitude and duration due to greater intensity of management (U.S.-EPA, 2010). Recently published data, however, suggest comparable N_2O emission rates between U.S. pasture and rangeland. Emission rates for pasture not subjected to grazing, haying, or fertilizer addition ranged from 0.9 to 7.7 g N_2O-N ha^{-1} d^{-1} in Michigan, Missouri, and Arkansas (Table 11.3). Addition of fertilizer N to pasture in Michigan had a negligible effect on N_2O emission (Ambus and Robertson, 2006), whereas application of poultry manure on a bermudagrass [*Cynodon dactylon* (L.) Pers.] pasture in Arkansas increased N_2O emission by 45% compared to pasture without manure (Sauer et al., 2009). Coupled evaluations of N_2O flux and soil NO_3^- dynamics in the study by Sauer et al. (2009) found the two variables to be positively associated, with increases in both parameters within 60 days after litter application. Interseeding annual ryegrass (*Lolium multiflorum*) L. Lam. on manure-amended soils decreased N_2O emission by 50% (Sauer et al., 2009). Grazing of a grass/legume pasture in Ohio increased N_2O emission by 51% compared to haying, though it is uncertain if the increase was associated with N addition from livestock waste and/or changes in near-surface soil physical properties due to trampling. Nitrous oxide flux dependence on soil physical properties was found by Nkongolo et al. (2010), where N_2O emissions from ungrazed smooth bromegrass were significantly correlated with soil thermal diffusivity, thermal conductivity, and soil temperature.

Greenhouse Gas Balance of U.S. Grasslands

The net exchange of CO_2, CH_4, and N_2O from an agroecosystem constitutes its net GWP (Robertson et al., 2000). To date, studies with inclusive GWP assessments specific to U.S. grasslands are rare. Liebig et al. (2010) concurrently evaluated GHG flux and SOC dynamics for three grazing management systems in North Dakota and found grazed native vegetation to serve as a net GHG sink at moderate and high stocking rates (-0.62 ± 0.08 and -0.78 ± 0.03 Mg CO_{2e} ha^{-1} yr^{-1}, respectively). Crested wheatgrass fertilized at 45 kg N ha^{-1} yr^{-1} and grazed at a high stocking rate in the same study was found to be a net GHG source (0.40 ± 0.23 Mg CO_{2e} ha^{-1} yr^{-1}). By way of an inclusive literature synthesis, Eagle et al. (2010) estimated fertilization and irrigation of grazinglands in the U.S. to result in average net GHG emissions of 0.63 and 0.83 Mg CO_{2e} ha^{-1} yr^{-1}, respectively. Conversely, net uptake of GHGs was found for management-intensive (rotational) grazing (-2.25 Mg CO_{2e} ha^{-1} yr^{-1}) and grazing-induced changes to native vegetation (-1.50 Mg CO_{2e} ha^{-1} yr^{-1}).

Estimates of GWP for U.S. pastureland have yet to be published. Research conducted in Europe, however, may suggest anticipated responses to management for some U.S. pasturelands. In a 2-year evaluation of nine sites, Soussana et al. (2007) found net uptake of GHGs not to differ significantly from zero (-0.85 ± 0.77 Mg CO_{2e} ha^{-1} yr^{-1}) for seeded, intensive permanent, and semi-natural grasslands. Near-neutral GWP estimates from additional European grassland sites highlight trade-offs in pasture management associated with sequestering soil C through fertilizer application, intensive grazing, and/or haying at the expense of increased soil–atmosphere N_2O flux and CH_4 emissions from ruminants (Leahy et al., 2004; Soussana et al., 2010).

Methane emissions via enteric fermentation from ruminant livestock are a significant contributor to the GHG balance of managed grasslands, although quantifiable estimates for management practices on rangeland and pastureland are lacking. Collectively, CH_4 emissions from ruminants accounted for approximately 25% of total U.S. CH_4 emissions in 2008 (U.S.-EPA, 2010). Of this, CH_4 emissions from enteric fermentation by beef and dairy cattle comprised 95% of CH_4 emissions from agricultural activities. Specific contributions of CH_4 emissions from ruminant livestock to net GWP of managed grasslands depend on many factors, most notably livestock type, diet quality, and feed intake (Westberg et al., 2001). Based on multiple experiments conducted throughout North America, beef and dairy cattle have been found to emit 200 ± 30 g CH_4 d^{-1} ($n = 15$) and 312 ± 121 g CH_4 d^{-1} ($n = 8$), respectively (Westberg et al., 2001; Kebreab et al., 2006). To date, assessments of CH_4 emissions from cattle in the U.S. have been conducted in isolation of grazing and/or pasture management practices where GHG dynamics from soil have been measured. Despite this limitation, published CH_4 emission rates from cattle have been applied to derive net GWP estimates for rangeland (Liebig et al., 2010), with results indicating CH_4 from enteric fermentation to negate between 12 and 33% of C sequestered in soil depending on stocking rate and length of grazing period.

Greenhouse gas emissions from livestock production are not limited to managed grasslands and associated livestock, but also include manure-associated emissions of CH_4 and N_2O (Steinfeld et al., 2006; U.S.-EPA, 2010). Prevalent GHG emissions from manure include direct emissions of CH_4 and N_2O and indirect emissions of N_2O (Soussana et al., 2010). Among direct emissions, liquid/slurry storage systems contribute mainly to CH_4 emissions due to a prevalence of anaerobic conditions, whereas dry/solid storage systems can be significant sources of both CH_4 and N_2O (Kebreab et al., 2006). Additionally, direct emissions from dung pats in grasslands averaged 0.96 g CH_4 d^{-1} cow^{-1} for grazing cattle (Yamulki et al., 1998). Indirect N_2O emissions are produced mainly from volatilized N redeposited on surrounding land, which can account for up to 65% of N fed to livestock (Todd et al., 2008). Collectively, estimates of CH_4 and N_2O emissions from manure management systems for the year 2008 in the U.S. were 45.0 and 17.1 Tg CO_{2e}, respectively, with the vast majority of emissions (49 to 77%) associated with beef and dairy cattle and sheep (U.S.-EPA, 2010). Based on a limited number of life-cycle assessments focused on beef and dairy production, direct and indirect CH_4 and N_2O emissions associated with manure management account for 12 to 30% of total GHG emissions generated over the cow–calf life-cycle (Ogino et al., 2007; Pelletier et al., 2010; de Vries and de Boer, 2010).

CONCLUSIONS AND RECOMMENDATIONS

Synthesis of NEE findings suggests most rangeland ecosystems in the U.S. are net CO_2 sinks, with assimilation rates ranging from 0.2 to 1.1 Mg C ha^{-1} yr^{-1}. However, high annual variability in NEE due to precipitation dynamics drives CO_2 loss/gain from these ecosystems, and therefore limits broad generalizations regarding their role as consistent CO_2 sinks. Rangeland ecosystems in the southwest U.S. appear to be net CO_2 sources (−0.9 to −1.4 Mg C ha^{-1} yr^{-1}), a finding that corroborates previous summaries where prevalence of soils high in inorganic C contributed to significant CO_2 efflux. Assessments of NEE in pasturelands exhibited a wide response range (−0.6 to 3.5 Mg C ha^{-1} yr^{-1}) and the few assessments ($n = 4$) limit generalizations regarding their role to affect CO_2 balance of U.S. grasslands.

Overall, managed grasslands in the U.S. exhibit a capacity to serve as minor CH_4 sinks, a finding that supports previous summaries addressing CH_4 flux for grasslands. Compiled observations from subtropical desert, northern mixed-grass rangelands, and pastures in the central U.S. show mean CH_4 uptake rates of -2.4 ± 1.1, -5.6 ± 1.0, and -3.7 ± 0.2 g CH_4-C ha^{-1} d^{-1}, respectively (Figure 11.3A). Management impacts (e.g. N application, grazing) on CH_4 uptake in grasslands are subtle to non-existent, suggesting an inherent resilience of CH_4 assimilation to management interventions. Sources of CH_4 from grasslands appear to be limited to regions affected by termite activity in subtropical deserts and wetland landscapes with hydrophytic vegetation. It is important to note, however, that conclusions regarding CH_4 sink capacity and source strength of U.S. grasslands continue to remain tenuous given the scarcity of findings.

Recent measurements of N_2O flux from U.S. grasslands suggest emission rates to be 47 to 102% greater than previously summarized (Franzluebbers and Follett, 2005). Among major rangeland ecosystems, mean N_2O emissions were 1.4 ± 0.4 g N_2O-N ha^{-1} d^{-1} in subtropical desert and 3.2 ± 0.8 g N_2O-N ha^{-1} d^{-1} in northern mixed-grass rangelands (Figure 11.3B). Compiled N_2O emissions from pastures and wetlands in the U.S. were 4.7 ± 1.2 and 5.1 ± 0.5 g N_2O-N ha^{-1} d^{-1}, respectively, suggesting the central role of soil moisture and N to enhance N_2O loss from grasslands. Based on recent findings, N_2O emissions were found to be elevated by woody plant encroachment, invasion of non-native grasses, and N fertilization, but not grazing intensity.

While significant progress has been made to better understand GHG dynamics in U.S. agroecosystems, deficiencies in understanding continue to undermine the development of robust estimates of net GWP for U.S. rangelands and pastures. Such deficiencies stem from a lack of measurements across all major grassland categories, as well as limitations associated with basic understanding of mechanistic processes related to GHG flux. Future assessments

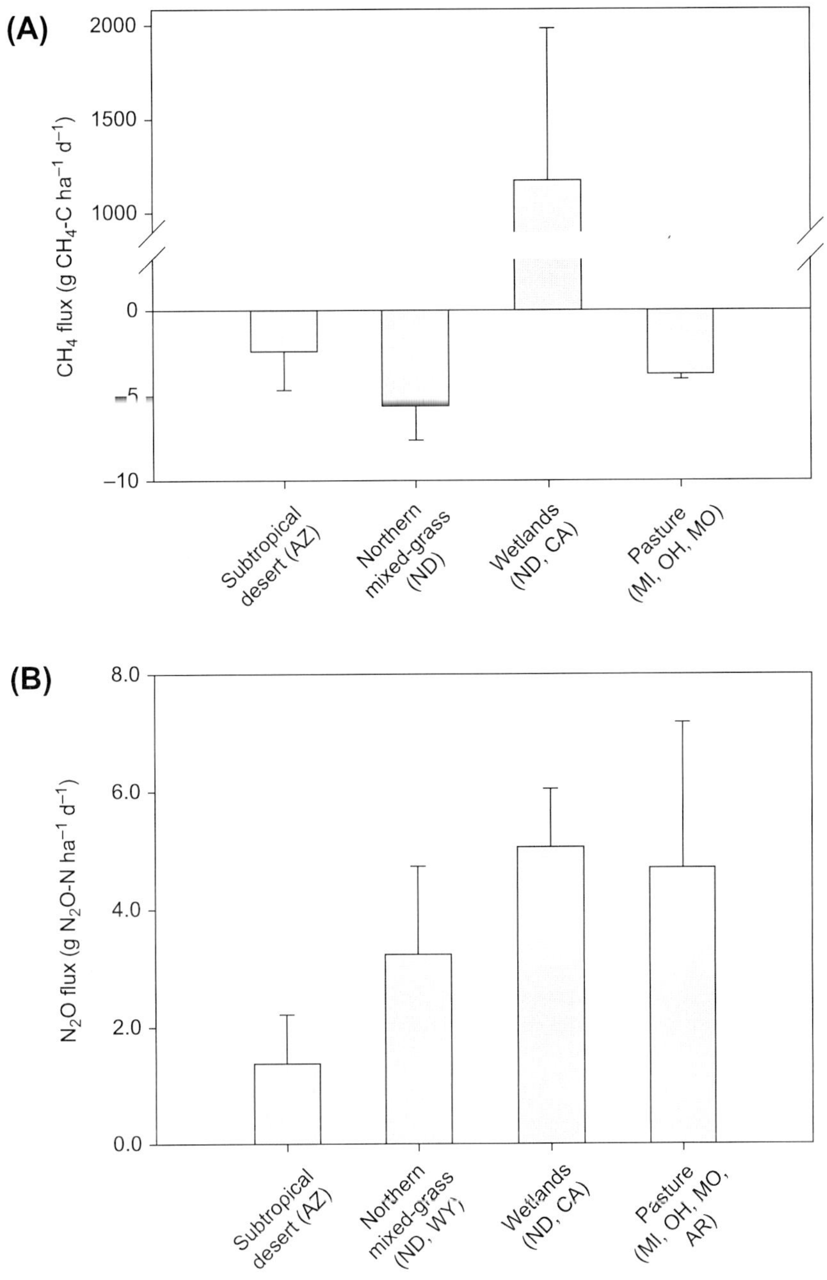

FIGURE 11.3
Mean CH_4 (A) and N_20 (B) flux within major grassland categories in the U.S. Error bars reflect ±1 standard error.

of GHG flux will no doubt develop and/or improve net GWP estimates for rangeland and pastures. In ourview, such assessments should take into consideration the following recommendations:

1. Rangeland CO_2 flux is strongly affected by the timing, frequency, and amount of precipitation. Moreover, precipitation received during periods of active growth is often more important than total precipitation received over the course of a year. Accordingly, accurate quantification of CO_2 dynamics following "rainfall pulses" is recommended to help identify precipitation thresholds necessary to achieve net annual CO_2 assimilation in rangelands.

2. Emission hotspots, spatial variation in micro-topography, and temporal variation of GHG flux in rangelands and pastures represent significant challenges in acquiring useful data from small chamber methods. Increased use of automated chambers and micrometeorological methods may help to address challenges associated with spatial and temporal variation in GHG flux from grasslands.
3. Climate change predictions for U.S. grasslands include shifts in temperature and precipitation patterns. Responses of microbial communities in grassland soils to warming and altered precipitation—and consequent fluxes of CO_2, CH_4, and N_2O from those communities—are not clear. Use of new, sophisticated molecular and biochemical tools to better understand microbial dynamics in soils should be coupled with year-round measurements of GHG fluxes to improve the prediction of feedback responses to climate change.
4. Concurrent assessments of GHG flux and SOC dynamics are needed in rangelands and pastures to more accurately ascertain net GWP of livestock production systems. Such assessments should be conducted over timescales necessary to encapsulate inherent variability associated with production practices. Where applicable, GHG emissions associated with upstream production processes should be included in assessments.

Acknowledgments

We thank Marty Schmer, Robert Gleason, Peter Stahl, and Alan Franzluebbers whose comments improved earlier drafts of the manuscript. This publication is based on work supported by the USDA Agricultural Research Service GRACEnet Project.

References

Adair, E.C., Reich, P.B., Hobbie, S.E., Knops, J.M.H., 2009. Interactive effects of time, CO_2, N, and diversity on total belowground carbon allocation and ecosystem carbon storage in a grassland community. Ecosyst 12, 1037–1052.

Allard, V., Soussana, J.-F., Falcimagne, R., Berbigier, P., Bonneford, J.M., Ceschia, E., D'hour, P., Hénault, C., Laville, P., Martin, C., Pinarès-Patino, C., 2007. The role of grazing management for the net biome productivity and greenhouse gas budget (CO_2, N_2O, and CH_4) of a semi-natural grassland. Agric. Ecosyst. Environ. 121, 47–58.

Allen, M.F., 2007. Mycorrhizal fungi: highways for water and nutrients in arid soils. Vadose Zone J. 6, 291–297.

Ambus, P., Robertson, G.P., 2006. The effect of increased N deposition on nitrous oxide, methane and carbon dioxide fluxes from unmanaged forest and grassland communities in Michigan. Biogeochem. 79, 315–337.

Archer, S., Boutton, T.W., Hibbard, K.A., 2001. Trees in grasslands: biogeochemical consequences of woody plant expansion. In: E-Schulze, D., Harrison, S., Heimann, M., Holland, E., Lloyd, J., Prentice, I., Schimel, D., D. (Eds.), Global Biogeochemical Cycles in the Climate System. Academic Press, San Diego, pp. 115–137.

Bouwman, A.F., 1990. Exchange of greenhouse gases between terrestrial ecosystems and the atmosphere. In: Bouwman, A.F. (Ed.), Soils and the Greenhouse Effect. John Wiley, New York, pp. 61–127.

Bowden, W.B., 1986. Gaseous nitrogen emissions from undisturbed terrestrial ecosystems—an assessment of their impacts on local and global nitrogen budgets. Biogeochem. 2, 249–279.

Bremer, D.J., Ham, J.M., 2010. Net carbon fluxes over burned and unburned native tallgrass prairie. Rangeland Ecol. Manage. 63, 72–81.

Bridgham, S.D., Megonigal, J.P., Keller, J.K., Bliss, N.B., Trettin, C., 2006. The carbon balance of North American wetlands. Wetlands 26, 889–916.

Chimner, R.A., Welker, J.M., 2005. Ecosystem respiration responses to experimental manipulations of winter and summer precipitation in a mixedgrass Prairie, WY, USA. Biogeochem. 73, 257–270.

Clark, J.S., Campbell, J.H., Grizzle, H., Acosta-Martinez, V., Zak, J.C., 2009. Soil microbial community response to drought and precipitation variability in the Chihuahuan Desert. Microb. Ecol. 57, 248–260.

Derner, J.D., Jin, V.L., 2012. Managing agricultural greenhouse gases: rangeland management. In: Liebig, M.A., Franzluebbers, A.J., Follett, R.F. (Eds.), Managing Agricultural Greenhouse Gases: Coordinated Agricultural Research through GRACEnet to Address our Changing Climate. Academic Press, San Diego, CA.

Derner, J.D., Schuman, G.E., 2007. Carbon sequestration and rangelands: a synthesis of land management and precipitation effects. J. Soil Water Conserv. 62, 77–85.

de Vries, M., de Boer, I.J.M., 2010. Comparing environmental impacts for livestock products: a review of life cycle assessments. Livestock Sci. 128, 1–11.

Dugas, W.A., Heuer, M.L., Mayeux, H.S., 1999. Carbon dioxide fluxes over bermudagrass, native prairie, and sorghum. Agric. For. Meterol. 93, 121–139.

Dunfield, P., Knowles, R., 1995. Kinetics of inhibition of methane oxidation by nitrate, nitrite, and ammonium in a humisol. Appl. Environ. Microbiol. 61, 3129–3135.

Eagle, A.J., Henry, L.R., Olander, L.P., Haugen-Kozyra, K., Millar, N., Robertson, G.P., 2010. Greenhouse gas mitigation potential of agricultural land management in the United States: a synthesis of the literature. Nicholas Institute for Environmental Policy Solutions. Duke University, Durham, NC.

Emmerich, W.E., 2003. Carbon dioxide fluxes in a semiarid environment with high carbonate soils. Agric. Forest. Meterol. 116, 91–102.

Falge, E., Baldocchi, D., Tenhunen, J., Aubinet, M., Bakwind, P., Berbigier, P., Bernhofer, C., Burba, G., Clement, R., Davis, K.J., Elbers, J.A., Goldstein, A.H., Grelle, A., Granier, A., Guðmundsson, J., Hollinger, D., Kowalski, A.S., Katul, G., Law, B.E., Malhi, Y., Meyers, T., Monsons, R.K., Mungert, J.W., Oechel, W., Paw, K.T., Pilegaard, U.K. Rannik, Ü, Rebmann, C., Suyker, A., Valentini, R., Wilson, K., Wofsy, S., 2003. Seasonality of ecosystem respiration and gross primary production as derived from FLUXNET measurements. Agric. For. Meterol. 113, 53–74.

Flechard, C.R., Ambus, P., Skiba, U., Rees, R.M., Hensen, A., van Amstel, A., van den Pol-van Dasselaar, A., Soussana, J.-F., Jones, M., Clifton-Brown, J., Raschi, A., Horvath, L., Neftel, A., Jocher, M., Ammann, C., Leifeld, J., Fuhrer, J., Calanca, P., Thelman, E., Pilegaard, K., Di Marco, C., Campbell, C., Nemitz, E., Hargreaves, K.J., Levy, P.E., Ball, B.C., Jones, S.K., van de Bulk, W.C.M., Groot, T., Blom, M., Domingues, R., Kasper, G., Allard, V., Ceschia, E., Cellier, P., Laville, P., Henault, C., Bizouard, F., Abdalla, M., Williams, M., Baronti, S., Berretti, F., Grosz, B., 2007. Effects of climate and management intensity on nitrous oxide emissions in grassland systems across Europe. Ag. Ecosys. Environ. 121, 135–152.

Follett, R.F., Reed, D.A., 2010. Soil carbon sequestration in grazing lands: societal benefits and policy recommendations. Rangeland Ecol. Manage. 63, 4–15.

Frank, A.B., 2002. Carbon dioxide fluxes over a grazed prairie and seeded pasture in the Northern Great Plains. Environ. Pollut. 116, 397–403.

Frank, A.B., Karn, J.F., 2005. Shrub effects on carbon dioxide and water vapor fluxes over grasslands. Rangeland Ecol. Manage. 58, 20–26.

Frank, A.B., Liebig, M.A., Tanaka, D.L., 2006. Management effects on soil CO_2 efflux in northern semiarid grassland and cropland. Soil Tillage Res. 89, 78–85.

Franzluebbers, A.J., Follett, R.F., 2005. Greenhouse gas contributions and mitigation potential in agricultural regions of North America: Introduction. Soil Tillage Res. 83, 1–8.

Fynn, A.J., Alvarez, P., Brown, J.R., George, M.R., Kustin, C., Laca, E.A., Oldfield, J.T., Schohr, T., Neely, C.L., Wong, C.P., 2010. Soil carbon sequestration in U.S. rangelands: issues paper for protocol development. Env. Defense Fund, New York.

Galbally, I.E., Kirstine, W.V., Meyer, C.P., Wang, Y.P., 2008. Soil-atmosphere trace gas exchange in semiarid and arid zones. J. Environ. Qual. 37, 599–607.

Gilmanov, T.G., Svejcar, T.J., Johnson, D.A., Angell, R.E., Saliendra, N.Z., Wylie, B.K., 2006. Long-term dynamics of production, respiration, and net exchange in two sagebrush-steppe ecosystems. Rangeland Ecol. Manage. 59, 585–599.

Gilmanov, T.G., Johnson, D.A., Saliendra, N.Z., 2003a. Growing season CO_2 fluxes in a sagebrush-steppe ecosystem in Idaho: Bowen ratio/energy balance measurements and modeling. Basic Appl. Ecol. 4, 167–183.

Gilmanov, T.G., Venna, S.B., Sims, P.L., Meyers, T.P., Bradford, J.A., Burba, G.G., Suyker, A.E., 2003b. Gross primary production and light response parameters of four Southern Plains ecosystems estimated using long-term CO_2-flux tower measurements. Global Biogeochem. Cycles 17 1071:40-1-40-16.

Gleason, R.A., Tangen, B.A., Browne, B.A., Euliss Jr., N.H., 2009. Greenhouse gas flux from cropland and restored wetlands in the Prairie Pothole Region. Soil Biol. Biochem. 41, 2501–2507.

Hamerlynck, E.P., Scott, R.L., Moran, M.S., Keefer, T.O., Huxman, T.E., 2010. Growing season ecosystem and leaf-level gas exchange of an exotic and native semiarid bunchgrass. Oecologia 163, 561–570.

Havstad, K., Peters, D., Allen-Diaz, B., Bartolome, J., Bestelmeyer, B., Briske, D., Brown, J., Brunson, M., Herrick, J., Huntsinger, L., Johnson, P., Joyce, L., Pieper, R., Svejcar, T., Yao, J., 2009. The western United States rangelands: a major resource. In: Wedin, W.F., Fales, S.L. (Eds.), Grassland: Quietness and Strength for a New American Agriculture. ASA-CSSA-SSSA, Madison, WI, pp. 75–93.

Haynes, R.J., Williams, P.H., 1993. Nutrient cycling and soil fertility in the grazed pasture ecosystem. Adv. Agron. 49, 119–199.

Jarvis, S.C., 1997. Emission processes and their interactions in grassland soils. In: Jarvis, S.C., Pain, B.F. (Eds.), Gaseous Nitrogen Emissions from Grasslands. CAB International, Wallingford, UK, pp. 1–17.

Kebreab, E., Clark, K., Wagner-Riddle, C., France, J., 2006. Methane and nitrous oxide emissions from animal agriculture: a review. Can. J. Anim. Sci. 86, 135–158.

Kjelgaard, J.F., Heilman, J.L., Mcinnes, K.J., Owens, M.K., Kamps, R.H., 2008. Carbon dioxide exchange in a subtropical, mixed C_3/C_4 grassland on the Edwards Plateau. Texas. Agric. For. Meterol. 148, 953–963.

Knutsen, G.A., Euliss Jr., N.H., 2001. Wetland restoration in the prairie pothole region of North America: a literature review, U.S. Department of the Interior, U.S. Geological Survey, Biological Science Report USGS/BRD/BSR 2001-0006, Reston, VA. p. 54.

Kool, D.M., Dolfing, J., Wrage, N., Willem Van Groenigen, J., 2011. Nitrifier denitrification as a distinct and significant source of nitrous oxide from soil. Soil Biol. Biochem. 43, 174–178.

Leahy, P., Kiely, G., Scalon, T.M., 2004. Managed grasslands: a greenhouse gas sink or source. Geophys. Res. Lett. 31 L20507, doi:10.1029/2044GL021161.

Leffler, A.J., Peek, M.S., Hipps, L., Ivans, S., Ryell, R.J., Caldwell, M.M., 2007. Potential contribution of respiration by *Anabrus simplex* (Mormon crickets) to net CO_2 exchange in three Great Basin ecosystems. West. N. Am. Nat. 67, 109–119.

Liebig, M.A., Gross, J.R., Kronberg, S.L., Phillips, R.L., Hanson, J.D., 2010. Grazing management contributions to net global warming potential: a long-term evaluation in the northern Great Plains. J. Environ. Qual. 39, 799–809.

Liebig, M.A., Kronberg, S.L., Gross, J.R., 2008. Effects of normal and altered cattle urine on short-term greenhouse gas flux from mixed-grass prairie in the northern Great Plains. Agric. Ecosyst. Environ. 125, 57–64.

Liebig, M.A., Morgan, J.A., Reeder, J.D., Ellert, B.H., Gollany, H.T., Schuman, G.E., 2005. Greenhouse gas contributions and mitigation potential of agricultural practices in northwestern USA and western Canada. Soil Tillage Res. 83, 25–52.

Martens, D.A., Emmerich, W., Jean McLain, E.T., Johnsen, T.N., 2005. Atmospheric carbon mitigation potential of agricultural management in the southwestern USA. Soil Tillage Res. 83, 95–119.

McCulley, R.L., Boutton, T.W., Archer, S.R., 2007. Soil respiration in a subtropical savanna parkland: response to water additions. Soil Sci. Soc. Am. J. 71, 820–828.

McLain, J.E.T., Ahmann, D.A., 2008. Increased moisture and methanogenesis contribute to reduced methane oxidation in elevated CO_2 soils. Biol. Fertil. Soils. 44, 623–631.

McLain, J.E.T., Martens, D.A., McClaran, M.P., 2008. Soil cycling of trace gases in response to mesquite management in a semiarid grassland. J. Arid Environ. 72, 1654–1665.

McLain, J.E.T., Martens, D.A., 2006. Moisture controls on trace gas fluxes in semiarid riparian soils. Soil Sci. Soc. Am. J. 70, 367–377.

McLain, J.E.T., Martens, D.A., 2005. N_2O production by heterotrophic N transformations in a semi-arid soil. Appl. Soil Ecol. 32, 253–263.

Niemuth, N.D., Quamen, F.R., Naugle, D.E., Reynolds, R.R., Esty, M.E., Shaffer, T.L., 2007. Benefits of the Conservation Reserve Program to grassland bird populations in the Prairie Pothole region of North Dakota and South Dakota USDA Farm Service Agency. RFA OS-IA-04000000–N34.

Nkongolo, N.V., Johnson, S., Schmidt, K., Eivazi, F., 2010. Greenhouse gas fluxes and soil thermal properties in a pasture in central Missouri. J. Environ. Sci-China 22, 1029–1039.

Norton, U., Mosier, A.R., Morgan, J.A., Derner, J.D., Ingram, L.J., Stahl, P.D., 2008. Moisture pulses, trace gas emissions and soil C and N in cheatgrass and native grass-dominated sagebrush-steppe in Wyoming, USA. Soil Biol. Biochem. 40, 1421–1431.

Novick, K.A., Stoy, P.C., Katul, G.G., Ellsworth, D.S., Siqueira, M.B.S., Juang, J., Oren, R., 2004. Carbon dioxide and water vapor exchange in a warm temperate grassland. Oecologia 138, 259–274.

Oates, L.G., Jackson, R.D., Allen-Diaz, B., 2008. Grazing removal decreases the magnitude of methane and the variability of nitrous oxide emissions from spring-fed wetlands of a California oak savanna. Wetlands Ecol. Manage. 16, 395–404.

Ogino, A., Orito, H., Shimada, K., Hirooka, H., 2007. Evaluating environmental impacts of the Japanese beef cow-calf system by the life cycle assessment method. Animal Sci. J. 78, 424–432.

Owensby, C.E., Ham, J.M., Auen, L.M., 2006. Fluxes of CO_2 from grazed and ungrazed tallgrass prairie. Rangeland Ecol. Manage. 59, 111–127.

Pelletier, N., Priog, R., Rasmussen, R., 2010. Comparative life cycle environmental impacts of three beef production strategies in the Upper Midwestern United States. Agric. Syst. 103, 380–389.

Phillips, R.L., Beeri, O., 2008. The role of hydropedologic vegetation zones in greenhouse gas emissions for agricultural wetland landscapes. Catena 72, 386–394.

Polley, H.W., Emmerich, W., Bradford, J.A., Sims, P.L., Johnson, D.A., Saliendra, N.Z., Svejcar, T., Angell, R., Frank, A.B., Phillips, R.L., Snyder, K.A., Morgan, J.A., Sanabria, J., Mielnick, P.C., Dugas, W.A., 2010. Physiological and environmental regulation of interannual variability in CO_2 exchange on rangelands in the western United States. Global Change Biol. 16, 990–1002.

Potter, C.S., Davidson, E.A., Verchot, L.V., 1996. Estimation of global biogeochemical controls and seasonality in soil methane consumption. Chemosphere 32, 2219–2246.

Robertson, G.P., Paul, E.A., Harwood, R.R., 2000. Greenhouse gases in intensive agriculture: contributions of individual gases to the radiative forcing of the atmosphere. Science 289, 1922–1925.

Sanderman, J., Amundson, R., 2010. Soil carbon dioxide production and climatic sensitivity in contrasting California ecosystems. Soil Sci. Soc. Am. J. 74, 1356–1366.

Sanderson, M.A., Wedin, D., Tracy, B., 2009. Grassland: definition, origins, extent, and future. In: Wedin, W.F., Fales, S.L. (Eds.), Grassland: Quietness and Strength for a New American Agriculture. ASA-CSSA-SSSA, Madison, WI, pp. 57–74.

Sauer, T.J., Compston, S.R., West, C.P., Hernandez-Ramirez, G., Gbur, E.E., Parkin, T.B., 2009. Nitrous oxide emissions from bermudagrass pasture: interseeded winter rye and poultry litter. Soil Biol. Biochem. 41, 1417–1424.

Scott, R.L., Huxman, T.E., Williams, D.G., Goodrich, D.C., 2006. Ecohydrological impacts of woody-plant encroachment: seasonal patterns of water and carbon dioxide exchange within a semiarid riparian environment. Global Change Biol. 12, 311–324.

Shrestha, R.K., Lal, R., Penrose, C., 2009. Greenhouse gas emissions and global warming potential of reclaimed forest and grassland soils. J. Environ. Qual. 38, 426–436.

Singer, J.W., Franzluebbers, A.J., Karlen, D.L., 2009. Grass-based farming systems: soil conservation and environmental quality. In: Wedin, W.F., Fales, S.L. (Eds.), Grassland: Quietness and Strength for a New American Agriculture. ASA-CSSA-SSSA, Madison, WI, pp. 121–136.

Singh, B.K., Bardgett, R.D., Smith, P., Reay, D.S., 2010. Microorganisms and climate change: terrestrial feedbacks and mitigation options. Nature Rev. Microbiol. 8, 779–790.

Skinner, R.H., 2008. High biomass removal limits carbon sequestration potential of mature temperate pastures. J. Environ. Qual. 37, 1319–1326.

Smith, K.A., Dobbie, K.E., Ball, B.C., Bakken, L.R., Situala, B.K., Hansen, S., Brumme, R., Borken, W., Christensen, S., Priemé, A., Fowler, D., MacDonald, J.A., Skiba, U., Klemedtsson, L., Kasmir-Klemedtsson, A., Degórska, A., Orlanski, P., 2000. Oxidation of atmospheric methane in Northern European soils, comparison with other ecosystems, and uncertainties in the global terrestrial sink. Global Change Biol. 6, 791–803.

Soussana, J.F., Allard, V., Pilegaard, K., Ambus, P., Amman, C., Campbell, C., Ceschia, E., Clifton-Brown, J., Czobel, S., Domingues, R., Flechard, C., Fuhrer, J., Hensen, A., Horvath, L., Jones, M., Kasper, G., Martin, C., Nagy, Z., Neftel, A., Raschi, A., Baronti, S., Rees, R.M., Skiba, U., Stefani, P., Manca, G., Sutton, M., Tuba, Z., Valentini, R., 2007. Full accounting of the greenhouse gas (CO_2, N_2O, CH_4) budget of nine European grassland sites. Agric. Ecosys. Environ. 121, 121–134.

Soussana, J.F., Tallec, T., Blanfort, V., 2010. Mitigating the greenhouse gas balance of ruminant production systems through carbon sequestration in grasslands. Animal 4, 334–350.

Steinfeld, H., Gerber, P., Wassenaar, T., Castel, V., Rosales, M., de Haan, C., 2006. Livestock's long shadow: environmental issues and options. Food and Agriculture Organization. Rome, Italy, p. 408.

Stursova, M., Crenshaw, C.L., Sinsabaugh, R.L., 2006. Microbial responses to long-term N deposition in a semiarid grassland. Microb. Ecol. 51, 90–98.

Suyker, A.E., Verma, S.B., 2001. Year-round observations of the net ecosystem exchange of carbon dioxide in a native tallgrass prairie. Global Change Biol. 7, 279–289.

Suyker, A.E., Verma, S.B., Burba, G.G., 2003. Interannual variability in net CO_2 exchange of a native tall-grass prairie. Global Change Biol. 9, 255–265.

Svejcar, T., Angell, R., Bradford, J.A., Dugas, W., Emmerich, W., Frank, A.B., Gilmanov, T., Haferkamp, M., Johnson, D.A., Mayeux, H., Mielnick, P., Morgan, J., Saliendra, N.Z., Schuman, G.E., Sims, P.L., Snyder, K., 2008. Carbon fluxes on North American rangelands. Rangeland Ecol. Manage. 61, 465–474.

Todd, R.W., Cole, A.N., Clark, N.R., Flesch, T.K., Harper, L.A., Baek, B.H., 2008. Ammonia emissions from a beef cattle feedyard on the southern High Plains. Atmos. Environ. 42, 6797–6805.

U.S. Census Bureau, 2011. The 2011 statistical abstract of the United States. U.S. Govt. Print. Office, Washington, DC. website—http://www.census.gov/compendia/statab/ (verified January 7, 2011).

USDA (U.S. Department of Agriculture), 2011. National Agriculture Statistics Service. Agricultural statistics database: livestock production totals for 2010. Available online at. http://www.nass.usda.gov/ (verified January 7, 2011).

USDA (U.S. Department of Agriculture), 2009. Summary Report: 2007 National Resources Inventory. NRCS, Washington, DC, and Center for Survey Statistics and Methodology, Iowa State Univ, Ames, IA, p 123.

U.S.-EPA (U.S. Environmental Protection Agency), 2010. Inventory of U.S. greenhouse gas emissions and sinks: 1990–2008. U.S. Environmental Protection Agency, Office of Atmospheric Programs, EPA 430-R-10-006, April 2010. Available online at www.epa.gov/climatechange/emissions/usinventoryreport.html (verified January 20, 2011).

Virginia, R.A., Jarrell, W.M., Francovizcaino, E., 1982. Direct measurement of denitrification in a *Prosopis* (mesquite) dominated Sonoran desert ecosystem. Oecologia 53, 120–122.

Wedin, W.F., Fales, S.L. (Eds.), 2009. Grassland: quietness and strength for a new American agriculture. ASA-CSSA-SSSA, Madison, WI.

Westberg, H., Lamb, B., Johnson, K.A., Huyler, M., 2001. Inventory of methane emissions from U.S. cattle. J. Geophys. Res. 106 (D12), 12633–12642.

Wolf, B., Zheng, X., Bruggemann, N., Chen, W., Dannenmann, M., Han, X., Sutton, M.A., Wu, H., Yao, Z., Butterbach-Bahl, K., 2010. Grazing-induced reduction of natural nitrous oxide release from continental steppe. Nature 464, 881–884.

Xu, L., Baldocchi, D.D., 2004. Seasonal variation in carbon dioxide exchange over a Mediterranean annual grassland in California. Agric. For. Meterol. 1232, 79–96.

Yamulki, S., Jarvis, S.C., Owen, P., 1998. Nitrous oxide emissions from excreta applied in a simulated grazing pattern. Soil Biol. Biochem. 30, 491–500.

Zhang, L., Wylie, B.K., Ji, L., Gilmanov, T.G., Tieszen, L.L., 2010. Climate-driven interannual variability in net ecosystem exchange in the Northern Great Plains grasslands. Rangeland Ecol. Manage. 63, 40–50.

CHAPTER 12

Mitigation Opportunities for Life-Cycle Greenhouse Gas Emissions during Feedstock Production across Heterogeneous Landscapes

Paul R. Adler[1], **Stephen J. Del Grosso**[2], **Daniel Inman**[3], **Robin E. Jenkins**[4], **Sabrina Spatari**[5], **Yimin Zhang**[6]

[1]USDA-Agricultural Research Service, Pasture Systems and Watershed Management Research Unit, University Park, PA
[2]USDA-Agricultural Research Service, Soil Plant Nutrient Research Unit, Fort Collins, CO
[3]National Bioenergy Center, National Renewable Energy Laboratory, Golden, CO
[4]DuPont Engineering Research and Technology, Wilmington, DE
[5]Civil, Architectural & Environmental Engineering, Drexel University, Philadelphia, PA
[6]Strategic Energy Analysis Center, National Renewable Energy Laboratory, Golden, CO

CHAPTER OUTLINE

NB: The views expressed in this chapter are those of the authors and do not necessarily reflect the views or policies of DuPont or its subsidraries. Mention of trade names or commercial products in this publication is solely for the purpose of providing specific information and does not imply recommendation or endorsement by the U.S. Department of Agriculture. USDA is an equal opportunity provider and employer.

Managing Agricultural Greenhouse Gases. DOI: 10.1016/B978-0-12-386897-8.00012-7
2012, Published by Elsevier Inc.

Abbreviations: bgy, billion gallons per year; CARB, California Air Resources Board; CH_4, methane; CO_2, carbon dioxide; CSBP, Council on Sustainable Biomass Production; EISA, Energy Independence and Security Act; GHG, greenhouse gas; GREET, Greenhouse Gases, Regulated Emissions, and Energy Use in Transportation; GWI, global warming intensity; ha, hectare; IPCC, Intergovernmental Panel on Climate Change; kg CO_2e, kilograms carbon dioxide equivalent; K, potassium; LCA, life-cycle assessment; LCFS, Low Carbon Fuel Standard; Mg, megagrams; MJ, megajoule; N, nitrogen; N_2O, nitrous oxide; NO_3, nitrate; NO_x, oxides of nitrogen; NREL, National Renewable Energy Laboratory; RFS, Renewable Fuel Standard; RSB, Roundtable on Sustainable Biofuels; U.S., United States

LIFE-CYCLE BIOFUEL COMPLIANCE POLICY STANDARDS

Lawmakers in the United States (U.S.) have passed legislation at both the federal and state levels to increase the use of renewable energy and reduce greenhouse gas (GHG) emissions (CARB, 2010; United States Government, 2007). The Renewable Fuel Standard (RFS) in the Energy Independence and Security Act (EISA) of 2007 is intended to prompt the growth of advanced and cellulosic biofuels industries through aggressive annual volumetric mandates. These mandates culminate in the year 2022 with a minimum annual production volume of 36 billion gallons per year (bgy) of renewable fuel, of which 16 bgy will be derived from cellulosic feedstocks, 4 bgy will be derived from "advanced biofuels", and 1 bgy will be biomass-based diesel (United States Government, 2007). There is general agreement that the remaining 15 bgy of renewable fuel will come from corn grain-based ethanol, because that industry is mature and well established with an annual capacity of nearly 13.5 billion gallons in the U.S., as of 2010.

In 2007, the California Air Resources Board (CARB) adopted the Low Carbon Fuel Standard (LCFS) (CARB, 2010), which seeks to incrementally reduce life-cycle GHG emissions of California's transportation fuels. Under this program, each fuel provider must ensure that the overall carbon intensity of its fuel pool meets the annual carbon-intensity target for a given year. Under the LCFS, a fuel produced from a lower-carbon feedstock will be of higher value than an identical fuel produced from a higher-carbon feedstock. For example, ethanol produced from forest residues, which have an estimated lower GHG intensity, would better help a fuel producer meet the LCFS requirements than would ethanol produced from corn, which has a higher GHG intensity. The LCFS program allows individual fuel producers to provide detailed documentation about the GHG intensity of a particular fuel pathway, thus rewarding innovation toward GHG emissions reductions. In contrast, the federal Renewable Fuel Standard (RFS) does not consider incremental differences in GHG emissions among production processes of biofuel feedstocks. Instead, the RFS specifies distinct GHG performance thresholds for each of the four different categories of biofuels (EPA, 2010). Each policy approach has its trade-offs: the LCFS incremental approach rewards innovations in the development of low-carbon fuel pathways (Turner et al., 2007). However, for fuel pathways that show great uncertainty in their estimated carbon intensity, it becomes difficult to measure and distinguish incremental improvements within a fuel pathway. In such cases, a threshold structure such as that of the RFS has advantages in its ability to separate the fuels that may qualify from those that are uncertain to meet low carbon goals.

In addition to the state and federal initiatives to increase renewable fuel use and reduce GHG emissions, two emerging voluntary sustainability standards are receiving significant attention in the U.S. and internationally. These voluntary standards would provide biomass growers the opportunity to have their feedstocks certified as "sustainably produced low-carbon feedstocks". It is important to observe and monitor what is included under the term "sustainably produced" to ensure that biofuels aimed at mitigating GHG emissions do not do so at the expense of other environmental and socioeconomic values and resources. This is addressed in many of the low-carbon and renewable fuel standards (see, e.g., CARB, 2010; Dehue et al., 2007; United States Government, 2007). The U.S.-based Council on Sustainable Biomass Production (CSBP) and the international Roundtable on Sustainable Biofuels (RSB) have developed sustainability criteria for both biomass and bioenergy production. Both CSBP and RSB consider similar environmental, social, and economic principles applicable to biomass feedstock producers as well as to biofuels producers, and both prioritize objective life-cycle GHG accounting, among other sustainability categories. Although research has identified many potential ecological risks associated with expanded biofuels production (see, e.g., Fargione et al., 2010), knowledge from ecological research has yet to be translated into policy. Because increased feedstock production is expected to play an important role in meeting biofuels mandates, there is a pressing need to examine its impacts in various ecological and socioeconomic ("sustainability") categories in isolation; to understand the contribution of this particular process to the overall biofuels production system; and to identify preferred management practices to improve the overall sustainability of feedstock production. There are numerous metrics to consider when assessing sustainability, but life-cycle global warming intensity (GWI) is widely regarded as one key measure of sustainability, and it is one we focus on in this chapter.

Biomass production is a significant contributor to the total life-cycle GHG emissions associated with bioenergy (Adler et al., 2007; Hsu et al., 2010). However, there are challenges in quantifying the GHG emissions associated with bioenergy feedstock production, because the GHG emission profile of a particular feedstock can vary considerably by both geography and management, as well as by the life-cycle methodology used to conduct the analysis and the approaches used to quantify certain GHG emissions (e.g. N_2O emissions from nitrogen fertilization). Currently, regulatory agencies develop generic GHG emission profiles for each feedstock and use them in the regulatory standards. There is growing recognition that product life-cycle models are improved when they incorporate geospatial specificity (Geyer et al., 2010). This is especially true for land-based resources such as agricultural feedstocks entering biofuel markets. As such, there is interest in developing technical guidelines for entity-level GHG emissions reporting that account for geographic variations.

In this study, we (1) demonstrate the importance of feedstock production to the overall fuel life-cycle by estimating the GWI [calculated using 2007 IPCC 100a weighting factors for the individual greenhouse gases (Forster et al., 2007) summed over the cradle to farm gate life-cycle] of six feedstock production systems: corn grain (*Zea mays* L.), winter barley (*Hordeum vulgare* L.), corn stover (*Zea mays* L.), winter wheat straw (*Triticum aestivum* L.), spring wheat straw (*Triticum aestivum* L.), and switchgrass (*Panicum virgatum* L.); (2) show that, in addition to the large GWI variations among feedstocks, which are recognized by the regulatory standards, there are large variations within each feedstock, which have largely gone unrecognized; (3) identify the components of feedstock production that have the greatest potential to improve the GHG emission profile; and (4) describe strategies for overcoming the greatest GHG challenges within feedstock production and discuss their GHG-mitigation potential. Understanding how agricultural inputs, including upstream effects and management practices, contribute to the GWI of biomass production will facilitate identification of opportunities for "decarbonizing" the largest contributors to biofuel life-cycle GWI. Very likely, region-specific practices will be necessary for the development of low-GHG feedstocks for bioenergy. Life-cycle profiles specific to regions and feedstocks will help focus agronomic research on developing management strategies that reduce the carbon footprint of biomass feedstock production systems, with the overall goal of minimizing the carbon footprint of biofuel production. Transferring such practices through effective incentive programs will be an essential step for developing sustainable biomass energy markets.

REVIEW AND ANALYSIS

We conducted a review and analysis of published life-cycle inventories for GHG emissions for six feedstocks that are either used or proposed for renewable, advanced, and cellulosic biofuel categories under the RFS. We reviewed both deterministic and stochastic models from literature, all of which use attributional LCA methods. We summarize the methods applied in the studies cited. The functional unit for all biomass feedstocks we compared is 1 Mg (megagram = 10^6 grams) dry weight.

Of the published studies we reviewed, we mined the complete profiles to disaggregate contributions to feedstock GHG emissions into the following categories, where possible: production and application of fertilizer (nitrogen, phosphorus, potassium), lime, and herbicides; energy for tillage, planting, and harvest; and direct and indirect nitrous oxide (N_2O) emissions. We also reviewed published life-cycle studies of various agricultural feedstocks. Results from the studies reviewed were not normalized or aligned in any way, the data presented herein are "as-is". For specific methods and assumptions of each individual study reviewed, we refer the reader to the specific literature (Hsu et al., 2010; Spatari and MacLean, 2010; Kim et al., 2009; Adler et al., 2007; Liska et al., 2009; Sheehan et al., 2004). All the studies reviewed, with the exception of Adler et al. (2007) and Kim et al. (2009), generalized the conditions of production sites when estimating direct and indirect N_2O emissions.

FEEDSTOCK PRODUCTION IS A SIGNIFICANT GWI COMPONENT OF THE FUEL LIFE-CYCLE

In the reference cases we reviewed, feedstock production contributed 46–68% of the total biofuel life-cycle GWI for ethanol derived from corn grain, corn stover, wheat straw, and switchgrass (Figure 12.1).

Overall, the cellulosic-based ethanol scenarios evaluated have much smaller carbon footprints than the corn grain-based ethanol scenario. Not only is the overall carbon footprint much

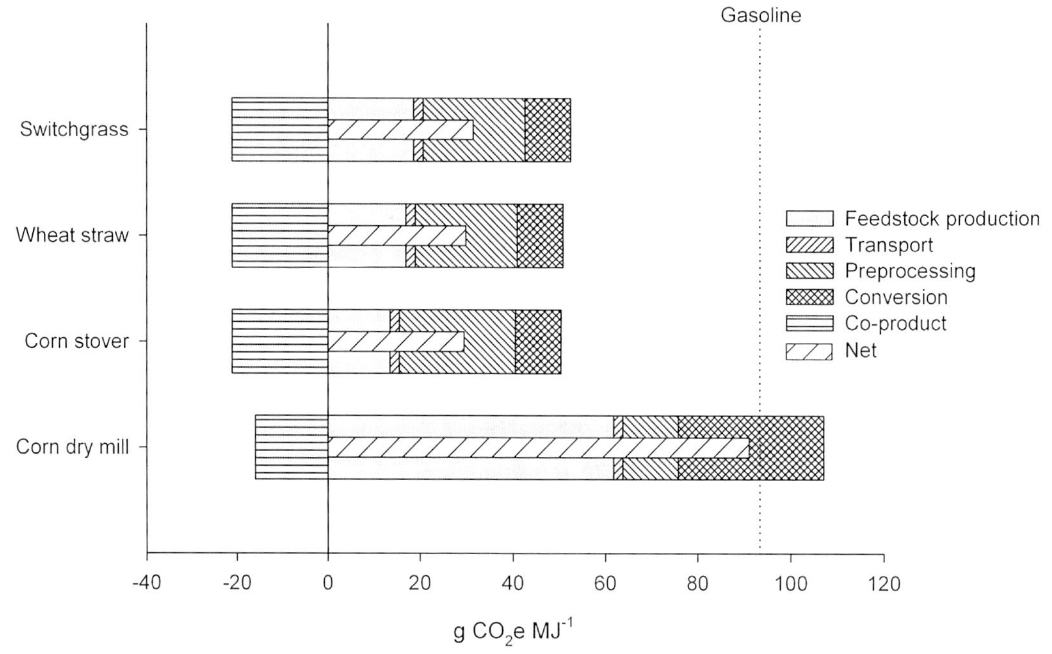

FIGURE 12.1
Global warming intensity (g CO_2-eq) of four feedstock-to-ethanol pathways on a per MJ of ethanol basis. Each stacked bar presents the relative contribution of the key seed-to-biorefinery exit gate unit processes. Feedstock data used in this figure are from Hsu et al. (2010). The gasoline reference line includes end use in a light-duty internal combustion vehicle (EPA, 2010).

greater for corn grain-based ethanol, the relative contribution of specific unit processes is quite different when compared to the cellulosic cases. The most salient difference when comparing the corn grain-based scenario to the cellulosic scenarios is the contribution of feedstock production to the total life-cycle GWI. For corn grain-based ethanol, the feedstock production phase accounts for over half of the life-cycle GWI (see Figure 12.1). In contrast, the relative contribution of preprocessing (e.g. drying, grinding, densifying) is greater for the cellulosic cases than for corn grain-based ethanol. It should be noted that the cellulosic cases presented in Figure 12.1 are calculated with the assumption that the biomass will be field-dried, baled, transported to an intermediate facility, dried further, and densified (pelletized) before being fed to a biorefinery. This "advanced" preprocessing step is included because transporting and handling biomass in bales and in loose forms for long distances may be economically prohibitive. Densifying the biomass allows it to be handled using existing grain handling infrastructure. Compared to cellulosic biomass, corn grain is energy dense out of the field, and is handled efficiently using current grain infrastructure.

GWI SIZE OF LIFE-CYCLE COMPONENTS VARY WITHIN AND BETWEEN FEEDSTOCKS

Many studies have reported ranges and point estimates of the GWI of producing a variety of biofuel feedstocks. Figure 12.2 presents data from a select group of studies that have reported the GWI of the feedstock production for a variety of crops and residues.

Overall, the median estimates of GWI span an order of magnitude from nearly 50 kg carbon dioxide equivalent (kg CO_2e) per Mg to approximately 500 kg CO_2e per Mg of feedstock produced. Although the overall range in estimates is large, the estimates within similar feedstocks (e.g. agricultural residues, grain, and dedicated energy crops) are loosely clustered. This is somewhat surprising, considering that these studies were all conducted and published independently. The data from Kim et al. (2009) are an exception, however. Their estimates for corn and corn stover are considerably higher and slightly lower, respectively, than estimates from the other studies.

The uncertainty represented in Figure 12.2 is the result of data variability, data quality, and LCA methodology differences. Uncertainty ranges reported for established and existing systems

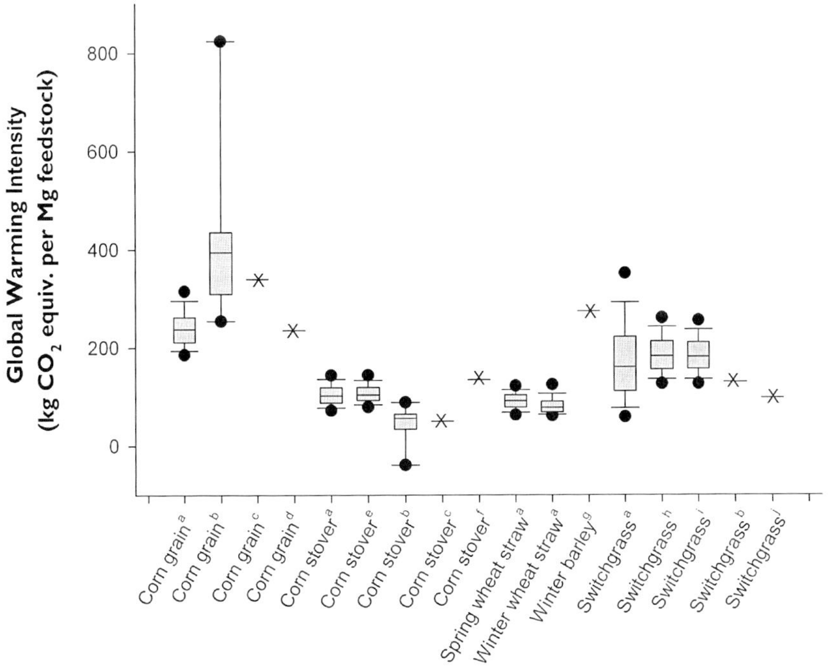

FIGURE 12.2
Box-and-whisker plots of the seed-to-gate global warming intensity (kg CO_2 equivalents per Mg of feedstock produced and harvested) with uncertainty ranges and point estimates for select feedstocks from literature. [a]Hsu et al., 2010; [b]Kim et al., 2009; [c]Argonne National Laboratory, 2009; [d]Liska et al., 2009; [e]Spatari and MacLean, 2010; [f]Sheehan et al., 2004; [g]Spatari and Stadel, 2011; [h]Spatari and MacLean, 2010. *Near-term estimates*; [i]Spatari and MacLean, 2010. *Mid-term estimates*; [j]Adler et al., 2007.

(e.g. corn grain, corn stover, and wheat straw) reflect the range of values reported for biomass yield, agricultural inputs, and management practices in the U.S. In contrast, the uncertainty ranges reported for switchgrass reflect the fact that switchgrass is not produced on a large scale in the U.S. and that certain assumptions regarding the distributions of variables such as yield and inputs must be made, which vary across studies. As more data are collected and more information regarding the probability distribution functions of the crop production parameters is acquired, the uncertainty ranges for switchgrass will most likely be reduced.

The relative contribution of farming activities and field emissions for producing a variety of biomass feedstocks is presented in Figure 12.3.

Farming and field operations vary in their GHG impacts. In general, nitrogen (N) fertilizer-related emissions, including N fertilizer input (GHGs released during the process of fertilizer production, as well as during transport) and field emission of N_2O are the largest contributors to the overall carbon footprint of the biomass feedstocks presented. Hence, proper accounting of N_2O emissions from soil is particularly important in the context of quantifying the net carbon footprint of any biomass cropping system. In the National Renewable Energy Laboratory (Hsu et al., 2010) and GREET cases, the N_2O emissions are calculated by applying an emission factor to the N fertilizer application rate, as per the International Panel on Climate Change (IPCC) Tier I Guidelines (de Klein et al., 2006). This approach does not take into consideration spatial and temporal variability as well as differences in crop, and management practices, which all influence the actual N_2O emissions from soil. To better quantify GHG fluxes of different biomass cropping systems under a range of environmental (climate, soil properties, etc.) and management (crop rotation, tillage) variables at a regional level, Adler et al. (2007) used DayCent, a biogeochemical model, to simulate carbon and nitrogen fluxes among the atmosphere, soil, and vegetation for five biomass production scenarios. Estimation of GHG emissions using the IPCC Tier I emission factors and biogeochemical models is often termed as a bottom-up approach. This is different from a top-down approach, which estimates the global N_2O emissions from the changes in atmospheric concentrations of N_2O and allocates the portion that was attributable to agricultural soils by subtracting N_2O emissions from other contributors (e.g. industrial sources).

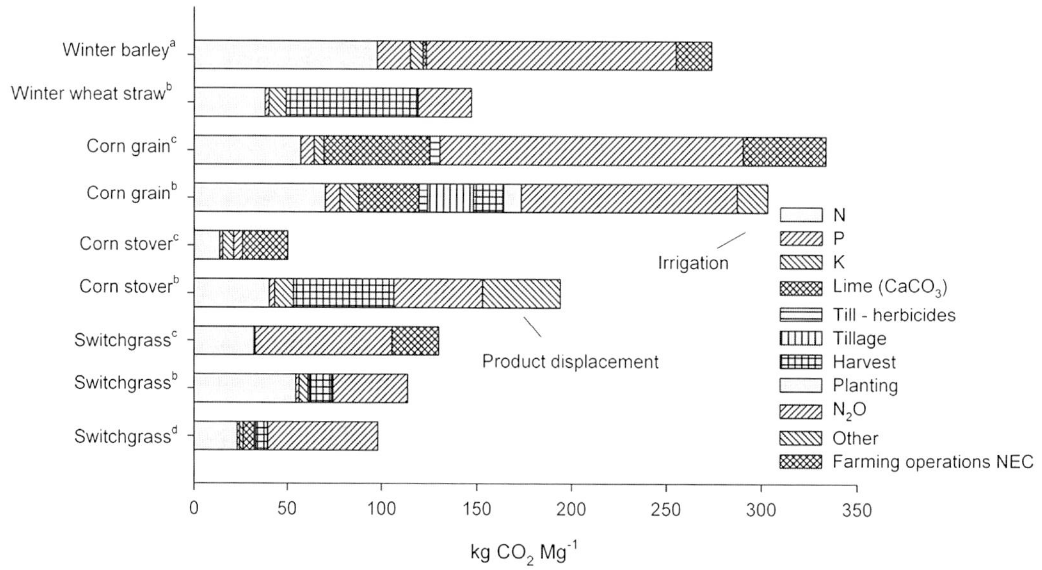

FIGURE 12.3

Estimated global warming intensity of specific farming operations and field emissions required to produce one Mg of biomass. [a]Spatari and Stadel, 2011; [b]Hsu et al., 2010; [c]Argone National Laboratory, 2009; Adler et al., 2007. Notes: NEC (not elsewhere categorized).

HETEROGENEITY OF N_2O EMISSIONS IN THE LANDSCAPE

In estimating the life-cycle GHG emissions of a biofuel, the relative importance of the components, their variability, and uncertainty in the magnitude of their estimates vary at different scales. Uncertainty describes how much an estimate differs from the actual value within a sample, while variability describes how much an estimator varies across space and through time. Uncertainty in estimates of the GHG footprint of the various LCA components is a function of the methods used and the availability of input data required by the respective methods. Estimates of emissions from farm operations are relatively certain because the input data (fuel type and amount) are often easily acquired, and the methods to calculate GHG emissions from combustion are highly reliable. Emissions from production and transport of farm inputs are not as certain, because chemicals applied and distance transported can vary substantially across agricultural regions and among individual farmers, so assumptions must be made to derive general estimates. Estimates of soil GHG fluxes are perhaps the most uncertain, as they are influenced not only by farm-management decisions, but also by how these decisions interact with environmental variables. Soil GHG fluxes are difficult to measure, because they are highly variable in space and time, and the spatial coverage and sampling frequency required to reliably measure emissions are not feasible beyond the plot level. Consequently, models of varying complexity are used to quantify these fluxes, but the models have limited reliability because of inherent uncertainty in the input data as well as structural uncertainty implicit in the models themselves.

Soils can be a source or sink of carbon dioxide (CO_2) and methane (CH_4) but are almost always a source of N_2O. We concentrate on soil N_2O emissions because in many (if not most) cases, these appear to be the largest contributor to the GHG footprint of feedstock production (see, e.g., Adler et al., 2007; Kim et al., 2009). Major factors that control soil GHG fluxes are quantity and quality of organic matter inputs to the soil, nutrient availability, redox potential, water stress on microbial activity, and temperature. The most important factor that controls soil N_2O emissions is N inputs. The IPCC Tier 1 methodology, the most commonly used methodology to estimate direct N_2O emissions (as a result of N application within the project boundary), assumes that on an annual basis, 1% of N inputs from fertilizers (synthetic and organic) and unharvested crop residues is emitted directly from soil as N_2O-N (de Klein et al., 2006). At large spatial and temporal scales, this methodology is reliable, but as scale decreases, errors can be large, and the 95% confidence interval for the IPCC Tier 1 emission factor is 0.3 to 3%. The reason for this wide uncertainty range is that, in addition to N inputs, emissions are influenced by other management practices, weather, and soil type. For example, greater than 2% of applied N fertilizer was observed to be lost as N_2O-N from rain-fed corn and soybean cropping in Iowa (Parkin and Kaspar, 2006), compared to less than 0.5% for irrigated corn in Colorado (Halvorson et al., 2010). To investigate how these factors interact to control emissions at the county scale, we used results from DayCent model simulations. DayCent, an ecosystem model of intermediate complexity, represents the processes (nitrification and denitrification) that lead to N_2O emissions, as well as other plant—soil processes (net primary productivity, decomposition, vertical water, and temperature flow). DayCent (Del Grosso et al., 2010) is currently used to generate N_2O emission estimates from agricultural soils for the annual Inventory of U.S. Greenhouse Gas Emissions and Sinks (EPA, 2011) and is described in detail in Chapter 14 of this volume.

We first analyze the impacts of weather and soil type on N_2O emissions across the U.S. at the county level. To minimize the impacts of management, we present results from leguminous hay simulations, because hay management is not as variable as management of other crops in terms of nitrogen addition rates. Results are 20-year means based on DayCent outputs reported in the U.S. Greenhouse Gas Inventory (EPA, 2011). On a per-unit-area basis, N_2O emissions vary at different scales across the country (Figure 12.4A). Emissions

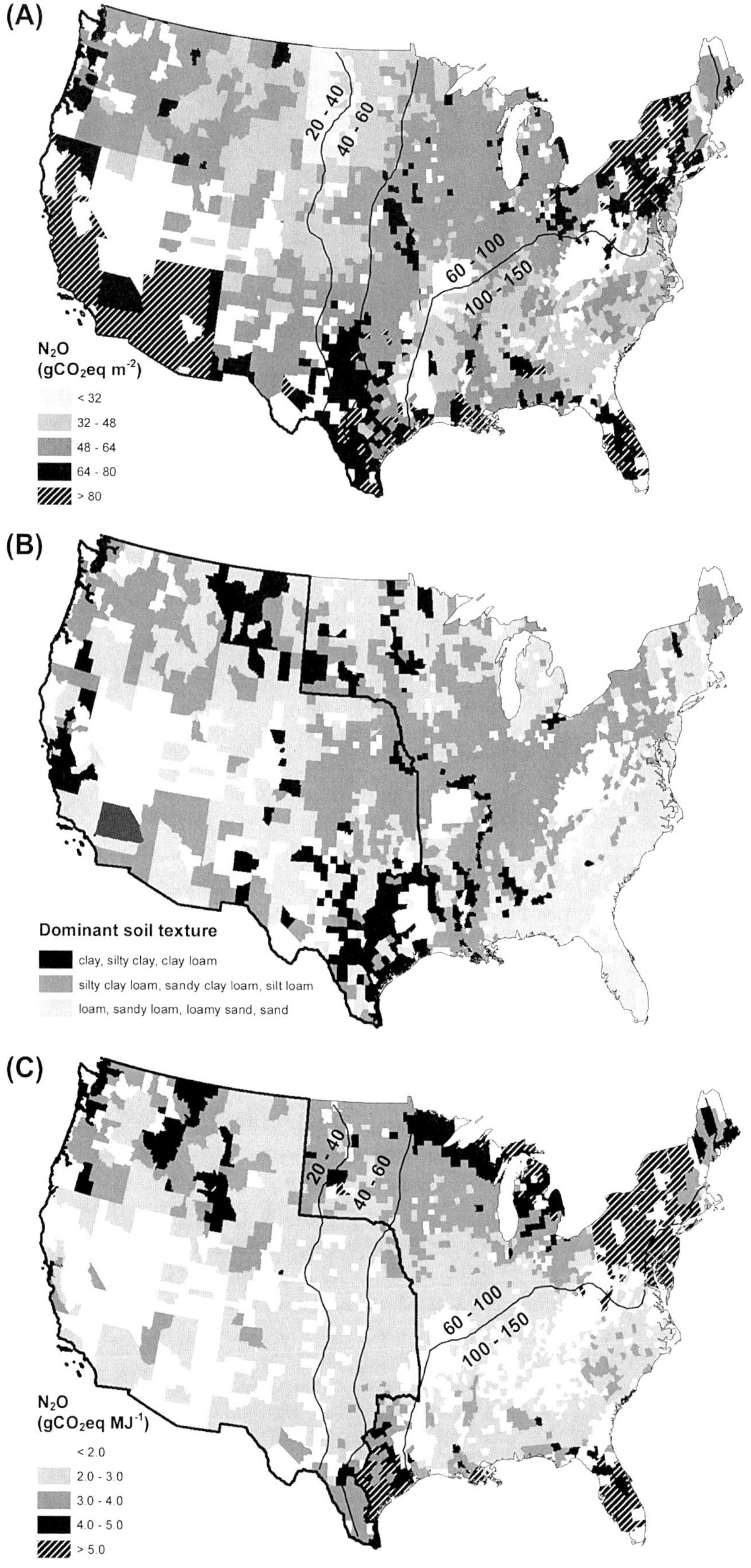

FIGURE 12.4

Impacts of weather and soil type on N_2O emissions from leguminous hay production across the United States. Results are 20-year means based on DayCent county level outputs reported in the U.S. GHG Inventory (EPA, 2011) A—N_2O emissions expressed on a unit area basis (g CO_2eq/m^2), B—dominant soil textures, and C—yield weighted N_2O emissions (g CO_2eq/MJ). To convert DayCent simulated hay yields from g C/m^2 to MJ/m^2, we assumed a biomass carbon concentration of 45% and a higher heating value (HHV) of 18 MJ/kg biomass.

tend to be highest in the northeast, the southwest, and parts of Texas. Irrigation in the western U.S. supports high growth and nitrogen fixation rates by alfalfa hay and, hence, high N_2O emissions in some counties in the west. But irrigation interacts with soil texture (Figure 12.4B) and other factors, so emissions are not uniformly high in the irrigated counties. In both the irrigated and non-irrigated regions, emissions from finer-textured soils are sometimes higher but not always. One reason for this inconsistent pattern is that finer textured soils tend to retain more water, which facilitates high denitrification rates, and high denitrification rates often result in high N_2O emissions. But as poorly drained soils become more anaerobic, N_2O is more likely to be further reduced to N_2 before diffusing from soil, thus leading to lower N_2O emissions even though denitrification rates are high. In the north-central region of the country, emissions tend to increase as precipitation increases, but in most counties in the southeast, where precipitation is highest, emissions are moderate to low. Coarse-textured soils in the southeast facilitate nitrate (NO_3) leaching and tend to have lower N_2O emissions, but some counties in Florida have high emissions even though soil texture is coarse. This highlights the importance of using spatially explicit process-based models to quantify emissions, because they have the potential to represent the interactions that control emissions, whereas models based on emission factors do not include these interactions.

Because biomass yield varies by locations, to facilitate direct comparisons, it is instructive to quantify emissions per unit of product (Figure 12.4C). By this metric, emissions per megajoule (MJ) of energy are moderate to low in the southwest but can be high, moderate, or low in most other parts of the country. This implies that biomass yields are much higher in the southwest, where the growing season is long, temperature limitation is minimal, and water supply is controlled. Controlled water supply is expected to increase N use efficiency, because minimal water stress optimizes plant N uptake, so less excess N is available for nitrification and denitrification. Emissions per MJ are also low for most counties in the southeast, where rainfall is high, and the growing season is long. But again, there are some exceptions, because rainfall and temperature interact with other factors such as soil texture, depth of profile, and soil organic matter levels to control emissions.

Similar to direct N_2O emissions, indirect N_2O emissions are driven by soil properties, weather, N additions, and other management practices. Indirect N_2O is from N that left the farm in a form other than N_2O (volatilized NO_x and NH_3 and leached NO_3) and was converted to N_2O offsite. On average, indirect N_2O emissions are estimated to total about 30% the amount of direct emissions from cropped soils across the U.S. (USDA, 2011). Similar to direct emissions, DayCent predicts higher indirect emissions in wetter areas and as N amendments increase. In contrast to direct N_2O emissions, DayCent model results show higher indirect emissions from coarse soils, compared to those from medium- and fine-textured soils, due to coarser textured soils having higher rates of NO_3 leaching and N volatilization. In sum, the patterns exhibited by the DayCent model are consistent with field observations and other models. Stehfest and Bouwman (2006) derived statistical models for N_2O and NO emission based on observations from 58 studies around the world and found higher N_2O emissions from fine-textured compared to medium- and coarse-textured soils. Similarly, field observations show that NO_3 leaching tends to be higher in coarse-textured soils (Delgado and Bausch, 2005). When total N_2O emissions (i.e. direct and indirect N_2O combined) are considered, DayCent shows that emissions increase with N additions and precipitation, and are highest with fine-textured soils and lower but variable for medium- and coarse-textured soils.

The results shown in Figures 12.4A and 12.4B are consistent with those of Rochette et al. (2008), who found that N_2O emissions factors varied regionally across Canada. Although crop management decisions are constrained by climate and soil fertility within regions, there is still expected to be high variability among farms within regions, due to specific crop requirements, management decisions, and environmental conditions (e.g. type, rate, and timing of fertilizer application and soil properties). Even within farms, spatial variability is expected to be high, due to differences in crop rotations, landscape position, and soil texture. Observations show that topography can significantly influence measured emissions (Grant and Patty, 2003). Unfortunately, models such as DayCent do not entirely account for variability due to landscape position, because DayCent is a one-dimensional model that does not account for topography and lateral transport of water and nutrients. However, these types of models can still be used to identify regional patterns of emissions and general impacts of management options on emissions and guide crop selection and other management decisions, which are within farmers' control.

MITIGATION OPPORTUNITIES TO REDUCE THE GWI OF FEEDSTOCK PRODUCTION

Understanding the major GHG contributors in the life-cycle of a biofuel product can help us identify feasible and cost-effective mitigation strategies, actionable at a farm level. As shown in Table 12.1, N fertilizer use, N_2O emissions, and harvest contribute significantly to the GWI across all feedstocks. The total contribution from N, including both N fertilizer use and soil N_2O emissions, accounted for 42–80% of the GWI for feedstock production in the cases we examined. The GWI contribution of total N from residues was lower than from dedicated feedstocks, since only the portion of N removed with residue harvest was allocated to these feedstocks. Lime, irrigation, and potassium (K) fertilizer are also significant, but only to a specialized group of feedstocks: lime and irrigation for corn grain, displaced forage for corn stover, and K fertilizer for winter and spring wheat.

Phosphorus, herbicides, crop planting, and tillage energy each account for less than 5% of the GWI associated with feedstock production (Table 12.2). In addition, K, lime, product displacement, and irrigation also typically account for less than 5% of the feedstock GWI, except for the cases noted in Table 12.1.

Although energy for tillage will not significantly impact the fuel life-cycle, tillage type can indirectly have a large impact on feedstock GWI through its effect on soil carbon, which we describe further in Table 12.3.

Only those practices that contribute more than 5% of the GWI of feedstock production are considered significant enough to evaluate alternatives for mitigation. Although the following

TABLE 12.1 Input Parameters with >5% Impact on Seed-to-Gate Global Warming Intensity (GWI) using IPCC 100a(v1.02) Weighting. Percent Contribution Towards the Total GWI for Feedstock Production is Shown for Each Parameter Detailed in Figure 12.3 from Feedstocks in Hsu et al. (2010)

	N fertilizer	N_2O	Lime	Harvesting[a]	Irrigation	K fertilizer	Forage displacement	Total N contribution
Feedstock	*Percent contribution (%)*							
Corn grain	18	39	10	5	5	–	–	57
Corn stover	17	25	–	27	–	–	21	42
Winter wheat Straw	25	19	–	47	–	6	–	44
Spring wheat Straw	26	20	–	48	–	6	–	46
Switchgrass	46	34	–	9	–	–	–	80

[a]Harvest includes baling for forage and residue feedstocks. All grain crops assumed a single pass harvester for both grain and residue. Notes: All data are based on feedstock life cycle inventories from Hsu et al. (2010).

TABLE 12.2 Categorization of Input Parameters by Contribution Towards the Total Global Warming Intensity (GWI) for Feedstock Production and Potential for Mitigation

	Percent contribution to GWI	
Practice	<5%	>5%
Standard practice		
	Phosphorus Herbicide Planting Tillage energy	Potassium Lime Harvest Forage displacement Irrigation
Mitigation opportunity		
		N fertilizer N_2O Tillage/SOC

Notes: All data are based on feedstock life-cycle inventories from Hsu et al. (2010). Seed-to-gate GWI was calculated using IPCC 100a (v1.02) weighting factors. The input parameters K, lime, product displacement, and irrigation typically account for less than 5% of the feedstock GWI, except for the cases noted in Table 12.1.

practices each contribute greater than 5% of the GWI for specific crops, there are probably minimal opportunities for GHG mitigation, for the reasons noted:

1. K and lime rates probably cannot be reduced without also affecting crop yield.
2. We consider the harvest technology employed across feedstocks to be mature, without alternatives for improvement.
3. Displacement of corn stover used for animal feed lacks a clear alternative for substitution.

Potential opportunities exist to reduce GWI of irrigation. Conceivably, alternative energy sources and alternative irrigation methods could reduce irrigation's impact on feedstock GWI. In practice, however, such measures would likely be very expensive and difficult to implement, because they would require overhauling existing irrigation systems and/or moving to irrigation systems that are more energy efficient but less water efficient, e.g. moving from center-pivot to flood irrigation.

N fertilizer use, N_2O emissions, and tillage impact on soil carbon are the most promising factors through which to reduce the GWI of feedstock production, with the latter being more important for annual crops. N source, application rate, application time, and placement are key strategies in precision agriculture to reduce N inputs. Of these, changing N source and application rate are the most broadly verified methods to reduce embodied energy and N_2O emissions from N management. The amount of energy required to produce fertilizer varies from one N source to another. Anhydrous ammonia is the first product of the Haber-Bosch process, requires the least amount of energy to produce, and has a lower overall GWI than urea-containing compounds. Urea-containing compounds, when used as a source of N, contribute to GWI in three ways: requiring energy during production and transportation, release of N_2O during nitrification and denitrification, and release of CO_2 from urea $[(NH_2)_2CO]$.

Another opportunity for GWI mitigation is the use of nitrification inhibitors. Nitrification inhibitors are imbedded in fertilizers to inhibit the conversion of ammonium ions to nitrates ions, which release N_2O and N_2 when broken down by denitrifying bacteria. Polymer-coated urea has potential to reduce indirect emissions of N_2O since plant uptake is better matched with N release thereby reducing leaching losses of N. Similarly, variable-rate application of N fertilizer allows for reduced overall N application, higher N-use efficiency by the crop, and reduced N loss to leaching and denitrification.

TABLE 12.3 Mitigation Strategies to Reduce the Global Warming Intensity (GWI) of Bioenergy Feedstock Production; a Case Study for Corn Grain

	Practice	Energy GWI[a]	SOC GWI[b]	N_2O reduction[c]	N fertilizer reduction[d]	Total GWI[e]
N management		**kg CO_2-e (kg N)$^{-1}$**	**kg CO_2-e (ha)$^{-1}$**	**%**	**%**	**kg CO_2-e (Mg)$^{-1}$**
	NH_3	1.9	–	–	–	95
	Urea	2.9	–	–	–	139
	UAN 32[f]	5.5	–	–	–	194
	PCU[g]	2.9	–	14–58	–	102–130
		kg CO_2-e kg^{-1}			–	
	Nitrification inhibitors[h]	4.7	–	31–44	20	54–149
	VRA N[i]	–	–	–	6–46	51–185
Tillage management		**kg CO_2-e (ha)$^{-1}$**			–	
	No-till[j]	0	920–1500	–	–	(–)99–(–)161[k]
	Conservation till	95.0	–	–	–	10
	Reduced till	140.0	280–830	–	–	(–)45–(–)104[k]
	Conventional till	305.0	0	–	–	33

[a]*Energy GWI: global warming intensity, as determined by using IPCC 100a v1.02 weighting factors for N_2O (Forster et al., 2007) and IPCC tier I methods for N_2O emissions and CO_2 emissions from urea, PCU, and UAN 32 (de Klein et al., 2006; Solomon et al., 2007); only the direct emissions associated with producing N fertilizer material (Swiss Center for Life Cycle Inventories, 2009), nitrification inhibitor, and/or performing tillage operations.*

[b]*SOC = soil organic carbon. Tillage impacts on soil carbon represent the projected change, in the top 30 cm, when switching from conventional tillage to a lower-intensity tillage system (reduced or no-till systems; conservation tillage was not an option). The projection is based on 80 studies (Ogle et al., 2005) and a reference SOC stock of 50 Mg C/ha. The range reported for total GWI reflects impacts from both energy and SOC GWI where applicable.*

[c]*Reduction in N_2O emissions reported for PCUs and nitrification inhibitors (Akiyama et al., 2010)*

[d]*N fertilizer reduction, the amount that nitrification inhibitors [nitrification inhibitor impacts on N fertilizer reduction based on DayCent model simulations of the central US (Stephen Del Grosso, unpublished data)] and VRA N [N fertilizer reduction associated with VRA N is from Khosla and Alley (1999)] have been reported to reduce N fertilizer input without negative yield consequences.*

[e]*Total GWI = [(energy GWI for N fertilizer material × N application rate × (1 – N fertilizer reduction factor/100)) + (GWI energy for nitrification inhibitor × nitrification inhibitor application rate) + (N application rate × (1 – N_2O reduction factor/100) × (1 – N fertilizer reduction factor/100) × N_2O emission factor × N_2O GWP weighting factor) + ((N application rate/(N content/100)) × carbon content × CO_2:C ratio)]. Where: the feedstock is corn grain, grain yield (15.5% moisture) = 9.3 Mg ha^{-1} (USDA-NASS, 2011); mean N fertilizer application rate = 157 kg N ha^{-1} (USDA-NASS, 2011); N_2O emission factor = 1.25% N applied is released as N_2O (de Klein et al., 2006); production-related GWI of fertilizer products (energy GWI) are from EcoInvent v 2.1 (2009); N_2O GWP weighting factor = 298 (Forster et al., 2007); N content = 46% (urea, PCU), 32% (UAN 32); carbon content = 20% (urea, PCU), 7% UAN 32; CO_2:C ratio = 44:12 = 3.67.*

[f]*UAN 32 = Urea ammonium nitrate 32% N.*

[g]*PCU = Polymer-coated urea.*

[h]*Dicyandiamide was used in this table as an example of one, of many, denitrification inhibitors. Application rate can vary from 0.2 to 2 kg ha^{-1} (Gowariker et al., 2009).*

[i]*VRA N = Variable-rate application of N fertilizer.*

[j]*Energy GWI for no-till was listed as 0 since no tillage operations were performed.*

[k]*The (–) sign represents the sequestration of atmospheric CO_2 as SOC.*

Excess N not incorporated into plant biomass can be converted to N_2O or lost from the system via other pathways such as leaching. Efficient mitigation strategies should aim to minimize excess N, while ensuring there is sufficient N to satisfy plant demand. These strategies include application of nitrification and urease inhibitors, time-released fertilizer, applying fertilizer during the growing season, strategically placing fertilizer in the rooting zone instead of uniformly across the soil surface, reducing the amount of N fertilizer applied, and growing winter season cover crops to scavenge excess N. The feasibility and effectiveness of these strategies vary regionally. Nitrification inhibitors appear to be consistently reliable in the western U.S. for both dryland wheat and irrigated corn. However, the efficacy of these chemicals in the central and eastern U.S. varies temporally, and during some years, N_2O emissions are not reduced compared to standard fertilizers (Parkin and Hatfield, 2010; Sistani et al., 2011; Curtis Dell, unpublished data).

Several issues should be weighed when considering the most cost-effective GHG mitigation options. First are emissions under business-as-usual conditions. Parcels of land where the environmental conditions and traditional management practices imply high emissions have a greater potential for emissions reductions than do areas where business-as-usual emissions are low. Second, the impacts of GHG mitigation strategies on crop yields and environmental impacts (e.g. NO_3 leaching) should be considered. Often, strategies intended to reduce GHG emissions also lessen other environmental impacts, but this is not always the case. Third, interactions among different GHG mitigation strategies can be important. For example, reduced tillage intensity usually results in increased carbon storage in soils, but no-till and other conservation tillage practices can increase or decrease N_2O emissions, depending on soil texture and hydrology. Conversion to no-till can initially increase N_2O but as soil structure develops and aeration increases emissions tend to decrease in the long term (Six et al., 2004). No-till is more likely to increase N_2O emissions in poorly drained soils partly because no-till tends to increase soil water content and enhance denitrification rates in these soils (Rochette, 2008). Fourth, the costs and benefits of different mitigation options need to be compared. Lastly, the ability of decision support tools to represent the impacts of different mitigation strategies must be assessed. Currently, decision support tools are limited, because they do not represent the impacts of all available mitigation options (e.g. fertilizer placement), and the options they do represent (e.g. nitrification inhibitors) have not been adequately tested.

A general strategy to mitigate N_2O emissions is to supply N only when plants need it, because it is the excess N that is eligible to be lost from the plant–soil system. Although continuously supplying N to exactly match plant demand is not feasible in field situations, there are strategies to make N supply more synchronous with plant demand. One is to employ multiple applications of N fertilizer during the growing season. The drawback with this approach is that each farm operation has economic and GHG emission costs associated with farm machinery operation. Consequently, fertilizers have been formulated to gradually release N as soil moisture and temperature increases or to stabilize N in the less mobile ammonium form. In theory, stabilized fertilizers should reduce N_2O emissions, but in practice their efficacy varies. The formulations are not perfect, nor is the knowledge of the future weather patterns that control plant N demand and N release after the fertilizer has been applied. Thus, the ability to choose a fertilizer that will release N over an optimal temporal duration is limited, and perfect synchrony between plant N demand and supply cannot be assured. In addition to improving temporal synchrony, placement of fertilizer within the rooting zone also provides a mitigation opportunity. Theoretically, targeting N fertilizer spatially where roots reside should reduce N_2O emissions, but as with stabilized fertilizers, experimental evidence for the efficacy of this option is mixed. One reason experimental results regarding the efficacy of N_2O mitigation options are mixed is because N can leave the plant–soil system in various forms, but typically only N_2O is measured. For example, an experiment may conclude that a stabilized fertilizer was not effective, because N_2O emissions were not reduced compared to a conventional fertilizer. But it is possible that the stabilized fertilizer reduced non-N_2O-N gas losses (NH_3, NO_x) that were not measured. This highlights the importance of accounting for gaseous N and NO_3 leaching losses that contribute to indirect N_2O emissions. In this regard, strategies that minimize NO_3 leaching in wet regions with sandy soils should be considered. Ammonium-based fertilizers with nitrification inhibitors should both reduce N_2O losses from nitrification as well as NO_3 leaching. Indirect emissions from both N gas volatilization and NO_3 are not as important in poorly drained clay, so efforts should be made to decrease direct N_2O emissions, particularly from denitrification. But to our knowledge, there are no commercially available fertilizers designed to inhibit denitrification.

PRECISION AGRICULTURE

Although on a global scale, carbon emissions related to land-use change are more important to control for, in terms of reducing agriculture's GHG footprint, than are the emissions of CH_4 and N_2O (Burney et al., 2010), from the standpoint of an individual producer, controlling and managing chemical inputs, especially N fertilizer, is actionable at the farm level. Meeting the world's food, feed, and fiber demands while reducing N_2O emissions without extensification is a lofty but potentially achievable goal. Only 30–50% of applied N fertilizer is taken up and used by the crop (Cassman et al., 2002; Smil, 1999); improving N use efficiency would reduce both the direct and indirect emissions of N_2O from farm fields. Precision agriculture does offer an opportunity to better control the quantity, timing, and placement of chemicals without reducing yields. Numerous studies have shown improved N management in various crops through precise placement and timing of fertilizer application. For example, variable rate application of N has been shown to reduce overall N application, as compared to recommended application rates, in corn by 22 kg N per hectare (ha) (Khosla and Alley, 1999) and 39 kg N per ha (Bausch and Diker, 2001) and by 42 kg N per ha for winter wheat (Mulla and Bhatti, 1997). Delayed N fertilization has also been shown to have little to no impact on yields in corn (Scharf et al., 2002). Delayed N application (i.e. applying N fertilizer after the start of the growing season) not only has potential to increase nitrogen use efficiency, but may also be effective for reducing denitrification by avoiding cooler temperatures and wetter conditions associated with spring application at the time of planting. Precision fertilizer management should be used in combination with other best management practices to maximize any GHG emissions reductions. For example, Snyder (2008) identifies best management practices in four key areas—N source, N rate, N timing, and N placement—that should be used to minimize GHG emissions related to N application. As discussed before, the efficacy of these options intended to reduce N_2O emissions is relatively uncertain, because results from experiments are not consistent. However, it is certain that crop yields per unit of N applied have increased in the past and are continuing to increase. By implication, GHG intensity (emissions per unit of product) is decreasing. As GHG intensity continues to decrease, accumulation of experimental evidence should improve our understanding of soil N transformations and fluxes. This understanding can be incorporated into more sophisticated models that better represent the factors that control emissions and can be used to identify best management practices that not only decrease GHG intensity, but decrease absolute emissions from agricultural systems as well. Although most precision agriculture studies are focused on conventional crops, there is nothing precluding its effectiveness and use in biomass cropping systems. For example, using in-season remote sensing and variable-rate application of N could be very effective at optimizing inputs in a cropping system that includes residue removal. As opposed to using an application rate uniformly based on the amount of nutrient contained in the biomass plus some yield goal, in-season remote sensing provides an opportunity to detect areas of a field that have nutrient deficiencies while variable rate application equipment allows these areas to be fertilized differentially. Likewise, similar precision approaches could be valuable to perennial cropping systems such as switchgrass.

SUMMARY AND CONCLUSIONS

Life-cycle assessment (LCA) is currently being used to quantify the GWI of biofuels expected to play an important role in meeting the mandated requirements such as those specified by the RFS under the EISA of 2007. To date, most LCAs of biofuels have largely taken broad spatial boundary assumptions and have not considered geography-specific variability in feedstock emissions characteristics. We conducted a review and analysis of studies that used deterministic and stochastic approaches to estimate the contributions of feedstock production to the life-cycle GWI of ethanol through point estimates and confidence intervals, respectively. Our review and analysis highlights the importance of considering specific agricultural management operations on the life-cycle GWI of a given feedstock-to-biofuel pathway. The GWI of growing ethanol feedstocks can vary significantly. Moreover, the relative contribution of specific management operations (e.g. fertilizer, tillage) is highly variable among cellulosic feedstocks. As this study shows, there are many opportunities for significant mitigation within the biomass production phase—mitigation opportunities that individual producers can enact. The two key field-level leverage points for mitigating the GWI of producing biomass are managing N fertilizer and N_2O emissions. Judicious choice of N fertilizer product as well as use of nitrification inhibitors offer the potential for substantial reductions in the carbon footprint of growing biomass. Further, incorporating precision nutrient management as well as reduced tillage intensity offers even more potential to minimize biomass production's carbon footprint.

POLICY IMPLICATIONS

Results of this study are a starting point from which more rigorous and perhaps regionally specific data on key leverage points and production systems should be collected to help guide biomass producers toward minimizing the GWI of their particular biomass production system. It should be noted, however, that modifying on-farm management and/or adopting new production methods and technologies necessarily comes with increased risk borne by the producers (i.e. farmers). This point should not be minimized when considering policies geared toward promoting low GWI biofuel production. In order to fully realize the potential of reducing the GWI of producing biofuels, attention needs to be paid to the on-farm activities that have the most impact on the overall GWI of a biofuel. To this end, on-farm mitigation strategies, such as those outlined in this study, should be considered in policies geared towards incentivizing low carbon biofuel production.

Including entity-level measurements of the GWI of feedstock production could incentivize feedstock producers to use lower GWI practices. Currently the site-specific GWI of feedstock production is not specifically accounted for within the LCFS. Compared to a biorefinery, measuring and monitoring production practices and inputs, at the farm level, has heretofore been considered onerous and unreasonable (Farrell, et al., 2007a, b). In addition to being easier to monitor, it has been suggested that the conversion stage has a larger range and more variability in GWI than feedstock production. Although there are many parameters that contribute to the GWI of feedstock production, much of the challenge in tracking and certifying the feedstock GWI can be overcome by focusing on the following high-impact leverage points: N fertilizer product, N_2O emissions, and tillage. Analogous monitoring and incentive programs currently exist. For example, there are programs to capture the credit of management practices on soil carbon (CCX, 2009) and N management strategies over a range of institutions from non-governmental organizations such as On-Farm Network (OFN, 2011), American Carbon Registry (ACR, 2011), and American Farmland Trust with their BMP Challenge program (AFT, 2011) to State and Federal (USDA-NRCS) nutrient management programs. Each program has a system in place to monitor the implementation of the program and therefore proven ability to monitor on-farm. Low-GWI feedstock production programs could effectively monitor N application rate, tillage practices, and crop yield through coordination with agencies such as the USDA-NRCS. However, quantifying N_2O emissions will require a model-based approach because measuring N_2O emissions in the field for program participants is simply impractical. Process-based models are currently used to quantify N_2O emissions from crop production for the U.S. Greenhouse Gas Inventory (EPA, 2011). Although the RSB is using IPCC methods to estimate N_2O emissions, the CSBP plans to develop a tool that biomass growers could use to calculate farm-level GHG emissions. The American Carbon Registry envisions that an aggregator will model many farms within a region, similar to what a biorefinery could do with all feedstock supplied to their facility. Such an approach could then create an incentive for the biorefinery to source the lowest GWI feedstocks within their collection radius.

References

Adler, P.R., Del Grosso, S.J., Parton, W.J., 2007. Life-cycle assessment of net greenhouse-gas flux for bioenergy cropping systems. Ecol. Appl. 17, 675–691.

Akiyama, H., Yan, X., Yagi, K., 2010. Evaluation of effectiveness of enhanced-efficiency fertilizers as mitigation options for N_2O and NO emissions from agricultural soils: meta-analysis. Global Change Biol. 16, 1837–1846.

American Carbon Registry. Available online at http://www.americancarbonregistry.org (verified November 11, August 25, 2011).

American Farmland Trust, BMP Challenge. Available online at http://www.farmland.org/programs/environment/solutions/nutrient-BMP-Challenge.asp (verified November 11, 2011).

Argonne National Laboratory, 2009. The Greenhouse Gases, Regulated Emissions, and Energy Use in Transportation (GREET) Model. Release v1.8c. Argonne National Laboratory, Argonne, IL.

Bausch, W.C., Diker, K., 2001. Innovative remote sensing techniques to increase nitrogen use efficiency of corn. Commun. Soil Sci. Plant Anal. 32, 1371–1390.

Burney, J.A., Davis, S.J., Lobell, D.B., 2010. Greenhouse gas mitigation by agricultural intensification. Proc. Natl. Acad. Sci. USA 107, 12052–12057.

CARB, 2010. Low Carbon Fuel Standard. Final Regulation Order. Available online at http://www.arb.ca.gov/regact/2009/lcfs09/lcfscombofinal.pdf (verified February 13, 2012).

Cassman, K.G., Dobermann, A., Walters, D.T., 2002. Agroecosystems, nitrogen-use efficiency, and nitrogen management. Ambio 31, 132–140.

Chicago Climate Exchange (CCX), 2009. Continuous Conservation Tillage and Conversion to Grassland Soil Carbon Sequestration Offset Project Protocol. https://www.theice.com/publicdocs/ccx/protocols/CCX_Protocol_Conservation_Tillage_and_Grassland.pdf (verified February 12, 2012).

de Klein, C., Novoa, R.S.A., Ogle, S., Smith, K.A., Rochette, P., Wirth, T.C., McConkey, B.G., Mosier, A., Rypdal, K., 2006. N_2O emissions from managed soils, and CO_2 emissions from lime and urea application. In: Eggleston, S. (Ed.), 2006 IPCC Guidelines for National Greenhouse Gas Inventories, Vol. 4. Agriculture, Forestry, and Other Land Use. IGES, Japan, pp. 1–54.

Dehue, B., Meyer, S., Hamelinck, C.N., 2007. Towards a harmonised sustainable biomass certification scheme Available at http://www.ecofys.com/com/publications/documents/HarmonisedSustainableBiomassScheme_final.pdf (verified August 09, 2011). Ecofys Netherlands BV, Ultrecht, Netherlands.

Del Grosso, S.J., Ogle, S.M., Parton, W.J., Breidt, F.J., 2010. Estimating uncertainty in N_2O emissions from U.S. cropland soils. Global Biogeochem. Cycles 24 doi:10.1029/2009GB003544.

Delgado, J.A., Bausch, W., 2005. Potential use of precision conservation techniques to reduce nitrate leaching in irrigated crops. J. Soil Water Conserv. 60, 379–387.

EPA., 2010. Renewable fuel standard program (RFS2) regulatory impact analysis. EPA-420-R-10–006. Assessment and Standards Division, Office of Transportation and Air Quality, Washington, DC.

EPA., 2011. Inventory of U.S. greenhouse gas emissions and sinks: 1990–2009. U.S. Environmental Protection Agency, Office of Atmospheric Programs, Washington, DC.

Fargione, J.E., Plevin, R.J. , Hill, J.D. 2010. The ecological impact of biofuels. Annu. Rev. Ecol. Evol. Syst. 41(1) 351–377.

Farrell, A.E., et al., 2007a. A Low-Carbon Fuel Standard for California, Part 1: Technical Analysis UC Berkeley and UC Davis, Available at http://www.energy.ca.gov/2007publications/UC-1000-2007-002/UC-1000-2007-002-PT1.PDF (verified February 12, 2012).

Farrell, A.E., et al., 2007b. A Low-Carbon Fuel Standard for California, Part 2: Policy Analysis UC Berkeley and UC Davis, Available at http://www.energy.ca.gov/2007publications/UC-1000-2007-002/UC-1000-2007-002-PT2.PDF (verified February 12, 2012).

Forster, P., Ramaswamy, V., Artaxo, P., Berntsen, T., Betts, R., Fahey, D.W., Haywood, J., Lean, J., Lowe, D.C., Myhre, G., Nganga, J., Prinn, R., Raga, G., Schulz, M., Van Dorland, R., 2007. Changes in atmospheric constituents and in radiative forcing. In: Solomon, S., et al. (Eds.), Climate Change 2007: The Physical Science Basis. Contribution of Working Group I to the Fourth Assessment Report of the Intergovernmental Panel on Climate Change. Cambridge University Press, Cambridge, United Kingdom and New York, NY, USA, pp. 129–234.

Geyer, R., Lindner, J., Stoms, D., Davis, F., Wittstock, B., 2010. Coupling GIS and LCA for biodiversity assessments of land use. Int J. Life Cycle Assess. 15 (7), 692–703.

Gowariker, V., Krishnamurthy, V.N., Gowariker, S., Dhanorkar, M., Paranjape, K., 2009. The Fertilizer Encyclopedia. John Wiley and Sons, Hoboken, NJ.

Grant, R.F., Patty, E., 2003. Modelling variability in N_2O emissions from fertilized agricultural fields. Soil Biol. Bioch. 35, 225–243.

Halvorson, A.D., Grosso, S.J.D., Alluvione, F., 2010. Nitrogen source effects on nitrous oxide emissions from irrigated no-till corn. J. Environ. Qual. 39, 1554–1562.

Hsu, D.D., Inman, D., Heath, G., Wolfrum, E.J., Mann, M.K., Aden, A., 2010. Life cycle environmental impacts of selected U.S. ethanol production and use pathways in 2022 supporting information. Environ. Sci. Technol. 44, 5289–5297.

Khosla, R., Alley, M.M., 1999. Soil-specific nitrogen management on Mid-Atlantic coastal plain soils. Better Crops Plant Food 83, 6–7.

Kim, S., Dale, B.E., Jenkins, R., 2009b. Life cycle assessment of corn grain and corn stover in the United States. Int. J. Life Cycle Assess. 14, 160–174.

Liska, A.J., Yang, H.S., Bremer, V.R., Klopfenstein, T.J., Walters, D.T., Erickson, G.E., Cassman, K.G., 2009. Improvements in life cycle energy efficiency and greenhouse gas emissions of corn-ethanol. J. Industr. Ecol. 13, 58–74.

Mulla, D.J., Bhatti, A.U., 1997. An evaluation of indicator properties affecting spatial patterns in N and P requirements for winter wheat yield. In: Stafford, J.D. (Ed.), Precision Agriculture '97. Bios Scientific Publications, Oxford, Herndon, VA, pp. 145–153.

Ogle, S.M., Breidt, F.J., Paustian, K., 2005. Agricultural management impacts on soil organic carbon storage under moist and dry climatic conditions of temperate and tropical regions. Biogeochemistry 72, 87–121.

On-Farm Network, 2011. Available online at http://www.agtechonfarm.net/ (verified November 11, 2011).

Parkin, T.B., Hatfield, J.L., 2010. Influence of nitrapyrin on N_2O losses from soil receiving fall-applied anhydrous ammonia. Agric. Ecosys. Environ. 136, 81–86.

Parkin, T.B., Kaspar, T.C., 2006. Nitrous oxide emissions from corn-soybean systems in the Midwest. J. Environ. Qual. 35, 1496–1506.

Rochette, P., 2008. No-till only increases N_2O emissions in poorly-aerated soils. Soil Tillage Res. 101, 97–100.

Rochette, P., Worth, D.E., Lemke, R.L., McConkey, B.G., Pennock, D.J., Wagner-Riddle, C., Desjardins, R.J., 2008. Estimation of N_2O emissions from agricultural soils in Canada. I. Development of a country-specific methodology. Can. J. Soil Sci. 88, 641–654.

Scharf, P.C., Wiebold, W.J., Lory, J.A., 2002. Corn yield response to nitrogen fertilizer timing and deficiency level. Agron. J. 94, 435–441.

Sheehan, J., Aden, A., Paustian, K., Killian, K., Brenner, J., Walsh, M., Nelson, R., 2004. Energy and environmental aspects of using corn stover for fuel ethanol. J. Industr. Ecol. 7, 117–146.

Sistani, K.R., Jn-Baptiste, M., Lovanh, N., Cook, K.L., 2011. Atmospheric emissions of nitrous oxide, methane, and carbon dioxide from different nitrogen fertilizers. J. Environ. Qual. 40, 1–9.

Six, J., Ogle, S.M., Breidt, F.J., Conant, R.T., Mosier, A.R., Paustian, K., 2004. The potential to mitigate global warming with no-tillage management is only realized when practised in the long term. Global Change Biol. 10, 155–160.

Smil, V., 1999. Nitrogen in crop production: an account of global flows. Global Biogeochem. Cycles 13, 347–662.

Snyder, C.S., 2008. Fertilizer Nitrogen BMPs to Limit Losses that Contribute to Global Warming. International Plant Nutrition Institute, Norcross, GA.

Solomon, S., Qin, D., Manning, M., Chen, Z., Marquis, M., Averyt, K.B., Tignor, M., Miller, H.L. (Eds.), 2007. Climate Change 2007: The Physical Science Basis. Contribution of Working Group I to the Fourth Assessment Report of the Intergovernmental Panel on Climate Change. Cambridge University Press, Cambridge, United Kingdom and New York, NY, USA, pp. 1–996.

Spatari, S., Stadel, A., 2011. Life Cycle Greenhouse Gas Emissions of Winter Barley. Drexel University, Philadelphia.

Spatari, S., MacLean, H.L., 2010. Characterizing model uncertainties in the life cycle of lignocellulose-based ethanol fuels. Environ. Sci. Technol. 44 (22), 8773–8780.

Stehfest, E., Bouwman, L., 2006. N_2O and NO emission from agricultural fields and soils under natural vegetation: summarizing available measurement data and modeling of global annual emissions. Nutr. Cycling Agroecosyst. 74, 207–228.

Swiss Center for Life Cycle Inventories, 2009. Ecoinvent. Release v.2.1. Swiss Center for Life Cycle Inventories. Duebendorf, Switzerland.

Turner, B.T., Plevin, R.J., O'Hare, M., Farrell, A.E., 2007. Creating Markets for Green Biofuels: Measuring and improving environmental performance. UCB-ITS-TSRC-RR-2007-1. UC Berkeley Transportation Sustainability Research Center, Berkeley. April 2007.

United States Government, 2007. Energy Independence and Security Act of 2007, Public Law 110-140, p. 311.

USDA, 2011. Agriculture and Forestry Greenhouse Gas Inventory: 1990–2008. Office of the Chief Economist. U.S. Department of Agriculture, Washington, DC.

USDA-NASS, 2011. National Agricultural Statistics Service Quick Stats [Online] Available by National Agricultural Statistics Service. http://www.nass.usda.gov/QuickStats (verified August 2, 2011).

CHAPTER 13

Greenhouse Gas Fluxes of Drained Organic and Flooded Mineral Agricultural Soils in the United States

Leon Hartwell Allen, Jr.
U.S. Department of Agriculture, Agricultural Research Service, Center for Medical, Agricultural, and Veterinary Entomology (CMAVE), Chemistry Research Unit, Gainesville, FL

CHAPTER OUTLINE

Abbreviations: CO_2, carbon dioxide; GWP, global warming potential; GHG, greenhouse gas; CO_2 equiv., greenhouse gas CO_2 equivalent based on equivalent calculated global warming potential; IPCC, Intergovernmental Panel on Climate Change; CH_4, methane; NO_3, nitrate; N_2O, nitrous oxide; U.S., United States; WFPS, water-filled pore space

Managing Agricultural Greenhouse Gases. DOI: 10.1016/B978-0-12-386897-8.00013-9

INTRODUCTION

Crop production on drained organic soils and flooded mineral soils represent only small areas of the agricultural landscape of the U.S. Undrained organic soils are typically flooded and anaerobic (anoxic), but the upper parts of the soil profile are converted to aerobic (oxic) conditions when drained. In comparison, mineral soils that originally are typically aerobic in the upper parts of the soil profile are converted to anaerobic conditions when flooded for rice culture. Both of these landscapes can have large contributions to GHG exchange, both on a unit land area basis and in total. Carbon dioxide emissions alone from drained organic soils represent a large fraction of agricultural GHG emissions and can offset nearly half of the beneficial sequestration of soil carbon (C) of other cropping systems. Cropping on flooded mineral soils, namely rice production, can lead to large emissions of CH_4 and N_2O depending on management practices. Because cropping systems, GHG exchanges, and potential management practices are different for drained organic soils and flooded mineral soil agriculture, these systems will be discussed in two separate sections. However, techniques for assessing flux amelioration will be discussed jointly toward the end of this chapter.

ORGANIC SOILS (HISTOSOLS AND PEATLANDS)

Organic Soils Background

DEFINITIONS OF ORGANIC SOILS

Couwenberg (2011) identified a number of common names of organic soils, many of which appear in national soil classifications under various names. Furthermore, the definition of Histosol is complex and not uniform. In USDA soil taxonomy, definitions have varied somewhat across time. Organic soil horizon materials are defined by having more organic carbon than contained in the definition of mineral soil materials (Soil Survey Staff, 1999). The organic materials are often referred to as peat or muck. A simple definition of organic soil material content in organic soils (Histosols) would be soil materials having more than 18% soil organic carbon when the mineral fraction contains 60% or more clay. With no clay content in the mineral fraction, soil layers having 12% or more organic carbon content are considered to be Histosols or organic soils. The defining percentage of soil organic carbon follows a linear relationship between these two points, so that an organic soil contains more than 12 to 18% soil organic carbon as the clay content of the mineral fraction increases from 0 to 60%. These definitions might be relaxed when these soils have been modified over long periods of time by drainage for cultivation with subsequent microbial oxidation that may gradually reduce the percentage of organic components, which would thereby increase the percentage of mineral matter content. Soil Survey Staff (1999), Buol et al. (2003), and Couwenberg (2011) provide more detailed definitions of Histosols.

The dominant suborders of Histosols in USDA soil taxonomy are folists, fibrists, hemists, and saprists (Soil Survey Staff, 1999). Folists are more or less freely drained Histosols derived mainly from leaves, twigs, and branches resting on bedrock or fragmental materials. Fibrists, hemists, and saprists are wet Histosols, distinguished by state of decomposition. Fibrists are little decomposed, hemists are moderately decomposed, and saprists are well decomposed. Most folists of the U.S. are found in Hawaii and Alaska, and most fibrists are found in southeastern Alaska. Most hemists are found in Minnesota and Alaska, whereas saprists are found largely in Michigan, Minnesota, Florida, Wisconsin, Louisiana, and parts of the eastern coastal areas (Soil Survey Staff, 1999). In addition to Histosols, the soil order Gelisols (soils over permafrost) include organic soils in the suborder Histels, which are all found in Alaska (Soil Survey Staff, 1999).

There are a number of terms and definitions associated with organic soils, peatlands, and associated organic matter. Some of these include peat, muck, fen, bog, pocosin, mull, mor, and moder (Collins and Kuehl, 2001). Terminology varies from country to country, and not all terms are used worldwide. In the U.S., peat refers to organic soil material in which the original plant materials are recognizable, whereas muck refers to organic soil materials in which the original plant materials are not recognizable. A fen is a peat accumulating wetland which has inflows and outflows, whereas a bog is a peat accumulating area that has no significant inflows or outflows. Furthermore, organic matter accumulation in peatlands can occur in a wide range of landscape and landform conditions (Richardson and Vepraskas, 2001; Rydin and Jeglum, 2006); however, discussion of those aspects is beyond the scope of this chapter.

FORMATION OF PEATLANDS AND ORGANIC SOILS

Peatland and organic soils were formed mainly as a result of flooded or waterlogged conditions that inhibited decomposition of plant materials because diffusion of oxygen was impeded by water. Decomposition is also inhibited by low temperatures in high northern (or low southern) latitudes. Peat material can be divided into three types (limnic organic sediments, fibrous, and woody) based in part on its biological origin associated with landforms and climate (e.g. Rydin and Jeglum, 2006). Limnic peat accumulates in relatively deep water and may be associated with ponds and shallow lakes in glaciated areas. Fibrous peats are formed from mosses such as sphagnum, and from sedges, reeds, or cattails. In northern areas, the sphagnum might be located in raised bogs, whereas sedges, reeds, and cattails would be associated with shallow flooded landscapes. Finally, woody peat is generally associated with deciduous or evergreen trees and undergrowth vegetation. Much of the tropical peatlands are associated with tropical woodland species (Couwenberg et al., 2010).

SPATIAL EXTENT OF PEATLANDS AND ORGANIC SOILS: U.S. AND GLOBAL

Rydin and Jeglum (2006) outlined the difficulty associated with providing accurate estimates of worldwide areas of peatlands because of the variability of the quality of land or soil survey information within various countries. Furthermore, each country defines its own land or soil terms that are governed by national information needs. Moreover, landscapes may gradually transition from peatland to non-peatland, which make boundaries difficult to define. Even within the U.S., state-by-state soil survey boundaries or aerial interpretations that define Histosols may not be exact.

Although organic soils cover about 10.00 Mha in the contiguous 48 states (Lucas, 1982; Stephens et al., 1984), only about 0.75 Mha are drained for agriculture (USDA, 2011). Bridgham et al. (2001) reported 10.00 Mha of peatlands for the lower 48 states plus Hawaii and Puerto Rico, with another 0.88 Mha of non-wetland folists for a total of 10.88 Mha of Histosols (Table 13.1). Recently, Joosten (2010) cited 9.18 Mha of peatland for the lower 48 states. The largest drained area is about 0.24 to 0.28 Mha in the Florida Everglades Agricultural Area (EAA) (Stephens and Johnson, 1951; Snyder, 2005; Allen, 2007) and about 0.10 to 0.15 Mha in the Sacramento-San Joaquin Delta of California (Rojstaczer and Deverel, 1993, 1995; Deverel and Rojstaczer, 1996; Mount and Twiss, 2005). Minnesota and Michigan have larger total areas of organic soils (Lucas, 1982) that are dispersed in many smaller units than the Florida EAA but less total area is drained. Other drained organic soils are found in Wisconsin, Indiana, Iowa, Ohio, Illinois, New York, North Carolina, South Carolina, and Louisiana. According to an assessment by Bridgham et al. (2001), 21 of the lower 49 states (includes Hawaii) have more than 0.05 Mha of Histosols. The rank of these states in total Histosol areas (from largest to smallest) is Minnesota, Michigan, Florida, Wisconsin, Louisiana, North Carolina, Maine, New York, Hawaii, Georgia, Indiana, Massachusetts, Maryland, Mississippi, New Hampshire, Alabama, Washington, New Jersey, South Carolina, California, and Virginia (Table 13.1). Eight states had no indicated Histosols (Arkansas, Arizona, Kansas, Kentucky, Oklahoma, South Dakota, Tennessee, and West Virginia) possibly because mapping units were too small for inclusion.

TABLE 13.1 Area of Organic Soils (Mha) in the United States by States within Regions (Adapted from Bridgham et al., 2001*). Adapted with Permission from CRC and the Taylor Francis Group

State	Rank	Area (Mha)	Data†	State	Rank	Area (Mha)	Data†
Midwest				**South**			
Illinois	24	0.0356	M	Alabama	17	0.0809	S
Indiana	12	0.1490	S	Arkansas		–	M
Iowa	29	0.0301	M	Florida	4	1.5943	S
Kansas		–	M	Georgia	11	0.1879	S
Michigan	3	1.6511	S	Kentucky		–	M
Minnesota	2	2.4345	S	Louisiana	6	0.9537	M
Missouri	37	0.0051	M	Mississippi	15	0.0908	S
Nebraska	38	0.0044	M	North Carolina	7	0.6339	S
North Dakota	42	0.0026	M	Puerto Rico	40	0.0028	M
Oklahoma		–	M	South Carolina	20	0.0650	S
South Dakota		–	M	Tennessee		–	M
Wisconsin	5	1.3476	S	Texas	36	0.0052	M
Total		5.6601		Virginia	22	0.0549	S
				Total		3.6693	
Northeast				**West**			
Connecticut	23	0.0434	S	Alaska	1	4.2460	S
Delaware	25	0.0356	S	Arizona		–	M
Maine	8	0.3965	S	California	21	0.0617	S
Maryland	14	0.0949	M	Colorado	26	0.0335	S
Massachusetts	13	0.1364	M	Hawaii	10	0.1920	M
New Hampshire	16	0.0899	M	Idaho	32	0.0236	S
New Jersey	19	0.0732	M	Montana	31	0.0260	S
New York	9	0.3131	S	Nevada	35	0.0074	S
Ohio	28	0.0309	S	New Mexico	43	0.0001	M
Pennsylvania	33	0.0163	M	Oregon	27	0.0329	S
Rhode Island	34	0.0119	S	Utah	41	0.0028	S
Vermont	30	0.0270	M	Washington	18	0.0790	M
West Virginia		–	M	Wyoming	39	0.0030	M
Total		1.2692		*Total*		4.7079	

Total Histosols = 15.3065 Mha
Total Wetland Histosols = 14.4270 Mha = Total Histosols - Folists (0.8795 Mha)
Total Peatlands = 23.1781 Mha = Total Histosols - Folists + Alaska Histels (8.7511 Mha)
Total Peatlands (lower 49 plus Puerto Rico) = 23.1781 - Alaska (4.4260 + 8.7511) = 10.001

**Soil Survey Staff (1998) Query for Histosol soil components in the National MUIR and STATSGO data sets 8/98. Natural Resource Conservation Service, Lincoln, NE and Statistical Laboratory, Iowa State University, Ames, IA.*
†S = STATSGO, M = MUIR. The highest Histosol area was taken from either STATSGO (State Soil Geographic database) or MUIR (Map Unit Interpretation Record database).
Folist and Histel areas were taken from STATSGO.

Northern Europe, especially Fenno-Scandinavia, has a significant proportion of natural peatlands with large areas that have been impacted by humans for many years (Jones et al., 2004, 2005; Montanarella et al., 2006; Maljanen et al., 2010). These peatlands are especially prevalent in Estonia, Finland, Ireland, Latvia, Netherlands, Norway, Poland, Sweden, and United Kingdom (mainly Scotland), with the greatest land areas in Finland, Sweden, Norway, Poland, United Kingdom, Germany, and Ireland. The actual estimated areas are variable, depending on whether estimates by soil mapping units are derived from the European Soils Database or the Map of Organic Carbon in Topsoils of Europe (Jones et al., 2004; Table 13.1 of Montanarella et al., 2006).

Recent worldwide peatland areas range from 381 to 416 Mha (Rydin and Jeglum, 2006; Joosten, 2010), while estimates for North America range from 136 to 188 Mha (Rydin and

TABLE 13.2 Peatland Areas of the World Summarized from Wetlands International (Adapted from Datasets Compiled by Joosten, 2010). Adapted with Permission from Marcel Silvius and Hans Joosten of Wetlands International

World peatland areas by continent, with listing of countries >2.0 Mha	Surface area Mha	Peatland area Mha	% Peatland	Peat C stock Gtonnes	Forest peatland area, Mha	Degrading peatland area, Mha	% Degraded
Africa total	3033	13.01	0.4	10.78	4.58	1.42	10.9
Sudan	251	2.99	1.2	1.98	0.00	0.10	3.3
America total	4386	154.44	3.5	201.11	39.31	2.27	1.5
Canada & USA (N. America)	2112	135.76	6.4	184.14	25.90	1.51	1.1
Canada	997	113.38	11.4	154.97	15.90	0.18	0.2
USA (Alaska)	152	13.20	8.7	15.50	7.00	0.01	0.1
USA (lower 48)	963	9.18	1.0	13.67	3.00	1.31	14.3
Latin Am/Carib (S. America)	2274	18.67	0.8	16.97	13.41	0.76	4.1
Brazil	855	5.47	0.6	5.44	5.00	0.30	5.5
Peru	128	5.00	3.9	1.00	4.00	0.01	0.2
Asia total	4566	154.57	3.4	184.60	26.53	9.51	6.2
Nontropical Asia							
Russia, Asian part	1360	117.63	8.7	117.61	2.00	0.90	0.8
China	957	3.35	0.3	3.22	0.05	2.71	81.0
Mongolia	157	2.63	1.7	0.75	0.00	1.51	57.4
Tropical Asia							
Indonesia	190	26.55	13.9	54.02	14.00	12.50	47.1
Malaysia	33	2.67	8.1	5.43	1.40	1.20	45.0
Papua New Guinea	46	5.99	12.9	5.98	4.00	0.50	8.3
Europe total	948	50.46	5.3	47.57	18.16	31.27	62.0
Russia, European part	348	19.94	5.7	19.95	5.00	6.26	31.4
Finland	34	7.94	23.5	5.29	6.00	6.33	79.6
Sweden	45	6.56	14.6	5.00	3.00	1.31	19.9
Norway	39	2.97	7.7	2.23	0.27	0.53	17.9
Belarus	21	2.24	10.8	1.31	0.70	1.81	80.8
World	15190	381.36	2.5	452.66	93.15	44.76	11.7
USA (lower 48)/World, %		2.40%		3.02%	3.22%	2.93%	
USA (Alaska)/World, %		3.46%		3.42%	7.51%	0.02%	

Jeglum, 2006; Joosten, 2008, 2010) (Table 13.2). The largest difference in worldwide estimates resides in estimates for the U.S. (especially the Alaska component) [e.g. approximately 62 Mha by Rydin and Jeglum (2006) and 22 Mha by Joosten (2010)]. An earlier estimate of peatland in Alaska was about 60 Mha (Kivinen and Pakarenin, 1981), while Joosten (2010) recently estimated only 13 Mha (Table 13.2). Decreases in worldwide peatland area appear to be based on data such as those reported by Bridgham et al. (2001, 2006, 2007a,b) for Alaska. The estimate in Table 13.1 for the contiguous 48 states is slightly more than 9 Mha, which compares well with an earlier estimate of 10 Mha (Lucas 1982; Stephens et al., 1984) and a historical estimate of about 11 Mha by Bridgham et al. (2001).

The country clusters with the largest peatland areas are Russia (139.0 Mha), Canada plus the U.S. (186.0 Mha), and tropical Southeast Asia (32.4 Mha) (Indonesia, 27.0 Mha; Malaysia, 2.5 Mha; and Papua New Guinea, 2.9 Mha) (Rydin and Jeglum, 2006). However, country clusters differ slightly in the information of Joosten (2010), namely 170.8 Mha for Russia, 135.8 Mha for Canada plus the U.S., and 35.2 Mha for tropical Southeast Asia (Indonesia, 26.5 Mha; Malaysia, 2.7 Mha; and Papua New Guinea, 6.0 Mha). The area "best estimates" of Page et al. (2011) for tropical Southeast Asia are less at 24.8 Mha. Differences among estimates of peatland areas will likely decrease as methods for assessment are refined and unified.

Although tropical peatlands and organic soils are beyond the major focus of this chapter, data will be presented here for comparisons because of their importance in the global carbon balance. The best estimate for total tropical peatland is 44.1 Mha, which is about 11% of the global peatland area (Page et al., 2011). The minimum estimated area is 38.7 Mha. Page et al. (2011) also estimated an area of 21.6 Mha for shallow Histosols, which totals 65.7 Mha for all tropical peatland and organic soils. Tropical region areas are 24.8 Mha for Southeast Asia (Indonesia, Malaysia, Papua New Guinea, Brunei, Thailand, Myanmar, Vietnam, and Philippines), 10.8 Mha for South America, 5.6 Mha for Africa, 2.3 Mha for Central America and Caribbean, and 0.6 Mha for other parts of tropical Asia (China, India, Bangladesh, and Sri Lanka).

Greenhouse Gas Flux from Drained Organic Soils

CARBON DIOXIDE

Carbon dioxide evolution from drained organic soils has been a concern for a number of years (Knipling et al., 1971; Volk, 1973; Armentano, 1980). Although drained organic soils represent less than 1% of the U.S. cropland, they can contribute a large amount to CO_2 emissions per unit land area because of microbial oxidation, which results in soil subsidence. Based on early subsidence measurements (Shih et al., 1979) of about 2.54 cm yr^{-1}, Allen (2007) estimated EAA soils of Florida emitted 25.4 Mg C ha^{-1} yr^{-1} (93.1 Mg CO_2 ha^{-1} yr^{-1}) to the atmosphere. With a land area of 0.240 Mha, this results in an annual efflux of 22.4 Tg CO_2. The most recent estimate of subsidence in the Florida EAA is 1.45 cm yr^{-1} (Shih et al., 1998) which calculates to 14.5 Mg C ha^{-1} yr^{-1} (53.2 Mg CO_2 ha^{-1} yr^{-1}) with an annual efflux of 12.7 Tg CO_2 (Allen, 2007). In a 1-year study, Elder and Lal (2008) reported CO_2 emissions from a drained organic soil in Ohio of 22.5, 18.9, and 20.8 Mg C ha^{-1} yr^{-1} for moldboard/disking, no tillage, and bare herbicide treatments, respectively, which are comparable to emissions in the Florida EAA. Other estimates of subtropical emissions ranged from 9.5 to 21.9 Mg C ha^{-1} yr^{-1} (Armentano and Menges, 1986; Joosten, 2010). The annual efflux estimates of Allen (2007) from the Florida EAA are about 42% of EPA (2011) estimates from 2000 to 2009 of total U.S. organic soil annual C emissions. EPA (2011) estimated organic soil emissions of 27.7 Tg CO_2 from cropland remaining as cropland plus 2.6 Tg CO_2 from land converted to cropland, totaling 30.3 Tg CO_2. This is the same as the USDA (2011) estimate of 30.3 Tg CO_2 for the same period and similar to the estimated 34.5 Tg CO_2 of Ogle et al. (2003) for 1982–1997. Furthermore, the total estimated CO_2 emissions from U.S. organic soils offsets nearly 40% of the average

2003–2008 annual gains in C sequestration (78.3 Tg CO_2) by U.S. grazed lands and cropland mineral soils, calculated from USDA (2011) data.

Globally, GHG emissions from developing tropical areas may be much more important than emissions in the U.S. Based on an emissions versus water table depth relationship of 91 Mg CO_2 ha^{-1} yr^{-1} per m depth, Hooijer et al. (2010) estimated annual CO_2 emissions from croplands on tropical Southeast Asia drained peatland to be 86 Mg CO_2 m^{-2} yr^{-1} (23.5 Mg C m^{-2} yr^{-1}). This rate is similar to early calculated rates of emissions from Florida EAA soils, but it is about 1.6-fold greater than recently calculated emission rates from Florida EAA soils (Allen, 2007) and about 2.5-fold greater than the calculated rates of emission from subtropical peatlands based on the default interpolated emission value of Joosten (2010). Furthermore, Hooijer et al. (2010) estimated total CO_2 fluxes from tropical Southeast Asia peatlands in 2006 to be 632 Tg of CO_2. Based on estimated annual total global emissions of CO_2 of 28,000 Tg, annual emissions of CO_2 from disturbed tropical Southeast Asia peatlands would be about 2.3% of total global emissions. Moreover, CO_2 emissions from tropical Southeast Asia peatlands would be 21- to 23-fold larger than estimated emissions from all U.S. Histosols. Based on the emission estimates of at least 50 Mg CO_2 ha^{-1} yr^{-1} from drained tropical peatlands by Couwenberg et al. (2010), the total annual CO_2 emissions might be twice as large as emissions based on estimates of Hooijer et al. (2010). Clearly, CO_2 emission from drained Histosols is a global phenomenon, and not solely a problem of agriculture on organic soils in the U.S.

NITROUS OXIDE AND METHANE FLUXES IN THE U.S. AND SOUTHERN CANADA

Drained organic soils are not only a source of CO_2, but they also emit N_2O. In a 2-year study of drained organic soils of the Florida EAA, annual N_2O emissions ranged from 7 to 48 kg N_2O-N ha^{-1} for sugarcane, 16 to 97 kg N_2O-N ha^{-1} for St. Augustinegrass [*Stenotaphrum secundatum* (Walt.) Kuntze], and 59 to 165 kg N_2O-N ha^{-1} for fallow (non-cropped) conditions (Duxbury et al., 1982). Lower emissions occurred during a low-rainfall year. For drained organic soils of New York, annual emissions were 72 to 85 kg N_2O-N ha^{-1} for an onion (*Allium cepa* L.) field and 76 to 152 kg N_2O-N ha^{-1} for a sweet corn (*Zea mays* L.) field. Elder and Lal (2008) found annual emissions of 96.9, 35.9, and 29.5 kg N_2O-N ha^{-1} for moldboard/disking, no tillage, and bare herbicide treatments, respectively, in Ohio organic soils. The crops were corn followed by winter wheat (*Triticum aestivum* L.) except for bare soil. The bare treatment in Ohio emitted the least N_2O in contrast to the fallow treatment in the Florida EAA, which emitted the most N_2O. However, the Florida EAA fallow was probably maintained by tillage, but this was not reported specifically. For comparison, annual CO_2 emissions were 22.5, 18.9, and 20.8 Mg CO_2-C ha^{-1} yr^{-1} for moldboard/disking, no tillage, and bare herbicide treatments, respectively. The CO_2 eq. of N_2O emissions measured by Elder and Lal (2008) was 20 to 33% of the CO_2 emissions. Annual CH_4 fluxes were 0.9, 7.6, and −4.5 kg CH_4-C ha^{-1} for these same respective treatments. These annual CH_4 fluxes were essentially negligible, although the maximum range of individual daily fluxes varied from about −300 (uptake) to +300 (emission) g CH_4-C ha^{-1} d^{-1} (Elder and Lal, 2008).

Emissions of GHGs from organic soils drained for agriculture in southern Canada should be similar to emissions in the northern U.S. Rochette et al. (2010) measured annual N_2O emissions of 8.1 kg N_2O-N ha^{-1} yr^{-1} (range of 3.6 to 11.8 kg N_2O-N ha^{-1} yr^{-1}) in 2004 and 25.5 kg N_2O-N ha^{-1} yr^{-1} (range of 13.0 to 40.2 kg N_2O-N ha^{-1} yr^{-1}) in 2005 for drained, cultivated organic soils south of Montreal, Canada. Water tables were higher in 2005 than 2004 for both irrigated and non-irrigated fields. There was no relationship of emissions to N fertilization but there was a negative linear relationship between cumulative N_2O emissions and mean water table depth over the range of 0.59 to 0.79 m. Cumulative emissions were about 30 kg N_2O-N ha^{-1} yr^{-1} with a 0.59 m mean annual water table depth and about 10 kg N_2O-N ha^{-1} yr^{-1} with a 0.79 m mean annual water table depth. Nitrous oxide emissions were also positively correlated with CO_2 emissions. This relationship implies both lower N_2O emissions and lower CO_2 emissions occurred for increasing mean water table depth over the range of 0.59 to 0.79 m. Increasing emissions of CO_2 are expected with increasing water table depths (e.g. Stephens et al., 1984). However, the relationships might have been different for shallow water tables over the range of 0 to 0.6 m mean water table depth.

NITROUS OXIDE AND METHANE FLUXES FROM NON-U.S. ORGANIC SOILS

There are limited sources of information on N_2O emissions from U.S. drained organic soils, but there are a number of studies in Europe for comparison. In a review of GHG emissions of farmed organic soils in Finland, Sweden, and the Netherlands, Kasimir-Klemedtsson et al. (1997) estimated N_2O emissions ranged from 2 to 38 kg N_2O-N ha^{-1} yr^{-1}, depending largely on water table depth and N-fertilizer applications. Their review also indicated low CH_4 emissions from farmed organic soils (0 to 1.6 kg CH_4 ha^{-1} yr^{-1}) compared to undrained peatland (70 to 200 kg CH_4 ha^{-1} yr^{-1}).

Based on 674 plot studies in various types of drained forest organic soils in Sweden, Ernfors et al. (2008) calculated emissions of 2.0 kg N_2O-N ha^{-1} yr^{-1} with a range of 0.06 to 28.1 kg N_2O-N ha^{-2} yr^{-1}. They found a linear relationship between ln N_2O emission (expressed in g N_2O m^{-2} yr^{-1}) and ln C:N ratio of the organic soil [ln (N_2O emission) = 8.69 − 3.26 ln (C:N); linear regression $r^2 = 0.87$; 95% confidence interval band in the reference], indicating that organic soils which were nitrogen-poor had lower emissions.

228 Flessa et al. (1998) measured N_2O and CH_4 fluxes weekly from four sites in southern Germany that had been drained for about 200 years. Two sites were meadow and two sites were cultivated fields. They found mean annual N_2O emissions (November 1993 through October 1994) of 4.2, 19.8, 15.6, and 56.4 kg N_2O-N ha^{-1} for a fertilized meadow, an unfertilized meadow, a fertilized field with potato (*Solanum tuberosum* L.) in 1993 and maize in 1994, and an unfertilized field with potato in 1993 and rye (*Secale cereale* L.) in 1993–1994, respectively. The average annual CH_4 flux (uptake) was −0.79, −1.22, −0.31, and −0.20 kg CH_4-C ha^{-1} for these same sites. The major N_2O emission rates from the unfertilized meadow occurred from mid-November to mid-December (0.066 g N_2O-N m^{-2} d^{-1}) and were related to high soil nitrate-N content (80 mg NO_3-N kg^{-1}). Likewise, the greatest N_2O emissions from the fertilized field (0.022 g N_2O-N m^{-2} d^{-1}) occurred in early January following a period of high soil NO_3-N content (~100 mg NO_3-N kg^{-1}). Furthermore, soil pH played a major role in annual N_2O emissions. Sites with soil pH values below 4.6 had a mean N_2O emission of 0.040 g N_2O−N m^{-2} d^{-1}. This emission rate was about three times greater than for a site that had a soil pH of ~5.0 and more than 10 times the rate of an average of five sites with soil pH between 6.2 and 6.7 (Flessa et al., 1998). These results suggest low pH is a major factor affecting N_2O emissions from organic soils. Accordingly, soil liming might be a mitigation option for decreasing N_2O emissions of drained organic soils in the U.S. However, the fact that emissions seem to peak in the early winter period might make this option more appropriate for drained organic soils in northern states than southern states and the Sacramento-San Joaquin Delta of California.

Klove et al. (2010) measured mean annual N_2O flux from drained organic soils under cultivation at the Bodin research farm, Northern Norway, and obtained an average of 1.51 kg N_2O-N ha^{-1} yr^{-1}, a value below the IPCC standard value of 8 kg N_2O-N ha^{-1} yr^{-1} for cultivated peat soils (IPCC, 2001) and in the lower range of values reported in Sweden, Germany, and Finland (Maljanen et al., 2003). Part of the reason for this low value might be that no N_2O emissions data were collected for the cold December–March period when high emission rates have often been observed in northern cultivated organic soils.

Maljanen et al. (2003) measured N_2O emissions from 8.3 to 11.0 kg N_2O-N ha^{-1} yr^{-1} from shallow (0.2 m) cultivated organic soils in Finland. Bare tilled plots had annual emissions ranging from 6.5 to 8.9 kg N_2O-N ha^{-1} yr^{-1}. One year of data for forested plots had an average annual emission of 4.2 kg N_2O-N ha^{-1} yr^{-1}. A sizeable portion of the annual N_2O emissions came during the winter (15–36% for the cultivated organic soil and ≤60% for the bare tilled soil). For cultivated soils, emissions tended to increase with water-filled pore space (WFPS) up to 90%. Emissions of N_2O were linearly related to net nitrification in the soil *in situ* from June through September with an $r^2 = 0.58$.

Martikanen et al. (1993) observed N_2O flux of 0.51 to 1.43 kg N_2O-N ha^{-1} yr^{-1} in drained bog and fen sites of Finland. Perhaps these values were low because winter measurements were few and emissions were low for those samplings, in contrast to reports with winter samplings (e.g. Flessa et al., 1998; Maljanen et al., 2003). Undrained minerotrophic fens and low N (0.4 to 0.5%N) ombrotrophic bogs exhibited emissions <0.04 kg N_2O-N ha^{-1} yr^{-1} (Martikanen et al., 1993).

A recent review of GHG balances of managed peatlands in the Nordic countries indicates agriculture on peat soil would release an average of 9.93 kg N_2O-N ha^{-1} yr^{-1} and 1.58 kg CH_4-C ha^{-1} yr^{-1} (Maljanen et al., 2010). Although there were more cases of CH_4 uptake than release, cases of emissions of CH_4 to the atmosphere tended to be larger. Also, the average CO_2 release was 4.88 Mg CO_2-C ha^{-1} yr^{-1}, which is somewhat less than CO_2 releases from drained organic soils in the U.S.

Recent research indicated that mean annual emission of N_2O-N was about 91 kg ha^{-1} for drained, fertilized agricultural peatland of tropical Southeast Asia, compared with about 6 kg N_2O-N ha^{-1} (with upper range of 57 kg N_2O-N) for fens drained for agriculture in temperate Europe (Couwenberg et al., 2010). Data extracted from several other recent sources provided mean annual emissions ranging from 1 to 25 kg N_2O-N ha^{-1} (Martikanen et al., 1993: Klove et al., 2010; Ernfors et al., 2008; Maljanen et al., 2003; Rochette et al., 2010). Moreover, graphs of daily emissions showed that effluxes of N_2O-N were highly variable over time.

Management of Cropped Organic Soils for Mitigating GHG Emissions

Stephens and Stewart (1976) and Stephens et al. (1984) pointed out that organic soil subsidence could be attributed to (1) densification, (2) soil removal by erosion (wind or water) or fire, or (3) microbial biochemical oxidation (Tate, 1980; Tate and Terry, 1980). Stephens et al. (1984) showed microbial oxidation was primarily governed by depth to water table and temperature. Within a given climatic zone, annual subsidence rates were linearly related to depth to water table (Stephens et al., 1984). Little can be done about the effect of soil temperature (other than the cooling effect of plant shading), but the water table can be controlled for production of certain crops, particularly rice. However, Shih et al. (1982) found that soil surface and soil profile temperatures were 1 to 2°C lower in crops of sugarcane than celery (*Apium graveolens* L.), sweet corn, and pasture. They predicted a 7 to 16% reduction in oxidative subsidence of soil in sugarcane based on reductions in soil temperatures due to plant shading. Glaz (1995) and Snyder (2005) explained that shallow water table management and crop selection could reduce microbial oxidation and ameliorate loss of organic soils. Research in Florida indicates shallow water tables for short durations and flooding durations of up to 2 weeks can be tolerated in sugarcane production without sacrificing productivity (Glaz et al., 2005, 2008; Morris, 2005; Glaz and Gilbert, 2006; Glaz and Morris, 2006; Allen, 2007; Gilbert et al., 2007, 2008; Glaz, 2007). Morris et al. (2004) reported that after 1 week of flooding, drainage to a water-table depth of 0.16 m was as effective as the 1-week flood in controlling rates of microbial oxidation, suggesting soil would not have to be continually flooded in order to conserve the Histosols in the Florida EAA. Also, Reddy et al. (1993) found *Typha* spp. growing in nutrient-enriched waters of the natural Florida Everglades could accumulate organic deposits at the rate of 1.1 cm yr^{-1}. Whiting and Chanton (1993) found *Typha* spp. in warm environments to be much more productive than wetland bog species in northern climates. Future research could focus on adapting systems for cropping under shallow water tables and periodic flooding, using strategies to decrease microbial oxidation, thereby potentially sequestering C. However, strategies intended to sequester C in soils can also impact the fluxes of N_2O and CH_4 (USDA, 2011).

Mean/low/high estimates of annual mitigation potentials for restoring (reflooding) organic soils have been calculated to be 73/7/139 Mg CO_2 ha^{-1} for warm climates and 37/4/70 Mg CO_2 ha^{-1} for cool climates (Smith et al., 2008). Estimates by Freibauer et al. (2004) for restoring European peatlands were up to 16.9 Mg CO_2-equiv. ha^{-1} yr^{-1}. For CH_4, mean/low/high estimates of annual mitigation potential by Smith et al. (2008) were −3.32/−0.05/−15.30 Mg CO_2-equiv. ha^{-1} for both warm and cool climates. The negative values indicate an increase of CH_4 emissions upon restoration (reflooding) of organic soil. With regard to N_2O, mean/low/high estimates of annual mitigation potential were 0.16/0.05/0.28 Mg CO_2-equiv. ha^{-1} for both warm and cool climates. The estimates of both Freibauer et al. (2004) and Smith et al. (2008) clearly indicate that restoration of drained organic soils should result in a net reduction in GWP with respect to these GHGs.

Modeling of whole system processes might help in decision making in managing drained organic soils. A few peatland models exist, such as those of Frolking et al. (2002) and St-Hilaire et al. (2010). However, more detailed models are needed to simulate biogeochemical processes affecting GHG exchange induced by drained and disturbed organic soils (e.g. van Huissteden et al., 2006, 2009).

FLOODED AGRICULTURE ON MINERAL SOILS

Rice as the Major Crop in Wetland Agriculture

Wetland agriculture implies mainly rice production systems. The total U.S. area cropped to rice in 2008, 2009, and 2010 was 1.21, 1.27, and 1.47 Mha, respectively, mainly in Arkansas, California, Louisiana, Mississippi, Missouri, and Texas (USDA Crop Production 2010 Summary, 2011). These rice producing areas represented about 1% of total U.S. cropped land. Flooded soils of rice cropping systems are anaerobic and have low redox potentials (Reddy et al., 1986; Yu et al., 2001). Anaerobic methanogenic bacteria metabolize digestible organic matter and release CH_4 in the soil (Mayer and Conrad, 1990). Methane is transported from soil to the atmosphere mainly by diffusion through air channels (aerenchyma) that exist from the roots through the stems and the leaf sheaths of rice plants (Cicerone and Shetter, 1981; Holzapfel-Pschorn and Seiler, 1986). Nouchi et al. (1990) and Mariko et al. (1991) demonstrated that rice plants have micropores on the leaf sheaths that allow CH_4 to escape from the plant to the atmosphere. Oxygen required for root function diffuses to the root via the same air channels (Allen, 1997). Some oxygen that diffuses from roots is used by methanotrophic bacteria to oxidize CH_4 before it escapes from the soil. Using a methyl fluoride CH_4 oxidation inhibitor technique, Epp and Chanton (1993) found oxidation percentages of 23 to 90% for *Sagitteria lancifolia* Willd. (broadleaf arrowhead) and *Pontederia cordata* L. (pickerelweed) grown in microcosms. Using potted *S. lancifolia* plants, Schipper and Reddy (1996) found an average CH_4 oxidation percentage of 65% using the methyl fluoride oxidation inhibitor technique and average CH_4 oxidation percentage of 79% using a mass balance technique. Lombardi et al. (1997) found lower CH_4 oxidation percentages of both *S. lancifolia* and *P. cordata* in field studies than in greenhouse studies, possibly because root densities were less for these plants in the field. Only about 10 to 20% of the CH_4 that is generated in flooded rice soil typically escapes to the atmosphere, based on rice microcosm studies of Holzapfel-Pschorn et al. (1985) and Frenzel et al. (1992).

Utilizing soil that had not been cropped for more than a year, and without any organic additions, Allen et al. (2003) found CH_4 emissions were essentially non-existent for the first 40 days of rice culture although the soil redox potential decreased drastically within 3 days of flooding and continued decreasing gradually for 30 more days. In the absence of metabolizable organic matter, CH_4 emissions were dependent on root exudates or root sloughing for carbon sources (Feldman, 1988; Minoda and Kimura, 1994; Minoda et al., 1996; Watanabe and Kimura, 1998).

Whiting and Chanton (1993) showed CH_4 emissions of wetland systems were linearly related to photosynthetic rates, which were generally higher in fertile, warm wetland systems [such as cattail (*Typha* spp.) and rice]. Later, Whiting and Chanton (2001) studied the balance of CH_4 emissions versus CO_2 uptake of wetland systems and concluded the combined fluxes produced a reduction in net GWP. Ziska et al. (1998) and Allen et al. (2003) found elevated CO_2 enhanced CH_4 emissions of rice via enhanced photosynthetic rates (Baker and Allen, 1993; Allen et al., 1995). Greater photosynthesis provides larger amounts of available nonstructural carbohydrates in rice leaves (Rowland-Bamford et al., 1990, 1996), which can increase translocation of photoassimilates to roots and increase root exudation as a substrate for methanogenesis (Minoda and Kimura, 1994; Minoda et al., 1996; Watanabe and Kimura, 1998). Since the work in China of Khalil et al. (1991), more than a thousand studies have been conducted worldwide on the effects of cultivars, management of organic matter inputs (including plant residues), types and timing of nitrogen fertilization, and water table controls for reducing CH_4 emissions from rice production (Web of Science-Thomson Reuters database search, December 5, 2011). Denier van der Gon (1999) reported declining use of organic inputs in rice farming (in China) was leading to a decrease in CH_4 emissions. Also, more than 200 studies have been conducted worldwide on the combination of the CH_4 and N_2O emissions by rice culture (Web of Science-Thomson Reuters database search, December 5, 2011). Nevertheless, targeted research is needed in the U.S. on strategies of cultivar selection, water table management, and fertility management for decreasing CH_4 and N_2O emissions in rice fields.

Management of Flooded Soil Agriculture for Mitigating GHG Emissions

Management practices to decrease CH_4 and N_2O emissions from U.S. rice systems may include: (1) avoiding application and incorporation of crop residues into the soil; (2) timing applications of mid-season drainage to minimize CH_4 emissions so that they do not coincide with high nitrogen in the soil that could lead to a burst of N_2O emissions; (3) managing nitrogen fertilizers to improve nitrogen-use efficiency through split fertilizer applications and use of nitrification inhibitors; and (4) selection of rice cultivars with low root exudation or root sloughing and with low CH_4 transport capacity (assuming that this would not interfere with oxygen transport capacity) (Majumdar, 2003). Severe or prolonged drainage should not be implemented at panicle initiation (Baker et al., 1997a,b) or flowering (Towprayoon et al., 2005) because this practice could reduce yields.

Cultivar selection affects total CH_4 emissions, apparently related to total biomass production of the crop (Lindau et al., 1995; Huang et al., 1997a,b). Huang et al. (1997b) also found a strong linear relationship between daily CH_4 emissions and both aboveground biomass and root biomass. Furthermore, in Texas rice field conditions, Huang et al. (1997a) generally found greater CH_4 emissions in clayey soils with a high sand content than in clayey soils with a low sand content. Huang et al. (1997a) also developed a relationship among a soil index, a variety index and grain yield for predicting CH_4 emissions. This relationship appears to be a good predictor, but a grain yield to methane emissions efficiency would be more useful for evaluating the best cultivars for sustaining grain yields while minimizing CH_4 emissions.

Research is needed to assure U.S. rice cultural practices minimize GWP through management of CH_4 and N_2O emissions. Modeling could help predict the flow of events influencing GHG emissions, thereby providing insights for designing improved management systems. Huang et al. (1998, 2004) and Sass et al. (2000) proposed a model for predicting CH_4 emissions from flooded rice fields in Texas. They indicated that key predictors of CH_4 emissions were organic matter additions, net productivity of rice, cultivar characteristics, soil texture, and soil temperature. These models have been applied to agricultural management alternatives and water management for predicting GHG emissions of rice in various regions of China (Li et al., 2005). Bossio et al. (1999) accurately predicted CH_4 emissions from a California rice field using a model by Nouchi et al. (1994) based on CH_4 concentration in soil water and temperature. Burning rice straw rather than incorporating it in the soil was found to decrease CH_4 emissions, but contributed to air quality problems and increased CO_2 emissions (Bossio et al., 1999). Harvesting a certain fraction of rice straw for other purposes (e.g. as source of biofuel) might be an option for decreasing CH_4 emissions if economically feasible.

Since emissions from flooded fields include both CH_4 and N_2O, conditions that decrease emissions of one GHG might increase emissions of the other (Bronson et al., 1997). Partial drainage can allow oxygen to penetrate the soil and reduce CH_4 emissions, but this practice can lead to N_2O emissions. In laboratory studies, Yu and Patrick (2003) found soil redox potentials for minimal GWP caused by CH_4 and N_2O production were between −150 and +180 mv at pH of 7.0. Accordingly, CH_4 production would occur at redox potentials below −150 mv and N_2O production would occur at redox potentials above +180 mv. This information could make possible appropriate management of irrigated rice fields for minimizing production of both N_2O and CH_4. Devising management practices for controlling soil redox potentials would require field data and verification. Jiao et al. (2006) found that the redox range for minimal emissions was between −100 mv and +100 mv in a rice field with a pH of 6.7. Furthermore, they found it would be difficult to control N_2O emissions during an experiment with four intermittent field drainage events during the cropping cycle, because redox increased to greater than +100 mv and led to N_2O emissions. Compared to a continuously flooded field, CH_4 emissions were reduced 24% but N_2O emissions were increased 24% by the intermittent drainage (Jiao et al., 2006). Thus, their experience suggests it would be difficult to achieve the goals of ameliorating GHG emissions with water management practices alone, because reductions in emissions of CH_4 might be more than offset by increases in emissions of N_2O, which would lead to greater GWP.

Roy et al. (1997) found a continuous increase in CH_4 production in soil slurries as oxidizing agents of nitrate (NO_3^1-), ferric iron [Fe (III)], and sulfate (SO_4^2-) were depleted. Although some CH_4 was produced in the presence of the oxidizing agents, the amounts became significant only after SO_4^2- was depleted. Lindau et al. (1998) studied the effects of gypsum and phosphogypsum applied at rates of 2.5, 5.0, and 10.0 $Mg\ ha^{-1}$ on CH_4 emissions in field studies. Compared to control plots, gypsum decreased CH_4 emissions 49, 52, and 66%, respectively, and phosphogypsum decreased emissions 47, 46, and 51%, respectively. Nitrous oxide emissions were negligible. Apparently the oxidizing effect of SO_4^2- decreased CH_4 emissions. Increases of ^{226}Ra activities found rice grain grown in phosphogypsum treated soil were small and deemed not a health risk. However, gypsum or phosphogypsum treatments would need to be evaluated in many regions over multiyear periods before they could be considered a management tool for ameliorating GHG emissions in flooded rice culture (Lindau et al., 1998).

ASSESSMENT OF FLUX AMELIORATION

Verification of management effects on GHG emissions from cultivated organic soils and wetland agriculture should include micrometeorological methods such as eddy covariance flux studies (Smith et al., 1994; Baldocchi, 2003; Edwards et al., 2003; Hsieh et al., 2005; Pattey et al., 2006; Kroon et al., 2007; Lohila et al., 2007; Denmead, 2008; Skinner and Wagner-Riddle, 2012). Such methodology could be built around the concepts of the AmeriFlux Network (2011), which shares similarities to systems for obtaining carbon balances over a peat soil in Finland (Lohila et al., 2007; Shurpali et al., 2009). Where applicable, chamber-based flux measurement techniques could complement eddy covariance measurements for assessing management effects on CO_2, CH_4, and N_2O flux (Lohila et al., 2007; Denmead, 2008). Full GHG balances have been reported in several studies involving peat soils, rice fields, and other agricultural systems (Hendriks et al., 2007; McMillan et al., 2007; Soussana et al., 2007), but currently are lacking in the U.S.

CONCLUSIONS AND RECOMMENDATIONS

There is a general lack of published work documenting the contributions of agricultural practices on GHG emissions from both U.S. organic soils and rice culture on mineral soils. However, there is a growing body of information from Europe (and Southeastern Asia) that could help in designing good experimental approaches for documenting the effects of modified agricultural practices (e.g. shallow water tables) on GHG exchanges from drained organic soils. Research in the Florida EAA has focused on adapting sugarcane cropping practices that could ameliorate GHG emissions and diminish organic soil subsidence.

Emission of CO_2 from microbial oxidation of drained organic soils offsets almost half of the carbon sequestered by other cropping systems in the U.S. Nitrous oxide emissions from drained organic soils also contribute to offsetting favorable GHG balances of other cropping systems. Research should be promoted to minimize microbial oxidation of organic soils used for agriculture, including adapting crops that can be grown in higher water table conditions.

Rice culture in flooded mineral soils can emit large amounts of CH_4. Field-scale research should be promoted to evaluate current cultural practices that could ameliorate CH_4 emissions while avoiding N_2O emissions.

Continuous eddy covariance measurements of CO_2, CH_4, and N_2O fluxes, in conjunction with other flux measurement methods, would provide verification of the effectiveness of management strategies for reducing the impact of agriculture on GHG emissions. The same technologies could be applied to both cropped organic soils and rice culture systems.

Acknowledgments

Thanks go especially to Barry Glaz whose comments and suggestions helped shape the content of this chapter particularly regarding drained organic soils, and to an anonymous reviewer. This publication is based in part on work supported by the USDA Agricultural Research Service GRACEnet Project.

References

Allen Jr., L.H., 1997. Mechanisms and rates of O_2 transfer to and from submerged rhizomes and roots via aerenchyma. Soil Crop Sci. Soc. Florida Proc. 56, 41–54.

Allen Jr., L.H., 2007. Carbon balance of sugarcane agriculture on Histosols of the Everglades Agricultural Area: review, analysis, and global energy perspectives. Soil Crop Sci. Soc. Florida Proc. 66, 7–14.

Allen Jr., L.H., Baker, J.T., Albrecht, S.L., Boote, K.J., Pan, D., Vu, J.C.V., 1995. Carbon dioxide and temperature effects on rice. In: Peng, S., Ingram, K.T., H-Neue, U., Ziska, L.H. (Eds.), Climate Change and Rice. Springer-Verlag, Berlin, pp. 258–277.

Allen Jr., L.H., Albrecht, S.L., Colón-Guasp, W., Covell, S.A., Baker, J.T., Pan, D., Boote, K.J., 2003. Methane emissions of rice increased by elevated carbon dioxide and temperature. J. Environ. Qual. 32, 1978–1991.

AmeriFlux Network. 2011. http://public.ornl.gov/ameriflux/ (verified December 9, 2011).

Armentano, T.V., 1980. Drainage of organic soils as a factor in the world carbon cycle. BioScience 30, 825–830.

Armentano, T.V., Menges, E.S., 1986. Patterns of change in the carbon balance of organic soil-wetlands of the temperate zone. J. Ecol. 74, 755–774.

Baker, J.T., Allen Jr., L.H., 1993. Contrasting crop species responses to CO_2 and temperature: rice, soybean and citrus. In: Rozema, J., Lambers, H., van de Geijn, S.C., Cambridge, M.L. (Eds.), CO_2 and Biosphere. Kluwer Academic Publ., Dordrecht, the Netherlands, pp. 239–260.

Baker, J.T., Allen Jr., L.H., Boote, K.J., Pickering, N.B., 1997a. Rice responses to drought under carbon dioxide enrichment. 1. Growth and yield. Global Change Biol. 3, 119–128.

Baker, J.T., Allen Jr., L.H., Boote, K.J., Pickering, N.B., 1997b. Rice responses to drought under carbon dioxide enrichment. 2. Photosynthesis and evapotranspiration. Global Change Biol. 3, 129–138.

Baldocchi, D.D., 2003. Assessing the eddy covariance technique for evaluating carbon dioxide exchange rates of ecosystems: past, present and future. Global Change Biol. 9, 479–492.

Bossio, D.A., Horwath, W.R., Mutters, R.G., van Kessel, C., 1999. Methane pool and flux dynamics in a rice field following straw incorporation. Soil Biol. Biochem. 31, 1313–1322.

Bridgham, S.D., Ping, C.L., Richardson, J.L., Updegraff, K., 2001. Soils of northern peatlands: Histosols and Gleisols. In: Richardson, J.L., Vepraskas, M.J. (Eds.), Wetland Soils: Genesis, Hydrology, Landscapes, and Classification. CRC Press, Boca Raton, FL, USA, pp. 343–370.

Bridgham, S.D., Megonigal, J.P., Keller, J.K., Bliss, N.B., Trettin, C., 2006. The carbon balance of North American wetlands. Wetlands 26, 889–916.

Bridgham, S.D., Megonigal, J.P., Keller, J.K., Bliss, N.B., Trettin, C., 2007a. Wetlands (Chapter 13). In: King, A.W., Dilling, L., Zimmerman, G.P., Fairman, D.M., Houghton, R.A., Marland, G., Rose, A.Z., Wilbanks, T.J. (Eds.), The First State of the Carbon Cycle Report (SOCCR): The North American Carbon Budget and Implications for the Global Carbon Cycle. A Report by the U.S. Climate Change Science Program and the Subcommittee on Global Change Research National Oceanic and Atmospheric Administration. National Climatic Data Center, Asheville, NC, USA, pp. 139–148. Available online at <http://www.climatescience.gov/Library/sap/sap2-2/final-report/default.htm> (verified November 5, 2011).

Bridgham, S.D., Megonigal, J.P., Keller, J.K., Bliss, N.B., Trettin, C., 2007b. Wetlands—Supplemental Materials (Appendix F). In: King, A.W., Dilling, L., Zimmerman, G.P., Fairman, D.M., Houghton, R.A., Marland, G., Rose, A.Z., Wilbanks, T.J. (Eds.), The First State of the Carbon Cycle Report (SOCCR): The North American Carbon Budget and Implications for the Global Carbon Cycle. A Report by the U.S. Climate Change Science Program and the Subcommittee on Global Change Research National Oceanic and Atmospheric Administration. National Climatic Data Center, Asheville, NC, USA, pp. 177–192. Available online at <http://www.climatescience.gov/Library/sap/sap2-2/final-report/default.htm> (verified November 5, 2011).

Bronson, K.F., Neue, H.-U., Singh, U., Abao Jr., E.B., 1997. Automated chamber measurements of methane and nitrous oxide flux in a flooded rice soil: I. Residue, nitrogen, and water management. Soil Sci. Soc. Am. J. 61, 981–987.

Buol, S.W., Southard, R.J., Graham, R.C., McDaniel, P.A., 2003. Soil Genesis and Classification, fifth ed. Iowa State Press, A Blackwell Pub. Co., Ames, Iowa, p. 494 ISBN 0-8138-2873-2.

Cicerone, R.J., Shetter, J.D., 1981. Sources of atmospheric methane: measurements in rice paddies and a discussion. J. Geophys. Res. 86, 7201–7209.

Collins, M.E., Kuehl, R.J., 2001. Organic matter accumulation and organic soils. In: Richardson, J.L., Vepraskas, M.J. (Eds.), Wetland Soils: Genesis, Hydrology, Landscapes, and Classification. CRC Press, Boca Raton, FL, USA, pp. 137–162.

Couwenberg, J., 2011. Greenhouse gas emissions from managed peat soils: is the IPCC reporting guidance realistic? Mires and Peat 8, 1–10. ISSN 1819-754X. available online at. http://www.mires-and-peat.net/ (verified November 5, 2011).

Couwenberg, J., Dommain, R., Joosten, H., 2010. Greenhouse gas fluxes from tropical peatlands in south-east Asia. Global Change Biol. 16, 1715–1732.

Denier van der Gon, H., 1999. Changes in CH_4 emissions from rice fields from 1960 to 1990s. 2. The declining use of organic inputs in rice farming. Global Biogeochem. Cycles 13, 1053–1062.

Denmead, O.T., 2008. Approaches to measuring fluxes of methane and nitrous oxide between landscapes and the atmosphere. Plant and Soil 309, 5–24.

Deverel, S.J., Rojstaczer, S., 1996. Subsidence of agricultural lands in the Sacramento-San Joaquin Delta, California: role of aqueous and gaseous carbon fluxes. Water Resources Res. 32, 2359–2367.

Duxbury, J.M., Bouldin, D.R., Terry, R.E., Tate III, R.L., 1982. Emissions of nitrous oxide from soils. Nature 298, 462–464.

Edwards, G.C., Thurtell, G.W., Kidd, G.E., Dias, G.M., Wagner-Riddle, C., 2003. A diode laser based gas monitor suitable for measurement of trace gas exchange using micrometeorological techniques. Agric. Forest Meteorol. 115, 71–89.

Elder, J.W., Lal, R., 2008. Tillage effects on gaseous emissions from an intensively farmed organic soil in North Central Ohio. Soil Tillage Res. 98, 45–55.

EPA, 2011. Inventory of U.S. greenhouse gas emissions and sinks: 1990–2009, EPA 430-R-11-005. U.S. Environmental Protection Agency, Office of Atmospheric Programs, Washington, DC. Available online at <http://epa.gov/climatechange/emissions/downloads11/US-GHG-Inventory-2011-Complete_Report.pdf> (Verified February 7, 2012).

Epp, M.A., Chanton, J.C., 1993. Rhizospheric methane oxidation determined by the methyl fluoride inhibition technique. J. Geophys. Res. 98, 18413–18422.

Ernfors, M., von Arnold, K., Stendahl, J., Olsson, M., Klemedtsson, L., 2008. Nitrous oxide emissions from drained organic forest soils—an up-scaling based on C: N ratios. Biogeochemistry 89, 29–41.

Feldman, L.J., 1988. The habits of roots. BioScience 38, 612–618.

Flessa, H., Wild, U., Klemisch, M., Pfadenhauer, J., 1998. Nitrous oxide and methane fluxes from organic soils under agriculture. Euro. J. Soil Sci. 49, 327–335.

Freibauer, A., Rounsevell, M.D.A., Smith, P., Verhagen, J., 2004. Carbon sequestration in the agricultural soils of Europe. Geoderma 122, 1–23.

Frenzel, P., Rothfuss, F., Conrad, R., 1992. Oxygen profiles and methane turnover in a flooded rice microcosm. Biol. Fertil. Soils. 14, 84–89.

Frolking, S., Roulet, N.T., Moore, T.R., Lafleur, P.M., Bubier, J.L., Crill, P.M., 2002. Modeling seasonal to annual carbon balance of Mer Bleue Bog, Ontario, Canada. Global Biogeochem. Cycles 16 (3) Art. Number: 1030. doi: 10.1029/2001GB001457.

Gilbert, R.A., Rainbolt, C.R., Morris, D.R., Bennett, A.C., 2007. Morphological responses of sugarcane to long-term flooding. Agron. J. 99, 1622–1628.

Gilbert, R.A., Rainbolt, C.R., Morris, D.R., McCray, J.M., 2008. Sugarcane growth and yield responses to a 3-month summer flood. Agric. Water Manage. 95, 383–291.

Glaz, B., 1995. Research seeking agricultural and ecological benefits in the Everglades. J. Soil Water Conserv. 50, 609–612.

Glaz, B., 2007. Sugarcane response to month and duration of preharvest flood. J. Crop Improve. 20, 137–152.

Glaz, B., Morris, D.R., 2006. Sugarcane morphological, photosynthetic, and growth responses to water-table depth. J. Sust. Agric. 28, 77–97.

Glaz, B., Gilbert, R.A., 2006. Sugarcane response to water table, periodic flood, and foliar nitrogen on organic soil. Agron. J. 98, 616–621.

Glaz, B., Comstock, J.C., Tai, P.Y.P., Edme, S.J., Gilbert, R., Miller, J.D., Davidson, J.O., 2005. Evaluation of new Canal Point sugarcane clones: 2003–2004 harvest season. USDA, ARS, ARS-165. p. 32.

Glaz, B., Reed, S.T., Albano, J.P., 2008. Sugarcane response to nitrogen fertilization on a Histosol with shallow water table and periodic flooding. J. Agron. Crop Sci. 194, 369–379.

Hendriks, D.M.D., van Huissteden, J., Dolman, A.J., van der Molen, M.K., 2007. The full greenhouse gas balance of an abandoned peat meadow. Biogeosciences 4, 411–424.

Holzapfel-Pschorn, A., Seiler, W., 1986. Methane emission during a cultivation period from an Italian rice paddy. J. Geophys. Res. 91, 11803–11814.

Holzapfel-Pschorn, A., Conrad, R., Seiler, W., 1985. Production, oxidation and emission of methane in rice paddies. FEMS Microbiol. Ecol. 31, 343–351.

Hooijer, A., Page, S., Canadell, J.G., Silvius, M., Kwadijk, J., Wösten, H., Jauhiainen, J., 2010. Current and future CO_2 emissions from drained peatlands in Southeast Asia. Biogeosciences 7, 1505–1514.

Hsieh, C.-I., Leahy, P., Kiely, G., Li, C., 2005. The effect of future climate perturbations on N_2O emissions from a fertilized humid grassland. Nutr. Cycl. Agroecosys. 73, 15–23.

Huang, Y., Sass, R.L., Fisher Jr., F.M., 1997a. Methane emission from Texas rice paddy soils. 1. Quantitative multi-year dependence of CH_4 emission on soil, cultivar and grain yield. Global Change Biol. 3, 479–489.

Huang, Y., Sass, R.L., Fisher Jr., F.M., 1997b. Methane emission from Texas rice paddy soils. 2. Seasonal contribution of rice biomass production to CH_4 emission. Global Change Biol. 3, 491–500.

Huang, Y., Sass, R.L., Fisher Jr., F.M., 1998. A semi-empirical model of methane emission from flooded rice paddy soils. Global Change Biol. 4, 247–268.

Huang, Y., Zhang, W., Zheng, X.H., Li, J., Yu, Y.Q., 2004. Modeling methane emission from rice paddies with various agricultural practices. J. Geophys. Res.-Atmospheres. 109 (D08113) doi:10.1029/2003JD004401.

IPCC, 2001. In: Houghton, J.T., Ding, 0, Griggs, D.J., Noguer, M., van der Linden, P.J., Dai, X., Maskell, K., Johnson, C.A. (Eds.), Climate change 2001: the scientific base, contribution of Working Group I to the third assessment report of the Intergovernmental Panel on Climate Change. Cambridge University Press, Cambridge, UK.

Jiao, Z.H., Hou, A.X., Shi, Y., Huang, G.H., Wang, Y.H., Chen, X., 2006. Water management influencing methane and nitrous oxide emissions from rice field in relation to soil redox and microbial community. Comm. Soil Sci. Plant Anal. 37, 1889–1903.

Jones, R.J.A., Hiederer, R., Rusco, E., Montanarella, L., 2005. Estimating organic carbon in the soils of Europe for policy support. European J. Soil Sci. 56, 655–671.

Jones, R.J.A., Hiederer, R., Rusco, E., Loveland, P.J., Montanarella, L., 2004. The map of organic carbon in topsoils in Europe, Version 1.2, September 2003: Explanation of Special Publication Ispra 2004 No.72 (S.P.I.04.72). European Soil Bureau Research Report No. 17, EUR 21209 EN, 26 p. and 1 map in ISO B1 format. Office for Official Publications of the European Communities, Luxembourg. http://eusoils.jrc.ec.europa.eu/esdb_archive/octop/OCtopMapBkLet76.pdf (verified November 30, 2011).

Joosten, H., 2008. Chapter 2: What are peatlands? In: Parish, F., Sirin, A., Charman, D., Joosten, H., Minayeva, T., Silvius, M., Stringer, L. (Eds.), Assessment on Peatlands, Biodiversity and Climate Change: Main Report. Global Environment Centre, Kuala Lumpur and Wetlands International, Wageningen, pp. 8–19.

Joosten, H., 2010. The Global Peatland CO_2 Picture. Wetlands International. Available online at http://www.wetlands.org/WatchRead/tabid/56/mod/1570/articleType/ArticleView/articleId/2418/The-Global-Peatland-CO_2-Picture.aspx (verified November 5, 2011).

Kasimir-Klemedtsson, A., Klemedtsson, L., Berglund, K., Martikainen, P., Silvola, J., Oenema, O., 1997. Greenhouse gas emissions from farmed organic soils: a review. Soil Use Manage. 13, 245–250.

Khalil, M.A.K., Rasmussen, R.A., M-Wang, X., Ren, L., 1991. Methane emissions from rice fields in China. Environ. Sci. Technol. 25, 979–981.

Kivinen, E., Pakarenin, E., 1981. Geographical distribution of peat resources and major peatland complex types in the world. Annals Academiae Scientanum Fennicae Series A 3, 1–28.

Klove, B., Sveistrup, T.E., Hauge, A., 2010. Leaching of nutrients and emission of greenhouse gases from peatland cultivation at Bodin, Northern Norway. Geoderma 154, 219–232.

Knipling, E.B., Schroder, V.N., Duncan, W.G., 1971. CO_2 evolution from Florida organic soils. Soil Crop Sci. Soc. Florida Proc. 30, 320–326.

Kroon, P.S., Hensen, A., Jonker, H.J.J., Zahniser, M.S., van 't Veen, W.H., Vermeulen, A.T., 2007. Suitability of quantum cascade laser spectroscopy for CH_4 and N_2O eddy covariance flux measurements. Biogeosciences 4, 715–728.

Li, C.S., Frolking, S., Xiao, X.M., Moore, B., Boles, S., Qiu, J.J., Huang, Y., Salas, W., Sass, R., 2005. Modeling impacts of farming management alternatives on CO_2, CH_4, and N_2O emissions: a case study for water management of rice agriculture of China. Global Biogeochem. Cycles 19 (3) Art. Number: GB3010. DOI: 10.1029/2004GB002341.

Lindau, C.W., Wickersham, P., DeLaune, R.D., Collins, J.W., Bollich, P.K., Scott, L.M., Lambremont, E.N., 1998. Methane and nitrous oxide evolution and ^{15}N and ^{226}Ra uptake as affected by application of gypsum and phosphogypsum to Louisiana rice. Agric. Ecosys. Environ. 68, 165–173.

Lindau, C.W., Bollich, P.K., DeLaune, R.D., 1995. Effect of rice variety on methane emission from Louisiana rice. Agric. Ecosys. Environ. 54, 109–114.

Lohila, A., Aurela, M., Regina, K., J-Tuovinen, P., Laurila, T., 2007. Wintertime CO_2 exchange in a boreal agricultural peat soil. Tellus 59B, 860–873.

Lombardi, J.E., Epp, M.A., Chanton, J.P., 1997. Investigation of the methyl fluoride technique for determining rhizospheric methane oxidation. Biogeochemistry 36, 153–172.

Lucas, R.E., 1982. Organic soils (Histosols): formation, distribution, physical and chemical properties and management for crop production. Research Report 435 from the Michigan State University Agricultural Experiment Station and Cooperative Extension Service, East Lansing, in cooperation with the Institute of Food and Agricultural Sciences. Agricultural Experiment Stations, University of Florida, Gainesville, pp. 77.

Majumdar, D., 2003. Methane and nitrous oxide emission from irrigated rice fields: proposed mitigation strategies. Current Sci. 84, 1317–1326.

Maljanen, M., Liikanen, A., Silvola, J., Martikainen, P.J., 2003. Nitrous oxide emissions from boreal organic soil under different land-use. Soil Biol. Biochem. 35, 1–12.

Maljanen, M., Sigurdsson, B.D., Guomundsson, J., Oskarsson, H., Huttunen, J.T., Martikainen, P.J., 2010. Greenhouse gas balances of managed peatlands in the Nordic countries—present knowledge and gaps. Biogeosciences 7, 2711–2738.

Mariko, S., Harazono, Y., Owa, N., Nouchi, I., 1991. Methane in flooded soil water and the emission through rice plants to the atmosphere. Environ. Exp. Bot. 31, 343–350.

Martikanen, P.J., Nykanen, H., Crill, P., Silvola, J., 1993. Effect of a lowered water table on nitrous oxide fluxes from northern peatlands. Nature 366, 51–53.

Mayer, H.P., Conrad, R., 1990. Factors influencing the population of methanogenic bacteria and the initiation of methane production upon flooding of paddy soil. FEMS Microbiol. Ecol. 73, 103–112.

McMillan, A.M.S., Goulden, M.L., Tyler, S.C., Stanley, C., 2007. Stoichiometry of CH_4 and CO_2 flux in a California rice paddy. J. Geophys. Res.-Biogeosciences 112 (G1) Art. Number: G01008. doi: 10.1029/2006JG000198.

Minoda, T., Kimura, M., 1994. Contribution of photosynthetic carbon to the methane emitted from paddy fields. Geophys. Res. Letters 21, 2007–2010.

Minoda, T., Kimura, M., Wada, E., 1996. Photosynthates as dominant source of CH_4 and CO_2 in soil water and CH_4 emitted to the atmosphere from paddy fields. J. Geophys. Res. 101, 21091–21097.

Montanarella, L., Jones, R.J.A., Hiederer, R., 2006. The distribution of peatland in Europe. Mires and Peat vol. 1 Art. 01. http://www.mires-and-peat.net/map01/map_1_1.pdf (verified November 30, 2011).

Morris, D.R., 2005. Dry matter allocation and root morphology of sugarcane, sawgrass, and St. Augustinegrass due to water-table depth. Soil Crop Sci. Soc. Florida Proc. 64, 80–86.

Morris, D.R., Glaz, B., Daroub, S.H., 2004. Organic soil oxidation potential due to periodic flood and drainage depth under sugarcane. Soil Sci. 169, 600–608.

Mount, J., Twiss, R., 2005. Subsidence, sea level rise, and seismicity in the Sacramento-San Joaquin Delta. San Francisco Estuary and Watershed Science vol. 3 (Issue 1) (March 2005), Article 5. Available online at http://www.barryyeoman.com/pdfs/mount2005.pdf (verified November 10, 2011).

Nouchi, I., Mariko, S., Aoki, K., 1990. Mechanism of methane transport from the rhizosphere to the atmosphere through rice plants. Plant Physiol. 94, 59–66.

Nouchi, I., Hosono, T., Aodi, K., Minami, K., 1994. Seasonal variation in methane flux from rice paddies associated with methane concentration in soil water, rice biomass and temperature and its modeling. Plant and Soil 161, 195–208.

Ogle, S.M., Breidt, F.J., Eve, M.D., Paustian, K., 2003. Uncertainty in estimating land use and management impacts on soil organic carbon storage for US agricultural lands between 1982 and 1997. Global Change Biol. 9, 1521–1542.

Page, S.E., Rieley, J.O., Banks, C.J., 2011. Global and regional importance of the tropical peatland carbon pool. Global Change Biol. 17, 798–818.

Pattey, E., Strachan, I.B., Desjardins, R.L., Edwards, G.C., Dow, D., MacPherson, J.I., 2006. Application of a tunable diode laser to the measurement of CH_4 and N_2O fluxes from field to landscape scale using several micrometeorological techniques. Agric. Forest Meteorol. 136, 222–236.

Reddy, K.R., Feijtel, T.C., Patrick Jr., W.H., 1986. Effect of soil redox conditions on microbial oxidation of organic matter. In: Chen, Y., Avnimelech, Y. (Eds.), The Role of Organic Matter in Modern Agriculture. Martinus Nijhoff Publ., Dordrecht, the Netherlands, pp. 117–156.

Reddy, K.R., DeBusk, W.F., DeLaune, R.D., Koch, M.S., 1993. Long-term nutrient accumulation rates in the Everglades. Soil Sci. Soc. Am. J. 57, 1147–1155.

Richardson, J.L., Vepraskas, M.J. (Eds.), 2001. Wetland Soils: Genesis, Hydrology, Landscapes, and Classification. CRC Press, Boca Raton, FL, USA.

Rochette, P., Tremblay, N., Fallon, E., Angers, D.A., Chantigny, M.H., MacDonald, J.D., Bertrand, N., L- Parent, E., 2010. N_2O emissions from an irrigated and non-irrigated organic soil in eastern Canada as influenced by N fertilizer addition. Euro. J. Soil Sci. 61, 186–196.

Rojstaczer, S., Deverel, S.J., 1993. Time dependence in atmospheric carbon inputs from drainage of organic soils. Geophys. Res. Letters 20, 1383–1386.

Rojstaczer, S., Deverel, S.J., 1995. Land subsidence in drained histosols and highly organic mineral soils of California. Soil Sci. Soc. Am. J. 59, 1162–1167.

Rowland-Bamford, A.J., Baker, J.T., Allen Jr., L.H., Bowes, G., 1996. Interactions of CO_2 enrichment and temperature on carbohydrate accumulation and partitioning in rice. Environ. Exp. Bot. 36, 111–124.

Rowland-Bamford, A.J., Allen Jr., L.H., Baker, J.T., Boote, K.J., 1990. Carbon dioxide effects on rice carbohydrate status and partitioning in rice. J. Exp. Bot. 41, 1601–1608.

Roy, R., Kluber, H.D., Conrad, R., 1997. Early initiation of methane production in anoxic rice soil despite the presence of oxidants. FEMS Microbiol. Ecol. 24, 311–320.

Rydin, H., Jeglum, J.K., 2006. The Biology of Peatlands. Oxford University Press, UK, p. 360.

Sass, R.L., Fisher Jr., F.M., Huang, Y., 2000. A process-based model for methane emissions from irrigated rice fields: experimental basis and assumptions. Nutr. Cycl. Agroecosys. 58, 249–258.

Schipper, L.A., Reddy, K.R., 1996. Determination of methane oxidation in the rhizosphere of *Sagittaria lancifolia* using methyl fluoride. Soil Sci. Soc. Am. J. 60, 611–616.

Shih, S.F., Glaz, B., Barnes Jr., R.E., 1998. Subsidence of organic soils in the Everglades Agricultural Area during the past 19 years. Soil Crop Sci. Soc. Florida Proc. 57, 20–29.

Shih, S.F., Stewart, E.H., Allen Jr., L.H., Hilliard, J.W., 1979. Variability of depth to bedrock in Everglades organic soil. Soil Crop Sci. Soc. Florida Proc. 38, 66–71.

Shih, S.F., Rahi, G.S., Ozaki, H.Y., Smajstrla, A.G., 1982. Effect of water table and crops on soil temperature. Soil Crop Sci. Soc. Florida Proc. 41, 47–54.

Shurpali, N.J., Hyvonen, N.P., Huttunen, J.T., Clement, R.J., Reichstein, M., Nykanen, H., Biasi, C., Martikainen, P.J., 2009. Cultivation of a perennial grass for bioenergy on a boreal organic soil—carbon sink or source? GCB Bioenergy 1, 35–50.

Skinner, R.H., Wagner-Riddle, C., 2012. Micrometeorological methods for assessing greenhouse gas flux. In: Liebig, M.A., Franzluebbers, A.J., Follett, R.F. (Eds.), Managing Agricultural Greenhouse Gases: Coordinated Agricultural Research through GRACEnet to Address our Changing Climate. Academic Press, San Diego, CA.

Smith, K.A., Clayton, H., Arah, J.R.M., Christensen, S., Ambus, P., Fowler, D., Hargreaves, K.J., Skiba, U., Harris, G.W., Wienhold, F.G., Klemedtsson, L., Galle, B., 1994. Micrometeorological and chamber methods for measurement of nitrous-oxide fluxes between soils and the atmosphere—overview and conclusions. J. Geophys. Res.-Atmospheres 99 (D8), 16541–16548.

Smith, P., Martino, D., Cai, Z., Gwary, D., Janzen, H., Kumar, P., McCarl, B., Ogle, S., O'Mara, F., Rice, C., Scholes, B., Sirotenko, O., Howden, M., McAllister, T., Pan, G., Romanenkov, V., Schneider, U., Towprayoon, S.,

Wattenbach, M., Smith, J., 2008. Greenhouse gas mitigation in agriculture. Phil. Trans. Royal Soc.-B (Biological Sciences) 363, 789–813.

Snyder, G.H., 2005. Everglades Agricultural Area soil subsidence and land use projections. Soil Crop Sci. Soc. Florida Proc. 64, 44–51.

Soil Survey Staff, 1999. Soil Taxonomy; A Basic System for Soil Classification for Making and Interpreting Soil Surveys, second ed. U.S. Department of Agriculture, Natural Resource Conservation Service. Agricultural Handbook number 436. Available online at http://soils.usda.gov/technical/classification/taxonomy/ (verified November 5, 2011).

Soussana, J.F., Allard, V., Pilegaard, K., Ambus, P., Amman, C., Campbell, C., Ceschia, E., Clifton-Brown, J., Czobel, S., Domingues, R., Flechard, C., Fuhrer, J., Hensen, A., Horvath, L., Jones, M., Kasper, G., Martin, C., Nagy, Z., Neftel, A., Raschi, A., Baronti, S., Rees, R.M., Skiba, U., Stefani, P., Manca, G., Sutton, M., Tuba, Z., Valentini, R., 2007. Full accounting of the greenhouse gas (CO_2, N_2O, CH_4) budget of nine European grassland sites. Agric. Ecosys. Environ. 121, 121–134.

Stephens, J.C., Stewart, E.H., 1976. Effect of climate on organic soil subsidence. Proceedings of the Second International Symposium on Land Subsidence, Anaheim, CA, 13–17 December 1976. IAHS-AISH Pub. No. 121, pp. 647–655.

Stephens, J.C., Johnson, L., 1951. Subsidence of organic soils in the upper Everglades region of Florida. Soil Crop Sci. Soc. Florida Proc. 11, 191–237.

Stephens, J.C., Allen Jr., L.H., Chen, E., 1984. Organic soil subsidence. In: Holzer, T.L. (Ed.), Reviews in Engineering Geology, vol. VI. Geological Soc.. of America, Boulder, Colorado, pp. 107–122.

St-Hilaire, F., Wu, J., Roulet, N.T., Frolking, S., Lafleur, P.M., Humphreys, E.R., Arora, V., 2010. McGill wetland model: evaluation of a peatland carbon simulator developed for global assessments. Biogeosciences 7, 3517–3530.

Tate III, R.L., Terry, R.E., 1980. Variations in microbial activity in Histosols and its relationship to soil moisture. Appl. Environ. Micro. 40, 313–317.

Tate III, R.L., 1980. Microbial oxidation of organic matter of Histosols. Chap. 5. In: Alexander, M. (Ed.), Advances in Microbial Ecology, vol. 4. Plenum Publishing Corp., New York, pp. 169–201.

Towprayoon, S., Smakgahn, K., Poonkaew, S., 2005. Mitigation of methane and nitrous oxide emissions from drained irrigated rice fields. Chemosphere 59, 1547–1556.

USDA Crop Production 2010 Summary, 2011. National Agricultural Statistics Service, U.S. Department of Agriculture, January 2011 ISSN: 1057–7823. Available online at http://usda.mannlib.cornell.edu/usda/current/CropProdSu/CropProdSu-01-12-2011_new_format.pdf (verified December 4, 2011).

USDA, 2011. U.S. Agriculture and Forestry Greenhouse Gas Inventory: 1990–2008. Climate Change Program Office, Office of the Chief Economist, U.S. Department of Agriculture, Tech. Bulletin No. 1930. 159 p. June, 2011. Available online at http://www.usda.gov/oce/climate_change/AFGG_Inventory/1990_2008/USDA_GHG_Inv_1990–2008_June2011.pdf (verified November 5, 2011).

Van Huissteden, J., Petruscu, A.M.R., Hendriks, D.M.D., Rebel, K.T., 2009. Sensitivity analysis of a wetland methane emission model based on temperate and arctic wetland sites. Biogeosciences 6, 3035–3051.

Van Huissteden, J., Van den Bos, R.M., Marticorena Alvarez, I., 2006. Modeling the effect of water-table management on CO_2 and CH_4 fluxes from peat soils. Netherlands J. Geosci. 85, 3–18.

Volk, B.G., 1973. Everglades Histosol subsidence: 1. CO_2 evolution as affected by soil type, temperature and moisture. Soil Crop Sci. Soc. Florida Proc. 32, 132–135.

Watanabe, A., Kimura, M., 1998. Factors affecting variation in CH_4 emission from paddy soils grown with different rice cultivars: a pot experiment. J. Geophys. Res. 103, 18947–18952.

Whiting, G.J., Chanton, J.P., 1993. Primary production control of methane emission from wetlands. Nature 364, 794–795.

Whiting, G.J., Chanton, J.P., 2001. Greenhouse carbon balance of wetlands: methane emission versus carbon sequestration. Tellus 53B, 521–528.

Yu, K., Patrick Jr., W.H., 2003. Redox range with minimum nitrous oxide and methane production in a rice soil under different pH. Soil Sci. Soc. Am. J. 67, 1952–1958.

Yu, K.W., Wang, Z.P., Vermoesen, A., Patrick Jr., W.H., van Cleemput, O., 2001. Nitrous oxide and methane emissions from different soil suspensions: effect of soil redox status. Biol. Fertil. Soils 34, 25–30.

Ziska, L.H., Moya, T.B., Wassmann, R., Namuco, O.S., Lantin, R.S., Aduna, J.B., Abao Jr., E., Bronson, K.F., Neue, H.U., Olszyk, D., 1998. Long-term growth at elevated carbon dioxide stimulates methane emission in tropical paddy rice. Global Change Biol. 4, 657–665.

SECTION 4

Modeling to Estimate Soil Carbon Dynamics and Greenhouse Gas Flux from Agricultural Production Systems

CHAPTER 14

DayCent Model Simulations for Estimating Soil Carbon Dynamics and Greenhouse Gas Fluxes from Agricultural Production Systems

Stephen J. Del Grosso[1], William J. Parton[2], Paul R. Adler[3], Sarah C. Davis[4], Cindy Keough[2], Ernest Marx[2]
[1]USDA, Agricultural Research Service, Fort Collins, CO
[2]Natural Resource Ecology Laboratory, Colorado State University, Fort Collins, CO
[3]USDA, Agricultural Research Service, Pasture Systems and Watershed Management Research Unit, University Park, PA
[4]Energy Biosciences Institute, University of Illinois at Urbana-Champaign, Urbana, IL

CHAPTER OUTLINE

INTRODUCTION

Models are needed to quantify soil greenhouse gas (GHG) fluxes, NO_3 leaching, and other carbon and nutrient flows at regional and larger scales because it is not feasible to implement the required measurement intensity. Models currently used for this purpose range from simple spreadsheet calculators which assume that, for example, N_2O emissions are proportional to N inputs, to more complex models that represent the processes that control emissions. In this context, the Intergovernmental Panel on Climate Change defined three approaches for calculating GHG emissions (IPCC, 2006). Briefly, Tier 1 methods use default emission factors, Tier 2 methods use country- or region-specific factors, and Tier 3 methods use complex

Managing Agricultural Greenhouse Gases. DOI: 10.1016/B978-0-12-386897-8.00014-0

simulation models or extensive monitoring systems. An example of the simple approach is IPCC (2006) default Tier 1 methodology which assumes that 1% of N inputs from fertilizer and crop residues are emitted as N_2O directly from soils on an annual basis.

DayCent is a more complex biogeochemical model (IPCC Tier 3 method) that is currently used to estimate N_2O emissions from cropped soils for the U.S. National GHG inventory (Del Grosso et al., 2010; EPA, 2011; USDA, 2011). Advantages of simple models include easy acquisition of input data, minimal computing requirements, and high transparency. However, confidence intervals for estimates based on these types of models tend to be large because they do not account for all the factors that control emissions and how they interact. In contrast, DayCent represents most of the soil and plant processes that control emissions as well as their interactions.

Compared to IPCC Tier 1 methodology, DayCent estimates of N fluxes almost always agree more closely with field measurements (Del Grosso et al., 2005, 2008a; 2008b). However, using complex models has disadvantages, which include: the need for highly resolved input data, state of the art computing infrastructure and programming expertise, and lack of transparency. Nonetheless, we advocate using more complex models when possible to estimate C and nutrient flows because they are likely to yield more reliable results and because they more realistically represent how land management practices interact with each other and environmental factors to control these fluxes. In particular, DayCent accounts for the interactions and feedbacks between management practices such as tillage intensity, irrigation, and fertilizer additions, as well as environmental factors such as soil type and weather.

Even though process-based models are more acculturate and precise than simple models, uncertainties of many quantities (e.g. N and C gas fluxes) are rather large. Similarly, process-based models do a better job of representing the impacts of land management strategies on C and N fluxes, but not all available technologies are well represented in the models. Consequently, it is necessary to continually test model results using field observations and to modify model algorithms to better represent the suite of available land management options. In this chapter we describe the history and current version of the DayCent model, recent model improvements and applications, model limitations and uncertainties, and planned future improvements.

DayCent MODEL HISTORY

DayCent (Del Grosso et al., 2001; Parton et al., 1998) is the daily time step version of the CENTURY model (Parton et al., 1994a). CENTURY was initially developed in the 1980s and was built using knowledge gained from previous models such as the ELM grassland ecosystem model. ELM (Innis, 1978) was a complex ecosystem model developed using data from the Short Grass Steppe Long Term Ecological Research program (SGS-LTER). ELM is computationally complex and uses phenology to control most of the plant growth processes. Plant growth is a function of species-specific quantum efficiencies and light saturation kinetics, phenological stage, temperature, water, and nitrogen and phosphorus limitations. ELM is difficult to parameterize because of limited knowledge regarding how these factors interact to control plant growth. Subsequent models, such as CENTURY, were not as detailed and were easier to use and parameterize. The original CENTURY model (Parton et al., 1987) was a generalized grassland ecosystem model that was developed to simulate biome level plant production, nutrient cycling, and soil organic matter dynamics in the Great Plains. CENTURY was one of the first models used to extrapolate climate change impacts on carbon cycling at regional and global scales (Schimel et al., 1991; Parton et al., 1994b, 1995). In the early 1990s, CENTURY was expanded to simulate different types of grasslands, common crops, forests, and savanna systems (Metherell, 1993).

As interest in greenhouse gas fluxes from ecosystems increased, DayCent was developed in the late 1990s. The monthly time step used in CENTURY is adequate to model plant growth and

SOM changes, but finer resolution is required to represent the processes that control fluxes of N_2O, CH_4, and other trace gases. One reason for this is that these processes often respond non-linearly to important drivers. For example, the majority of N_2O fluxes for a given month can occur during the 2 or 3 days following a rainfall event. A model operating on a monthly time step would not simulate the short-term increase in soil water content following rainfall events that wet soils above field capacity and enhance denitrification rates and N_2O emissions. However, a model operating on a daily time step has the ability to represent the short-term environmental conditions that lead to pulses of trace gas emissions.

The original DayCent model (Parton et al., 1998) simulated soil processes (decomposition, nitrification, denitrification, water flow, temperature flow) on a daily or finer time step while plant production was at a weekly time step. Land management events were scheduled monthly, as in CENTURY. The nitrification and denitrification submodels were refined in the early 2000s (Del Grosso et al., 2000a; Parton et al., 2001) and a CH_4 oxidation submodel was added (Del Grosso et al., 2000b). DayCent was applied to investigate the impacts of climate and land-use change on net soil GHG fluxes (CO_2, CH_4, N_2O) at both the site and regional levels (Del Grosso et al., 2002, 2005; Pepper et al., 2005). Beginning in 2005, DayCent simulations have been used to calculate N_2O emissions from agricultural soils for the annual Inventory of U.S. Greenhouse Gas Emissions and Sinks compiled by the EPA and reported to the United Nations Framework Convention on Climate Change (Del Grosso et al., 2006; EPA, 2011). Model driver and testing data from networks such as GRACEnet (Jawson et al., 2005) and TRAGNET (Ojima et al., 2000) were crucial for model development and testing before DayCent could be applied to simulate regional and national scale GHG emissions for the U.S. National GHG Inventory.

DayCent MODEL OVERVIEW

DayCent simulates exchanges of C, nutrients, and trace gases among the atmosphere, soil, and vegetation. The model operates on a daily time step and requires daily maximum/minimum temperature and precipitation data, and soil texture, bulk density, and hydraulic properties must be specified for each soil layer. Ammonium is assumed to be confined to the top 15 cm layer(s) of soil while nitrate is mobile and can leach out the bottom of the profile. Users can specify soil horizination, but typically, soil layers are 5 cm or thinner near the surface and 15 or 30 cm thick for deeper layers. Common management and disturbance events (plowing, burning, harvesting, fertilizing, irrigating, etc.) can readily be implemented in DayCent.

The model simulates decomposition and nutrient mineralization of plant litter and soil organic matter, plant growth and senescence, and soil water and temperature fluxes (Figure 14.1). Flows of C and nutrients are controlled by the amount of C in the various pools, the C/N and lignin/N ratios of the pools, soil water/temperature factors, and soil physical properties related to texture. Soil organic matter (SOM) is divided into three pools based on decomposition rates (Parton et al., 1994a). Surface organic matter is divided into two pools, active and slow. Decomposition of SOM and external nutrient additions supply the mineral N pool which is available for plant growth and microbial processes that result in N transformations and trace gas fluxes. Plant growth is controlled by vegetation-specific maximum growth and temperature/water response parameters, nutrient availability, and 0-1 multipliers that reflect shading, water, and temperature stress. Net primary productivity (NPP) is allocated among leafy, woody, and root compartments as a function of plant type, phenology, soil water content, and nutrient availability (Metherell et al., 1993). The land surface submodel of DayCent simulates water flow and evapotranspiration for the plant canopy, litter and soil profile, as well as soil temperature throughout the profile (Parton et al., 1998; Eitzinger et al., 2000).

Soil C levels are a function of C inputs from plant litter and organic matter amendments and C losses from decomposition of litter and organic matter. High quality litter and organic matter decompose more quickly and contribute less than low quality litter and organic matter to the

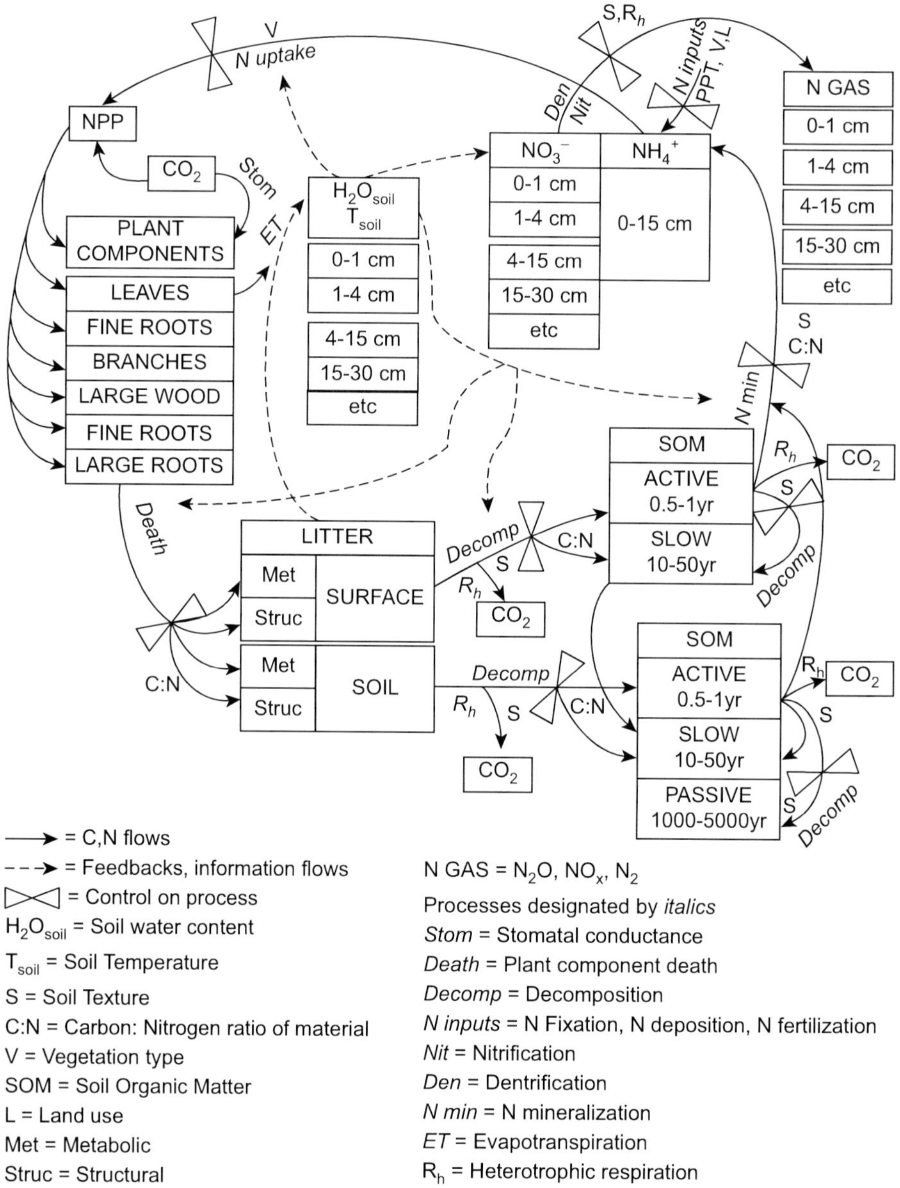

FIGURE 14.1
DayCent model flow diagram.

stable soil C pools. Litter quality is a function of N concentration and the lignin/N ratio of material. Carbon losses from decomposition are a function of soil texture and cultivation. As soils become coarser textured, a higher portion of C in material is assumed to be lost as CO_2 during each stage of the decomposition process. Cultivation events affect decomposition by incorporating surface biomass and litter into the soil and by enhancing decomposition rates. Operationally, users define the portions of aboveground litter and biomass that will be mixed into the soil, the extent to which disturbance events enhance decomposition rates of the SOM pools, and the time period during which decomposition will be enhanced.

The nitrification submodel in DayCent simulates N_2O and NO_x emissions as a function of soil NH_4^+, water content, temperature, pH, and texture (Parton et al., 2001). Nitrification is limited by moisture stress when soil water-filled pore space (WFPS) is too low and by O_2 availability when WFPS is too high. Optimum WFPS for nitrification is approximately 55%, with a higher optimum for clay than sandy soils. The denitrification submodel simulates N_2O, N_2, and NO_x

emissions as a function of soil NO_3, water content, labile C availability, and soil physical properties that influence gas diffusion rates (Del Grosso et al., 2000a). Denitrification, an anaerobic process, does not occur until WFPS exceeds 50–60%, then it increases exponentially as WFPS increases and levels off as soils approach saturation. Users can also simulate application of fertilizers containing nitrification inhibitors. Operationally, the user specifies an amount by which nitrification rates will be reduced and the number of weeks after application that the inhibitor will be effective. In addition to the direct impact on nitrification, the inhibitors also indirectly impact denitrification and NO_3 leaching rates because with the inhibitor, applied N remains in the ammonium form longer, so less NO_3 is available to be denitrified or leached below the rooting zone.

The ability of DayCent to simulate grain and forage yields, N_2O emissions, and NO_3 leaching has been tested for various systems around the world (e.g. Stehfest et al., 2007; Li et al., 2006; Jarecki et al., 2008; Smith et al., 2008; Hartman et al., 2007; Abdalla et al., 2010). Although DayCent does not always show good agreement with measured N_2O values on a daily time scale (Del Grosso et al., 2008a), the model usually shows better agreement with annual mean N_2O emissions than default IPCC (2006) Tier 1 methodology (Del Grosso et al., 2005, 2008b).

RECENT MODEL IMPROVEMENTS

As mentioned above, plant production was at a weekly time step in the original version of DayCent. Also, management events could only be scheduled monthly. For example, more than one fertilizer addition during the same month could not be simulated. Monthly scheduling also limited the ability to simulate growing cover crops because the cover crop could not be planted until the month after the previous crop was harvested. The current version allows daily scheduling of management events and plant growth is calculated daily. This allows more realistic representation of the impacts of management and environmental factors on plant production, microbial processes, and other factors that influence C, nutrient, and GHG fluxes.

RECENT MODEL TESTING

The new DayCent version that includes daily plant growth and event scheduling was recently tested using crop yield data from the National Agricultural Statistics Service (NASS) for three mid-Atlantic counties where rotations involving food, feed, and bioenergy crops are becoming more popular. Data required for model inputs were acquired for Richmond and Nottaway counties in Virginia and Queen Anne County in Maryland. Specifically, average planting and harvest dates and fertilizer addition rates for corn (*Zea mays*), winter wheat (*Triticum aestivum*), winter barley (Hordeum vulgare), full season soybean (*Glycine max*), soybean after winter wheat, and soybean after winter barley were compiled using data from NASS. Soils and weather were acquired by locating the geographical center of cropped land in each county, and the dominant STATSGO soil map unit and DAYMET cell that intersected the center were identified. The STATSGO (Soil Survey Staff, 2011) database includes surface soil texture for association units and DAYMET (Thornton et al., 1997; Thornton and Running, 1999) has daily weather at one square km resolution across the U.S. Before simulating modern crop rotations, pre-settlement vegetation and historical cropping practices were simulated to initialize SOM pools based on potential native vegetation maps and historical records as described in Del Grosso et al. (2006). Modern rotations of summer corn and soybean and winter barley or wheat were conducted for 10 years and average crop yields were calculated. DayCent properly showed highest yields in Queen Anne County and lowest for Nottaway County for all crops, but yields in Queen Anne County tended to be overestimated (Figure 14.2). The model also showed lower soybean after wheat yields compared to full season beans and lower soybean after barley than after wheat, except for Queen Anne County, where the model incorrectly shows higher soybean yields for soybean after barley than after wheat.

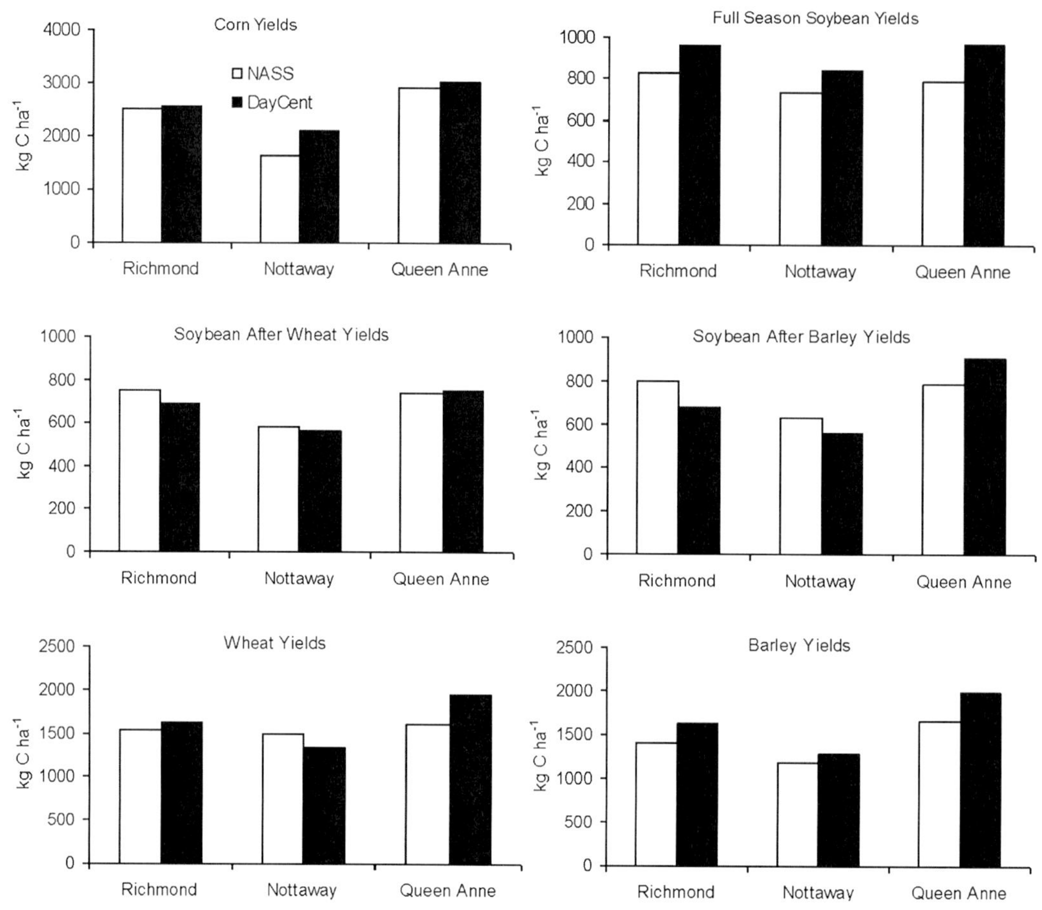

FIGURE 14.2
DayCent simulated crop yields compared with county level crop yield data from the National Agricultural Statistics Service (NASS) for Queen Anne county in Maryland and Richmond and Nottaway counties in Virginia. Values are 10-year averages (2001–2010).

RECENT MODEL APPLICATIONS

DayCent is currently used to calculate N_2O emissions from soils used to grow major crops and non-federal grasslands for the annual Inventory of U.S. Greenhouse Gas Emissions and Sinks (Del Grosso et al., 2010; EPA, 2011). Although model simulations were conducted at the county level for the inventory, results are presented at the agricultural/economic regional level as defined by McCarl et al. (1993). Most states correspond to agricultural/economic regions except some states (e.g., CA, IA, TX) are divided into two or more regions. Briefly, each of the major crops were simulated in each county using county-specific weather data from DAYMET (Thornton et al., 1997; Thornton and Running, 1999; http://www.daymet.org) and soils information from STATSGO database (Soil Survey Staff, 2011). Nitrous oxide emissions were summed for each major crop and aggregated to the regional or state level because key land management data needed to drive the model (crop rotation, fertilizer amount) were only available at this resolution.

Previous analyses have shown that total emissions tend to be high in regions where a large portion of the land area is used for intensive cropping (Del Grosso et al., 2010). On a per unit area basis, emissions tend to be high in finer textured soils that have high SOM, and in some northern regions where soils are subject to freeze/thaw cycles during the spring season (EPA, 2011). Here, we investigate how the portion of N inputs that are emitted from soil as N_2O varies regionally to facilitate comparison with the default IPCC emission factor used to calculate direct soil N_2O emissions. To be consistent with the default IPCC Tier 1 methodology which assumes that 1% of N inputs are converted to direct soil N_2O emissions annually, we divided regional level DayCent generated N_2O emissions by N inputs from synthetic fertilizer amendments, manure amendments, and unharvested crop residues.

DayCent emission factors averaged over 1990–2009 ranged from less than 0.7% to almost 4% (Figure 14.3). An interesting pattern is that most of the high emission factors are in regions receiving relatively high rainfall that have fine textured soils with high SOM (e.g. central and east Texas). In contrast, emission factors are low in the southeast and mid-Atlantic states even though rainfall is high because the soils tend to be coarse textured and low in organic matter, where NO_3 leaching is the major N loss pathway. One exception is Michigan, where the emission factor is high but soils are coarse. In this case, the high emission factor may result from spring freeze/thaw cycles. Surface soils can thaw on top of frozen sub-surface layers and create ponding of water which facilitates denitrification events. In most of the Corn Belt, emissions factors are close to or a little higher than the default IPCC value. When evaluating how water affects the patterns in Figure 14.3, note that the simulations assumed that east of the 98th meridian no irrigation was used, while west of the 98th parallel, all crops besides winter wheat and sorghum were assumed to be irrigated.

Of course, many other factors in addition to soil texture, water content, and SOM influence N_2O emissions. Consequently, correlations with any single controlling variable tend to be poor. This highlights the utility of process-based models that have the potential to represent how these and other factors (e.g. plant N demand, timing of fertilizer application) interact to control emissions, whereas simple methods such as the default IPCC methodology only consider the effect of N inputs. The process-based models can identify regions and time periods where and when emissions are high and mitigation efforts should be targeted. Although the default IPCC Tier 1 methodology cannot be used to identify hot spots and emission pulses at small scales, when estimates are aggregated to sufficiently large scales, this methodology showed good agreement with more complex models and with emissions estimated using atmospheric-based top-down methods (Del Grosso et al., 2008b).

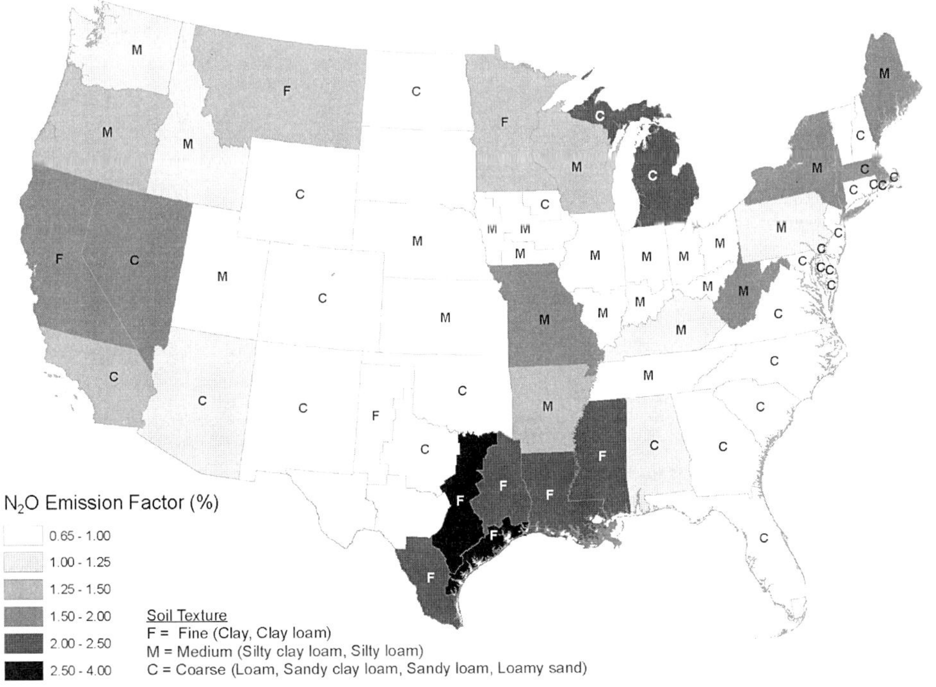

FIGURE 14.3
DayCent generated soil N_2O emission factors as a % of N inputs from fertilizers, manure, and crop residues for major crops in the U.S. For comparison, the default IPCC Tier 1 emission factor for N_2O is 1% of N inputs.

In addition to the U.S. National GHG Inventory, DayCent has also been recently applied to investigate the impacts of biofuel cropping at the regional scale. Simulations suggest that converting land in the central U.S. currently used for corn ethanol cropping to miscanthus (*Miscanthus x giganteus*) for cellulosic feedstock production would reduce GHG by more than 400% and converting to fertilized switchgrass (*Panicum virgatum*) would decrease GHG by about 100% (Davis et al., 2011). Reductions were higher for miscanthus because fast growth increased organic matter inputs and soil C sequestration and N_2O emissions were relatively low. Nitrous oxide emissions were low because no N fertilizer was applied to miscanthus because it appears to facilitate symbiotic N fixation (Davis et al., 2010). As part of biofuel life cycle analysis, DayCent has recently been used to quantify the changes in soil GHG fluxes from introducing winter barley for ethanol production into corn soybean rotations in the mid-Atlantic region (P. Adler, unpublilshed data).

MODEL LIMITATIONS AND PLANNED IMPROVEMENTS

Model limitations can be placed into two broad categories: model output error and inability to represent land management strategies and currently available technologies. Errors in model output derive from uncertainty in model drivers and imperfections in model algorithms and parameterizations (Del Grosso et al., 2010). Errors associated with uncertainty in model drivers can be assessed by performing a series of Monte Carlo simulations in which model drivers are acquired by drawing from probability distribution functions for each model driver. Errors resulting from imperfections in model algorithms and parameters can be quantified by comparing model outputs with observed values from field experiments. After accounting for both of these sources of error, confidence intervals in regional and larger scale GHG emissions calculated using process-based models such as DayCent are narrower than those calculated using simpler models (Del Grosso et al., 2010). However, uncertainty is still rather large compared to estimates of other sources of anthropogenic GHG emissions and tends to increase as spatial and temporal scales decrease (Del Grosso et al., 2010; Ogle et al., 2010).

Recent analysis showed the majority of error in DayCent model outputs is due to imperfections in model algorithms and parameters, as opposed to uncertainty in model drivers (Del Grosso et al., 2010). This suggests efforts to improve the model should concentrate on comparing model outputs with observations for various C and N vectors (grain yields, SOM changes, NO_3 leaching, N_2O, NO_x, NH_3 emissions) from field experiments to identify weaknesses and potential resolutions of model shortcomings. Further testing and improvement of model algorithms and parameters is expected to increase accuracy and reliability of model outputs at various scales.

In addition to errors associated with model outputs, another limitation is that not all available GHG mitigation strategies are well represented in the models. For example, fertilizer placement and type (except for nitrification inhibitors) are not represented in DayCent. Urea reduced N_2O emissions compared to anhydrous ammonia for corn in Minnesota (Venterea et al., 2010), while Halvorson et al. (2009) observed lower emissions with UAN applied to irrigated corn in Colorado compared to urea. In addition, Halvorson et al. (2010) showed that stabilized N sources (i.e. poly coated urea, nitrification inhibitors) can substantially reduce emissions compared to urea for irrigated corn in Colorado, especially when combined with no-till cultivation. The ability to represent nitrification inhibitors has been implemented in DayCent (Del Grosso et al., 2008a), but this feature has not been extensively tested. In addition to fertilizer type, placement has also been shown to impact emissions (e.g. Venterea et al., 2005). Currently, DayCent cannot distinguish surface application from sub-surface banding because fertilizer N is assumed to be uniformly distributed in the top 15 cm soil layer upon application. Future versions of the model will include the ability to distinguish different types of N fertilizer, as well as placement.

References

Abdalla, M., Jones, M., Yeluripati, J., Smith, P., Burke, J., Williams, M., 2010. Testing DayCent and DNDC model simulations of N_2O fluxes and assessing the impacts of climate change on the gas flux and biomass production from a humid pasture. Atmos. Environ. 44 2961-2297.

Davis, S.C., Parton, W.J., Dohleman, F.G., Gottel, N.R., Smith, C.M., Del Grosso, S.J., Kent, A.D., DeLucia, E.H., 2010. Comparative biogeochemical cycles of bioenergy crops reveal nitrogen-fixation and low greenhouse gas emissions in a Miscanthus x giganteus agro-ecosystem. Ecosystems 13, 144–156. doi: 10.1007/s10021-009-9306-9.

Davis, S.C., Parton, W.J., Del Grosso, S.J., Keough, C., Marx, E., Adler, P.R., DeLucia. E.H., 2011. Impact of second-generation biofuel agriculture on greenhouse gas emissions in the corn-growing regions of the U.S. Frontiers in Ecology and the Environment. (e-View) doi:10.1890/110003.

Del Grosso, S.J., Mosier, A.R., Parton, W.J., Ojima, D.S., 2005. DAYCENT model analysis of past and contemporary soil N_2O and net greenhouse gas flux for major crops in the USA. Soil Tillage Res. 83, 9–24.

Del Grosso, S.J., Ogle, S.M., Parton, W.J., Breidt, F.J., 2010. Estimating uncertainty in N_2O emissions from U.S. cropland soils. Global Biogeochem. Cycles 24, GB1009.

Del Grosso, S.J., Parton, W.J., Mosier, A.R., Hartman, M.D., Brenner, J., Ojima, D.S., Schimel, D.S., 2001. Simulated interaction of carbon dynamics and nitrogen trace gas fluxes using the DAYCENT model. In: Schaffer, M., Ma, L., Hansen, S. (Eds.), Modeling Carbon and nitrogen Dynamics for Soil Management. CRC Press, Boca Raton, Florida, pp. 303–332.

Del Grosso, S.J., Ojima, D.S., Parton, W.J., Mosier, A.R., Peterson, G.A., Schimel, D.S., 2002. Simulated effects of dryland cropping intensification on soil organic matter and greenhouse gas exchanges using the DAYCENT ecosystem model. Environ. Poll. 116, S75–S83.

Del Grosso, S.J., Parton, W.J., Mosier, A.R., Ojima, D.S., Kulmala, A.E., Phongpan, S., 2000a. General model for N_2O and N_2 gas emissions from soils due to denitrification. Global Biogeochem. Cycles 14, 1045–1060.

Del Grosso, S.J., Parton, W.J., Mosier, A.R., Ojima, D.S., Potter, C.S., Borken, W., Brumme, R., Butterbach-Bahl, K., Crill, P.M., Dobbie, K., Smith, K.A., 2000b. General CH_4 oxidation model and comparisons of CH_4 oxidation in natural and managed systems. Global Biogeochem. Cycles 14, 999–1019.

Del Grosso, S.J., Mosier, A.R., Parton, W.J., Ojima, D.S., 2005. DAYCENT model analysis of past and contemporary soil N_2O and net greenhouse gas flux for major crops in the USA. Soil Tillage Res. 83, 9–24.

Del Grosso, S.J., Parton, W.J., Mosier, A.R., Walsh, M.K., Ojima, D.S., Thornton, P.E., 2006. DAYCENT national-scale simulations of N_2O emissions from cropped soils in the USA. J. Environ. Qual. 35, 1451–1460.

Del Grosso, S.J., Halvorson, A.D., Parton, W.J., 2008a. Testing DAYCENT model simulations of corn yields and nitrous oxide emissions in irrigated tillage systems in Colorado. J. Environ. Qual. 37, 1383–1389.

Del Grosso, S.J., Wirth, T., Ogle, S.M., Parton, W.J., 2008b. Estimating agricultural nitrous oxide emissions. EOS. 89, 529–530.

Eitzinger, J., Parton, W.J., Hartman, M.D., 2000. Improvement and validation of a daily soil temperature submodel for freezing/thawing periods. Soil Science 165, 525–534.

EPA., 2011. Inventory of U.S. greenhouse gas emissions and sinks. Off. of Atmos. Programs, Washington, DC. Available online at http://www.epa.gov/climatechange/emissions/usinventoryreport.html (verified February 24, 2011)1990–2009.

Halvorson, A.D., Del Grosso, S.J., Alluvione, F., 2010. Tillage and nitrogen source effects on nitrous oxide emissions from irrigated cropping systems. Soil Sci. Soc. Am. J. 74, 436–445.

Halvorson, A.D., Del Grosso, S.J., Alluvione, F., 2009. Nitrogen rate and source effects on nitrous oxide emissions from irrigated cropping systems in Colorado. Better Crops Plant Food 93, 16–18.

Hartman, M.D., Baron, J.S., Ojima, D.S., 2007. Application of a coupled ecosystem-chemical equilibrium model, DayCent-Chem, to stream and soil chemistry in a Rocky Mountain watershed. Ecol. Model. 200, 493–510.

Innis, G.S., 1978. Grassland Simulation Model: Ecological Studies, 26. Springer-Verlag, New York.

IPCC, 2006. IPCC guidelines for national greenhouse gas inventories. vol. 4: Agriculture, forestry and other land use. In: De Klein, C., et al. (Eds.). IGES, Japan, pp. 11.1–11.54.

Jarecki, M.K., Parkin, T.B., Chan, A.S., Hatfield, J.L., Jones, R., 2008. Comparison of DAYCENT-simulated and measured nitrous oxide emissions from a corn field. J. Environ. Qual. 37, 1685–1690.

Jawson, M.D., Shafer, S.R., Franzluebbers, A.J., Parkin, T.B., Follett, R.F., 2005. GRACEnet: greenhouse gas reduction through agricultural carbon enhancement network. Soil Tillage Res. 83, 167–172.

Li, X.Y., Meixner, T., Sickman, J.O., Miller, A.E., Schimel, J.P., Melack, J.M., 2006. Decadal-scale dynamics of water, carbon and nitrogen in a California chaparral ecosystem: DAYCENT modeling results. Biogeochemistry 77, 217–245.

McCarl, B.A., Chang, C.C., Atwood, J.D., Nayda, W.I., 1993. Documentation of ASM: the U.S. agricultural sector model, Technical Report TR-93, Agricultural Experimental Station. Texas, College Station.

Metherell, A.K., Harding, L.A., Cole, C.V., Parton, W.J., 1993. CENTURY soil organic matter model environment, technical documentation, Agroecosystem version 4.0. Great Plains System Research Unit Technical Report No. 4. USDA-ARS. Colorado, Fort Collins.

Ogle, S.M., Breidt, F.J., Easter, M., Williams, S., Killian, K., Paustian, K., 2010. Scale and uncertainty in modeled soil organic carbon stock changes for US croplands using a process-based model. Glob. Change Biol. 16, 810–822.

Ojima, D.S., Mosier, A.R., Del Grosso, S.J., Parton, W.J., 2000. TRAGNET analysis and synthesis of trace gas fluxes. Global Biogeochem. Cycles 14, 995–997.

Parton, W.J., Schimel, D.S., Cole, C.V., Ojima, D.S., 1987. Analysis of factors controlling soil organic matter levels in Great Plains grasslands. Soil Sci. Soc. Am. J. 51, 1173–1179.

Parton, W.J., Hartman, M.D., Ojima, D.S., Schimel, D.S., 1998. DAYCENT: its land surface submodel: description and testing. Global and Planetary Change 19, 35–48.

Parton, W.J., Holland, E.A., Del Grosso, S.J., Hartman, M.D., Martin, R.E., Mosier, A.R., Ojima, D.S., Schimel, D.S., 2001. Generalized model for NO_x and N_2O emissions from soils. J. Geophys. Res. 106, 17403–17420.

Parton, W.J., Ojima, D.S., Cole, C.V., Schimel, D.S., 1994a. A general model for soil organic matter dynamics: sensitivity to litter chemistry, texture and management. In: Quantitative Modeling of Soil Forming Processes. SSSA, Spec. Pub. 39, Madison, WI, pp. 147–167.

Parton, W.J., Schimel, D.S., Ojima, D.S., 1994b. Environmental change in grasslands: assessment using models. Clim. Change 28, 111–141.

Parton, W.J., Scurlock, J.M.O., Ojima, D.S., Schimel, D.S., Hall, D.O., Coughenour, M.B., Garcia Moya, E., Gilmanov, T.G., Kamnalrut, A., Kinyamario, J.I., Kirchner, T., Kittel, T.G.F., Menaut, J.C., Sala, O.E., Scholes, R.J., van Veen, J.A., 1995. Impact of climate change on grassland production and soil carbon worldwide. Glob. Change Biology 1, 13–22.

Pepper, D.A., Del Grosso, S.J., McMurtrie, R.E., Parton, W.J., 2005. Simulated carbon sink response of shortgrass steppe, tallgrass prairie and forest ecosystems to rising [CO_2], temperature and nitrogen input. Global Biogeochem. Cycles 19, GB1004.

Schimel, D.S., Kittel, T.G.F., Parton, W.J., 1991. Terrestrial biogeochemical cycles: global interactions with the atmosphere and hydrology. Tellus. 43AB, 188–203.

Smith, W.N., Grant, B.B., Desjardins, R.L., Rochette, P., Drury, C.F., Li, C., 2008. Evaluation of two process-based models to estimate N_2O emissions in eastern Canada. Can. J. Soil Sci. 15, 31–51.

Soil Survey Staff, 2011. Natural Resources Conservation Service, United States Department of Agriculture. U.S. General Soil Map (STATSGO2). Available online at http://soildatamart.nrcs.usda.gov (verified February 24, 2011).

Stehfest, E., Heistermann, M., Priess, J.A., Ojima, D.S., Alcamo, J., 2007. Simulation of global crop production with the ecosystem model DayCent. Ecol. Model. 209, 203–219.

Thornton, P.E., Running, S.W., 1999. An improved algorithm for estimating incident daily solar radiation from measurements of temperature, humidity, and precipitation. Agri. Forest Meteorol. 93, 211–228.

Thornton, P.E., Running, S.W., White, M.A., 1997. Generating surfaces of daily meteorology variables over large regions of complex terrain. J. Hydrol. 190, 214–251.

U.S. Department of Agriculture, 2011. U.S. Agriculture and Forestry Greenhouse Gas Inventory: 1990–2008. In: Del Grosso, S.J., Eve, M. (Eds.), Technical bulletin 1930. Available online at http://www.usda.gov/oce/global_change/AFGGInventory1990_2008.htm (verified February 24, 2011).

Venterea, R.T., Dolan, M.S., Ochsner, T.E., 2010. Urea decreases nitrous oxide emissions compared with anhydrous ammonia in a Minnesota corn cropping system. Soil Sci. Soc. Am. J. 74, 407–418.

Venterea, R.T., Burger, M., Spokas, K.A., 2005. Nitrogen oxide and methane emissions under varying tillage and fertilizer management. J. Environ. Qual. 34, 1467–1477.

CHAPTER 15

COMET2.0—Decision Support System for Agricultural Greenhouse Gas Accounting

Keith Paustian[1,2], Jill Schuler[2], Kendrick Killian[2], Adam Chambers[3], Steven Del Grosso[4], Mark Easter[2], Jorge Alvaro-Fuentes[2], Ram Gurung[2], Greg Johnson[3], Miles Merwin[5], Stephen Ogle[2], Carolyn Olson[6], Amy Swan[2], Steve Williams[2], Roel Vining[7]

[1]Department of Soil and Crop Sciences, Colorado State University, Fort Collins, CO
[2]Natural Resource Ecology Laboratory, Colorado State University, Ft. Collins, CO
[3]USDA, Natural Resources Conservation Service, National Air Quality and Atmospheric Change Team, Portland, OR
[4]USDA, Agricultural Research Service, Soil Plant Nutrient Research Unit, Fort Collins, CO
[5]Ridgeback Tree Farm, Portland, OR
[6]USDA, Natural Resources Conservation Service, Office of the Deputy Chief, Washington, DC
[7]USDA, Natural Resources Conservation Service, Information Technology Center, Fort Collins, CO

CHAPTER OUTLINE

Abbreviations: AIC, Akaike Information Criterion; ARI, average relative increment; CCX, Chicago Climate Exchange; CO_2, carbon dioxide; CO_2 eq, carbon dioxide equivalents; CB, Corn Belt; CTIC, Conservation Technology Information Center; FIA, Forest Inventory and Analysis; GP, Great Plains; GHG, greenhouse gas; IPCC, Intergovernmental Panel on Climate Change; LRR, Land Resource Region; MLRA, Major Land Resource Area; CH_4, methane; NASS, National

Managing Agricultural Greenhouse Gases. DOI: 10.1016/B978-0-12-386897-8.00015-2

Agricultural Statistics Service; NRI, Natural Resources Inventory; NI, nitrification inhibitor; N_2O, nitrous oxide; SOC, soil organic carbon; SSURGO, Soil Survey Geographic Database

INTRODUCTION

The potential of the agriculture sector for greenhouse gas (GHG) mitigation in the U.S. has been assessed by various authors over the past decade (CAST, 2004; Follett, 2001; Lal et al., 1999; Morgan et al., 2010; Sperow et al., 2003). While estimates vary—depending on assumptions and data used—there is a broad consensus that the technical potential is quite significant, on the order of 100–200 Tg CO_2eq yr^{-1} (Paustian et al., 2006). Similarly, the economic potential for agricultural mitigation, i.e. the portion of the technical potential that might be achieved given economic constraints, has been assessed using economic models combined with biophysical analysis of the technical potentials (Antle et al., 2007, McCarl and Schneider, 2001; Murray et al., 2005). While these analyses invariably show lower emission reduction potentials when factoring in economic considerations, they generally suggest a significant and low-cost mitigation potential that can be accomplished with existing technologies. Moreover, the additional environmental benefits in terms of soil, water, and air quality from many of the conservation practices suggested for carbon sequestration and GHG reduction have been widely acknowledged (CAST, 2004; Morgan et al., 2010).

While there is not currently a mandatory GHG reduction policy in the U.S. (and there is uncertainty about the role that agriculture might play in a future policy), there is growing interest in the agricultural sector within voluntary GHG offset markets (e.g. VCS, 2011a; Reicosky et al., 2012). Furthermore, marketers and distributors of agricultural food products are becoming interested in reducing the GHG footprint of their supply chain, for which on-farm production of food and fiber represents the front end. In addition, GHG emission reduction efforts are increasingly being considered in USDA farm programs (Vilsack, 2010).

Workable policies, market initiatives, and sound science and tools to support decision-making and program implementation will be required to empower U.S. farmers and ranchers to contribute to GHG mitigation efforts by adopting improved management practices that store more carbon in the soil and biomass and reduce nitrous oxide (N_2O) and methane (CH_4) emissions. One of the key issues with respect to implementation of agricultural GHG reduction activities is the need to reliably quantify emissions and emissions reductions in ways that are practical and cost effective. Moreover, it is essential that the reductions claimed are consistent with real reductions in GHGs added to the atmosphere.

Agricultural GHG emissions are difficult to quantify for a variety of reasons, including their dispersed, non-point nature, the complexity of the underlying processes involved, the number of different factors that influence emission processes, and their considerable spatial and temporal heterogeneity (CAST, 2004). To deal with these issues, there are several attributes of an effective quantification system. First, it needs to provide acceptable accuracy and precision with minimal bias and with robust estimates of uncertainty. Second, quantification approaches need to be locally specific and at the same time be applicable to all the major agricultural regions and systems in the country. Spatial variability in climate conditions and soils properties are major reasons for differences in the response of soil C stocks and/or N_2O fluxes to a particular management change, such as changes in tillage, crop rotation or nutrient management. Analyses of mitigation policy design have shown that the efficiency of the mitigation policy increases sharply as more of the spatial heterogeneity in management responses is accounted for in the estimation method (Antle et al., 2003). Third, estimation systems should be able to account for the interaction of different components of a management system in terms of their effects on soil GHG fluxes. Changes in a management system for the purposes of reducing GHG emissions often involve more than one element. For example, in the Great Plains (GP) region, changes in tillage may be linked to a change in crop rotation such as in introducing no-till along with a more continuous cropping sequence (i.e. reduced bare fallow) (Peterson et al., 1998). Similarly, the addition of legume cover crops in an annual crop rotation could be linked with reductions in mineral N fertilizer rates. Estimation systems that can incorporate these interactions are better able to represent a comprehensive GHG accounting. Finally, quantifying emissions needs to be done in a cost-effective manner relative to the value of the mitigation itself. For many if not most of the GHG sources and sinks related to agriculture, direct measurement of fluxes requires sophisticated instruments and expertise that are largely confined to a research environment and thus would be prohibitively expensive and impractical to implement for routine measurement and monitoring of agricultural mitigation projects. Although direct measurement of soil C stock changes is more feasible to implement for a large aggregated project area (Mooney et al., 2007), it would be very expensive if applied at the individual farm or small project scale (Subak, 2000). Hence, most ongoing efforts for quantifying GHG emissions for mitigation activities in agriculture have focused on practice-based approaches using models.

Practice-based approaches can utilize either empirical models, often formulated as simple "emission factors" for a particular management or land- use practice, or they can employ dynamic process-based models. Many of the practice-based methods developed to date use an empirical emission factor approach, such as in the former Chicago Climate Exchange (CCX) agricultural offset program. However, a dynamic modeling approach has the advantage of being able to integrate interacting management factors as well as local soil and climate conditions (as described below) and thus can provide a more flexible and general basis for quantifying agricultural GHG mitigation activities. Examples of practice-based systems that include a dynamic simulation model as part of the quantification procedures include the COMET-VR (CarbON Management and Evaluation Tool-Voluntary Reporting) system (Paustian et al., 2009) for soil C emissions or removals under the US 1605b voluntary emissions reporting program, a soil C methodology using the RothC model (Coleman and Jenkinson, 1999) under approval for the Verified Carbon Standard registry (VCS, 2011b), and the DNDC model (Li et al., 1992) for N_2O emissions estimates in the American Carbon Registry (Pearson and Salas, 2010).

In this paper, we describe the COMET2.0 (CarbON Management and Evaluation Tool 2.0) system, which represents a further development of the COMET-VR system. Improvements include a wider range of management options for cropland and grazing land systems, as well as new modules for perennial crops (i.e. orchards and vineyards) and agroforestry. Estimates of woody biomass C stock changes have been added as a source/sink category along with a module to estimate N_2O emissions from soils. New features in the COMET2.0 system will be described along with examples of system performance.

SYSTEM OVERVIEW

User Interface

The COMET2.0 system is available online at www.comet2.colostate.edu and runs directly on the web, requiring no downloads. One of the main objectives of the COMET2.0 system is to enable agricultural producers, consultants, land managers, agency personnel, and others to easily use the system to produce an estimate of their GHG footprint using advanced models, without the need for prior training or experience with GHG emission and carbon sequestration calculations. The user is only required to provide information on their land-use and management practices. For ease of use, inputs are structured as a series of pull-down menus. The system has extensive embedded help routines that provide definitions for terms used. From the user selections, the appropriate data (e.g. climate, crop sequences, and management input files) to run the models are retrieved from a database server.

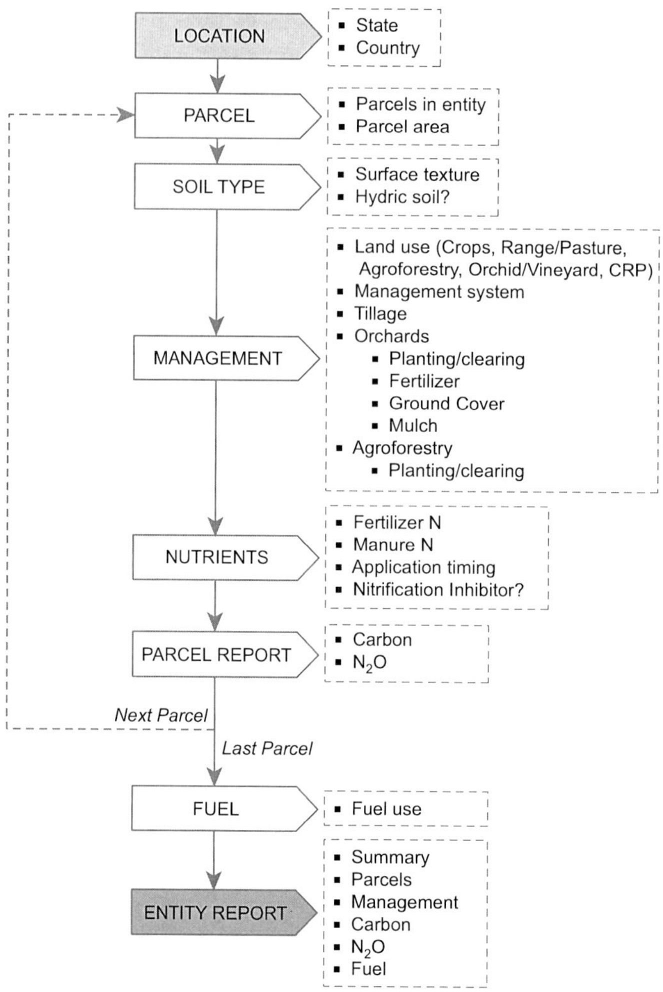

FIGURE 15.1
Flow chart of data entry and report results for the COMET2.0 system.

Figure 15.1 gives an overview of the work flow for entering data into the system. After providing information on location (state and county), the user specifies the number of individual parcels to include in the analysis. Soil characteristics and management are assumed to be homogeneous within a parcel; hence a parcel can be an individual field or portion of a field (e.g. if soil type varies strongly within a field, the field can be subdivided). An enhancement in the COMET2.0 system when compared with COMET-VR is that any number of parcels can be specified within a work session and a summary report for all parcels combined is generated. The system also allows the user to "redo" estimates for parcels that have previously been completed in the analysis.

For each parcel, the user specifies soil properties (surface texture type and whether the soil is classified as hydric or non-hydric), followed by a series of menus to define specific management practices for each parcel. Management inputs for soil C estimates include selections for both current and past management. The inclusion of information on historic land use is used to capture the current trajectory of soil C change, which is influenced by land-use changes (e.g. plow out of grassland) that may have occurred several decades ago. A new feature in COMET2.0 is that each land-use type (i.e. annual crop rotations, continuous hay, pasture/range, agroforestry, orchard/vineyard and Conservation Reserve Program set-aside) is organized under a separate tab (Figure 15.2). Each tab has specific choices for management systems

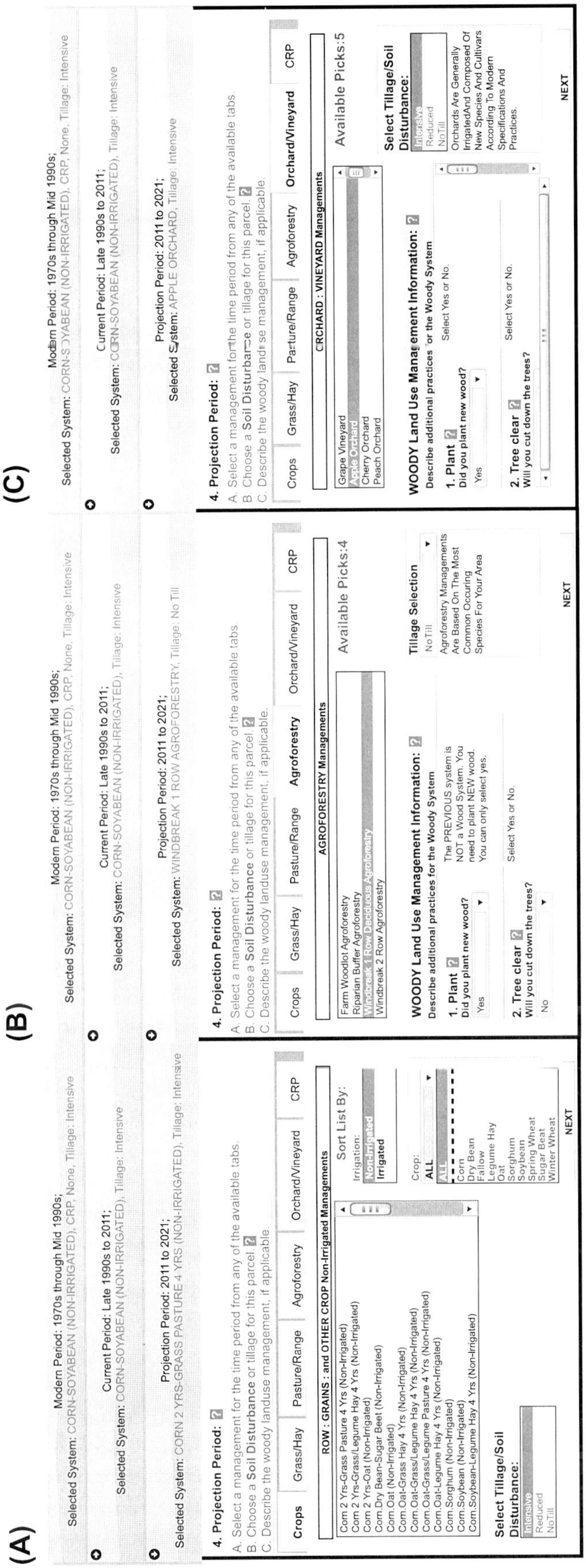

FIGURE 15.2
Example of menu selections and sorting functions for management practices for (A) annual crops, (B) agroforestry, and (C) orchards and vineyards.

in pull-down menus. The system also provides the capability to sort the menu of management options by irrigated vs. non-irrigated systems or by specific crop type, to make it easier to find and select the appropriate system. For each time period, the type of tillage system, classified as intensive, reduced or no-tillage, is also selected. For orchard/vineyard systems, additional options for mulching and cover crops are available as well as planting and clearing options for both orchard/vineyard and agroforestry systems (described in more detail below). Finally, nutrient management options include rate of N fertilizer or manure addition, method of application, timing of application and whether nitrification inhibitors are used. The system then computes soil C and woody C stock changes and N_2O flux rates which are displayed in a "parcel report". If several parcels have been specified then the process described above is repeated for each parcel. After all parcels have been specified, the fuel usage page is displayed and the user has the option of inputting actual fuel consumption for field operations (i.e. including diesel, natural gas, electricity—e.g. for irrigation pumping) or using the fuel consumption rates estimated from the Natural Resources Conservation Service (NRCS) Energy Tool (NRCS, 2011) which are automatically computed in COMET2.0. Finally, results from all parcels are displayed in a summary report, including all emissions, fuel use, specified management variables, and parcel attributes. The summary report can be downloaded as an Excel™ spreadsheet or as a text file.

An additional change from the COMET-VR version is that COMET2.0 computes both a baseline scenario as well as a projection scenario for each parcel. The projection scenario is typically specified as a change in current management aimed at sequestering C and/or reducing GHG emissions, which is projected forward for a 10-year period. The baseline scenario is taken as the current management (i.e. no changes), also projected forward for 10 years. Results for both baseline and projection scenarios and an overall estimate of the net change in GHG emissions, i.e. emissions under the projection scenario minus the baseline, are provided in the report.

Model Inputs Database

Many factors influence the response of soil C to management changes, including climate, soil properties, and previous land management. A MySQL™ database server houses a large amount of geo-referenced data that, depending on user inputs, are selected to run the Century ecosystem model (Parton et al., 1994). Climate data consist of county-average monthly temperature and precipitation, derived from PRISM (Parameter-elevation Regressions on Independent Slopes Model) data (Daly et al., 2002). Land-use and management data are derived from a variety of sources to provide the detailed information needed to run the simulation model. Information on crop rotations and pasture management systems were based largely on USDA-NRCS National Resources Inventory (NRI), which records land use and crop types on ca. 400,000 inventory locations on U.S. agricultural lands since 1982. Information on historical crop management systems for major regions in the U.S. were gathered from a variety of journal publications and agricultural yearbooks. Additionally, a supplemental land-use and management survey instrument, called the Carbon Sequestration Rural Appraisal (Brenner et al., 2002), was administered by USDA personnel in 2005 for each of the Land Resources Regions in the 48 states. Fertilizer practices by crop and region were obtained primarily from the USDA Economic Research Service's 1995 Cropping Practices Survey (ERS, 1997). Tillage practices are grouped into three categories: intensive, moderate, and no till. Intensive tillage is defined as multiple tillage operations every year, including significant soil inversion (e.g. plowing, deep disking) and low surface residue coverage. This definition corresponds to the intensive tillage and "reduced" tillage systems as defined by Conservation Technology Information Center (CTIC) (1998). No tillage was defined as not disturbing the soil except through the use of fertilizer and seed drills and where no till is applied to all crops in the rotation. Moderate tillage made up the remainder of the cultivated area, including mulch tillage and ridge tillage as defined by CTIC and intermittent no till. The specific tillage implements and applications used for different crops, rotations, and regions to represent the three tillage classes were derived from the 1995 Cropping Practices Survey by the Economic Research Service (ERS, 1997).

Simulation Model and Run Controls

The calculations of soil C stock changes are done by the Century ecosystem model (Metherell et al., 1993; Parton et al., 1994). The Century model simulates C (and also N, P, and S), soil temperature, and water dynamics for cropland, grassland, forest, and savanna (mixed forest-grassland) systems, thus it is well suited to represent land-use and management changes. In the COMET2.0 implementation, only C and N dynamics are simulated.

User inputs entered on the website are interpreted by a set of run control programs to select the proper data records from the database server and build the input files necessary to run the Century model and the N_2O emission model (described below). By running the model "on-the-fly" for each set of user inputs, the system is able to represent 100s of millions of unique combinations of climate and management practices across the country.

NEW MODULES

In addition to changes to the user interface and a greatly expanded selection of management options, major enhancements of the COMET2.0 system when compared with COMET-VR are the inclusion of orchard/vineyard and agroforestry practices and the inclusion of N_2O emissions into the system, as outlined below.

Orchard/Vineyard Component

Orchard and vineyard land-use systems have been incorporated into COMET2.0 and changes in woody biomass C (for both orchard/vineyard and agroforestry) have been added as an emission source/sink. To simulate woody biomass and soil C for orchard and vineyard systems, several changes were made to the tree growth submodel in Century, which was originally designed for forests, to account for distinct structural, physiological and management attributes of woody perennial crops. The tree submodel in Century allocates carbon and nutrients to five biomass pools: leaves, fine roots, fine branches, large wood, and coarse roots. In modern orchard and vineyard plants, a large fraction of annual production is allocated to fruit (Lakso and Flore, 2003). To accommodate orchard and vineyard crops, an additional pool was added to Century for fruits and nuts. As with the other biomass pools, the new fruit/nut pool is parameterized by the C:N ratio, lignin content, monthly death rates, and growth allocation rate. An event was also added to specify tree flowering date, which initiates allocation to the fruit pool.

The most common management practices that affect C cycling in orchards include irrigation, fertilization, organic matter amendments, groundcover management, pruning of branches, thinning of young fruit, and harvest and removal of mature fruit. No modifications to Century were required for irrigation, fertilization, organic matter amendments, or groundcover management. New management options were added to simulate tree planting, woody removal through pruning, and harvest and removal of fruit. Typical modern orchards often manage the orchard floor with grass alleys between tree rows, and herbicide applied within the tree rows. Killing herbaceous cover near the tree root zone reduces competition between the trees and herbaceous cover for water and nutrients (Tworkoski and Glenn, 2001; Welker and Glenn, 1990). Additionally, irrigation and N fertilizer are most often applied only within the tree row. In order to allocate fertilizer N and soil available N between the tree and grass crops, a new algorithm was created that apportions both fertilizer N additions and soil available N according to the relative areas of each.

Management of orchards varies greatly by species, region, and between conventional and organic agriculture. We used county-level National Agricultural Statistics Service (NASS) statistics to determine which perennial crops were the most dominant crops in each Land Resource Region in the U.S. (USDA-NASS, 2008; USDA-NRCS, 2006). We evaluated management practices by species and region by reviewing university and government extension literature, grower association publications, published experimental data, and personal communications with horticultural scientists. We focused on management practices that were common, had the greatest impact on GHG emissions, and could be simulated and validated in the model. Some sources of variation in management, such as species cultivar or rootstock may have a significant impact on GHG emissions, but could not be assessed due to limited data. In COMET2.0, users are only allowed to select orchard and vineyard crops that are commonly grown in their region. Additionally, some management practices are fixed in the system and others can be chosen by the user. Management options that are fixed include irrigation, timing of management events, pruning amounts, and N fertilizer amounts. Users can select tillage type, planting and removal of orchards, ground cover type, N addition type, and mulch additions.

Agroforestry Module

Wind breaks, shelterbelts, riparian woodland, and farm woodlots are available in COMET2.0 as management options. Agroforestry practices, as defined by the structure and species present, are based on the NRCS agroforestry practice standards. The user can select the desired practice from a pull-down menu, similar to other land-use types, and changes in woody biomass C stocks and soil C are calculated by the Century model. Model-based calculations are the default option in the system.

The system also includes an empirical model for making projections of woody biomass C increment for existing agroforestry operations. This option can only be used if the user has field measurements for their agroforestry systems. The tree survey model uses an individual tree growth algorithm that follows work of Lessard et al. (2000, 2001) in which average relative increment (ARI) is calculated over the 10-year projection period as,

$$ARI = a \cdot exp(-bx) \cdot x^c \qquad [1]$$

where x is the average diameter at breast height (dbh), and a, b and c are empirical constants. Carbon pools are then derived from diameter-based allometric equations that predict total aboveground biomass components for 10 broad species groups in the U.S. (Jenkins et al., 2003, 2004). Both published and unpublished data for the U.S. Forest Service Forest Inventory and Analysis (FIA) program were used to develop the growth increment model. Only records for trees that were not overtopped by adjacent trees were selected. Locations of tree records were recorded and aggregated by Major Land Resource Area (MLRA). Data were then fitted by non-linear regression for each combination of MLRA and species. Because of the short (10-year) projection period in COMET2.0, tree mortality or changes in inter-tree competition were assumed to be negligible. From a telephone survey of USDA-NRCS foresters and literature sources, a total of 84 agroforestry prescriptions were developed and grouped by Land Resource Region (LRR; a superset of MLRAs) and used as menu selections for agroforestry practices in COMET2.0. More details of the agroforestry module are provided by Merwin et al. (2009) and Merwin and Townsend (2007).

N_2O Emission Submodel

Meta-models for direct and indirect N_2O emissions from soil were derived using results from DAYCENT (Del Grosso et al., 2012, 2005; Parton et al., 1998) simulations. Direct soil N_2O emissions are those that occur "on-farm", i.e. from the field being managed, whereas indirect emissions are the result of N lost via volatilization and leaching and subsequently deposited in another location, where it is subject to loss as N_2O. The meta-models use soil N_2O emissions from DAYCENT for estimating direct emissions. Indirect emissions are estimated from

TABLE 15.1 Cropland Model Goodness of fit (R^2) Values for FASOMGHG (Forest and Agricultural Sector Optimization Model with Greenhouse Gases) regions (Murray et al., 2005), i.e. Corn Belt (CB), Northern Plains (NP), Lake States (LS), North East (NE), Pacific Northwest (PNW), Pacific Southwest (PSW), Rocky Mountains (RM), South Central (SC), South East (SE) and South West (SW)

	CB	NP	LS	NE	PNW	PSW	RM	SC	SE	SW
N_2O flux	0.831	0.856	0.851	0.893	0.923	0.915	0.909	0.893	0.897	0.928
Ammonia volatilization	0.917	0.898	0.947	0.948	0.918	0.918	0.906	0.916	0.917	0.913
Nitrate leaching	0.763	0.498	0.816	0.858	0.492	0.451	0.309	0.690	0.939	0.848

TABLE 15.2 Grassland Model's Goodness of Fit (R^2) Values for FASOMGHG Regions, see Table 15.1 for Regional Definitions

	CB	NP	LS	NE	PNW	PSW	RM	SC	SE	SW
N_2O flux	0.800	0.909	0.778	0.878	0.943	0.963	0.942	0.919	0.935	0.956
Ammonia volatilization	0.981	0.976	0.979	0.984	0.989	0.985	0.985	0.982	0.986	0.980
Nitrate leaching	0.462	0.496	0.531	0.590	0.716	0.822	0.709	0.857	0.592	0.727

simulated N volatilization and leaching from DAYCENT, together with IPCC (2006) emission factors for the fractions of volatilized and leached N that are lost as N_2O.

Simulations were conducted to capture a wide range of management practices and environmental variation (i.e. edaphic and climate) that occur in U.S. cropland and grassland systems. Separate models were developed for each of 10 agricultural regions (Murray et al., 2005) to account for broad-scale impacts of the interactions between climate, soil, and management on emission rates. Using the DAYCENT results, a total of 60 separate meta-models were produced using a linear mixed-effect modeling approach (Pinheiro and Bates, 2000) in which the emissions were predicted as a function of nitrogen inputs, timing of fertilizer application, use of nitrification inhibitors, mean temperature, mean precipitation:potential evapotranspiration ratio (using the county-averaged climate data from PRISM), soil texture, crop/grass type, residue removal rate, and tillage practice. A random effect was included to address correlation in output from the same model simulation. Model selection used the Akaike Information Criterion (AIC). In addition to the 60 meta-models, seven logistic models were created to determine the probability of nitrate leaching in semiarid/arid regions, as part of the calculation of indirect N_2O emissions. Furthermore, 10 more statistical models were created to capture N_2O emissions during fallow years due to the fertilizer applied in the previous cropping season. The goodness of fit (R^2) statistics calculated for the cropland and grassland meta-models are shown in Tables 15.1 and 15.2.

SYSTEM TESTING AND APPLICATIONS

Cropland Management Practices

Estimates of soil C stock changes associated with changing crop rotations and tillage practices were simulated with COMET2.0 and compared to change rates from field measurements in several long-term experiments in the central U.S., including both row crop systems in the Corn Belt (CB) and small grain cropping systems in the Great Plains (GP). Corn Belt sites included well-published long-term field experiments in Dixon Springs, IL, Hoyteville, South Charleston and Wooster, OH, Lancaster, WI, Lexington, KY, and Nashua, IA, and GP experiments included Mandan, ND, Sidney, NE, and Sterling, Stratton, and Walsh, CO. Field experiment treatments used in the analysis included intensive and no-till treatments for different crop rotations, mainly continuous corn (*Zea mays* L.) and corn–soybean (*Glycine max* L.) rotations in the

Corn Belt and wheat (*Triticum aestivum* L.)—fallow and reduced fallow and continuous cropping in the Great Plains.

To obtain the results in COMET2.0, state and county locations for the experiments were entered along with the reported soil surface texture and hydric conditions at the site and the management treatments used in the experiments. Where reported, the management conditions prior to the start of the experiment were entered or, if not reported, were assumed to be the most common cropping practices for the region (i.e. corn—soybean in the CB and winter wheat—fallow in the GP). Management treatments that most closely matched the rotation and tillage practices used in the experiments were selected from the available menu choices.

Since soil C change rates for most of the long-term experiments are inferred from comparison of C stocks across treatments, we used intensive tillage treatments (mainly moldboard plow) as the baseline condition and used the corresponding no-till treatment as the projected mitigation option in COMET2.0. Thus stock changes predicted over the 10-year projection period in COMET2.0 were computed as the net difference between annual soil C stock changes in the projection and baseline scenario. In some cases, the baseline scenario of the intensive tillage treatment (and in one case for the mitigation scenario) in COMET2.0 was a predicted decrease in C stocks, so the net C accumulation rates represent a measure of the net soil C *increase relative to the baseline.* These were compared to the corresponding measured soil C stock differences between treatment, divided by experimental time period, to derive an inferred net annual rate, which is also a measure of the difference relative to the baseline (i.e. intensive till treatment).

For comparison, we also estimated soil C change rates for no-till adoption using the (Intergovernmental Panel on Climate Change) IPCC Tier 1 methods (IPCC, 2006) and the CCX protocol (CCX, 2009) for each of the long-term sites. For the IPCC Tier 1 calculations we used mean soil C stocks reported for the intensive tillage treatment (i.e. baseline condition) and then applied the Tier 1 land-use, tillage and input factors (IPCC, 2006) to compute predicted soil C stocks with no-till adoption. The default IPCC stock change factors were derived for a 20-year time frame (IPCC, 2006) and so annual change rates were computed as the difference between no-till and intensive till soil C stocks, divided by 20. All sites in the CB are within the temperate moist climate designation in the IPCC method and all GP sites in the temperate dry climate, so where stock change factors differ by climate type, the appropriate factors were used. Values for the input factor (which relates to type of crop rotation and external organic amendments) were set to 1 ("medium" category) for all treatments, except for wheat—summer fallow rotations which were classified in the "low" category, with an input factor value of 0.95, as per IPCC methodology guidance (IPCC, 2006). For the CCX protocol, a fixed annual rate of soil C increase for continuous "conservation" tillage (e.g. no-till, strip till) is specified according to broad climate zones, where the CB sites are within CCX Zone A, the Mandan, ND, site is in Zone B and the other GP sites are in Zone D (CCX, 2009).

Overall, COMET2.0 gave a reasonable representation of the direction and magnitude of net soil C change rates when compared to the field experiment results (Figure 15.3A). However, the COMET2.0 system tended to underestimate net (relative to the baseline) C stock increase rates under no till for some CB sites, in particular for the three sites in Ohio. The measured mean rate of soil C increase for the mitigation scenarios at CB sites was 0.3 Mg C ha^{-1} yr^{-1} while the corresponding mean for COMET2.0 estimates was 0.19 Mg C ha^{-1} yr^{-1}. Estimates for the GP sites were better (mean estimate from field measurements of net soil C increase across GP mitigation scenarios was 0.16 vs. 0.15 Mg C ha^{-1} yr^{-1} for COMET2.0) and, in particular, the interaction of tillage and crop rotation at Mandan, ND, was well represented, where changing from intensive tillage to no-till had little or no benefit in the wheat—fallow rotation (−0.008 Mg C yr^{-1} measured, +0.08 Mg C yr^{-1} modeled), but substantially increased soil C (+0.38 Mg C yr^{-1} measured, + 0.37 Mg C yr^{-1} modeled) when no-till

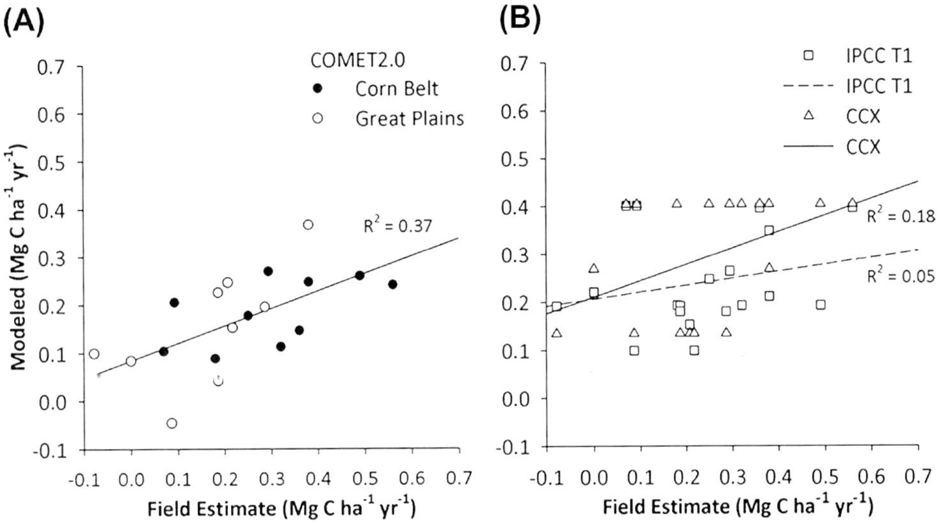

FIGURE 15.3
(A) Estimated mean net soil C stock changes (0–20 cm depth) for a 10-year period following adoption of mitigation practices [including no-till adoption and changes in crop rotation (see text)], based on field measurements and COMET2.0. Field experiment sites included Dixon Springs, IL (Kitur et al., 1994), Hoyteville, South Charleston and Wooster, OH (Collins et al., 2000; Dick et al., 1997), Lancaster, WI (Karlen et al., 1994), Lexington, KY (Ismail et al., 1994; Collins et al., 2000), Nashua, IA (Karlen et al., 1991), Mandan, ND (Black and Tanaka, 1997), Sidney, NE (Lyon et al., 1997), and Sterling, Stratton and Walsh, CO (Sherrod et al., 2003). (B) Estimated soil C stock changes following adoption of no-till and changes in crop rotation (see text) for field sites listed in (A) using the IPCC (2006) Tier 1 method and the CCX (2009) protocol.

adoption was coupled with a continuous cropping rotation. The COMET2.0 estimates accounted for about 36% of the total variation across the full data set (Figure 15.3A); model estimates were better for the GP sites ($R^2 = 0.46$) than for CB sites ($R^2 = 0.32$) (separate regressions not shown).

Both IPCC and CCX methods (Figure 15.3B) tended to overestimate C gains for many of the sites and accounted for much less of the observed variability across sites, compared with COMET2.0. The CCX method accounted for more of the total variation across the range of sites ($R^2 = 0.18$) than the IPCC Tier 1 method ($R^2 = 0.05$). Since both the IPCC and CCX approaches are based on broad-scale and highly aggregated data sources, their relative lack of sensitivity to more local-scale effects on soil C dynamics is to be expected.

Even with the much more disaggregated approach that is possible with the use of a process-based model in COMET2.0 system, a major issue in assessing system performance for individual field experimental data is that much of the input data in COMET2.0 represent general conditions derived from county and MLRA aggregate data sources, which only approximate the specific conditions at a particular field site. This adds an additional source of error and uncertainty that is independent of the accuracy of the underlying model. Capabilities to more precisely represent field-scale management and soil conditions would help improve the accuracy of the model-based estimates (see Concluding Remarks).

Management Impacts on N_2O Emissions

Nitrogen management scenarios representative for a Corn Belt location (Hamilton Co., IA) were run using COMET2.0 to examine the relationship between N_2O flux and N addition rate and to compare flux rates for spring vs. fall application and with application of nitrification inhibitors. To simplify the analysis for individual crops, COMET2.0 was run for either continuous corn (at annual fertilization rates of 0, 80, 160, 240, and 360 kg N ha^{-1}) or continuous spring wheat (at annual rates of 0, 80, and 160 kg N ha^{-1}), applied either at planting (spring) or after harvest and fall tillage. Currently, the model does not distinguish

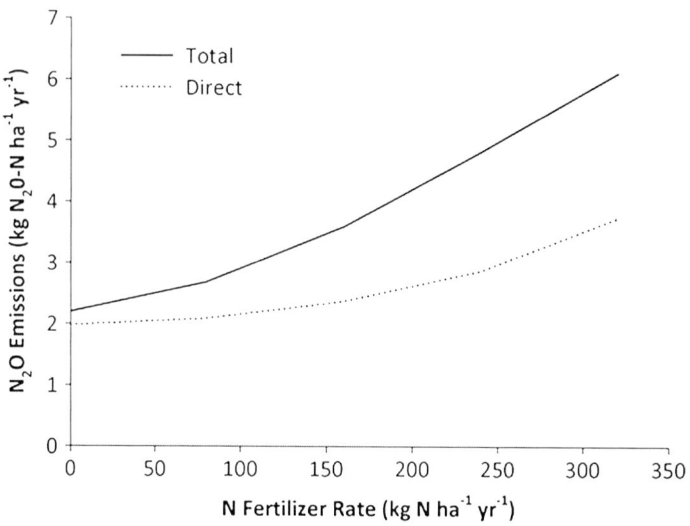

FIGURE 15.4
Response of direct and total (direct + indirect) N_2O emissions to annual rate of N fertilizer addition (0, 80, 160, 240, and 320 kg N ha^{-1} yr^{-1}) for simulated intensive tillage, continuous corn on a non-hydric silt loam soil in Hamilton Co., Iowa, as estimated by the COMET2.0 system. Lines are linear interpolations between the five N addition rates.

between different types of mineral fertilizers, although manure and inorganic fertilizer additions are treated separately.

Both direct and total (direct + indirect) N_2O emissions increased in a curvilinear fashion as a function of increasing fertilizer rates, so that for each additional increment of N fertilizer added, the fraction lost as N_2O increased (Figure 15.4). A similar non-linear pattern has been reported from field studies (Bouwman et al., 2002; Ma et al., 2009; McSwiney and Robertson, 2005) and most recently by Hoben et al. (2011) for fertilizer rate experiments in Michigan. In the Hoben et al. (2011) study, direct emissions at the highest fertilizer rate for corn (225 kg N ha^{-1} yr^{-1}), over five farms and 2 years, averaged around 3 kg N_2O-N ha^{-1} yr^{-1}, with a range of 1–7 kg N_2O-N ha^{-1} yr^{-1}. Direct emissions at that N rate from the continuous corn example in COMET2.0 were about 2.7 kg N_2O-N ha^{-1} yr^{-1} (Figure 15.4). In our Iowa example, implied direct emission factors for corn, i.e. the percent of fertilizer applied lost as N_2O, were 2.6% at the lowest N rate (80 kg N ha^{-1} yr^{-1}) and between 1.2 and 1.4% for the higher N rates. These are somewhat higher than the mean default (global) IPCC direct emission factor of 1% but well within the range of reported emission factors (1–3%) (IPCC, 2006). Total predicted emissions from wheat were approximately 60–70% lower than for corn at the same fertilizer rates (Figure 15.5). This is consistent with greater N use efficiency in wheat as compared to corn in the Midwest as reported by Syswerda et al. (2011). The fraction of total emissions comprised of indirect emissions (i.e. offsite emissions from N that is volatilized or leached and re-deposited elsewhere) increased with increasing fertilizer addition, from a low of 10% with no added fertilizer to as high as 30–40% of total emissions at the highest rates (Figure 15.5). Although we have no direct measurements of indirect N_2O emissions *per se* to compare to, this pattern is consistent with greater losses of fertilizer N, particularly via leaching, as fertilizer addition rates increase.

Aside from the effects of variable fertilizer rates, COMET2.0 predicted somewhat higher N_2O losses with fall compared to spring fertilizer application, between 6 and 17% higher, with the greater difference between spring and fall occurring at the highest N addition rates (Figure 15.5). It is generally assumed that N_2O emissions will be higher for fall vs. spring applied fertilizers (Snyder et al., 2007) due to the longer time period that added N is vulnerable to losses prior to crop uptake. Field studies in Canada have shown higher losses for fall vs. spring applications (Hao et al., 2001; Hultgreen and Leduc, 2003), although Millar et al. (2010) point out that similar studies for the Midwest U.S. are currently lacking.

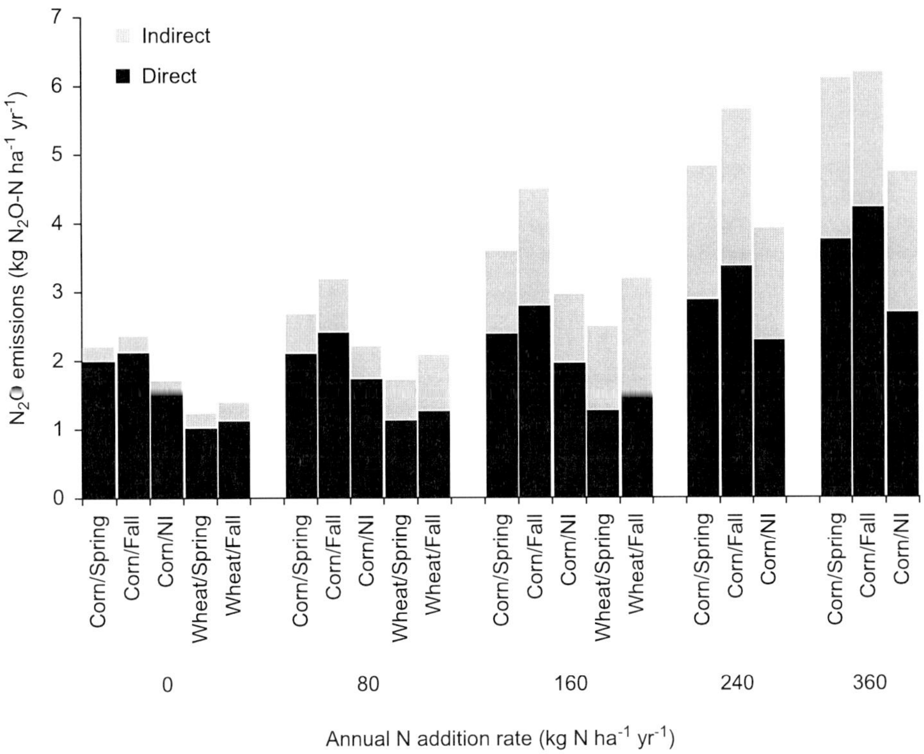

FIGURE 15.5
Annual N_2O emissions, direct and indirect, estimated by COMET2.0 for some alternative N management practices (i.e. spring vs. fall fertilization for continuous corn or continuous spring wheat and use of nitrification inhibitors for spring fertilized continuous corn).

With application of N inhibitors (NI), COMET2.0 predicted a decrease in N_2O fluxes of around 20% (18–22%) for total emissions and 18–29% for direct emissions, with the greatest percentage decrease in direct emissions at the highest N application rate (Figure 15.5). Although field results on the effect of nitrification inhibitors on N_2O are variable, the relative magnitude of the reduction in fluxes corresponded to the meta-analysis by Akiyama et al. (2010) who reported a range of 26–43% reductions in direct N_2O emissions for upland cropland with NI application. The authors also noted increasing NI effectiveness with increasing N addition rates which is consistent with the results from COMET2.0.

Biomass and Soil C in Orchard Systems

The new orchard submodel in Century was tested for experimental sites where detailed management and woody biomass measurements were available. Most orchard and vineyard studies do not give woody biomass data, so there were a very limited number of sites used in testing. A total of seven studies, with eight sites and nine unique treatments were tested with the model: vineyards in Oakville, CA (Williams and Smith, 1991), apple in Fletcher, NC (Barden et al., 2002), pecan in Sparks, OK, and Madill, OK (Smith and Wood, 2006), and Las Cruces, NM (Andales et al., 2006), citrus in Israel (Feigenbaum et al., 1987), almonds in Shafter, CA (Esparza et al., 1999), and peach in Silver Springs, MD (Miller and Walsh, 1988) (Figure 15.6). The comparison of modeled vs. measured woody biomass C demonstrate that the model did reasonably well simulating growth of various orchard species. A linear regression of the points showed that the model slightly overestimated growth, with a slope of 1.2 ($R^2 = 0.65$). The largest overestimations of growth relative to total measured woody biomass C occurred in pecan orchards in Madill, OK (Smith and Wood, 2006), and citrus in Israel (Feigenbaum et al., 1987). Differences between modeled and measured biomass C may be due

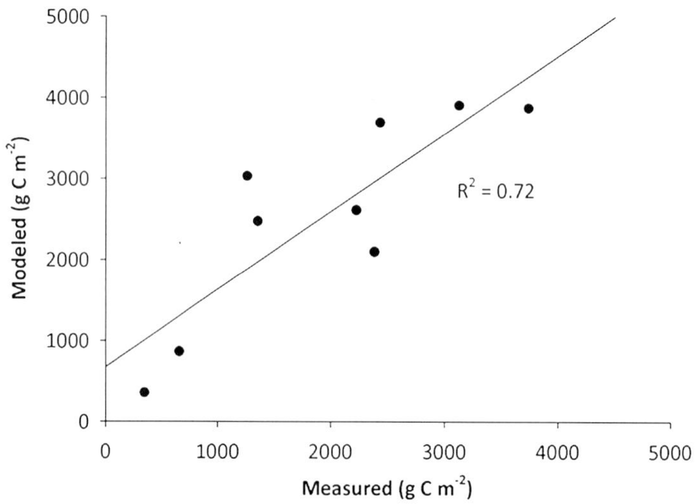

FIGURE 15.6
Century model results for woody biomass C (g C m^{-2}) for orchard crops from experimental sites. The experimental sites from left to right across the X-axis: vineyards in Oakville, CA (Williams and Smith, 1991), apple in Fletcher, NC (Barden et al., 2002), pecan in Madill, OK (Smith and Wood, 2006), citrus in Israel (Feigenbaum et al., 1987), almonds in Shafter, CA (Esparza et al., 1999), peach in Silver Springs, MD (Miller and Walsh, 1988), pecans in Las Cruces, NM (two treatments) (Andales et al., 2006), and pecan in Sparks, OK (Smith and Wood, 2006).

to cultivar types, management practices or local conditions not captured by the model. More research is needed on woody biomass C accumulation in orchards to further refine modeling capabilities.

There were even fewer studies that evaluated soil organic C (SOC) in orchard soils, so very little validation was done for modeled SOC estimates. One study, from an apple orchard in Ithaca, NY, that tested different groundcover treatments, provided complete soil data for validation (Figure 15.7) (Merwin and Stiles, 1994; Merwin et al., 1994). The model captured the trends of SOC change in both treatments, with a loss of SOC in the tilled treatment and an increase in SOC in the sod treatment.

The orchard module of the COMET2.0 system was further tested across all regions for various combinations of crop types, cropping histories, and management practices. The results were evaluated for anomalous woody biomass or soil carbon values. For further validation, testing was done in a few counties where experimental sites were located and cropping histories and management practices were chosen to approximate that of the experiments. Similar to the evaluations described earlier for soil C changes in annual crop systems using experimental field data, soil and management attributes were selected from the system menu options that most closely aligned with the conditions reported in the experiment. Figure 15.8 shows two examples for experimental sites in Fletcher, NC (Barden et al., 2002; Parker, 2008), and Sparks, OK (Smith and Wood, 2006). Although time series data to evaluate the COMET2.0 predictions were not available, the system was able to reasonably reflect differences between orchard types in apparent growth rates and biomass stocks over the first 4–5 years after planting.

These results illustrate the need for studies and measurements of both biomass and C stocks and N_2O flux rates in orchard/vineyard and agroforestry systems. Although, in terms of land area, these systems make up a relatively small portion of the agricultural area in the U.S., they may offer significant opportunities with respect to both GHG mitigation and other environmental benefits in agricultural landscapes. However, a limitation to a broader inclusion of perennial crops and agroforestry in agricultural mitigation policies is the paucity of empirical data to assess impacts of different management alternatives on GHG emissions.

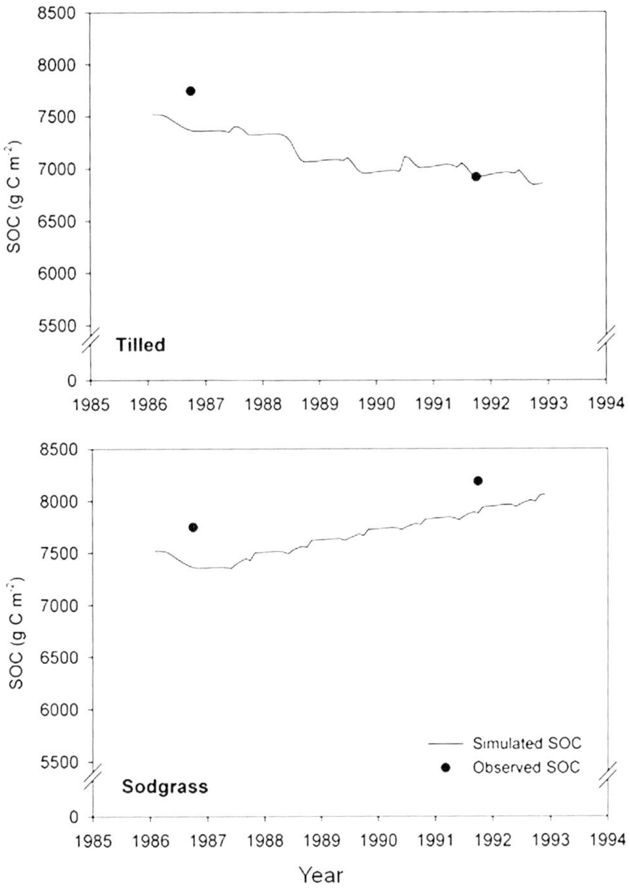

FIGURE 15.7
Measured (circles) and simulated (lines) SOC stocks (g C m^{-2} for 0–20 cm) for two groundcover treatments in an apple orchard in Ithaca, NY (Merwin and Stiles, 1994; Merwin et al., 1994). Adapted from Alvaro-Fuentes (unpubl.).

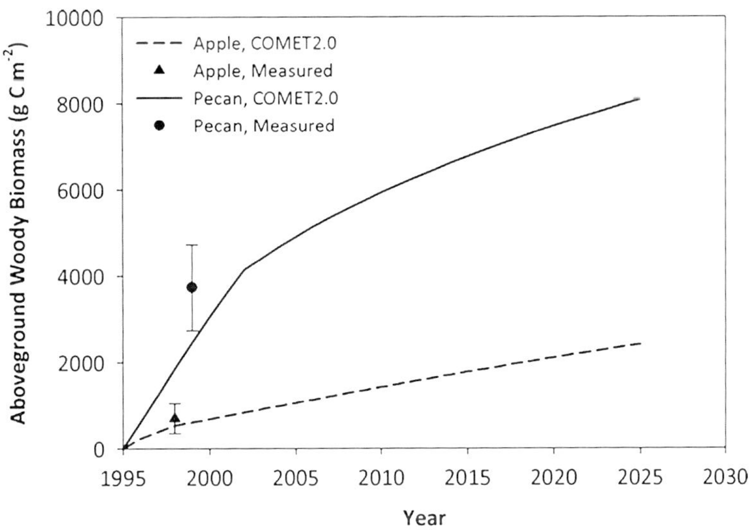

FIGURE 15.8
Projections of woody aboveground biomass C with COMET2.0 (solid lines) and field measurements (squares, showing means with standard deviations), using approximate management histories from experimental sites for pecans in Lincoln County, OK (Smith and Wood, 2006), and apples in Henderson County, NC (Barden et al., 2002; Parker, 2008).

CONCLUDING REMARKS

For U.S. agriculture to participate in the emerging private-sector GHG offset market, or to implement efficient government policies aimed at reducing the GHG footprint of agricultural activities, cost-effective and robust quantification approaches are crucial. As outlined previously, direct on-farm measurement is either impractical or too costly for most if not all agricultural GHG sources, which necessitates the use of practice-based approaches for policy implementation. Advantages of a practiced-based approach employing dynamic process-based models, such as used in the COMET2.0 system, are that multiple interacting processes and local conditions can be integrated for a more accurate and complete GHG accounting. By making the underlying processed-based models accessible to non-expert users through a web-based interface, it enables the primary "actors" involved—the land managers—to evaluate different scenarios and trade-offs to determine the options that are best suited for their particular conditions.

However, to further improve our capabilities to quantify GHG emissions and removals with this type of approach, several needs can be identified. Although COMET2.0 is able to account for some of the local-scale impacts of climate, soil, land-use history, and specific management practices, the menu-provided options cannot precisely reflect the management conditions for a given field. In addition, some key interactions, such as between user-specified variation in N management practices and their effects on soil organic matter are not currently coupled in COMET2.0 because the N management options and N_2O estimates are done using the DAYCENT-based metamodel, while the soil C balance is estimated with real-time Century simulations using region-specific (i.e. MLRA and county) information in the management practice database. Also, while the most important soil and biomass sources are accounted for in COMET2.0, emissions and mitigation options from livestock management are not included.

To address these and other issues, a new version of the system, called COMET-Farm™, is presently under development and will be released in 2012. Unlike COMET2.0, COMET-Farm™ has a fully spatial interface which allows specific fields to be delimited by the user and fields are automatically stratified by major soil types using Soil Survey Geographic Database (SSURGO) from Web Soil Survey (http://websoilsurvey.nrcs.usda.gov/app/HomePage.htm). To capture the true field-level management information, the user can specify crops and management practices on a year-by-year and seasonal basis starting in 2000, and they can develop similar year and seasonally resolved mitigation scenarios. The user can also save their information for future use. Both soil carbon dioxide (CO_2) and N_2O dynamics are simulated in a fully integrated way using real-time simulation of the DAYCENT model. The livestock module includes emission estimates for both enteric CH_4 as well as CH_4 and N_2O emissions from manure storage and management. The field and livestock modules are linked to account for manure applications to land as well as import or export of manure off-farm. An expanded energy module accounts for emissions and emission reductions from energy use for livestock as well as field operations and includes accounting for on-farm energy production.

A major opportunity with this type of web-based information and analysis system is that it provides a framework whereby individual producers, participating in a larger aggregate GHG mitigation project, can do detailed self-reporting and thereby greatly reduce transactional and measurement costs for large aggregated projects. Added facilities for documentation of management records could be easily added to facilitate project verification requirements.

Finally, any practice-based system is only as good as the information and models that go into it. Additional research is needed to improve basic understanding and model capabilities to predict the dynamics of soil GHGs and other agricultural GHG sources with improved quantification of uncertainties. Long-term experiments with coordinated designs and protocols, such as in the GRACEnet Project described in this volume, will play a key role in further improvements in models and systems to quantify agricultural GHG mitigation options.

Acknowledgments

Support from a cooperative agreement with USDA-NRCS (68-3A75-4-106) is gratefully acknowledged.

References

Akiyama, H., Yan, X., Yagi, K., 2010. Evaluation of effectiveness of enhanced-efficiency fertilizers as mitigation options for N_2O and NO emissions from agricultural soils: meta-analysis. Glob. Change Biol. 16, 1837–1846.

Andales, A., Wang, J., Sammis, T., Mexal, J., Simmons, L., Miller, D., Gutschick, V., 2006. A model of pecan tree growth for the management of pruning and irrigation. Agri. Water Manag. 84, 77–88.

Antle, J.M., Capalbo, S.M., Mooney, S., Elliott, E.T., Paustian, K., 2003. Spatial heterogeneity, contract design and the efficiency of carbon sequestration policies for agriculture. J. Environ. Econ. Manag. 46, 231–250.

Antle, J.M., Capalbo, S.M., Paustian, K., Ali, M.K., 2007. Estimating the economic potential for agricultural soil carbon sequestration in the Central United States using an aggregate econometric-process simulation model. Clim. Change. 80, 145–171.

Barden, J.A., Cline, J.A., Kushad, M.M., Parker, M.L., 2002. Various measures of tree vigor, yield, and yield efficiency of apple trees in the 1990 NC-140 systems trial as influenced by location, cultivar, and orchard system. J.Amer. Pomological Soc. 56, 208–214.

Black, A.L., Tanaka, D.L., 1997. A conservation tillage-cropping systems study in the Northern Great Plains of the United States. In: Paul, E.A., Paustian, K., Elliott, E.T., Cole, C.V. (Eds.), Soil Organic Matter in Temperate Agroecosystems. CRC Press, Inc., Boca Raton, FL, pp. 335–342.

Bouwman, A.F., Boumans, L.J.M., Batjes, N.H., 2002. Emissions of N_2O and NO from fertilized fields: summary of available measurement data. Glob. Biogeochem. Cycles 16 (4), 1058–1070.

Brenner, J., Paustian, K., Bluhm, G., Killian, K., Cipra, J., Dudek, B., Williams, S., Kautza, T., 2002. Analysis and reporting of carbon sequestration and greenhouse gases for conservation districts in Iowa. In: Kimble, J.M., Lal, R., Follett, R.F. (Eds.), Agriculture Practices and Policies for Carbon Sequestration in Soil. Lewis Publishers, CRC Press, Boca Raton, FL, pp. 127–140.

CAST., 2004. Climate Change and Greenhouse Gas Mitigation: Challenges and Opportunities for Agriculture. Council for Agricultural Science and Technology, Ames, IA, 120 p.

Coleman, K., Jenkinson, D.S., 1999. RothC-26.3. A Model for the Turnover of Carbon in Soils. Herts, Rothamsted Research. Harpenden, Hertfordshire, UK.

Collins, H.P., Elliott, E.T., Paustian, K., Bundy, L.C., Dick, W.A., Huggins, D.R., Smucker, A.J.M., Paul, E.A., 2000. Soil carbon pools and fluxes in long-term corn belt agroecosystems. Soil Biol. Biochem. 32, 157–168.

CCX., 2009. Chicago Climate Exchange. Continuous Conservation Tillage and Conversion to Grassland—Soil Carbon Sequestration Offset Project Protocol. Available online at https://www.theice.com/publicdocs/ccx/protocols/CCX_Protocol_Conservation_Tillage_and_Grassland.pdf (verified November 8, 2011).

CTIC., 1998. 1998 Crop Residue Management Executive Summary. Conservation Technology Information Center, West Lafayette, IN.

Daly, C., Gibson, W.P., Taylor, G.H., Johnson, G.L., Pasteris, P.A., 2002. A knowledge-based approach to the statistical mapping of climate. Clim. Res. 22, 99–113.

Del Grosso, S.J., Mosier, A.R., Parton, W.J., Ojima, D.S., 2005. DAYCENT model analysis of past and contemporary soil N_2O and net greenhouse gas flux for major crops in the USA. Soil Tillage Res. 83, 9–24.

Del Grosso, S.J., Parton, W.J., Adler, P.R., Davis, S.C., Keough, C., Marx, E., 2012. DayCent model simulations for estimating soil carbon dynamics and greenhouse gas fluxes from agricultural production systems. In: Liebig, M.A., Franzluebbers, A.J., Follett, R.F. (Eds.), Managing Agricultural Greenhouse Gases: Coordinated Agricultural Research through GRACEnet to Address our Changing Climate. Academic Press, San Diego CA.

Dick, W.A., Edwards, W.M., McCoy, E.L., 1997. Continuous application of no tillage to Ohio soils: changes in crop yields and organic matter-related soil properties. In: Paul, E.A., Paustian, K., Elliott, E.T., Cole, C.V. (Eds.), Soil Organic Matter in Temperate Agroecosystems. CRC Press, Inc., Boca Raton, FL, pp. 171–182.

ERS., 1997. Cropping Practices Survey Data—1995. Economic Research Service, United States Department of Agriculture. Available online at http://www.ers.usda.gov/Data/ARMS/CropOverview.htm (verified September 1, 2011).

Esparza, G., DeJong, T.M., Grossman, Y.L., 1999. Modifying "peach" to model the vegetative and reproductive growth of almonds. Acta Hort. 499, 91–98.

Feigenbaum, S., Bielorai, H., Erner, Y., Dasberg, S., 1987. The fate of ^{15}N labelled nitrogen applied to mature citrus trees. Plant Soil. 97, 179–187.

Follett, R.F., 2001. Soil management concepts and carbon sequestration in cropland soils. Soil Tillage Res. 61, 77–92.

Hao, X., Chang, C., Carefoot, J.M., Janzen, H.H., Ellert, B.H., 2001. Nitrous oxide emissions from an irrigated soil as affected by fertilizer and straw management. Nutr. Cycling Agroecosyst. 60 (1), 1–8.

Hoben, J.P., Gehl, R.J., Robertson, G.P., Millar, N., Grace, P.R., 2011. Nonlinear nitrous oxide (N_2O) response to nitrogen fertilizer in on-farm corn crops of the US Midwest. Glob. Change Biol. 17, 1140–1152.

Hultgreen, G., Leduc, P., 2003. The effect of nitrogen fertilizer placement, formulation, timing, and rate on greenhouse gas emissions and agronomic performance. Saskatchewan Department of Agriculture and Food. Saskatchewan, Canada, Regina. Final Report Project No. 5300G, ADF#19990028.

IPCC., 2006. 2006 IPCC guidelines for national greenhouse gas inventories, prepared by the National Greenhouse Gas Inventories Programme. In: Eggleston, H.S., Buendia, L., Miwa, K., Ngara, T., Tanabe, K. (Eds.), 2006 IPCC guidelines for national greenhouse gas inventories, prepared by the National Greenhouse Gas Inventories Programme. Publisher: IGES, Japan.

Ismail, I., Blevins, R.L., Frye, W.W., 1994. Long-term no-tillage effects on soil properties and continuous corn yields. Soil Sci. Soc. Am. J. 58, 193–198.

Jenkins, J.C., Chojnacky, D.C., Heath, L.S., Birdsey, R.A., 2003. National-scale biomass estimators for United States tree species. For. Sci. 49 (1), 12–35.

Jenkins, J.C., Chojnacky, D.C., Heath, L.S., Birdsey, R.A., 2004. Comprehensive database of diameter-based biomass regressions for North American tree species. Gen. Tech. Rep. NE-319. U.S. Dept. of Agriculture, Forest Service. Northeastern Research Station, Newtown Square, PA.

Karlen, D.L., Berry, E.C., Colvin, T.S., 1991. Twelve-year tillage and crop rotation effects on yields and soil chemical properties in northeast Iowa. Commun. Soil Sci. Plant Anal. 22, 1985–2003.

Karlen, D.L., Wollenhaupt, N.C., Erbach, D.C., Berry, E.C., Swan, J.B., Eash, N.S., Jordahl, J.L., 1994. Long-term tillage effects on soil quality. Soil Tillage Res. 32, 313–327.

Kitur, B.K., Phillips, S.R., Olson, K.R., Ebelhar, S.A., 1994. Tillage effects on selected chemical properties of Grantsburg silt loam. Commun. Soil Sci. Plant Anal. 24, 1509–1527.

Lakso, A.N., Flore, J.A., 2003. Carbohydrate partitioning and plant growth. In: Baugher, T.A., Singha, S. (Eds.), Concise Encyclopedia of Temperate Tree Fruit. Hawthorn Press, Binghamton, NY.

Lal, R., Follett, R.F., Kimble, J., Cole, C.V., 1999. Managing U.S. cropland to sequester carbon in soil. J. Soil Water Conserv. 54, 374–381.

Lessard, V.C., 2000. Updating Indiana annual forest inventory and analysis plot data using eastern broadleaf forest diameter growth models. In: Reams, G.A., McRoberts, R.E., van Deusen, P.C. (Eds.), Proceedings of the Second Annual Forest Inventory and Analysis Symposium, Salt Lake City, UT, October 17–18, 2000. Gen. Tech. Rep. SRS-47. U.S. Dept. of Agriculture, Forest Service, Southern Research Station, Asheville, NC, p. 143.

Lessard, V.C., McRoberts, R.E., Holdaway, M.R., 2001. Diameter growth models using Minnesota forest inventory and analysis data. For. Sci. 47, 301–310.

Li, C.S., Frolking, S., Frolking, T.A., 1992. A model of nitrous-oxide evolution from soil driven by rainfall events. 1. Model structure and sensitivity. J. Geophys. Res. (Atmos.) 97 (D9), 9759–9776.

Lyon, D.J., Monz, C.A., Brown, R.E., Metherell, A.K., 1997. Soil organic matter changes over two decades of winter wheat fallow cropping in western Nebraska. In: Paul, E.A., Paustian, K., Elliott, E.T., Cole, C.V. (Eds.), Soil Organic Matter in Temperate Agroecosystems. CRC Press, Inc., Boca Raton, FL, pp. 343–351.

Ma, B.L., Wu, T.Y., Tremblay, N., Deen, W., Morrison, M.J., Mclaughlin, N.B., Gregorich, E.G., Stewart, G., 2009. Nitrous oxide fluxes from corn fields: on-farm assessment of the amount and timing of nitrogen fertilizer. Glob. Change Biol. 16 (1), 156–170.

McCarl, B.A., Schneider, U.A., 2001. The cost of greenhouse gas mitigation in U.S. agriculture and forestry. Science 294, 2481–2482.

McSwiney, C.P., Robertson, G.P., 2005. Nonlinear response of N_2O flux to incremental fertilizer addition in a continuous maize (*Zea mays* L.) cropping system. Glob. Change Biol. 11 (10), 1712–1719.

Merwin, I.A., Stiles, W.C., 1994. Orchard groundcover management impacts on apple tree growth and yield, and nutrient availability and uptake. J. Amer. Soc. Hort. Sci. 119, 209–215.

Merwin, I.A., Stiles, W.C., van Es, H.M., 1994. Orchard groundcover management impacts on soil physical properties. J. Am. Soc. Hort. Sci. 119, 216–222.

Merwin, M.L., Townsend, L.R., 2007. Online tool for estimating carbon storage in agroforestry practices. In: Oliver, A., Campeau, S. (Eds.), When Trees and Crops Get Together: Economic Opportunities and Environmental Benefits from Agroforestry: Proc. 10th North American Agroforestry Conf. Université Laval, Québec, Canada, Quebec City, Canada, pp. 523–556. 10–13 June 2007.

Merwin, M.L., Easter, M., Townsend, L.R., Vining, R.C., Johnson, G.L., 2009. Estimating carbon stock change in agroforestry and family forestry practices. In: Gold, M.A., Hall, M.M. (Eds.), Agroforestry Comes of Age: Putting Science into Practice: Proc. 11th North American Agroforestry Conf, Columbia, MO, USA, pp. 17–24. May 31–June 3, 2009. http://www.centerforagroforestry.org/pubs/proceedings.pdf (verified September 1, 2011).

Metherell, A.K., Harding, L.A., Cole, C.V., Parton, W.J., 1993. CENTURY Soil organic matter model environment. Agroecosystem version 4.0. Technical documentation. GPSR Tech. Report No. 4, USDA/ARS, Ft, Collins, CO.

Millar, N., Robertson, G.P., Grace, P.R., Gehl, R.J., Holben, J.P., 2010. Nitrogen fertilizer management for nitrous oxide (N_2O) mitigation in intensive corn (maize) production: an emissions reduction protocol for US Midwest agriculture. Mitig. Adapt. Strateg. Glob. Change 15, 185–204.

Miller, A.N., Walsh, C.S., 1988. Growth and seasonal partitioning of dry matter in eight-year-old "Loring" peach trees. J. Amer. Soc. Hort. Sci. 113, 309–314.

Mooney, S., Gerow, K., Antle, J., Capalbo, S., Paustian, K., 2007. Reducing standard errors by incorporating spatial autocorrelation into a measurement scheme for soil carbon credits. Clim. Change 80, 55–72.

Morgan, J.A., Follett, R.F., Allen, L.H., Del Grosso, S., Derner, J.D., Dijkstra, F., Franzluebbers, A., Fry, R., Paustian, K., Schoeneberger, M.M., 2010. Carbon sequestration in agricultural lands of the United States. J. Soil Water Conserv. 65 (1), 6A–13A.

Murray, B.C., Sommer, A.J., Depro, B., Sohngen, B.L., McCarl, B.A., Gillig, D., de Angelo, B., Andrasko, K., 2005. Greenhouse Gas Mitigation Potential in US Forestry and Agriculture. EPA Report 430-R-05-006, Washington, DC.

NRCS., 2011. Energy Tools: Animal Housing, Irrigation, Nitrogen, Tillage. United States Department of Agriculture. Natural Resources Conservation Service. Available online at http://energytools.sc.egov.usda.gov/ (verified November 8, 2011).

Parker, M.L., 2008. Personal communication.

Parton, W.J., Ojima, D.S., Cole, C.V., Schimel, D.S., 1994. A general model for soil organic matter dynamics: sensitivity to litter chemistry, texture and management. In: Quantitative Modeling of Soil Forming Processes. Special Publication 39, SSSA, Madison, WI, pp. 147–167.

Parton, W.J., Hartman, M., Ojima, D., Schimel, D., 1998. DAYCENT and its land surface submodel: description and testing. Global Planetary Change 19, 35–48.

Paustian, K., Antle, J.M., Sheehan, J., Paul, E.A., 2006. Agriculture's Role in Greenhouse Gas Mitigation. Pew Center on Global Climate Change, Washington, DC, 76 p.

Paustian, K., Brenner, J., Easter, M., Killian, K., Ogle, S., Olson, C., Schuler, J., Vining, R., Williams, S., 2009. Counting carbon on the farm: reaping the benefits of carbon offset programs. J. Soil Water Conserv. 64, 36A–40A.

Pearson, T., Salas, W., 2010. The American Carbon Registry Methodology for N2O Emission Reductions through Changes in Fertilizer Management. Winrock International, Little Rock, Arkansas. Available online at http://www.americancarbonregistry.org/carbon-accounting/ (verified September 1, 2011).

Peterson, G.A., Halvorson, A.D., Havlin, J.L., Jones, O.R., Lyon, D.J., Tanaka, D.L., 1998. Reduced tillage and increasing cropping intensity in the Great Plains conserves soil C. Soil Tillage Res. 47, 207–218.

Pinheiro, J.C., Bates, D.M., 2000. Mixed-effect Models in S and S-Plus. Springer, New York.

Reicosky, D.C., Goddard, T., Enerson, D., Chan, A.S.K., Liebig, M.A., 2012. Agricultural greenhouse gas trading markets in North America. In: Liebig, M.A., Franzluebbers, A.J., Follett, R.F. (Eds.), Managing Agricultural Greenhouse Gases: Coordinated Agricultural Research through GRACEnet to Address our Changing Climate. Academic Press, San Diego, CA.

Sherrod, L.A., Peterson, G.A., Westfall, D.G., Ahuja, L.R., 2003. Cropping intensity enhances soil organic carbon and nitrogen in a no-till agroecosystem. Soil Sci. Soc. Am. J. 67 (5), 1533–1543.

Smith, M.W., Wood, B.W., 2006. Pecan tree biomass estimates. Hort. Sci. 41, 1286–1291.

Snyder, C.S., Bruulsema, T.W., Jensen, T.L., 2007. Greenhouse gas emissions from cropping systems and the influence of fertilizer management—a literature review. International Plant Nutrition Institute, Norcross, Georgia, USA. Available at www.ipni.net/ipniweb/portal.nsf/0/D27FE7F63BC1FCB3852573CA0054F03E (verified November 14, 2011).

Sperow, M., Eve, M.D., Paustian, K., 2003. Potential soil C sequestration on U.S. agricultural soils. Clim. Change 57, 319–339.

Subak, S., 2000. Agricultural soil carbon accumulation in North America: considerations for climate policy. Global Environ. Change 10, 185–195.

Syswerda, S.P., Basso, B., Hamilton, S.K., Taurig, J., Robertson, G.P., 2011. Long-term nitrate loss along an agricultural intensity gradient in the upper Midwest USA. J. Environ. Qual In press.

Tworkoski, T.J., Glenn, D.M., 2001. Yield, shoot and root growth, and physiological responses of mature peach trees to grass competition. Hort. Sci. 36, 1214–1218.

USDA-NASS., 2008. County Data—Crops, 1980–2008. U.S. Department of Agriculture, National Agriculture Statistics Service, Washington, DC.

USDA-NRCS., 2006. Land Resource Regions and Major Land Resource Areas of the United States, the Caribbean, and the Pacific Basin. U.S. Department of Agriculture Handbook 296.

VCS., 2011a. Agriculture, Forestry and Other Land Use (AFOLU) Requirements. Verified Carbon Standard Version 3. Available online at http://www.v-c-s.org/sites/v-c-s.org/files/AFOLU%20Requirements%2C%20v3.0.pdf (verified September 1, 2011).

VCS., 2011b. Adoption of Sustainable Agricultural Land Management (SALM). Verified Carbon Standard. Available online at http://www.v-c-s.org/methodologies/adoption-sustainable-agricultural-land-management-salm (verified September 1, 2011).

Vilsack, T., 2010. Agriculture Secretary Vilsack announces new steps to meet the challenge of climate change. Remarks to UN Climate Change Conference, Cancun, Mexico. 9 Dec. 2010. USDA news release no. 0642.10.

Welker, W.V., Glenn, D.M., 1990. Peach-tree growth as influenced by grass species used in a killed-sod planting system. Hort. Sci. 25, 514–515.

Williams, L.E., Smith, R.J., 1991. The effect of rootstock on the partitioning of dry weight, nitrogen and potassium, and root distribution of Cabernet Sauvignon grapevines. Am. J. Enol. Vitic. 42, 118–122.

CHAPTER 16

CQESTR Simulations of Soil Organic Carbon Dynamics

Hero T. Gollany[1], Ronald F. Follett[2], Y. Liang[3]
[1]USDA-ARS, Columbia Plateau Conservation Research Center, Pendleton, OR
[2]USDA-ARS, Soil Plant Nutrient Research Center, Fort Collins, CO
[3]Department of Biological & Agricultural Engineering, University of Arkansas Division of Agriculture, Fayetteville, AR

CHAPTER OUTLINE

Abbreviations: CO_2, carbon dioxide; CQESTR, process-based carbon sequestration model; CS, conservation tillage; DT, conventional disking; H, hay (clover); LTEs, long-term field experiments; M, maize; MM, continuous maize; MP, conventional moldboard plow tillage; NT, no-tillage; NT-25, no-tillage with 25% of crop residue removed; NT-50, no-tillage with 50% of crop residue removed; NT-75, no-tillage with 75% of crop residue removed; NT-90, no-tillage with 90% of crop residue removed; O, oat; SO, soybean; SOC, soil organic carbon; SOM, soil organic matter; ST, stubble-mulch tillage; WF, wheat–fallow rotation; WW, continuous winter wheat

INTRODUCTION

Concerns about carbon dioxide (CO_2) emissions from fossil fuels, global climate change, and a desire to mitigate the greenhouse effect have prompted interest in using soil to store C as soil organic matter (SOM). Furthermore, concerns about fossil fuel supplies have revitalized the

Managing Agricultural Greenhouse Gases. DOI: 10.1016/B978-0-12-386897-8.00016-4

search for renewable bioenergy (Hill et al., 2006) to reduce net emissions of CO_2 into the atmosphere (Lynd et al., 1991; Kim and Dale, 2005; Farrell et al., 2006). Agricultural management in the U.S. is undergoing rapid change in response to these concerns. Other pressures facing the U.S. are its dependence on fossil fuels, high costs of energy and fertilizer, food quality demands by consumers, localized water shortages, environmental concerns about chemical pollutants and soil erosion, and most critically maintaining soil productivity (Lal, 2006; Paul et al., 1997).

Soil organic carbon (SOC) is about 58% of the total SOM fraction (Nelson and Sommers, 1982). Soil organic C and SOM are related terms and can be expressed as SOM = SOC/0.58. While using crop residue for bioenergy has the potential to decrease SOC and affect regional SOC balances, many of the management techniques designed to address soil conservation issues have also demonstrated a potential to increase soil C stocks above those maintained under past management. These practices include: (1) reduced cultivation, (2) direct seeding or no-tillage (NT), (3) reduction in the frequency of fallow in semiarid regions, (4) elimination of residue burning, (5) utilization of cover crops, green manure, animal manure, and/or municipal and food processing wastes, (6) the use of high-yielding crop varieties, (7) incorporation of biomass crops in a rotation, (8) the use of agro-forestry, and (9) reversion to grassland or forest vegetation to increase biomass on marginal land (Lal et al., 1998; Paul et al., 1997). Although there has been some evaluation of how these shifts in management or land use have affected regional C balances, only recently have process-based models been used to predict future changes on SOC dynamics and greenhouse gas (GHG) emissions.

The amount of SOC in an agricultural soil is the net difference between inputs of organic C and outputs through mineralization, loss and deposition by erosion, and translocation of dissolved organic C through the soil profile (Campbell et al., 1996; Mertens et al., 2007). Jenkinson (1981) defined the SOC turnover rate as the SOC accrual/loss rate divided by annual source C inputs, which is basically the same as the C sequestration efficiency defined by Follett et al. (2005). Gollany et al. (2006) calculated SOC storage index as a ratio of final SOC divided by the sum of source C added over a 52-year period for nine treatments in a long-term experiment. The estimated SOC storage index ranged from 0.22 (manure treatments with wheat residue returned) to 0.69 (wheat residue burned and no manure added). The storage index doubled with the wheat residue returned (0.45) compared to when organic amendments (0.23) were added.

Predictions of SOC dynamics when converting from one management practice to another are becoming increasingly important policy and management tools. A number of models simulating C cycling and SOM have been proposed during the last three decades. Process-based models (e.g. CENTURY, Parton et al., 1988, 1996; Paustian et al., 1992; CQESTR, Rickman et al., 2002; Liang et al., 2009; DAYCENT, Del Grosso et al., 2005; EPIC, Izaurralde et al., 2006) have been developed and utilized by scientists and government agencies to estimate the potential for C storage in soils. These models require rigorous testing. Model validation and future predictions concerning SOC management can only be conducted using a long-term SOC database. Fortunately, the necessary data exists from several LTEs. In this chapter we will use CQESTR (sequester), a process-based simulation model, to estimate management effects on SOC using data from LTEs located in three distinct regions of the U.S.

CQESTR MODEL DESCRIPTION

The CQESTR model is a biophysical process-based model that simulates SOC dynamics in agricultural management systems. It has been in continuous development by Agricultural Research Service (ARS) scientists at the Pendleton Research Center since 2000 (Liang et al., 2009; Rickman et al., 2001). The CQESTR model is used for the field-scale evaluation of SOC, operates on a daily time step, and performs long-term (100-year) simulations. The major inputs include crop rotation, yields (including aboveground and belowground biomass), organic amendments (e.g. manure, municipal, and organic industrial wastes), tillage information, and weather data (i.e. mean monthly air temperature and monthly precipitation). Additional required input data include N content of crop residues and organic amendments, root distribution characteristics of the crops, the number and thickness (depth) of soil layers, the initial SOM or SOC content, and soil bulk density of each layer. Soil layers used in CQESTR are user defined and are not necessarily delineated by soil characteristics or morphology. Soil layers are defined by the depth increments of available SOM or SOC observations. A main consideration in the development of CQESTR is its use of easily obtainable data inputs instead of detailed physical or chemical fractionations of C sources (Rickman et al., 2001).

The basic model structure and decomposition module are illustrated in Figure 16.1. Each organic residue addition is tracked separately, without partitioning, according to its placement relative to soil, either on the surface or buried in the soil. The total SOC budget can be represented by Eq. [1], using units of dry weight per unit area within each soil layer,

$$\mathrm{TOC} = (C_{SOM} - C_{DOM}) + \sum_{l=1}^{u}(C_{S,l} - C_{DS,l}) + \sum_{m=1}^{v}(C_{R,m} - C_{DR,m}) + \sum_{n=1}^{w}(C_{A,n} - C_{DA,n}) \quad [1]$$

where TOC = total soil organic carbon; C_{SOM} = carbon in the stable soil organic matter; C_{DOM} = decomposed organic matter lost as CO_2; $C_{S,\,l}$ = carbon in shoot residue l; $C_{DS,l}$ = carbon lost as CO_2 from decomposed shoot residue l; $C_{R,m}$ = carbon in root residue m; $C_{DR,m}$ = carbon lost as CO_2 from decomposed root residue m; $C_{A,n}$ = carbon in organic amendment n; $C_{DA,n}$ = carbon lost as CO_2 from decomposed amendment n; and u, v, w = all applications of organic materials from the initial time to the current day (Liang et al., 2009).

Compared with the original version of CQESTR, several improvements have been made (Liang et al., 2009). Major modifications to the model include: (1) the terminating point for organic residue decomposition and residue-to-SOM transfer was changed from calendar-based time to thermal time; (2) separation of surface residue into two compartments, each with a different water coefficient, based on the non-uniform degradation of accumulated surface

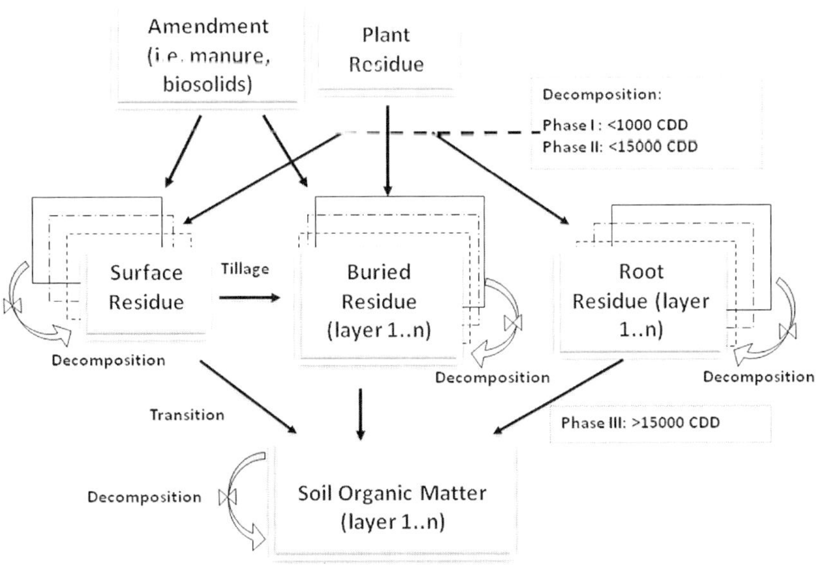

FIGURE 16.1
Flow diagram for the CQESTR model. The empty arrows represent the computed change in mass on a daily basis. The solid arrows depict biomass additions to the residue pools or transfers from the residue pools to the stable SOM pool. CDD is the cumulative degree-days or thermal time. *(Adapted after Liang et al., 2009.)*

residue in an NT cropping system as a result of annual residue layering; (3) the quantification of drainage and soil texture effects using calibration datasets with a wide range of edaphic conditions; (4) the inclusion of partially decomposed residue with stable simulated SOC when comparing with SOC observations; and (5) the inclusion of maximum SOC capacity or soil carbon saturation index.

EXPERIMENTAL SITES DESCRIPTION

Agricultural LTEs ranging from 23 to 134 years in duration, with a wide range of mean annual precipitation (420 to 1109 mm), temperature (11 to 20°C), surface soil textures (i.e. loamy sand, loam, silt loam), and soil drainage classes (i.e. somewhat poorly to well drained) were selected in three regions (Figure 16.2; Table 16.1). The selected LTEs included Akron, CO (Akron Plots), Champaign, IL (Morrow Plots), Florence, SC (Pee Dee Plots), Pendleton, OR (Pendleton Plots), and Columbia, MO (Sanborn Field). These LTEs were selected to represent a wide range of edaphic conditions, fertility management, crop rotations, and crop residue removal practices (i.e. harvest or burning).

The Akron Plots at Central Great Plains Research Station, CO, had been cultivated for 60 years before tillage treatments (stubble-mulch tillage, reduced tillage and no-tillage) were established in 1967 (Peterson et al., 1998) and modified in 1989 to include differences in cropping intensity (Halvorson et al., 2002a). The soil is a Weld silt loam (fine, smectic, mesic Aridic Paleustoll) developed from loess parent material under short-grass prairie and is moderately well drained to well drained. The cropping systems selected for model simulations were fallow–wheat (*Triticum aestivum* L.), (WF) under conventional stubble-mulch tillage (ST) and NT. The top 0 to 5 cm and 5 to 10 cm depths were used for SOC simulations.

The Morrow Plots at Champaign, IL, were established in 1876 (Darmody and Peck, 1997). The soil is a Flanagan silt loam (fine, smectitic, mesic Aquic Argiudoll), developed in loess over loamy glacial till under tallgrass prairie vegetation. Tile-drainage lines were installed in the plot borders in 1904 to improve soil drainage. Cropping systems were continuous maize (*Zea mays* L.) (MM), maize–oat (*Avena sativa* L.) (M–O), and maize–oat–clover (*Trifolium pratense* L.) (M–O–H) rotations. In 1967, soybean (*Glycine max* L.) replaced oat in the maize–oat rotation. All plots were under conventional moldboard plow tillage (MP), and fertilizer treatments were subplots of the cropping system treatments. Control (no fertilizer) or manure treatments for each rotation were assigned to two or three subplots. From 1876 until 1955, all aboveground biomass (grain, straw, stover, and hay) was removed. Starting in 1955, crop residues were returned to selected subplots, and in 1967 to all subplots. We selected all the above cropping systems for SOC simulation in the 0 to 15 cm depth.

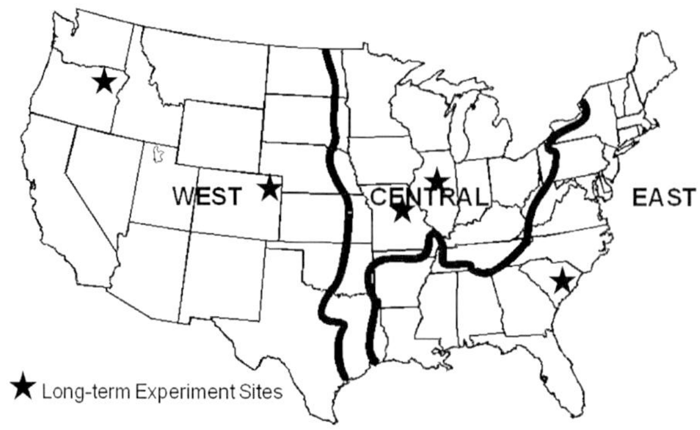

FIGURE 16.2
Approximate location of the five long-term experiment sites in relation to regional delineations for syntheses used in this volume.

TABLE 16.1 Location, Mean Annual Precipitation (MAP) and Temperature (MAT), Experiment Duration and Sampling Period, Crop Rotation, Management Practices, Soil Sampling Depth Used in Simulation, and Soil Texture Classes for the Selected Long-Term Experiment Sites

Site[†]	Akron, CO (Akron Plots)	Champaign, IL (Morrow Plots)	Columbia, MO (Sanborn Field)	Florence, SC (Pee Dee Plots)	Pendleton, OR (Pendleton Plots)
Location	40° 09' N 103° 09' W	40° 06' N 88° 13' W	38° 57' N 93° 20' W	34° 18' N 79° 44' W	45° 44' N 118° 37' W
MAP (mm)	420	940	973	1109	420
MAT (°C)	19.1	1[illegible]	13	20	11
Initiation year	1967, 1989	1876	1888	1979	1931
Duration (years)[‡]	23	134	122	25	79
Earliest sample[§]	1982	1904	1915	1979	1931
Latest sample	2005	1993	1986	2004	2005
Crop rotation[¶]	WF	MM M–O M–O–H/SO	MM WW	M–W, C(SO)	WF
Management[#]	ST NT	MP Residue removal before 1950 2 fertility levels	MP Residue removal before 1955 Fertilizer Manure (13.4 Mg ha^{-1} yr^{-1})	DT CS	MP Residue burning 90 kg N ha^{-1} Manure (11 Mg ha^{-1} yr^{-1})
Sampling depth (cm)	0–5 5–10	0–15	0–20	0–5 5–10	0–30 30–60
Soil texture[††]	SiL	SiL	SiL and L	LS	SiL

[†] *CO, Colorado; IL, Illinois; MO, Missouri; SC, South Carolina; OR, Oregon.*
[‡] *Refers to the duration time period for reported data used in simulation.*
[§] *Earliest soil sampling does not necessarily occur at the initiation of an experiment.*
[¶] *Crop rotation abbreviations: F, fallow; M, maize; MM, continuous maize; O, oat; H, clover; C(SO), cotton or soybean; W, wheat; WW, continuous wheat. "–", separates years.*
[#] *Tillage abbreviations: CS, conservation tillage; DT, disking; MP, conventional moldboard plow; ST, stubble-mulch.*
[††] *Soil texture abbreviations: L, loam; S, sand; Si, silt.*

The Pendleton Plots at the Columbia Basin Agricultural Research Center, Pendleton, OR, were established in 1931 on a site broken from a shrub–grassland or sagebrush-grassland steppe in the mid-1880s, and had already been farmed for ~50 years (Rasmussen and Albrecht, 1998). The soil is a Walla Walla silt loam (coarse–silty, mixed, superactive, mesic Typic Haploxeroll) developed in loess overlying basalt and is well drained. The crop residue plots under MP were selected for SOC simulations in the top 0 to 30 cm and 30 to 60 cm depths. Three wheat–fallow (WF) treatments selected were: (1) manure—22 Mg ha^{-1} wet manure added before plowing in the spring of the fallow year; (2) fertilizer—90 kg N ha^{-1} applied to each wheat crop, with straw plowed in the spring of the summer fallow; and (3) fall burn—no fertilizer added, with straw burned in the fall and the soil plowed in the spring before the fallow summer for minimum residue return (Machado, 2011; Rasmussen and Albrecht, 1998).

The Pee Dee Plots in the middle-Coastal Plain region of South Carolina at Florence, SC, were established in 1979 on a Norfolk loamy sand (fine-loamy, kaolinitic, thermic Typic Kandiudult) and the soil is well drained (Karlen et al., 1996). Primary tillage used at this site consisted of conventional tillage (disking, DT) to a depth of 15 cm. Conservation tillage (CS) eliminated surface tillage and used only in-row subsoiling (Hunt et al., 1996; Novak et al., 2007). Both tillage treatments received in-row subsoiling to fracture a root restrictive layer. Each plot was double-cropped with maize and winter wheat followed either by cotton (*Gossypium hirsutum* L.) or soybean. We are reporting SOC for the top 0 to 5 cm depth, although there were SOC measurements for three soil depths that were reported previously (Gollany et al., 2010; Novak et al., 2007).

Sanborn Field at Columbia, MO, was established in 1888 (Buyanovsky et al., 1997; Miles and Brown, 2011). The soil is a Mexico silt loam (fine, smectitic, mesic Vertic Epiaqualf), developed in loess over glacial till under tallgrass prairie, transitioning into a Lindley loam (fine-loamy, mixed, superactive, mesic Typic Hapludalf), developed in loess over glacial till under hardwood forest. Most of Sanborn Field consists of somewhat poorly drained soils. Cropping systems used in these simulations were continuous maize (MM), and continuous winter wheat (WW), with fertilizer (full treatment) or manure. Primary tillage consisted of fall MP followed by secondary disk tillage and a harrow in spring. Crop residues were removed by hand from the plots prior to 1950, and only standing stubble and root residues were returned to the soil. After 1950, straw and stover from wheat and maize plots were retained. Simulations of SOC in the top 0 to 20 cm were carried out for wheat and maize in manure and fertilized treatments under MP and NT with 90% straw or stover removed prior to 1950.

THE CQESTR MODEL APPLICATIONS

The CQESTR model was first used to simulate SOM dynamics under MP at Pendleton, OR, and MP and NT at Lexington, KY (Rickman et al., 2001). The model has been used since then for various field applications in temperate agroecosystems across North America (Gollany et al., 2011), for tillage comparisons in tropical regions of Brazil (Leite et al., 2009), and Saharan Steppe of Africa (Gollany et al., unpublished). The CQESTR applications range from comparison of SOC dynamics under prevalent agricultural practices, including tillage (MP, NT, ridge-till, sweep tillage, and stubble-mulch tillage), nutrient management (e.g. manure vs. synthetic fertilizer), residue management (e.g. burning, removal, retention), and cropping system diversity (Gollany et al., 2011, 2010; Liang et al., 2008; Rickman et al., 2002).

Comparison of Conventional Tillage and Conservation Tillage

Agronomic practices such as fallow, crop residue removal, and intensive tillage have contributed to accelerated SOC losses during the first 50 years after converting native prairie or forest to production agriculture (Follett et al., 1997; Huggins et al., 1998; Janzen et al., 1998; Johnson et al., 2005; Liebig et al., 2005; Paustian et al., 1998; Rasmussen and Albrecht, 1998). Tillage affects SOC in three fundamental ways: (1) through the physical disturbance and exposure of soil aggregates to disruptive forces that make physically protected soil C in the aggregate available for decomposition; (2) through effects of tillage on the soil microclimate, the influence of surface residue coverage on soil temperature, water interception and infiltration, and the effects of tillage-induced changes in porosity on soil aeration and water relations (Paustian et al., 1997); and (3) through stimulation of biotic activity. Conservation tillage practices were introduced to U.S. agriculture in the late 1940s to increase surface residue and soil cover.

The initial impetus for conservation tillage was to increase surface residue for erosion control. The NT management system is a conservation tillage practice which gained widespread acceptance in the 1980s primarily as a soil conservation measure; however, its value in preserving and possibly increasing SOC stocks was recognized in the 1990s. Long-term experiment results often support a recommendation to increase adoption of NT to preserve or increase SOC (Reicosky et al., 1995; Lal et al., 1998). Differences in SOC between NT and conventional tillage are most extreme near the surface, primarily due to differences in the distribution of C inputs. Benefits of NT for SOC sequestration varied with soil, climate, cropping system, and depth of measurement (West and Post, 2002), which may help explain apparent conflicts in the literature (Baker et al., 2007). The CQESTR model simulation for conventional tillage (moldboard plowing, disking, or stubble-mulch tillage) and NT represent opposite ends of the tillage spectrum at each location and would be expected to show the greatest differences in SOC. We will focus our discussion in this section on the model simulation results of these two tillage systems for the three regions of the U.S.

Since NT management was not present at all the sites, the actual management practices excluding tillage were used for CQESTR simulations. The NT was simulated for the plots with fertilizer or manure since the inception of the experiment. Actual annual grain and crop residue yields for each treatment at the site were used in the NT simulation (assuming comparable yields throughout the period of simulation) with the resulting harvest of crop residue under each treatment being equivalent to each harvest rate. Although NT reduces N availability due to N-immobilization (Andraski and Bundy, 2008) and potentially lowers N-mineralization rates, no yield reduction was assumed since the simulations were for long periods of time (e.g. several decades) and no severe consequences due to the residue harvest were imposed on any of the NT simulations. The NT simulation also assumed no change in soil bulk density due to tillage, although changes in soil bulk density are possible in some soils and both increases and decreases under NT have been reported in the literature (Gantzer and Blake, 1978; Lal et al., 1989; Tollner et al., 1984).

WESTERN REGION (OR)

Many of the agricultural practices that have traditionally been used in the Western region (e.g. moldboard plowing, cereal crop monoculture cropping, fallowing with no cover crop, and removal or burning of crop residues) represent "worst-case" scenarios for SOC maintenance. Currently, conservation tillage practices such as direct seeding or NT have been adopted in the region. We simulated NT scenarios for the long-term crop residue experiment at Pendleton to examine the effect of tillage on SOC at two soil depths (Figure 16.3A, OR). During the 73 years of MP, the observed SOC decreased at rates of 0.022 and 0.025 Mg C ha^{-1} yr^{-1} in the 0 to 30 cm and 30 to 60 cm depths, respectively. The observed decline in SOC under MP with fertilization (90 kg N ha^{-1}) was also predicted by the CQESTR model. Depending upon soil depth, decreases of SOC of up to 40% after the start of cultivation have been documented previously in the semiarid Western region and other parts of the U.S. (Follett, 2009; Follett et al., 1997; Janzen et al., 1998; Peterson et al., 1998). The predicted SOC in the 0 to 30 cm depth increased from 47.2 to 49.0 Mg C ha^{-1} under NT during the 73-year period, which could be attributed to an increase in the labile and slowly oxidized SOC pools that responded to the change in tillage practice. Increased C inputs from improved crop yields under NT with

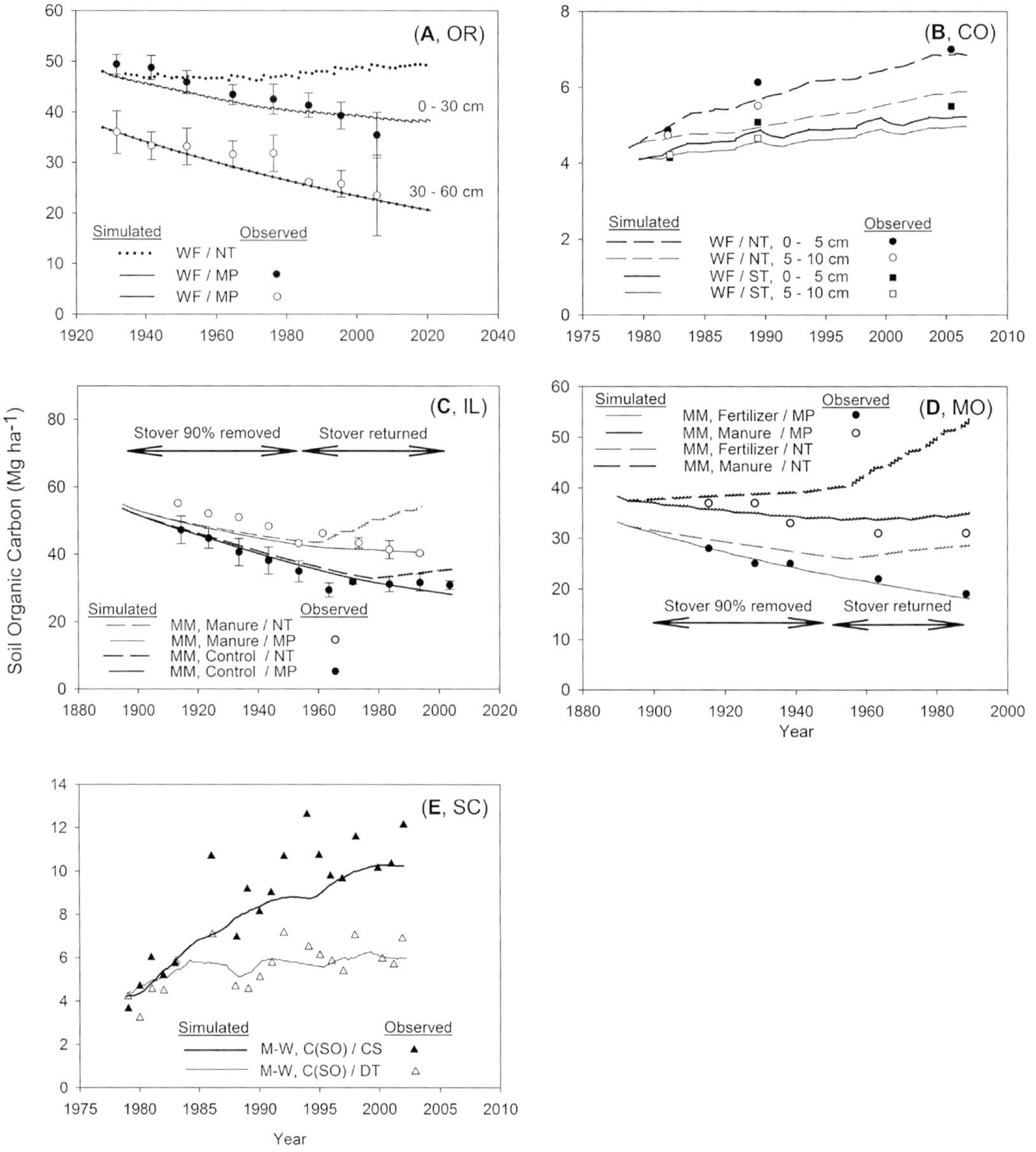

FIGURE 16.3
Simulated and observed soil organic carbon dynamics for winter wheat–fallow (WF), continuous winter wheat (WW), continuous maize (MM), maize–wheat–cotton, or soybean (M–W, C(SO)); (A, OR) in the 0–30 and 30–60 cm depths, for fertilized treatment under moldboard plow tillage (MP) or simulated no-till (NT) at Pendleton Plots, OR; (B, CO) in the 0–5 and 5–10 cm depths, under stubble-mulch tillage (ST) or NT at Akron Plots, CO; (C, IL) in the 0–15 cm depth for control (no fertilizer) and manure treatments under MP or simulated NT at Morrow Plots, IL; (D, MO) in the 0–20 cm depth for fertilized and manure treatments under MP or simulated NT at Sanborn Field, MO; and (E, SC) in the 0–5 cm depth under disk tillage (DT) or conservation tillage (CS) at Pee Dee Plots, SC. Wheat straw was removed during harvest before 1950 at Sanborn Field, and stover was removed during harvest before 1955 at Morrow and retained afterwards. The symbols are measured values and the lines are simulations from the CQESTR model. *(Adapted after Gollany et al., 2010.)*

adequate fertilization and weed control under limited precipitation have been reported (Paul et al., 1997). After 44 years of conservation tillage in a dryland WF system at Pendleton, sweep tillage increased SOC by 14% compared to MP tillage (Gollany et al., 2005). Decomposition of surface-placed residue such as under NT usually occurs more slowly than that of buried residues (Douglas et al., 1980; Ghidey and Alberts, 1993).

WESTERN REGION (CO)

The conventional tillage winter wheat–summer fallow is the predominant cropping system at Akron Plots in the Central Great Plains. Stubble-mulch (ST), reduced tillage, and NT were initiated in 1967, with SOC measured at various soil depths in 1982, 1989, and 2005 (Halvorson et al., 1997; Mikha et al., 2010). Soil bulk density was not measured in 1989 therefore the bulk density values reported for 1982 were used to calculate the observed SOC. The CQESTR model was used to simulate SOC in the 0 to 5 cm and 5 to 10 cm soil depths for the ST (100% disturbance) and NT treatments under WF cropping system (Figure 16.3B, CO). From 1982 to 2005, observed SOC increased at rates of 0.06 and 0.09 Mg ha^{-1} yr^{-1} in the top 0 to 5 cm under ST and NT, respectively. The slow rates of SOC sequestration under NT and ST were well simulated by the CQESTR model. The simulated SOC in the 0 to 5 cm depth increased from 4.9 to 6.8 Mg C ha^{-1} under NT over 23 years. The greater SOC for the NT soil is likely due to a lower C oxidation rate in these soils (Peterson et al., 1998), since the residue and grain productivity of the tillage treatments were similar. Reduced tillage and NT systems generally increase SOC, biological activity, aggregate stability, porosity, and water infiltration in the surface soil layers (Doran 1980a, b; Liebig et al., 2004; Puget et al., 1999). Mikha et al. (2010) reported a greater amount of soil macro-aggregates at this site and attributed it to tillage elimination.

CENTRAL REGION (IL)

The continuous maize and maize–soybean rotations are the predominant cropping systems in the North Central region. We simulated SOC for continuous maize (MM) in two long-term treatments, manure and control, under MP and NT cropping systems at Morrow Plots of the Central region. The observed SOC in the 0 to 15 cm soil depth decreased for the MM in the manure and control treatments under MP due to cultivation and 90% stover removal from 1876 until 1955 (Figure 16.3C, IL). Cultivation decreased SOC because of the greater opportunity for both erosion and microbial oxidation. The observed SOC in the control plot decreased from 44.3 to 32.1 Mg C ha^{-1} (27%) between 1915 and 2004, and this was also simulated by CQESTR. The observed and simulated SOC followed the same pattern for the manure treatment, although the total SOC losses were less pronounced than the control treatment. The total observed SOC losses were 4.6 and 12.2 Mg C ha^{-1} for the manure and control plots, respectively, during the 89 years of MM under MP. The decline in SOC from long-term stover removal under MP has been reported for this region in several studies (Allmaras et al., 2004; Blanco-Canqui et al., 2006; Larson et al., 1979; Vitosh et al., 1997). The simulated SOC in the control plots increased by 2.0 Mg C ha^{-1} at a rate of 0.006 Mg C ha^{-1} yr^{-1} from the late 1970s to 2004, with residue retention under the NT system. With stover retention under the NT system, the simulated SOC in the manure plots increased by 8.5 Mg C ha^{-1} from the late 1970s to 2004. This is most likely the result of increased N, delayed residue decomposition, improved soil physical properties and reduced erosion, or a combination of these factors. Wander et al. (1998) reported that tillage impacted the depth distribution of SOC in three Illinois soils. Generally, NT increased SOC and particulate organic matter (POM) in the surface 5 cm, but decreased values were found in the 5 to 17.5 cm depth. Losses of SOC could be reversed with stover retention and manure application under MM with NT cropping systems in this region.

CENTRAL REGION (MO)

The loss of SOC after converting native tallgrass prairie to MM cropping systems under the MP system and two fertilization treatments (fertilizer and manure) at Sanborn Field was very well simulated by the CQESTR model (Figure 16.3D, MO). Continuous loss of SOC under MP was observed in the fertilized treatments. The rapid decline in SOC was caused by land-use conversion and 90% stover removal at harvest before 1950. Observed and simulated SOC decreased with fertilizer or manure treatment until stover was retained. By 1988, observed SOC in the 0 to 20 cm depth for the fertilized and manure plots under MP were 19.0 and 31.0 Mg C ha^{-1}, respectively. The predicted SOC in the fertilized plots were 18.2 and 28.6 Mg C ha^{-1} under the MP and the NT scenarios, respectively, indicating an increase in the SOC for the top 20 cm under NT. The predicted SOC for manure plots under NT increased by 0.3 Mg C ha^{-1} yr^{-1} to 53.0 Mg C ha^{-1} with stover return for 38 years. In an NT system aboveground crop residue will primarily stay above ground. Only limited amounts of crop residue will be transported below ground by macro-fauna activity and leaching of dissolved C into the soil. In this region, CQESTR simulation results indicate that SOC could be maintained with manure application under an MM cropping system. This is most likely due to large biomass input as the result of high net primary biomass production and manure addition.

EASTERN REGION (SC)

At the Pee Dee Plots in the Eastern region, we simulated SOC dynamics for two long-term tillage treatments (Figure 16.3E, SC). The CQESTR model captured temporal changes in SOC for the 0 to 5 cm soil depth under conventional disking (DT) and conservation tillage (CS). The mean and standard error of observed SOC in the 0 to 5 cm depth for DT was 4.2 ± 0.5 Mg C ha^{-1} in 1979 (Novak et al., 2007). An increase of 1.6 Mg C ha^{-1} in the 0 to 5 cm depth was observed under DT from 1979 to 1983. During the same 4-year period, CQESTR also predicted an increase in SOC of 1.3 Mg C ha^{-1} under DT. In the late 1980s, observed SOC decreased because of management changes after 1986, and this decrease was also simulated by CQESTR. Another increase in SOC occurred when paratill substituted for in-row subsoiling after 1996. During the first 10 years, observed SOC increased by 5.4 Mg C ha^{-1} under CS. The mean and standard error of the observed SOC for CS was 3.6 ± 0.1 Mg C ha^{-1} in 1979 (Novak et al., 2007). Simulated SOC stocks followed the same trend as observed values. The CQESTR model predicted an increase in SOC from 4.1 to 8.7 Mg C ha^{-1} during the first 10 years. During 23 years, the simulated SOC stocks for the CS system increased from 4.1 to 10.2 Mg C ha^{-1}, while observed values increased from 3.6 to 10.5 Mg C ha^{-1}. In the late 1990s, CQESTR underestimated some high SOC values for the CS system. The model predicted, on average, increases of 0.07 and 0.27 Mg C ha^{-1} yr^{-1} for DT and CS, respectively. This is in agreement with the reported near doubling of mean SOC, in the 0 to 5 cm depth, for CS (1.2%) compared to DT (0.72%) for this site, after 14 years of conservation tillage with a cropping system that included maize, soybean, wheat, and cotton (Hunt et al., 1996). Similar results were found in other long-term tillage comparisons of continuous maize and a maize–soybean rotation under MP in Ohio (Lal et al., 1994; Dick and Durkalski, 1997; Dick et al., 1998). This increase in SOC for CS was attributed to the low initial SOC of these soils and POM in the surface 5 cm compared to DT. Small changes in SOC can substantially change soil properties. An increase in SOC occurred for both DT and CS systems presumably because of high levels of crop management resulting in improved yields when most of the residues were returned to the experimental site (Karlen et al., 1989). Conservation tillage in this region with low inherent levels of SOM resulted in great stratification of SOC among soil depths and could lead to greater soil functional benefit, such as soil water retention and nutrient content, with adoption of conservation tillage (Franzluebbers, 2004). These results showed the importance of soil type, soil texture, and tillage interactions on SOC accretion.

Comparison of Fertilizer and Manure

Fertilization and manure addition can play a key role in C cycling and can act as a source or sink depending on management. Addition of organic C, whether from animal manure, municipal or industrial waste, not only provides micro- and macro-nutrients needed for crop production, but can contribute to long-term SOC increase if managed properly. Application of manure has been promoted as a management practice to sustain or increase SOC (Lal et al., 1998; Tester, 1990; Eghball, 2002; Edmeades, 2003). In eastern Oregon, application of cattle manure at 11.2 MT ha^{-1} yr^{-1} over a 56-year period resulted in an increase in SOC of 0.02 Mg C ha^{-1} yr^{-1} (Liebig et al., 2005).

The increase in biological activity following manure application is the result of an additional food source for soil biota, as well as improved soil structure, which increases soil water holding capacity, water movement, and allows better air exchange. Manure addition can improve soil physical properties such as available water holding capacity (Estevez et al., 1996; Hudson, 1994; Rawls et al., 2003; Olness and Archer, 2005). Soil enzyme activities are usually greater in manure plots compared to inorganic fertilizer plots (Fauci and Dick, 1994; Martens et al., 1992). However, application of manure with high N content may increase N_2O and CH_4 emissions (Cates and Keeney, 1987; Paul et al., 1993; Coyne et al., 1995; Chang et al., 1998), thus reducing GHG mitigation potential. Improved management of animal manures, such as optimizing the timing of application to synchronize with crop uptake and avoiding excess application, will ensure the most positive effects of manure additions on SOC storage while reducing GHG emissions (Johnson et al., 2005).

In general, a proper N fertility program in continuous cropped systems increases crop yield and SOC and has a potential to mitigate GHG emissions (Gollany et al., 2004; Snyder et al., 2009). Conversely, loss of SOC occurs even with a high N fertilization rate in fallow systems under MP (Halvorson et al., 2002b). Increasing N application rates above the baseline is unlikely to have a major impact on SOC sequestration (Rassmussen and Parton, 1994) and could lead to increased N_2O and CO_2 emissions (Mosier et al., 2006) if its application is not synchronized with nutrient uptake by crops. In eastern Oregon, under a semiarid environment, loss of SOC occurred in a WF rotation across a range of N rates (Rassmussen and Parton, 1994). In contrast, Gregorich et al. (1996) showed a 13% increase in SOC after 32 years of continuous maize with N fertilization under MP. The SOC changes were dependent on the aboveground and root biomass response to N fertilization (Huggins and Fuchs, 1997).

WESTERN REGION (OR)

Decreases in SOC were observed in the 0 to 30 cm and 30 to 60 cm soil depths and were well predicted by the CQESTR model in the fertilized (90 kg N ha^{-1}) wheat–fallow treatments under MP at Pendleton (Figure 16.4A, OR). Observed SOC decreased by 14.0 and 12.5 Mg C ha^{-1} over 73 years in the 0 to 30 cm and 30 to 60 cm depths in the fertilized plots, respectively. This indicates that N fertilizer had a limited effect on SOC through residue production in this semiarid environment. The primary effect of N fertilizer is to increase crop vegetative biomass while concurrently creating a source of C inputs to the soil. Fallow systems under dryland conditions are not conducive to SOC accumulation. After 44 years of conservation tillage, application of 180 kg N ha^{-1} increased SOC storage only 3% above that found under the 45 kg N ha^{-1} application rate (Gollany et al., 2005).

Manure application for 73 years in a wheat–fallow cropping system slightly increased SOC in the top 30 cm depth. Simulated SOC followed the same trend as observed for both soil depths. In 2005, the predicted and observed SOC in the topsoil (0–30 cm) of the manure plots were the same (46.3 Mg C ha^{-1}). A slight increase in SOC with manure application is because of increased enzyme activity and microbial biomass. Both enzyme activity and microbial biomass were correlated with the total amounts of organic C added as manure in these plots (Collins et al., 1992). An increase in microbial population can potentially improve soil physical properties and increase the physical protection of SOC. Manure application improved soil structure and increased fungal and bacterial populations (N'Dayegamiye and Angers, 1990). Declines in SOC at rates of 0.18 Mg C ha^{-1} yr^{-1} in the fertilized plots and of 0.15 Mg C ha^{-1} yr^{-1} with manure application for the 30–60 cm soil depth were predicted by CQESTR. Continuous loss of SOC in the 30 to 60 cm depth was due either to the use of fallow, or perhaps as a result of loss of native grasses, which have a greater net biomass because of their larger belowground crown and root biomass compared to the cultivated wheat crop.

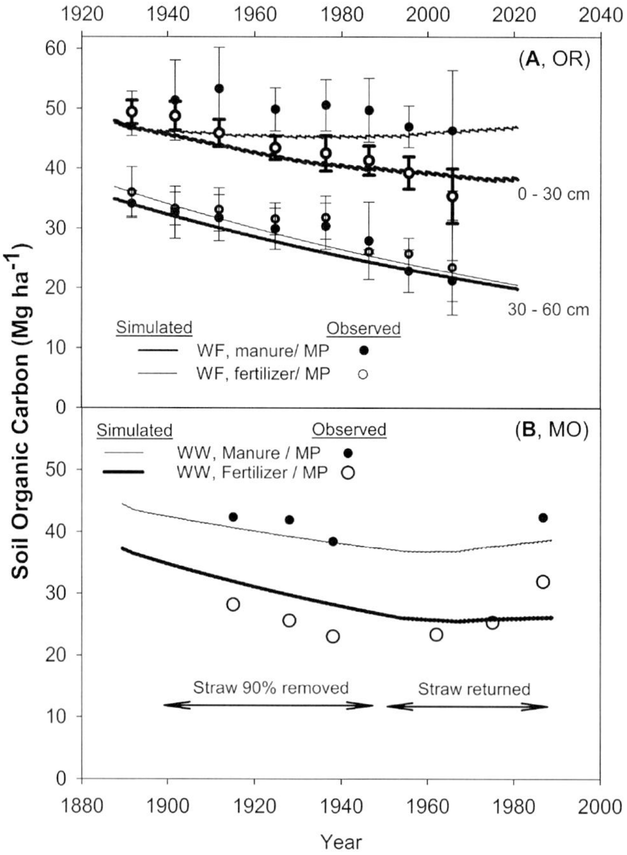

FIGURE 16.4
Simulated and observed soil organic carbon dynamics for winter wheat—fallow (WF) and continuous winter wheat (WW); (A, OR) in the 0–30 and 30–60 cm depths for fertilized and manure treatments under moldboard plow tillage (MP) at Pendleton Plots, OR; and (B, MO) in the 0–20 cm depth for fertilized and manure treatment under MP at Sanborn Field, MO, where wheat straw was removed during harvest before 1950 and retained afterwards. The symbols are measured values and the lines are simulations from the CQESTR model.

CENTRAL REGION (MO)

Continuous decrease in SOC was observed when 90% of straw was harvested in the fertilized plots before 1950 at Sanborn Field in the South Central region (Figure 16.4B, MO). During the first 23 years of MP, observed SOC in the 0 to 20 cm depth decreased by 3.9 Mg C ha^{-1} with manure and by 4.8 Mg C ha^{-1} with fertilizer application. After the 1950s, when the straw was retained, SOC increased and the upward trend in SOC for both manure and fertilized treatments during this period was simulated by CQESTR. An increase in SOC at a rate of 0.08 Mg C ha^{-1} yr^{-1} in manure plots was most likely due to both enzyme activity and straw retention after 1950. Researchers in California found that animal manures, as well as green manures and crop residues added to soils, increase enzyme activities within a few weeks, compared to unamended soils (Martens et al., 1992). The simulated and observed SOC in 1988 were 26.2 and 32.0 Mg C ha^{-1} for fertilized plots under MP, respectively. The underestimated SOC at this sampling date was attributed to either overestimation of decomposition rates of SOM by CQESTR in this environment, or possible error in sample collection or sieving and processing, since the predicted SOC in the fertilized plots under NT scenario (Figure 16.6C) approximated the observed SOC in 1988 under MP (Gollany et al., 2011; Liang et al., 2008). It is possible that the sample contained POM that was measured as total SOC, which CQESTR does not account for in simulated stable SOC pool. The simulated SOC loss in the manure plot, with 90% straw removal until 1950, was 3.5 Mg C ha^{-1} under MP. After

71 years, SOC stocks reached a steady-state condition with manure application under WW cropping (Gollany et al., 2011; Miles and Brown, 2011).

Comparison of Monoculture and Crop Rotation

Crop rotation can influence SOC. In the southeastern U.S., increased cropping intensity and organic C inputs led to increased SOC under both conventional and conservation tillage systems (Franzluebbers, 2005). The SOC sequestration (0.53 Mg C ha^{-1} yr^{-1}) with a cover crop and NT was about two times greater than without a cover crop (0.28 Mg C ha^{-1} yr^{-1}), and it was suggested SOC could be increased by 0.22 Mg C ha^{-1} yr^{-1} with increasing cropping system complexity irrespective of tillage management (Franzluebbers, 2005). After 35 years, legume-based cropping systems had 20 Mg C ha^{-1} yr^{-1} greater SOC than monoculture maize (Gregorich et al., 2001). Most of the additional SOC was found deeper in the soil profile in the legume system presumably because of quantity and quality of root inputs. Wander et al. (1994) showed long-term cover crop use reduced soluble C and increased retention of total C and POM, while manure applications increased C mineralization and soluble C.

CENTRAL REGION (IL)

At Morrow Plots, observed SOC decreased at a rate of 0.004 Mg C ha^{-1} yr^{-1} in the 0 to 15 cm soil depth with manure application, which was lower than the 0.006 Mg C ha^{-1} yr^{-1} loss for the control plot in a maize–oat (M–O) rotation under MP cropping system until 1955 (Figure 16.5A, IL). The CQESTR model simulated the decline in SOC with 90% crop residue removal from 1876 to 1955. Inclusion of oat increased SOC compared to MM (Figure 16.3C, IL) and was more pronounced in the manure plots than in the control plot. Addition of manure has shown to increase SOC even when maize silage was removed (Vitosh et al., 1997).

Observed SOC in the control and the manure plots declined with 90% crop residue removal in maize–oat–hay (M–O–H) rotation under MP at Morrow (Figure 16.5B, IL); however, the rate of decline was less than in MM (Figure 16.3C, IL) or maize–oat rotation (Figure 16.5A, IL), especially when manure was applied. During the first 49 years, observed SOC decreased by 4.6 and 8.3 Mg C ha^{-1} in the 0 to 15 cm depth of the manure and control plots, respectively. The CQESTR model simulated the SOC dynamic and decline in SOC. Simulated SOC decreased by 11.3 Mg C ha^{-1} in the control plots under MP, but increased by 13.3 Mg C ha^{-1} in the manure plots during 89 years. The maintenance of high SOC under perennial crops is associated with a relative increase in POM and an increase in the physical protection of SOC (Wander et al., 1994). Aggregate stability increases rapidly after perennials are established because of the characteristics of the belowground biomass of most perennial grasses. These characteristics include a dense, fibrous root system, rhizodeposits, and fungal hyphae and microbially derived polysaccharides and gums, all of which are effective at bonding mineral and SOM particles.

Comparison of Residue Removal Rates

Crop residues are the major component of organic C entering a soil system. Roots and rhizodeposits are also an important source for SOC and play a critical role in the C cycle (Johnson et al., 2006). Many attempts have been made to quantify belowground C inputs (Balabane and Balesdent, 1992; Bolinder et al., 1999; Bottner et al., 1999; Flessa et al., 2000). However, belowground biomass is seldom measured because of lack of appropriate techniques to measure total C inputs and quantification of belowground biomass is highly variable. The CQESTR model computes belowground biomass from the amount of aboveground biomass added. It is estimated that one-third of the C derived from crop residues remains in the soil after 1 year (Angers and Chenu, 1977) and 80 to 90% decomposes after 2 years (Broder and Wagner, 1988; Buyanovsky and Wagner, 1997). Only a small fraction of the crop residue added is converted to stable SOC. Soil C in its stable form as SOM responds gradually to changes in

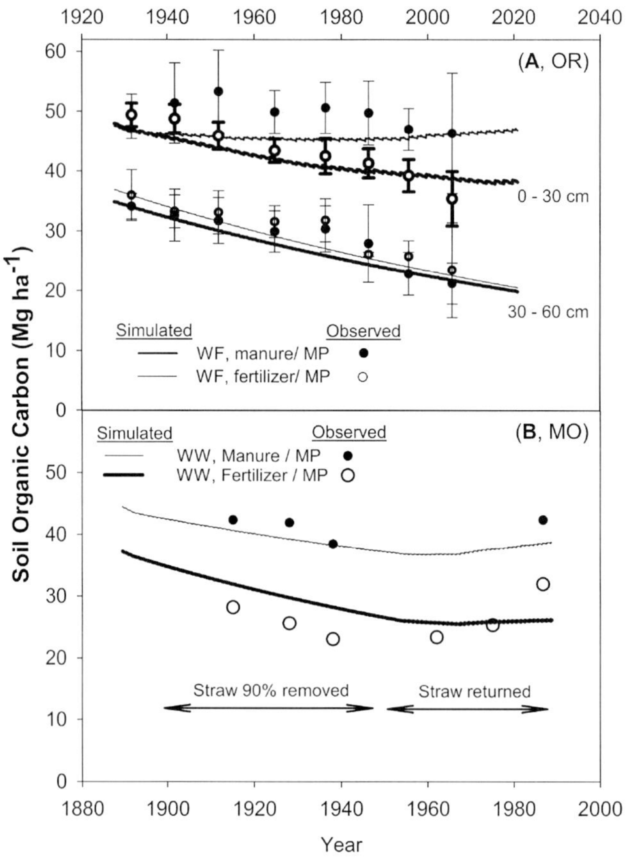

FIGURE 16.5
Simulated and observed soil organic carbon dynamics for: (A, IL) maize–oat (M–O); and (B, IL) maize–oat–clover (M–O–H) in the 0–15 cm depth for control (no fertilizer) and manure treatments under conventional moldboard plow (MP) at Morrow Plots, IL. Stover was removed during harvest before 1955 and retained afterwards. The symbols are measured values and the lines are simulations from the CQESTR model.

agricultural management. Because of this slow process and complex interactions between climate, crop, soil, and management practices, mechanistic models have been used to describe and predict short- and long-term effects of crop residue removals on SOC dynamics and turnover (Gollany et al., 2011; Liang et al., 2008; Lemke et al., 2010).

We examined the effect of crop type and residue harvest on SOC while implementing four crop residue removal scenarios (i.e. 25, 50, 75, and 90% stover or straw removal). Simulations for residue removal scenarios were carried out under NT assuming no change in soil bulk density due to tillage or residue removal, since empirical data on soil bulk density changes due to residue removal were not reported in the literature for these sites.

CENTRAL REGION (IL)

Cultivation and 90% stover harvest from 1876 until 1955 decreased observed SOC under an MP cropping system at the Morrow Plots by 11.4 and 12.2 Mg C ha^{-1} for maize–oat (M–O) rotation in the manure and control treatments, respectively (Figure 16.6A, IL). The predicted SOC increased by 2.0 Mg C ha^{-1} at a rate of 0.002 Mg C ha^{-1} yr^{-1} with residue retention in the late 1950s and continued through the 1990s under the NT scenario with manure application. The CQESTR model captured the decline in SOC prior to the mid-1950s, although CQESTR underestimated SOC in the manure plot. There was a lag time in CQESTR response to return of crop residue in the early 1950s presumably due to overestimation of decomposition rates of SOM in this environment. Alternatively, CQESTR may have underestimated the

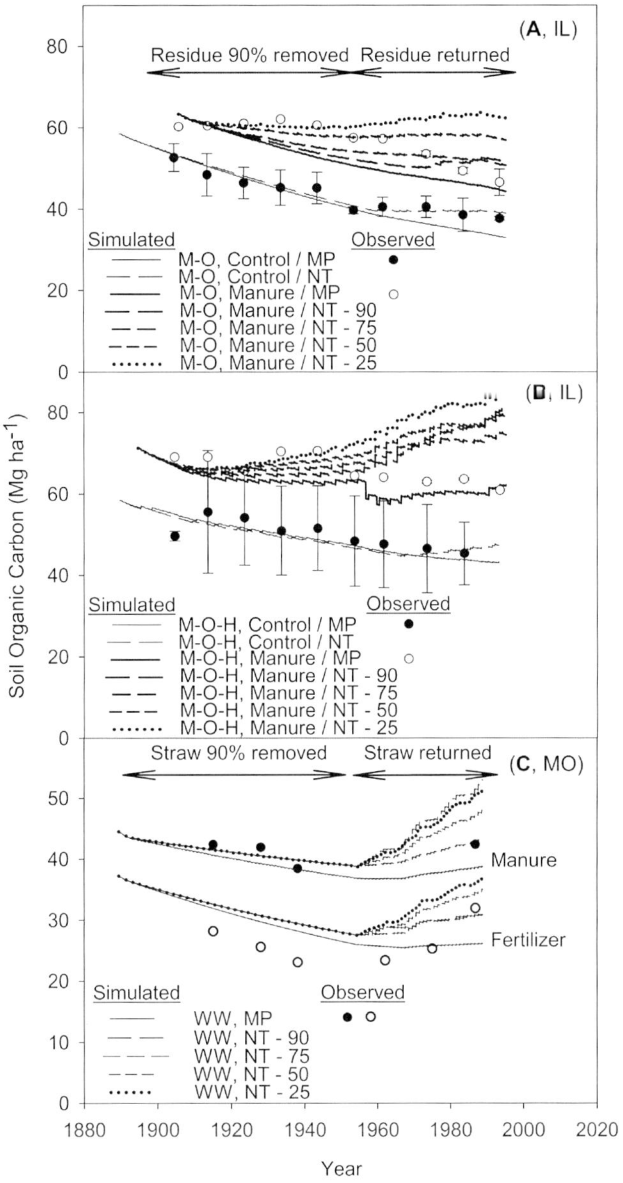

FIGURE 16.6

Simulated and observed soil organic carbon dynamics for: (A, IL) maize–oat (M–O); (B, IL) maize–oat–clover (M–O–H) in the 0–15 cm depth for control (no fertilizer) and manure treatments under conventional moldboard plow (MP) and simulated no-till (NT) at Morrow Plots, IL; (C, MO) in the 0–20 cm depth for fertilized and manure treatments under MP or simulated NT at Sanborn Field, MO; with stover or straw removal at 25 (NT-25), 50 (NT-50), 75 (NT-75), and 90% (NT-90); stover was removed during harvest before 1955 at Morrow Plots, before 1950 at Sanborn Field, and retained afterwards. The symbols are measured values and the lines are simulations from the CQESTR model.

contribution of oat biomass to SOC before 1960s because specific site parameters were unknown and some values, such as N content of the oat residue and residue or root to shoot ratio, were only estimated (Gollany et al., 2011). Huggins et al. (1998) suggested factors other than C inputs to these plots were controlling management-induced variation in SOC (i.e. SOC decay rates). The CQESTR model predicted SOC well after 1967, when oat was replaced by soybean, most likely because appropriate estimates of plant parameters were used for soybean in the model. Simulations for residue removal scenarios were carried out under NT for the manure plots only since the control plots were still losing SOC after stover returned. Simulated losses of SOC under NT management were 1.0, 6.5, 11.5, and 12.7 Mg C ha^{-1} with 25, 50, 75, and 90% stover removal scenarios, respectively. The CQESTR model predicted that it would take more than 89 years for the SOC in the maize–oat rotation with manure application to return to pre-cultivation levels even with NT management under minimal residue removal. This indicates restocking stable SOC through the inclusion of grass may occur more slowly than the rapid losses of physically stabilized SOC following residue removal. A decrease in SOC can reduce nutrient availability and consequently reduce crop yield and source C input into the soil system if supplemental fertilizer is not applied to replace nutrients lost with residue removal (Blanco-Canqui and Lal, 2009; Larson et al., 1979; Wilhelm et al., 2004).

The CQESTR model captured the SOC dynamic and decline in SOC with 90% crop residue removal from 1876 until 1955 in the maize–oat–hay (M–O–H) rotation under MP at Morrow Plots (Figure 16.6B, IL). The observed SOC in the M–O–H rotation under MP decreased during the first 49 years, by 4.6 Mg C ha^{-1} in manure, and by 8.3 Mg C ha^{-1} in control plots. The rate of SOC decline under the M–O–H rotation with crop residue removal was less severe, especially for the manure treatment, than the other rotations at this site (Figures 16.3C, IL and 16.6A, IL). The simulated SOC also decreased at a rate of 0.006 Mg C ha^{-1} yr^{-1} during 89 years in the control plots under the MP cropping system. The predicted SOC increased by 13.2 Mg C ha^{-1} at a rate of 0.004 Mg C ha^{-1} yr^{-1} during 89 years, with 90% residue removal up to the mid-1950s, and residue retention through the 1990s under the NT scenario only with manure amendment. Most of the increase came after leaving residue in the field in the late 1950s. The predicted SOC increased by 7.4, 11.5, 13.1, and 17.4 Mg C ha^{-1} with the 90, 75, 50, and 25% residue removal scenarios, respectively, under NT and manure application since 1905 in the M–O–H cropping system (Figure 16.6B, IL). Other studies under North Central region conditions have examined the complex interactions between stover harvest, N fertilization, and SOC dynamics. In a 13-year-old experiment in Minnesota, Clapp et al. (2000) found a substantial increase (14%) in SOC in both 0 to 15 cm and 15 to 30 cm depths where maize stover was returned under chisel plow compared to continuous NT maize plots, where SOC remained nearly unchanged over time. They also found that the half-life for original or relic SOC was lengthened when stover was not harvested and N fertilization was partially mixed with the stover during chiseling, which is more important for C sequestration and long-term soil health.

CENTRAL REGION (MO)

The CQESTR model was used to simulate SOC dynamics for continuous winter wheat (WW) with fertilizer or manure treatments at the Sanborn Field, and simulated NT management with straw removal at 25, 50, 75, and 90% for potential bioenergy production.

Wheat straw harvests by hand and removal from the plots until the 1950s was well simulated by CQESTR. Simulated SOC under NT increased by 4.1, 8.6, 12.0, and 13.4 Mg C ha^{-1} in the manure plots, and by 3.0, 3.2, 6.9, and 9.2 Mg C ha^{-1} in the fertilized plots with straw removal of 90, 75, 50, and 25%, respectively, relative to MP with fertilizer management over 33 years (Figure 16.6C, MO). This indicates that NT adoption can reverse the trend of SOC loss, even under moderate residue removal rates at least in these types of soils and under climatic conditions similar to those in the South Central region. However, residue removal rates greater than 50% elsewhere have not maintained SOC.

Comparison of Residue Burned and Residue Retained

Monocultures or simple crop rotation, inclusion of fallow, intensive tillage and cultivation operations, and residue burning to reduce weeds and disease and promote clean tillage, were dominant management practices in the cereal-producing semiarid Northwest region for more than a century and are still practiced by some farmers in the region. These practices were

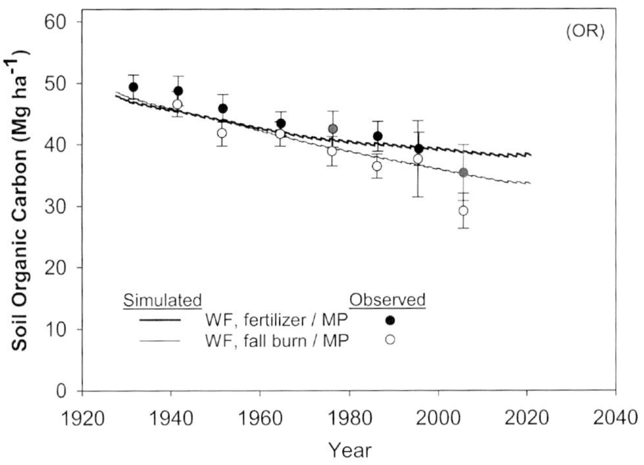

FIGURE 16.7
Simulated and observed soil organic carbon dynamics for winter wheat–fallow rotation in the 0–30 cm depth for fall burned and fertilized treatment, under conventional moldboard plow tillage (MP) at Pendleton, OR. The symbols are measured values and the lines are simulations from the CQESTR model.

generally detrimental to the maintenance of SOC. Rasmussen and Albrecht (1998) showed accelerated soil C loss and reduced microbial activity in burned plots for over 20 years.

We simulated SOC change for fertilized and burned management systems at the Pendleton long-term plots. The observed SOC stocks were well predicted by the CQESTR model (Figure 16.7, OR). Observed SOC decreased by 17.4 Mg C ha^{-1} for the straw burned treatment, which was greater than 13.4 Mg C ha^{-1} lost in the 0 to 30 cm depth of the fertilized treatment. Gollany et al. (2011) reported a decline of 0.24 Mg SOC ha^{-1} yr^{-1} in the 30 to 60 cm depth for burned plots over 64 years. Simulated SOC followed the same trend for both soil depths. Repeated, long-term residue burning can have a negative effect on overall soil health, including reduced C input to the soils, reduced microbial activity (Ladd et al., 2002), and decreased water-stable aggregates (Wuest et al., 2005).

CONCLUSIONS

The CQESTR model captured spatial and temporal changes in SOC under most agronomic practices within the five LTEs evaluated. In general, observed and predicted SOC losses were greater under conventional tillage in fertilized cropping systems than manure or intensified cropping management systems. The CQESTR model predicted an increase in SOC under conservation tillage for all sites without crop residue removal. The SOC losses decreased after crop residue was retained, while manure application maintained SOC when straw was retained at Pendleton Plots (OR) and Sanborn Field (MO), or when stover was retained under continuous maize at Sanborn Field, or under maize–oat–hay rotation at Morrow Plots (IL) under conventional tillage. An increase in observed and predicted SOC at Akron (CO) and Pee Dee (SC) Plots for stubble-mulch, disking, and conservation tillage practices indicate that high levels of crop and soil management have improved SOC. Furthermore, SOC appears not to have reached steady-state equilibrium at these sites under current management systems (although it is approaching equilibrium under conventional tillage). The model predicted SOC will not be maintained in the maize–oat rotation with stover removal even with manure application under NT management scenarios for 89 years; however, in the maize–oat–hay rotation it is possible to maintain SOC with stover removal and manure application under NT management at Morrow Plots. At Sanborn Field, the model predicted SOC maintenance is possible with a residue removal rate of 50% or lower in the fertilized continuous wheat plots, and SOC can be maintained under straw removal with manure addition. This indicates NT adoption can reverse the trend of SOC loss, even under moderate straw removal rates, at least in soils and the climatic condition similar to those in the South Central region.

The SOC stocks of a particular field are directly influenced by several factors, including initial SOC, crop yield, crop rotation intensity, field management practices (e.g. tillage type, crop residue removal, manure application, fertilization, etc.), edaphic conditions, and climate. Agronomic management such as tillage, fertilization, and manure addition can play a key role in C cycling. The SOC might be increased by reducing tillage and organic amendment application (e.g. crop residue or organic waste), thus mitigating GHG emissions. More long-term (>25 year) data would be useful to validate modeling of SOC, because stable SOC changes occur very slowly and require several decades to reach equilibrium.

Acknowledgments

The authors thank Dr. C. Walthall for his support to the modeling effort, and Dr. M. Liebig, ARS-Mandan and reviewers for helpful comments and suggestions. This publication is based on work supported by the U.S. Department of Agriculture-Agricultural Research Service under the ARS GRACEnet Project. USDA is an equal opportunity provider and employer.

References

Allmaras, R.R., Linden, D.R., Clapp, C.E., 2004. Corn-residue transformations into root and soil carbon as related to nitrogen, tillage and stover management. Soil Sci. Soc. Am. J. 68, 1366–1375.

Andraski, T.W., Bundy, L.G., 2008. Corn residue and nitrogen source effects on nitrogen availability in no-till corn. Agron. J. 100, 1274–1279.

Angers, D.A., Chenu, C., et al., 1977. Dynamics of soil aggregation and C sequestration. In: Lal, R. (Ed.), Soil Processes and the Carbon Cycle. CRC Press, Boca Raton, FL, pp. 199–206.

Baker, J.M., Ochsner, T.E., Venterea, R.T., Griffis, T.J., 2007. Tillage and soil carbon sequestration—what do we really know? Agri. Ecosys. Environ. 118, 1–5.

Balabane, M., Balesdent, J., 1992. Input of fertilizer-derived labeled N to soil organic matter during a growing season of maize in the field. Soil Biol. Biochem. 24, 89–96.

Blanco-Canqui, H., Lal, R., 2009. Corn stover removal for expanded uses reduces soil fertility and structural stability. Soil Sci. Soc. Am. J. 73, 418–426.

Blanco-Canqui, H., Lal, R., Post, W.M., Izaurralde, R.C., Owens, L.B., 2006. Rapid changes in soil carbon and structural properties due to stover removal from no-till corn plots. Soil Sci. 171, 468–482.

Bolinder, M.A., Angers, D.A., Giroux, M., Laverdiere, M.R., 1999. Estimating C inputs retained as soil organic matter from corn (Zea mays L.). Plant Soil 215, 85–91.

Bottner, P., Pansu, M., Sallih, Z., 1999. Modelling the effect of active roots on soil organic matter turnover. Plant Soil 216, 15–25.

Broder, M.W., Wagner, G.H., 1988. Microbial colonization and decomposition of corn, wheat and soybean residue. Soil Sci. Soc. Am. J. 52, 112–117.

Buyanovsky, G.A., Brown, J.R., Wagner, G.H., et al., 1997. Sanborn field effect of 100 years of cropping on soil parameter. In: Paul, E.A. (Ed.), Soil Organic Matter in Temperate Agroecosystems: Long-Term Experiments in North America. CRC Press, Boca Raton, FL, pp. 205–226.

Buyanovsky, G.A., Wagner, G.H., 1997. Crop residue input to soil organic matter on Sanborn Field. In: Paul, E.A., et al. (Eds.), Soil Organic Matter in Temperate Agroecosystems: Long-Term Experiments in North America. CRC Press, Boca Raton, FL, pp. 73–83.

Campbell, C.A., McConkey, B.G., Zentner, R.P., Dyck, F.B., Selles, F., Curtin, D., 1996. Long-term effects of tillage and crop rotations on soil organic C and total N in a clay soil in southwestern Saskatchewan. Can. J. Soil Sci. 76, 395–401.

Clapp, C.E., Allmaras, R.R., Layese, M.F., Linden, D.L., Dowdy, R.H., 2000. Soil organic carbon and ^{13}C abundance as related to tillage, crop residue, and nitrogen fertilization under continuous corn management in Minnesota. Soil Tillage Res. 55, 127–142.

Cates Jr., R.L., Keeney, D.R., 1987. Nitrous oxide production throughout the year from fertilized and manured maize fields. J. Environ. Qual. 16, 443–447.

Chang, C., Cho, C.M., Janzen, H.H., 1998. Nitrous oxide emission from long-term manured soils. Soil Sci. Soc. Am. J. 62, 677–682.

Collins, H.P., Rasmussen, P.E., Douglas Jr., C.L., 1992. Crop rotation and residue management effects on soil carbon and microbial dynamics. Soil Sci. Soc. Am. J. 56, 783–788.

Coyne, M.S., Villalba, A., Blevins, R.L., 1995. Nitrous oxide loss from poultry manure-amended soil after rain. J. Environ. Qual. 24, 1091–1096.

Darmody, R.G., Peck, T.R., 1997. Soil organic changes through time at the University of Illinois Morrow Plots. In: Paul, E.A., et al. (Eds.), Soil Organic Matter in Temperate Agroecosystems: Long-Term Experiments in North America. CRC Press, Boca Raton, FL, pp. 161–169.

Del Grosso, S.J., Mosier, A.R., Parton, W.J., Ojima, D.S., 2005. DAYCENT model analysis of past and contemporary soil N_2O and net greenhouse gas flux for major crops in the USA. Soil Tillage Res. 83, 9–24.

Dick, W.A., Durkalski, J.T., et al., 1997. No-till production agriculture and carbon sequestration in a Typic Fragiudalf soil of northeastern Ohio. In: Lal, R. (Ed.), Management of Carbon Sequestration in Soil. CRC Press, Boca Raton, FL, pp. 59–71.

Dick, W.A., Blevins, R.L., Frye, W.W., Peters, S.E., Christenson, D.R., Pierce, F.J., Vitosh, M.L., 1998. Impacts of agricultural management practices on C sequestration in forest-derived soils of the eastern Corn Belt. Soil Tillage Res. 47, 235–244.

Doran, J.W., 1980a. Microbial changes associated with residue management with reduced tillage. Soil Sci. Soc. Am. J 44, 518–524.

Doran, J.W., 1980b. Soil microbial and biochemical changes associated with reduced tillage. Soil Sci. Soc. Am. J 44, 765–771.

Douglas Jr., C.L., Allmaras, R.R., Rasmussen, P.E., Ramig, R.E., Roager Jr., N.C., 1980. Wheat straw composition and placement effects on decomposition in dryland agriculture of the Pacific Northwest. Soil Sci. Soc. J. 44, 833–837.

Edmeades, D.C., 2003. The long-term effects of manures and fertilisers on soil productivity and quality: a review. Nutr. Cycl. Agroecosyst. 66, 165–180.

Estevez, B., N'Dayegamiye, A., Coderre, D., 1996. The effect on earthworm abundance and selected soil properties after 14 years of solid cattle manure and NPKMg fertilizer application. Can J. Soil Sci. 76, 351–355.

Eghball, B., 2002. Soil properties as influenced by phosphorus- and nitrogen-based manure and compost applications. Agron. J. 94, 128–135.

Fauci, M.F., Dick, R.P., 1994. Soil microbial dynamics: short and long-term effects of inorganic and organic nitrogen. Soil Sci. Soc. Am. J. 58, 801–806.

Farrell, A.E., Plevin, R.J., Turner, B.T., Jones, A.D., O'Hare, M., Kammen, D.M., 2006. Ethanol can contribute to energy and environmental goals. Science 311, 506–508.

Flessa, H., Ludwig, B., Heil, B., Merbach, W., 2000. The origin of soil organic C, dissolved organic C and respiration in a long term maize experiment in Halle, Germany, determined by ^{13}C natural abundance. J. Plant Nutr. Soil Sci. 163, 157–163.

Follett, R.F., 2009. US agriculture's relationship to soil carbon. J. Soil Water Conserv. 64 (6), 159A–165A.

Follett, R., Paul, E.A., Leavitt, S.W., Halvorson, A.D., Lyon, D., Peterson, G.A., 1997. Carbon isotope ratios of Great Plains soils in wheat-fallow systems. Soil Sci. Soc. Am. J. 61, 1068–1077.

Follett, R.F., Shafer, S.R., Jawson, M.D., Franzluebbers, A.J., 2005. Research and implementation needs to mitigate greenhouse gas emissions from agriculture in the USA. Soil Tillage Res. 83, 159–166.

Franzluebbers, A.J., 2005. Soil organic carbon sequestration and agricultural greenhouse gas emissions in the Southeastern USA. Soil Tillage Res. 83, 120–147.

Franzluebbers, A.J., 2004. Tillage and residue management effects on soil organic matter. In: Magdoff, F., Weil, R.R. (Eds.), Soil Organic Matter in Sustainable Agriculture. CRC Press, Boca Raton, FL, pp. 227–268.

Gantzer, C.J., Blake, G.R., 1978. Physical characteristics of Le Sueur clay loam soil following no-till and conventional tillage. Agron. J. 70, 853–857.

Ghidey, F., Alberts, E., 1993. Residue type and placement effects on decomposition: field study at model evaluation. Trans. ASAE 36, 1611–1617.

Gregorich, E.G., Drury, C.F., Baldock, J.A., 2001. Changes in soil carbon under long-term maize in monoculture and legume-based rotation. Can. J. Soil Sci. 81, 21–31.

Gregorich, E.G., Ellert, B.H., Drury, C.F., Liang, B.C., 1996. Fertilization effects on soil organic matter turnover and corn residue placement. Soil Sci. Soc. Am. J 60, 472–476.

Gollany, H.T., Rickman, R.W., Liang, Y., Albrecht, S.L., Machado, S., Kang, S., 2011. Predicting agricultural management influence on long-term soil organic carbon dynamics: Implications for biofuel production. Agron. J. 103, 234–246.

Gollany, H.T., Novak, J.M., Liang, Y., Albrecht, S.L., Rickman, R.W., Follett, R.F., Wilhelm, W.W., Hunt, P.G., 2010. Simulating soil organic carbon dynamics with residue removal using the CQESTR model. Soil Sci. Soc. Am. J. 74, 372–383.

Gollany, H.T., Allmaras, R.R., Copeland, S.M., Albrecht, S.L., Douglas Jr., C.L., 2006. Incorporated source carbon and nitrogen fertilization effects on carbon storage and soluble silica in a Haploxeroll. Soil Sci. 171, 585–597.

Gollany, H.T., Allmaras, R.R., Copeland, S.M., Albrecht, S.L., Douglas Jr., C.L., 2005. Tillage and nitrogen fertilizer influence on carbon and soluble silica relation in a Pacific Northwest Mollisol. Soil Sci. Soc. Am. J. 69, 1102–1109.

Gollany, H.T., Molina, J.A.E., Clapp, C.E., Allmaras, R.R., Layese, M.F., Baker, J.M., Cheng, H.H., 2004. Nitrogen leaching and denitrification in continuous corn as related to residue management. Environ. Manage. vol. 33 (Supplement 1), S289–S298.

Halvorson, A.D., Peterson, G.A., Reule, C.A., 2002a. Tillage system and crop rotation effects on dryland crop yields and soil carbon in the Central Great Plains. Agron. J. 1429–1436.

Halvorson, A.D., Vigil, M.F., Peterson, G.A., Elliot, E.T., 1997. Long-term tillage and crop residue management study at Akron, Colorado. In: Paul, E.A., et al. (Eds.), Soil Organic Matter in Temperate Agroecosystems. Long-term Experiments in North America. CRC Press, Boca Raton, FL, pp. 361–370.

Halvorson, A.D., Wienhold, B.J., Black, A.L., 2002b. Tillage, nitrogen, and cropping system effects on soil carbon sequestration. Soil Sci. Soc. Am. J. 66, 906–912.

Hill, J., Nelson, E., Tilman, D., Polasky, S., Tiffany, D., 2006. Environmental, economic, and energetic costs and benefits of biodiesel and ethanol biofuels. Proc. Natl. Acad. Sci. 103, 11206–11210.

Huggins, D.R., Buyanovsky, G.A., Wagner, G.H., Brown, J.R., Darmody, R.G., Peck, T.R., Lesoing, G.W., Vanotti, M.B., Bundy, L.G., 1998. Soil organic C in the tallgrass prairie-derived region of the Corn Belt: effects of long-term crop management. Soil Tillage Res. 47, 219–234.

Huggins, D.R., Fuchs, D.J., 1997. Long-term N management effects on corn yield and soil C of an Aquic Haplustoll in Minnesota. In: Paul, E.A., et al. (Eds.), Soil Organic Matter in Temperate Agroecosystems: Long-Term Experiments in North America. CRC Press, Boca Raton, FL, pp. 121–128.

Hudson, B.D., 1994. Soil organic matter and available water capacity. J. Soil Water Conserv. 49, 189–194.

Hunt, P.G., Karlen, D.L., Matheny, T.A., Quisenberry, V.L., 1996. Changes in carbon content of a Norfolk loamy sand after 14 years of conservation or conventional tillage. J. Soil Water Conserv. 51, 255–258.

Izaurralde, R.C., Williams, J.R., McGill, W.B., Rosenberg, N.J., Quiroga Jakas, M.C., 2006. Simulation soil C dynamics with EPIC: model description and testing against long-term data. Ecol. Model. 192, 362–384.

Janzen, H.H., Campbell, C.A., Izaurralde, R.C., Ellert, B.H., Juma, N.G., McGill, W.B., Zentner, R.P., 1998. Management effects on soil C storage on the Canadian prairies. Soil Tillage Res. 47, 181–195.

Jenkinson, D.S., 1981. The fate of plant and animal residues in soil. In: Greenland, D.J., Hayes, M.H.B. (Eds.), The Chemistry of Soil Processes. John Wiley and Sons, Chichester, UK, pp. 505–561.

Johnson, J.M.F., Allmaras, R.R., Reicosky, D.C., 2006. Estimating source carbon from crop residues, roots and rhizodeposits using the national grain-yield database. Agron. J. 98, 622–636.

Johnson, J.M.F., Reicosky, D.C., Allmaras, R.R., Sauer, T.J., Venterea, R.T., Dell, C.J., 2005. Greenhouse gas contributions and mitigation potential of agriculture in the central USA. Soil Tillage Res. 83, 73–94.

Karlen, D.L., Berit, W.R., Hunt, P.G., Matheny, T.A., 1989. Soil-test values after eight years of tillage research on a Norfolk loamy sand. Commun. Soil Sci. Plant Anal. 32, 1803–1812.

Karlen, D.L., Hunt, P.G., Matheny, T.A., 1996. Fertilizer ^{15}Nitrogen recovery by corn, wheat, and cotton grown with and without pre-plant tillage on Norfolk loamy sand. Crop Sci. 36, 975–981.

Kim, S., Dale, B., 2005. Life cycle assessment of various cropping systems utilized for producing biofuels: bioethanol and biodiesel. Biomass Bioenergy 29, 426–429.

Ladd, J.N., Amato, M., Zhou, L., Schultz, J.E., 2002. Differential effects of rotation, plant residue and nitrogen fertilizer on microbial biomass and organic matter in an Australian Alfisol. Soil Biol. Biochem. 26, 82–1831.

Lal, R., 2006. Managing soils for feeding a global population of 10 billion. J. Sci. Food Agric. 86, 2273–2284.

Lal, R., Kimble, J.M., Follett, R.F., Cole, C.V., 1998. The Potential of U.S. Cropland to Sequester Carbon and Mitigate the Greenhouse Effect. Sleeping Bear Press, Ann Arbor, USA.

Lal, R., Logan, T.J., Fausney, N.R., 1989. Long-term tillage and wheel traffic effects on a poorly drained Mollic Ochraqualf in Northwest Ohio. 1. Soil physical properties, root distribution and grain yield of corn and soybean. Soil Tillage Res. 14, 341–358.

Lal, R., Mahboubi, A.A., Fausey, N.R., 1994. Long-term tillage and rotation effects on properties of a central Ohio soil. Soil Sci. Soc. Am. J. 58, 517–522.

Larson, W.E., 1979. Crop residue: energy production or erosion control? J. Soil Water Conserv. 34, 74–76.

Lemke, R.L., VandenBygaart, A.J., Campbell, C.A., Lafond, G.P., Grant, B., 2010. Crop residue removal and fertilizer N: effect on soil organic carbon in a long-term crop rotation experiment on a Udic Boroll. Agric. Ecosyst. Environ. 135, 42–51.

Liang, Y., Gollany, H.T., Rickman, R.W., Albrecht, S.L., Follett, R.F., Wilhelm, W.W., Novak, J.M., Douglas Jr., C.L., 2008. CQESTR Simulation of management practice effects on long-term soil organic carbon. Soil Sci. Soc. Am. J. 72, 1486–1492.

Liang, Y., Gollany, H.T., Rickman, R.W., Albrecht, S.L., Follett, R.F., Wilhelm, W.W., Novak, J.M., Douglas Jr., C.L., 2009. Simulating soil organic matter with CQESTR (v. 2.0): model description and validation against long-term experiments across North America. Ecol. Model. 220 (4), 568–581.

Liebig, M.A., Morgan, J.A., Reeder, J.D., Ellert, B.H., Gollany, H.T., Schuman, G.E., 2005. Greenhouse gas contributions and mitigation potential of agricultural practices in northwestern USA and western Canada. Soil Tillage Res. 83, 25–52.

Liebig, M.A., Tanaka, D.L., Wienhold, B.J., 2004. Tillage and cropping effects on soil quality indicators in the northern Great Plains. Soil Tillage Res. 78, 131–141.

Leite, L.F.C., Doraiswamy, P.C., Causarano, H.J., Gollany, H.T., Milak, S., Mendonca, E.S., 2009. Modeling organic carbon dynamics under no-tillage and plowed systems in tropical soils of Brazil using CQESTR. Soil Tillage Res. 102, 118–125.

Lynd, L.R., Cushman, J.H., Nichols, R.J., Wyman, C.F., 1991. Fuel ethanol from cellulosic biomass. Science 231, 1318–1323.

Machado, S., 2011. Soil organic carbon dynamics in the Pendleton long-term experiments: implication for biofuel production in Pacific Northwest. Agron. J. 103, 253–260.

Martens, D.A., Johanson, J.B., Frankenberger Jr., W.T., 1992. Production and persistence of soil enzymes with repeated addition of organic residues. Soil Sci. 153, 53–61.

Mertens, J., Vanderborght, J., Kasteel, R., Pütz, T., Merckx, R., Feyen, J., Smolders, E., 2007. Dissolved organic carbon fluxes under bare soil. J. Environ. Qual. 36, 597–606.

Mikha, M.M., Benjamin, J.G., Vigil, M.F., Nielsen, D.C., 2010. Cropping intensity impacts on soil aggregation and carbon sequestration in the Central Great Plains. Soil Sci. Soc. Am. J. 74, 1712–1719.

Miles, R.J., Brown, J.R., 2011. The Sanborn Field experiment: implications for long-term soil organic carbon levels. Agron. J. 103, 267–278.

Mosier, A.R., Halvorson, A.D., Reule, C.A., Liu, X.J., 2006. Net global warming potential and greenhouse gas intensity in irrigated cropping systems in northeastern Colorado. J. Environ. Qual. 35, 1584–1598.

N'Dayegamiye, A., Angers, D.A., 1990. Effects of long-term cattle manure application on physical and biological properties of Newbois silt loam cropped to corn. Can. J. Soil Sci. 70, 259–262.

Nelson, D.W., Sommers, L.E., et al., 1982. Total carbon, organic carbon, and organic matter. In: Page, A.L. (Ed.), Methods of Soil Analysis. Part 2: Chemical and Microbiological Properties, second ed.). ASA, Madison, WI, pp. 539–579. Agronomy 9.

Novak, J.M., Bauer, P.J., Hunt, P.G., 2007. Carbon dynamics under long-term conservation and disk tillage management in a Norfolk loamy sand. Soil Sci. Soc. Am. J. 71, 453–456.

Olness, A., Archer, D., 2005. Effect of organic carbon on available water in soil. Soil Sci. 170, 90–101.

Parton, W.J., Stewart, J.B.W., Cole, C.V., 1988. Dynamic of C, N, P and S in grassland soils: a model. Biogeochem 5, 109–131.

Parton, W.J., Ojima, D.S., Schimel, D.S., 1996. Models to evaluate soil organic matter storage and dynamics. In: Carter, M.R., Stewart, B.A. (Eds.), Structure and Organic Matter Storage in Agricultural Soils. CRC Press, Lewis Publishers, New York, pp. 421–448.

Paul, E.A., Paustian, K., Elliott, E.T., Cole, C.V., 1997. In: Paul, E.A., et al. (Eds.), Soil Organic Matter in Temperate Agroecosystems: Long-Term Experiments in North America. CRC Press, Boca Raton, FL.

Paul, J.W., Beauchamp, E.G., Zhang, X., 1993. Nitrous and nitric oxide emissions during nitrification and denitrification from manure-amended soil in the laboratory. Can. J. Soil. Sci. 73, 539–553.

Paustian, K., Cole, C.V., Sauerbeck, D., Sampson, N., 1998. CO_2 mitigation by agriculture: an overview. Climate Change 40, 135–162.

Paustian, K., Collins, H.P., Paul, E.A., 1997. Management controls on soil carbon. In: Paul, E.A., et al. (Eds.), Soil Organic Matter in Temperate Agroecosystems: Long-Term Experiments in North America. CRC Press, Boca Raton, FL, pp. 15–49.

Paustian, K., Parton, W.J., Persson, J., 1992. Modeling soil organic matter in organic-amended and nitrogen-fertilized long-term plots. Soil Sci. Soc. Am. J. 56, 476–488.

Peterson, G.A., Halvorson, A.D., Havlin, J.L., Jones, O.R., Lyon, D.J., Tanaka, D.L., 1998. Reduced tillage and increased cropping intensity in the Great Plains conserves soil C. Soil Tillage Res. 47, 207–218.

Puget, P., Angers, D.A., Chenu, C., 1999. Nature of carbohydrates associated with water-stable aggregates of two cultivated soils. Soil Biol. Biochem. 31, 55–63.

Rasmussen, P.E., Parton, W.J., 1994. Long-term effects of residue management in wheat-fallow: I. Inputs, yield, and soil organic matter. Soil Sci. Soc. Am. J. 58, 523–530.

Rasmussen, P.E., Albrecht, S.L., 1998. Crop management effects on organic carbon in semi-arid Pacific Northwest soils. In: Lal, R., et al. (Eds.), Management of Carbon Sequestration in Soil. CRC Press, Boca Raton, FL, pp. 209–219.

Rawls, W.J., Pachepsky, Y.A., Ritchie, J.C., Sobecki, T.M., Bloodworth, H., 2003. Effect of soil organic carbon on soil water retention. Geoderma 116, 61–76.

Reicosky, D.C., Lindstrom, M.J., 1995. Impact of fall tillage on short-term carbon dioxide flux. In: Lal, R., et al. (Eds.), Soils and Global Change. CRC Press, Boca Raton, FL, pp. 177–187.

Rickman, R., Douglas, C., Albrecht, S., Berc, J., 2002. Tillage, crop rotation, and organic amendment effect on changes in soil organic matter. Environ. Pollut. 116, 405–411.

Rickman, R.W., Douglas Jr., C.L., Albrecht, S.L., Bundy, L.G., Berc, J.L., 2001. CQESTR: a model to estimate carbon sequestration in agricultural soils. J. Soil Water Conserv. 56, 237–242.

Snyder, C.S., Bruulsema, T.W., Jensen, T.I., Fixen, P.E., 2009. Review of greenhouse gas emissions from crop production systems and fertilizer management effects. Agric. Ecosyst. Environ. 133, 247–266.

Tester, C.F., 1990. Organic amendment effects on physical and chemical properties of a sandy soil. Soil Sci. Soc. Am. J. 54, 827–831.

Tollner, E.W., Hargrove, W.L., Langdale, G.W., 1984. Influence of conventional and no-till practices on soil physical properties. J. Soil Water Conserv. 39, 73–76.

Vitosh, M.L., Lucas, R.E., Silva, G.H., 1997. Long-term effects of fertilizer and manure on corn yield, soil carbon and other soil chemical properties in Michigan. In: Paul, E.A., et al. (Eds.), Soil Organic Matter in Temperate Agroecosystems: Long-Term Experiments in North America. CRC Press, Boca Raton, FL, pp. 129–139.

Wander, M.M., Bidart, M.G., Aref, S., 1998. Tillage impacts on the depth distribution of total and particulate organic matter in three Illinois soils. Soil Sci. Soc. Am. J. 62, 1704–1711.

Wander, M.M., Traina, S.J., Stinner, B.R., Peters, S.E., 1994. Organic and conventional management effects on biologically active soil organic matter pools. Soil Sci. Soc. Am. J. 58, 1130–1139.

West, T.O., Post, W.M., 2002. Soil organic carbon sequestration rates by tillage and crop rotation: a global data analysis. Soil Sci. Soc. Am. J. 66, 1930–1946.

Wilhelm, W.W., Johnson, J.M.F., Hatfield, J.L., Voorhees, W.B., Linden, D.R., 2004. Crop and soil productivity response to corn residue removal: a literature review. Agronomics J. 96, 1–17.

Wuest, S.B., Caesar-TonThat, T.C., Wright, S.F., Williams, J.D., 2005. Organic matter addition, N, and residue burning effects on infiltration, biological, and physical properties of an intensively tilled silt-loam soil. Soil Tillage Res. 84, 154–167.

CHAPTER 17

Development and Application of the EPIC Model for Carbon Cycle, Greenhouse Gas Mitigation, and Biofuel Studies

R.C. Izaurralde[1], W.B. McGill[2], J.R. Williams[3]
[1]Joint Global Change Research Institute, Pacific Northwest National Laboratory/University of Maryland, College Park, MD
[2]Ecosystem Science and Management, University of Northern British Columbia, Prince George, BC, Canada
[3]Texas Agri-Life Extension Service, Texas A&M University, Temple, TX

CHAPTER OUTLINE

Managing Agricultural Greenhouse Gases. DOI: 10.1016/B978-0-12-386897-8.00017-6

Abbreviations: EPIC, original meaning: Erosion Productivity Impact Calculator; current meaning: Environmental Policy Integrated Climate; DNDC, DeNitrification DeComposition; SOCRATES, Soil Organic Carbon Reserves And Transformations in agro-EcoSystems; EPIC, agricultural policy environmental extender; NPP, net primary productivity; NEE, net ecosystem exchange

INTRODUCTION

Agricultural models for describing crop productivity and biophysical processes started to be developed during the 1970s and 1980s due to the confluence of, at least, three major factors: (1) the existence of a quantitative understanding of the effects of plant growth factors on yield, (2) the need to quantify aspects of productivity and environmental quality, and (3) the increasing availability of main frame and personal computers. The EPIC (Erosion Productivity Impact Calculator) model (Williams et al., 1989) was one such model. The original purpose of the development of EPIC was the need of the U.S. Department of Agriculture to have a tool for evaluating the status of natural resources and, specifically, for quantifying reductions in crop productivity due to soil erosion (Williams et al., 2008b).

The rationale behind the development of the EPIC model was simple: construct a model capable of describing the interactions between soil erosion and crop productivity using readily available inputs and computational resources (Williams et al., 2008b). The original EPIC model included modules for simulating weather, hydrology, erosion–sedimentation, nutrient cycling, plant growth, tillage, soil temperature, management, and economics (Williams et al., 1984). The first major application of EPIC was during the 1985 Resource Conservation Act national appraisal of soil erosion impacts (Putnam et al., 1988). Subsequently, the EPIC model has undergone many enhancements thanks to the adoption and feedback of a large, worldwide, user community (Gassman et al., 2005). Model enhancements have included, for example, submodels to simulate pesticide application and fate, nitrification–volatilization, salinity, CO_2 fertilization effects on crop growth and water use, wind erosion and sediment yield with new methods, and coupled carbon–nitrogen (C:N) cycling (Williams et al., 2008a). All of these capabilities plus others in progress have rendered an EPIC model capable of simulating a full suite of terrestrial ecosystem processes in both managed and non-managed lands for producing results relevant to science and policy applications. In order to reflect these new capabilities, the meaning of the well-known acronym EPIC was recently changed from "Erosion Productivity Impact Calculator" to "Environmental Policy Integrated Climate" (Williams et al., 2008b).

As mentioned above, a recent improvement of the EPIC model entailed the implementation of a coupled C:N submodel to simulate terrestrial carbon (C) dynamics as affected by environmental and management factors (Izaurralde et al., 2006). The objective of this chapter is to describe the latest developments and examples of applications of the EPIC model for C cycle, greenhouse gas mitigation, and biofuel studies.

DESCRIPTION OF THE EPIC MODEL AND ITS CARBON/NITROGEN MODEL

Brief Description of the EPIC Model

In its current version (EPIC version 1102, October 2011), the EPIC model can simulate the growth and development of over 100 plant species including all major crops, grasses, legumes, and some trees. These crops can be grown in monoculture or as intercrops, in complex rotations, and under a wide range of management operations including tillage, irrigation, fertilization, and liming (Williams, 1995). EPIC uses the concept of radiation-use efficiency (Monteith, 1977) by which a fraction of daily photosynthetically active radiation is intercepted by the plant canopy and converted into plant biomass. Daily gains in plant biomass are affected by vapor pressure deficits and atmospheric CO_2 concentration (Stockle et al., 1992a, b). Plant phenology is controlled via heat-unit calculations where each crop/plant species has base and optimal air temperatures for growth. Potential daily gains in biomass are affected by environmental stresses such as water, temperature, nutrients (primarily N and P), and aeration. All stresses are calculated every day during the simulation, but only the value of the most severe stress during each day is used to reduce potential plant growth and crop yield. Stress factors for soil strength, temperature, and aluminum toxicity are also calculated daily and used to adjust potential root growth (Jones et al., 1991). EPIC is driven by daily weather consisting of solar radiation, air temperature, precipitation, wind speed, and relative humidity, which is either predicted from weather parameters or read from input files. Processes are modified by topographical (e.g. slope gradient and length) characteristics and field/watershed dimensions, soil layer properties [e.g. layer depth (layers or horizons), bulk density, C and N contents, pH] and management information (e.g. cropping systems, planting, fertilization, irrigation, harvesting), as inputs to run EPIC (Gassman et al., 2005; Williams, 1995; Zhang et al., 2010).

Early in its development, the EPIC model lacked an explicit representation of the C cycle. Previous versions of EPIC used the PAPRAN model of Seligman and van Keulen (1981) to calculate N transformations and dynamics. In turn, soil organic C was calculated daily for each soil layer as a fixed fraction of the soil organic nitrogen using a C/N ratio of 10. Although simple, the approach produced realistic results of soil C dynamics and also explained lateral fluxes of C through eroded sediments. In a model comparison conducted in Canada, Izaurralde et al. (2001) compared the performance of six models: Century (Parton et al., 1987); DNDC (DeNitrification DeComposition; Li et al., 1994), *ecosys* (Grant, 1996a, b), RothC (Jenkinson and Coleman, 1994), SOCRATES (Soil Organic Carbon Reserves And Transformations in agro-EcoSystems; Grace and Ladd, 1995); and EPIC (Williams, 1995). Models were selected for comparison based on their representation of soil C dynamics, past testing, performance, and accessibility. Experimental data for the comparison were selected from long-term experiments in Alberta and Saskatchewan (Izaurralde et al., 2001). SOCRATES best mimicked long-term C change with Century being a close second. The EPIC model fared well in its ability to predict long-term soil C trends and to link erosion impacts on net C storage. Notwithstanding EPIC's ability to deal with long-term C dynamics, this comparison revealed a clear need to improve the representation of the coupled C:N cycle in EPIC. This is discussed in the following section.

Implementing and Testing of a Coupled Carbon–Nitrogen Model in EPIC

The development of the coupled carbon–nitrogen C:N model in EPIC (Izaurralde et al., 2006) followed a daily timescale implementation of the approach used in the monthly timescale Century model (Parton et al., 1987, 1993, 1994; Vitousek et al., 1994). The modeling approach distributed the C and N between "litter" and "soil organic matter". The C and N in soil organic matter is distributed among three pools or compartments: active (microbial), slow, and passive (Figure 17.1). These pools differ in size and function while their turnover times range from days to hundreds of years. Litter consists of organic materials added to soil as plant residues, roots, and manures, which are partitioned into two litter compartments: metabolic and structural (Figure 17.1) according to their N and lignin contents. Plant lignin concentration is modeled as a sigmoidal function of plant age. The distribution of C and N among these five pools is implemented in every soil layer up to a total of 15, with the exception that the surface litter fraction has a slow compartment in addition to the metabolic and structural pools (Izaurralde et al., 2006).

Actual daily coupled C and N transformations are based on coupled potential transformations adjusted for supply of electron acceptors (O_2, NO_3^-, NO_2^-, and N_2O; inclusion of Fe^{+3}, SO_4^{-2}, and CO_2 is under development). In turn potential coupled transformations are based on daily maximum C transformations adjusted for N supply.

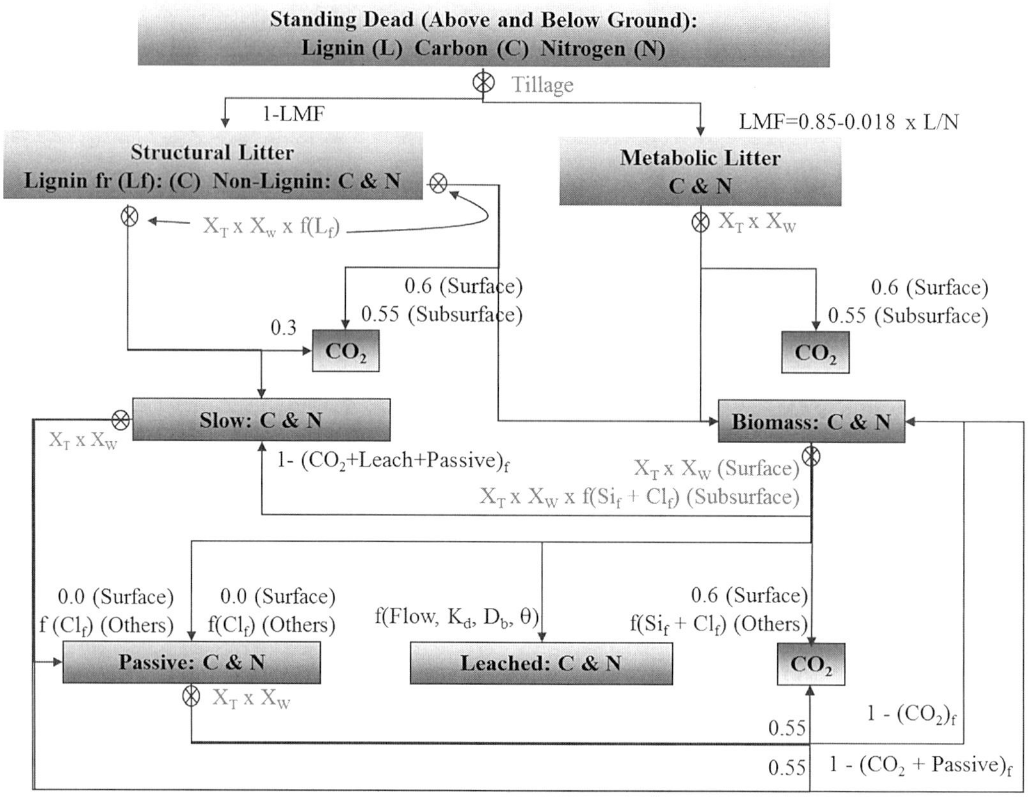

FIGURE 17.1
Diagram of soil C and N pools and flows in EPIC. Numbers indicate the fraction of C that goes to CO_2. Symbols: XW and XT are moisture and temperature controls on soil biological processes; LMF = fraction of litter that is metabolic (kg kg^{-1}); Lf = fraction of structural litter that is lignin (kg kg^{-1}); lower case f = "function of", and subscript f = "fraction"; Si_f = fraction of soil mineral component that is silt; Cl_f = fraction of soil mineral component that is clay; K_d = distribution coefficient of organic compounds between soil solid and liquid phases; DB = soil bulk density; θ = soil volumetric water content. *(Redrawn from Izaurralde et al., 2006.)*

Daily maximum C transformations are calculated as regulated by soil moisture, temperature, tillage, and depth within the profile. These transformations are considered maxima because they reach completion only when sufficient quantities of organic and inorganic N are available to allow coupled C and N transformations. The N supply is calculated using the N/C ratio of each compartment and the maximum flow of C from it combined with the amount of mineral N present in the soil. The demand for N is established by the N/C ratio of the receiving compartment and the maximum flow of C to it. The coupled potential daily flows of C and N are constrained by the total supply of N when N supply is less than combined N demands associated with C-driven flows. Net mineralization of N occurs when N supply exceeds N demand associated with C-driven flows. For more detailed explanations of the method used to calculate coupled potential C and N transformations see Izaurralde et al. (2006). The coupled C and N model in EPIC has been implemented in the watershed version of EPIC, called Agricultural Policy Environmental Extender (APEX) (Williams and Izaurralde, 2006).

At least two factors need to be considered to ensure good model performance for predicting soil organic matter dynamics (Smith et al., 1997). The first is to be able to predict, with an acceptable degree of accuracy, the plant productivity responsible for determining the C inputs that arise from the non-harvested plant components such as crop residues and decaying roots. A second factor is a realistic initialization of the soil C pools. This is a critical factor to consider, especially in short-term simulation experiments. The recommended approach has been to

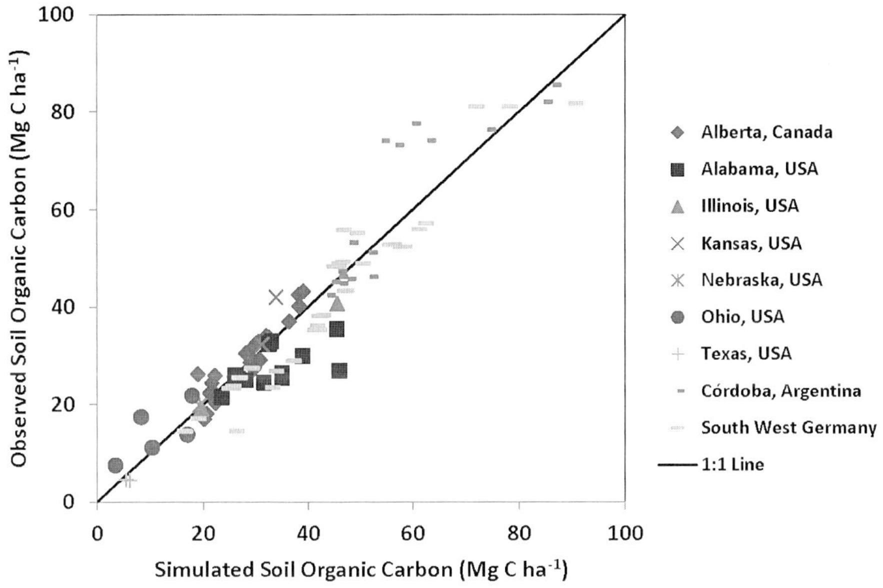

FIGURE 17.2
Comparison between simulated and observed soil organic C for a range treatment × site combinations ranging from 10 to 90 Mg C ha^{-1}. Fitted linear regression: OBS = 0.9SIM + 4.16; $R^2 = 0.907$, $p < 0.01$, n = 109). Please see color plate section at the back of the book.

"spin up" the model for a long-term period that captures the climatic, edaphic, topographic, geomorphic, and management effects on the distribution of the "kinetic" pools of soil organic matter. In EPIC, there are at least three ways to initialize the microbial and passive C pools at the beginning of a simulation: (1) spin up the model using past environmental and management information, (2) let the model estimate an initial fractionation of the soil C pools based on the number of years of cultivation, and (3) input, if known, the observed amounts and depth distribution of C in the two litter and three C pools. The rationale for the second option is that soils experience a decline in the slow C pool when cultivated. The fraction of the passive C pool varies between 40 and 70%. The default value of the active (microbial) C pool is 4%. The initial value of the slow C pool is obtained by difference.

Initial testing of the coupled soil C and N module in EPIC was reported by Izaurralde et al. (2006) using two datasets; one reported soil C accrual in soils under the Conservation Reserve Program while the other described soil C dynamics in two long-term crop rotations at Breton, Alberta. Further testing of the coupled C and N model in EPIC has been conducted so far using datasets from the U.S. (Causarano et al., 2007; Izaurralde et al., 2007; Wang et al., 2005; He et al., 2006), Canada (Izaurralde et al., 2006), Argentina (Apezteguía et al., 2009), and Germany (Gaiser et al., 2008). Figure 17.2 shows an overall comparison between simulated and observed soil organic C for a range of 10–90 Mg C ha^{-1}. A recent simulation study by Farina et al. (2011) found EPIC to be able to reproduce soil organic C within 10% of observed values in two long-term crop rotations [maize (*Zea mays* L.)—durum wheat (*Triticum durum* L.); sunflower (*Helianthus annuus* L.)—durum wheat] under conventional and no tillage.

Denitrification

As modeled, denitrification is driven by C oxidation in roots and microbes to generate electrons. Oxygen is the primary electron acceptor. Upon depletion of O_2 competition among NO_3^-, NO_2^- and N_2O for residual electrons leads to denitrification as follows:

$$5\ CH_2O + 5\ HOH \rightarrow 5\ CO_2 + 20\ H^+ + 20\ e^- \qquad [1]$$

$$4\ NO_3 + 8\ H^+ + 8\ e^- \rightarrow 4\ NO_2^- + 4\ HOH \qquad [2]$$

$$4\ NO_2 + 8\ H^+ + 8\ e^- + 4\ H^+ \rightarrow 2\ N_2O + 6\ HOH \qquad [3]$$

$$2\,N_2O + 4\,H^+ + 4\,e^- \rightarrow 2\,N_2 + 2\,HOH \quad [4]$$

$$\text{Overall}: \; 5\,CH_2O + 4\,NO_3 + 4\,H^+ \rightarrow 5\,CO_2 + 2\,N_2 + 7\,HOH + \text{energy} \quad [5]$$

The potential supply of electrons is calculated daily and based on coupled potential daily flows of C and N. Fate of the potential supply of electrons is calculated hourly. Electrons are first passed to O_2, based on O_2 concentration at the surface of both plant roots and soil microorganisms to form CO_2. Michaelis-Menten uptake kinetics is used to evaluate electron transfer to O_2.

If the potential supply of electrons exceeds those accepted by O_2, and if oxides of N (NO_3^-, NO_2^- or N_2O) are present, then denitrification occurs. Uptake of electrons by organisms reducing oxides of N is estimated via Michaelis-Menten kinetics (Grant and Pattey, 1999). Complete reduction of one mole of NO_3^- consumes five moles of electrons, compared to three moles of electrons per mole of NO_2^- and one mole of electrons per mole of N_2O. Accordingly, the concentrations of each of the oxides of N are weighted to account for the variation in numbers of moles of electrons that each species accepts. Competition has been reported for electron donors among nitrate reducers, ferric iron reducers, sulfate reducers, and methanogens (Achtnich et al., 1995), and between nitrate and nitrite reduction in denitrification (Almeida et al., 1995). Consequently, the Michaelis-Menten expression contains terms for competitive inhibition such that NO_3^- inhibits reduction of NO_2^-, and both inhibit reduction of N_2O.

Denitrification Feedback to Decomposition

If the potential supply of electrons is matched by the total accepted by O_2 plus oxides of N then decomposition equals potential decomposition. In this scenario variables are updated and calculations start again for the next day. If, however, potential supply of electrons exceeds those accepted by O_2 plus oxides of N, then decomposition is reduced sufficiently such that actual decomposition produces an electron supply equal to the total of electrons that can be accepted by O_2 plus oxides of N.

Gas and Solute Movement

Movement of solutes (NO_3^-, NO_2^-) and dissolved gases (O_2, CO_2, and N_2O) is the product of their concentration in soil water and the volume of water moving vertically and laterally. Water flows from a soil layer when soil water content exceeds field capacity. Water drains from the layer daily as a function of layer storage and saturated conductivity until the storage returns to field capacity (Williams et al., 2008a, 25).

Movement of O_2, CO_2, and N_2O also occurs in the gas phase in EPIC. Within each day, each gas is transferred on an hourly basis within the soil profile, between the profile and the atmosphere above or between the profile and the material below. The profile is divided into computational layers such that soil layer thicknesses do not exceed 10 cm. Properties of the soil profile layers are interpolated among the computational layers. Redistribution within the profile is calculated using the Crank-Nicholson procedure (Press et al., 1992, 840), although the implicit and explicit procedures are also available in EPIC. Gas diffusion in soil is modified from that in air to account for tortuosity based on total porosity and proportion of pore space filled with water according to Millington and Quirk (1961). To avoid the layer below the soil profile becoming an infinite sink or source due to arbitrary values for the lower boundary layer, the gas lower boundary layer is set to be impermeable. Each gas is redistributed hourly between the gas and liquid phases using Henry's law. Flux of each gas into or out of the profile is calculated from the surface boundary term of the diffusion equation at each diffusion time step and is accumulated into daily fluxes.

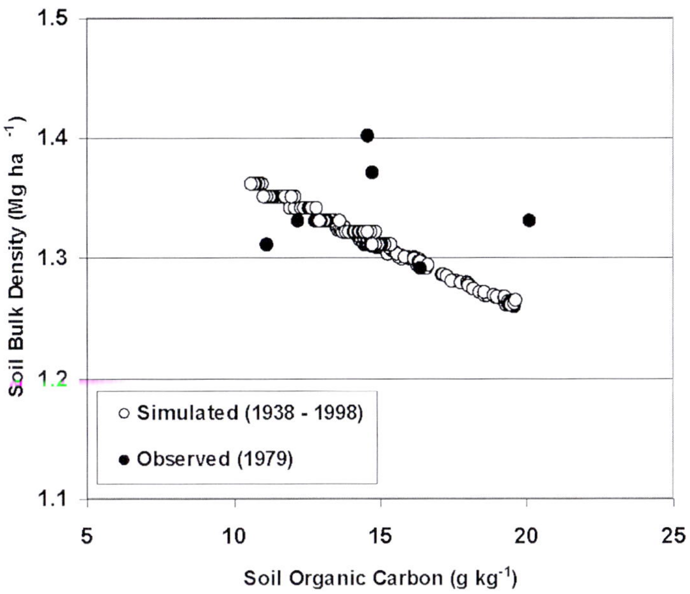

FIGURE 17.3
Simulated and observed relationship between soil organic C concentration and soil bulk density. *(Adapted from Izaurralde et al., 2006.)* Please see color plate section at the back of the book.

Soil Bulk Density and Soil Organic Matter

In EPIC, changes in soil organic matter excerpt a feedback on soil properties. The strongest feedback is given by the inverse relationship between soil organic matter content and soil bulk density. The equation proposed by Adams (1973) is used in EPIC to estimate, annually, the effect of soil organic matter on soil bulk density (Izaurralde et al., 2006). Figure 17.3 shows a general comparison between simulated and observed soil bulk density as affected by soil organic matter concentration. Provided there is no erosion, the EPIC model also recalculates, annually, the thickness of the soil layer required to accommodate the soil mass in each layer (Apezteguía et al., 2009; He et al., 2006). When soil organic matter increases, soil bulk density decreases, and this decrease may bring about a suite of changes in biophysical properties such as soil porosity, infiltration, aeration, greenhouse gas fluxes, etc. There has been limited testing of these attributes of EPIC; thus, there is opportunity to examine these assumptions in controlled experiments.

Ecosystem Carbon Balance and Fluxes

Soil C changes are simulated as a function of climatic conditions, soil properties, and management practices. Gains in C occur via litter inputs and organic amendments. Losses of C occur via heterotrophic respiration (CO_2), water erosion (particulate and soluble C), wind erosion (particulate C), and leaching (soluble C). Leaching of soluble C is modified by sorption, as soluble material moves down through subsurface layers.

Originally, the EPIC model tracked C in vegetation in an implicit manner, i.e. assuming a value of 42% C in plant material. With the incorporation of the coupled soil C and N model into EPIC, there was the need to address ecosystem C more explicitly in order to describe C cycling at an ecosystem scale (Figure 17.4):

$$\partial \mathrm{Soil}C/\partial t + \partial \mathrm{Veg}C/\partial t = C_P - C_{Ra} - C_{Rh} + C_{\mathrm{Added}} - C_{\mathrm{Subtracted}} \quad [6]$$

where $\partial \mathrm{Soil}C/\partial t$ is mass change of soil C per unit area (to a given depth) per unit time, $\partial \mathrm{Veg}C/\partial t$ is mass change of vegetation C per unit time, C_P is C captured via photosynthesis, C_{Ra} is C respired via autotrophic respiration, C_{Rh} is C respired via heterotrophic respiration, C_{Added} is

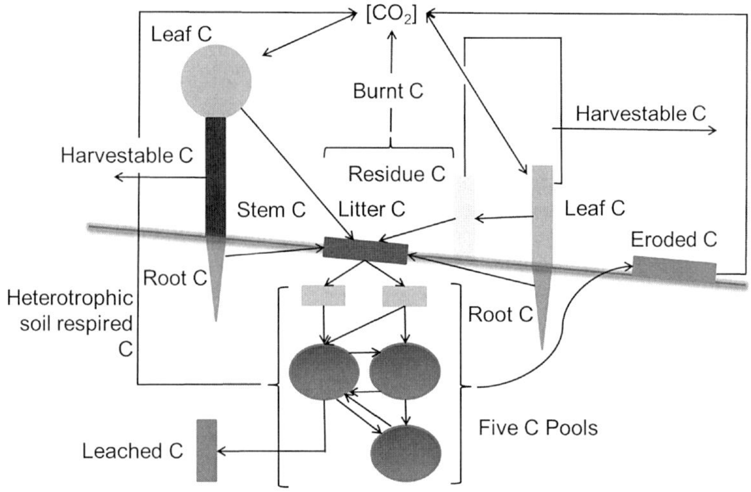

FIGURE 17.4
Diagram representing the ecosystem carbon balance in EPIC.

C added via manure, sedimentation, or a given technology (e.g. biochar), and $C_{Subtracted}$ is C subtracted (removed) via harvested products, erosion, leaching, or vegetation burning. Left-hand equation terms represent changes in ecosystem C stocks while right-hand equation terms represent C fluxes. Since EPIC does not deal explicitly with C_P and C_{Ra}, these terms in Eq. [6] are replaced by the computed value of Net Primary Productivity or NPP_C (i.e. the difference between photosynthesis and autotrophic respiration, in C units). The difference between C_{Rh} and NPP_C yields the vertical exchange of C between the land and the atmosphere, known as Net Ecosystem Exchange, or NEE_C:

$$NEE_C = C_{Rh} - NPP_C \quad [7]$$

In EPIC, calculations of NEE_C are reported as in Eq. [7] but the calculation of the vertical exchange of C between the land and the atmosphere can be refined by including the influence of C added or subtracted via lateral flows such as manure or erosion (Eq. [7]), which are terms explicitly computed by the model.

Carbon Dioxide Emissions from Fossil Fuels

A book-keeping algorithm is included in EPIC to calculate CO_2-C emissions associated with tillage, planting, fertilization, pest control, and other field operations (Zhang et al., 2010). The C-emission factors include emissions associated with the production, distribution, and use of materials, both man-made (e.g. pesticides and fertilizers) and natural (e.g. seeds and water for irrigation) based on summaries and tabulations documented by West and Marland (2002) and Lal (2004).

EXAMPLES OF APPLICATIONS

Soil Carbon Sequestration Studies

Many cultivated soils worldwide have experienced considerable losses of organic C as CO_2 during oxidation of organic matter and sediment transport during erosion. Global losses of 55 Pg C from cultivation of mineral soils have been estimated to have occurred since ~1800, of which approximately two-thirds have been estimated to be recoverable (i.e. recaptured from the atmosphere) during 50–100 years by implementing improved agronomic practices (e.g. nutrient management, crop rotations, and advanced tillage practices) (Cole et al., 1997). Well-tested simulation models containing modules of soil organic matter dynamics are called to play a fundamental role in soil C sequestration monitoring schemes (Brown et al., 2010) (Figure 17.5).

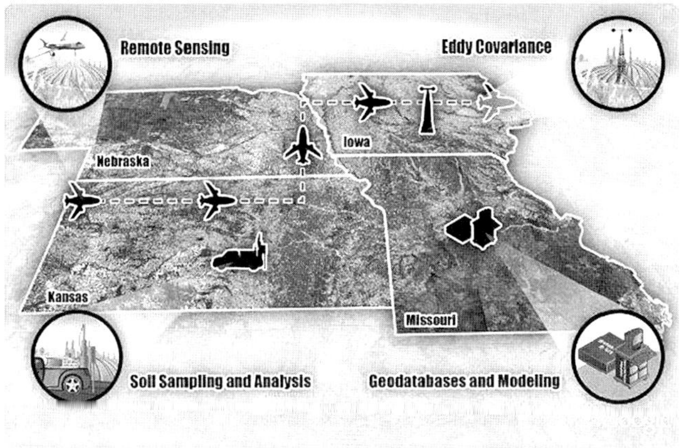

FIGURE 17.5
A vision for employing high-resolution biophysical modeling with other techniques to monitor soil carbon over large areas. *(Adapted from Brown et al., 2010.)*

The EPIC modules of soil organic matter dynamics have been tested by an increasingly larger user community. Three examples are given here. The first is from Gaiser et al. (2010) who tested EPIC for central European conditions using data from 13 representative sites in the German State Baden-Württemberg. Model calibration steps included adjustments of N depositions, N_2-fixation by bacteria during fallow, nutrient content of organic fertilizers, mixing efficiency of tillage implements, and four crop parameters. Predicted average soil C sequestration rates of 0.341 Mg C ha^{-1} yr^{-1} under reduced tillage compared well with observations (0.343 Mg C ha^{-1} yr^{-1}). In the second, Apezteguía et al. (2009) tested EPIC in its ability to capture the effects of land-use change on soil organic C content. Data from a 40-year chronosequence describing the conversion of a temperate shrubland forest to cultivated land in semiarid Córdoba, Argentina, were used to drive the EPIC model. In the chronosequence, a soil organic C loss of 38.4 Mg C ha^{-1} to a depth of 40 cm was estimated after 40 years of cultivation while the 44.1 C Mg ha^{-1} loss simulated by EPIC was 15% greater than observed. The model also captured the effects of three levels of tillage intensity (moldboard plow, chisel, no tillage) on soil organic C content. In the last example, Causarano et al. (2007) used an automated parameter optimization procedure to calibrate the EPIC model for a site-specific experiment in the Coastal Plain of central Alabama. After selecting optimal plant and soil C parameters, EPIC was used to simulate maize and cotton (*Gossypium hirsutum* L.) yields and soil organic C dynamics on different soil landscape positions (summit, sideslope, and drainageway) during the first 5 years of conservation tillage adoption. Overall, EPIC captured 88% of the measured yield variation performing well on drainageway positions and poorly on sideslope positions. Across landscape positions, EPIC explained ~40% of the total variation in soil organic C content in the top 20 cm showing the lowest errors on the sideslope and summit positions.

Applications of the EPIC model in soil C sequestration studies at regional scales are emerging. Doraiswamy et al. (2007) applied the EPIC model to assess the potential for C sequestration in degraded soils of sub-Saharan Africa. High and low resolution imagery was used to classify land use on a 64 km^2 area near Omarobougou, Mali. Small fields and low-level tillage technology predominate in the area. The area also is characterized by rainfall variability, low crop yields, and degradation problems. Modeled results suggested that application of ridge tillage could help control erosion and lead to soil C sequestration due largely to increased water infiltration. The combination of modeling results and land-use classification led to the conclusion that it would be possible to sequester ~54 kg C ha^{-1} yr^{-1} with large-scale adoption of ridge tillage, improved nutrient-use efficiency, and residue retention.

Another example of a large-scale application of EPIC in soil C sequestration studies is from Causarano et al. (2008). These authors used the EPIC model to simulate the impacts of crop and soil practices on the soil organic C content of Iowa croplands. Using publicly available climate, soil, crop management, fertilizer rates, and tillage data, EPIC simulations were able to simulate state crop-yield averages during 1970–2005 ($R^2 = 0.87$). Moreover, simulated soil organic C explained 75% of the variation in measured soil organic C. Assuming the persistence of current trends in conservation tillage adoption, the model predicted total soil C stock of 506 Tg C by 2019 in the top 20 cm, representing an increase of 28 Tg compared to 1980. However, simulation results also suggested the need to investigate further the dynamics of soil organic C in subsurface horizons.

There is also the need to improve our understanding of climate change impacts on future soil organic C stocks. Thomson et al. (2006) employed EPIC to simulate the impacts of climate change on agricultural productivity and soil C sequestration in the Huang-Hai Plain in northeastern P.R. China. The simulation study examined the extent to which increased cropping intensity and reduced tillage could affect soil C sequestration under two climate change scenarios (A2 and B2) from the HadCM3 global climate model at two periods in the 21st century (2015–2045 and 2070–2099). Temperature increases larger than 5°C and increases in annual precipitation of up to 300 mm are projected for this region by the end of the century. Simulated soil organic C levels under no-till management were higher than under conventionally tilled management. Simulated soil C sequestration rates for continuous wheat systems averaged 0.4 Mg C ha^{-1} yr^{-1} under no-till during the first future period and 0.2 Mg C ha^{-1} yr^{-1} during the second. Under wheat–corn cropping, no-till increased soil organic C at rates of 0.8 and 0.4 Mg C ha^{-1} yr^{-1}. Based on the modeling results, Thomson et al. (2006) suggested that a shift from conventional to no tillage in continuous wheat practiced over 16 million hectares of the Huang-Hai Plain could offset 240 Tg C during the first future period and 180 Tg C during the second. Corresponding C offsets for wheat–corn cropping systems were estimated at 675 and 495 Tg C.

Net Ecosystem Carbon Balance Studies

Carbon flows in intensively managed ecosystems are substantially different from less managed or unmanaged ecosystems. In croplands and forest plantations, a good fraction of the net C fixed by plants (i.e. economic yield) is exported from the place of origin and emitted or stored in products elsewhere. In pasturelands or grasslands, there is more recycling of fixed C, but a portion of the C is also removed by livestock as a component of meat products. Carbon can also be added to agricultural systems via manure or biochar additions, which sometimes may represent a net addition (Izaurralde et al., in review). Construction of net ecosystem C balances either through spreadsheet calculations or simulations is a very useful exercise to understand the magnitude and uncertainty associated with the various C-flux components (see Eq. [6]). Izaurralde et al. (2007) applied EPIC to evaluate the role of erosion–deposition processes on the net ecosystem C balance (NECB) and soil C changes of three small watersheds (~1 ha) located at the USDA North Appalachian Experimental Watershed near Coshocton, OH. Thirty-six- to 49-year long simulations of ecosystem processes including plant growth, runoff, and water erosion were used to quantify sediment C yields. Interpretation of simulated results suggested eroded C re-deposited inside (30–212 kg C ha^{-1} yr^{-1}) or outside (73–179 kg C ha^{-1} yr^{-1}) watershed boundaries was similar in magnitude to a simulated soil C sequestration rate of 225 kg C ha^{-1} yr^{-1} and to literature values. Analyses of simulated NECB suggested that a watershed under a plow till system (W128) was a source of C to the atmosphere while two other watersheds under no-till (W118 and W188) behaved as C sinks for atmospheric CO_2 (Table 17.1). These results demonstrated a clear need for accounting for the proportion of eroded soil C that could be transported outside watershed boundaries and the proportion that is lost to the atmosphere as CO_2.

TABLE 17.1 Simulated Carbon Balance of Three Watersheds as Affected by Topography, Soil Properties, and Long-Term Management at the USDA North Appalachian Experiment Station Near Coshocton, OH. Modified from Izaurralde et al. (2007)

	Watershed number[§] and length of simulation		
	No. 118—49 years	*No. 128—36 years*	*No. 188—36 years*
		kg C ha^{-1} yr^{-1}	
Initial C in soil profile	71.7	94.7	94.7
Added C	290.6	211.5	212.5
Plant C	279.4	209.1	211.2
Manure C	11.2	2.5	1.2
Subtracted C	270.5	220.7	198.3
Respired C	265.5	215.3	190.8
Sediment C in runoff	1.9	3.0	4.7
Soluble C in runoff	1.7	1.7	1.7
Leached C	1.3	0.8	1.1
Final C in soil profile	91.8	85.3	108.8

[§]*Watershed 118: maize–soybean, no-till; watershed 128: continuous corn, conventional till; and watershed 188: continuous corn, no till.*

Biofuel Studies

Model preparation for simulation of biomass production and environmental outcomes represents the latest development in the EPIC model (Zhang et al., 2010). Simulation of crop yields of annual crops and cropping systems has been at the core of EPIC's capability. Simulation of biofuel crops grown for cellulosic ethanol production has added new challenges in model development, testing, and application during recent years. Three examples are given here. The first is the fact that EPIC utilizes a single algorithm to simulate growth and development of many plant species; thus, the effort so far has concentrated on the parameterization and testing of known and potential biofuel species [e.g. Miscanthus (*Miscanthus x giganteus*)]. In particular, the correct and dynamic partitioning of plant biomass above and below ground has represented a challenge for the modeling community as alerted by Zhang et al. (2011). The second relates to the transition between live and dead vegetation as the main interest (i.e. the economic yield) resides maximizing cellulose yield while minimizing water and nutrient contents. The third example is computational; i.e. how to make efficient computations of the EPIC model across many types of soils, especially marginal soils. In the case of growing perennial crops for cellulosic ethanol production, the objective is to locate soils of suitable production capacity that do not compete with land used for food and feed production (Zhang et al., 2010).

Upcoming Improvements

BIOCHAR

Biochar is a product of low temperature heating of plant residues or other lignocellulosic organic material in the presence of low O_2 concentration. Under such conditions solid residue (biochar), liquids, and gases are produced. The liquids and gases can be used for fuels or industrial feedstock (Baoteng et al., 2010). Biochar has a high C content (up to 80% depending on pyrolysis conditions; Antal and Grønli, 2003), which appears to be relatively recalcitrant (Kimetu and Lehmann, 2010); it normally has a high pH, low bulk density, and may be concentrated in nutrients from the original plant residues (Laird et al., 2010). Biochar is potentially a soil amendment and may be used both to store C due in part to recalcitrant C, and to improve soil aeration, water-holding, and nutrient-retention properties (Joseph et al., 2010). Many soils treated with biochar support enhanced plant growth due to improved physical and chemical properties (Lehmann, 2007).

As part of a program to develop and simulate climate change adaptation options for agriculture in the central U.S., a module is being added to EPIC to accommodate addition of biochar to soils. Specifically, the module will follow the change in pH, soil C and N, soil bulk density, and ecosystem C balance following addition and subsequent transformations of biochar in soils in EPIC (Lychuk, Izaurralde, and Hill, personal communication). The module will be tested to simulate biochar impacts on Oxisol soil parameters and maize yield, compared to a 4-year study reported by Major et al. (2010). Upon successful implementation, simulations of $\geq$20 years will be possible, thereby allowing long-term evaluation of biochar additions on crops, soil properties, and ecosystem C balances. Unique to this approach is the progression in bulk density, CEC, and pH over time following biochar addition.

SUBSOIL CARBON SORPTION

Access to C substrates drives most heterotrophic biological processes and associated emission of trace gases in EPIC. Sorption of substrates regulates their supply to microbes and the rate of leaching within and through soil profiles. Transformation products of biomass C in subsurface layers are the only entities subjected to leaching in EPIC (Izaurralde et al., 2006). Leaching equations currently implemented in EPIC use a linear partition coefficient and soil water content to calculate movement of organic materials from surface litter to subsurface layers as modified by sorption. The subsurface Allocation of Biomass Leached (ABL; kg ha^{-1}) follows the treatment of Williams (1995) in EPIC:

$$ABL = BMC \times (1.0 - \exp(-QV/(0.01 \times ST + 0.1 \times KdBM \times BD)) \qquad [8]$$

where ABL is the amount of C leached (kg ha^{-1}), BMC is the weight of C in soil microbial biomass and associated products (kg ha^{-1}), QV is the vertical flow (mm), ST is the volume of water stored in the soil layer (mm), $KdBM$ is the linear adsorption coefficient for biomass (m^3 Mg^{-1}), and BD is the soil layer bulk density (Mg m^{-3}) (Williams et al., 2008a, 191).

Recently, sorption data were developed at the Oak Ridge National Laboratory using 250 samples from B and C horizons from the Ultisol, Mollisol, and Alfisol soil orders in the U.S. (Mayes et al., 2010). The parameters for the Langmuir isotherm:

$$RE = \left[k \times Q_m \times X_f/(1 + k \times X_f)\right] - b \qquad [9]$$

where RE is amount of organic C sorbed or desorbed (mg kg^{-1}), X_f is the final equilibrium concentration of dissolved organic C (mg L^{-1}), k is binding affinity, Q_m is maximum adsorption capacity (mg kg^{-1}), and b is the desorption potential (mg kg^{-1}). These parameters were related to soil pH, clay content, Fe, and C for each soil order. Based on these data, EPIC is being updated to treat subsoil sorption, and hence leaching, of organics on a more soil-specific basis (LeDuc and Izaurralde; personal communication).

NITRIFICATION

Nitrification ($NH_4^+ \rightarrow NO_2^- \rightarrow NO_3^-$) in EPIC is based on the first-order kinetic rate equation of Reddy et al. (1979), with modifications for soil temperature, moisture, and pH (Williams et al., 2008a, 50). Volatilization of NH_3 is calculated concurrently with nitrification, but N_2O production is not associated with it. A revised treatment of nitrification is being developed. It includes the influence of O_2 availability on nitrification and associated production of N_2O. The approach is based on the hypothesis that under insufficient O_2 supply ammonium-oxidizing bacteria use NO_2^- as an electron acceptor thereby producing N_2O (Schwab, Izaurralde, McGill, Williams, Schmid; personal communication).

METHANE

Methane production is currently not modeled in EPIC. As with nitrification, revisions are being implemented to allow EPIC to deal with this important greenhouse gas to complement the treatment of denitrification. As with N_2O, CH_4 may either escape to the atmosphere or be further metabolized within the soil.

Mechanisms of Methanogenesis

Two mechanisms exist for producing CH_4: acetogenic methanogenesis (AM; Eq. [11]) and hydrogenic methanogenesis (HM; Eq. [12]). Acetogenic methanogenesis is a disproportionation reaction in which one C atom in acetate gains an e^- to yield CH_4 and one C atom loses an e^- to yield CO_2 resulting in equimolar amounts of CH_4 and CO_2 with no net release or consumption of electrons. Hydrogenic methanogenesis entails reduction of CO_2 as terminal electron acceptors (TEA) yielding CH_4 ($8\ e^-$/mol CH_4 formed) in the absence of O_2 or mineral TEAs.

Basic Relationships when Mineral Electron Acceptors have been Depleted

$$\text{Fermentation: } 4\ CH_2O + 2\ HOH \rightarrow CH_3COOH + 8\ H^+ + 8\ e^- + 2CO_2 \quad [10]$$

$$\text{Acetogenic methanogenesis: } CH_3COOH \rightarrow CH_4 + CO_2 \quad [11]$$

$$\text{Hydrogenic methanogenesis: } 8\ H^+ + 8\ e^- + CO_2 \rightarrow CH_4 + 2\ HOH \quad [12]$$

$$\text{Summary: } CH_2O \rightarrow 2\ CH_4 + 2\ CO_2 \quad [13]$$

In the overall summary (Eq. [13]) eight moles of e^- are involved for two moles of CO_2 and two moles of CH_4 combined (4 mol e^-/mol CO_2).

Methanogenesis in EPIC (under Development)

Revisions under way in EPIC calculate mol e^- available from potential CO_2 respiration at four mol e^-/mol CO_2 potentially formed. Electrons are accepted by TEAs according to relative free energy changes associated with their reductions. Consequently, NO_3^-, Fe (and Mn) and $SO_4^=$ compete with CO_2 for electrons in the absence of O_2 (Achtnick et al., 1995; Yao and Conrad, 2000). Eventually, all O_2 and mineral TEAs may be exhausted if biological activity is high and O_2 diffusion is restricted (e.g. high HOH content). In such instances, fermentation (Eq. [10]) and methanogenesis (Eqs [11] and [12]) take over. In EPIC, methanogenesis is represented as the least favorable destination for e^- and becomes important only after O_2 and mineral TEAs are exhausted. The total e^- remaining for methanogenesis is electrons available for methanogenesis, or ESM (mol e^- m^{-2}).

The treatment of methanogenesis being developed for EPIC is based on the connection of e^- generation to potential CO_2 respiration. Electrons available for methanogenesis arose from potential CO_2 respiration and are reassigned to CO_2 during methanogenesis to produce CH_4 in HM (Eq. [12]).

During HM in EPIC flow of e^- is to CO_2, thereby converting some CO_2 to CH_4. The eight mol e^- produced in association with two mol CO_2 (Eq. [10]) are consumed in reduction of one mol CO_2 to CH_4 during HM (Eq. [12]). The C balance reduces CO_2 emitted by one mole and augments CH_4 by one mole for each eight e^- used. Alternatively, during AM, one mole each of CH_4 and CO_2 is produced from disproportionation of acetate (Eq. [11]) as part of the suite of reactions leading to HM, which in turn is driven by total e^- supply. The C balance is adjusted to include the additional loss of CH_4 that is not re-oxidized before leaving the soil.

SUMMARY

This chapter has attempted to provide a comprehensive—albeit incomplete—review of the EPIC model in relation to C cycle, greenhouse gas mitigation, and biofuel applications. From its original capabilities and purpose (i.e. quantify the impacts of erosion on soil productivity), the EPIC model has evolved into a comprehensive terrestrial ecosystem model for simulating with more or less process-level detail many ecosystem processes such as weather, hydrology, plant growth and development, C dynamics (including erosion), nutrient cycling, greenhouse gas emissions, and the most complete set of manipulations that can be implemented on a parcel of land (e.g. tillage, harvest, fertilization, irrigation, drainage, liming, burning, pesticide application).

This chapter has also provided details and examples of the latest efforts in model development such as the coupled C:N model, a microbial denitrification model with feedback to the C decomposition model, updates on calculation of ecosystem C balances, and CO_2 emissions from fossil fuels. The chapter has included examples of applications of the EPIC model in soil C sequestration, net ecosystem C balance, and biofuel studies.

Finally, the chapter has provided the reader with an update on upcoming improvements in EPIC such as the addition of modules for simulating biochar amendments, sorption of soluble C in subsoil horizons, nitrification including the release of N_2O, and the formation and consumption of CH_4 in soils. Completion of these model development activities will render an EPIC model with a thorough representation of biogeochemical processes and the capability of simulating the dynamic feedback of soils to climate and management in terms not only of transient processes (e.g. soil water content, heterotrophic respiration, N_2O emissions) but also of fundamental soil properties (e.g. soil depth, soil organic matter, soil bulk density, water limits).

Acknowledgments

This work was funded in part by the DOE Carbon Sequestration in Terrestrial Ecosystems, CSiTe (DOE BER Office of Science DE-AC05-76RL01830), the DOE Great Lakes Bioenergy Research Center (DOE BER Office of Science DE-FC02-07ER64494), and the DOE OBP Office of Energy Efficiency and Renewable Energy (DE-AC05-76RL01830).

References

Achtnich, C., Bak, F., Conrad, R., 1995. Competition for electron donors among nitrate reducers, ferric iron reducers, sulfate reducers, and methanogens in anoxic paddy soil. Biol. Fert. Soils 19, 65–72.

Adams, W.A., 1973. The effect of organic matter on the bulk and true densities of some uncultivated podzolic soils. J. Soil Sci. 24, 10–17.

Almeida, J.S., Reis, M.A.M., Carrondo, M.J.T., 1995. Competition between nitrate and nitrite reduction in denitrification by Pseudomonas fluorescens. Biotech. Bioengineer. 46, 476–484.

Antal, M.J., Grønli, M., 2003. The art, science, and technology of charcoal production. Industrial Eng. Chem. Res. 42, 1619–1640.

Apezteguía, H.P., Izaurralde, R.C., Sereno, R., 2009. Simulation study of soil organic matter dynamics as affected by land use and agricultural practices in semiarid Córdoba, Argentina. Soil Tillage Res. 102, 101–108.

Boateng, A.A., Mullen, C.A., Goldberg, N.M., Hicks, K.B., Devine, T.E., Lima, I.M., McMurtrey, J.E., 2010. Sustainable production of bioenergy and biochar from the straw of high-biomass soybean lines via fast pyrolysis. Environ. Progress Sustainable Energy 29, 175–183. doi: 10.1002/ep.10446.

Brown, D.J., Hunt Jr., E.R., Izaurralde, R.C., Paustian, K.H., Rice, C.W., Schumaker, B.L., West, T.O., 2010. Soil organic carbon change monitored over large areas. Eos 91, 441–442.

Causarano, H.J., Shaw, J.N., Franzluebbers, A.J., Reeves, D.W., Raper, R.L., Balkcom, K.S., Norfleet, M.L., Izaurralde, R.C., 2007. Simulating field-scale soil organic carbon dynamics using EPIC. Soil Sci. Soc. Am. J. 71, 1174–1185.

Causarano, H.J., Doraiswamy, P.C., McCarty, G.W., Hatfield, J.L., Milak, S., Stern, A.J., 2008. EPIC modeling of soil organic carbon sequestration in croplands of Iowa. J. Environ. Qual. 37, 1345–1353.

Cole, C.V., Duxbury, J., Freney, J., Heinemeyer, O., Minami, K., Mosier, A., Paustian, K., Rosenberg, N., Sampson, N., Sauerbeck, D., Zhao, Q., 1997. Global estimates of potential mitigation of greenhouse gas emissions by agriculture. Nutrient Cycling Agroecosys 49, 221–228.

Doraiswamy, P.C., McCarty, G.W., Hunt Jr., E.R., Yost, R.S., Doumbia, M., Franzluebbers, A.J., 2007. Modeling soil carbon sequestration in agricultural lands of Mali. Agric. Syst. 94, 63–74.

Farina, R., Seddaiu, G., Orsini, R., Steglich, E., Roggero, P.P., Francaviglia, R., 2011. Soil carbon dynamics and crop productivity as influenced by climate change in a rainfed cereal system under contrasting tillage using EPIC. Soil Tillage Res. 112, 36–46.

Gaiser, T., Stahr, K., Billen, N., Mohammad, M.A.-R., 2008. Modeling carbon sequestration under zero tillage at the regional scale. I. The effect of soil erosion. Ecol. Model 218, 110–120. doi: /10.1016/j.ecolmodel.2008.06.025.

Gassman, P.W., Williams, J.R., Benson, V.W., Izaurralde, R.C., Hauck, L.M., Jones, C.A., Atwood, J.D., Kiniry, J.R., Flowers, J.D., 2005. Historical development and applications of the EPIC and APEX models. CARD Working Paper 05-WP. Center for Agricultural and Rural Development. Iowa State University, Ames, Iowa, USA. Available at http://www.card.iastate.edu/publications/synopsis.aspx?id=763 (verified November 20, 2011).

Grace, P.R., Ladd, J.N., 1995. SOCRATES v2.00 User Manual. Cooperative Research Centre for Soil and Land Management, PMB 2 Glen Osmond 5064, South Australia.

Grant, R.F., 1996a. Ecosys. In: Global Change and Terrestrial Ecosystems Focus 3 Wheat Network. Model and Experimental Meta. Data, 2nd Edition. GCTE Focus 3 Office, NERC Centre for Ecology and Hydrology, Wallingford, Oxon, U.K., pp. 65–74.

Grant, R.F., 1996b. Ecosys. In: Global Change and Terrestrial Ecosystems Task 3.3.1 Soil Organic Matter Network (SOMNET): 1996 Model and Experimental Metadata. GCTE Focus 3 Office, NERC Centre for Ecology and Hydrology, Wallingford, Oxon, U.K., pp. 19–24.

Grant, R.F., Pattey, E., 1999c. Mathematical modeling of nitrous oxide emissions from an agricultural field during spring thaw. Global Biogeochem. Cycles 13, 679–694.

He, X., Izaurralde, R.C., Vanotti, M.B., Williams, J.R., Thomson, A.M., 2006. Simulating long-term crop productivity and soil organic carbon dynamics with the EPIC model using data from Arlington, WI. J. Environ. Qual. 35, 1608–1619.

Izaurralde, R.C., Williams, J.R., McGill, W.B., Rosenberg, N.J., Quiroga Jakas, M.C., 2006. Simulating soil C dynamics with EPIC: model description and testing against long-term data. Ecol. Modelling 192, 362–384.

Izaurralde, R.C., Williams, J.R., Post, W.M., Thomson, A.M., McGill, W.B., Owens, L.B., Lal, R., 2007. Long-term modeling of soil C erosion and sequestration at the small watershed scale. Climatic Change 80, 73–90.

Izaurralde, R.C., Haugen-Kozyra, K.H., Jans, D.C., McGill, W.B., Grant, R.F., Hiley, J.C., 2001. Soil organic carbon dynamics: measurement, simulation and site to region scale-up. In: Lal, R., Kimble, J.M., Follett, R.F., Stewart, B.A. (Eds.), Assessment Methods for Soil Carbon. Lewis Publishers, Boca Raton, FL, pp. 553–575.

Izaurralde, R.C., McGill, W.B., Williams, J.R., Link, R.P., Manowitz, D.H., and Schwab, D.E. Integrating microbial denitrification into EPIC: model description and initial testing. Environ. Sci. Technol. In preparation.

Izaurralde, R.C., Post, W.M., and West, T.O., 2012. Managing carbon: ecological limits and constraints. In Land Use and the Carbon Cycle: Science and Applications in Coupled Natural-Human Systems. Cambridge Univ. Press, UK. In press.

Jenkinson, D.S., Coleman, K., 1994. Calculating the annual input of organic matter to soil from measurements of total organic carbon and radiocarbon. European J. Soil Sci. 45, 167–174.

Jones, C.A., Dyke, P.T., Williams, J.R., Kiniry, J.R., Benson, V.W., Griggs, R.H., 1991. EPIC: an operational model for evaluation of agricultural sustainability. Agric. Syst. 37, 341–350.

Joseph, S.D., Camps-Arbestain, M., Lin, Y., Munroe, P., Chia, C.H., Hook, J., Van Zwieten, L., Singh, B.P., Lehmann, J., 2010. An investigation into the reactions of biochar in soil. Australian J. Soil Res. 48, 501–515.

Kimetu, J.M., Lehmann, J., 2010. Stability and stabilisation of biochar and green manure in soil with different organic carbon contents. Australian J. Soil Res. 48, 577–585. DOI: 10.1071/SR10036.

Laird, D.A., Pierce, F., Davis, D., Horton, R., Wang, B., Douglas, K., 2010. Impact of biochar amendments on the quality of a typical Midwestern agricultural soil. Geoderma 158, 443–449.

Lal, R., 2004. Carbon emission from farm operations. Environ. Int. 30, 981–990.

Lehmann, J., 2007. Bioenergy in the black. Front. Ecol. Environ. 5, 381–387.

Li, C., Frolking, S., Harriss, R., 1994. Modeling carbon biogeochemistry in agricultural soils. Global Biogeochem. Cycles 8, 237–254.

Major, J., Rondon, M., Molina, D., Riha, S.J., Lehmann, J., 2010. Maize yield and nutrition during 4 years after biochar application to a Colombian savanna oxisol. Plant Soil 333, 117–128.

Mayes, M.A., Heal, K.R., Brandt, C., Jagadamma, S., Jardine, P.M., 2010. Dependence of sorption of dissolved organic carbon on soil order classification. http://a-c-s.confex.com/crops/2010am/recordingredirect.cgi/id/9997 (accessed October 6, 2011).

Millington, R.J., Quirk, J.P., 1961. Permeability of porous solids. Trans. Faraday Soc. 57, 1200–1207.

Monteith, J.L., 1977. Climate and the efficiency of crop production in Britain. Philos. Trans. R. Soc. London B 281, 277–294.

Parton, W.J., Ojima, D.S., Cole, C.V., Schimel, D.S., 1994. A general model for soil organic matter dynamics: sensitivity to litter chemistry, texture and management. In: Quantitative Modeling of Soil Forming Processes. SSSA Spec. Public No. 39, Madison, WI, pp. 147–167.

Parton, W.J., Scurlock, J.M.O., Ojima, D.S., Gilmanov, T.G., Scholes, R.J., Schimel, D.S., Kirchner, T., Menaut, J.-C., Seastedt, T., Garcia Moya, E., Kamnalrut, A., Kinyamario, J.I., 1993. Observations and modelling of biomass and soil organic matter dynamics for the grassland biome worldwide. Global Biogeochem. Cycles 7, 785–809.

Parton, W.J., Schimel, D.S., Cole, C.V., Ojima, D.S., 1987. Analysis of factors controlling soil organic matter levels in Great Plains grasslands. Soil Sci. Soc. Am. J 51, 1173–1179.

Press, W.H., Flannery, B.P., Teukolsky, S.A., Vetterling, W.T., 1992. Numerical Recipes in FORTRAN: The Art of Scientific Computing. Cambridge Univ. Press, 963 p.

Putnam, J., Williams, J., Sawyer, D., 1988. Using the erosion productivity calculator (EPIC) model to estimate the impact of soil erosion for the 1985 RCA appraisal. J. Soil Water Conserv. 43, 321–326.

Reddy, K.R., Khaleel, R., Overcash, M.R., Westerman, P.W., 1979. A nonpoint source model for land areas receiving animal wastes: II. Ammonia volatilization. Trans. ASAE 22, 1398–1404.

Seligman, N.G., van Keulen, H., 1981. PAPRAN: a simulation model of annual pasture production limited by rainfall and nitrogen. In: Frissel, M.J., van Veen, J.A. (Eds.), Simulation of Nitrogen Behaviour of Soil–Plant Systems. PUDOC, Wageningen, The Netherlands, pp. 192–221.

Smith, P., Smith, J.U., Powlson, D.S., McGill, W.B., Arah, J.R.M., Chertov, O.G., Coleman, K., Franko, U., Frolking, S., Jenkinson, D.S., Jensen, L.S., Kelly, R.H., Klein-Gunnewiek, H., Komarov, A.S., Li, C., Molina, J.A.E., Mueller, T., Parton, W.J., Thornley, J.H.M., Whitmore, A.P., 1997. A comparison of the performance of nine soil organic matter models using datasets from seven long-term experiments. Geoderma 81, 153–225.

Stockle, C.O., Williams, J.R., Jones, C.A., Rosenberg, N.J., 1992a. A method for estimating the direct and climatic effects of rising atmospheric carbon dioxide on growth and yield of crops. I. Modification of the EPIC model for climate change analysis. Agric. Syst. 38, 225–238.

Stockle, C.O., Dyke, P.T., Williams, J.R., Jones, C.A., Rosenberg, N.J., 1992b. A method for estimating the direct and climatic effects of rising atmospheric carbon dioxide on growth and yield of crops. II. Sensitivity analysis at three sites in the Midwestern USA. Agric. Syst. 38, 239–256.

Thomson, A.M., Izaurralde, R.C., Rosenberg, N.J., He, X., 2006. Climate change impacts on agriculture and soil carbon sequestration potential in the Huang-Hai Plain of China. Agric. Ecosystems Environ. 114, 195–209.

Vitousek, P.M., Turner, D.R., Parton, W.J., Sanford, R.L., 1994. Litter decomposition on the Mauna Loa environmental matrix, Hawaii: patterns, mechanisms, and models. Ecology 75, 418–429.

Wang, X., He, X., Williams, J.R., Izaurralde, R.C., Atwood, J.D., 2005. Sensitivity and uncertainty analyses of crop productivity and soil organic carbon simulated with EPIC. Trans. ASAE 48, 1041–1054.

West, T.O., Marland, G., 2002. A synthesis of carbon sequestration, carbon emissions, and net carbon flux in agriculture: comparing tillage practices in the United States. Agric. Ecosyst. Environ. 91, 217–232.

Williams, J.R., 1995. The EPIC model. In: Singh, V.P. (Ed.), Computer Models of Watershed Hydrology. Water Resources Publications, Highlands Ranch, CO, pp. 909–1000.

Williams, J.R., Izaurralde, R.C., 2006. The APEX model. In: Singh, V.P., Frevert, D.K. (Eds.), Watershed Models. CRC Press, Boca Raton, FL, pp. 437–482.

Williams, J.R., Jones, C.A., Dyke, P.T., 1984. A modeling approach to determining the relationship between erosion and soil productivity. Trans. ASAE 27, 129–144.

Williams, J.R., Jones, C.A., Kiniry, J.R., Spanel, D.A., 1989. The EPIC crop growth model. Trans. ASAE 32, 497–511.

Williams, J.W., Izaurralde, R.C., Steglich, E.M., 2008a. Agricultural Policy/Environmental eXtender Model Theoretical Documentation Version 0604. BREC Report # 2008–17.

Williams, J.R., Arnold, J.G., Kiniry, J.R., Gassman, P.W., Green, C.H., 2008b. History of model development at Temple. Texas. Hydrol. Sci. 53, 948–960.

Yao, H., Conrad, R., 2000. Electron balance during steady-state production of CH_4 and CO_2 in anoxic rice soil. European J. Soil Sci. 51, 369–378.

Zhang, X., Izaurralde, R.C., Manowitz, D., West, T.O., Post, W.M., Thomson, A.M., Bandaru, V.P., Nichols, J., Williams, J.R., 2010. An integrative modeling framework to evaluate the productivity and sustainability of biofuel crop production systems. Global Change Biol.—Bioenergy 2, 258–277.

Zhang, X., Izaurralde, R.C., Arnold, J.G., Sammons, N.B., Manowitz, D.H., Thomson, A.M., Williams, J.R., 2011. Comment on "Modeling Miscanthus in the Soil and Water Assessment Tool (SWAT) to Simulate Its Water Quality Effects As a Bioenergy Crop". Environ. Sci. Tech. 45, 6211–6212.

CHAPTER 18

The General Ensemble Biogeochemical Modeling System (GEMS) and its Applications to Agricultural Systems in the United States

Shuguang Liu[1], Zhengxi Tan[2], Mingshi Chen[2], Jinxun Liu[3], Anne Wein[4], Zhengpeng Li[2], Shengli Huang[2], Jennifer Oeding[3], Claudia Young[5], Shashi B. Verma[6], Andrew E. Suyker[6], Stephen Faulkner[7], Gregory W. McCarty[8]

[1]U.S. Geological Survey (USGS) Earth Resources Observation and Science (EROS) Center, Sioux Falls, SD

[2]ASRC Research and Technology Solutions, Contractor to USGS EROS Center, Sioux Falls, SD

[3]Stinger Ghaffarian Technologies (SGT) Inc., Contractor to the USGS EROS Center, Sioux Falls, SD

[4]U.S. Geological Survey, Menlo Park, CA

[5]Earth Resources Technology (ERT), Inc., Contractor to the USGS EROS Center, Sioux Falls, SD

[6]School of Natural Resources, University of Nebraska-Lincoln, Lincoln, NE

[7]Aquatic Ecology Laboratory, USGS Leetown Science Center, Kearneysville, WV

[8]U.S. Department of Agriculture (USDA)—Agricultural Research Station (ARS) Hydrology and Remote Sensing Laboratory, Beltsville, MD

CHAPTER OUTLINE

Managing Agricultural Greenhouse Gases. DOI: 10.1016/B978-0-12-386897-8.00018-8

Abbreviations: GEMS, the General Ensemble Biogeochemical Modeling System; EDCM, Erosion-Deposition-Carbon-Model; GHG, greenhouse gas; IPCC, Intergovernmental Panel on Climate Change (IPCC); NPP, net primary production; SOC, soil organic carbon; SSURGO, Soil Survey Geographic Database; NRCS, Natural Resources Conservation Service; NRI, National Resources Inventory; CTIC, Conservation Technology Information Center

GENERAL ENSEMBLE BIOGEOCHEMICAL MODELING SYSTEM (GEMS)

The General Ensemble Biogeochemical Modeling System (GEMS) (Liu, 2009; Liu et al., 2004c) was developed to integrate well-established ecosystem biogeochemical models with various spatial databases for the simulations of biogeochemical cycles over large areas. Figure 18.1 shows the overall structure of the GEMS. Some of the key components are described below.

Multiple Underlying Biogeochemical Models

To avoid biases from individual models and to quantify the uncertainty of model outputs, GEMS simultaneously uses multiple site-scale biogeochemical models to simulate ecosystem dynamics over time and space. Previous applications of GEMS (Liu, 2009; Liu et al., 2004a, b, c, 2008; Tan et al., 2005, 2010, 2009b; Zhao et al., 2009) included the application of the CENTURY (Parton et al., 1994, 1987) and Erosion-Deposition-Carbon-Model (EDCM; see Liu et al., 2003). We are in the process of incorporating more models into GEMS including a spreadsheet model to account for carbon storage and greenhouse gas (GHG) emissions using the

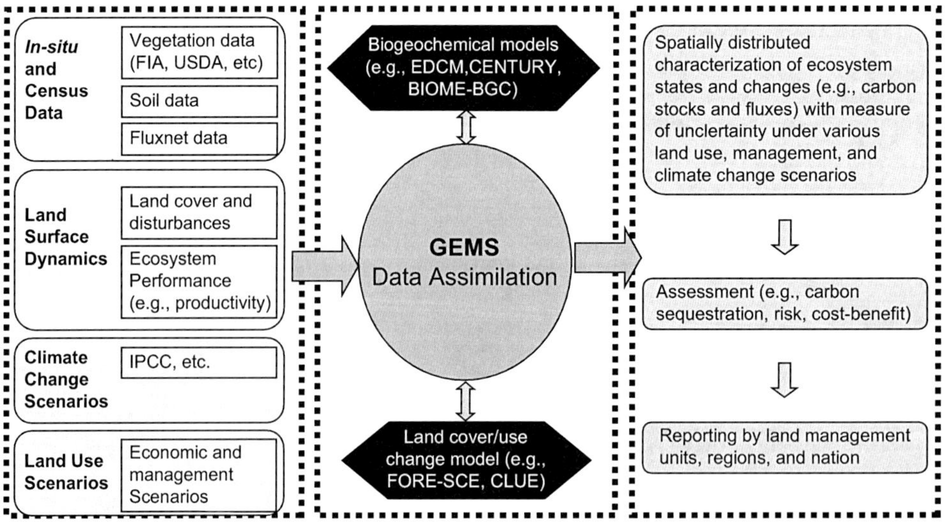

FIGURE 18.1
Structure and major components of the General Ensemble Modeling System (GEMS)

Intergovernmental Panel on Climate Change (IPCC) approach (IPCC, 2003), wetland biogeochemical models, and other models such as DeNitrification-DeComposition model (DNDC) (Li, 2000).

Monte Carlo Simulations

In addition to addressing model structure uncertainties using model ensemble, algorithms are implemented in GEMS to address the transfer and impacts of input data uncertainty (Liu, 2009). Monte Carlo ensemble simulations of each simulation unit (one site/pixel or group of sites/pixels with similar biophysical conditions) are used to incorporate the uncertainties and variability (as measured by variances and covariance) of state and driving input variables. Consequently, GEMS can provide uncertainty estimates of the predicted variables in time and space.

Model Inputs: Management Practices and Others

GEMS is designed to upscale carbon stocks and fluxes from sites to regions with a spatially explicit, dynamic consideration for land cover and land use change. Major driving variables include land cover and land use, climate, soils, disturbances, and management history. GEMS has the capability of modeling the impacts of land-surface disturbances and management practices; these include land use and land cover change, fertilization, cultivation, fallow, crop composition, crop rotation, manure addition, tillage practices, grazing, harvesting options (or residue management), wildfire incidents, and hurricane events (Liu et al., 2008, 2004a, b, c; Tan et al., 2005, 2009a, 2010; Zhao et al., 2009).

In general, channeling all management practices into biogeochemical modeling systems over large areas is a critical challenge because of the complexity, diversity, and spatial and temporal changes of management practices. In addition, most of the management practices cannot be detected using remote sensing techniques; the only data available are agricultural census data at the county, state, or resource management level (including NRI data). Stochastic approaches have been implemented in GEMS to downscale census data to site/pixel level. For example, if all crops are mapped into one category (i.e. cropland) as is often seen in land cover maps, GEMS would use county-level crop composition data (fractions or probabilities of all crops in a county) from an agricultural census to downscale the aggregated class cropland into different crops. In addition, if the land cover maps are snapshots with a time interval longer than one year, GEMS would create the missing annual land cover maps using crop rotation probabilities, which can be obtained from agricultural census data or expert knowledge. Additional examples can be found in Liu (2009).

Model Outputs

While different biogeochemical models in GEMS have different output variables, their common output variables include gross and net primary productivity, autotrophic and heterotrophic respiration, grain production, dynamics of carbon pools of vegetation and soils, and methane (CH_4) and nitrous oxide (N_2O) fluxes for agricultural systems. At the regional scale, outputs are in standardized file formats (e.g. network Common Data Form (NetCDF)) to facilitate sharing, analysis, and visualization.

Data Assimilation

Several data assimilation approaches were applied with GEMS for two purposes: (1) to understand and quantify the dynamics of model parameters, and (2) to detect model structure deficiencies (Chen et al., 2008; Liu et al., 2008; Zhao et al., 2010). This capability becomes useful when various observations are applied to calibrate the models at the site to regional scales.

Simulation of Agricultural Practices: EDCM as an Example

GEMS can drive a number of biogeochemical models to simulate carbon dynamics and GHG emissions. It is beyond the scope of this chapter to discuss all the algorithms behind individual models in GEMS because model formulations and algorithms are model specific and diverse. Instead in this section, we concentrate on the description of the EDCM, the core, and the first model that was coupled with GEMS.

NET PRIMARY PRODUCTION (NPP) AND IMPROVEMENTS IN CROP GENETICS AND AGRONOMICS

Representing the net amount of carbon fixed through photosynthesis into an ecosystem, NPP directly regulates the storage and rates of change of organic carbon in vegetation and soil. The prediction of the spatial and temporal change of NPP is critical for the simulation of carbon dynamics for a site or region.

The algorithms for NPP calculation follow the procedures that are well documented in the literature (Metherell et al., 1993; Parton et al., 1993). Calculation algorithms for NPP use the concept of potential primary productivity (PPP) and the limiting effects of moisture, temperature, and nutrients (Liu et al., 2003). Accordingly, PPP is the optimal primary productivity a system can reach without limitation from controlling variables; this is because the limiting factors change over time, as does NPP.

Grain yield and harvest index have increased dramatically since the 1940s at different rates in the U.S. for almost every crop (Hay, 1995), indicating that NPP of these crops must have changed as well. To account for these changes, in addition to using the land use and land cover change data (specifically crop rotation or transition), EDCM incorporates temporal changes in grain yield and the harvest index of crops in simulations. These temporal change patterns were derived from long-term U.S. agricultural census data, thereby allowing for improvements of crop genetics and management practices to be represented in the model. Details of the accounting formula and procedures can be found in Liu et al. (2003).

SOIL CARBON DYNAMICS

EDCM uses up to 10 soil layers to simulate the dynamics of soil organic carbon (SOC) in the profile. The thickness of the surface soil layer is fixed at the plowing depth at 20 or 30 cm, while the thicknesses of other layers are flexible. Five SOC pools (i.e. metabolic, structural, fast, slow, and passive) in each soil layer are used in EDCM to characterize the quantity and quality of SOC, which follows the practice used by the CENTURY model for the surface soil depth (Metherell et al., 1993; Parton et al., 1987, 1993). The SOC dynamics in each of the layers were simulated as a result of the interactions of the following processes: erosion or deposition, litter input, decomposition, and leaching.

Litter Input: Harvesting and Residue Management

Plant residue input directly regulates net carbon flux into the soil and, therefore, the amount of SOC storage. The amount of plant residue input varies over time and space, depending on a variety of factors including NPP and harvesting practices. Higher NPP usually means higher residue return to the soil for a given harvesting practice. In practice, the fraction of non-grain biomass removed from the site has important implications to the maintenance of SOC and site fertility. EDCM explicitly tracks the amount of biomass removed from the site as grain and straw, and the amount returned to the soil using NPP, harvest index, and the allocation of biomass in the crop (e.g. grain, aboveground, and belowground) (Liu et al., 2003).

In addition to litter input from the soil surface, soil receives litter input from root mortality in the soil profile. EDCM uses species-specific rooting characteristics (e.g. rooting depth and root vertical distribution) to track the growth and death of roots in each soil layer.

Finally, the decomposition of plant residues is simulated following the CENTURY 4.0 model according to residue quality indexes (e.g. C/N ratio and lignin content) and environmental conditions of the soil (Metherell et al., 1993; Parton et al., 1993).

Soil Carbon Decomposition in Soil Profile

EDCM simulates the decomposition of SOC in each SOC pool in each layer, calculated by using a pool-specific maximum decomposition rate, layer-specific soil moisture, soil temperature, and soil aeration. The approach is consistent with the CENTURY 4.0 model (Metherell et al., 1993; Parton et al., 1993; Paustian et al., 2012). The effects of soil texture on SOC turnover and lignin content of structural material on SOC decomposition are also considered.

For the simulation of plant growth and SOC decomposition, it is necessary to predict the temporal change of soil moisture in the soil profile. EDCM uses an innovative statistically based approach to simulate the dynamics of soil moisture using monthly precipitation observations, and has been tested successfully in dramatically different climate regions (Li, 2000; Liu et al., 2003).

Soil aeration has a strong impact on SOC decomposition (Li, 2000; Renault and Sierra, 1994; Renault and Stengel, 1994). Several studies (Bouwman, 1989; Van Dam et al., 1997; Voroney et al., 1981) indicate SOC decomposition in subsurface soil horizons is slower than can be explained by soil moisture, temperature, and soil texture, which are usually sufficient for the prediction of SOC dynamics in the surface layer. To our knowledge, no effective physically based modeling approach for the dynamics of soil aeration has been proposed. In EDCM, we hypothesize that soil aeration decreases with soil depth and we model its effect on decomposition using an aeration factor analogous to other factors included in the CENTURY model's treatment of decomposition (Liu et al., 2003, 2010). This approach is based on the assumption that the diffusion of oxygen to deep layers becomes increasingly difficult as depth increases.

Surface and internal drainage has been recognized as a major force driving SOC dynamics in cropland (Baker et al., 2007). In general, poorly drained environments favor SOC accumulation and well-drained environments enhance the soil organic matter decomposition and C emissions (Tan et al., 2004). Improvement in the drainage conditions through an internal tile drainage system within poorly drained soils (such as hydric soils, organic soils, and peatland) can promote crop root development and increase crop biomass (both above and below ground) and grain yields (Kanwar et al., 1988). Since the 1970s, a massive tile drainage system was developed in the Corn Belt (e.g. Iowa, Illinois, Ohio, etc.) to convert native prairies and other hydric soils to highly productive croplands. In order to evaluate the effect of tile drainage on SOC budgets, Liu et al. (2010) added an empirical equation to the EDCM to define drainage conditions at any depth in a soil profile in a tile-drained system.

IMPACTS OF SOIL EROSION AND DEPOSITION

Erosion and deposition of soil, carbon, and nutrients are important processes affecting carbon balance and GHG emissions (Harden et al., 1999; Liu et al., 2003; McCarty and Ritchie, 2002; Stallard, 1998; Van Oost et al., 2007). A suite of management practices and disturbances impacts soil erosion and deposition. EDCM treats the impacts of soil erosion and deposition on ecosystem productivity, SOC, and GHG fluxes in detail (Liu et al., 2003). It models the evolution of the soil profile as it is altered by soil erosion and deposition processes. Soil properties (e.g. soil texture and bulk density) and processes (e.g. moisture, temperature, and SOC decomposition) are explicitly tracked or simulated in each layer.

CH_4 AND N_2O FLUXES

In EDCM, CH_4 oxidation from agricultural systems is simulated according to the algorithms presented in Del Grosso et al. (2012). Nitrous oxide emissions are simulated as a function of fertilization rate, methods of fertilization, and N mineralization rate in the soil (Liu et al.,

1999). The model uses the algorithm based on previous work by Cao et al. (1996) and Zhang et al. (2002) and modified to fit in EDCM's monthly step prediction. In wetland systems and rice fields, flooding time and water depth are required inputs to calculate the CH_4 emission. Methane production is calculated as the function of temperature and carbon substrate in the flooded soil.

STUDY AREAS AND MODELING DESIGN

Study Areas

NEBRASKA EDDY FLUX TOWER SITES

Three flux tower sites were used to calibrate and test the model at the plot scale through data assimilation. The study sites are located at the University of Nebraska Agricultural Research and Development Center near Mead, NE (Figure 18.2). One site (#1: 41°09′54.2″ N, 96°28′35.9″ W, 361 m) is equipped with center pivot irrigation and was planted as continuous corn (*Zea mays* L.). The second site (#2: 41°09′53.5″ N, 96°28′2.3″ W, 362 m) is also equipped with center pivot irrigation and was planted to a corn–soybean (*Glycine max.* L.) rotation. The third site (#3: 41°10′46.8″ N, 96°26′22.7″ W, 362 m) relies on rainfall and is planted in corn–soybean rotation. Soil at the sites are deep silty clay loams, typical of eastern Nebraska, consisting of four soil series: Yutan (fine-silty, mixed, superactive, mesic Mollic Hapludalfs),

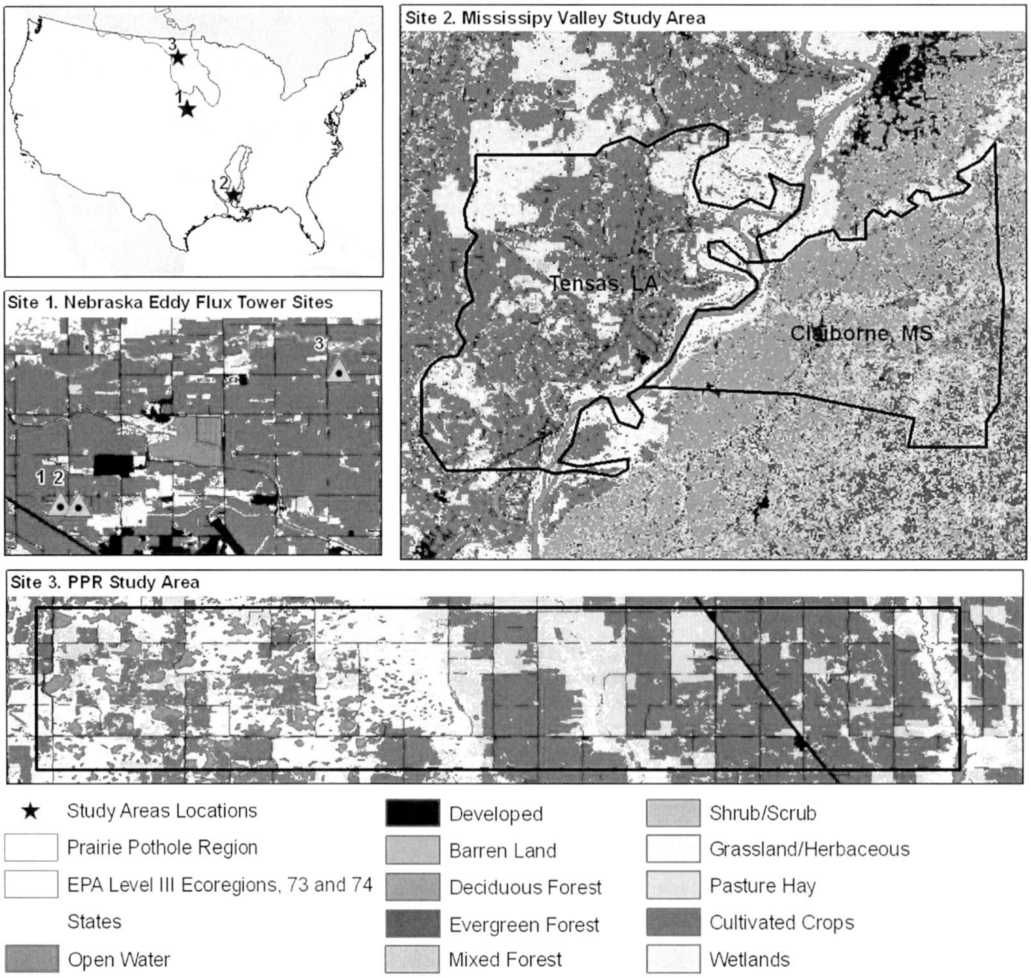

FIGURE 18.2
Study area locations. Please see color plate section at the back of the book.

Tomek (fine, smectitic, mesic Pachic Argialbolls), Filbert (fine, smectitic, mesic Vertic Argialbolls), and Filmore (fine, smectitic, mesic Vertic Argialbolls).

Prior to initiation of the study, these study sites had a variable cropping history. All three sites were uniformly tilled by disking prior to initiation of the study in 2001 to homogenize the top 0.1 m of soil and incorporate fertilizers as well as previously accumulated surface residues. The sites have been in no-till since 2001 (except Site 1, where conservation plow was used in the autumn of 2005). Results from the first 4 years documented declining yields with continuous irrigated maize (Site 1) because of difficulties in achieving uniform and adequate plant population due to a heavy litter layer. To address these constraints conservation-plow operations were employed at Site 1, resulting in partial inversion of topsoil layers.

Fluxes of CO_2, water vapor, and sensible heat were measured employing eddy covariance systems at all three sites. Details of measurements and analyses of these fluxes and supporting variables are provided in Verma et al. (2005).

REGIONAL APPLICATIONS: MISSISSIPPI VALLEY AND PRAIRIE POTHOLES

Three counties were selected as representative of the land cover change characteristics in the three regions. Tensas Parish, LA, is located in the Mississippi Alluvial Plain (MAP, EPA Level III Ecoregion 73), a roughly 14 million ha lowland valley shaped by the Mississippi River that extends from southern Illinois to the Gulf of Mexico. The meander belts, valley trains, and back swamps are comprised of fine-textured and poorly drained clay and silt soils (EPA-Western Ecology Division, 2010). Originally, the MAP was dominated by bottomland hardwood forests, but flood control levees have reduced the natural historic floodplain to 10% of its original extent (Mac et al., 1998). This facilitated the large-scale conversion of the original forest to agricultural cropland, which now covers nearly 59% of the MAP (Faulkner et al., 2011). In 2001, Tensas Parish land cover was primarily cropland (54%) followed by wetlands (33%), forests (3%), and other (water, developed, etc., 10%) (Zhu et al., 2010). Nearly two-thirds of the MAP is dominated by Sharkey or Tensas clay soils, with the remaining soils consisting of silt loams or silty clay loams (USDA-NRCS, 2006).

Claiborne County, MS, is located in the Mississippi Valley Loess Plain (MVLP, EPA Level Three Ecoregion 74). The MVLP lies adjacent to the eastern edge of the MAP, extending from western Kentucky to Louisiana. It consists of bluff hills, loess plains, and southern rolling plains with loess (wind-blown silt) soils. Vegetation is primarily upland forests dominated by oak (*Quercus*), hickory (*Carya*), and southern yellow pine (*Pinus*) (EPA-Western Ecology Division, 2010). As of 2001, Claiborne County was dominated by forests (73%, consisting of 47% deciduous, 6% evergreen, 9% mixed, and 11% anthropogenic disturbances), followed by wetlands (10%) and croplands (10%, includes hay and pasture), and other land cover classes (7%) (Zhu et al., 2010). All soils in the county are classified as silt loam (USDA-SCS, 1963).

A 256 km^2 block in Stutsman County, North Dakota, was selected as an area for GEMS applications in the Prairie Pothole Region of the United States (Figure 18.2). The study area is characterized by a dynamic continental climate with a mean annual precipitation of approximately 440 mm (Carroll et al., 2005). Native vegetation within the study area was mixed grass prairie. However, the landscape has been substantially altered, and the majority of the prairie grasslands have been converted to agricultural croplands.

Modeling Design

In this study, GEMS was applied at two spatial scales to illustrate its capability in simulating the consequences of various management practices on soil carbon dynamics and GHG emissions. As an example, GEMS was used to model the impacts of crop rotation, fertilization, crop residue management, irrigation, and tillage at the site scale (i.e. the Nebraska flux tower corn–soybean rotation site). Table 18.1 lists the scenario setups for these simulations. GEMS

TABLE 18.1 Modeling Experiment Setup for Simulating Impacts of Management Practices at the Nebraska Flux Tower Sites. Results are Presented in Figure 18.3

Case (see Figure 18.3)	Crop rotation (yearly)	Cultivation types	Harvest types	Fertilization (g N/m^2)	Irrigation (cm)
Base	Corn–soybean	CONV–CONV	G–G	26–5	34–34
Corn–corn	Corn–corn	CONV–CONV	G–G	26–5	34–34
Tillage	Corn–soybean	NT–NT	G–G	26–5	34–34
Residue	Corn–soybean	CONV–CONV	RED–RED	26–5	34–34
Fertilization	Corn–soybean	CONV–CONV	G–G	15–0	34–34
Irrigation	Corn–soybean	CONV–CONV	G–G	26–5	0–0

Note: CONV: conventional tillage. NT: no tillage.
G: 100% grain harvested, and other plant materials are left on site.
RED: 100% grain harvested, 50% aboveground non-grain biomass removed, and roots not removed.

was calibrated using the eddy flux measurements of carbon exchanges between the cropland and the atmosphere, grain production, and other measurements before being applied to simulate the carbon dynamics and GHG fluxes under various scenarios listed in Table 18.1. In order to address the long-term impacts of management practices, our simulations ran from 2000 to 2050.

To illustrate the applicability of GEMS in incorporating high-resolution remotely sensed land cover and land use change information at the regional scale, we applied the model at the selected area in the Prairie Pothole region from 1972 to 2008. A land cover change database from 2000 to 2008 was created based on the USDA Crop Data Layer derived from the Indian Advanced Wide Field Sensor (AWiFS), which has a spatial resolution of 56 m (USDA-National Agricultural Statistics Service, 2011). It was assumed that there was no land cover change before 2000 because of lack of information.

In addition to illustrating the wide range of geographic areas suitable for GEMS application, we also applied the model to the Mississippi Valley to simulate the potential of carbon sequestration and reduction of GHG emissions under future climate change and land use scenarios. Annual land cover change scenarios (R: "reference land use, land cover, and land management"; L: "enhanced land use and land cover with reference land management") from 2001 to 2050 were predicted using the Forecasting Scenario (FORE-SCE) model (Sohl et al., 2007); those results are presented in Zhu et al. (2010). Monthly climate data were from IPCC SRES (Special Report on Emissions Scenarios) A2 scenario, which is at the higher end of the SRES emissions scenarios characterized by high regional heterogeneity.

For both regional applications, soil information was from the national Soil Survey Geographic (SSURGO) Database. Model simulations were constrained by grain yields for crops from the USDA agricultural census data (USDA-National Agricultural Statistics Service, 2010) and forest growth curves from the USDA Forest Service Forest Inventory and Analysis (FIA) (USDA-Forest Service, 2010).

RESULTS

Impacts of Management Practices on SOC at Site Scale

Figure 18.3 shows the impacts of various management practices on SOC dynamics at the corn–soybean rotation site in Nebraska. Apparently, all management activities affected SOC dynamics but the impacts differed. Crop residue management was the most effective practice affecting SOC dynamics directly. A 50% removal of the residue from the field would reduce SOC by about 840 g C m^{-2} (or 17.3%) in the 50-year simulation. Figure 18.3 shows that this decrease will continue after 2050. Of course, the magnitude of the decrease depends on the

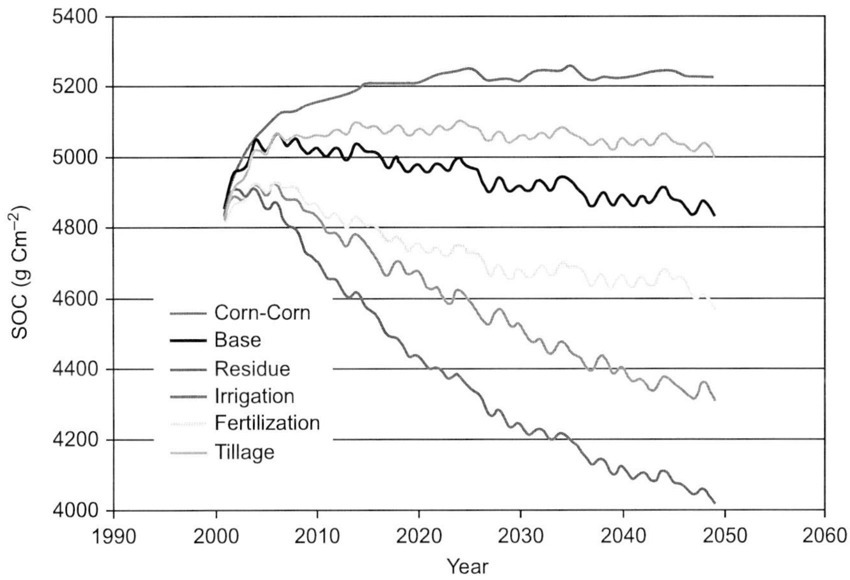

FIGURE 18.3
Soil carbon dynamics under various management practices. The scenarios are specified in Table 18.1. Please see color plate section at the back of the book.

fraction of residue removed from the site. High removal rates would accelerate and increase the reduction of SOC (Gollany et al., 2012).

Replacing corn–soybean rotation by continuous corn sequestered 371 g C m^{-2} (or 7.7%) of SOC during the 50-year simulation. Implementing no-till instead of the conventional tillage would increase SOC by 148 g C m^{-2} (or 3.1%). Irrigation and fertilization are important practices for maintaining high SOC content as well. Without irrigation and fertilization, SOC would decrease by 546 and 280 g C m^{-2}, respectively. Of course, those changes should be interpreted with caution. First, the relative fast increases of SOC in the initial years might be due to artifacts of improper initialization of the SOC pools. We did not spin-up (i.e. start the model runs years earlier than the intended starting date of model simulations to make sure that the model runs had enough time to stabilize and the states of the models were close to reality at the starting date) the model runs because of the difficulties in prescribing the details of management practices before the installation of the flux tower at the sites. Second, the changes are very small if they are represented on the basis of total soil mass, and very difficult to detect in the field. For example, replacing the corn–soybean rotation with continuous corn resulted in a 7.7% increase in SOC, which is equivalent to the standard errors of the field measurements of SOC (about 5 to 8%) (Verma et al., 2005). Therefore, given the challenges in measuring small SOC change in the field, these modeling results do not conflict with field observations of C neutrality at these sites (Verma et al., 2005).

Quantification of Regional Carbon Stocks and GHG Fluxes

PRAIRIE POTHOLE REGION

Figure 18.4 shows the simulated spatial details of SOC change in the study area of the Prairie Pothole Region. On average, this system lost SOC at a rate of 5 g C m^{-1} yr^{-1} from 1998 to 2007. However, the spatial variability of SOC change was high varying from strong sources (<−60 g C m^{-1} yr^{-1}) to strong sinks (>60 g C m^{-1} yr^{-1}). Variability in SOC responses was mainly caused by the spatial variability of management practices (e.g. crop rotation) and the existing SOC storage. The C sources mainly occurred in cropping systems with high levels of baseline SOC, which tend to be C sources following the conversion of grassland to cropland (Bellamy et al., 2005; Liu et al., 2010; Tan et al., 2006a, b, 2007). Accordingly, land use change

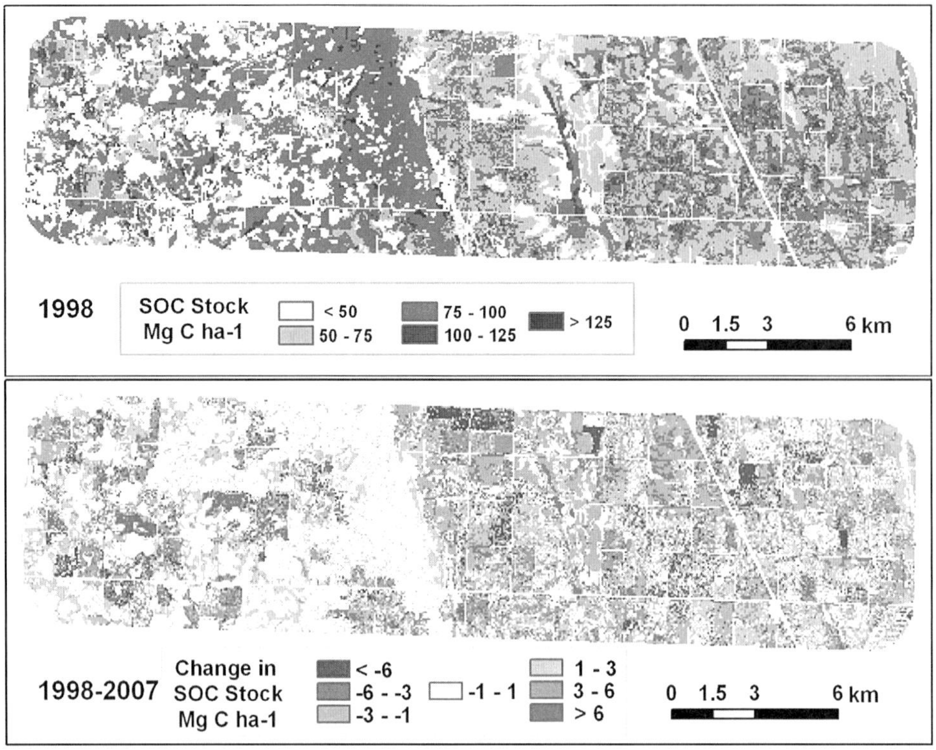

FIGURE 18.4
GEMS simulated changes in soil organic carbon in the 0–100 cm depth within the study area of the Prairie Pothole region. Please see color plate section at the back of the book.

was a major factor driving SOC dynamics. Planted areas for barley (*Hordeum vulgare* L.), spring wheat (*Triticum aestivum* L.), and sunflower (*Helianthus annuus* L.) declined sharply from 1998 to 2007, while the areas of corn and soybean expanded from 0.5 to 13.5% and from 1.0 to 14.6%, respectively. In fact, the SOC loss rate in the region has become smaller since the mid-1980s due mainly to an expansion of conservation tillage and restoration of grassland from croplands in the last decade (Follett et al., 2009).

TABLE 18.2 GEMS Simulated Changes in Total Carbon Stock (Tg, terragram, or 10^{12} gram), and Cumulative and Additional Carbon Sequestration in two counties (Tensas Parish, LA, and Claiborne County, MS) of the Mississippi Valley, Calculated using the Specified Method, and using the "Reference Land Use, Land cover, and Land Management" (R) and "Enhanced Land use and Land Cover with Reference Land Management" (L) Scenarios. Values Represent the Amount at the End of the Given Year in the Top 20 cm Layer

	Total carbon Stocks, by method, in Tg[1]			Cumulative carbon sequestration, by method, in Tg[1]			Additional carbon sequestration, by method, in Tg[2]		
Year	*GEMS-spreadsheet*	*GEMS-Century*	*GEMS–EDCM*	*GEMS-spreadsheet*	*GEMS-Century*	*GEMS–EDCM*	*GEMS-spreadsheet*	*GEMS-Century*	*GEMS–EDCM*
2001	40.91	34.22	43.30						
2010	43.45	38.37	42.56	2.54	4.15	−0.74	0.30	0.47	0.02
2020	45.57	42.11	43.71	4.67	7.90	0.41	0.52	0.54	0.15
2030	47.32	45.88	45.24	6.41	11.66	1.94	0.90	0.95	0.39
2040	48.48	49.14	46.70	7.58	14.92	3.39	1.27	1.42	0.82
2050	49.36	51.89	48.07	8.45	17.67	4.76	1.64	1.75	1.08

[1] *Values were evaluated using the "enhanced land use and land cover with reference land management" (L) scenario.*
[2] *Values represent the difference between the L scenario and the "reference land use and land cover and land management" (R) scenario.*

MISSISSIPPI VALLEY

Table 18.2 shows the dynamics of total carbon stocks as simulated by GEMS using GEMS-spreadsheet, GEMS-Century, and GEMS–EDCM methods in the two counties in Mississippi Valley. Note that the results presented for the Mississippi Valley included not only agricultural lands but also forests, wetlands, and other land cover.

The initial conditions of carbon stocks in vegetation and soils were not synchronized among these models. The purpose was to preserve the uncertainty and mimic the general observation that different initial conditions are used by different modelers. Although the GEMS-Century method began with a lower carbon stock value in 2001, it reached a higher carbon stock value in 2050 than the other two models (GEMS-spreadsheet and EDCM) because of a higher carbon sequestration rate during the study period. From 2001 to 2050, the total carbon sequestration (the net change in carbon stocks) calculated using the GEMS-CENTURY method (17.67 Tg) was much higher than that calculated using the GEMS-Spreadsheet (8.45 Tg) and GEMS–EDCM methods (4.77 Tg) (Table 18.2). The corresponding annual rates of carbon sequestration were 0.35, 0.17, and 0.1 Tg C yr^{-1} from the GEMS-CENTURY, GEMS-Spreadsheet, and GEMS–EDCM methods, respectively.

The differences shown here might be attributed to differences in the input data sources, initial parameter values, and simulation algorithms of each model, especially between the GEMS-CENTURY and GEMS–EDCM methods. For example, a higher rate of carbon sequestration from the GEMS-CENTURY method might have been caused by the lower initial biomass carbon values, faster biomass accumulation (compared to the GEMS-Spreadsheet method), and SOC accumulation. In contrast, the lower carbon-sequestration estimate from the GEMS–EDCM method can be attributed to lower biomass accumulation (compared to the GEMS-Spreadsheet method) and SOC loss. Further study to reconcile the differences among the modeling approaches within GEMS should be conducted.

All three methods estimated significantly higher ecosystem carbon stocks for the "enhanced land use and land cover with reference land management" (L) scenario, indicating additional carbon sequestration of 1.64, 1.75, and 1.08 Tg from the GEMS-Spreadsheet, GEMS-CENTURY, and GEMS–EDCM methods, respectively, relative to the "reference land use, land cover, and land management" (R) scenario. These amounts represented an additional 20%, 10%, and 23% increase, respectively, above the carbon-sequestration values calculated using the R scenario (Table 18.2). The result suggests that these models, rather consistently, are capable of quantifying additional carbon sequestration from enhanced changes in land use and land cover activities such as the Wetland Reserve Program (NRCS, 2011), although their initialization and performance on the absolute estimates of C stocks were quite different.

Table 18.3 lists major differences in CH_4 and N_2O emissions between the GEMS-spreadsheet and GEMS–EDCM methods (no results were generated from the GEMS-CENTURY method). Table 18.3 revealed the following: (1) the GEMS-spreadsheet method estimated an annual CH_4-emission rate on wetlands more than double that of the GEMS–EDCM method; (2) estimates of N_2O emissions demonstrated opposite temporal trends, although both methods produced similar N_2O-emission rates; and (3) the GEMS-spreadsheet method showed small increases in annual emission rates of CH_4 and N_2O, whereas the GEMS–EDCM method showed decreasing trends. Field studies in this region suggested both CH_4 and N_2O emission rates were greatly affected by soil moisture, temperature, and substrate availability, and thus varied considerably depending on site conditions. For example, CH_4 emissions from rice paddies ranged from 2 to 1642 kg C ha^{-1} yr^{-1} (Lindau et al., 1990). The complexity of ecosystems and the management practices in the region makes estimation of N_2O and CH_4 fluxes challenging. Additional work is needed to address discrepancies among different modeling approaches. For CH_4 and N_2O emissions, we found that uncertainty of the CH_4 and N_2O emission factors using the GEMS-spreadsheet method was very high. Reducing the uncertainty relies heavily on certainty

TABLE 18.3 Annual Emission Rates of Methane and Nitrous Oxide (Gg, gigagram, or 10^9 gram) and their Total Differences (between 2001 and 2050), for the "Reference Land use, Land cover, and Land Management" (R) and the "Enhanced Land Use and Land Cover with Reference Land Management" (L) Scenarios

	CH_4 from wetland (Gg C)				N_2O from all land (Gg N)			
Year	*GEMS-spreadsheet*		*GEMS–EDCM*		*GEMS-spreadsheet*		*GEMS–EDCM*	
	L	R	L	R	L	R	L	R
2001	28.47	28.42	15.50	15.47	2.74	2.74	2.77	2.76
2010	28.88	28.53	13.32	13.20	2.78	2.77	1.98	1.99
2020	29.26	28.36	12.66	12.45	2.82	2.76	1.91	1.92
2030	29.80	28.24	13.57	13.27	2.87	2.77	1.86	1.89
2040	30.43	28.10	13.04	12.65	2.92	2.77	1.74	1.77
2050	31.01	27.94	12.92	12.42	2.96	2.76	1.73	1.77
Difference	2.54	−0.48	−2.58	−3.05	0.22	0.02	−1.04	−0.99

of field observations of CH_4 and N_2O fluxes at the regional scale. At present, field observations demonstrate a high uncertainty in GHG fluxes in the Mississippi Valley (Zhu et al., 2010).

DISCUSSION

Many site-level biogeochemical models have been developed and tested extensively over the past three decades. With proper calibration and validation, they can be used to quantify the impacts of various management practices on SOC and GHG fluxes in agricultural systems at the field scale. We have demonstrated this capability for GEMS–EDCM at a site in Nebraska.

Although many site-scale models have been applied to regional and global studies, the appropriateness and efficacy of such model extrapolation are not well addressed and tested in the literature (Parton et al., 1994). For example, to our knowledge, few models have the capability of systematically simulating the impacts of agricultural management practices over large areas. GEMS incorporates information from different sources into the modeling processes. In addition, the use of the model ensemble in GEMS makes it ideal to address uncertainties in model structure, model parameters, and input data. Results from the Mississippi Valley indicated that the differences among the models within GEMS (specifically the biases and errors in the individual models) are significant in the estimation of carbon dynamics and GHG fluxes. On the development or technical side of GEMS, procedures should be put in place to address issues across models within GEMS such as consistent initialization and automated schemes for constraining model simulations with observations from various sources and different spatial and temporal scales. GEMS is the biogeochemical modeling system for the U.S. Geological Survey's assessment of national potentials for biological C sequestration and reduction of CH_4 and N_2O emissions (Zhu et al., 2010). Existing algorithms will be tested and improved and new algorithms will be added, if needed.

In addition to the challenges in model development, a major difficulty is obtaining information about the multitude of agricultural practices that affect SOC and GHG fluxes (see Eagle et al. (2010) for an exhaustive list of agricultural land management practices). First, to our knowledge, there is no common data repository for sharing agricultural management practice data, and each project locates available data from various sources, collects, compiles, and uses it to prepare agricultural practice inputs. For example, the Forest and Agricultural Sector Optimization Model—Green House Gas version (FASOM-GHG) (Adams et al., 2005) has accomplished national data compilations for various cropland mitigation strategies including changing crop composition, rice acreage reduction, crop fertilizer rate reduction, crop tillage alteration, grassland conversion, and irrigated/dryland conversion for 63 U.S. production regions. Second, subtle but important relationships among practice data are not captured. For example, USDA Natural Resources Conservation Service (NRCS, 2002) observed that the Conservation Technology Information Center (CTIC) reported the area in various tillage systems by individual crops on an annual basis; however, it did not differentiate between long-term no-till practices versus intermittent or "rotational no till" (e.g. tilled corn—no-tilled soybean rotations). Third, there are uncertainties inherent in survey-based data such as sampling design. Fourth, some agricultural practices are not routinely monitored. For example, information about cover crop practices is scattered. Finally, these problems are exacerbated across local, regional and global scales of analysis. For example, the downscaling of agricultural practice projections from the Integrated Model to Assess the Global Environment (IMAGE, 2006) is limited to crop composition and fertilizer and manure use. These data challenges are opportunities to improve the analysis of potential SOC and GHG fluxes in agricultural systems.

Acknowledgments

The research was funded by the LandCarbon Project, Geographic Analysis and Monitoring (GAM) Program, and the Global Change Research and Development Program (R&D) of the U.S. Geological Survey. Work of Zhengxi Tan, Mingshi Chen, Zhengpeng Li, and Shengli Huang (ARTS) was performed under USGS contract G08PC91508, and work of Jinxun Liu and Jennifer Oeding (SGT) and Claudia Young (ERT) was performed under USGS contract G10PC00044. Any use of trade, firm, or product names is for descriptive purposes only and does not imply endorsement by the U.S. government.

References

Adams, D., Alig, R., McCarl, B.A., Murray, B.C., 2005. FASOMGHG conceptual structure, and specification: documentation. Texas A&M Univ., College Station, TX. Available at http://agecon2.tamu.edu/people/faculty/mccarl-bruce/papers/1212FASOMGHG_doc.pdf (verified March 16, 2011).

Baker, J.M., Ochsner, T.E., Venterea, R.T., Griffis, T.J., 2007. Tillage and soil carbon sequestration: what do we really know? Agric. Ecosyst. Environ. 118, 1–5.

Bellamy, P.H., Loveland, P.J., Bradley, R.I., Lark, R.M., Kirk, G.J.D., 2005. Carbon losses from all soils across England and Wales 1978–2003. Nature 437, 245–248.

Bouwman, A.F., 1989. Modelling soil organic matter decomposition and rainfall erosion in two tropical soils after forest clearing for permanent agriculture. Land Degrad. Dev. 1, 125–140.

Cao, M., Marshall, S., Gregson, K., 1996. Global carbon exchange and methane emissions from natural wetlands: application of a process-based model. J. Geophys. Res. Atmos 101, 14399–14414.

Carroll, R., Pohll, G., Tracy, J., Winter, T., Smith, R., 2005. Simulation of a semipermanent wetland basin in the Cottonwood Lake area, east-central North Dakota. J. Hydrol. Eng. 10, 70–84.

Chen, M., Liu, S., Tieszen, L.L., Hollinger, D.Y., 2008. An improved state-parameter analysis of ecosystem models using data assimilation. Ecol. Modell. 219, 317–326.

Del Grosso, S.J., Parton, W.J., Adler, P.R., Davis, S.C., Keough, C., Marx, E., 2012. DayCent model simulations for estimating soil carbon dynamics and greenhouse gas fluxes from agricultural production systems (this volume). In: Liebig, M.A., Franzluebbers, A.J., Follett, R.F. (Eds.), Managing Agricultural Greenhouse Gases: Coordinated Agricultural Research through GRACEnet to Address our Changing Climate. Elsevier Publ., San Diego, CA.

Eagle, A.J., Henry, L.R., Olander, L.P., Haugen-Kozyra, K., Millar, N., Robertson, G.P., 2010. Greenhouse gas mitigation potential of agricultural land management in the United States: a synthesis of the literature. Technical Working Group on Agricultural Greenhouse Gases (T-AGG) Report; Companion Report to Assessing Greenhouse Gas Mitigation Opportunities and Implementation Options for Agricultural Land Management in the United States, NI R 10–04. Nicholas Inst. for Environ. Policy Solutions, Duke Univ., Durham, NC. Available at http://nicholasinstitute.duke.edu/ecosystem/land/TAGGDLitRev (verified February 23, 2010).

EPA-Western Ecology Division, 2010. US Level III Descriptions, Level III and IV Ecoregions of the Continental United States. EPA, Corvallis, OR. Available at http://www.epa.gov/wed/pages/ecoregions/level_iii_iv.htm#Level%20III (verified March 14, 2011).

Faulkner, S., Barrow, W., Keeland, B., Walls, S., Telesco, D., 2011. Effects of conservation practices on wetland ecosystem services in the Mississippi Alluvial Valley. Ecol. Appl. 21 (Suppl. 3), S31–S48.

Follett, R.F., Kimble, J.M., Pruessner, E.G., Samson-Liebig, S., Waltman, S., 2009. Soil organic carbon stocks with depth and land use at various U.S. sites. In: Lal, R., Follett, R.F. (Eds.), Soil Carbon Sequestration and the Greenhouse Effect, second ed.). SSSA Spec. Publ. 57. ASA-CSSA-SSSA, Madison, WI, pp. 29–46.

Gollany, H.T., Follett, R.F., Liang, Y., 2012. CQESTR simulations of soil organic carbon dynamics (this volume). In: Liebig, M.A., Franzluebbers, A.J., Follett, R.F. (Eds.), Managing Agricultural Greenhouse Gases: Coordinated Agricultural Research through GRACEnet to Address our Changing Climate. Elsevier Publ., San Diego, CA.

Harden, J.W., Sharpe, J.M., Parton, W.J., Ojima, D.S., Fries, T.L., Huntington, T.G., Dabney, S.M., 1999. Dynamic replacement and loss of soil carbon on eroding cropland. Global Biogeochem. Cycles 13, 885–901.

Hay, R.K.M., 1995. Harvest index: a review of its use in plant breeding and crop physiology. Ann. Appl. Biol. 126, 197–216.

IMAGE, 2006. Integrated modelling of global environmental change. An overview of IMAGE 2.4. © Netherlands Environ. Assess. Agency (MNP), Bilthoven Oct. 2006, MNP publication number 500110002/2006.

IPCC, 2003. Good practice guidance for land use, land-use change and forestry. Inst. for Global Environ. Strategies for the IPCC, Hayama, Kanagawa, Japan.

Kanwar, R.S., Baker, J.L., Baker, D.G., 1988. Tillage and split N-fertilization effects on subsurface drainage water quality and crop yields. Trans. Am. Soc. Agric. Eng. 31, 453–461.

Li, C.S., 2000. Modeling trace gas emissions from agricultural ecosystems. Nutr. Cycling Agroecosyst. 58, 259–273.

Lindau, C.W., Patrick, W.H., Delaune, R.D., Reddy, K.R., 1990. Rate of accumulation and emission of N2, N2O and CH4 from a flooded rice soil. Plant and Soil 129, 269–276.

Liu, J., Liu, S., Loveland, T.R., Tieszen, L.L., 2008. Integrating remotely sensed land cover observations and a biogeochemical model for estimating forest ecosystem carbon dynamics. Ecol. Modell 219, 361–372.

Liu, S., 2009. Quantifying the spatial details of carbon sequestration potential and performance. Geophys. Monogr. 183, 117–128.

Liu, S., Bliss, N., Sundquist, E., Huntington, T.G., 2003. Modeling carbon dynamics in vegetation and soil under the impact of soil erosion and deposition. Global Biogeochem. Cycles 17, 1074.

Liu, S., Kaire, M., Wood, E., Diallo, O., Tieszen, L.L., 2004a. Impacts of land use and cimate change on carbon dynamics in south-central Senegal. J. Arid Environ 59, 583–604.

Liu, S., Liu, J., Loveland, T.R., 2004b. Spatial-temporal carbon sequestration under land use and land cover change. In: Bridging the Pacific and Atlantic, Proc., Geospatial Information Research, Int. Conf. on Geoinformatics, 12th, Univ. of Gävle, Sweden. 7–9 June 2004. Gävle Univ. Press, pp. 525–532.

Liu, S., Loveland, T.R., Kurtz, R.M., 2004c. Contemporary carbon dynamics in terrestrial ecosystems in the southeastern plains of the United States. Environ. Manage. 33, S442–S456.

Liu, S., Reiners, W.A., Keller, M., Schimel, D.S., 1999. Model simulation of changes in N_2O and NO emissions with conversion of tropical rain forests to pastures in the Costa Rican Atlantic Zone. Global Biogeochem. Cycles 13, 663–677.

Liu, S., Tan, Z., Li, Z., Zhao, S., Yuan, W., 2010. Are soils of Iowa USA currently a carbon sink or source? Simulated changes in SOC stock from 1972 to 2007. Agric. Ecosyst. Environ. 140, 106–112.

Mac, M.J., Opler, P.A., Puckett Haeker, C.E., Doran, D.P., 1998. Status and trends of the nation's biological resources. Vol. 1-2. U.S. Geol. Survey, Reston, VA.

McCarty, G.W., Ritchie, J.C., 2002. Impact of soil movement on carbon sequestration in agricultural ecosystems. Environ. Pollut. 116, 423–430.

Metherell, A.K., Harding, L.A., Cole, C.V., Parton, W.J., 1993. CENTURY soil organic matter model environment: technical documentation, Agroecosystem Version 4.0. USDA, Agric. Res. Serv., Great Plains System Res. Unit, Fort Collins, CO.

NRCS, 2011. Wetlands Reserve Program. USDA, Washington, DC. Available at http://www.nrcs.usda.gov/programs/wrp (verified June 22, 2011).

NRCS, 2002. Quantifying the change in greenhouse gas emissions due to natural resource conservation practice application in NE, "The Nebraska Carbon Storage Project" USDA, Report to the Nebraska Conservation Partnership, March 2002. http://dnr.ne.gov/Carbon/NebraskaNRCS.pdf (accessed April 21, 2011).

Parton, W.J., Ojima, D.S., Schimel, D.S., 1994. Environmental change in grasslands: assessment using models. Clim. Change 28, 111–141.

Parton, W.J., Schimel, D.S., Cole, C.V., Ojima, D.S., 1987. Analysis of factors controlling soil organic-matter levels in Great Plains grasslands. Soil Sci. Soc. Am. J. 51, 1173–1179.

Parton, W.J., Scurlock, J.M.O., Ojima, D.S., Gilmanov, T.G., Scholes, R.J., Schimel, D.S., Kirchner, T., Menaut, J.C., Seastedt, T., Garcia Moya, E., Kamnalrut, A., Kinyamario, J.I., 1993. Observations and modeling of biomass and soil organic matter dynamics for the grassland biome worldwide. Global Biogeochem. Cycles 7, 785–809.

Paustian, K., Schuler, J., Killian, K., Chambers, A., Del Grosso, S., Easter, M., Alvaro-Fuentes, J., Gurung, R., Johnson, G., Merwin, M., Ogle, S., Olson, C., Swan A,, Williams, S., Vining, R. COMET2.0 – Decision Support System for Agricultural Greenhouse Gas Accounting. In M.A. Liebig, A.J. Franzluebbers, and R.F. Follet (ed.) Managing Agricultural Greenhouse Gases. Academic Press - Elsevier, Amsterdam. In Press.

Renault, P., Sierra, J., 1994. Modeling oxygen diffusion in aggregated soils: II. Anaerobiosis in topsoil layers. Soil Sci. Soc. Am. J. 58, 1023–1030.

Renault, P., Stengel, P., 1994. Modeling oxygen diffusion in aggregated soils: I. Anaerobiosis inside the aggregates. Soil Sci. Soc. Am. J. 58, 1017–1023.

Sohl, T.L., Sayler, K.L., Drummond, M.A., Loveland, T.R., 2007. The FORE-SCE model: a practical approach for projecting land cover change using scenario-based modeling. J. Land Use Sci. 2, 103–126.

Stallard, R.F., 1998. Terrestrial sedimentation and the carbon cycle: coupling weathering and erosion to carbon burial. Global Biogeochem. Cycles 12, 231–257.

Tan, Z., Johnston, C.A., Liu, S., Loveland, T.R., Tieszen, L.L., Kurtz, R., Liu, J., 2005. Soil organic carbon dynamics as related to land use history in the northwestern Great Plains. Global Biogeochem. Cycles 19 GB3011.

Tan, Z., Lal, R., Liu, S., 2006a. Using experimental and geospatial data to estimate regional carbon sequestration potential under no-till management. Soil Sci. 171, 950–959.

Tan, Z., Lal, R., Smeck, N.E., Calhoun, F.G., 2004. Relationships between surface soil organic carbon pool and site variables. Geoderma 121, 187–195.

Tan, Z., Liu, S., Johnston, C.A., Liu, J., Tieszen, L.L., 2006b. Analysis of ecosystem controls on soil carbon source-sink relationships in the northwest Great Plains. Global Biogeochem. Cycles 20 GB4012.

Tan, Z., Liu, S., Li, Z., Loveland, T.R., 2007. Simulated responses of soil organic carbon stock to tillage management scenarios in the northwest Great Plains. Carbon Balance Manage 2, 7.

Tan, Z., Liu, S., Tieszen, L.L., Tachie-Obeng, E., 2009a. Simulated dynamics of carbon stocks driven by changes in land use, management and climate in a tropical moist ecosystem of Ghana. Agric. Ecosyst. Environ. 130, 171–176.

Tan, Z., Tieszen, L.L., Liu, S., Tachie-Obeng, E., 2010. Modeling to evaluate the response of savanna-derived cropland to warming-drying stress and nitrogen fertilizers. Clim. Change 100, 703–715.

Tan, Z., Tieszen, L.L., Tachie-Obeng, E., Liu, S., Dieye, A.M., 2009b. Historical and smulated ecosystem carbon dynamics in Ghana: land use, management, and climate. Biogeosciences 6, 45–58.

USDA-Forest Service, 2010. Forest inventory and analysis national program. USDA, Arlington, VA. Available at http://www.fia.fs.fed.us/tools-data (verified January 25, 2011).

USDA-National Agricultural Statistics Service, 2010. The census of agriculture. USDA, Washington, DC. Available at http://www.agcensus.usda.gov/ (verified December 18, 2010).

USDA-National Agricultural Statistics Service, 2011. Cropland data layer—examples. USDA, Washington, DC. Available at http://www.nass.usda.gov/research/Cropland/sarsexample3.html (verified February 10, 2011).

USDA-NRCS, 2006. Soil survey of Tensas Parish, Louisiana, Louisiana Online Soil Survey Manuscripts. USDA, Washington, DC. Available at http://soils.usda.gov/survey/online_surveys/louisiana/index.html (verified March 16, 2011).

USDA-SCS, 1963. Soil survey, Claiborne County, Mississippi, Mississippi Online Soil Survey Manuscripts. USDA, Washington, DC. Available at http://soils.usda.gov/survey/online_surveys/mississippi/index.html (verified March 16, 2011).

Van Dam, D., Veldkamp, E., Van Breemen, N., 1997. Soil organic carbon dynamics: variability with depth in forested and deforested soils under pasture in Costa Rica. Biogeochemistry 39, 343–375.

Van Oost, K., Quine, T.A., Govers, G., De Gryze, S., Six, J., Harden, J.W., Ritchie, J.C., McCarty, G.W., Heckrath, G., Kosmas, C., Giraldez, J.V., da Silva, J.R.M., Merckx, R., 2007. The impact of agricultural soil erosion on the global carbon cycle. Science 318, 626–629.

Verma, S.B., Dobermann, A., Cassman, K.G., Walters, D.T., Knops, J.M., Arkebauer, T.J., Suyker, A.E., Burba, G.G., Amos, B., Yang, H., 2005. Annual carbon dioxide exchange in irrigated and rainfed maize-based agroecosystems. Agric. For. Meteorol. 131, 77–96.

Voroney, R.P., Van Veen, J.A., Paul, E.A., 1981. Organic C dynamics in grassland soils. 2. Model validation and simulation of the long-term effects of cultivation and rainfall erosion. Can. J. Soil Sci. 61, 211–224.

Zhang, Y., Li, C., Zhou, X., Moore III, B., 2002. A simulation model linking crop growth and soil biogeochemistry for sustainable agriculture. Ecol. Modell. 151, 75–108.

Zhao, S., Liu, S., Li, Z., Sohl, T.L., 2009. Ignoring detailed fast-changing dynamics of land use overestimates regional terrestrial carbon sequestration. Biogeosciences 6, 1647–1654.

Zhao, S., Liu, S., Li, Z., Sohl, T.L., 2010. Federal land management, carbon sequestration, and climate change in the southeastern US: a case study with Fort Benning. Environ. Sci. Technol. 44, 992–997.

Zhu, Z., Bergamaschi, B., Bernknopf, R., Clow, D., Dye, D., Faulkner, S., Forney, W., Gleason, R., Hawbaker, T., Liu, J., Liu, S., Prisley, S., Reed, B., Reeves, M., Rollins, M.G., Sleeter, B., Sohl, T.L., Stackpoole, S., Stehman, S., Striegl, R., Wein, A., 2010. A method for assessing carbon stocks, carbon sequestration, and greenhouse-gas fluxes in ecosystems of the United States under present conditions and future scenarios. U.S. Geol. Survey Sci. Investigations Rep. 2010-5233. Reston, VA.

SECTION 5

Measurements and Monitoring: Improving Estimates of Soil Carbon Dynamics and Greenhouse Gas Flux

CHAPTER 19

Quantifying Biases in Non-Steady-State Chamber Measurements of Soil—Atmosphere Gas Exchange

Rodney T. Venterea[1], Timothy B. Parkin[2]
[1]U.S. Department of Agriculture, Agricultural Research Service, Soil and Water Management Research Unit, St. Paul, MN
[2]U.S. Department of Agriculture, Agricultural Research Service, National Laboratory for Agriculture and the Environment, Ames, IA

CHAPTER OUTLINE

Abbreviations: CBC, chamber bias correction; CEAT, chamber error assessment tool; DP, deployment period; *F*, flux; GHG, greenhouse gas; *H*, height or volume-to-area ratio; HM, Hutchinson and Mosier (1981); HMR, revised Hutchinson and Mosier (1981); LR, linear regression; NDFE, non-linear diffusive flux estimator; NSS, non-steady state; PDF, pre-deployment flux; *TFU*, theoretical flux underestimation

INTRODUCTION

Chamber techniques offer the best currently available tool for statistically evaluating the effects of management practices and potential mitigation strategies on emissions of important trace and greenhouse gases (GHGs). However, limitations of non-steady-state (NSS) chamber methods for determining soil-to-atmosphere trace gas exchange rates have

Managing Agricultural Greenhouse Gases. DOI: 10.1016/B978-0-12-386897-8.00019-X been recognized for several decades (Hutchinson and Mosier, 1981). Of these limitations, the so-called "chamber effect" is one of the most challenging to overcome. The chamber effect can be defined as the inherent tendency for NSS gas flux chamber methods to produce a biased estimate of the actual pre-deployment flux (PDF), where the PDF is defined as the flux occurring immediately prior to placement of the chamber on the soil surface (Mathias et al., 1978; Healy et al., 1996; Livingston et al., 2006; Kutzbach et al., 2007; Kroon et al., 2008; Pederson et al., 2010). This effect can be particularly important for GHGs like nitrous oxide (N_2O) which often require more prolonged chamber deployment periods in order to facilitate analytical measurement. Despite widespread recognition of this limitation, there is little consensus regarding practical approaches to either estimating or reducing the magnitude of this effect. Rochette and Eriksen-Hamel (2008) showed that there is wide variation in method protocols used to determine soil N_2O flux, details of which can directly affect chamber-induced bias. Recent analysis has shown that intra- and inter-site flux comparisons can be confounded by chamber-induced artifacts (Venterea, 2010). Since NSS chambers generally tend to underestimate the actual PDF, this raises the likelihood that current emissions assessments at the regional, national, and global scale are negatively biased. This has important implications for effective management and mitigation of GHG emissions. The main objectives of this chapter are to (1) describe the physical basis of the chamber effect and how measurement methods and soil properties influence this effect, (2) describe currently available options for quantifying and minimizing flux estimation errors, and (3) briefly discuss the path toward solutions to this problem.

PHYSICAL BASIS FOR THE CHAMBER EFFECT

Most if not all researchers using NSS chambers are well aware of the general problem posed by the chamber effect. Its fundamental basis is that gas exchange between soil and the atmosphere is driven in most circumstances by diffusion (Hutchinson and Livingston, 2002). We know from Fick's law that the diffusive flux (F, with units[a] of $M\ L^{-2}$ soil T^{-1}) is proportional to the concentration gradient (dC/dz) at the soil–atmosphere interface, where the gradient is defined as the derivative of the gas concentration (C, $M\ L^{-3}$ gas) in the direction of the flux, or, under steady-state conditions

$$F = D_s \frac{dC}{dz} \quad [1]$$

where z indicates vertical distance (L) and D_s (L^3 gas L^{-1} soil T^{-1}) is the soil gas diffusivity. NSS chamber methods rely on placement of an enclosure on the soil surface, within which gases generated in the soil are allowed to accumulate for a given deployment period (DP). This accumulation of gas within the chamber alters the concentration gradient at the soil–atmosphere interface and therefore the soil–atmosphere flux. Thus, we are confronted with the situation where attempting to measure a quantity immediately alters its magnitude, which is a classical problem for other physical measurements such as temperature (Griffiths, 1926). Following chamber deployment, gas concentrations will not only change within the chamber itself, but also within the air-filled pores of the soil beneath the chamber. Depending on the insertion depth of the chamber walls into the soil, this may create horizontal gradients in trace gas concentrations in soil directly under and immediately adjacent to the chamber. This may induce lateral trace gas diffusion within the soil. This effect, as well as leakage of gas to the outside atmosphere from an improperly sealed chamber, may also alter chamber gas concentration dynamics and cause negatively biased estimates (Pedersen et al., 2010).

[a] Unit dimensions are indicated by M for mass, L for length, and T for time. Where appropriate, dimensions are also specified with respect to the quantity described by the unit, i.e. soil, gas, H_2O.

While the soil-to-atmosphere flux depends directly on the concentration gradient, which is a spatial derivative (i.e. (dC/dz) in Eq. [1]), in practice the flux using NSS chambers is calculated based on estimation of a time derivative. That is, flux is determined from the change in trace gas concentration over time in samples of the chamber contents taken at multiple times during the DP. In the simplest case, F can be estimated from

$$F = \frac{dC}{dt}\frac{V}{A} \quad [2]$$

where dC/dt is the linear regression (LR)-based slope of the chamber gas concentration versus time (t), V is the internal chamber volume (L^3 gas), and A is the soil surface area (L^2 soil) enclosed by the chamber. It is also convenient to define the chamber height (H) as the ratio of V to A, having units of L^3 gas L^{-2} soil which are commonly simplified to units of L (e.g. cm).

In cases where the direction of gas flux is from soil to atmosphere, which is usually considered to be a flux in the "positive" direction, gas concentration will increase in the chamber over time. This will cause the vertical gradient at the soil–atmosphere interface, (dC/dz) in Eq. [1], and therefore the flux (F) to decrease over time. This in turn will cause the time derivative, (dC/dt) in Eq. [2], obtained from measurements of chamber gas concentration, to decrease with time leading to the type of curvature shown in Figure 19.1 for a positive flux. Conversely, in cases where the flux is from atmosphere to soil (i.e. a "negative" initial flux), gas concentration will decrease in the chamber with time, thereby causing (dC/dz) at the soil–atmosphere interface and F to increase (i.e. become less negative) over time. This in turn causes (dC/dt) in Eq. [2] to increase with time, leading to the type of curvature shown in Figure 19.1 for a negative flux.

The remainder of this chapter will focus on the case of positive flux, which is generally found for carbon dioxide (CO_2) and N_2O. The approaches to be discussed here are not expected to apply to negative fluxes, which are commonly observed for gases including methane and nitric oxide. Such negative fluxes are influenced by biologically mediated gas consumption within the soil. Emissions of N_2O can also be influenced by biological transformation, i.e. denitrification. Using process-based modeling, Venterea et al. (2009) showed that biological consumption of N_2O in the soil profile is not likely to affect chamber N_2O concentration dynamics, except under extreme conditions. However, the significance of negative N_2O fluxes is currently a topic of research (Chapuis-Lardy et al., 2007). Senevirathna et al. (2007) developed a detailed numerical model that may provide some guidance for addressing the influence of biological consumption on chamber dynamics and flux calculations.

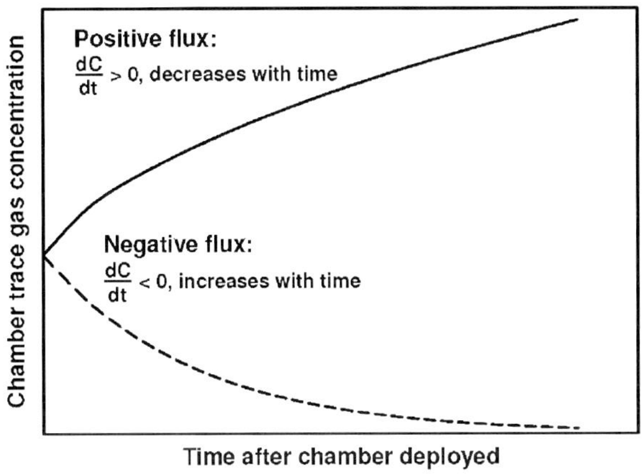

FIGURE 19.1
Hypothetical, generalized shapes of chamber time series data as affected by alteration of the concentration gradient at the soil–atmosphere interface for the case of positive and negative fluxes.

It is logical to expect that the degree of bias in a flux estimate based on Eq. [1] will increase when longer DPs are used, since dC/dt will be expected to change to a greater degree as time increases (Figure 19.1). Also, for a given DP, a smaller value of H would be expected to result in a greater increase in chamber concentration for a given flux, due to less dilution by the air initially present in the smaller chamber. Thus, it is also logical to expect that the degree of bias in a flux estimate will increase as H decreases. While these two general guidelines (i.e. bias increases with longer DP and smaller H) can be deduced based on a qualitative understanding of the chamber effect, they do not provide any quantitative information or practical guidance regarding the magnitude of the bias associated with any particular set of deployment conditions. They also do not indicate which set of measurement conditions will provide a reasonable or acceptable degree of bias.

EXPERIMENTAL APPROACHES

Experimental approaches have been used to attempt to directly measure the degree of chamber-induced bias. The main challenge of an experimental approach is that, at least under field conditions, it is generally impossible to have knowledge of the actual PDF, since any attempt to measure it will necessarily modify it. Thus, laboratory investigations have attempted to utilize various experimental systems designed to create a flux of known magnitude through a porous medium (Ney et al., 1994; Pumpanen et al., 2004; Widen and Lindroth, 2003; Butnor and Johnsen, 2004; Martin et al., 2004; Butnor et al., 2005). Limitations of this approach have been two-fold: (1) it is problematic to create an experimental system which maintains a known and steady flux for sufficient duration while also maintaining diffusion as the dominant transport process (i.e. minimizing pressure gradients); and (2) any relationships obtained between the measured flux and the PDF will likely be limited in applicability to the particular set of measurement conditions evaluated (i.e. DP, H, and media properties).

NON-LINEAR FLUX CALCULATION SCHEMES

Due to the limitations described above with regard to experimental methods, theoretical and numerical approaches have been used to develop bias correction techniques. Currently, the most commonly used approach for bias correction is to apply a flux calculation (FC) scheme that estimates the PDF by fitting a non-linear function to the chamber time series data. The available non-linear FC schemes include what we will refer to as (1) Type I, or entirely empirical approaches, which include polynomial functions which are not derived from any particular theoretical considerations, e.g. the quadratic model of Wagner et al. (1997); (2) Type II, or semi-theoretical approaches based on relatively simplified gas transport theory, in particular the model of Hutchinson and Mosier (1981) (HM), and a revised version of the HM model (HMR) of Pedersen et al. (2010); and (3) Type III, or theoretical approaches that are based on more rigorous gas transport theory, e.g. Livingston et al. (2006) and Venterea (2010). Type I and II schemes use various means to estimate the value of (dC/dt) at the moment of chamber deployment ($t = 0$), and then use this estimate in Eq. [1] to calculate F, while Type III methods do not calculate (dC/dt) directly, but utilize different numerical approaches to estimate the PDF.

Prior to the study by Livingston et al. (2006), it was widely assumed that Type I and II FC schemes provided reasonably effective correction of chamber-induced bias. The major advance of Livingston et al. (2006) was derivation of an analytical solution to a soil-gas transport equation describing one-dimensional (1D) diffusive flux of a trace gas from the soil into a closed chamber, assuming certain physical conditions (described below). Under these conditions, Livingston et al. (2006) clearly showed that previously developed Type I and II FC schemes do not fully account for chamber-induced biases.

Livingston et al. (2006) also presented a new (Type III) FC scheme called the non-linear diffusive flux estimator (or NDFE, software available at http://arsagsoftware.ars.usda.gov)

which, as mentioned above, is based on soil-gas transport theory with the following assumptions: (1) flux is driven by diffusion; i.e. pressure gradients are minimal, (2) diffusive flux is primarily 1D in the vertical direction; i.e. horizontal transport is minimal, (3) the chamber atmosphere is homogeneously mixed, (4) leakage out of the chamber is negligible, (5) irreversible consumption of gas in the soil (e.g. biological uptake) or in the chamber (e.g. gas-phase or surface reaction) is negligible, and (6) soil physical properties (i.e. total and air-filled porosity) are vertically uniform. Livingston et al. (2006) concluded that with a properly designed chamber, i.e. having appropriately sized vent tube and chamber insertion depth (Hutchinson and Livingston, 2002), assumptions (1) through (4) are generally valid, and therefore that the NDFE scheme is superior to all others. As mentioned above, assumption (5) appears to be generally valid for CO_2 and N_2O.

However, even when the above assumptions are valid, the NDFE scheme is relatively inefficient to apply to large datasets. The method requires non-linear iterative regression, will often arrive at multiple solutions for a given data set (Venterea and Baker, 2008), and can deliver flux estimates that are much greater than expected (Pedersen et al., 2010). Due to these issues, the NDFE model has seen limited application. Also, the general validity of some of the underlying assumptions have been questioned by Venterea and Baker (2008) (assumption 6) and Pedersen et al. (2010) (assumption 2). In spite of these limitations, the work of Livingston et al. (2006) has turned out to be useful in developing methods that provide at least an approximate estimate of the degree of chamber-induced bias.

BIAS ESTIMATION TECHNIQUES

Venterea et al. (2009) utilized the analytical solution of Livingston et al. (2006) to develop a spreadsheet-based Chamber Error Assessment Tool (CEAT) to estimate the degree of chamber-induced bias which may be useful in designing chambers and selecting DPs. The CEAT spreadsheet is available online at http://www.ars.usda.gov/pandp/docs.htm?docid=18991 or upon request from the authors. Given basic information regarding *H*, DP, and soil properties, CEAT generates theoretical chamber concentration time series data as well as the theoretical flux underestimation (*TFU*, %) of the PDF, where

$$TFU = 100\frac{(PDF - F)}{PDF} \quad [3]$$

The *TFU* is one way to express the bias of the flux estimate. By this definition, a positive value of *TFU* indicates that the *F* has been *underestimated.* For each set of conditions, CEAT estimates *TFU* for *F* calculated using LR as well as using Type I (quadratic) and Type II (HM) non-linear FC schemes.

It is important to note two critical findings of Livingston et al. (2006), i.e. the *TFU* calculated using Eq. [3] is independent of: (1) the PDF, and (2) the vertical distribution of trace gas production in the soil. Both of these findings may be counterintuitive, since it might be expected that, following chamber deployment, (1) a greater PDF would lead to greater and more rapid suppression of the inter-facial concentration gradient, and (2) trace gas production that is located closer to the soil surface may result in more deformation of the soil–atmosphere concentration gradient after chamber placement. For (1), since *TFU* is calculated relative to the PDF, it turns out that *TFU* is independent of the PDF. Finding (2) is more difficult to explain without a mathematical presentation, but was confirmed by Venterea and Baker (2008) using numerical modeling methods under a range of assumed production profiles. Findings (1) and (2) above greatly simplify bias estimation techniques. The result is that when *TFU* is calculated as a function of *H*, DP, and soil physical properties, the result can be generalized to varying flux magnitudes independent of the vertical distribution of gas production within the soil profile. Examples of CEAT output are given in Figure 19.2. These results illustrate that Type I and II non-linear schemes do not completely eliminate

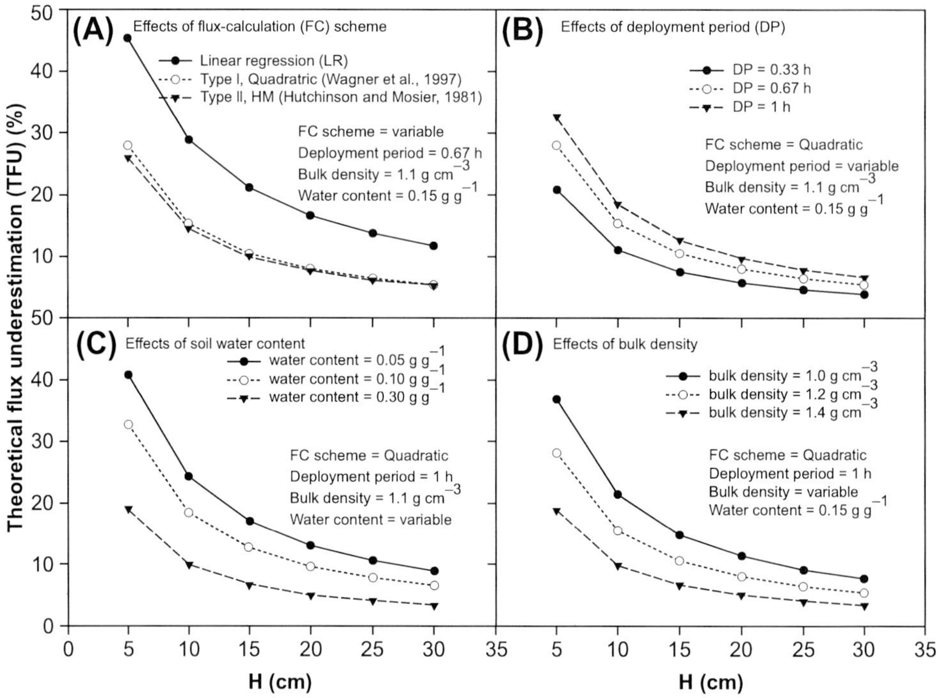

FIGURE 19.2
Theoretical flux underestimation (***TFU***) determined using the Chamber Error Assessment Tool (CEAT) as a function of the chamber volume to area ratio (***H***) and varying (A) flux calculation (FC) schemes, (B) deployment periods (DP), (C) gravimetric water content, and (D) bulk density.

chamber-induced biases. They also show how *TFU* increases with decreasing *H*, increasing DP, and differences in soil properties.

SOIL PROPERTY EFFECTS

As illustrated in Figure 19.2, soil physical properties will influence the magnitude of diffusion-driven, chamber-induced bias. These effects have been noted by Hutchinson et al. (2000), Venterea and Baker (2008) and Venterea et al. (2009). Gas transport theory predicts that increased soil air-filled porosity (ε) leads to increased non-linearity in chamber time series data, which results in increased *TFU*. Thus, if we compare soil profiles that differ with respect to ε but have the same PDF, the profile with greater ε will be expected to result in chamber time series data with more curvature and therefore a greater magnitude of bias (i.e. a greater *TFU*).

This effect can be illustrated with the CEAT spreadsheet tool which can be used to generate simulated chamber time series for different soil profiles assumed to have the same PDF but different physical properties. For example, in Figure 19.3, hypothetical soil profiles 1 and 2 are both assumed to have a PDF of 100 μg N m^{-2} h^{-1} and the same gravimetric water content (0.15 g g^{-1}), but with bulk densities of 1.0 and 1.4 g cm^{-3}, respectively, resulting in ε values of 0.47 and 0.26 cm^3 cm^{-3}, respectively. Comparing the chamber time series data for profiles 1 and 2 with $H = 5$ cm (Figure 19.3, upper curves), profile 1 (with greater ε) displays more non-linearity in the data and thus will lead to greater flux underestimation than profile 2. The CEAT-calculated *TFU*s for these same conditions for DP = 1 h are illustrated in Figure 19.2D. This effect diminishes with increased *H* and decreased DP. Thus, the chamber time series for profiles 1 and 2 tend to converge for $H = 30$ cm and DP < 0.5 (Figure 19.3, lower curves).

Since the effect of soil physical properties on chamber dynamics is not widely recognized, it is worth exploring its physical basis and practical implications. Fundamentally, what is

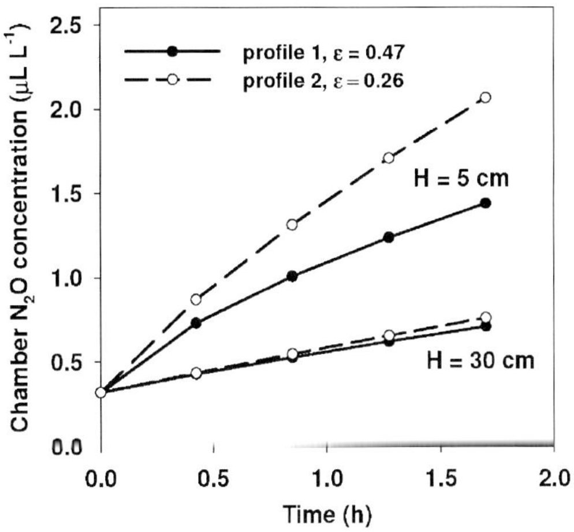

FIGURE 19.3
Theoretical chamber time series data generated using the Chamber Error Assessment Tool (CEAT) for two uniform soil profiles (1 and 2) assuming the same pre-deployment flux (100 μg N_2O-N m^{-2} h^{-1}) and varying chamber volume to area ratios (*H*). Profiles 1 and 2 were assumed to have the same gravimetric water content (0.15 g H_2O g^{-1}) but varying bulk density (1.0 and 1.4 g cm^{-3}, respectively), resulting in varying air-filled porosity (ε).

happening is that following chamber deployment, the soil with greater ε will accumulate more trace gas in the soil pores, therefore allowing less gas to diffuse into the chamber, compared to the soil with lower ε. This effect can be illustrated by dynamic modeling of soil-gas concentration profiles below the chamber following deployment using numerical techniques (e.g. Venterea and Baker, 2008). First of all, the soil with greater ε (in the above example, profile 1) will also have greater soil-gas diffusivity (D_s), since D_s increases with increasing ε (Rolston and Moldrup, 2002). Therefore, prior to chamber deployment, according to Fick's law (Eq. [1]), the soil with greater ε and D_s will have a proportionally smaller gradient (dC/dz) for a given PDF compared to the soil with smaller ε. This results in different initial concentration gradients for profiles 1 and 2, as illustrated in Figure 19.4 for $t = 0$ h.

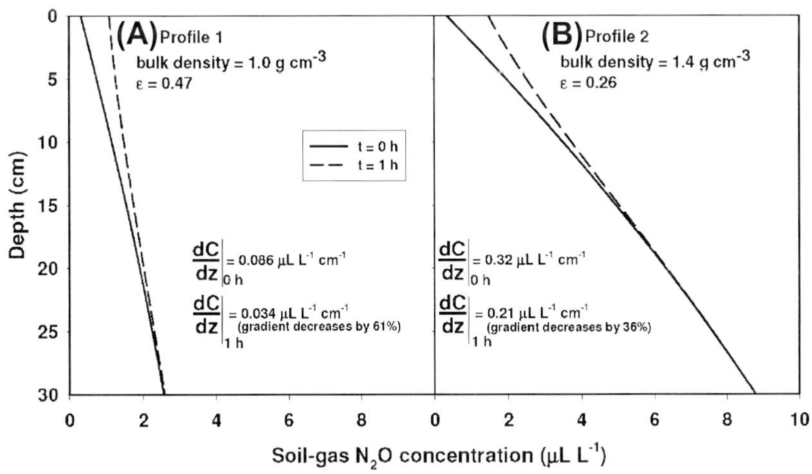

FIGURE 19.4
N_2O soil-gas concentration profiles prior to ($t = 0$ h) and after ($t = 1$ h) chamber deployment for (A) profile 1 and (B) profile 2, assuming pre-deployment flux of 100 μg N_2O-N m^{-2} h^{-1} and chamber volume to area ratio (*H*) of 5 cm. Values of the vertical gradient at the soil–atmosphere interface (dC/dz) were calculated from the linear regression slope over the 0 to 5 cm depth. *Data were generated using the model of Venterea and Baker (2008).*

When the chamber is deployed, gas accumulates at the soil surface (i.e. in the chamber) causing the soil–atmosphere concentration gradients (dC/dz) in both soils to be deformed (decreased) from their original pre-deployment state. Diffusion theory predicts that the initially smaller gradient present in the less dense, high-ε soil will be deformed to a greater extent (profile 1, decreased by 61%) compared to the gradient in the more dense, low-ε soil (profile 2, decreased by only 21%), as illustrated in Figure 19.4 for $t = 1$ h. Due both to the greater degree of deformation and the greater ε of the less dense soil (profile 1), the net result is that more gas accumulates within the soil pores of profile 1 compared with profile 2 over a given period of time. Because gas is produced at the same rate in both soils, since they have the same PDF, this means that less gas is available to diffuse into the chamber above profile 1. For the current example, the model calculates that 57% of the N_2O produced in the soil over a 1 h period following chamber deployment accumulates within the soil for profile 1, allowing only 43% of the N_2O produced to diffuse into the chamber. In contrast, 37% of the N_2O produced in the soil accumulates within the soil for profile 2, and 63% diffuses into the chamber. These differences in soil-gas storage and diffusion rates lead to the differences in chamber gas concentration data shown in Figure 19.3.

The effect described above can lead to two potentially important experimental artifacts. First, it can confound the comparison of fluxes measured within the same soil profile at different times due to varying water content. That is, since drier conditions will result in greater ε, assuming a relatively constant bulk density over time, greater non-linearity in chamber data and greater potential for underestimating the actual PDF will be expected under drier conditions. As described by Venterea (2010), this effect could lead to an apparent positive correlation between soil water content and gas flux within the same soil, when in fact the observation is an artifact of chamber-induced bias. Second, the soil property effect can confound the comparison of fluxes among soils having different physical properties resulting from either natural variation or experimental manipulation, for example when gas fluxes are compared among soils treated with different tillage or water management regimes (e.g. Omonode et al., 2011), or soils amended with different types or amounts of organic materials (e.g. Rochette et al., 2008). In this case, chamber-induced biases could lead to the mistaken conclusion that soils with greater bulk-density have greater gas fluxes. For the current example illustrated in Figures 19.3–19.4, assuming $H = 5$ cm and DP $= 1$ h, applying the quadratic FC scheme of Wagner et al. (1997) to the chamber data for profile 1 would generate a flux estimate of 63 μg N m^{-2} h^{-1} compared with 81 μg N m^{-2} h^{-1} for profile 2. Thus, the less dense soil appears to have a flux that is 22% less than the more dense soil, while in reality their fluxes are the same. The magnitude of this effect will be influenced by chamber geometry and deployment time. For example, for chamber data using $H = 30$ cm and DP $= 1$ h, the quadratic FC scheme estimates fluxes of 92 and 97 μg N m^{-2} h^{-1} for profiles 1 and 2, respectively, a difference of only 5%.

CHAMBER BIAS CORRECTION (CBC)

The artifacts of soil property effects described above will be most pronounced when LR is used; however, the artifacts do not disappear when Type I or II non-linear FC schemes are used. In theory, the NDFE scheme eliminates these artifacts for a completely uniform soil profile. However, practical and theoretical limitations of the NDFE scheme (described above) are problematic. Venterea (2010) developed a chamber bias correction (CBC) technique that overcomes some of the limitations of the NDFE method. The CBC method is based on the same theory as Livingston et al. (2006), but is readily adapted to spreadsheets, delivers a single result, and reduces the frequency of suspiciously large flux estimates. An outline of the CBC method is given below, and full details for its application are provided in the Appendix. The CBC method can be used in one of two ways: (1) a graphical option that can be used for approximating the degree of bias associated with a particular chamber method applied to a particular soil, similar to CEAT; or (2) a fully numerical option can be used for making

TABLE 19.1 Input Parameters Required for Applying the Chamber Bias Correction (CBC) Method, and Parameter Values for Example Soil Profiles

Parameter	Symbol	Units	Example profiles 1	2	3[‡]
Bulk density	ρ	g dry soil cm^{-3} bulk soil	1.0	1.4	1.19
Particle density	ρ_p	g dry soil cm^{-3} soil particle	2.65	2.65	2.65
Water content	θ	cm^3 H_2O cm^{-3} soil	0.15	0.21	0.14
Soil temperature	T_s	°C	20	20	20
Clay fraction	CF	g clay g^{-1} soil	0.22	0.22	0.22
Soil pH[†]	pH	—	na	na	na
Volume-to-area ratio	H	cm^3 gas cm^{-2} soil	5	5	10
Deployment period	DP	h	1.0	1.0	0.85, 1.7[§]

[†] *Soil pH is needed for CO_2 only (see Appendix).*
[‡] *Profile 3 has non-uniform soil properties (see Figure 19.5). Bulk density and water content values in the table are for samples taken from 0–10 cm depth interval.*
[§] *Two different values of DP (0.85 and 1.7 h) are examined for profile 3.*

quantitative bias corrections. In the latter case, since many of the required properties (Table 19.1) will vary over time, the most accurate results will be obtained if the properties are measured with an appropriate degree of temporal resolution to capture this variation. Additionally, some of these properties (e.g. bulk density, water content, temperature) also tend to vary with depth. This issue of vertical non-uniformity in soil properties is addressed in the examples given below.

There are three general steps involved in applying the CBC method for both the graphical and numerical options:

1. The initial flux estimate (F) is calculated in the conventional manner using either LR, the Type I quadratic method (Wagner et al., 1997), or the Type II HM method (Hutchinson and Mosier, 1981).
2. The degree of bias associated with the initial flux estimate is determined and expressed as *TFU* (% basis)
3. The PDF is determined by rearranging Eq. [3] and solving for PDF knowing F and *TFU*.

Step 2 (determining *TFU*) consists of calculating values of two scaling parameters E_1 and E_2, which are then used together to calculate *TFU*. The parameter E_1 accounts for soil property effects, while the parameter E_2 accounts for the effects of H and DP. The difference between the more approximate graphical option and the more rigorous numerical option is that with the graphical option the user determines E_1 visually using nomograms (see Venterea, 2010), while the numerical option uses physically based calculations. While the numerical method for determining E_1 is mathematically straightforward, it involves several calculation steps. These steps are presented in the Appendix, and a spreadsheet template containing the calculations is available for downloading at http://www.ars.usda.gov/pandp/people/people.htm?personid=31831 or upon request from the authors. For the graphical technique, once E_1 is obtained, the steps starting with Step 2c and Eq. [A8] in the Appendix can be followed.

As examples, the chamber time series data given in Figure 19.3 for the vertically uniform soil profiles (1 and 2) used for the previous example, as well as an additional, vertically non-uniform soil profile (profile 3), were analyzed using the CBC method. Profile 3 has vertically variable bulk density and water content (and therefore variable ε), as shown in Figure 19.5A, based on high-resolution measurements in a soil used for corn/soybean production under a moldboard plow tillage regime (Venterea and Baker, 2008). Chamber concentration time series date for profile 3 generated using the numerical model of

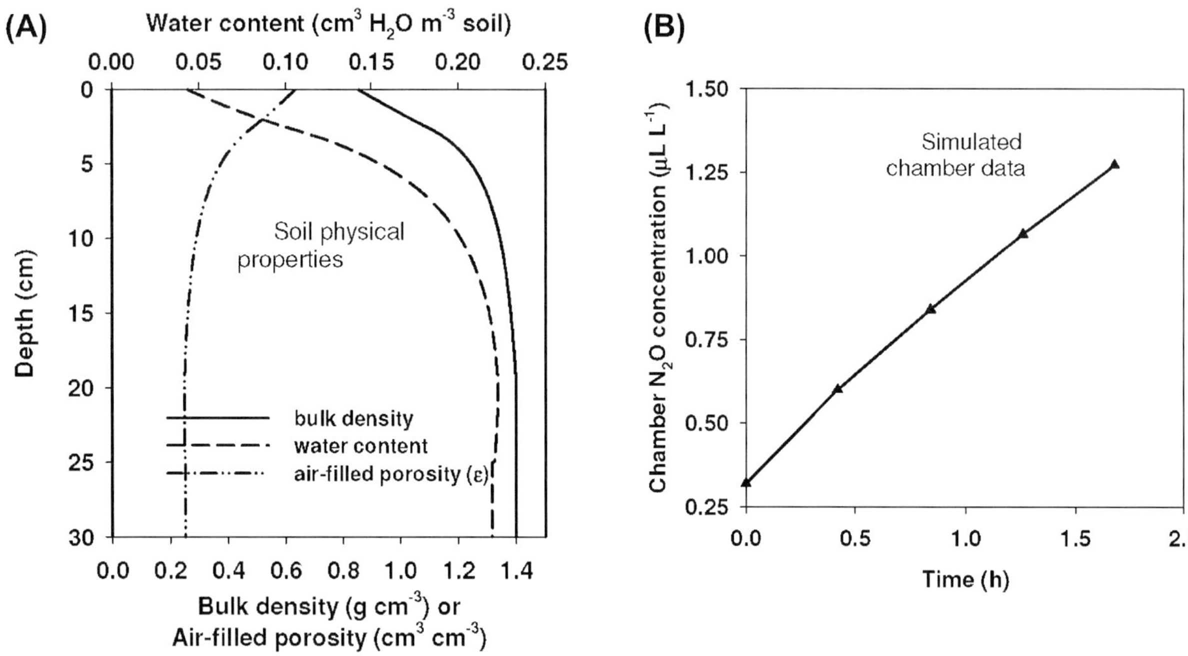

FIGURE 19.5
(A) Vertical profiles of bulk density, water content and air-filled porosity (ε) for profile 3; (B) chamber concentration data for profile 3 simulated using the model of Venterea and Baker (2008) assuming a pre-deployment flux (PDF) of 100 g m^{-2} h^{-1} and chamber volume-to-area ratio (*H*) of 10 cm. Lines in (A) were obtained by non-linear regression of measured bulk density and water content in a soil used for corn/soybean production under a moldboard plow tillage regime *(Venterea and Baker, 2008).*

Venterea and Baker (2008) are shown in Figure 19.5B, assuming a PDF of 100 μg N m^{-2} h^{-1} and $H = 10$ cm.

Input parameter values required by the CBC method for each of the three profiles are given in Table 19.1. Obviously, for a non-uniform soil (e.g. profile 3), values of the required input parameters will depend on the sampling depth considered. Analysis of this and other non-uniform soil profiles has indicated that input parameters based on a 10 cm sampling depth will produce accurate *TFU*-correction (Venterea, 2011, unpublished data). Thus, bulk density and water content values that would be obtained for samples taken from 0–10 cm depth interval were used as input parameters for profile 3 (Table 19.1). Values of *F* estimated using the quadratic method of Wagner et al. (1997), and the calculated values of E_1, E_2, *TFU* and PDF using the CBC method for all three soil profiles are given in Table 19.2. The estimated PDF values (last column of Table 19.2) are very close to the theoretical value (100 μg N m^{-2} h^{-1})

TABLE 19.2 Values of the Uncorrected Flux (*F*), the Two Scaling Parameters (*E*1 and *E*2) and the Calculated Pre-Deployment flux (PDF) for Three Different Soil Profiles and Measurement Conditions

Profile	H cm	DP h	F[†] μg N m^{-2} h^{-1}	E_1[‡] cm^6 gas cm^{-4} soil h^{-1}	E_2 —	TFU %	Calculated PDF μg N m^{-2} h^{-1}
1	5	1.0	63.0	53.7	−0.764	36.4	99.0
2	5	1.0	81.0	9.51	0.966	17.8	98.6
3	10	0.85	83.4	36.7	1.177	16.2	99.6
		1.7	77.9	36.7	0.4836	21.9	100.9

[†] *F values determined using the quadratic method of Wagner et al. (1997).*
[‡] *See Appendix for discussion about units for* E_1.

that was used to generate the time series data. In contrast, the F values calculated using the quadratic method underestimated the PDF by 16 to 37% (Table 19.2).

LIMITATIONS OF THE CBC METHOD

One limitation of the CBC method is that it requires values for soil properties listed in Table 19.1 as inputs. Thus, errors in estimation of these properties would necessarily lead to errors in the final flux estimates. Like measurement of any other variable soil property, these errors can be minimized by increasing the spatial and temporal intensity of sampling. Of course, this will increase the time and expense required for making flux estimates. To date, there has not been a sensitivity assessment of how errors in estimating these properties will translate into errors in flux estimation using the CBC method. This needs to be addressed in future studies.

Another potentially important limitation is the assumption of 1D vertical diffusion (assumption 2) within the theory underlying the CBC method. Livingston et al. (2006) concluded that the 1D assumption was reasonable in the case where chambers have wall-insertion depths that are properly sized with respect to soil physical properties. That is, soils expected to have greater ε values should use chambers with deeper insertion depths. Guidelines put forth by Hutchinson and Livingston (2001, 2002) recommend wall depths of at least 5 cm, and as much as 20 cm, depending on soil properties and DPs. Pedersen et al. (2010) raised the possibility that, at least under certain conditions, the 1D assumption may not be reasonable. Pedersen et al. (2010) pointed out that the theory underlying the NDFE and CBC methods would predict chamber gas concentrations will continue to increase infinitely with extended DPs, but that this prediction is not consistent with empirical observation. Pedersen et al. (2010) also put forth a modification of the Hutchinson and Mosier (1981) FC scheme (the HMR model) which attempts to account for the possibility of both lateral (i.e. 2D) diffusion and leakage from an imperfectly sealed chamber (search HMR at http://www.r-project.org/). Future studies are required to evaluate the robustness of the HMR model.

BIAS VERSUS PRECISION

This chapter has focused on *bias* estimation and correction, but another important consideration with regard to flux estimation errors is the *precision* of the method. While bias is the degree to which a flux estimate differs from the PDF, precision is the degree to which repeated estimates agree with each other (Figure 19.6). Practices that decrease the bias of flux estimates will also generally decrease the precision of those estimates (Venterea et al., 2009). Reducing the deployment period (DP), increasing the chamber height (H), and utilizing non-linear FC schemes (relative to LR) all serve to reduce the bias of flux estimates. However, these same practices also reduce the precision of the flux estimates. The degree to which each of these practices will decrease the precision of the overall flux estimates will depend upon the precision of the sampling and analytical procedure used to determine the chamber headspace gas concentration. Nonlinear models such as the quadratic, HM and HMR methods are inherently more sensitive to small deviations in headspace concentration data. Thus, an inherent trade-off exists between the bias and precision of flux estimates obtained using a particular chamber configuration, deployment time, and FC scheme in a given soil (Venterea et al., 2009; Parkin and Venterea, 2010).

CURRENT RECOMMENDATIONS

Gas transport theory predicts that variations in chamber methods and soil physical properties will affect not only the absolute magnitude of flux estimates, but also the relative difference in fluxes among soils that differ with regard to physical properties. This factor, together with the inherent trade-off between bias and precision described above, make it impossible to

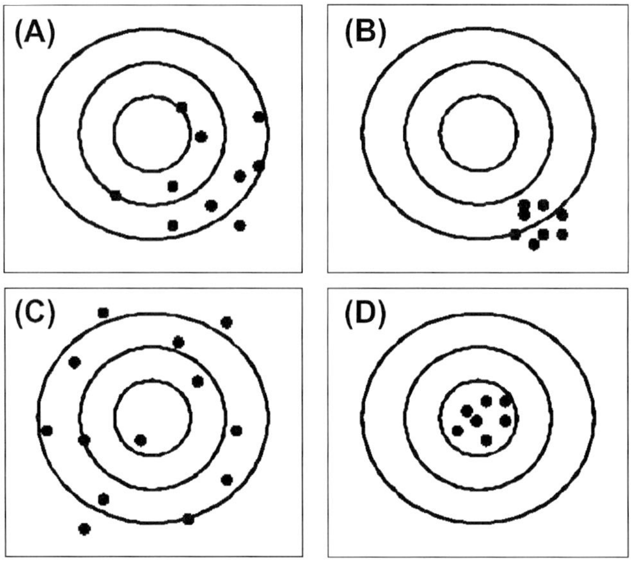

FIGURE 19.6
Diagram illustrating bias and precision of chamber methods. Bias is illustrated by how close the flux estimates (points) cluster around the target bull's-eye, which represents the pre-deployment flux (PDF), and precision is illustrated by the degree of agreement in flux estimates. The following scenarios are illustrated: (A) high bias/low precision, (B) high bias/high precision, (C) low bias/low precision, and (D) low bias/high precision (the ideal outcome).

recommend any single, uniform approach with regard to chamber geometry, deployment time, or flux calculation that will be optimal across all measurement conditions. However, based on our current understanding, the following practices and considerations relative to chamber-induced biases are recommended:

1. In cases where soil property effects would be expected to confound data interpretation (as discussed above), or in cases where estimates of the *absolute* flux magnitude is of primary importance, measures described above to evaluate the degree of bias (e.g. using the CEAT method, especially when initially designing chambers) or correct for bias (e.g. using the CBC method) should be considered. Due to natural variation in soil water content, this will include many if not the vast majority of situations.
2. While many chamber versus time data sets appear to be highly linear, gas transport theory has shown that even data displaying r^2 values greater than 0.99 can result in substantial flux underestimation using LR. Thus, while it is tempting to apply LR, it is not recommended as a primary method unless combined with bias correction techniques. Several researchers have used the approach of making preliminary or periodic evaluations of linearity to justify using LR and also to justify using a reduced number of sampling points based on the assumption of linearity. However, because soil conditions (e.g. water content) can directly affect the degree of linearity, this approach is also not recommended. Thus, non-linear FC schemes are recommended, using four time points at a minimum.
3. While we cannot at this time recommend any one specific FC scheme, analysis to date indicates that the quadratic FC scheme (Wagner et al., 1997) generates flux estimates with lower bias than LR but greater precision than the HM method, when both the quadratic and HM schemes are applied with a reasonable model failure criteria (Venterea et al., 2009). Thus, an argument can be made that the quadratic method best balances the bias and precision trade-off. An additional benefit is that the quadratic method does not require equally spaced time points. Preliminary analysis has indicated that selection of the optimal FC scheme based on the mean square error (which incorporates both bias and precision) may vary depending on measurement conditions (i.e. H, DP, soil properties) (Parkin and

Venterea, 2010). Ongoing analysis is evaluating the benefits of the newly developed HMR scheme, which also does not require equally spaced time points (Parkin et al., unpublished data).

4. The precision of the sampling and analytical method used to determine chamber gas concentrations should be quantified by making repeated (i.e. $\geq$20) measurements of samples of standard gases that are collected and analyzed as similarly as possible to actual chamber samples. The sampling and analytical precision can then be used to determine (1) the overall precision of the flux estimate accounting for DP, H, and soil properties, using the method of Venterea et al. (2009), and (2) the minimum detectable flux using the methods of Parkin et al. (2012).

In addition to the above recommendations, the reader is also directed to Parkin and Venterea (2010) and Hutchinson and Livingston (2002).

ULTIMATE SOLUTIONS

The bias-precision trade-off problem would disappear if sampling and analytical error were eliminated, in which case very short DPs and large H values together with Type II non-linear FC schemes could largely eliminate chamber-induced bias. While it is not realistic to expect complete error elimination, advances in measurement instrumentation have the potential to greatly reduce errors. Relatively low-cost instrumentation is available that can measure near-ambient CO_2 concentrations with relatively high precision. That is, real-time infrared analyzers can be used in conjunction with NSS chambers to make flux estimates using DPs less than 5 min which can largely eliminate chamber-induced biases with acceptable precision (e.g. Davidson et al., 2002). While high-precision, high-sensitivity analyzers for use with N_2O flux measurements have been introduced and have potential to achieve the same result (e.g. Laville et al., 2011), the reliability of some of these analyzers has been questioned (e.g. Flechard et al., 2007; Yamulki and Jarvis, 1999) and will require further testing. Also, high-precision analyzers generally require real-time, in-situ analysis, in contrast with more commonly used techniques such as gas chromatography which require collection of discrete gas samples for subsequent analysis. While real-time analysis has its advantages, it also has limitations in that only one chamber location can be measured at one time per analyzer in contrast with discrete sample collection techniques which can utilize rotational sampling regimes to sample multiple locations at one time. Use of high-precision N_2O analyzers in combination with automated chamber technology could partly overcome this limitation although the degree of replication that is practical with automated chambers is generally less than can be achieved with manual chambers. While micrometeorological methods for N_2O flux determination avoid chamber-related issues altogether, they have their own set of limitations (Hutchinson and Livingston, 2002). In addition, the practical utility of micrometeorological methods for use in replication plot studies aimed at detecting statistically significant treatment effects is limited.

References

Butnor, J.R., Johnsen, K.H., 2004. Calibrating soil respiration measures with a dynamic flux apparatus using artificial soil media of varying porosity. Eur. J. Soil Sci. 55, 639–647.

Butnor, J.R., Johnsen, K.H., Maier, C.A., 2005. Soil properties influence estimates of soil CO_2 efflux from three chamber-based measurement systems. Biogeochemistry 73, 283–301.

Chapuis-Lardy, L., Wrage, N., Metay, A.L., Chotte, J., Bernoux, M., 2007. Soils, a sink for N_2O? A review. Global Change Biol. 13, 1–17.

Davidson, E.A., Savage, K., Verchot, L.V., Navarro, R., 2002. Minimizing artifacts and biases in chamber-based measurements of soil respiration. Agr. Forest Meteorol. 113, 21–37.

Flechard, C., Ambus, P., Skiba, U., Rees, R., Hensen, A., van Amstel, A., et al., 2007. Effects of climate and management intensity on nitrous oxide emissions in grassland systems across Europe. Agricult. Ecosys. Environ. 121, 135–152.

Fuller, E.N., Schettler, P.D., Giddings, J.C., 1966. A new method for prediction of binary gas-phase diffusion coefficients. Ind. Eng. Chem. 58, 19–27.

Griffiths, E., 1926. Methods of Measuring Temperature, second ed. Charles Griffin and Co, London, p. 203.

Healy, R.W., Striegl, R.G., Russell, T.F., Hutchinson, G.L., Livingston, G.P., 1996. Numerical evaluation of static-chamber measurements of soil–atmosphere gas exchange: identification of physical processes. Soil Sci. Soc. Am. J. 60, 740–747.

Hutchinson, G.L., Livingston, G.P., Healy, R.W., Striegl, R.G., 2000. Chamber measurement of surface-atmosphere trace gas exchange: numerical evaluation of dependence on soil, interfacial layer, and source/sink properties. J. Geophys. Res. Atm. 105 (D7), 8865–8875.

Hutchinson, G.L., Mosier, A.R., 1981. Improved soil cover method for field measurement of nitrous oxide fluxes. Soil Sci. Soc. of Am. J. 45, 311–316.

Hutchinson, G.L., Livingston, G.P., 2001. Vents and seals in nonsteady-state chambers used for measuring gas exchange between soil and the atmosphere. Eur. J. Soil Sci. 52, 675–682.

Hutchinson, G.L., Livingston, G.P., 2002. Soil-atmosphere gas exchange. In: Dane, J.H., Topp, G.C. (Eds.), Methods of Soil Analysis. Part 4. SSSA Book Ser. 5. SSSA, Madison, WI, pp. 1159–1182.

Hutchinson, G.L., Rochette, P., 2003. Non-flow-through steady-state chambers for measuring soil respiration: numerical evaluation of their performance. Soil Sci. Soc. Am. J. 67, 166–180.

Kroon, P.S., Hensen, A., van den Bulk, W.C.M., Jongejan, P.A.C., Vermeulen, A.T., 2008. The importance of reducing the systematic error due to non-linearity in N_2O flux measurements by static chambers. Nutrient Cycl. Agroecosys 82, 175–186.

Kutzbach, L., Schneider, J., Sachs, T., Giebels, M., Nykänen, H., Shurpali, N.J., Martikainen, P.J., Alm, J., Wilmking, M., 2007. CO_2 flux determination by closed-chamber methods can be seriously biased by inappropriate application of linear regression. Biogeosci. Disc. 4, 2279–2328.

Laville, P., Lehuger, S., Loubet, B., Chaumartin, F., Cellier, P., 2011. Effect of management, climate and soil conditions on N_2O and NO emissions from an arable crop rotation using high temporal resolution measurements. Agr. Forest Meteorol 151, 228–240.

Livingston, G.P., Hutchinson, G.L., Spartalian, K., 2006. Trace gas emission in chambers: a non-steady-state diffusion model. Soil Sci. Soc. Am. J. 70, 1459–1469.

Martin, J.G., Bolstad, P.V., Norman, J.M., 2004. A carbon dioxide flux generator for testing infrared gas analyzer-based soil respiration systems. Soil Sci. Soc. Am. J. 68, 514–518.

Matthias, A.D., Yarger, D.N., Weinback, R.S., 1978. A numerical evaluation of chamber methods for determining gas fluxes. Geophys. Res. Lett. 5, 765–768.

Ney, M.S., Mattson, K.G., Bormann, B.T., 1994. Biases of chamber methods for measuring soil CO2 efflux demonstrated with a laboratory apparatus. Ecology 75, 2460–2463.

Omonode, R.A., Smith, D.R., Gál, A., Vyn, T.J., 2011. Soil nitrous oxide emissions in corn following three decades of tillage and rotation treatments. Soil Sci. Soc. Am. J. 75, 152–163.

Parkin, T.B., Venterea, R.T., 2010. USDA-ARS GRACEnet Project Protocols. Chapter 3. Chamber-Based Trace Gas Flux Measurements (replaces 2003 version). Available at http://www.ars.usda.gov/SP2UserFiles/Program/212/Chapter%203.%20GRACEnet%20Trace%20Gas%20Sampling%20Protocols.pdf.

Parkin, T.B., Venterea, R.T., Hargreaves, S.K., 2012. Calculating the detection limits of chamber-based soil greenhouse gas flux measurements. J. Environ. Qual. (In press).

Pedersen, A.R., Petersen, S.O., Schelde, K., 2010. A comprehensive approach to soil atmosphere trace-gas flux estimation with static chambers. Eur.. J. Soil Sci. 61, 888–902.

Pumpanen, J., Kolari, P., Ilvesniemi, H., Minkkinen, K., Vesala, T., Niinistö, S., Lohila, A., et al., 2004. Comparison of different chamber techniques for measuring soil CO_2 efflux. Agricult. For. Meteorol 123, 159–176.

Rochette, P., Eriksen-Hamel, N.S., 2008. Chamber measurements of soil nitrous oxide flux: are absolute values reliable? Soil Sci. Soc. Am. J. 72, 331–342.

Rochette, P., Angers, D.A., Chantigny, M.H., Gagnon, B., Bertrand, N., 2008. N_2O fluxes in soils of contrasting textures fertilized with liquid and solid dairy cattle manures. Can. J. Soil Sci. 88, 175–187.

Rolston, D.E., Moldrup, P., 2002. Gas diffusivity. In: Dane, J.H., Topp, G.C. (Eds.), Methods of Soil Analysis. Part 4. SSSA Book Ser. 5. SSSA, Madison, WI, pp. 1113–1139.

Sander, R., 1999. Compilation of Henry's Law Constants for Inorganic and Organic Species of Potential Importance in Environmental Chemistry (Version 3). http://www.henrys-law.org (confirmed August 27, 2009).

Senevirathna, D.G.M., Achari, G., Hettiaratchi, J.P.A., 2007. A mathematical model to estimate errors associated with closed flux chambers. Environ. Model. Assess. 12, 1–11.

Snoeyink, V.L., Jenkins, D., 1980. Water Chemistry. John Wiley & Sons, New York.

Venterea, R.T., 2010. Simplified method for quantifying theoretical underestimation of chamber-based trace gas fluxes. J. Environ. Qual. 39, 126–135.

Venterea, R.T., Baker, J.M., 2008. Effects of soil physical nonuniformity on chamber-based gas flux estimates. Soil Sci. Soc. Am. J. 72, 1410–1417.

Venterea, R.T., Spokas, K.A., Baker, J.M., 2009. Accuracy and precision analysis of chamber-based nitrous oxide gas flux estimates. Soil Sci. Soc. Am. J. 73, 1087–1093.

Wagner, S.W., Reicosky, D.C., Alessi, R.S., 1997. Regression models for calculating gas fluxes measured with a closed chamber. Agron. J. 89, 279–284.

Widen, B., Lindroth, A., 2003. A calibration system for soil carbon dioxide-efflux measurement chambers: description and application. Soil Sci. Soc. Am. J. 67, 327–334.

Yamulki, S., Jarvis, S.C., 1999. Automated chamber technique for gaseous flux measurements: evaluation of a photoacoustic infrared spectrometer trace gas analyzer. J. Geophys. Res. 104, 5463–5469.

APPENDIX: STEPS AND CALCULATIONS USED IN CBC METHOD

Note: A spreadsheet that can be used as a template for the calculations given below is available for downloading at http://www.ars.usda.gov/pandp/people/people.htm?personid=31831 or upon request from the authors.

Note about units: The required parameter inputs with units and symbols are listed in Table A1. Units selected for the soil input parameters need to be consistent with the units used for *H* and DP. For example, in Table A1, all *L* units are in cm and *T* units are in h. Thus[b], h must be used as the time unit for DP and cm as the length unit for *H*. This will result in a dimensionless term inside the parentheses and for E_2 in Eq. [A8]. Any set of units can be used to calculate the initial uncorrected flux estimate (*F*), i.e. units for *F* do not have to be consistent with units for E_1, *H*, or DP, and the corrected PDF will always have the same units as *F*.

Note about high-organic or high-clay soils: For soils with clay content >40% or organic matter >5%, Eq. [A7] may need to be modified for greater accuracy. For these cases, the reader is referred to the theory section of Venterea (2010).

Steps for using the chamber-bias correction technique of Venterea (2010):

1. Calculate *F* in the conventional manner using either linear regression, the Type I quadratic method of Wagner et al. (1997), or the Type II method of Hutchinson and Mosier (1981) together with Eq. [1].

2a. Calculate the following intermediate parameters needed for determining E_1:
Volumetric water content (θ) (if only gravimetric water content θ_g is known):

$$\theta = \theta_g \rho \tag{A1}$$

Total porosity (φ):

$$\varphi = \left(1 - \frac{\rho}{\rho_p}\right) \tag{A2}$$

Henry's law gas–liquid partitioning coefficient (*K*):

$$K = K_{25} \exp\left[\chi\left(\frac{1}{T_s + 273.15} - \frac{1}{298.15}\right)\right] \tag{A3}$$

where K_{25} is the Henry's constant at 25°C, and χ is a temperature response factor (Sander, 1999). Values for K_{25} and χ are listed in Table A1.

[b] More rigorously, the actual units of *H* are not cm but really cm^3 gas cm^{-2} soil. Thus, the actual units of E_1 in the examples are cm^6 gas cm^{-4} soil h^{-1} (Table 19.2).

TABLE A1 Parameter Values for use in Eqs. [A3] and [A4]

	K_{25}[†]	K[†]	D_{25}[‡]
Gas	cm^3 gas cm^{-3} H_2O	K	$cm^2\ h^{-1}$
N_2O	0.6116	2600	511.7
CO_2	0.8318	2400	652.3

[†] *Sander (1999).*
[‡] *Fuller et al. (1966); Healy et al. (1996).*

Gas diffusivity in free air (D):

$$D = D_{25}\left[\frac{273.15 + T_s}{298.15}\right]^{1.72} \quad \text{[A4]}$$

where D_{25} is the diffusivity at 25°C (Rolston and Moldrup, 2002). Values for D_{25} are listed in Table A1.

Campbell soil pore-size distribution parameter (b):

$$b = 13.6 \quad CF + 3.5 \quad \text{[A5]}$$

where CF is the clay fraction ($0 < CF < 1$) (Rolston and Moldrup, 2002).

Correction factor for pH (applicable to CO_2 only) (β):

$$\beta = 1 + 10^{(pH - pK_a)} + 10^{(2pH - pK_a - pK_b)} \quad \text{[A6]}$$

where $pK_a = 6.42$ and $pK_b = 10.43$ are the equilibrium constants for dissociation of carbonic acid and bicarbonate, respectively (values at 25°C per Snoeyink and Jenkins, 1980). If desired, values of pK_a and pK_b at varying temperature can be found in Table 4-7 of Snoeyink and Jenkins (1980). The β parameter accounts for the formation of soluble carbonate species from dissolved CO_2 which will also influence chamber CO_2 dynamics (Hutchinson and Rochette, 2003). Since pH is not expected to affect N_2O gas–liquid partitioning, β should be set equal to 1 for N_2O. For CO_2, β reduces to 1 for pH less than approximately 5.0.

2b. Once the above intermediate terms are calculated, E_1 can be calculated from

$$E_1 = [\varphi + \theta(\beta K - 1)]D\varphi^2(1 - \theta/\varphi)^{(2+3/b)} \quad \text{[A7]}$$

2c. E_2 is then calculated from

$$E_2 = \ln\left(\frac{H^2}{E_1 DP}\right) \quad \text{[A8]}$$

TABLE A2 Regression Coefficients Required for Calculating the Theoretical Flux Underestimation (TFU) in Eq. [A9] for Different Flux-Calculation Schemes

	Regression coefficient			
Flux-calculation scheme	*a*	*b*	*c*	*d*
LR	44.3456	−5.5105	0.1799	0.0363
HM[†]	25.0140	−3.2561	0.2772	0.0439
Quadratic[‡]	26.8575	−3.5666	0.2814	0.0471

[†] *Model of Hutchinson and Mosier (1981).*
[‡] *Model of Wagner et al. (1997).*

2d. TFU is then determined from

$$TFU = \frac{a + bE_2}{1 + cE_2 + dE_2^2} \quad [A9]$$

where E_2 is dimensionless, TFU is expressed as a percentage, and a, b, c, and d are regression coefficients specific to each FC scheme, as shown in Table A2.

3. And finally, the PDF is determined from

$$PDF = \frac{F}{\left(1 - \frac{TFU}{100}\right)} \quad [A10]$$

where PDF has the same units as F.

CHAPTER 20

Advances in Spectroscopic Methods for Quantifying Soil Carbon

James B. Reeves, III[1], Gregory W. McCarty[2], Francisco Calderon[3], W. Dean Hively[4]
[1]USDA-Agricoltural Research Service, Environmental Management and Byproduct Utilization Laboratory, Beltsville, MD
[2]USDA-Agricoltural Research Service, Hydrology and Remote Sensing Laboratory, Beltsville, MD
[3]USDA-Agricoltural Research Service, Central Great Plains Research Station, Akron, CO
[4]USGS Eastern Geographic Science Center, Reston, VA

CHAPTER OUTLINE

NB: Mention of a trade name, proprietary product, or specific equipment does not constitute a quarantine or warranty by the U.S. Department of Agriculture or by the U.S. Geological Survey and does not imply its approval to the exclusion of any other products that may be suitable.

Abbreviations: ATR, attenuated total reflection; CAI, cellulose absorption index; DRIFTS, diffuse reflectance infrared (IR) Fourier transform (FT) spectroscopy; mid-IR, mid-infrared (2500−25,000 nm); MLR, multi-linear regression; NDVI, normalized difference vegetative index; NIR, near-infrared (1100−2498 nm); NIRS, NIR spectroscopy; NN, neural nets; PCR, principal component regression; PLS, partial least squares; SW-NIR, shortwave NIR (400−1098 nm); SOC, soil organic carbon; SOM, soil organic matter; VIS, visible (400−700 nm).

INTRODUCTION

The objective of this discussion is to cover the basics and present state of using spectroscopy to determine the content and composition of soil carbon (SC). For the purposes of this discussion near-infrared (NIR) will include any spectral data from 400 to 2500 nm (25,000 to

Managing Agricultural Greenhouse Gases. DOI: 10.1016/B978-0-12-386897-8.00020-6

$4000\ cm^{-1}$). This includes the visible region (400 to 700 nm; 25,000 to $14,300\ cm^{-1}$), the region from 700 to 1100 nm (14,300 to $9090\ cm^{-1}$) sometimes named the shortwave NIR or near-NIR, and the 1100 to 2500 nm (9090 to $4000\ cm^{-1}$) region, which is often referred to by the remote sensing community as the shortwave-infrared and others as the NIR. The region from 2500 to 25,000 nm (4000 to $400\ cm^{-1}$) will be considered the mid-infrared (mid-IR). Although most published research studies define the wavelengths associated with their terminology, many advertisements for instrumentation do not, thus just saying that an instrument covers the NIR tells one little. Finally this chapter covers both proximal sensing in the laboratory and field and remote sensing from planes, satellites, etc., but only from the standpoint of carbon measurements. Remote sensing has many other applications related to assessment of greenhouse gas emissions from agricultural ecosystems other than direct measurement of soil carbon, but those are the subject of another chapter. This chapter does not attempt to be a comprehensive review of all the work done over the last several decades on qualitative measurement of organic and mineral fractions of soil using primarily mid-IR spectroscopy and more recently quantitative analysis using both NIR and mid-IR. For general references on NIR spectroscopy (NIRS) readers are referred to several texts (Burns and Ciurczak, 1992, 2001, 2008; Roberts et al., 2004; Williams and Norris, 1987, 2001). For reviews specific to soils see: Bellon-Maurel and McBratney, 2011; chapter 26 (Malley et al., 2004) in the book by Roberts et al., 2004; Reeves, 2010 and Stenberg et al., 2010.

While NIRS for the determination of the composition of agricultural products such as grains, forages, foods, and soils has been in use for at least four decades (Burns and Ciurczak, 1992, 2001, 2008; Roberts et al., 2004; Williams and Norris, 1987, 2001), there is still a great deal of misconception by new practitioners of what is required for accurate results. Many of the same considerations derived from NIRS research will apply if mid-IR and remote sensing applications are to be successful, i.e. wisdom must guide the separation of signal from noise, the use of sensors, and the choice of a robust set of calibration data. The understanding of these basic concepts and characterization of current advances is of much value and the description of the requirements and subtleties of such an understanding is the objective of this chapter.

First and foremost, it must be stated that except for rare occasions, the use of spectroscopic analysis represents a secondary method as understood by soil analytical chemists. By this we mean that for all C content or compositional determinations by NIR or mid-IR, reference data are needed to formulate a predictive calibration. Ideally, this set of soil samples, for which both analytical and spectral data are obtained, will include a robust sampling of the full range of C contents and soil composition that is found in the unknown samples for which analytical values will be predicted. The calibration data can be related to the spectral data using any number of mathematical procedures, which for the most part involve some method of regression analysis where the spectra are the independent or X data and the analytical values are the dependent predicted (Y) data.

Within the research community this approach has come to be known as chemometrics, which can be applied using specific wavelengths selected by stepwise regression methods, or can incorporate entire high-resolution spectra using robust statistical procedures such as partial least squares (PLS) regression, neural networks and many others (Naes et al., 2002; Westerhaus et al., 2004). The development of a calibration between spectral data and analytical data is a considerably more nuanced process than the calibration of a carbon combustion analyzer, where a single sensor element is used to determine the CO_2 content of an oxidized sample (Franks et al., 2001, Nelson and Sommers, 1996). The importance of taking a careful approach to this process cannot be overstated.

NIR versus Mid-IR Spectra

Chemometric methods, which extract quantitative information from spectral reflectance data, have traditionally been applied to NIR spectra, and more recently to mid-IR spectra as well

(Reeves, 2010). With few exceptions, soil properties that calibrate well to NIR tend to calibrate well, and often better, in the mid-IR spectral region (McCarty et al., 2002; Reeves et al., 2010), although the two regions differ greatly at the theoretical level. The mid-IR spectral region consists of fundamental and some overtone bands of energy absorbed by numerous functional groups, e.g. C—H, N—H, O—H, C=C, C—N, etc., while the NIR spectral region, with a few exceptions, consists of overtones, harmonics, and muted combination bands of C—H, N—H, and O—H. As an example of the chemical structures that give rise to radiation absorbance is the simple case of a water molecule. Any complex oscillation can be decomposed into three normal modes of oscillation and result in three fundamental absorption features and overtones thereof (Figure 20.1). Additional spectral absorption features can be generated by combinations of overtones or fundamental absorptions (Figure 20.2) resulting from formation of doubly excited vibrational states as well as features created by combinations of these normal modes. The fundamental (first harmonic) resonance frequencies for oscillating chemical bonds occur only within the mid-IR region whereas the NIR spectral region contains higher order harmonics (overtones) and combination features (Workman and Weyer, 2008) and does not contain fundamental harmonics.

Comparison of mid-IR and NIR spectra (Figure 20.3) for soils reveals the marked differences in character of information contained within each spectrum. The mid-IR region is dominated with strong signals at the fundamental frequencies and the NIR region has muted and broad overtone and combination adsorption features (McCarty et al., 2002; Mardari et al., 2005; Reeves, 2010; Workman and Weyer, 2008). Also the overall absorption intensity for the NIR is a factor of 10 lower than the mid-IR. Although both spectral regions may contain similar information rising from the chemical bonds contained within a sample, it may be expected that the strength of the information content may vary between spectral regions. This conclusion is supported by the finding that soil properties that calibrate well or poorly in mid-IR also do so in NIR (i.e. a strong correlation in goodness of fit) (Siebielec et al., 2004; McCarty and Reeves, 2006, Reeves, 2010). Moreover, it has generally been found that stronger soil property calibrations can be developed in the mid-infrared region (McCarty et al., 2002; McCarty and Reeves, 2006; Mardari et al., 2005; Reeves, 2010; Viscarra Rossel et al., 2006).

Carbon in soils is present in many forms: humic substances, inorganic C in the form of carbonates, charred C, microbial cell walls, fine roots, non-degraded or partially degraded plant residues, etc., and both the exact composition of these fractions and the relative ratios between them vary from sample to sample. In addition, for most agricultural soils the organic fraction is a relatively small part of the entire soil (often $<5\%$) with the remainder being mineral in nature, and this has profound effects, which are also spectral region dependent, on calibration development and spectral interpretation.

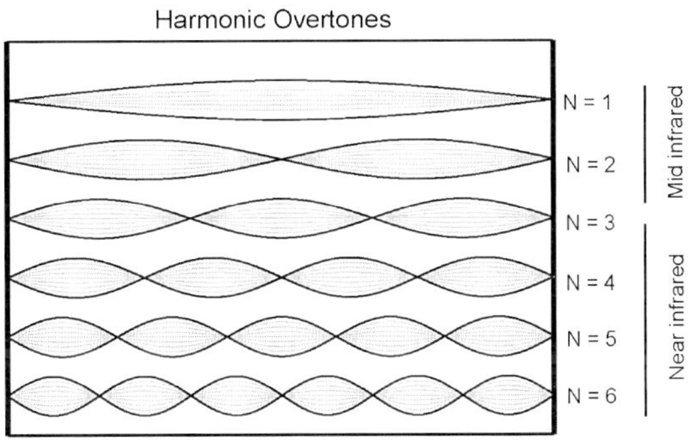

FIGURE 20.1
Demonstration of origin of overtone bands in near- and mid-infrared spectral regions.

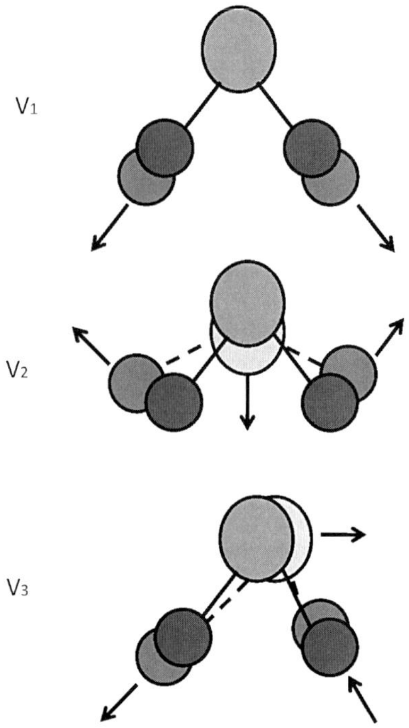

FIGURE 20.2
Demonstration of origin of combination bands in water.

The results of this can be seen in Figure 20.4. As demonstrated, while the mid-IR spectrum of a soil has many defined peaks, interpreting these peaks as if the soil was a simple mixture of organic fractions is made largely impossible by the dominating effect on the spectra of the inorganic fractions such as silica, the heterogeneity of the soil matrix, and the abundant complexity of the organic components of soils.

Accurate calibrations for soil C are possible using both the NIR or mid-IR spectral range and the calibration process is basically the same, but there are differences that need discussion. Spectral interpretation as previously discussed is generally considered much easier for the mid-IR than for the NIR due to the nature of the bands in the spectra. Also, while decades of work in the mid-IR have produced large spectral databases such as the Sadtler IR spectral libraries (Bio-Rad, www.bio-rad.com) and texts on spectral interpretation (Smith, 1999; Socrates, 2001; and many others), spectral databases and publications for the NIR are much more limited. Although several texts have included chapters on NIR interpretation (Williams and Norris, 1987, 2001), the first full text was published in 2008 (Workman and Weyer, 2008). It should finally be noted that texts are largely devoted to interpreting the spectra of pure organic compounds. Interpreting the spectra of complex multi-component materials, such as soils or forages, is more complex and difficult (Reeves et al., 1999).

Spectral Subtraction

Many researchers have used spectral subtraction, i.e. subtracting a spectrum of ashed soil from each original soil spectrum, to reduce noise associated with background soil reflectance and accentuate the signature of the organic component of soils (Cox et al., 2000). However, this may be risky in the 1500 to 400 cm^{-1} (6666 to 25,000 nm) spectral region due to physical changes that occur in some clays during the ashing process, the nature of which varies with the type of clay present, thus leading to erroneous interpretation. In support of this general conclusion that heat treatments designed to oxidize soil organic matter can have strong influence on mineralogy, McCarty et al. (2010b) found that the

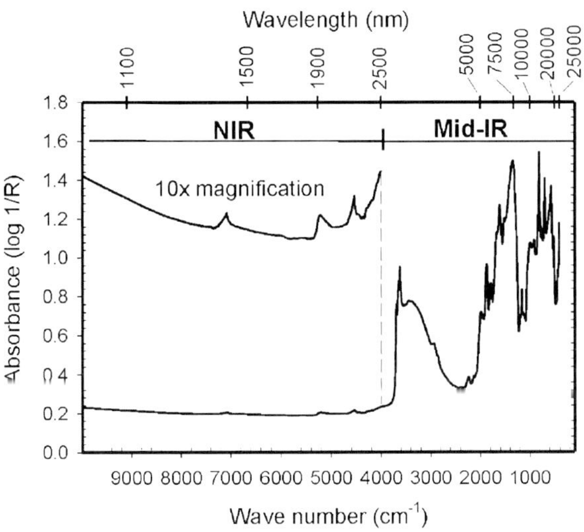

FIGURE 20.3
Average spectra for 405 soil samples collected from the eastern shore of Maryland: top (10,000 to 4000 cm^{-1}) = expanded near-infrared, average spectrum × 10; bottom = average near-Infrared shifted by 0.05 to match mid-infrared spectrum (4000 to 400 cm^{-1}) at 4000 cm^{-1} due to differences in detectors, etc. between the two spectral ranges.

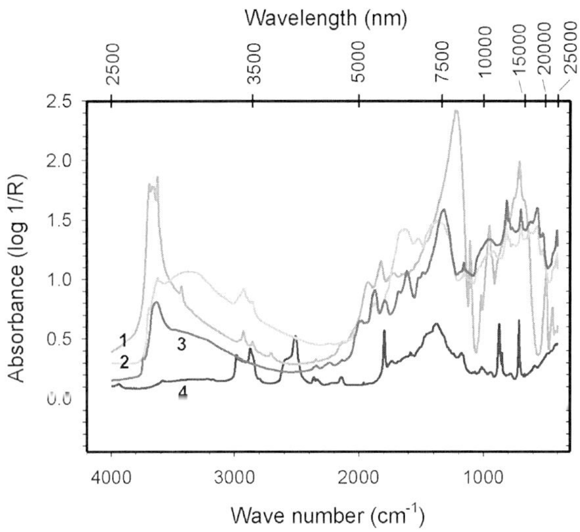

FIGURE 20.4
Mid-infrared spectra of: kaolinite clay (1); soil with 9.80% organic carbon (0.59% carbonate) before (2) and after (3) ashing at 550°C, and 5% calcium carbonate diluted in KBr (4). Please see color plate section at the back of the book.

mid-IR difference spectra of oven dried (105°C) soil before and after heat treatment (400°C) contained features related to the water spectrum. This finding indicated that the 400°C heat treatment commonly used for loss of ignition determination of soil organic matter can liberate bound water or dehydrate minerals and thereby cause bias estimates of soil C by this procedure.

While mid-IR spectra contain fundamental and some overtone bands due to C—H, N—H, O—H, C=O, aromatic rings, silica, clays, etc. (Smith, 1999), as seen in Figure 20.5A, the NIR contains bands due to OH in clays, but silica absorbs very little, which explains why quartz windows and beam splitters are used in NIR spectrometers. It should also be noted that the peaks at ~5200 and 7100 cm^{-1} are due to water. The large upward spectral slope from 6000 to 10,000 cm^{-1} is related to sensible color differences. As ashing removes the color contribution from the soil spectrum, this sloping baseline in the NIR makes spectral subtraction using spectra of ashed soils as shown for the mid-IR (Figure 20.4) useless in the NIR unless a curved baseline correction is applied to the spectra. The results of acidification can be seen in Figure 20.5B, where the carbonate band at ~4150 cm^{-1} is clearly absent in the spectrum of the acidified sample.

In Figures 20.6 and 20.7 are mid-IR and NIR spectra of high and low C soils with little inorganic C. As can be seen for the mid-IR, except for the bands at 2930 and 2860 (aliphatic C—H), the spectral signature of mineral soil overwhelms any organic bands. Unfortunately, the spectral signature of inorganic C has bands either close (2980 cm^{-1}) or at virtually the same (2870 cm^{-1}) location. This may be the reason why, as was previously noted, the presence of inorganic C in soils degrades mid-IR calibrations for organic C (McCarty et al., 2002). In the NIR (Figure 20.5A), the adsorption band for inorganic C (centered on 2400 nm) is distinct from those for organic C (centered around 2200 nm) and this degradation does not occur. Finally, the lower overall absorption levels previously noted for the NIR spectral range permits the incident radiation to penetrate deeper into the sample, which can provide better sample representation especially with unground soil samples. A quote to keep in mind when applying spectral subtraction is: "And our old friend spectral subtraction is virtually a seismograph. It will detect not only whether your sample is changed by 0.1% at some point in time, but will

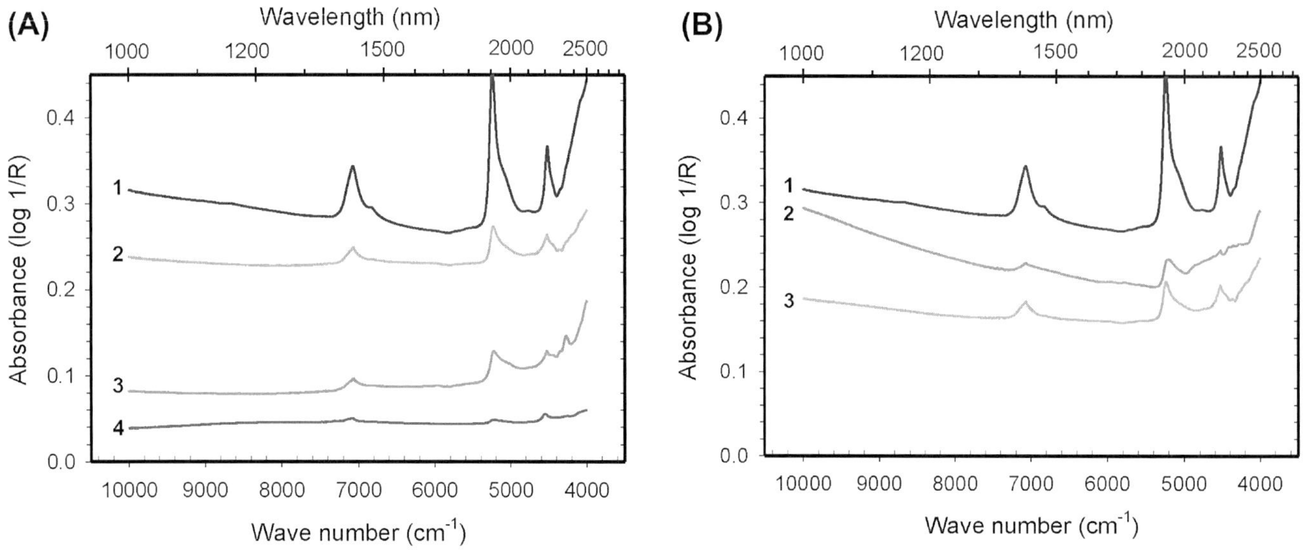

FIGURE 20.5

(A) Near-infrared spectra of: (1) southern bentonite clay, (2) acidified soil with 6.54% carbonate C after and before acidification (3) and (4) ground silica. (B) Near-infrared spectra (arbitrarily scaled for viewing). Top to bottom: (1) southern bentonite clay, (2) soil (scaled for viewing), with 0.09% organic carbon content after acidification, and (3) soil with and 9.80% organic carbon after acidification.

also seem to detect the phases of the moon and the mood you were in while you were measuring the data" (Hirschfeld, 1984).

CHEMOMETRICS

Chemometrics is the area of statistics that involves relating spectra to analyte values. Over the last several decades researchers have investigated virtually every old and newly developed regression-based technique for use in predicting soil constituents. Early efforts were concentrated on multi-linear regression (MLR) techniques, which selected a limited number of spectral wavelengths found to best predict the analyte of interest (Westerhaus et al., 2004). Over time and with the development of faster and more accessible computer power, chemometric methods, which utilized the entire spectral range such as principal components regression (PCR), partial least squares regression (PLS), neural networks (NN), and many other methods, have become the dominant force in chemometrics (Westerhaus et al., 2004, Naes et al., 2002). However, recent discussions such as those on an NIR Discussion Website (http://www.impublications.com/cgi-bin/discus/discus.cgi) indicate that the move from specific wavelengths methods such as MLR to whole spectrum methods may have had as much to do with marketing and sales as science. Considering that the overall body of work for soils and NIRS or mid-IR spectroscopy is extremely small compared to other agricultural products, considerable research on the best chemometric methods still needs to be done as also recently concluded by Viscarra Rossel and Behrens (2010).

CALIBRATION DEVELOPMENT

One of the most important chemometric questions still to be answered for soils is just how wide a range of soil types and/or varying organic matter composition can be encompassed within a single calibration which may, or may not, be convolved with the method used for calibration development (Viscarra Rossel and Behrens, 2010). While considerable work on the question of local versus global calibrations has been carried out (e.g. Brown, 2007; Cobo et al., 2010; Guerrero et al., 2010; Madari et al., 2005; Minasny et al., 2009; Reeves and Smith, 2009; Sankey et al., 2008; Terhoeven-Urselmans et al., 2010; Wetterlind and Stenberg, 2010), the exact basis upon which such decisions can be made is still unknown. For example, grains and

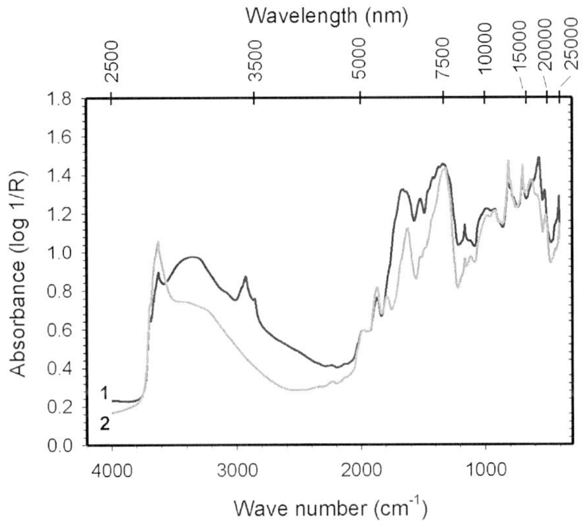

FIGURE 20.6
Mid-IR spectra of soils with low carbonate content carbonates (acidified) and respective organic carbon contents at 9.80% (1) and 0.09% (2), respectively. Please see color plate section at the back of the book.

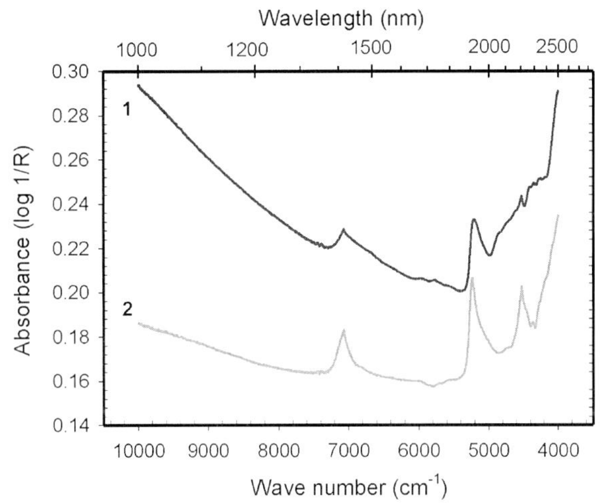

FIGURE 20.7
Near-IR spectra of soils with low carbonate content and respective organic carbon contents of C at 0.09% (1) and 9.80% (2), respectively.

oil seed calibrations are now designed to cover all the variations due to varietal differences, seasons, etc. in one universal calibration using neural networks. But such calibrations require tens of thousands of samples collected and analyzed over decades according to a properly designed protocol. This approach has so far not been tested with soils except for much smaller sample sets. The largest sample set analyzed by neural networks, known to the authors, are sets of ~950–1100 samples (Janik et al., 2009; Viscarra Rossel and Behrens, 2010), a number many chemometricians would still consider insufficient. Futhermore, the composition of soils surely differs more than that of grains and oilseeds across varieties and years.

The biggest error committed by many researchers is using too few samples for calibrations. For example, neural network calibrations based on sample numbers less than a few thousand are of little value, yet efforts are often made using as few as 50 samples. Even with other methods such as PLS regression, researchers often use a few dozen samples at most. As demonstrated in recent work by Delwiche and Reeves (2010), PLS regression is subject to overprediction to the extent that, with a sufficient number of wavelengths (700) and relatively few samples (30–58), it is possible to "predict" random numbers with considerable accuracy. Considering that modern Fourier transform instruments produce NIR and IR spectra that can easily contain twice as many spectral data points as this, one can see where basing calibrations on sample sets of a few dozen or even perhaps 100 samples can lead to erroneous conclusions through overfitting and random fitting of data. If the number of samples is too small to develop a robust local calibration, one alternative is to calibrate based upon a "global" calibration derived from large spectral libraries, with the incorporation of a small number of "local" samples to improve accuracy (Brown, 2007; Christy, 2008; Guerrerro et al., 2010; Wetterlind and Stenberg, 2010,). This approach can avoid overfitting of datasets while also potentially reducing the number of local samples required to derive accurate predicted values.

The accuracy of any calibration is also dependent on the specific spectral information available. In Figure 20.8, NIR spectra obtained on several different types of spectrophotometers are shown. One can easily see that the resulting spectra are different. While the spectra shown are not all for the same fields, many of these differences result from differences in spectral resolution (number of data points collected over a specific spectral range) and differences among configurations, i.e. sampling method, data format, etc. All other things being equal, the higher

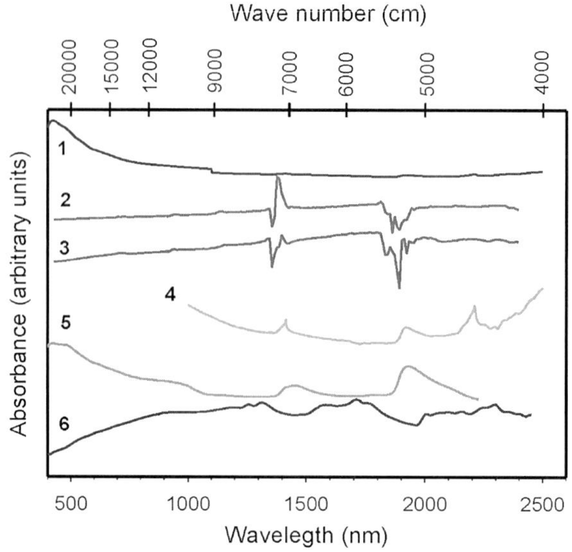

FIGURE 20.8
Near-infrared spectra of soils including: (1) pseudo-absorbance (PA) NIRSystems 6500 scanning monochromator; (2) reflectance (R) from Hyperion satellite at 4.5% and (3) 80% ground cover; (4) PA from an FT-NIR laboratory instrument; (5) PA from a tillage-shank VIS-NIR Veris scanning system; and (6) airborne reflectance collected using a hyperspectral VIS-NIR imaging spectrometer (HyperSpecTIR®). All X units converted to nm and all spectra normalized to between 0 and 1 and shifted for display. Please see color plate section at the back of the book.

the resolution of the spectra, the lower the signal to noise ratio, i.e. higher resolution may contain more information but is noisier. Co-added multiple scans can reduce this noise. Additionally averaging data points can decrease the noise but also decrease the amount of information in the spectrum. For spectra such as those from a Fourier transformation NIR sensor with >3000 data points, averaging reduces the noise but does not reduce calibration accuracy due to the high information content. On the other hand, the coarser resolution spectra associated with satellite sensors, with each data point representing the sum of reflectance within wavebands spanning tens of nm, may not always contain sufficient spectral information to support accurate soil C determinations even when obtained at a high signal to noise level.

CALIBRATION TRANSFER

Calibration transfer among instruments has received little attention with regard to soil analysis. This is most likely due to the fact that the interest in chemometric analysis of soils has really only taken off in the last decade. To date, there is still no large-scale operational use of chemometrics for soil C determinations as compared to more simple materials such as grains. Government organizations such as the U.S. Grain Inspection Packers and Stockyards Administration have taken the role in developing long-term stable and robust calibrations for grain products that are now used operationally (GIPSA, 2006). Due to the complex nature of soils, which are highly variable in mineralogical and organic composition as reflected by soil classification schemes (http://soils.usda.gov/technical/classification/), the problems encountered are likely to be even more complex than those found with comparatively uniform plant materials such as grains.

One advantage of spectrometric methods is that one is predicting new/future samples based on older existing data, thus libraries of existing samples offer a quick and less costly method to obtain calibrations even if the C values need to be re-determined due to changes in methods (e.g. the literature contains 14 different variants in the dichromate method for soil carbon assessment). One obvious desire in using libraries would be to avoid having to rescan calibration samples on each spectrometer that is in use. Any differences between any two spectrometers or even between the same spectrometer over time due to environmental conditions or changes with time or repair (i.e. replacement of a source or changes in the motion of a grating, etc.) can produce inaccurate results if the calibration is not properly transferred. Probably the easiest difference to envision is a slight discrepancy in the exact wavelengths scanned due to differences between gratings or grating position. Differences between instrument types, e.g. grating versus FT or diode array instruments, are even greater. To overcome these differences and to avoid the problem of having to rescan samples, a process called calibration transfer has been developed in which one spectrometer is mapped to another (Inge and Hurburgh, 2008). This is one of the most important efforts required for large-scale use of calibrations across multiple instruments.

In the same manner as for spectra, the analytical results must be consistent and replicable across the variety of samples used in a calibration. Unfortunately, just as any differences in spectrometers affect a calibration, so do differences in samples and resampling methods. While it may be acceptable within a specific study for all samples to be measured as slightly higher or lower than the true values for any given analyte (a systematic bias), since relative within-site differences remain accurate, a baseline shift of this kind can cause havoc if one is interested in differences among diverse soil sample datasets, especially if samples from many different studies/libraries/instruments are being combined. A prime example of this effect can be found in early NIR work on cotton where calibrations were failing when developed using reference data obtained on different instruments of the same make and model, which were later determined to produce disparate spectral values for the same analyte (Montalvo et al. 2004). Recent tests in the laboratory of one of the authors (Reeves) have shown that two combustion units purchased about 10 years apart, but of the same basic make and model, show biases relative to Kjeldhal N values for forages, but one positive and one negative. Systematic differences in analyte or spectral values due to changes in instrumentation are not acceptable when combining data into one dataset for calibration development or transfer. Failure to recognize these effects can lead to failed calibrations or even failure to accurately ascertain whether certain calibrations are possible. A great amount of effort can be wasted by heedlessly combining spectra obtained from different sources or using library compiled without taking into account spectral differences due to instrumental effects.

INSTRUMENTATION

There are too many makes and models of spectrometers in operation, both being made today and older but still operational units, to say which specific spectrometers produce acceptable calibrations and which do not. However, from personal experience, personal communications, and review of the literature, some general statements can be made. As shown in Table 20.1, successful calibrations for soil C do not require any specific type of spectrometer, i.e. FT, scanning monochromator, etc. Calibrations developed using identical soil samples (McCarty et al., 2002) scanned on various FT mid-IR spectrometers, ranging from old non-dynamically aligned units such as the Perkin Elmer PE2000 and the Digilab FTS60 to new dynamically aligned units (Digilab FTS-7000) and even the portable Surface Optics SOC-400 unit, have all been excellent (unpublished data and Table 20.1). Unfortunately, due to differences among sample characteristics and ranges of observed values, which greatly affect calibration results, it is not possible to use calibration results published in the literature to make statements about any specific instrument, but generally, any well-built, research quality spectrometer with the proper spectral range (VNIR or MIR with good spectral resolution) and accessories (standardized and controlled light source and sample handling equipment) appears to be capable of developing accurate calibrations for soils.

TABLE 20.1 Comparison of Calibration Results Using set of 201 Samples on Different Instruments Under Varying Processing and Methods

	Dried and sieved		Dried and ground		Soil moist	
NIR	**RMSD**	**R2**	**RMSD**	**R2**	**RMSD**	**R2**
Monochromator	0.153	0.931	0.148	0.935	0.155	0.928
FTNIR	0.171	0.913	0.145	0.938		
FTIR	0.175	0.909	0.145	0.938		
SOC 400	0.175	0.909	0.140	0.942	0.208	0.871

RMSD = Relative Mean Squared Deviations; R^2 = Calibration R^2.

One technique that has advanced the use of NIR sensors is the incorporation of fiber optics to transmit the reflectance signal from the sampling area to the detectors. Many spectrometers can be equipped with fiber optic accessories, i.e. FOSS NIRSystems model 6500 (FOSS NIRSystems, Laurel, MD) or they come designed specifically with fiber optic probes, i.e. ASD FieldSpec® 3 Portable Spectroradiometer (ASD, Inc., Boulder, CO). The Veris shank-mounted unit (Christy, 2008) has a light source in the probe head, but uses fiber optics to return the reflectance to the sensor. In contrast, most laboratory devices use an external light source or use fiber optics to both deliver the light and collect the reflected energy. Fiber optic probes offer flexibility in sampling and spectrometer design, but are much more common in the NIR than in the mid-IR. Due to the presence of some hydroxyl groups in even the best silica-based probes even NIR spectra become noisy in the region beyond 1900 nm (5350 cm^{-1}) and such probes are useless in the mid-IR as silica absorbs the mid-IR radiation. As a result fiber optic probes in the mid-IR are much less common due to their expense and fragility. The fibers are made of exotic glasses such as chalcogenide (As-S) or AgCl:AgBr [two probes are needed to cover entire mid-IR spectral range (Ingram, 2011) and are both expensive and more fragile than those used in the NIR]. More recently, hollow wave guides have been developed that allow the entire mid-IR spectral range to be scanned with one probe (PIKE Technologies, 2011).

Other methods widely used in the mid-IR, but not in the NIR, which have been used for soils include attenuated total reflection (ATR) (Linker et al., 2005, 2006; Jahn et al., 2006) and photo-acoustic spectroscopy (Du et al., 2007, 2009; Du and Shaviv, 2008; Du and Zhou, in press). When light reflects within a crystal it actually leaves the crystal for a short distance called the evanescent wave and can penetrate samples placed on the crystal, resulting in a reflectance spectrum (Coleman, 1993). This requires very close contact with the crystal and thus high pressure for solid samples using diamond anvils or extremely finely ground samples (75 μm sieve to avoid scratching the crystal or being pressed into non-diamond crystals such as ZnSe); the setup is expensive.

Photo-acoustic mid-IR spectroscopy is based on the sound waves created in a sealed compartment when a sample is heated by the mid-IR radiation (Coleman, 1993). The results of Du et al. (2007, 2009), Du and Shaviv (2008), and Du and Zhou (in press) indicate that this method could be used to determine soil C and other analytes, and could provide better spectra than ATR due to the use of dry samples rather than the pastes that are used with the ATR.

PARTICLE SIZE CONSIDERATIONS

1. Other effects of particle size include reduced ability to acquire representative spectra because mid-IR radiation cannot penetrate a sample as deeply as can NIR radiation, and the sample area scanned may be very small in an FT spectrometer. For example, that area viewed is often only a mm or two in diameter. Thus, large particles could be a problem for either Fourier transformation NIR or mid-IR spectroscopy. Our experience has indicated

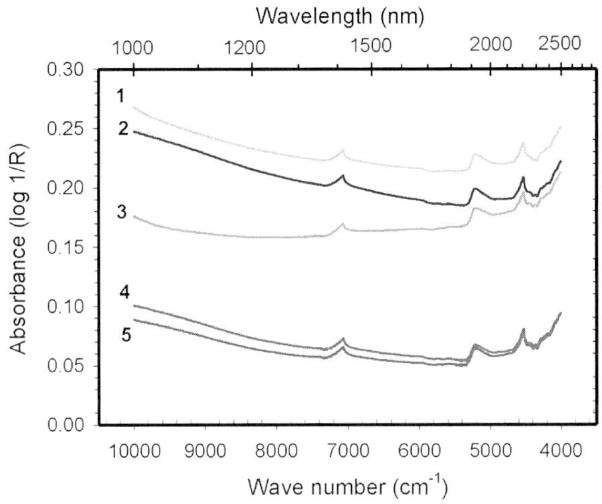

FIGURE 20.9
Near-infrared spectra of five soil samples from the eastern shore of Maryland demonstrating differences in overall soil brightness as reflected by overall absorbance differences, and scattering effects as reflected by overall slope differences, due to particle size differences among soils. Please see color plate section at the back of the book.

that grinding samples to pass a 20-mesh screen is sufficient to avoid these problems when using diffuse reflectance infrared FT spectroscopy (DRIFTS; Reeves et al., 2010 and unpublished data).

2. There are two other effects that need to be mentioned. In the NIR, systematic differences in particle sizes between samples results in overall shifts in baselines between spectra due to greater trapping of light within coarser soils. Larger particle size allows deeper penetration of the NIR radiation resulting in an overall higher absorption compared to similar samples with smaller particle sizes, which perhaps explains differences in soil brightness seen in Figure 20.9. Note also that for the top two spectra, the overall slope of the baseline of the spectra varies. This variation is caused by an effect called multiplicative scatter due to different scattering effects at different wavelengths. Spectral pretreatments such as derivatives and multiplicative scatter correction or standard normal variate correction (Naes et al., 2002) can be applied to the spectra prior to regression analysis to reduce or eliminate these effects. Finally, in addition to the specular distortion effects previously discussed for mid-IR spectra, work by Salisbury et al. (1987, 1988) demonstrated that the spectra of minerals in the mid-IR can vary greatly with the particle size. While a single chapter does not allow a discussion of all the ramifications of these and other spectral effects, this information is provided both as a warning and as a starting point for those who may be interested in applying mid-IR or NIR in new and innovating ways, such as using probes to scan earthworm burrows (Leue at al., 2010), soil aggregates, or intact soil cores where the spectra of the native soil, containing organic debris and coarse rock fragments, may be very different than uniformly ground soils.

SCALE AND SAMPLING INTENSITY

Broad-scale techniques relying on large spectral libraries perform best when properly seeded with local samples. Otherwise accuracy is often severely limited due to differences between the composition of unknown soils and those that make up the calibration dataset. The most accurate results are obtained in local studies where the calibration data are collected in close proximity to the unknowns and the range is well sampled. This limits the utility of predictions to well-sampled areas, although the number of samples relative to the abundance of predicted output is certainly favorable, particularly for areas in which a large amount of information on spatial variability of soil C is desired.

The cost of collecting samples and analyzing them by conventional procedures, such as combustion analysis for C, is much greater than the cost of obtaining spectra for the same samples. But as shown in Table 20.2, for any method the real costs come to the labor costs involved in obtaining the samples. A scanning monochromator using sealed cells can easily scan 150 samples per day, but it would take at least 2 days to collect and process the samples prior to scanning if cores to 30 cm were taken. Table 20.2 shows that substantially more samples (300 to 360 samples) can be easily scanned on an FT instrument with an autosampler. Whenever large numbers of soil samples are being collected and chemically analyzed, such as within private, university or government laboratories that serve the agricultural or environmental community, implementing routine collection of spectral reflectance data for all processed samples could be used to quickly and cheaply develop large datasets for chemometric model development and calibration.

PREDICTION ACCURACY

Many studies have used some form of hyperspectral reflectance data combined with chemometric approaches to measure soil C content in soils. Reviews can be found in Viscarra Rossel et al. (2006), and Ladoni et al. (2010). Numerous spectral features have been associated with soil C, with an overview provided in Ladoni et al. (2010). Prediction accuracies (R^2) for soil C have varied considerably (e.g. 0.00 to 0.98; Viscarra Rossel et al., 2006) depending on the sensor, soil conditions, mathematical treatments, and selection of calibration data, although the majority of studies are now obtaining prediction accuracies >0.8. In the most successful cases, which often occurred under laboratory conditions, accuracy and repeatability (residual mean squared errors) have approached those found in analytical laboratory studies, indicating that under the best conditions a well-calibrated spectral assessment can provide a rapid, cheap, and accurate alternative to traditional laboratory analysis for determination of soil C content. However, the extreme variability of results indicates the continuing importance of advancing research into the development and application of best chemometric methods for sample selection, spectral calibration, and prediction of unknowns (Viscarra Rossel and Behrens, 2010).

PROXIMAL AND REMOTE SENSING

With recent advances in aircraft geopositioning technology, it has become possible to produce raster maps of reflectance spectra collected using a wide variety of airborne NIR imaging spectrophotometers (e.g. AHS-160, http://www.argonst.com/docs/AHS-160.pdf; HyMap, http://www.hyvista.com/wp_11/wp-content/uploads/2011/02/EARSEL98_HyMap.pdf and SpecTIR, http://www.spectir.com). These sensors usually include about 200 bands in 10 nm increments from 400 to 2500 nm, although airborne NIR sensors have also been used. With proper application of chemometric methods such as PLS regression, these hyperspectral images can be translated into high-resolution (0.5–5 m) distributed raster maps of soil properties (Ben-Dor et al., 2009; McCarty et al., 2010a; Selige et al., 2006). While results can be good, prediction models using airborne imagery tend to be weaker than those derived in the laboratory, due to a reduced spectral information content that stems from the interference of ambient light conditions and atmospheric water vapor adsorption features (Ben-Dor et al., 2009; Stevens et al., 2008). Still, several studies have shown the potential for airborne sensors to assist in high-resolution mapping of soil constituents such as C even at relatively low (1.5 to 3%) observed ranges of C content (Ben-Dor et al., 2009; Hively et al., 2011). In soils with low C concentrations, the soil C signal can be masked by other soil characteristics (Ben-Dor et al., 2009; Stevens et al., 2008).

Additional sources of interference in airborne spectroscopy include soil moisture, which may vary across the mapped landscape, and bidirectional reflectance distribution resulting from sun angle interactions with surface roughness. Spatial averaging of adjacent pixels can help

TABLE 20.2 Characterizations of Different Spectral Modes and Methods for Obtaining Soil C Spectral Data

	Proximate sensing						
	Near-infrared			*Mid-infrared*		*Remote sensing*	
	FTNIR	*Monochromator*	*Fiber optic*	*Bench top*	*Portable*	*Plane*	*Satellite*
Range	10,000–4000 cm^{-1}	400–2500 nm (25,000–400 cm^{-1})	Cut-off at ~2300 nm	4000–400 cm^{-1}	4000–400 cm^{-1} (600 with ZnSe BS)	400–2500	400–2500
Spectral resolution	4 to 8 cm^{-1}	10 nm		4 to 8 cm^{-1}	4 to 8 cm^{-1}	10 nm	10 nm
Sample area	<.1 cm^2	>.1 cm^2	1 to 100s cm^2	<.1 cm^2	<.1 cm^2 and up	1 to 100s km^2/day	100s– 1000s km^2/day
Pixel area	–	–	–	–	–	1–3 m^2	30 m^2
Time per sample	1 min	1–3 min		1 min	1 min	–	–
Samples/day	300	150	150–300	300	300	–	–
Effect of moisture	Low	Low	Low	High	High	Cloud dependent	Cloud dependent

overcome noise (Ben-Dor et al., 2009; Hively et al., 2011). The spectral information content of airborne hyperspectral imagery is generally less than that of laboratory methods. Collection of soil samples for an adequate calibration dataset can be challenging due to discrepancies in the scale of analytical chemistry (sampling ~10 g soil) versus image-based spectral analysis (sampling ~10 m^2 soil surface). The use of chemometric analysis of airborne hyperspectral imagery to map soil C content is naturally limited to areas of bare soils, such as newly tilled agricultural fields. For those areas, the resulting high-resolution raster maps of measured soil properties can contribute substantially to the understanding of field-scale agricultural processes, including accumulation and redistribution of soil C (Hively et al., 2011). In more complex landscapes, masks based on calculated indices that correlate well with green vegetation (such as Normalized Difference Vegetative Index, NDVI), crop residue (such as Cellulose Absorption Index, CAI) , soil moisture, and atmospheric water vapor, can help to remove pixels that exhibit non-soil characteristics from the analysis.

It is also possible to develop maps of soil properties using proximal measurements, such as driving through the landscape and frequently collecting sensor measurements along transects or on gridded patterns (Adamchuk and Viscarra Rossel, 2010; de Gruijter et al., 2010, Stevens, 2008). Although this may be a quicker and cheaper method than gridded soil sampling, this method has similar limitations in the number of sample points collected, and associated errors in interpolation of soil characteristics. One such method employs an NIR spectral sensor pulled through the soil on a tillage shank, with an internal light source and signal return provided using a fiber optic cable protected with a quartz window (Christy, 2008). By driving transects through the field this method is simpler than gridded sampling, and also overcomes the interference from surface vegetation and residue that prevents the use of aerial imagery for estimating soil C content of no-till soils. An efficient calibration protocol for tillage shank measurements is employed by the Vertis Corporation (McCarty et al., 2010a) where the transects are first driven, and then spectral variation is analyzed by the computer and 10 soil sampling locations are identified within the field that represent observed clusters of spectral variation. In this way, likelihood is increased that the calibration data will fully represent the range of unknown (spectrally predicted) soil conditions.

A few satellites (e.g. Hyperion) provide imagery with sufficient spectral definition to support hyperspectral chemometric analysis, and future space technology missions are expected to improve spectral and spatial resolution (e.g. HyspIRI). Currently, imagery from the Hyperion satellite can provide spectra with 198 usable wavebands from 400 to 2500 nm, covering a 7 km swath image width at 30 m spatial resolution. At this large pixel size, it becomes difficult to collect soil samples that accurately represent the 900 m^2 spectral sampling locations. However, it has been used with success to predict soil C content in several studies including Cheng et al. (2009), Gomez et al. (2008), and Zhang et al. (2009). Similar to airborne image collection, satellite hyperspectral imagery provides the opportunity to limit analysis to areas with little crop residue through the use of cellulose adsorption indices (Daughtry et al., 2006) and crop residue mapping (Bannari et al., 2007). Studies have also applied regression analysis to multi-spectral satellite data to predict soil C with varying degrees of success (Ladoni et al., 2010). Derivation of chemometric predictions from satellite imagery presents additional challenges to overcome noise associated with atmospheric adsorptions and large scales of analysis.

WATER EFFECTS AND SPECTROSCOPY

Water has varied effects on soil reflectance, depending on the spectral range and scanning method used as well as the properties of the soil. Many laboratory studies control this variable by drying soils prior to spectral analysis, although it has also been suggested that controlled rewetting prior to analysis will improve calibrations with soil C content (Stenberg, 2010). As shown in Figure 20.10, with pure water a path length of even 0.5 mm results in total

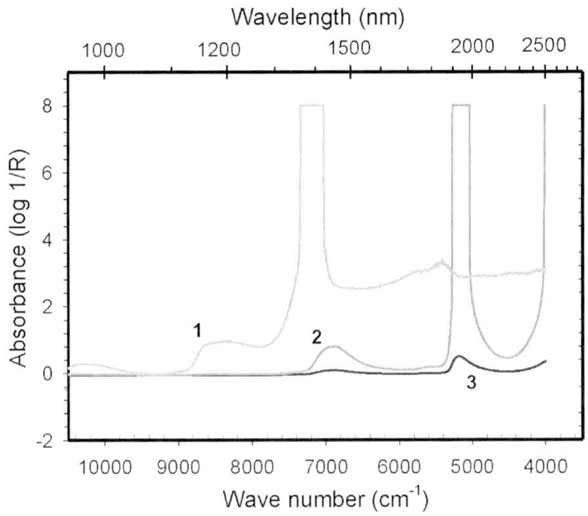

FIGURE 20.10
Near-infrared spectrum of water with a path length of 2 mm (1), 0.5 mm (2), and 0.1 mm (3); all spectra zeroed to baseline. Please see color plate section at the back of the book.

absorption at 5180 cm^{-1}, while 1 mm is possible at 6900 cm^{-1}, and 20 mm or more at 8300 cm^{-1} and beyond (data not shown). For the mid-IR, the maximum path length is more on the order of tenths of a nm (data not shown), leading to strong adsorption and interference from ambient water. Another point of consideration is to note that while the water absorption peak is at 5180 cm^{-1}, the water obscures the entire spectral region from ~7400 to 4000 cm^{-1}. The absorption shows as a noisy line due to analog to digital conversion problems (Figure 20.10). While reflectance spectroscopy on field-moist samples performed using a fiber optic probe placed directly on the soil or a sealed cell does not require transmission through mm of soil to obtain a spectrum, the same cannot be said for the mid-IR as presently used (non-sealed cells) or for remote sensing methods where substantial amounts of water vapor between soil and the detector may be present. In the vapor state, both the NIR and mid-IR spectra of water consist of many sharp bands (Figure 20.11A), which can have the appearance of noise (Figure 20.11B) and degrade the ability of mid-IR to develop robust calibrations for soil C. As a rule, collection of reflectance spectra from dry samples will yield the best predictions of soil characteristics.

PRESENT STATUS AND FUTURE RESEARCH NEEDS

The success with NIRS applications for assessing many soil parameters has resulted in a rapidly evolving research field in which many new approaches are being tested. Each NIR spectrum integrates information about soil mineral and organic composition (Stenberg et al., 2010). In their reviews, Reeves (2010) and Stenberg et al. (2010) cite a number of studies where NIRS is used to characterize or predict a wide variety of soil C attributes beyond total C, such as alkyl and aromatic functional groups (Terhoeven-Urselmans et al., 2006), microbial biomass (Cécillon et al., 2008), and organic matter in soil aggregates (Velasquez et al., 2007). Hartmann and Appel (2006) and Cabassi et al. (2008) used NIRS to study the decomposition of cellulose, manure, or plant material in soils. Especially important is developing methods to measure soil C quality. Vasques et al. (2009) found that visible-NIRS is sensitive to total, recalcitrant, and mineralizable C. Thomsen et al. (2009) were able to correlate NIR spectra to labile soil C. NIR also correlated with resistant C, but Thomsen et al. (2009) explained that this was possibly due to correlation with total soil C. The potential problems with C forms can be seen in the research by Hammes et al. (2007) where they demonstrated the many forms C may take and the varied methods used to measure it. Ongoing and future research needs also to

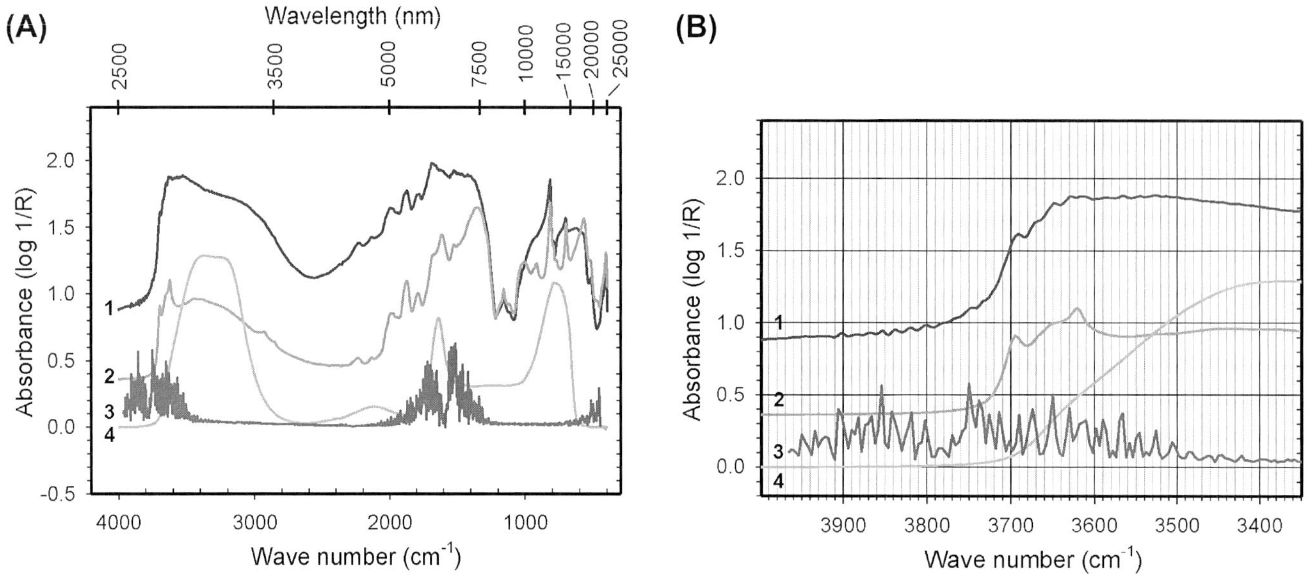

FIGURE 20.11

(A) Average infrared (IR) spectra of 201 soils scanned field moist in open cells (1), scanned dried (2), water vapor by pyrolysis of $CuSO_4 \cdot (H_2O)_5$ (3), and liquid water by attenuated total reflection using 9 bounce ZnSe crystal (4). (B) Average near infra-red (NIR) of: 201 soils scanned field moist in open cells (1), scanned dried (2) water vapor by pyrolysis of $CuSO_4 \cdot (H_2O)_5$ (3), liquid water by attenuated total reflection using 9 bounce ZnSe crystal (4). Please see color plate section at the back of the book.

assess calibrations for volumetric C content rather than C per unit mass basis in order to meet C credit requirements (Bellon-Maurel and McBratney, 2011).

Mid-IR has always been deemed more amenable to spectral interpretation than NIR spectroscopy. However, the fact that multivariate statistics can use NIR spectral information to calibrate for different soil organic matter (SOM) qualities indicates that there is information contained in NIR that can potentially be used for interpretation purposes. Viscarra Rossel et al. (2006) showed that qualitative interpretations using PLS loadings are better in the mid-IR than NIR because of the lower intensity, lower specificity, and overlap of the NIR bands. Spectral overlap is particularly marked in diffuse reflectance spectra (Stenberg et al., 2010), which happens to be a preferred mode of scanning soils. Nevertheless, the potential exists for self-deconvolution of diffuse reflectance NIR data to enhance bands and aid in mining the NIR spectra for interpretable patterns (McClure, 2008). Even with these limitations, the literature has instances where researchers attribute NIR frequencies to soil attributes. It is feasible to identify NIR regions explained by water, mineral, and organic matter in soil (Viscarra Rossel and Behrens, 2010), but their location can shift because real molecules do not act fully harmonically. Because of this, interpreting soil NIR spectra can be challenging, especially when the species of interest is present in small amounts, such as soil C.

This poor specificity can be compounded by scattering due to soil structure or quartz content (Stenberg, 2010). Clay minerals can absorb in the NIR because of metal—OH bending and O—H stretching combinations (Viscarra Rossel et al., 2006). Stenberg (2010) explains that clay features around 1400, 1900 and 2200 nm (7140, 5250, and 4545 cm^{-1}) are characteristic in soil, although they are affected by water content. Viscarra Rossel et al. (2006) observed high PLS loadings for soil organic C (SOC) at these water bands. Viscarra Rossel and Behrens (2010) used feature selection to identify bands (in nm) useful in SOC calibrations: iron oxides (650; 1540 cm^{-1}), C—H (825; 1212 cm^{-1}), C—H (1852, 5400 cm^{-1}), water (1915, 5220 cm^{-1}), C=O (1930, 5180 cm^{-1}), C=O (2033, 4920 cm^{-1}), N—H (2060, 4850 cm^{-1}), and Al—OH kaolin (2208, 4530 cm^{-1}). Bands near 1100, 1700—1800, and 2200—2400 nm (9100, 5880—5560, and 4545—4170 cm^{-1}) were identified as important for SOC calibration by

Stenberg (2010), and could be attributed to overtones of the C—H stretch fundamentals near 3400 nm (2900 cm^{-1}). Viscarra Rossel and Behrens (2010) identified bands useful for clay calibrations (nm) at: iron oxides (650, 1540 cm^{-1}), goethite (920, 1087 cm^{-1}), O—H (930, 1075 cm^{-1}), water (940, 1064 cm^{-1}), O—H kaolin (1395, 7168 cm^{-1}), water (1380, 7246 cm^{-1}), water (1915, 5200 cm^{-1}), Al—OH kaolin (2160 and 2208; 4630 and 4530 cm^{-1}), AlFe—OH smectite (2230, 4485 cm^{-1}), and illite (2450, 4080 cm^{-1}). With identification of relevant bands in soil spectra, such absorption bands may be useful for calibrations with a select number of frequencies instead of full spectra.

Mineral effects on NIR spectra are important because they affect calibration development. NIR should therefore have an advantage over mid-IR for soil C determinations because NIR is not sensitive to quartz, an important soil mineral (Reeves, 2010). Non-organics have been considered to contribute relatively little to the NIR spectrum (Reeves et al., 2005). Nevertheless, mid-IR often outperforms NIR on soil C calibrations (Reeves, 2010). As previously discussed, ashing and spectral subtraction can be used to further remove mineral signals from soil spectra, but it has caveats. Besides clays (Stenberg et al., 2010), carbonates (Bellon-Maurel and McBratney, 2011) also have weak absorption peaks in the NIR but more research is needed to determine the best way to calibrate for carbonates. Variations in carbonate content in large sample sets can produce non-linearity in soil C calibrations (Reeves and Smith, 2009). A carbonate peak can be observed in NIR spectra at 2381 nm or 4200 cm^{-1} (Reeves and Smith, 2009).

An important issue with NIR calibration research is the occurrence of surrogate calibrations where the correlation between other predicted analytes and the analyte of interest (total soil C) explains the observed variation (Stenberg et al., 2010). This effect may be especially pronounced in soils low in C, due to decreased signal from the soil C component. It may be even debatable whether calibrations for total N result from intercorrelations with C, because a large portion of soil N is organic and thus related to soil C (Chang et al., 2001; Martin et al., 2002). One good example is the determination of soil δ13C by NIRS, which may be explained by correlation of isotope ratios with carbon pools, and not directly by the spectral properties of the isotopes (Reeves et al., 2006). The wavelengths chosen by a PLS model can be used to determine if there is a correlation effect and thus avoid surrogate effects (Kusumo et al., 2011).

Calibration development for recalcitrant black C or char is another important research question. The dark color of chars may pose an advantage when using visible-NIR instrumentation to measure and predict char content, but the black carbon compounds can also mask the reflectance signal associated with additional soil constituents that may be desirable to predict. The first hurdle in developing an NIR calibration for char is the fact that many different methods are used to determine the analyte values, not all measuring the same char attribute (Reeves, 2010). This results in possible bias when combining datasets. Furthermore, depending on the pyrolysis temperature, char can have many of the same C forms already present in SOM, so calibrating for char specifically may be challenging (Reeves et al., 2008).

The issue of methods development is a recurring theme in soil science, with different laboratories trying different procedures to fit their capacity, soil types, and goals. It would be beneficial to arrive at a convention for how total soil C and soil fractions are analyzed so that studies can be combined into larger datasets for more robust calibrations (Reeves, 2010). This issue particularly affects more recently developed C variables like mineralizable C (Vasques et al., 2009; Thomsen et al., 2009) and char content, because many different laboratory methods or incubation conditions are being used to obtain the data.

Another issue that is being researched by several laboratories is sample preparation, which should be kept to a minimum in order to exploit the simplicity and rapidity of NIRS (Stenberg et al., 2010). The most common soil pretreatment is drying, sieving, and grinding to manage water and particle size scattering effects. However, Stenberg (2010) showed that standardized volumetric rewetting before scanning should be considered because it can reduce SOC prediction errors by up to 30%. Stenberg (2010) hypothesizes that rewetting changes the absorbance of clays, improving C calibrations.

Another issue is the pre-processing of the spectral data. Some common preprocessing treatments include derivatives, multiplicative scatter correction, baseline correction, and the standard normal variate (Stenberg et al., 2010). Each dataset may need different combinations of these in order to perform best.

ADVANCES IN INSTRUMENTATION

Dr. Howard Mark (Near Infrared Research Corporation) has often commented that spectrometers are inherently no more complicated than a good stereo, but the problem is that they are not made in sufficient volume to reduce prices. Also, the costs of components for spectrometers, and even the range of components that are available without unlimited funding, are based more on the broad industrial market than a niche market for spectrometers. Thus, high grade fiber optics, which could be used in the 1800 to 2500 nm spectral region, became available when the telecommunications industry needed low hydroxyl fibers and not because of their desirability for spectroscopy. At the International Diffuse Reflectance Conference in 2006, Dr. Jerry Workman (Thermo Electron Corporation) gave a keynote talk on the potential role of the telecommunications industry on the future of spectroscopy (no proceedings for meeting). He stated that there were a lot of people with ideas and products from the telecommunications industry that he predicted would change the face of spectroscopy. Increasing demands for fiber optic performance within the telecommunications industry have pushed rapid development of new diode lasers and tunable quantum cascade lasers with wavelengths further into the NIR and mid-IR regions. These new lasers are expected to lead to compact and rugged array type sensors that can fully span these spectral regions. In addition compact and rugged monochromators and Fourier transformation spectrometers, such as the PHAZIR (Figure 20.12) which is based on micro-moveable mirrors (Gomez, 2003) and Exoscan (Figure 20.13) a totally self-contained battery-operated mid-infrared spectrometer (http://www.a2technologies.com/exoscan_handheld.html), hold promise as more compact, field-ready instruments that are useful for in situ measurements. Satellite- and aircraft-based hyperspectral sensors with greater spectral and spatial resolution are

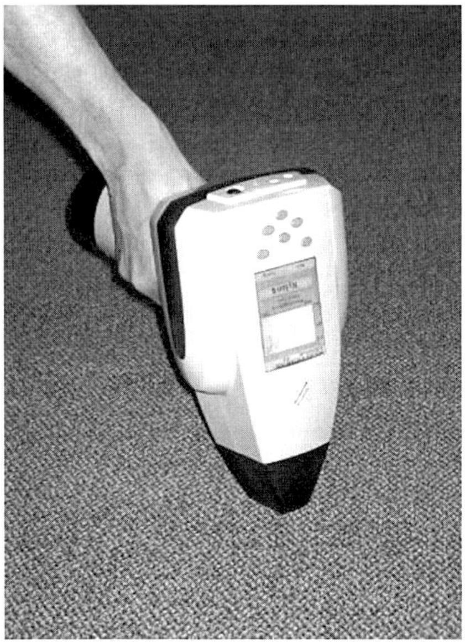

FIGURE 20.12
Phazir handheld near-infrared spectrometer.

FIGURE 20.13
EXOSCAN handheld Fourier transform mid-infrared spectrometer.

evolving rapidly, increasing the suitability of remote sensing applications for mapping terrestrial carbon in the landscape. Most of the information on these newer developments is present in industrial publications such as "LaserFocusWorld" (http://www.optoiq.com/index/404.html) and the reader is advised to consult this source if interested in future developments.

References

Adamchuk, V.I., Viscarra Rossel, R.A., 2010. Development of On-the-Go Proximal Soil Sensor Systems. In: Viscarra Rossel, R.A., McBratney, A., Minasny, B. (Eds.), Proximal Soil Sensing. Springer, NY, pp. 15–28.

Bannari, A., Staens, K., Khurshid, K.S., 2007. Remote sensing of crop residue using Hyperion (EO-1) data. IEEE 9, 2795–2799.

Bellon-Maurel, V., McBratney, A., 2011. Near-infrared (NIR) and mid-infrared (MIR) spectroscopic techniques for assessing the amount of carbon stock in soils. Critical review and research perspectives. Soil Biol. Biochem. 43, 1398–1410.

Ben-Dor, E., Chabrillat, S., Dematte, J.A.M., Taylor, G.R., Hill, J., Whiting, M., Sommer, S., 2009. Using imaging spectroscopy to study soil properties. Remote Sensing Environ. 113, 538–555.

Brown, D.J., 2007. Using a global VNIR soil-spectral library for local soil characterization and landscape modeling in a 2nd-order Uganda watershed. Geoderma 140, 444–453.

Burns, D.A., Ciurczak, E.W., 1992. Handbook of Near-Infrared Analysis. Marcel Dekker, Inc., New York, NY.

Burns, D.A., Ciurczak, E.W., 2001. Handbook of Near-Infrared Analysis, second ed. Marcel Dekker, Inc., New York, NY.

Burns, D.A., Ciurczak, E.W., 2008. Handbook of Near-Infrared Analysis, third ed. CRC Press, Taylor & Francis Group, Boca Raton, FL.

Cabassi, G., Gallina, P.M., Barzaghi, S., Cattaneo, T.M.P., Bechini, L., 2008. Near infrared monitoring of mineral-isation of liquid dairy manure in agricultural soils. J. Near Infr. Spectrosc. 16, 59–69.

Cécillon, L., Cassagne, N., Czarnes, S., Gros, R., Brun, J., 2008. Variable selection in near infrared spectra for the biological characterization of soil and earthworm casts. Soil Biol. Biochem. 40, 1975–1979.

Chang, C.W., Laird, D.A., Mausbach, M.J., Hurburgh, C.R., 2001. Near-infrared reflectance spectroscopy-principal components regression analyses of soil properties. Soil Sci. Soc. Am. J. 65, 480–490.

Cheng, P., Jian, W., Dajun, L., 2009. Quantitative prediction of soil organic matter content using hyperspectral remote sensing and geo-statistics. Trans. CSAE 25 (3), 142–147 (in Chinese with English abstract).

Christy, C.D., 2008. Real-time measurement of soil attributes using on-the-go near infrared reflectance spectroscopy. Comp. Electron. Agr. 61 (1), 10–19.

Cobo, J.G., Dercon, G., Yekeye, T., Chapungu, L., Kadzere, C., Murwira, A., Delve, R., Cadisch, G., 2010. Integration of mid-infrared spectroscopy and geostatistics in the assessment of soil spatial variability at landscape level. Geoderma 158, 398–411.

Coleman, P.B., 1993. Practical Sampling Techniques for Infrared Analysis. CRC Press, Inc., Boca Raton, FL.

Cox, R.J., Peterson, H.L., Young, J., Cusik, C., Espinoza, E.O., 2000. The forensic analysis of soil organic by FTIR. Forensic Sci. Inter. 108, 107–116.

Daughtry, C.S.T., Doraiswamy, P.C., Hunt Jr., E.R., Stern, A.J., McMurtrey III, J.E., Prueger, J.H., 2006. Remote sensing of crop residue cover and soil tillage intensity. Soil Tillage Res. 91, 101–108.

de Gruijter, J.J., McBratney, A.B., Taylor, J., 2010. Sampling for high-resolution soil mapping. In: Viscarra Rossel, R.A., McBratney, A., Minasny, B. (Eds.), Proximal Soil Sensing. Springer, NY, pp. 3–14.

Delwiche, S.R., Reeves III, J.B., 2010. A graphical method to evaluate spectral preprocessing in multivariate regression calibrations: example with Savitzky-Golay filters and partial least squares regression. Applied Spectrosc. 64, 73–82.

Du, C., Zhou., J., 2008. Evaluation of soil fertility using infrared spectroscopy: a review. Environmental Chemistry Letters. 7, 97–113.

Du, C., Zhou, J., Wang, H., Chen, X., Zhu, A., Zhang, J., 2009. Determination of soil properties using Fourier transform mid-infrared photoacoustic spectroscopy. Vibr. Spectrosc. 49, 32–37.

Du, C., Linker, R., Shaviv, A., 2007. Characterization of soils using photoacoustic mid-infrared spectroscopy. Applied Spectrosc. 61 (10), 1063–1067.

Du, C., Linker, R., Shaviv, A., 2008. Identification of agricultural Mediterranean soils using mid-infrared photo-acoustic. Geoderma 143, 85–90.

Franks, C.D., Kimble, J.M., Samson-Liebig, S.E., Sobecki, T.M., 2001. Organic carbon methods, microbial biomass, root biomass, and sampling design under development by Natural Resources Conservation Service (NRCS). Reprinted from Assessment Methods for Soil Science. Chapter 8. In: Lal, R., Kimble, J.M., Follett, E.F., Stewart, B.A. (Eds.), Lewis Publishers, Washington, DC.

GIPSA., 2006. USDA Near-infrared Transmittance Handbook (NIRT). USDA Grain Inspection, Packers & Stockyards Administration, Washington, DC. <http://www.gipsa.usda.gov/GIPSA/webapp?area=home&subject=lr&topic=hb-nirt>

Gomez, P., 2003. State-of-the-art of MEMS Technology for Design and Fabrication of Micro-mirrors. <http://web.eng.fiu.edu/~pgomez/papers/mems-mirrors.pdf>

Guerrero, C., Zornoza, R., Gómez, I., Mataix-Beneyto, J., 2010. Spiking of NIR regional models using samples from target sites: effect of model size on prediction accuracy. Geoderma 158 (1–2), 66–77.

Hammes, K., Schmidt, M.W.I., Smernik, R.J., Currie, L.A., Ball, W.P., et al., 2007. Comparison of quantification methods to measure fire-derived (black/elemental) carbon in soils and sediments using reference materials from soil, water, sediment and the atmosphere. Global Biogeochemical Cycles 21 doi:10.1029/2006GB00291.

Hartmann, H.P., Appel, T., 2006. Calibration of near infrared spectra for measuring decomposing cellulose and green manure in soils. Soil Biol. Biochem. 38, 887–897.

Hirschfeld, T., 1984. Computerized infrared: the need for caution. In: McClure, G.L. (Ed.), Computerized Quantitative Infrared Analysis. STP 934, ASTM, Philadelphia, PA, pp. 169–179.

Hively, W.D., McCarty, G.W., Reeves III, J.B., Lang, M., Oesterling, T., Delwiche, S., 2011. Use of airborne hyperspectral imagery to map soil properties in tilled agricultural fields, Appl. Environ. Soil Sci. [In press].

Inge, B., Hurburgh Jr., C.R., 2008. Standardisation of near infrared spectrometers: evaluation of some common techniques for intra- and inter-brand calibration. J. Near Infra. Spectrosc. 16 (6), 539–550.

Ingram, J., 2011. Comparison of CIR- & PIR-Fibers. <http://www.jtingram.com/sitebuildercontent/sitebuilderfiles/Comparison.pdf>.

Jahn, B.R., Linker, R., Upadhyaya, S.K., Shaviv, A., Slaughter, D.C., Shimulevich, I., 2006. Mid-infrared spectroscopic determination of soil nitrate content. Biosys. Engineer. 94 (4), 505–515.

Janik, L.J., Forrester, S.T., Rawson, A., 2009. The prediction of soil chemical and physical properties from mid-infrared spectroscopy and combined partial least-squares regression and neural networks (PLS-NN) analysis. Chemometr. Intelligent Lab. Systems 97, 179–188.

Ladoni, M., Bahrami, H.A., Alvaipanah, S.K., Norouzi, A.A., 2010. Estimating soil organic carbon from soil reflectance; a review. Precision Agri. 11, 82–99.

Leue, M., Ellerbrock, R.H., Gerke, H.H., 2010. DRIFT mapping of organic matter composition at intact soil aggregate surfaces. Vadose Zone J. 9, 317–324.

Linker, R., Shmulevich, I., Kenny, A., Shaviv, A., 2005. Soil identification and chemometrics for direct determination of nitrate in soils using FTIR-ATR mid-infrared spectroscopy. Chemosphere 60 (5), 652–658.

Linker, R., Weiner, M., Shmulevich, I., Shaviv, A., 2006. Nitrate determination in soil pastes using attenuated total reflectance mid-infrared spectroscopy: improved accuracy via soil identification. Biosys. Engineer. 94, 111–118.

Madari, B.E., Reeves III, J.B., Coelho, M.R., Machado, P.L.O.A., De-Polli, H., Coelho, R.J., Benites, V.M., Souza, L.F., McCarty, G.W., 2005. Mid- and near-infrared spectroscopic determination of carbon in a diverse set of soils from the Brazilian national soil collection. Spectrosc. Lett. 38, 721–740.

Malley, D.F., Martin, P.D., Den-Dor, E., 2004. Application in analysis of soils. In: Roberts, C.A., Workman, Jr., J., Reeves III, J.B. (Eds.), Near-Infrared Spectroscopy in Agriculture. American Society of Agronomy, Madison, WI, pp. 729–784.

Martin, P.D., Malley, D.F., Manning, G., Fuller, L., 2002. Determination of soil organic carbon and nitrogen at the field level using near-infrared spectroscopy. Can. J. Soil Sci. 82, 413–422.

McCarty, G.W., Reeves III, J.B., 2006. Comparison of near infrared and mid infrared diffuse reflectance spectroscopy for field-scale measurement of soil fertility parameters. Soil Sci. 171, 94–102.

McCarty, G.W., Reeves III, J.B., Yost, R., Doraiswamy, P.C., Doumbia, M., 2010b. Evaluation of methods for measuring soil organic carbon in West African soils. African J. Agri. Res. 5 (16), 2169–2177.

McCarty, G.W., Reeves III, J.B., Reeves, V.B., Follett, R.F., Kimble, J.M., 2002. Mid-infrared and near-infrared diffuse reflectance spectroscopy for soil carbon measurement. Soil Sci. Soc. Amer. J. 66, 640–646.

McCarty, G.W., Hively, W.D., Reeves III, J.B., Lang, M., Lund, E., Weatherbee, O., 2010a. Infrared sensors to map soil carbon in agricultural ecosystems. In: Viscarra Rossel, R.A., McBratney, A., Minasny, B. (Eds.), Proximal Soil Sensing. Springer, NY, pp. 165–176.

McClure, W.F., 2008. Analysis using Fourier transforms. In: Burns, D.A., Ciurczak, E.W. (Eds.), Handbook of Near-Infrared Analysis, third ed. CRC Press, Taylor & Francis Group, Boca Raton, FL.

Minasny, B., Tranter, G., McBratney, A.B., Brough, D.M., Murphy, B.W., 2009. Regional transferability of mid-infrared diffuse reflectance spectroscopic prediction for soil chemical properties. Geoderma 153, 155–162.

Naes, T., Isaksson, T., Fearn, T., Davies, T., 2002. A user-friendly Guide to Multivariate Calibration and Classification. NIR Publications, Chichester, West Sussex, UK.

Nelson, D.W., Sommers, L.E., 1996. Total carbon, organic carbon, and organic matter. In: Methods of Soil Analysis. Part 3. Chemical Methods. Book Series no. 5. Soil Science Society of America/American Society of AgronomySoc. Am, Madison, WI.

PIKE Technologies, 2011. FlexIR Hollow Waveguide Accessory—for Remote and Flexible Sampling in the Mid-Infrared Spectral Region. <http://www.piketech.com/products/product-documentation-pdfs/FlexIR_HWG_PDS.pdf#zoom=100%>

Reeves III, J.B., Smith, D.B., 2009. The potential of mid- and near-infrared diffuse reflectance spectroscopy for determining major- and trace-element concentrations in soils from a geochemical survey of North America. Appl. Geochem. 24, 1472–1481.

Reeves III, J.B., McCarty, G.W., Rutherford, D.W., Wershaw, R.L., 2008. Mid-infrared diffuse reflectance spectroscopic examination of charred pine wood, bark, cellulose, and lignin: implications for the quantitative determination of charcoal in soils. Appl. Spectrosc. 62, 182–189.

Reeves III, J.B., 2010. Near- versus mid-infrared diffuse reflectance spectroscopy for soil analysis emphasizing carbon and laboratory versus on-site analysis: where are we and what needs to be done? Geoderma 158, 3–14.

Reeves III, J.B., Francis, B.A., Hamilton, S.K., 2005. Specular reflection and diffuse reflectance spectroscopy of soils. Appl. Spectrosc. 59 (1), 39–46.

Reeves III, J.B., Zapf, C.M., Simkovic, I., Delwiche, S.R., 1999. Mid-infrared versus near-infrared diffuse reflectance spectroscopy for quantitative and qualitative analysis of agricultural products. Recent Res. Develop. Agric. Food Chem. 3, 201.

Reeves III, J.B., McCarty, G.W., Hively, W.D., 2010. Mid- versus near-infrared spectroscopy for on-site analysis of soils. In: Viscarra Rossel, R.A., McBratney, A., Minasny, B. (Eds.), Proximal Soil Sensing, Springer, NY, pp. 133–142.

Reeves III, J.B., Follett, R.F., McCarty, G.W., Kimble, J.M., 2006. Can near- or mid-infrared diffuse reflectance spectroscopy be used to determine soil carbon pools? Comm. Soil Sci. Plant Anal. 37 (15), 2307–2325.

Roberts, C.A., Workman, Jr., J., Reeves III, J.B. (Eds.), 2004. Near-Infrared Spectroscopy in Agriculture. Volume 44 in the series Agronomy. American Societies of Agronomy, Crop and Soil Science, Madison, WI.

Salisbury, J.W., Walter, L.S., D'Aria, D., 1988. Mid-infrared (2.5 to 13.5 mm) spectra of igneous rocks. USGS Open Report, 88–686. <http://pubs.er.usgs.gov/djvu/OFR/1987/ofr_87_263.djvu>

Salisbury, J.W., Walter, L.S., Vergo, N., 1987. Mid-infrared (2.1–25 μm) spectra of minerals. USGS Open Report, 87–263. <http://pubs.er.usgs.gov/usgspubs/ofr/ofr87263>

Sankey, J.B., Brown, D.J., Bernard, M.L., Lawrence, R.L., 2008. Comparing local vs. global visible and near-infrared (VisNIR) diffuse reflectance spectroscopy (DRS) calibrations for the prediction of soil clay, organic C and inorganic C. Geoderma 148, 149–158.

Selige, T., Bohner, J., Schmidhalter, U., 2006. High resolution topsoil mapping using hyperspectral image and field data in multivariate regression modeling procedures. Geoderma 136, 235–244.

Siebielec, G., McCarty, G.W., Stuczynski, T., Reeves III, J.B., 2004. Near- and mid-infrared diffuse reflectance spectroscopy for measuring soil metal content. J. Environ. Qual. 33, 2056–2069.

Smith, B., 1999. Infrared Spectral Interpretation: A Systematic Approach. CRC Press, Taylor & Francis Group, New York, NY.

Socrates, G., 2001. Infrared and Raman characteristic Group Frequencies: Tables and Charts, third ed. John Wiley & Sons, Inc., Hoboken, NJ.

Stenberg, B., 2010. Effects of soil sample pretreatments and standardised rewetting as interacted with sand classes on Vis-NIR predictions of clay and soil organic carbon. Geoderma 158 (1–2), 15–22.

Stenberg, B., Viscarra Rossel, R.A., Mouazen, A.M., Wetterlind, J., 2010. Visible and near infrared spectroscopy in soil science. Adv. Agron 107, 163–215 (Chapter 5).

Stevens, A., van Wesemael, B., Bartholomeus, H., Rosillon, D., Tychon, B., Ben-Dor, E., 2008. Laboratory, field and airborne spectroscopy for monitoring organic carbon content in agricultural soils. Geoderma 144, 395–404.

Terhoeven-Urselmans, T., Michel, K., Helfrich, M., Flessa, H., Ludwig, B., 2006. Nearinfrared spectroscopy can predict the composition of organic matter in soil and litter. J. Plant Nutr. Soil Sci. 169, 168–174.

Terhoeven-Urselmans, T., Vagen, T., Spaargaren, O., Shepherd, K.D., 2010. Prediction of soil fertility properties from a globally distributed soil mid-infrared spectral library. Soil Sci. Soc. America J. 74, 1792–1799.

Thomsen, I.K., Bruun, S., Jensen, L.S., Christensen, B.T., 2009. Assessing soil carbon lability by near infrared spectroscopy and NaOCl oxidation. Soil Biol. Biochem. 41, 2170–2177.

Vasques, G.M., Grunwald, S., Sickman, J.O., 2009. Modeling of soil organic carbon fractions using visible–near-infrared spectroscopy. Soil Sci. Soc. America J. 73, 176–184.

Velasquez, E., Pelosi, C., Brunet, D., Grimaldi, M., Martins, M., Rendeiro, A.C., Barrios, E., Lavelle, P., 2007. This ped is my ped: visual separation and near infrared spectra allow determination of the origins of soil macroaggregates. Pedobiologia 51, 75–87.

Viscarra Rossel, R.A., Behrens, T., 2010. Using data mining to model and interpret soil diffuse reflectance spectra. Geoderma 158, 46–54.

Viscarra Rossel, R.A., Walvoort, D.J.J., McBratney, A.B., Janik, L.J., Skjemstad, J.O., 2006. Visible, near infrared, mid infrared or combined diffuse reflectance spectroscopy for simultaneous assessment of various soil properties. Geoderma 131, 59–75.

Westerhaus, M., Workman Jr., J., Reeves III, J.B., Mark, H., 2004. Quantitative analysis. In: Roberts, C.A., Workman, Jr., J., Reeves III, J.B. (Eds.), Near-Infrared Spectroscopy in Agriculture. American Society of Agronomy, Madison, WI, pp. 133–174.

Wetterlind, J., Stenberg, B., Söderström, M., 2010. Increased sample point density in farm soil mapping by local calibration of visible and near infrared prediction models. Geoderma 156, 152–160.

Williams, P., Norris, K. (Eds.), 2001. Near-Infrared Technology in the Agricultural and Food Industries, second ed. Amer. Assoc. of Cereal Chemists, Inc., St. Paul, MN.

Williams, P., Norris, K. (Eds.), 1987. Near-Infrared Technology in the Agricultural and Food Industries. Amer. Assoc. of Cereal Chemists, Inc., St. Paul, MN.

Workman Jr., J., Weyer, L., 2008. Practical Guide to Interpretive Near-Infrared Spectroscopy. CRC Press, Taylor & Francis Group, New York, NY.

Zhang, T., Li, L., Zheng, B., 2009. Partial least squares modeling of Hyperion image spectra for mapping agricultural soil properties. In: Gao, W., Jackson, T.J. (Eds.), Remote Sensing and Modeling of Ecosystems for Sustainability VI Proc. of SPIE 7454 74540P-1.

CHAPTER 21

Micrometeorological Methods for Assessing Greenhouse Gas Flux

R. Howard Skinner[1], Claudia Wagner-Riddle[2]
[1]USDA-ARS Pasture Systems & Watershed Management Research Unit, University Park, PA
[2]School of Environmental Sciences, University of Guelph, Guelph, Ontario, Canada

CHAPTER OUTLINE

Abbreviations: BREB, Bowen ration energy balance; CP, closed-path carbon dioxide sensor; EC, eddy covariance; FG, flux gradient, FTIR, Fourier-transform infra-red spectroscopy; GPP, gross primary productivity; H, sensible heat; λE, latent heat; M-BR, modified Bowen ratio; NEE, net ecosystem exchange; OP, open-path carbon dioxide sensor; QCLAS, quantum cascade laser-based absorption spectrometer; Re, ecosystem respiration; TDLAS, tunable diode laser absorption spectroscopy

INTRODUCTION

Micrometeorological methods have been used for more than 50 years to measure CO_2 and water vapor exchange between the atmosphere and crop canopy, although it has only been in the last couple of decades that advances in instrumentation have allowed for routine, continuous flux measurements from multiple sites on a global scale. Continued advances now also allow for micrometeorological measurement of other greenhouse gases including N_2O and CH_4. These techniques have the advantage that they can measure fluxes over large spatial and temporal scales without the limitations imposed by enclosed chamber methods. Although they have many advantages, as with any measurement tool, problems and limitations exist that must be recognized and accounted for. The purpose of this chapter is to review the

Managing Agricultural Greenhouse Gases. DOI: 10.1016/B978-0-12-386897-8.00021-8
2012, Published by Elsevier Inc.

FIGURE 21.1
Eddy covariance (A, C) and flux gradient (B, D) systems used for year-round micrometeorological measurement of CO_2 and N_2O. Note other instrumentation needed for flux-gradient measurements such as sonic and cup anemometers are not shown in pictures depicting the air sampling intakes for flux-gradient.

instrumentation and methods involved in micrometeorological measurement of CO_2 and N_2O flux, identify limitations to the techniques, outline solutions for those limitations, and describe insights gained from micrometeorological techniques that have enhanced our understanding of trace gas fluxes. We focus on CO_2 and N_2O because of the importance of these gases in the greenhouse gas budget of agroecosystems. We start with a brief review of the two most deployed micrometeorological methods, flux-gradient and eddy covariance (Figure 21.1), then review issues common to both CO_2 and N_2O flux measurements using these methods, and summarize specific studies for each of these greenhouse gases.

METHODS BASED ON FLUX-GRADIENT THEORY

Using an analogy to the molecular diffusion process, the turbulent vertical flux of a gas is proportional to its vertical time-averaged concentration gradient. The proportionality factor is called the eddy diffusivity (K) and this approach is termed gradient-transport or K-theory (Stull, 1988). According to flux-gradient (FG) theory the flux of a gas (F_g in $\mu g\,m^{-2}\,s^{-1}$) is given by:

$$F_g = -K_g(z)\,\frac{\partial \bar{\rho}_g}{\partial z} \qquad [1]$$

where $K_g(z)$ is the eddy diffusivity for the gas at height z ($m^2\,s^{-1}$) and , $\partial \bar{\rho}_g\,/\partial z$ is the vertical gradient in density of the gas ($\mu g\,m^{-3}$). The $K_g(z)$ is assumed equal to the eddy diffusivity of other scalars such as heat (Businger, 1986). Note that expressing the gradient in terms of the gas density in Eq. [1], rather than mixing ratio, requires considerations of changes in air density induced by temperature or pressure, and application of density corrections (Webb et al., 1980) as discussed below.

Businger (1986) stated that the flux-gradient method was "by far the most used method in micrometeorology". Developments in eddy covariance instrumentation over the last 25 years have decreased the popularity of FG-based methods; situations in which these methods are still applied will be discussed here. Traditionally, application of FG-theory involved measurements at several heights, consequently also termed the profile method

(Kanemasu et al., 1979). Approaches based on measurements at two levels currently used in agriculture applications are the Bowen ratio energy balance (BREB), modified Bowen ratio (M-BR), and aerodynamic methods (Meyers and Baldocchi, 2005). These approaches all rely on the measurement of gas density or mixing ratio differences between two heights above a canopy, but differ in how the eddy diffusivity is estimated as detailed in Foken (2008). Briefly, in the BREB approach the eddy diffusivity is obtained from either sensible (H) or latent heat flux (λE) divided by the difference in potential temperature or water vapor mixing ratio measured between two heights, respectively. This approach also requires measurements of net radiation and soil heat flux to calculate H and λE using the energy budget equation and the measured Bowen ratio. The M-BR follows the same principles as BREB except that H or λE is directly measured using the eddy covariance method.

In the aerodynamic method, the eddy diffusivity for a gas is expressed according to Monin-Obukhov similarity theory (Denmead, 2008):

$$K_g(z) = \frac{k(z-d)u_*}{\varphi_g(\zeta)} \quad [2]$$

where k is the von Kármán constant, u_* is the friction velocity (m s^{-1}), z is the measurement height (m), d is the zero plane displacement height (m) and $\varphi_g(\zeta)$ is an empirical stability function specific to the gas in question. The dimensionless parameter ζ is equal to $(z-d)/L$, where the Obukhov length, L (m), is defined as (Foken, 2008):

$$L = -\frac{u_*^3}{k\frac{g}{\theta_v}\overline{w'\theta_v'}} \quad [3]$$

and g/θ_v is the buoyancy parameter (m s^{-2} K^{-1}) with g equal to acceleration due to gravity, θ_v the virtual potential air temperature, and $\overline{w'\theta_v'}$ is the kinematic heat flux (K m s^{-1}).

Empirical expressions derived from intensive measurement campaigns relating H obtained from eddy covariance and vertical air temperature gradients have yielded many forms for the stability correction for heat ($\varphi_h(\zeta)$), termed universal functions (Foken, 2008). These are assumed to be valid for other scalars such as gas density (Prueger and Kustas, 2005).

Inserting Eq. [2] into the flux-gradient expression (Eq. [1]) and applying it to two measurement heights z_2 and z_1, requires the use of the integrated form:

$$F_g = \frac{u_* k(\rho_2-\rho_1)}{\left[\ln\left(\frac{z_2-d}{z_1-d}\right) - \psi_{h_2} + \psi_{h_1}\right]} \quad [4]$$

where ρ_2 and ρ_1 are the gas densities at the measurement heights, and ψ_{h_2} and ψ_{h_1} are the integrated stability functions at z_2 and z_1, respectively, resulting from integration of $\phi_h(\varsigma)$. Various forms for ψ_h dependent on atmospheric stability conditions have been fitted to experimental data; the most commonly used functions are (Arya, 1988):

- if $(z-d)/L > 0$ (stable case) then

$$\psi_h = -\beta\frac{(z-d)}{L} \quad [5]$$

- if $(z-d)/L = 0$ (neutral case) then $\psi_h = 0$
- if $(z-d)/L < 0$ (unstable case) then

$$\psi_h = 2\ln\left(\frac{1+x^2}{2}\right) \quad [6]$$

where$x = (1-\gamma(z-d)/L)^{1/4}$. The values for the constants β and γ vary depending on the experiment from which the stability functions were derived (Prueger and Kustas, 2005).

General values of $\beta = 5$ and $\gamma = 16$ are often used (Denmead, 2008). Foken (2008) recommends universal functions $\varphi_h(\zeta)$ according to Businger et al. (1971) and recalculated by Högström (1988) using the now accepted von Kármán constant of 0.40; these yield $\beta = 7.8$ and $\gamma = 12$.

EDDY COVARIANCE METHOD

Eddy covariance (EC) systems monitor three-dimensional wind velocity and vertical flux densities of λE, and H and scalars including CO_2, water vapor and other greenhouse gases. In the case of CO_2, vertical flux is calculated as the covariance between fluctuations in vertical wind velocity and CO_2 concentration. Similar fluxes can be calculated for others. The general equation for describing eddy flux as it applies to CO_2 can be written as follows:

$$F_C = \overline{w'\rho'} \qquad [7]$$

where, $F_C = CO_2$ flux, $w =$ vertical wind velocity, $\rho = CO_2$ density, prime indicates instantaneous fluctuations about the mean, and the overbar denotes a time average. One governing constraint of flux measurements is that the mean vertical flux of dry air should be zero (Webb et al., 1980). Several adjustments to the general flux equation are routinely carried out and will be discussed later.

Use of the EC technique was first proposed in the early 1950s as a means to measure the vertical flux of CO_2, water vapor and energy between ecosystems and the atmosphere (Swinbank, 1951). A historical review of micrometeorological techniques (Baldocchi et al., 2001) indicated that the earliest reported use of micrometeorological techniques to measure CO_2 flux occurred in the late 1950s and early 1960s, when flux gradient methods were employed to monitor fluxes over agricultural crops during the growing season. The routine application of EC methods was delayed until the 1980s when technological advances in instrumentation permitted flux measurements for short periods during the peak growing season. Additional advances in the late 1980s and early 1990s enabled researchers to make accurate flux measurements for extended periods of months or years. Since that time, the number of EC towers has grown worldwide until, as of 2011, more than 500 tower sites from about 30 regional networks across five continents were operating on a long-term basis (The FLUXNET Project: http://daac.ornl.gov/FLUXNET/fluxnet.shtml).

Widespread use of EC measurements of CO_2 flux required technological advancements in the design, robustness, and speed of sensors for measuring wind velocity, atmospheric humidity, CO_2 concentration and other atmospheric constituents (Moore, 1986). Wind velocity and temperature fluctuations are typically measured with a three-dimensional sonic anemometer, and CO_2 and water vapor fluctuations with either an open- or closed-path infrared gas analyzer. Sensors are sampled at frequencies of 10 Hz or faster.

ISSUES AFFECTING FLUX MEASUREMENTS

Energy Balance Closure

The term energy balance closure refers to the agreement between available energy (i.e. net radiation minus soil heat flux and storage terms) and the turbulent fluxes of H and λE. Energy balance closure is frequently not observed with an average underestimation of turbulent fluxes of 20% compared to the available energy (Wilson et al., 2002). This issue has been reported for turbulent energy fluxes in the literature for many years. Twine et al. (2000) were the first to identify its implications for EC CO_2 fluxes showing that they were underestimated by the same proportion as turbulent energy fluxes. Foken et al. (2010) concluded that "the missing energy is most likely transported into the atmosphere by processes that cannot be captured by a single-station EC measurement". Fluxes obtained using the aerodynamic flux-gradient method are likely underestimated as well because the empirical stability functions were derived from EC measurements. The underestimate of N_2O fluxes using the micrometeorological methods has not been discussed in the literature, except for an arbitrary 1.3 correction applied by Wagner-Riddle et al. (1996b, 1997) and Maggiotto et al. (2000).

Density Correction

Increased interest in measurements of CO_2 fluxes led Webb et al. (1980) to develop corrections to account for apparent CO_2 fluxes induced by fluctuations in water vapor and temperature. These are widely known as WPL corrections and can be of the same magnitude as the uncorrected flux for trace gases with a large gas density to flux ratio such as CO_2 and N_2O (Pattey et al., 1992). Leuning (2004) provides a step-by-step guide for calculating corrections when using open-path and closed-path gas analyzers. Open-path analyzers require corrections for both H and λE fluxes. For closed-path analyzers, only the λE flux is usually considered (Meyers and Baldocchi, 2005) as temperature fluctuations are usually negligible by the time sampled air has traveled through tubing and reached the analyzer. Leuning (2004) recommends that closed-path measured trace gas concentrations be converted to mixing ratio using high-frequency temperature and pressure data (20 Hz) to avoid errors associated with this assumption. If closed-path air samples are dried then the λE correction is not needed as shown by Webb et al. (1980). Interestingly, what motivated Webb et al. (1980) to develop these corrections was the suggestion expressed in the literature at the time that "a correction should be applied to recover the original CO_2 density of the moist air" after pre-drying to avoid water vapor interference.

Flux Footprint Analysis

Flux footprint analysis is commonly used to quantify the contributing source area for flux measurements or to determine fetch requirements when devising N_2O and CO_2 flux experiments (Hsieh et al., 2000). A general "rule of thumb" is that a 100 to 1 fetch-to-sensor height ratio is necessary to adequately contain the source area contributing to measured flux. However, fetch-to-height ratio, footprint peak position and footprint magnitude can all change with surface roughness, thermal stability and sensor height, such that the general "rule of thumb" greatly underestimates fetch requirements above smooth surfaces, under stable conditions, or with high sensor placement (Leclerc and Thurtell, 1990; Hsieh et al., 2000). Conversely, strongly convective conditions lead to relatively small footprints with a roughly 10 to 1 fetch-to-height ratio (Davis et al., 2003). Detailed descriptions of flux footprint calculation procedures are beyond the scope of this review. Readers are referred to papers by Kljun et al. (2004), Hsieh et al. (2000), and Leclerc and Thurtell (1990) for examples of footprint calculation procedures. An excellent review of respective strengths and weaknesses of footprint modeling approaches is provided by Schmid (2002).

During long-term measurements, study areas are exposed to a large number of different atmospheric conditions, and, thus, require numerous footprint calculations to adequately describe the area being sampled. Such calculations are generally computationally intensive and, therefore, not well suited for routine application to large data sets (Kljun et al., 2004). To facilitate more accessible footprint estimates, simplified models have been devised whereby footprint estimates can be related to observation height, surface roughness, and atmospheric stability, allowing for quick but precise footprint estimations (Hsieh et al., 2000; Kljun et al., 2004).

MICROMETEOROLOGICAL MEASUREMENTS OF NITROUS OXIDE FLUXES

Concerns about stratospheric ozone destruction due to nitric oxide derived from N_2O prompted early studies on N_2O emissions from agricultural soils under field conditions

(Hutchinson and Mosier, 1979). In the decade that followed, evidence of the potential contribution of N_2O to global warming added to this concern, and the number of direct measurements of N_2O emissions increased to 104, as summarized by Eichner (1990). Although the pioneering study by Hutchinson and Mosier (1979) used a micrometeorological method (FG) combined with gas chromatography (GC) for measurements of high N_2O fluxes, the studies that followed during the 1980s were limited to the closed-chamber method. The main limitation for micrometeorological measurements of N_2O fluxes during that time was the lack of analytical instrumentation with the precision to detect small differences in N_2O concentration required for FG, or with the time response necessary for EC deployment. Indeed, Hutchinson and Mosier (1979) reported only 5 hourly measurements during periods of very high fluxes due to limitations of the GC technique (1–2 ppb minimum detectable difference).

The development of fast response and high-precision analyzers based on tunable diode laser absorption spectroscopy (TDLAS) and Fourier-transform infra-red spectroscopy (FTIR) in the late 1980s allowed for the application of micrometeorological methods in N_2O emission studies from agricultural soils (Wagner-Riddle et al., 2005; Table 21.1). Early studies were of short duration and focused on determining if the newly developed instruments had the resolution and time response to be used with FG or EC methods (Wienhold et al., 1994; Hargreaves et al., 1996; Wagner-Riddle et al., 1996a; Laville et al., 1997). These were followed by investigations using TDLAS or FTIR, mostly in FG applications over relatively long time periods focused on evaluating potential mitigation practices (Table 21.1). More recently, the development of continuous wave quantum cascade laser-based absorption spectrometers (QCLAS), which do not require cryogenic cooling, has spurred research evaluating this sensor in EC measurements (Kroon et al. 2007; Neftel et al., 2010). The current state-of-the-art is that sensors based on TDLAS and QCLAS with the required precision and time response are available for meeting the stringent requirements of the EC method (Edwards et al. 2003; Kroon et al., 2007). Fourier-transform infra-red analyzers suitable for FG but not EC are also available, providing the advantage of multi-gas measurements (Griffith et al., 2002).

Nitrous oxide production in the soil is controlled by the complex interplay of microbiological processes and soil conditions such as oxygen, carbon and nitrogen content, temperature and pH (Granli and Bøckman, 1994). Soil tillage, crop type, and the application of nitrogen fertilizers influence the physical and hydrological conditions of the soil and the timing and distribution of nutrient inputs. These factors combined regulate the N_2O dynamics in the soil profile, and the timing and magnitude of emissions. Nitrous oxide fluxes are highly episodic with peak emissions occurring after wetting of dry soil (Davidson, 1992), addition of nitrogen and carbon (Mosier et al., 1982), and freeze/thaw cycles (Christensen and Tiedje, 1990). Therefore, Mosier et al. (1996) suggested that field studies conducted throughout the calendar year are needed to improve flux estimates.

The geographical distribution and duration of N_2O flux studies has been limited (Table 21.1) particularly in comparison to CO_2 flux research. While a range of crops has been studied, a systematic approach considering typical pasture and cropping systems is needed. Micrometeorological methods do not interfere with soil conditions and hence can be used quasi-continuously (at hourly to half-hourly intervals) to capture the highly intermittent nature of N_2O emission episodes. In addition, tower-based micrometeorological measurements yield flux values integrated over a relatively large area depended on the height of the measuring tower, atmospheric and surface conditions, covering upwind distances ranging from 10 m to 1 km (Asman et al., 1999). Areal integration provided by micrometeorological methods is an advantage compared to chamber methods, as the large spatial variability of N_2O fluxes (Yates et al., 2006) dictates the need for a large number of chambers to obtain similar spatial integration. The ability to provide quasi-continuous temporal and spatially integrated

TABLE 21.1 Micrometeorological N_2O Flux Measurements from Agricultural Soil Reported in the Literature. Measurement Method, Analytical Technique, Duration of Study, Number of Observations, Location, Treatments, and Crops Studied are Listed for Each Investigation

	Method	Analysis	Duration (nobs[1])	Location	Treatment	Crop(s)	Purpose
Hutchinson and Mosier (1979)	FG[2]	GC[5]	May–Sep 1978 (5 h)	CO, USA	NH_3 fertilization	Irrigated maize	Determine fraction of N fertilizer lost as N_2O
Wienhold et al. (1994)	FG and EC[3]	TDLAS[6]	Aug 1992	Scotland	–	Grass	Compare methods
Hargreaves et al. (1996)	FG	GC TDLAS FTIR[7]	Aug 1993 (5 d)	Denmark	–	Harvested wheat	Compare methods
Wagner-Riddle et al. (1996b)	FG	TDLAS	Aug–Oct 1992, Apr–Sep 1992 (7 mo.)	ON, Canada	Irrigation, fertilizer, C addition	Bare soil	Evaluate management practices
Wagner-Riddle et al. (1997)	FG	TDLAS	Jul–Nov 1992 Mar 1993–Feb 1995 (2445 d)	ON, Canada	NH_4NO_3, manure addition	Alfalfa, bare soil, grass, maize	Estimate seasonal and annual N_2O emissions
Laville et al. (1997)	FG and EC	TDLAS	Oct–Nov 1995 (11 d)	France	NH_4NO_3 fertilization	Plowed wheat	Compare methods
Wagner-Riddle and Thurtell (1998)	FG	TDLAS	Mar 1993–Apr 1996 (12 mo.)	ON, Canada	Fall-plowing and manure addition	Alfalfa, barley, canola, soybean, grass, maize	Quantify spring thaw emissions and management effects
Grant and Pattey (1999)	FG	TDLAS	Mar–Apr 1996 (60 d)	ON, Canada	–	Barley stubble	Model N_2O emissions induced by spring thaw
Laville et al. (1999)	FG and EC	TDLAS	June 1996 (14 d)	France	Anhydrous ammonia fertilization	Maize	Compare methods
Maggiotto et al. (2000)	FG	TDLAS	Sep–Oct 1995, May–Oct 1996, Jun–Nov 1997 (353 d)	ON, Canada	Urea, slow-release urea, NH_4NO_3	Turfgrass	Determine effect of N fertilizer type on N_2O emissions
Grant and Pattey (2003)	FG	TDLAS	June 1998 (1454 30 min)	ON, Canada	Urea fertilization rate	Maize	Model N_2O emissions after fertilization
Scanlon and Kiely (2003)	EC	TDLAS	Jul 2002–Mar 2003 (8 mo.)	Ireland	Fertilizer and manure addition	Grazed grassland	Understand factors that control dynamics of N_2O flux
Phillips et al. (2007)	FG	TDLAS	July 2003–June 2005 (629 d)	Australia	Fertilization, grazing, irrigation	Dairy pasture	Improve N_2O emission factor
Wagner-Riddle et al. (2007)	FG	TDLAS	Jan 2000–Apr 2005 (2195 d)	ON, Canada	Tillage, fertilization rate and timing	Maize–soybean–wheat rotation	Evaluate N_2O emission reduction due to BMP practices

Continued

TABLE 21.1 Micrometeorological N_2O Flux Measurements from Agricultural Soil Reported in the Literature. Measurement Method, Analytical Technique, Duration of Study, Number of Observations, Location, Treatments, and Crops Studied are Listed for Each Investigation—continued

	Method	Analysis	Duration (nobs[1])	Location	Treatment	Crop(s)	Purpose
Harvey et al. (2008)	FG	TDLAS GC	October 2006 (20 d)	New Zealand	Pre- and post-grazing	Grass–clover paddock	Demonstrate technique for top-down N_2O emission verification
Pattey et al. (2008)	FG	TDLAS	Apr–Sep 2003 Mar–May 2004 (173 d)	QC, Canada	Urea, beef manure and cover crop addition	Edible peas	Quantify annual emissions
Denmead et al. (2010)	FG	FTIR	Oct 2005–Sep 2006 (634 d)	Australia	Soil type and residue management	Sugarcane	Provide data for assessing biofuels
Desjardins et al. (2010)	FG and REA[4]	TDLAS	Mar–Apr 2000, 2001, 2003 Apr–Jun 2004	ON, Canada		Maize–soybean rotation	Compare tower- and aircraft-based flux measurements
Kim et al. (2010)	EC	TDLAS	2003–2008	Ireland	Fertilizer and manure addition	Grazed grassland	Understand how dry periods affect N_2O flux
Kroon et al. (2010)	EC	QCLAS[8]	Apr 2006–Oct 2008 (20,957–30 min)	Netherlands	Fertilization	Dairy-managed peatland	Obtain annual N_2O balance
Neftel et al. (2010)	EC	QCLAS	June–Sep 2008 (~45 d)	Switzerland	NH_4NO_3 and and cattle slurry fertilization	Grass–clover non-grazed pasture	Evaluate analyzer
Metivier et al. (2009)	FG	TDLAS	May–Jul 2004	ON, Canada	Urea fertilization	Canola	Testing model to simulate N_2O fluxes at the field scale

[1]*nobs = number of observations: months, days, hours as/if reported.*
[2]*FG = flux-gradient.*
[3]*EC = eddy covariance.*
[4]*REA = relaxed eddy accumulation.*
[5]*GC = gas chromatography.*
[6]*TDLAS = tunable diode laser spectroscopy.*
[7]*Fourier-transform infra-red spectroscopy.*
[8]*QCLAS = quantum cascade laser spectroscopy.*

measurements makes micrometeorological methods specifically suited for long-term measurement of N_2O fluxes from agroecosystems, and for the study of management effects on annual emissions.

Despite the advantages discussed above, only a few multi-year micrometeorological N_2O emission studies have been conducted on agricultural sites up to now (Wagner-Riddle et al., 1997; Wagner-Riddle and Thurtell, 1998; Phillips et al., 2007; Wagner-Riddle et al., 2007; Denmead et al., 2008; Kim et al., 2010; Kroon et al., 2010). Cost of monitoring equipment and training of personnel in the use of new measurement technologies are possible reasons for this. In addition, there is a mismatch in the requirement of large plots for micrometeorological methods, and the typical agronomic experimental setup on multiple small plots. Several of the pressing research topics related to potential N_2O flux reduction through management practices (e.g. effect of N fertilizer timing, rate and type) require factorial experiments which can only be monitored with chamber methods. A combined approach spanning the spatial scales of chambers, to identify the most promising mitigation measures, and micrometeorological methods, to evaluate these over multiple years, would be beneficial. An example of the latter is the study of Wagner-Riddle et al. (2007), where a combination of best management practices (BMP), such as N rate tailored to crops' needs and applied as side-dress and no-tillage to a corn–soybean–wheat rotation were studied over 5 years. Year-round micrometeorological flux measurements showed mean annual N_2O emission decreased by 0.79 kg N ha^{-1} when compared to emissions for conventional practices of 2.19 kg N ha^{-1}. Large variation in emissions were observed within and between years, with 80% in the N_2O reduction occurring during the non-growing season (Nov–Apr) compared to 20% for the growing season, emphasizing the need for year-round measurements.

When considering which micrometeorological method (FG or EC) to deploy, preference is often given to EC since this method measures flux directly without relying on empirical functions as is the case for the FG method (Denmead, 2008). Most flux-gradient measurements listed in Table 21.1 have relied on a combination of eddy covariance measurement of friction velocity and sensible heat, and N_2O density differences between two or more heights. Lack of fast response sensors is commonly cited as a reason for using the FG technique. However, in current practice, fast response analyzers are often used with FG for N_2O flux measurement (Grant and Pattey, 2003; Phillips et al., 2007; Pattey et al., 2008; Desjardins et al., 2010; Wagner-Riddle et al., 2010). Fast response sensors (10 Hz) achieve high resolution in concentration difference measurements, due to the increased samples per measurement period and overall reduction of noise levels (Wagner-Riddle et al., 2005). In addition, long tube lengths and setup of multi-plot (2–4) experiments are possible with the FG method as high-frequency response is not required (Wagner-Riddle et al., 2005). A multi-plot setup allows the side-by-side comparison of management practices, which is extremely important for a trace gas with emissions that are highly variable in time. Nitrous oxide fluxes are driven by environmental events such as rainfall and spring thaw, so that management effects are confounded with weather events on a short time scale. This is illustrated in Figure 21.2, where applying N fertilizer according to pre-plant soil N test recommendations at the 6th leaf stage in 2000 (Figure 21.2C) significantly reduced emissions over 20 days following fertilization compared to applying the recommended rate at corn planting (Figure 21.2A). However, a similar reduction was not observed in 2003 when a timely rainfall promoted a large N_2O flux for the BMP treatment (Figure 21.2D).

Micrometeorological N_2O flux datasets can make a valuable contribution to model testing and validation. Development of the model *ecosys* for simulation of N_2O emissions associated with freeze/thaw and N fertilization has benefited from availability of high temporal resolution N_2O flux data (Grant and Pattey, 1999, 2003; Metivier et al., 2009). Recently, Kariyapperuma et al. (2011) showed that the mechanism underlying freeze/thaw induced N_2O flux in the DNDC model (i.e. release of ice-trapped N_2O) was not correct. They suggested this

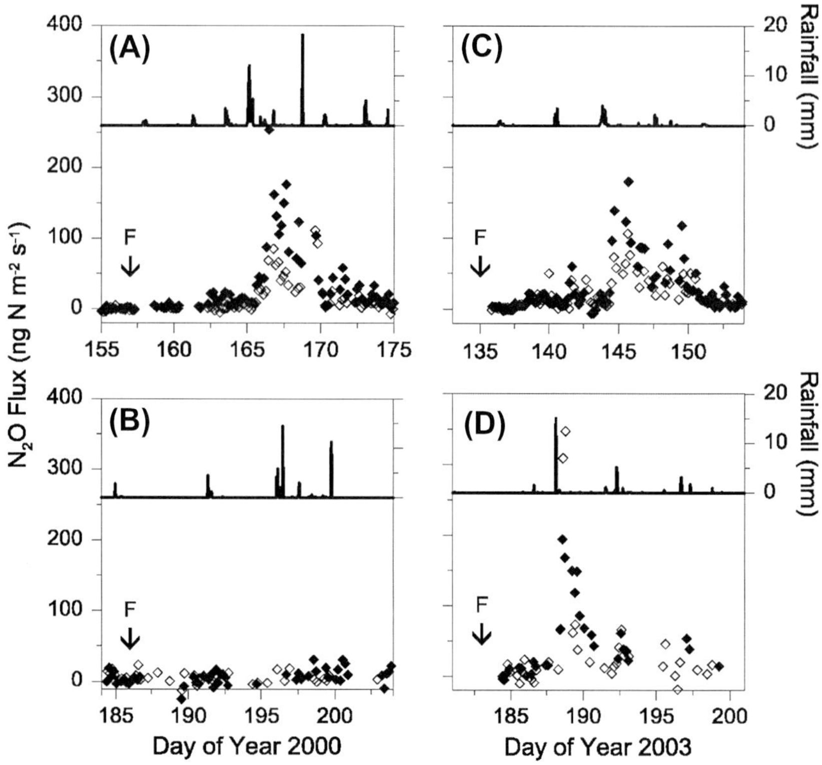

FIGURE 21.2
Hourly N_2O fluxes after corn fertilization under conventional practices (A, C) and best management practices (B, D) in 2000 (left graphs) and 2003 (right graphs). Fluxes for replicate 1.5 ha plots (plot 1, closed symbol; plot 2 open symbol) were measure using the flux-gradient method (Wagner-Riddle et al., 2007). The X-axis on each graph starts two days before the nitrogen fertilizer application indicated by an arrow, and ends 20 days after. Conventional practice consisted of fall and spring tillage, and N fertilizer rate of 150 kg N ha^{-1} applied as urea at planting. Best management practices consisted of no-tillage, and N fertilizer rate of 50 (in 2000) and 60 kg N ha^{-1} (in 2003) applied as urea-ammonium nitrate solution at the 6th leaf stage.

misconception may have been overlooked in previous studies due to the lack of continuously measured N_2O fluxes during winter and spring thaw conditions.

MICROMETEOROLOGICAL MEASUREMENTS OF CO_2 FLUX

Eddy covariance is the most commonly used technique for micrometeorological measurements of CO_2 flux, and a global network of EC sites (FLUXNET) has been organized for compiling, archiving, and distributing flux and associated data to ensure the inter-comparability of regional data (Baldocchi et al., 2001). However, differences in CO_2 measurement instrumentation, i.e. open- vs. closed-path systems, and the occasional use of alternative techniques, such as Bowen ratio methods, could potentially complicate comparisons among sites and global estimates of CO_2 flux. For example, a comparison of fluxes from 19 European grasslands included 13 sites that used open-path systems, five that employed closed-path systems, and one that used both (Gilmanov et al., 2007). Another analysis of worldwide grasslands included both eddy covariance and Bowen ratio data (Gilmanov et al., 2010).

Concerns over differences between EC and Bowen ration systems arise primarily because of differences in partitioning of energy fluxes, with EC systems typically favoring λE and Bowen ratio systems favoring H flux (Brotzge and Crawford, 2003). However, a recent analysis of a shortgrass steppe by Wolf et al. (2008) suggested that when all relevant corrections are applied to EC data, including signal asynchrony (either from spatial separation of the sensors or from delays introduced by the instruments) and frequency domain corrections which are not universally employed, there was no bias for H and λE between EC and Bowen ratio systems. The EC systems tended to vary from expected CO_2 flux values under low light and during the night, whereas Bowen ratio systems tended to show occasional higher than expected daytime CO_2 uptake (Wolf et al., 2008). In a forested system, Fredeen et al. (2007) also found that EC flux estimates were less than Bowen ratio estimates when fluxes were high, but were greater than Bowen ratio estimates when fluxes were low. Their analysis was unable to determine which system was most correct. Currently missing are studies comparing EC and Bowen ratio flux estimates for the entire growing season or for complete annual cycles to determine how differences between systems affect long-term estimates of net C exchange.

Several differences exist between open-path (OP) and closed-path (CP) EC systems which could potentially affect both ease of operation and flux estimates. Open-path systems typically consume less power and require less maintenance than CP systems, but tend to produce more erroneous data when increased turbidity of the sensor window occurs due to the presence of precipitation, dew, or dirt (Haslwanter et al., 2009). Differences in low-pass filtering, energy balance closure, and random flux uncertainty can also be observed between systems. Despite the differences, Haslwanter et al. (2009) found that annual net ecosystem exchange differed by only 25 $g\,C\,m^{-2}\,yr^{-1}$ between the two systems, and concluded that the use of CP vs. OP systems did not represent a serious source of bias for their data set.

Although comparisons between OP and CP systems often give substantially similar results, a growing body of evidence has shown that OP systems often indicate nighttime and off-season CO_2 uptake when it would not be physiologically reasonable to expect it (Burba et al., 2008). The unexpected uptake results from environmental and/or electronic surface heating of the CO_2 analyzer resulting in sensible heat fluxes inside the open-path cell which are different from those in ambient air (Burba et al., 2006). The erroneous uptake can be avoided by enclosing the OP sensor to attenuate temperature fluctuations, measuring the integrated temperature within the path itself, or applying modeled corrections to previously collected data based on standard weather measurements (Burba et al., 2008). Such corrections may make only small differences where net annual fluxes are concerned (Haslwanter et al., 2009), but can be crucial for physiological studies such as investigations into low-temperature limits to CO_2 uptake (Skinner, 2007).

Although one of the primary benefits of EC systems is the ability to continuously measure CO_2 flux across diurnal, seasonal, and annual scales, gaps frequently appear in the data for a number of reasons including equipment and power failures, system maintenance, obstruction of sensors by precipitation or dew, and during periods of weak turbulent mixing. In a survey of FluxNet sites, 9 to 65% of all observations were rejected or missing, with an average of 35% (Falge et al., 2001). In the majority of cases a greater percentage of nighttime vs. daytime observations were missing. Because data gaps do not occur randomly, specific gap-filling techniques are required for daytime and nighttime periods and for gaps during specific seasons of the year before accurate annual flux estimates can be determined. No universal methods have emerged for filling gaps, and the most appropriate method may be unique to the conditions at a given site. All published studies on gap-filling methods have studied CO_2 flux time series, with no reported studies on gap filling specific to N_2O flux. Falge et al. (2001) compared three major gap-filling methodologies against a set of 28 unique site-year combinations. Filling methods included: (1) mean diurnal variation wherein a missing observation is replaced by the mean for that time period from usually 4 to 15 adjacent days; (2) the use of look-up tables based on the environmental conditions associated with the missing data; and (3) use of nonlinear regression relationships between flux components and site- and season-specific associated controlling factor such as light and temperature. They concluded that errors introduced by gap filling did not differ among methods but were proportional to the number of gaps being filled.

In a separate analysis of daytime and nighttime fluxes, Stoy et al. (2006) found that gap filling had little effect on daytime data, but significantly altered nighttime results. Measuring nighttime respiration can be particularly troublesome because of calm and stable atmospheric conditions which often exist during the night, typically leading to underestimation of nighttime respiration (Wohlfahrt et al., 2005). Discrimination between calm and windy conditions is usually based on friction velocity (u_*) where some minimum value for u_* is usually designated, below which measured fluxes are rejected and replaced using gap-filling procedures

(Wohlfahrt et al., 2005; Moureaux et al., 2006; Acevedo et al., 2009). Moureaux et al. (2006) also suggested that data above a maximum u_* be discarded because of potential pressure pumping effects. Although widely used, no standard value exists for the u_*-correction which depends on site characteristics (the minimum threshold is generally higher for forested compared with crop or grassland sites) and individual researcher's judgment. However, a threshold value for u_* of about 0.2 m s^{-1} is typical (Davis et al., 2003). As an alternative, Acevedo et al. (2009) showed that the relationship between u_* and turbulence is not necessarily straightforward, and proposed that the standard deviation of vertical velocity fluctuations (σ_w) be used instead of u_*. They suggest that use of σ_w provides for easier determination of the threshold for data filtering.

Direct output from EC systems provides measurements of net ecosystem exchange (NEE), but there is much interest in partitioning daytime NEE into its component fluxes, gross primary productivity (GPP), and ecosystem respiration (Re) in order to evaluate the processes that control NEE (Stoy et al., 2006). Many of the methods used to fill missing NEE data can also be used to estimate GPP and Re. Most common methods employ some version of either using nighttime NEE to estimate daytime Re, or using light response curves with daytime Re derived from the intercept. Ecosystem respiration is then subtracted from NEE to obtain GPP. Ecosystem models are also often used to estimate GPP and Re. Desai et al. (2008) compared 23 different methods over 10 site-years of temperate forest data and found that most methods differed by less than 10% in their estimates of annual GPP and Re with a mean deviation of 9.8% for Re and 7.0% for GPP. On a diurnal basis, large variation among methods occurred for estimates of hourly Re, primarily because of differences from using air temperature vs. soil temperature for controlling respiration. However, strong correspondence existed among methods for estimating GPP due to the strong link between GPP and incoming radiation. In addition to GPP, such analysis can provide estimates of other photosynthetic parameters such as quantum yield, radiation use efficiency, and maximum gross photosynthesis (Gilmanov et al., 2005, 2010).

Gilmanov et al. (2010) partitioned eddy covariance NEE data into GPP and Re for 118 grassland, cropland, shrubland, savanna, and wetland ecosystems from around the world. They found the greatest GPP and Re in intensively managed grasslands, followed by croplands, shrublands and savannas, wetlands, and extensively managed grasslands. Interestingly, even though extensively managed grasslands had the lowest mean GPP, the highest GPP in the study occurred on an extensively managed tropical grassland in Brazil. Comparisons with selected forest sites found that maximum annual GPP was generally higher for forests, but the highest daily rate of photosynthetic uptake was for intensive maize crops in the Midwestern U.S. Annual GPP was typically higher in forests because of longer production periods, especially in tropical forests.

Unlike the limited application of micrometeorological techniques for N_2O measurements, the number of sites measuring CO_2 fluxes is large, as evidenced by the more than 500 locations participating in the FLUXNET network mentioned earlier. Early on, studies involving forests and C_3 crops dominated, whereas limited information was available for other ecosystems (Buchmann, 1999). Even though the number of non-forested sites has increased as the network has expanded, forested sites still tend to dominate. For example, in a study examining ecosystem respiration from 104 FLUXNET sites, 58% were forested compared with 15% grassland, 11% cropland, 9% shrubland, 5% savanna, and 3% wetland (Migliavacca et al., 2011). Even given the predominance of forests, data from enough non-forested sites now exist to provide comprehensive estimates of CO_2 flux from a number of ecosystems. As an example, Gilmanov et al. (2010) were able to provide worldwide GPP and Re estimates for 316 site-years from 118 locations representing non-forested ecosystems. Of those, 45% were extensively managed grasslands, 24% croplands, 15% intensively managed grasslands, 9% shrublands and savanna, and 8% wetlands. Such widespread coverage of the world's ecosystems will allow robust estimates of climate and management effects on CO_2 fluxes.

The continuous, large-scale flux measurements by EC systems can provide unique insights into mechanisms controlling CO_2 fluxes that would not be possible with more limited data collection systems. Of particular value is the ability to measure fluxes without the interference and climate alteration of enclosed chambers, the integration of fluxes over large areas, and the ability to capture flux data during periods that might otherwise be missed using non-continuous sampling methods. The need to measure photosynthetic fixation by undisturbed field crops was well recognized by the early 1960s (Lemon, 1960), and the ability to measure flux, free from the climate altering effects of chambers, remains one of the most attractive features of micrometeorological methods. Another is the ability to monitor CO_2 flux over large areas rather than being constrained by the small footprint of chamber measurements since the EC footprint is only constrained by the height of the sensors. As an extreme example, Davis et al. (2003) instrumented a 447 m tall TV tower with EC systems at 30, 122, and 396 m, providing flux footprints of hundreds to thousands of meters depending on sensor height and atmospheric conditions.

Several examples can be provided of the types of information that can be obtained due to the continuous nature of EC measurements. First, EC data can capture unique phenomena that might otherwise be missed. Aerosols from the 1991 Mount Pinatubo eruption decreased global solar radiation but greatly increased diffuse radiation for 2 years following the eruption. Using EC data collected at Harvard Forest, Gu et al. (2003) showed that the increased diffuse radiation enhanced noontime photosynthesis by 23% in 1992 and 8% in 1993. They argued that this increase in photosynthesis enhanced the terrestrial C sink and contributed to the decline in the growth rate of atmospheric CO_2 that occurred after the eruption. In a related study, Law et al. (2002) found that the highest solar radiation occurs on partly cloudy days due to increased diffuse radiation. They also showed that net C uptake was greater for a given irradiance level on cloudy days suggesting the photosynthesis was enhanced under diffuse radiation.

Second, EC data can provide CO_2 flux data for periods, such as wintertime, when chamber measurements are difficult or impossible to make. In a study on humid temperate pastures in the northeastern U.S., Skinner (2007) was able to show that in the absence of snow cover photosynthetic uptake occurred at air temperatures as low as $-4°C$, and that the transition from net C sink to net C source in the autumn was determined more by available solar radiation and the amount of live plant biomass than by temperature. In an examination of the transition from a wintertime C source to a springtime sink, Monson et al. (2005), determined that CO_2 assimilation during the snow-melt period represented a significant proportion of annual NEE, and that while warm, early-season temperatures were sufficient to initiate limited CO_2 uptake, the availability of liquid water from snow melt along with warmer air temperatures were required for full recovery of the capacity for ecosystem C uptake.

Third, complete annual CO_2 flux budgets across a number of locations can provide insights into the mechanisms that control NEE. In a comparison of 15 European forests, Valentini et al. (2000) found that GPP was largely independent of latitude, whereas NEE significantly increased with decreasing latitude. They concluded that Re rather than GPP was the primary determinant of NEE. Skinner and Adler (2010) also found that NEE for a newly established switchgrass (*Panicum virgatum* L.) stand was controlled by Re rather than GPP.

CONCLUSIONS

Micrometeorological methods have been found to be valuable resources for providing long-term, spatially relevant measurements of CO_2 and N_2O fluxes. In particular, they are crucial for capturing periodic high-flux events that have disproportional effects on total annual flux, but whose occurrence is difficult to predict. Large global networks of CO_2 flux monitoring sites are providing much needed information on ecosystem contributions to the global C budget.

A similar expansion of N_2O monitoring sites is needed to fully understand environmental and management effects on emissions of this important greenhouse gas.

References

Acevedo, O.C., Moraes, O.L.L., Degrazia, G.A., Fitzjarrald, D.R., Manzi, A.O., Campos, J.G., 2009. Is friction velocity the most appropriate scale for correcting nocturnal carbon dioxide fluxes? Agric. Forest Meteorol. 149, 1–10.

Arya, S.P., 1988. Introduction to micrometeorology. Academic Press, London.

Baldocchi, D., Falge, E., Gu, L., Olson, R., Hollinger, D., Running, S., Anthoni, P., Bernhofer, C., Davis, K., Evans, R., Fuentes, J., Goldstein, A., Katul, G., Law, B., Lee, X., Malhi, Y., Meyers, T., Munger, W., Oechel, W., Paw U, K.T., Pilegaard, K., Schmid, H.P., Valentini, R., Verma, S., Vesala, T., Wilson, K., Wofsy, S., 2001. FLUXNET: a new tool to study the temporal and spatial variability of ecosystem-scale carbon dioxide, water vapor, and energy flux densities. B. Am. Meteorol. Soc. 82, 2415–2434.

Brotzge, J.A., Crawford, K.C., 2003. Examination of the surface energy budget: a comparison of eddy correlation and Bowen ratio measurement systems. J. Hydrometeor. 4, 160–178.

Buchmann, N., 1999. Net CO_2 and H_2O fluxes of terrestrial ecosystems. Global Biogeochem. Cycles 13, 751–760.

Burba, G.G., Anderson, D.J., Xu, L., McDermitt, D.K., 2006. Correcting apparent off-season CO_2 uptake due to surface heating of an open-path gas analyzer: progress report of an ongoing experiment. Proceedings of the 27th Annual Conference of Agricultural and Forest Meteorology, San Diego, CA, p. 13.

Burba, G.G., McDermitt, D.K., Grelle, A., Anderson, D.J., Xu, L., 2008. Addressing the influence of instrument surface heat exchange on the measurements of CO_2 flux from open-path gas analyzers. Global Change Biol. 14, 1854–1876.

Businger, J.A., 1986. Evaluation of the accuracy with which dry deposition can be measured with current micrometeorological techniques. J. Climate Appl. Meteorol. 25, 1100–1124.

Businger, J.A., Wyngaard, J.C., Izumi, Y., Bradley, E.F., 1971. Flux-profile relationships in atmospheric surface layer. J. Atmos. Sci. 28, 181–187.

Christensen, S., Tiedje, J.M., 1990. Brief and vigorous N_2O production by soil at spring thaw. J. Soil Sci. 41, 1–4.

Davidson, E.A., 1992. Sources of nitric oxide and nitrous oxide following wetting of dry soil. Soil Sci. Soc. Am. J. 56, 95–102.

Davis, K.J., Bakwin, P.S., Yi, C., Berger, B.W., Zhao, C., Teclaw, R.M., Isebrands, J.G., 2003. The annual cycles of CO_2 and H_2O exchange over a northern mixed forest as observed from a very tall tower. Global Change Biol. 9, 1278–1293.

Denmead, O.T., 2008. Approaches to measuring fluxes of methane and nitrous oxide between landscapes and the atmosphere. Plant and Soil 309, 5–24.

Denmead, O.T., Macdonald, B.C.T., Bryant, G., Naylor, T., Wilson, S., Griffith, D.W.T., Wang, W.J., Salter, B., White, I., Moody, P.W., 2010. Emissions of methane and nitrous oxide from Australian sugarcane soils. Agric. Forest Meteorol. 150, 748–756.

Desai, A.R., Richardson, A.D., Moffat, A.M., Kattge, J., Hollinger, D.Y., Barr, A., Falge, E., Noormets, A., Papale, D., Reichstein, M., Stauch, V.J., 2008. Cross-site evaluation of eddy covariance GPP and RE decomposition techniques. Agric. Forest Meteorol. 148, 821–838.

Desjardins, R.L., Pattey, E., Smith, W.N., Worth, D., Grant, B., Srinivasan, R., MacPherson, J.I., Mauder, M., 2010. Multiscale estimates of N_2O emissions from agricultural lands. Agric. Forest Meteorol. 150, 817–824.

Edwards, G.C., Thurtell, G.W., Kidd, G.E., Dias, G.M., Wagner-Riddle, C., 2003. A diode laser based gas monitor suitable for measurement of trace gas exchange using micrometeorological techniques. Agr. For. Meteorol 115, 71–89.

Eichner, M.J., 1990. Nitrous oxide emissions from fertilized soils—summary of available data. J. Environ. Qual. 19, 272–280.

Falge, E., Baldocchi, D., Olson, R., Anthoni, P., Aubinet, M., Bernhofer, C., Burba, G., Ceulemans, R., Clement, R., Dolman, H., Granier, A., Gross, P., Grunwald, T., Hollinger, D., Jensen, N.-O., Katul, G., Keronen, P., Kowalski, A., Lai, C.T., Law, B.E., Meyers, T., Moncrieff, J., Moors, E., Munger, J.W., Pilegaard, K., Rannik, U., Rebmann, C., Suyker, A., Tenhunen, J., Tu, K., Verma, S., Vesala, T., Wilson, K., Wofsy, S., 2001. Gap filling strategies for defensible annual sums of net ecosystem exchange. Agric. Forest Meteorol. 107, 43–69.

Foken, T., 2008. Micrometeorology. Springer, Berlin, p 308.

Foken, T., Mauder, M., Liebethal, C., Wimmer, F., Beyrich, F., Leps, J.P., Raasch, S., DeBruin, H.A.R., Meijninger, W.M.L., Bange, J., 2010. Energy balance closure for the LITFASS-2003 experiment. Theoret. Appl. Climatol. 101, 149–160.

Fredeen, A.L., Waughtal, J.D., Pypker, T.G., 2007. When do replanted sub-boreal clearcuts become net sinks for CO_2? Forest Ecol. Manage. 239, 210–216.

Gilmanov, T.G., Aires, L., Barcza, Z., Baron, V.S., Belelli, L., Beringer, J., Billesbach, D., Bonal, D., Bradford, J., Ceschia, E., Cook, D., Corradi, C., Frank, A., Gianelle, D., Gimeno, C., Gruenwald, T., Guo, H., Hanan, N., Haszpra, L., Heilman, J., Jacobs, A., Jones, M.B., Johnson, D.A., Kiely, G., Li, S., Magliulo, V., Moors, E., Nagy, Z., Nasyrov, M., Owensby, C., Pinter, K., Pio, C., Reichstein, M., Sanz, M.J., Scott, R., Soussana, J.F., Stoy, P.C., Svejcar, T., Tuba, Z., Zhou, G., 2010. Productivity, respiration, and light-response parameters of world grassland and agroecosystem derived from flux-tower measurements. Rangeland Ecol. Manage. 63, 16–39.

Gilmanov, T.G., Soussana, J.F., Aires, L., Allard, V., Ammann, C., Baozarolo, M., Barcza, Z., Bernhofer, C., Campbell, C.L., Cernusca, A., Cescatti, A., Clifton-Brown, J., Dirks, B.O.M., Dore, S., Eugster, W., Fuhrer, J., Gimeno, C., Gruenwald, T., Haszpra, L., Hensen, A., Ibrom, A., Jacobs, A.F.G., Jones, M.B., Lanigan, G., Laurila, T., Lohila, A., Manca, G., Marcolla, B., Nagy, Z., Pilegaard, K., Pinter, K., Pio, C., Raschi, A., Rogiers, N., Sanz, M.J., Stefani, P., Sutton, M., Tuba, Z., Valentini, R., Williams, M.L., Wohlfahrt, G., 2007. Partitioning European grassland net ecosystem CO_2 exchange into gross primary productivity and ecosystem respiration using light response function analysis. Agric. Ecosys. Environ. 121, 93–120.

Gilmanov, T.G., Tieszen, L.L., Wylie, B.K., Flanagan, L.B., Frank, A.B., Haferkamp, M.R., Meyers, T.P., Morgan, J.A., 2005. Integration of CO_2 flux and remotely-sensed data for primary production and ecosystem respiration analyses in the northern Great Plains: potential for quantitative spatial extrapolation. Blobal Ecol. Biogeogr. 14, 271–292.

Granli, T., Bøckman, O.C., 1994. Nitrous oxide from agriculture. NorW. J. Agric. Sci 12, 1–127.

Grant, R.F., Pattey, E., 1999. Mathematical modeling of nitrous oxide emissions from an agricultural field during spring thaw. Global Biogeochem. Cycles 13, 679–694.

Grant, R.F., Pattey, E., 2003. Modelling variability in N_2O emissions from fertilized agricultural fields. Soil Biol. Biochem. 35, 225–243.

Griffith, D.W.T., Leuning, R., Denmead, O.T., Jamie, I.M., 2002. Air-land exchanges of CO_2, CH_4 and N_2O measured by FTIR spectrometry and micrometeorological techniques. Atm. Environ 36, 1833–1842.

Gu, L., Baldocchi, D.D., Wofsy, S.C., Munger, J.W., Michalsky, J.J., Urbanski, S.P., Boden, T.A., 2003. Response of a deciduous forest to the Mount Pinatubo eruption: enhanced photosynthesis. Science 299, 2035–2038.

Hargreaves, K.J., Wienhold, F.G., Klemedtsson, L., Arah, J.R.M., Beverland, I.J., Fowler, D., Galle, B., Griffith, D.W.T., Skiba, U., Smith, K.A., Welling, M., Harris, G.W., 1996. Measurement of nitrous oxide emission from agricultural land using micrometeorological methods. Atmos. Environ. 30, 1563–1571.

Harvey, M., Pattey, E., Saggar, S., Bromley, T., Dow, D., Kotkamp, M., Martin, R., Moss, R., Singh, J., 2008. Verification techniques for N_2O emission at the paddock scale in New Zealand: FarmGas2006. Austr. J. Exp. Agric. 48, 138–141.

Haslwanter, A., Hammerle, A., Wohlfahrt, G., 2009. Open-path vs. closed-path eddy covariance measurements of the net ecosystem carbon dioxide and water vapour exchange: a long-term perspective. Agric. Forest Meteorol. 149, 291–302.

Högström, U., 1988. Non-dimensional wind and temperature profiles in the atmospheric surface layer—a re-evaluation. Bound.-Layer Meteorol. 42, 55–78.

Hsieh, C.-I., Katul, G., Chi, T., 2000. An approximate analytical model for footprint estimation of scalar fluxes in thermally stratified atmospheric flows. Adv. Water Resour. 23, 765–772.

Hutchinson, G.L., Mosier, A.R., 1979. Nitrous oxide emissions from an irrigated corn field. Science 205, 1125–1127.

Kanemasu, E.T., Wesely, M.L., Hicks, B.B., Heilman. J.L. 1979. Techniques for calculating energy and mass fluxes. In B.J. Barfield and J.F. Gerber (eds.), Modification of the Aerial Environment of Plants. ASAE Monograph no. 2, p. 156–182.

Kariyapperuma, K.A., Wagner-Riddle, C., Furon, A.C., Li, C., 2011. Assessing winter and spring thaw nitrous oxide fluxes simulated by the DNDC model for agricultural soils in Ontario, Canada. Soil Sci. Soc. Am. J. 75, 678–690.

Kim, D.G., Mishurov, M., Kiely, G., 2010. Effect of increased N use and dry periods on N_2O emission from a fertilized grassland. Nutrient Cycl. Agroecosys. 88, 397–410.

Kljun, N., Calanca, P., Rotach, M.W., Schmid, H.P., 2004. A simple parameterization for flux footprint predictions. Bound.-Layer Meteorol. 112, 503–523.

Kroon, P.S., Schrier-Uijl, A.P., Hensen, A., Veenendaal, E.M., Jonker, H.J.J., 2010. Annual balances of CH_4 and N_2O from a managed fen meadow using eddy covariance flux measurements. Eur. J. Soil Sci. 61, 773–784.

Kroon, P.S., Hensen, A., Jonker, H.J.J., Zahniser, M.S., van 't Veen, W.H., Vermeulen, A.T., 2007. Suitability of quantum cascade laser spectroscopy for CH_4 and N_2O eddy covariance flux measurements. Biogeosciences 4, 715–728.

Laville, P., Jambert, C., Cellier, P., Delmas, R., 1999. Nitrous oxide fluxes from a fertilised maize crop using micrometeorological and chamber methods. Agric. Forest Meteorol. 96, 19–38.

Laville, P., Henault, C., Renault, P., Cellier, P., Oriol, A., Devis, X., Flura, D., Germon, J.C., 1997. Field comparison of nitrous oxide emission measurements using micrometeorological and chamber methods. Agronomie. 17, 375–388.

Law, B.E., Falge, E., Gu, L., Baldocchi, D.D., Bakwin, P., Berbigier, P., Davis, K., Dolman, A.J., Falk, M., Fuentes, J.D., Goldstein, A., Granier, A., Grelle, A., Hollinger, D., Janssens, I.A., Jarvis, P., Jensen, N.O., Katul, G., Mahli, Y., Matteucci, G., Meyers, T., Monson, R., Munger, W., Oechel, W., Olson, R., Pilegaard, K., Paw U, K.T., Thorgeirsson, H., Valentini, R., Verma, S., Vesala, T., Wilson, K., Wofsy, S., 2002. Environmental controls over carbon dioxide and water vapor exchange of terrestrial vegetation. Agric. Forest Meteorol. 113, 97–120.

Leclerc, M.Y., Thurtell, G.W., 1990. Footprint prediction of scalar fluxes using a Markovian analysis. Bound.-Layer Meteorol. 52, 247–258.

Lemon, E.R., 1960. Photosynthesis under field conditions. II. An aerodynamic method for determining the turbulent carbon dioxide exchange between the atmosphere and a corn field. Agron. J. 52, 697–703.

Leuning, R., 2004. Measurements of trace gas fluxes in the atmosphere using eddy covariance: WPL corrections revisited. In: Lee, X., Massman, W., Law, B. (Eds.), Handbook of Micrometeorology: a Guide for Surface Flux Measurement and Analysis. Ktuwer, Dordrecht, pp. 133–160.

Maggiotto, S.R., Webb, J., Wagner-Riddle, C., Thurtell, G.W., 2000. Nitrous and nitrogen oxide emissions from turfgrass fertilized with ammonium nitrate, urea and slow-release urea. J. Environ. Qual. 29, 621–630.

Metivier, K.A., Pattey, E., Grant, R.F., 2009. Using the *ecosys* mathematical model to simulate temporal variability of nitrous oxide emissions from a fertilized agricultural soil. Soil Biol. Biochem. 41, 2370–2386.

Meyers, T.P., Baldocchi, D.D., 2005. Current micrometeorological flux methodologies with applications in agriculture. In: Hatfield, J.L., Baker, J.M. (Eds.), Micrometeorology in Agricultural Systems, Agronomy series no. 47, ASA-CSSA-SSSA, Madison, WI.

Migliavacca, M., Reichstein, M., Richardson, A.D., Colombo, R., Sutton, M.A., Lasslop, G., Tomelleri, E., Wohlfahrt, G., Carvalhais, N., Cescatti, A., Mahecha, M.D., Montagnani, L., Papale, D., Zaehle, S., Arain, A., Arneth, A., Black, T.A., Carrara, A., Dore, S., Gianelle, D., Helfter, C., Hollinger, D., Kutsch, W.L., Lafleur, P.M., Nouvellon, Y., Rebmann, C., da Rocha, H.R., Rodeghiero, M., Roupsard, O., M-Sebastia, T., Seufert, G., Soussana, J.-F., van der Molen, M.K., 2011. Semiempirical modeling of abiotic and biotic factors controlling ecosystem respiration across eddy covariance sites. Global Change Biol. 17, 390–409.

Monson, R.K., Sparks, J.P., Rosenstiel, T.N., Scott-Denton, L.E., Huxman, T.E., Harley, P.C., Turnipseed, A.A., Burns, S.P., Backlund, B., Hu, J., 2005. Climatic influences on net ecosystem CO_2 exchange during the transition from wintertime carbon source to springtime carbon sink in a high-elevation, subalpine forest. Oecologia. 146, 130–147.

Moore, C.J., 1986. Frequency response corrections for eddy correlation systems. Bound.-Layer Meteor. 37, 17–35.

Mosier, A.R., Duxbury, J.M., Freney, J.R., Heinemeyer, O., Minami, K., 1996. Nitrous oxide emissions from agricultural fields: assessment, measurement and mitigation. Plant and Soil 181, 95–104.

Mosier, A.R., Hutchinson, G.L., Sabey, B.R., Baxter, J., 1982. Nitrous-oxide emissions from barley plots treated with ammonium-nitrate or sewage-sludge. J. Env. Qual. 11, 78–81.

Moureaux, C., Debacq, A., Bodson, D., Heinesch, B., Aubinet, M., 2006. Annual net ecosystem carbon exchange by a sugar beet crop. Agric. Forest Meteorol. 139, 25–39.

Neftel, A., Ammann, C., Fischer, C., Spirig, C., Conen, F., Emmenegger, L., Tuzson, B., Wahlen, S., 2010. N_2O exchange over managed grassland: application of a quantum cascade laser spectrometer for micrometeorological flux measurements. Agric. Forest Meteorol. 150, 775–785.

Pattey, E., Desjardins, R.L., Boudreau, F., Rochette, P., 1992. Impact of density-fluctuations on flux measurements of trace gases—implications for the relaxed eddy accumulation technique. Bound.-Layer Meteorol. 59, 195–203.

Pattey, E., Blackburn, L.G., Strachan, I.B., Desjardins, R., Dow, D., 2008. Spring thaw and growing season N_2O emissions from a field planted with edible peas and a cover crop. Can. J. Soil Sci. 88, 241–249.

Phillips, F.A., Leuning, R., Baigenta, R., Kelly, K.B., Denmead, O.T., 2007. Nitrous oxide flux measurements from an intensively managed irrigated pasture using micrometeorological techniques. Agric. Forest Meteorol. 143, 92–105.

Prueger, J.H., Kustas, W.P., 2005. Aerodynamic methods for estimating turbulent fluxes. In: Hatfield, J.L., Baker, J.M. (Eds.), Micrometeorology in Agricultural Systems. Agronomy series no. 47. ASA-CSSA-SSSA, Madison, WI, pp. 407–436.

Scanlon, T.M., Kiely, G., 2003. Ecosystem-scale measurements of nitrous oxide fluxes for an intensely grazed, fertilized grassland. Geophys. Res. Lett. 30.

Schmid, H.P., 2002. Footprint modeling for vegetation atmosphere exchange studies: a review and perspective. Agric. Forest Meteorol. 113, 159–183.

Skinner, R.H., 2007. Winter carbon dioxide fluxes in humid-temperate pastures. Agric. Forest Meteorol. 144, 32–43.

Skinner, R.H., Adler, P.R., 2010. Carbon dioxide and water fluxes from switchgrass managed for bioenergy production. Agric. Ecosys. Environ. 138, 257–264.

Stoy, P.C., Katul, G.G., Siqueira, M.B.S., Juang, J-Y., Novick, K.A., Uebelherr, J.M., Oren, R., 2006. An evaluation of models for partitioning eddy covariance-measured net ecosystem exchange into photosynthesis and respiration. Agric. Forest Meteorol. 141, 2–18.

Stull, R., 1988. An Introduction to Boundary Layer Meteorology. Kluwer Academic, Norwell, PA.

Swinbank, W.C., 1951. The measurement of vertical transfer of heat and water vapor by eddies in the lower atmosphere. J. Meteor. 8, 135–145.

Twine, T.E., Kustas, W.P., Norman, J.M., Cook, D.R., Houser, P.R., Meyers, T.P., Prueger, J.H., Starks, P.J., Wesely, M.L., 2000. Correcting eddy-covariance flux underestimates over a grassland. Agric. Forest Meteorol. 103, 279–300.

Valentini, R., Matteucci, G., Dolman, A.J., E-Schulze, D., Rebmann, C., Moors, E.J., Granier, A., Gross, P., Jensen, N.O., Pilegaard, K., Lindroth, A., Grelle, A., Bernhofer, C., Gurnwald, T., Aubinet, M., Ceulemans, R., Kowalski, A.S., Vesala, T., Rannik, U., Berbigier, P., Loustau, D., Gudmundsson, J., Thorgeirsson, H., Ibrom, A., Morgenstern, K., Clement, R., Moncrieff, J., Montagnani, L., Minerbi, S., Jarvis, P.G., 2000. Respiration as the main determinant of carbon balance in European forests. Nature 404, 861–865.

Wagner-Riddle, C., Thurtell, G.W., Kidd, G.E., Edwards, G.C., Simpson, I.J., 1996a. Micrometeorological measurements of trace gas fluxes from agricultural and natural ecosystems. Infrared Phys. Technol. 37, 51–58.

Wagner-Riddle, C., Thurtell, G.W., King, K.M., Kidd, G.E., Beauchamp, E.G., 1996b. Nitrous oxide and carbon dioxide fluxes from a bare soil using a micrometeorological approach. J. Environ. Qual. 25, 898–907.

Wagner-Riddle, C., Thurtell, G.W., Kidd, G.K., Beauchamp, E.G., Sweetman, R., 1997. Estimates of nitrous oxide emissions from agricultural fields over 28 months. Can. J. Soil Sci. 77, 135–144.

Wagner-Riddle, C., Thurtell, G.W., 1998. Nitrous oxide emissions from agricultural fields during winter and spring thaw as affected by management practices. Nutr. Cycl. Agroecosys. 52, 151–163.

Wagner-Riddle, C., Thurtell, G.W., Edwards, G.C., 2005. Trace gas concentration measurements for micrometeorological flux quantification. In: Hatfield, J.L., Baker, J.M. (Eds.), Micrometeorology in Agricultural Systems. Agronomy series no. 47. ASA-CSSA-SSSA, Madison, WI, pp. 321–344.

Wagner-Riddle, C., Furon, A., McLaughlin, N.L., Lee, I., Barbeau, J., Jayasundara, S., Parkin, G., Von Bertoldi, P., Warland, J., 2007. Intensive measurement of nitrous oxide emissions from a corn-soybean-wheat rotation under two contrasting management systems over 5 years. Global Change Biol. 13, 1722–1736.

Wagner-Riddle, C., Rapai, J., Warland, J., Furon, A., 2010. Nitrous oxide fluxes related to soil freeze and thaw periods identified using heat pulse probes. Can. J. Soil Sci. 90, 409–418.

Webb, E.K., Pearman, G.I., Leuning, R., 1980. Correction of flux measurements for density effects due to heat and water vapor transfer. Quart. J. R. Met. Soc. 106, 85–100.

Wienhold, F.G., Frahm, H., Harris, G.W., 1994. Measurements of N_2O fluxes from fertilized grassland using a fast-response tunable diode-laser spectrometer. J. Geophys. Res.-Atmos. 99, 16557–16567.

Wilson, K., Goldstein, A., Falge, E., Aubinet, M., Baldocchi, D., Berbigier, P., Bernhofer, C., Ceulemans, R., Dolman, H., Field, C., Grelle, A., Ibrom, A., Law, B.E., Kowalski, A., Meyers, T., Moncrieff, J., Monson, R., Oechel, W., Tenhunen, J., Valentini, R., Verma, S., 2002. Energy balance closure at FLUXNET sites. Agric. Forest Meteorol. 113, 223–243.

Wohlfahrt, G., Anfang, C., Bahn, M., Haslwanter, A., Newesely, C., Schmitt, M., Drosler, M., Pfadenhauer, J., Cernusca, A., 2005. Quantifying nighttime ecosystem respiration of a meadow using eddy covariance, chambers and modeling. Agric. Forest Meteorol. 128, 141–162.

Wolf, A., Saliendra, N., Akshalov, K., Johnson, D.A., Laca, E., 2008. Effects of different eddy covariance correction schemes on energy balance closure and comparisons with the modified Bowen ratio system. Agric. Forest Meteorol. 148, 942–952.

Yates, T.T., Si, B.C., Farrell, R.E., Pennock, D.J., 2006. Probability distribution and spatial dependence of nitrous oxide emission: temporal change in hummocky terrain. Soil Sci. Soc. Am. J. 70, 753–762.

CHAPTER 22

Remote Sensing of Soil Carbon and Greenhouse Gas Dynamics across Agricultural Landscapes

Craig S.T. Daughtry[1,5], E. Rayymond Hunt, Jr.[1], Peter C. Beeson[1], Sushil Milak[2], Megan W. Lang[3], Guy Serbin[4], Joseph G. Alfieri[1], Gregory W. McCarty[1], Ali M. Sadeghi[1]
[1]USDA-ARS Hydrology and Remote Sensing Laboratory, Beltsville, MD
[2]Science Systems and Applications, Inc., Lanham, MD
[3]USDA Forest Service, Northern Research Station, Beltsville, MD
[4]InuTeq, LLC, Washington, DC

CHAPTER OUTLINE

Abbreviations: AgRISTARS, Agriculture and Resources Inventory Surveys through Aerospace Remote Sensing; ASTER, Advanced Spaceborne Thermal Emission and Reflection radiometer; AVHRR, Advanced Very High Resolution Radiometer; BRDF, bidirectional reflectance distribution function; CAI, cellulose absorption index; CDL, Cropland Data Layer; CEAP, Conservation

Managing Agricultural Greenhouse Gases. DOI: 10.1016/B978-0-12-386897-8.00022-X
2012, Published by Elsevier Inc.

Effects Assessment Project; DEM, digital elevation models; EPIC, Environmental Policy Integrated Climate model; ET, evapotranspiration; EVI, Enhanced Vegetation Index; fAPAR, fraction of absorbed PAR; GOES, NOAA Geostationary Operational Environmental Satellites; GPP, gross primary productivity; LACIE, Large Area Crop Inventory Experiment; LAI, leaf area index; Landsat TM, Landsat Thematic Mapper; LiDAR, Light Detection and Ranging; MODIS, Moderate Resolution Imaging Spectroradiometer; NASA, National Aeronautics and Space Administration; NASS, USDA National Agricultural Statistical Service; NDTI, Normalized Difference Tillage Index; NDVI, Normalized Difference Vegetation Index; NEXRAD, Next-Generation Radar; NIR, near infrared (700–1200 nm); NOAA, National Oceanic and Atmospheric Administration; NPP, net primary productivity; NRCS, USDA Natural Resources Conservation Service; PAR, photosynthetically active radiation (400–700 nm); PRI, Photochemical Reflectance Index; PROSPECT, leaf optical properties model; SAIL, Scattering by Arbitrarily Inclined Leaves; SMAP, Soil Moisture Active Passive; SOC, soil organic carbon; SPOT, Satellite Pour l'Observation de la Terre; SSURGO, Soil Survey Geographic; STARFM, Spatial and Temporal Adaptive Reflectance Fusion Model; STATSGO2, U.S. General Soil Map; SWAT, Soil and Water Assessment Tool; SWIR, shortwave infrared (1200–2500 nm); TIR, thermal infrared (8–14 μm); USDA, United States Department of Agriculture; Vis, visible (400–700 nm); ε, conversion efficiency

INTRODUCTION

The overall goal of GRACEnet is to develop strategies for managing soil C sequestration and reducing emissions of greenhouse gases for agricultural lands. The geographical distribution of GRACEnet sites and use of common protocols provide an excellent sampling of U.S. agricultural conditions (Jawson et al., 2005; Walthall et al., 2012). Simulation models have been developed or modified to predict the impacts of crop and soil management practices on soil organic carbon (SOC), greenhouse gas emissions, and water quality. However, in order to assess the overall impact of the adoption of these management practices, the data acquired from small plots or fields must be extended to regional and national scales. The spatial scaling task is nontrivial because mechanisms controlling carbon, water, and energy exchanges are nonlinear and interact with each other.

While *in situ* measurements accurately describe the temporal variability of carbon, water, and energy fluxes for small areas, remote sensing offers a practical method to account for the spatial variability inherent across agricultural landscapes. Many of the biophysical characteristics of vegetation and soils needed by process models (field or watershed scales) can be derived directly or indirectly from remotely sensed data. Thus, synergies of *in situ* and remotely sensed measurements can potentially provide frequent, spatially explicit information about agricultural landscapes.

This chapter briefly reviews (1) fundamental spectral properties of vegetation and soils, (2) remote sensing products for assessing soil organic carbon and greenhouse gases, (3) synergies of remote sensing and process models to determine carbon dynamics, and then (4) discusses a case study that used simulation models to address effects of crop management decisions on soil carbon and water quality.

HISTORICAL BACKGROUND

One of the earliest agricultural applications of remotely sensed data was the use of aerial photographs to map the boundaries of soil series, to characterize soil properties, and to delineate soil erosion. Aerial photographs became the standard base maps for the national cooperative soil surveys (Baldwin et al., 1938) and since the 1950s, color-infrared aerial photography has been used to infer vegetation health and potential growth. The acquisition of color-infrared photographs is still an important component for many management applications.

The modern era of civilian satellite remote sensing began in 1972 with the launch of the Earth Resources Technology Satellite, later renamed Landsat-1. The Landsat program has evolved over time and is currently operated to systematically survey the earth and preserve historical observation records (Goward et al., 2009). Other nations have also developed the capability to launch and operate land observation satellites. Currently more than 30 moderate spatial resolution (10–100 m pixel size) satellites are in orbit and another 27 involving 23 nations are planned (Goward et al., 2009). Table 22.1 briefly summarizes key characteristics of the satellite sensors discussed in this chapter.

Satellite sensors with coarse (100–1000 m) and very coarse (>1000 m) spatial resolutions generally have swath widths that provide daily or nearly daily coverage. Images from the NOAA Advanced Very High Resolution Radiometer (AVHRR) and the NASA Moderate Resolution Imaging Spectroradiometer (MODIS) have been analyzed to track temporal changes in vegetation and climate over regional to global scales. The spectral vegetation indices from these sensors are the most widely used remotely sensed products (Justice and Tucker, 2009). Successes of AVHRR and MODIS have led to programs by other nations to provide global data at a coarse resolution.

TABLE 22.1 Characteristics of Various Satellite Sensors (Goward et al., 2009; Justice and Tucker, 2009)

Sensor[1]	Number of bands[2]	Spatial resolution[3]	Temporal resolution[4]	Swath km	CVGC[5]
Moderate Spatial Resolution					
Landsat TM	6 Vis/NIR/SWIR	30 m	16 days	185	3 months
	1 TIR	90 m			
ASTER	3 Vis/NIR	15 m	16 days nadir	60	n/a[7]
	6 SWIR	30 m	5 days off-nadir		
	5 TIR	90 m			
SPOT	3 Vis/NIR	10 m	26 days nadir	60	n/a[7]
	1 SWIR	20 m	2–3 days off-nadir		
	1 pan	5 m			
Hyperion	220 Vis/NIR/SWIR	30 m	16 days nadir	7.5	n/a[7]
			7–9 days off-nadir		
AWiFS	4 Vis/NIR/SWIR	56 m	5 days	737	1–2 months
Coarse Spatial Resolution					
AVHRR	5 Vis/NIR/TIR	1100 m	daily	2048	2–4 weeks
MODIS	2 Red/NIR	250 m	daily	2048	2–4 weeks
	5 Vis/SWIR	500 m			
	29 Vis/NIR/TIR	1000 m			
Very Coarse Spatial Resolution					
GOES	4 Vis/NIR/TIR	8 km	0.25 h[6]		<1 week

[1]Landsat TM = Landsat Thematic Mapper, USA; ASTER = Advanced Spaceborne Thermal Emission and Reflection, Japan and USA; SPOT = Satellite Pour l'Observation de la Terre, France; AWiFS = Advanced Wide-Field Sensor on ResourceSat-1, India; AVHRR= Advanced Very High Resolution Radiometer, USA; MODIS = Moderate Resolution Imaging Spectroradiometer, USA; GOES = Geostationary Operational Environmental Sensor, USA.
[2]Vis = Visible (400–700 nm); NIR = Near infrared (700–1200 nm); SWIR = Shortwave infrared (1200–2500 nm); TIR = thermal infrared (8–14 μm).
[3]Spatial resolution at nadir.
[4]Pointing capability facilitates rapid revisits, but is not designed for global coverage.
[5]CVGC = Clear view global coverage, based on orbital repeat and probability of clear view (no clouds).
[6]Images acquired every 15 minutes of western hemisphere from geostationary orbit.
[7]n/a., not applicable.

Remote sensing cannot directly assess soil carbon, except at the soil surface (Reeves et al., 2012). However, recent advances in remote sensing of vegetation and soils can provide some of the biophysical variables, including soil and crop management practices and local topography, needed by process models to predict soil carbon and greenhouse gas dynamics across agricultural landscapes. With spatially explicit inputs, these models can also provide accurate estimates of soil erosion, a major component of both carbon dynamics and water quality.

FUNDAMENTAL SPECTRAL PROPERTIES OF VEGETATION AND SOILS

Vegetation

When solar radiation interacts with matter, various proportions of the incident photons may be reflected, transmitted, or absorbed (Gates, 1980; Knipling, 1970). Absorption is determined by the molecular composition of the material while reflectance is influenced by both the index of refraction and the absorption spectrum. The spectral reflectance of vegetation canopies is determined primarily by (1) optical properties of leaves and stems, (2) amount and geometric arrangement of leaves and stems, (3) background (soil or residue) reflectance, (4) illumination and view angles, and (5) atmospheric transmittance (Bauer, 1985).

Reflectance of solar radiation from a typical green leaf is characterized by three distinct regions: visible, near infrared, and shortwave infrared (Figure 22.1). In the visible wavelength region (400–700 nm), the spectral properties of leaves are determined primarily by the concentrations of chlorophyll and other leaf pigments (Thomas and Gausman, 1977). In the near-infrared (NIR) wavelength region (700–1200 nm), leaf properties are determined by mesophyll structure and the cell wall–air interfaces (Slaton et al., 2001). Finally, in the shortwave-infrared (SWIR) wavelength region (1200–2500 nm), reflectance is affected primarily by the amount of water in the leaves (Hunt and Rock, 1989; Yilmaz et al., 2008). As leaves expand, mature, and senesce, physiological and morphological changes occur that affect their spectral properties. Various stresses, including nutrient deficiencies, water deficits, and damage by insects and diseases, also affect the optical properties of leaves (Walter-Shea and Biehl, 1990).

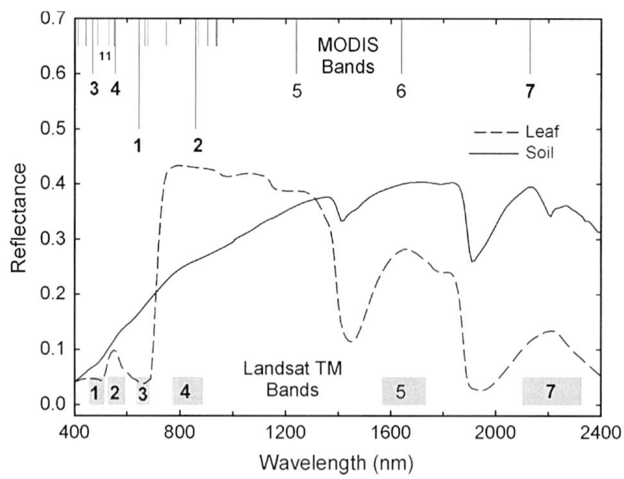

FIGURE 22.1

Reflectance spectra of a typical green corn leaf and a loamy soil (Barnes, a fine-loamy, mixed, superactive, frigid Calcic Hapludolls) from Morris, MN. Along the figure bottom, the positions of the Landsat TM bands are shown (Band 6 is in the thermal infrared, 10.4–12.5 μm). Positions of the MODIS bands are shown along the figure top. MODIS bands 1 and 2 have 250 m pixel resolution, MODIS bands 3 to 7 have 500 m pixel resolution, and MODIS bands 8 to 19 have 1000 m pixel resolution. MODIS band 11 is used for the Photochemical Reflectance Index (PRI), an estimate of radiation use efficiency.

When vegetation density is low, the reflectance of the soil and crop residue (plant litter) background significantly influences the overall scene (canopy) reflectance. When vegetation density is high, leaves are the primary scattering (i.e. reflecting and transmitting) elements and the background contributes little to overall canopy reflectance (Hatfield and Prueger, 2010; Kollenkark et al., 1982). The reflectance of vegetation canopies can be simulated by the Scattering by Arbitrarily Inclined Leaves (SAIL) model (Verhoef, 1984). The SAIL model requires leaf spectral reflectance and transmittance as inputs. Much of the variation observed in leaf spectral properties associated with changes in pigments, cellular structure, and water can be simulated with the PROSPECT model (Jacquemoud et al., 2009). One of the important outputs of the SAIL model is the changes in reflectance due to changes in sun angles and view angles. Simulations using all sun and sensor view angles create bidirectional reflectance distribution functions (BRDF) that can be useful for mission planning (Walthall et al., 2000).

Soils

The spectral reflectance of soils is determined by physical factors quite different from those of vegetation. Soil reflectance generally increases monotonically with increasing wavelength (Figure 22.1). The relative contributions of moisture content, iron-oxide content, organic matter content, particle size distribution, mineralogy, and soil structure to reflectance of naturally occurring soils have been thoroughly reviewed (Baumgardner et al., 1985; Ben-Dor, 2002). In perhaps the most comprehensive early study of soil reflectance, Stoner and Baumgardner (1981) defined five general classes of soil reflectance spectra. Organic matter content and iron oxide content were the primary factors determining shape of the reflectance spectra of soils. In general, soil reflectance increased as soil moisture content, particle size, surface roughness, organic matter content, and iron oxide content decreased. Spectral reflectance of soil is strongly correlated with soil organic matter among soils from the same parent materials; however, the relationship is sensitive to changes in iron and manganese oxides in soils from different parent materials (Henderson et al., 1992). Various clays and other soil minerals may have specific absorption features that are evident in the reflectance spectra of soils near 2200 nm (Brown et al., 2006; Serbin et al., 2009b,c).

The synoptic view of soils in the landscape and the tonal variations in the aerial photographs have long been used to delineate soil boundaries and identify inclusions within the predominant soil series (Baldwin et al., 1938). More recently, multispectral and hyperspectral images have also been used to aid soil survey, soil inventory, and soil management (Baumgardner et al., 1985; Ben-Dor et al., 2009).

Maps of soil properties at field and landscape scales have also been generated from airborne and satellite hyperspectral imaging spectrometers (Chen et al., 2008). Information on surface soil carbon has been extracted from hyperspectral images of bare or nearly bare soils using neural network, multivariate regression, or chemometric approaches (Gomez et al., 2008; McCarty et al., 2010; Selige et al., 2006). These spectroscopic approaches typically involved field sampling and laboratory spectral measurements or proximal in-field spectral measurements. Reeves et al. (2012) discussed recent advances in spectroscopic methods for quantifying soil carbon. Thus, remote sensing holds great promise for SOC measurements with intrinsic spatial information; however, optimization of the data analysis procedures is required to improve prediction ability.

Vegetation Indices

A distinguishing spectral characteristic of green vegetation is the step-like transition from low reflectance and low transmittance of the visible (400–700 nm) wavelength region where pigments strongly absorb to the high reflectance and high transmittance of the near-infrared (700–1200 nm) wavelength region where absorption is low (Figure 22.1). Senesced vegetation (e.g. crop residues) and soils lack this spectral feature (Figure 22.2). Recognition of these

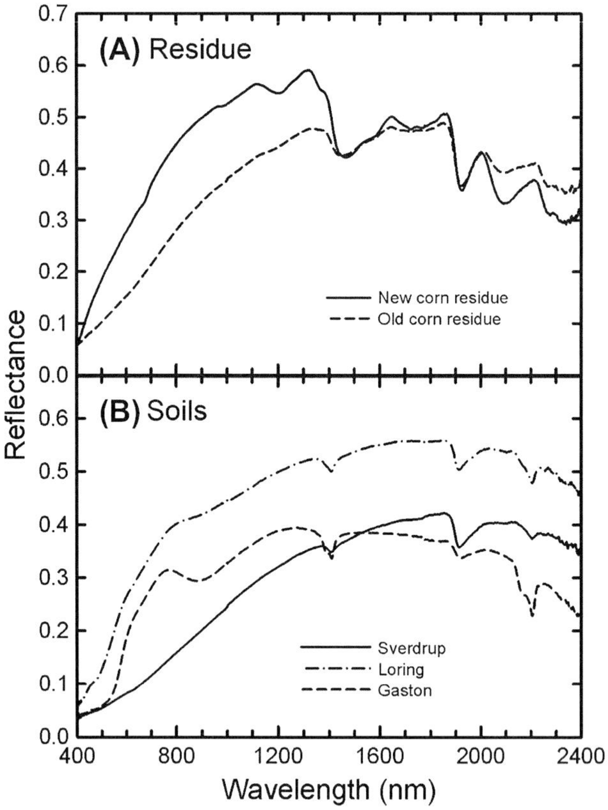

FIGURE 22.2
Reflectance spectra of (A) corn residue 1 week after harvest (fresh) and 7 months after harvest (old) and (B) three diverse agricultural soils. Sverdrup sandy loam is a sandy, mixed, frigid Typic Hapludolls from Morris, MN; Loring silt loam is a fine-silty, mixed, active, thermic Oxyaquic Fragiudalfs from Como, MS; and Gaston sandy clay loam is a fine, mixed, active, thermic Humic Hapludults from Salisbury, NC.

fundamental spectral differences led to the development of a multitude of spectral vegetation indices that are particularly sensitive to green vegetation. The Normalized Difference Vegetation Index (NDVI) is the *de facto* standard vegetation index:

$$\mathrm{NDVI} = (\mathrm{NIR} - \mathrm{Red})/(\mathrm{NIR} + \mathrm{Red}) \quad [1]$$

where NIR is the near-infrared (e.g. 760–900 nm) reflectance and Red is the red (e.g. 630–690 nm) reflectance (Rouse et al., 1974; Tucker, 1979). The NDVI works well because the red and NIR reflectances of vegetation are readily distinguishable from those of soils (Figure 22.1). Plant cover, biomass, leaf area index (LAI), and fraction of absorbed photosynthetically active radiation (fAPAR) are all highly correlated with NDVI during the vegetative growth phase of crops (Asrar et al., 1989; Daughtry et al., 1992; Hatfield and Prueger, 2010).

Numerous additional vegetation indices have been developed using various combinations of visible, near-infrared, and shortwave infrared wavebands. For example, the Optimized Soil Adjusted Vegetation Index (OSAVI) minimized the effects of soil color, roughness, and moisture content better than NDVI at low vegetation densities (Rondeaux et al., 1996). The Enhanced Vegetation Index (EVI), was designed to improve sensitivity to high biomass conditions and to minimize influence of atmospheric aerosols (Huete et al., 2002).

Although changes in leaf chlorophyll concentrations produce significant differences in leaf reflectance and transmittance spectra, the transition from leaf reflectance spectra to plant canopy reflectance spectra is complicated. Spatial and temporal variations in background reflectance and LAI often confound detection of the relatively subtle changes in canopy reflectance due to differences in leaf chlorophyll concentrations (Daughtry et al., 2000). Spectral vegetation indices sensitive to leaf chlorophyll concentrations but insensitive to background (soil or residue) reflectance are needed. Leaf and canopy reflectance simulation models, e.g. PROSPECT and SAIL (Jacquemoud et al., 2009), can also be used to develop and evaluate vegetation indices based on the theoretical understanding of radiation scattering by canopies (Goward and Huemmrich, 1992). A wide range of crop conditions and potential vegetation indices can be simulated and evaluated before testing with field data. Vegetation indices sensitive to leaf chlorophyll concentration and less sensitive to LAI and soil reflectance have been identified for nutrient management (Eitel et al., 2009; Haboudane et al., 2002; Hunt et al., 2011). Two-dimensional or planar indices that use pairs of indices to estimate leaf chlorophyll over a broad range of soil types have also had some success (Eitel et al., 2008; Fitzgerald et al., 2010). Nevertheless, robust vegetation indices that accurately assess N requirements of crops remain elusive.

REMOTE SENSING PRODUCTS FOR ASSESSING SOIL ORGANIC CARBON AND GREENHOUSE GASES

Crop selection, tillage, and fertilization are dynamic crop management practices that strongly influence the spatial distribution of biomass production and thus the organic matter inputs to the soil. These management practices also interact with climate, soil properties, and topography to influence SOC dynamics. Agricultural management practices may also include draining excess water from wet soils and irrigating dry soils.

Crop Identification, Area Estimation, and Condition Assessment

Crop type and phenology (developmental stage) are important for monitoring crop production and interpreting seasonal changes in greenhouse gas fluxes. Crop identification and area estimation have long been major thrusts of U.S. agricultural remote sensing programs, such as the Corn Blight Watch Experiment (MacDonald et al., 1972), the Large Area Crop Inventory Experiment (LACIE) (MacDonald and Hall, 1980), and the Agriculture and Resources Inventory Surveys through Aerospace Remote Sensing (AgRISTARS) (Bauer, 1985). These programs firmly established the feasibility of using multispectral scanner data from aircraft and satellites and digital analysis techniques to identify and estimate the areal extent of crops. These programs also recognized the importance of multi-temporal remotely sensed data for consistent, accurate crop identification and area estimation and laid the foundations for the USDA National Agricultural Statistical Service's (NASS) Cropland Data Layer (CDL), a crop-specific land cover classification product encompassing the entire conterminous U.S. (Johnson and Mueller, 2010).

Key phenological stages of crops can be detected by fitting temporal profile models to various vegetation indices calculated from images acquired throughout a growing season (Badhwar and Henderson, 1981; Sakamoto et al., 2010). Attributes of the temporal profile (e.g. shape, widths, and inflection points) contain useful information for identifying crop type and assessing crop conditions. However, to quantitatively monitor changes in vegetation over the growing season, images from sensors, such as the Landsat Thematic Mapper (TM), must be atmospherically corrected to land-surface reflectance (Liang et al., 2002; Masek et al., 2006; Moran et al., 1992). Additional processing of the remotely sensed imagery may be required to extract accurate information about the land surface. For example, south-facing slopes in the northern hemisphere receive more solar radiation and have higher apparent reflectance than north-facing slopes. With digital elevation data, slope and aspect may be calculated for each pixel and reflectance corrected to obtain better information about vegetation type, cover, LAI, and biomass (Riaño et al., 2003).

Clouds are also major obstacles in developing multi-date datasets. Composite images can be formed from the cloud-free portions of multiple images acquired over several days. Satellites with short revisit times (<5 days) are most suitable for composite images of agricultural crops (Table 22.1). Multiple MODIS images are routinely composited into 8-day and 16-day products, e.g. surface reflectance, NDVI, LAI, and fAPAR.

An ideal remote sensing system for monitoring agricultural crops and soils would produce images with moderate to high spatial resolution and frequent coverage for all land areas. Current sensors with fine to moderate spatial resolutions typically have long revisit times (e.g. Landsat, Table 22.1) which limit their ability to detect rapid changes associated with phenology or stress. However, sensors with frequent revisit times typically have coarse spatial resolutions (e.g. MODIS, Table 22.1) which limit their ability to monitor heterogeneous landscapes with small fields. A spatial and temporal adaptive reflectance fusion model (STARFM) that combined Landsat TM and MODIS data for predicting daily surface reflectance at Landsat TM spatial resolution and MODIS temporal frequency has been proposed and refined (Gao et al., 2006; Zhu et al., 2010). This technique should be especially useful for frequent monitoring of crop conditions within small fields or after episodic events, such as floods, droughts, or diseases.

Crop Residue and Tillage

Crop residues (or plant litter) on the soil surface decrease soil erosion, increase soil organic matter, improve soil quality, increase water infiltration, and reduce amounts of nutrients and pesticides that reach streams and rivers (Delgado, 2010; Lal et al., 1999). After harvest, crop residues often completely cover the soil surface, but when the soil is tilled, residue cover decreases.

Three useful categories of tillage intensity based on crop residue cover after planting have been defined: intensive tillage has <15% residue cover; reduced tillage has 15–30% residue cover; conservation tillage has >30% residue cover. Thus, quantification of crop residue cover is required to evaluate the effectiveness and extent of conservation tillage practices. For agricultural fields, the standard technique for measuring crop residue cover used by the USDA Natural Resources Conservation Service (NRCS) is visual determination of the presence of residue at selected points along a line (Morrison et al., 1993). Reviews of crop residue measurement techniques illustrate the problems that current techniques have addressing spatial variability of crop residue cover (Corak et al., 1993; Morrison et al., 1993; Thoma et al., 2004).

Shortly after harvest, crop residues are frequently much brighter than the soil, but as the residues weather and decompose, they may be either brighter or darker than the soil (Daughtry et al., 2010; Nagler et al., 2000). The reflectance spectra of both soils and crop residue lack the unique spectral signature of green vegetation in the 400 to 1000 nm wavelength region (Figures 22.1 and 22.2). Crop residues and soils are often spectrally similar and differ only in amplitude at a given wavelength (Aase and Tanaka, 1991; Baird and Baret, 1997).

A robust approach for discriminating crop residues from soil is based on detecting an absorption feature near 2100 nm that appears in all compounds possessing alcoholic –OH groups, such as cellulose (Murray and Williams, 1988). The absorption feature near 2100 nm is clearly evident in the reflectance spectra of the dry crop residues and is absent in the spectra of soils and live green vegetation (Elvidge, 1990; Nagler et al., 2000). The relative intensity of this absorption feature was defined as the cellulose absorption index (CAI). Crop residue cover is linearly related to CAI for diverse soils and crop residues (Serbin et al., 2009a).

Efforts to identify tilled fields using changes in surface reflectance have had mixed success, e.g. Baird and Baret, 1997; Thoma et al., 2004. Although tillage frequently roughens the soil surface and decreases soil reflectance, the effect is often short-lived and reflectance may increase as the soil surface is smoothed by rain or subsequent tillage. Nevertheless, supervised classification methods can often discriminate among a few (2–3) classes of residue cover, but require scene-specific training statistics (McNairn and Protz, 1993; Streck et al., 2002; van

Deventer et al., 1997). Advanced classification techniques, such as linear spectral unmixing analysis that exploit the information from all available spectral bands, have had success where soil and crop residues differ sufficiently in reflectance (Pacheco and McNairn, 2010). Few studies have successfully linked spectral and temporal domains to assess the timing and intensity of tillage operations. Zheng et al. (2012) used multi-temporal Landsat TM images to calculate normalized difference tillage index (NDTI) and showed that minimum NDTI values reliably indicated changes in surface conditions associated with tillage or planting operations in agricultural fields.

If tillage intensity is defined by the amount of residue cover, it is possible to use CAI to identify tillage categories across diverse agricultural landscapes (Daughtry et al., 2006; Serbin et al., 2009a). However, current hyperspectral sensors have a very narrow swath (e.g. Hyperion, Table 22.1) and do not have the capacity to map large areas in a timely manner (Goward et al., 2009). Therefore, the challenge is how to best use a few hyperspectral images and many multispectral images to produce regional surveys and maps of crop residue cover and tillage intensity. For example, a simple robust method for quantitatively mapping the fractions of photosynthetic vegetation, non-photosynthetic vegetation, and bare soil was developed using Hyperion and MODIS data (Guerschman et al., 2009). Hyperion hyperspectral data were used to establish a method for unmixing each of the three fractions using NDVI and CAI. A surrogate for CAI was identified in MODIS surface reflectance data by regression analysis. NDVI and the surrogate for CAI were then used to successfully assess the fractions of photosynthetic vegetation, non-photosynthetic vegetation, and bare soil with daily MODIS data over large areas of Australian savannas (Guerschman et al., 2009). Nevertheless, the relatively coarse spatial resolution of MODIS data (i.e. 250–1000 m, Table 22.1) would be a limitation for assessing crop residue cover and tillage intensity in agricultural regions with many fields and diverse crops.

Topography

Spatial assessment of SOC within agricultural landscapes is important for understanding carbon dynamics within agricultural ecosystems. However, current approaches require intensive measurements and errors in the spatial assessment of SOC distribution can result from inadequate or biased sampling of landscapes and from analytical errors in SOC measurements. The largest errors often result from the sampling procedure, which may provide a poorly stratified reflection of landscape variability or be based on a sample size that inadequately covers the actual range of SOC content.

Soil properties are usually best represented as continua as opposed to discrete or abrupt transitions as they are typically represented on soil maps. Both kriging and correlation approaches have been explored for predicting spatial structure of soil properties, such as SOC. Kriging depends on interpolation of the measured values and therefore predictions cannot exceed the sampling range (McBratney and Pringle, 1999). In contrast, the correlation approach relates the variable of interest to a landscape property that is conveniently measured (McKenzie and Ryan, 1999) and thereby allows extrapolation based on collinearity of the properties. Optimized sampling schemes to predict surface SOC have used ancillary variables, such as soil series, relative elevation, slope, conductivity, and soil reflectance (Simbahan and Doberrnann, 2006). Digital Elevation Models (DEMs) are often used to derive primary topographic metrics, such as slope and aspect, and compound or secondary metrics, which are based on the relationship between multiple primary metrics. Secondary topographic metrics, e.g. topographic wetness indices, have been used to determine the spatial distribution of key soil processes, such as soil saturation, and have also been useful topographic parameters to scale soil property measurements (Terra et al., 2004).

Although topographic information is an obvious covariant property, until very recently fine resolution data have not been readily available for large areas of interest. In the past, the spatial resolution of commonly available digital topographic data for the United States (vertical accuracies of 1-10 m) was insufficient to map the subtle variations in topography that often control soil erosion, sedimentation, the accumulation of soil organic matter, and the emission of greenhouse gases in many agricultural systems. However, DEMs derived from LiDAR data provide superior vertical accuracy ($\leq$0.15 m) and horizontal (typically 0.5 to 2 m) resolution. Additional information regarding the National LiDAR Dataset is available at the Center for LiDAR Information Coordination and Knowledge (CLICK) website at http://lidar.cr.usgs.gov.

The ability to resolve fine topographic variations is particularly vital in areas of low topographic variability, the Mid-Atlantic Coastal Plain, where subtle differences in topography can lead to divergent soil types and processes. As illustrated in Figure 22.3 only a fine resolution DEM derived from LiDAR data contained field scale topographic information related to the distribution of soil carbon at a study site located in this coastal plain landscape. Additionally, on the Eastern Shore of Maryland historic wetlands that have been drained and are currently farmed (prior converted croplands) are evident as shallow topographic depressions on agricultural fields (Lang and McCarty, 2009). Although most of these areas are difficult to discern using multispectral imagery, these areas are readily apparent in LiDAR-based DEMs. These converted croplands (e.g., historic Delmarva bays), although drained, are often wetter than surrounding areas and these wet conditions coupled with an ample supply of nitrogen fertilizer often lead to the emissions of greenhouse gases, such as nitrous oxide, and are potential zones for high nutrient losses to nearby surface water resources.

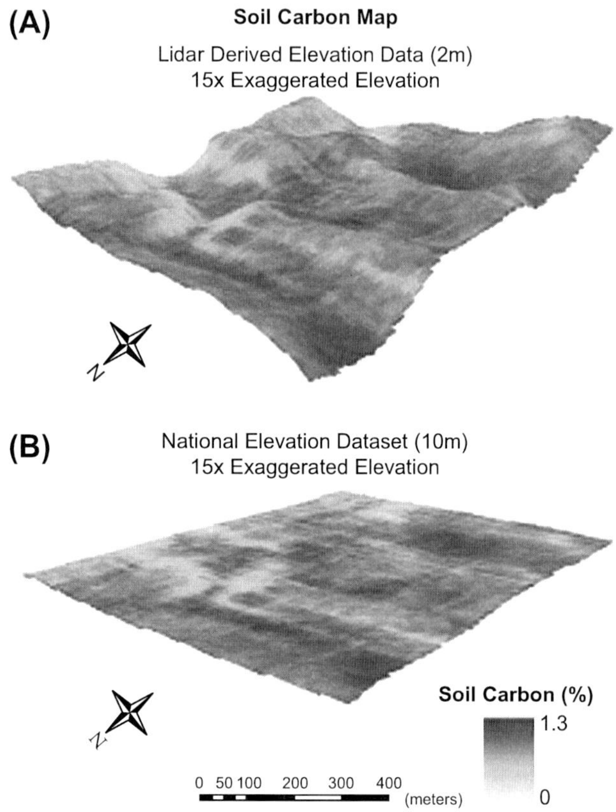

FIGURE 22.3
Soil organic carbon (SOC) map produced from aircraft hyperspectral data over a field in Maryland, draped on DEM products generated by: (A) high resolution LiDAR (2 m resolution) and (B) National Elevation Dataset (NED, 10 m resolution). Elevation within the field associated with the LiDAR data ranged from 11.1 to 17.3 m above sea level (ASL) and the NED data ranged from 11.9 to 14.2 m ASL. Although the general slope of the field is represented to some degree by the NED data, it contains essentially no information related to SOC distributions in the field as a function of topographic relief. By contrast, visual inspection shows that the LiDAR product is rich in information related to field-scale distribution of SOC from analysis of hyperspectral data. Please see color plate section at the back of the book.

SYNERGIES OF REMOTE SENSING AND PROCESS MODELS TO DETERMINE SOIL CARBON DYNAMICS

Although crop yields may be directly correlated with remotely sensed spectral indices (e.g. NDVI), the relationships are rarely extendable to other regions (Chang et al., 2003). Robust approaches use crop growth models that simulate biophysical processes in the soil—-crop—atmosphere system and provide continuous descriptions of crop growth and development. Remotely sensed data may be used to periodically update growth models or to predict crop characteristics (e.g. LAI, biomass) by inverting radiation transfer models (Doraiswamy et al., 2004, 2005).

There is a strong synergy between remotely sensed data and soil process-based simulation models. Remote sensing may be used to regionally determine a large number of spatially explicit inputs that are required by process models to estimate carbon dynamics (Brown et al., 2010). Remotely sensed estimates of vegetative production can determine the spatial distribution of carbon inputs into the soil and thus increase the accuracy of belowground carbon dynamics from soil-process simulation models. Nevertheless, there are still large uncertainties and both techniques require ground-based data for validation.

Soil Moisture and Temperature

The carbon flux from soil is the sum of root autotrophic respiration and soil heterotrophic respiration. Soil moisture and temperature control much of the variability in soil respiration (Mahecha et al., 2010; Reichstein et al., 2003). Predictions of soil temperature and moisture from remote sensing could help estimate soil carbon fluxes over large areas (Kimball et al., 2009).

Thermal (8 to 12 μm) and microwave (0.01 to 0.5 m) radiation are passively emitted according to the Stefan-Boltzmann law, where the amounts of radiation emitted are functions of surface temperature, emissivity, and wavelength. For thermal wavelengths, the emissivity is high for water and other materials, so passively emitted radiation is used to measure surface temperature (the basis for infrared thermometers). In the microwave region, dry soil has a high emissivity whereas water has a low emissivity, so passively emitted radiation is used to estimate soil moisture content (Jackson et al., 1999, 2010).

Water lost to the atmosphere through plant transpiration or soil water evaporation (i.e. evapotranspiration, ET) cools the earth's surface. As the surface soil layer (0—5 cm) dries, soil temperature increases rapidly, even in partially vegetated landscapes. As moisture deficiencies in the root zone develop, stomata close, transpiration is reduced, and canopy temperatures increase. Thermal infrared sensors can measure and map land surface temperatures which are indicators of both ET and surface moisture status (Anderson and Kustas, 2008; Moran, 2003). Empirical models have linked land surface temperature and NDVI to monitor drought conditions over large areas but have significant limitations (Kustas and Anderson, 2009). Robust, physically-based surface energy balance approaches have linked surface energy balance models with atmospheric boundary layer models to map daily ET at continental scales using thermal imagery from NOAA's Geostationary Operational Environmental Satellites (GOES; Table 22.1). These daily ET fluxes were disaggregated and fused with moderate-spatial resolution imagery from polar orbiting satellites (e.g. Landsat TM, ASTER, MODIS, AVHRR) to produce daily ET maps at sub-field scales (Anderson et al., 2011).

One of NASA's priorities for earth-observation satellites is the Soil Moisture Active Passive (SMAP) mission (Entekhabi et al., 2010). SMAP will merge active L-band radar backscatter with the passive L-band radiometer data to measure soil moisture in the top 0.1 m at a spatial resolution of about 10 km. This SMAP product promises to be a major advance because the upper horizons of soils are major sources of carbon respired by soil organisms (Reichstein et al., 2003). Furthermore, using data assimilation techniques and a water-balance model, root-zone moisture content may be inferred (Crow et al., 2008). Broadly applicable approaches to hydrologic modeling will likely involve assimilation of both thermal and microwave estimates of soil moisture (Anderson et al., 2011).

Surface Flux Measurements and Spatial Scaling

Measurements of surface fluxes of carbon, water, and energy from both individual field sites and regional networks, such as Ameriflux, provide invaluable data characterizing the local environment. However, scaling these data from fields to regions or continents remains a nontrivial task (Anderson et al., 2007). The mechanisms regulating the exchange of energy, water, and carbon between the land surface and atmosphere are highly nonlinear and interact with local surface and atmospheric conditions.

Two diverse approaches for scaling surface flux measurement have been developed. First, statistical approaches typically use a covariant, such as carbon inventories or soil maps, to upscale fluxes from field to regions (Izaurralde et al., 2012; Paustain et al., 2012; West et al., 2010). Second, aggregation approaches use the surface measurements to develop, parameterize, calibrate, and evaluate numerical models that can determine the turbulent fluxes on local to regional scales (Skinner and Wagner-Riddle, 2012). The numerical models used for aggregation may be subdivided into four categories: (1) canopy vegetation models (Baldocchi and Meyers, 1998; Lai et al., 2000), (2) surface energy balance models (Allen et al., 2007; Anderson et al., 1997), (3) vegetation index models (Glenn et al., 2010; Xiao et al., 2008), and (4) land surface models (Bonan et al., 2002; Ek et al., 2003).

Regardless of the approach for scaling fluxes, remote sensing offers the only practical method to account for the spatial and temporal variability of factors controlling land–atmosphere exchanges (Kalma et al., 2008). However, current remote sensing systems suitable for scaling surface fluxes are constrained by spatial and temporal resolution issues (see discussion in "Crop identification, area estimation, and condition assessment", above). Data fusion combines remotely-sensed data collected at differing spatial and temporal scales to produce continuous high-resolution datasets of synthetic observations with fine-scale variability, e.g. the STARFM algorithm (Gao et al., 2006). Data assimilation incorporates data from a broad range of sources, including remote sensing and in-situ observations, to fully describe the state of the land–atmosphere system and constrain the model output accordingly (Dorigo et al., 2007).

Process Models of Carbon Fluxes

COMPLEX MODELS AND REMOTE SENSING

Complex models such as EPIC (initially called the Erosion-Productivity Impact Calculator and later redefined as the Environmental Policy Integrated Climate version) simulate both plant and soil carbon fluxes, but require numerous site-specific data and parameters (Causarano et al., 2008; Izaurralde et al., 2006; Williams et al., 1984). Much of the site-specific data can be obtained spatially from remote sensing (e.g. crop type, LAI) and soil type databases (e.g. the Soil Survey Geographic, SSURGO or the U.S. General Soil Map, STATSGO2) (USDA-NRCS, 2011).

Small errors in model parameters or input data may be trivial for predicting carbon dynamics over short periods of time or over small areas, but those errors accumulate and become large when accumulated over longer time periods or larger areas. This limitation may be overcome by updating the process models with spatially explicit information on crop conditions. In Figure 22.4, LAI was used as the central variable for updating models (Doraiswamy et al., 2004, 2005). Red and near-infrared reflectances from Landsat or MODIS were used to predict LAI with an inversion of the SAIL model. Predicted LAI values were verified with measured values

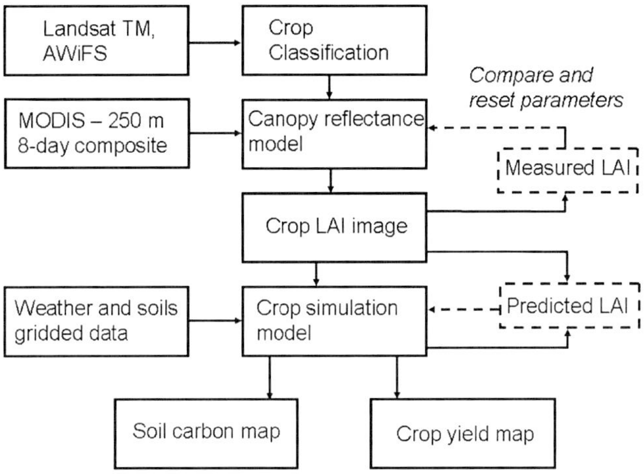

FIGURE 22.4
Mapping Leaf Area Index (LAI), soil carbon content and crop yields using remote sensing and simulation modeling. Two simulation models are used, (1) a canopy reflectance model to predict LAI from MODIS data and (2) a crop simulation model predicting LAI, soil carbon content and crop yields from weather and soils data. The key step for the first model is the comparison of the crop LAI image to measured data to test any site-specific parameters of the canopy reflectance model. For the second model, predicated LAI is compared to the crop LAI image to determine if any site-specific parameters need to be adjusted.

and site-specific parameters used by the SAIL model were adjusted as needed. After the comparison, predicted LAI was determined for each image in a time series. With gridded input data, the EPIC model was used to predict LAI over time, and spatial predictions were compared to the final crop LAI images (Figure 22.4).

With rigorously measured LAI data for different crops in a small region, the final predicted yield and soil carbon maps have higher accuracies than the same models without the LAI measurements. Furthermore, the combination of remote sensing and complex process models lose accuracy when applied over larger areas, because differences in crop genotypes and farming practices (e.g. tillage practices, planting dates, fertilizer rates, etc.) affect numerous site-specific parameters in these process models.

GLOBAL-SCALE MODELS AND REMOTE SENSING

Gross primary productivity (GPP) is the rate at which vegetation captures and stores energy as carbohydrates. Major factors affecting ecosystem GPP are leaf area index, physiological capacity of the plants (i.e. type of plants), and meteorological conditions. Net primary productivity (NPP) is the difference between GPP and the energy used for vegetation growth and maintenance respiration. Monteith (1972) proposed NPP could be estimated as a function of incident solar radiation and a series of conversion efficiencies (ε). Remotely-sensed canopy reflectance and vegetation indices can provide reliable estimates of the fraction of incident photosynthetically active radiation (PAR) absorbed by the plant canopy (fAPAR) (Asrar et al., 1989; Hatfield and Prueger, 2010; Kumar and Monteith, 1981). Because photosynthesis and plant respiration respond somewhat differently to environmental conditions, radiation-use efficiency for NPP is actually quite variable (Hunt and Running, 1992; Prince, 1991). Micrometeorological flux measurements of CO_2 show GPP is directly related to the amount of absorbed PAR.

The MODIS data product (MOD17) estimates gross primary production (GPP) for biomes based on the PAR-use efficiency model:

$$\text{GPP} = \varepsilon \sum (\text{fAPAR} \cdot \text{PAR}) \qquad [2]$$

where the summation is made over an 8-day compositing period (Heinsch et al., 2006; Reeves et al., 2005; Running et al., 2004). Reductions in PAR-use efficiency by cold temperatures and drought have been described using functions of minimum temperature and vapor pressure deficit (Heinsch et al., 2006). Other functions are used to estimate aboveground NPP, crop yield, carbon sequestration, and available forage (Hunt and Miyake, 2006; Hunt et al., 2004; Reeves et al., 2005). Equation [2] has also been combined with the Century model to estimate soil carbon fluxes at regional and global scales (Potter et al., 2006).

Certain leaf carotenoids track changes in photosynthesis rate and can be used to determine radiation-use efficiency by remote sensing (Drolet et al., 2005; Garbulsky et al., 2011; Middleton et al., 2009). When meteorological conditions inhibit photosynthesis, these carotenoids dissipate energy to avoid damaging the chloroplasts. This physiological process changes the leaf reflectance at 551 nm wavelength. The photochemical reflectance index (PRI) uses this wavelength, MODIS band 11 (Figure 22.1), to estimate the ratio of energy used for photosynthesis to energy dissipated, from which radiation-use efficiency may be calculated. Direct estimates of radiation-use efficiency using MODIS will increase the accuracy of global scale models in order to predict ecosystem carbon dynamics and the amount of carbon inputs into the soil.

CASE STUDY: SOUTH FORK WATERSHED

One of the primary goals of GRACEnet was to develop agricultural management strategies to enhance soil C sequestration, reduce greenhouse gas emissions, and improve environmental quality (Jawson et al., 2005; Walthall et al., 2012). Many management strategies have been examined at diverse locations. However, the question remains, how can the successful management strategies of GRACEnet be extended beyond original test sites? As reviewed in this chapter, remote sensing can provide important spatial and temporal data that will allow agricultural system models to be applied across landscapes. These models, when parameterized with site-specific data and calibrated to the current state, can evaluate management scenarios that will affect soil carbon, greenhouse gasses, and water quality at field and watershed scales. For example, corn (*Zea mays* L.) residues are potentially major feedstocks for bioenergy production. Recent studies have suggested that some removal of corn residue is sustainable (Graham et al., 2007). However, the impacts of shifting to continuous corn and removing the crop residues are not uniform for all fields or all crop management systems (Andrews, 2006; Johnson and Novak, 2012; Wilhelm et al., 2004). For the balance of this chapter, we will link diverse databases, remotely sensed data, and simulation models to address the environmental impact of selected crop management practices on soil erosion, soil carbon content, and water quality at field and watershed scales.

Description of the Watershed and Databases

The South Fork watershed of the Iowa River (Figure 22.5) covers about 788 km^2 and is part of the USDA Conservation Effects Assessment Project (CEAP) (Duriancik et al., 2008). The watershed is dominated by pothole depressions and artificial subsurface tile drainage needed to drain the hydric soils that cover 54% of the watershed (Tomer and James, 2004). The watershed is 84% cropland and the rest is mostly pasture or forest with very limited urban areas. Corn and soybean (*Glycine max* L. Merr.) are grown on more than 99% of the cropland area.

Temporal and geospatial data for the watershed include soil properties, climate, crop classification, and management practices. Soil properties for all fields were extracted from the SSURGO database and slopes were estimated using a moderate resolution (10 m) digital elevation model (USDA-NRCS, 2011). Climate data were assembled from first order and cooperative observer program stations in and around the watershed. Rainfall measured

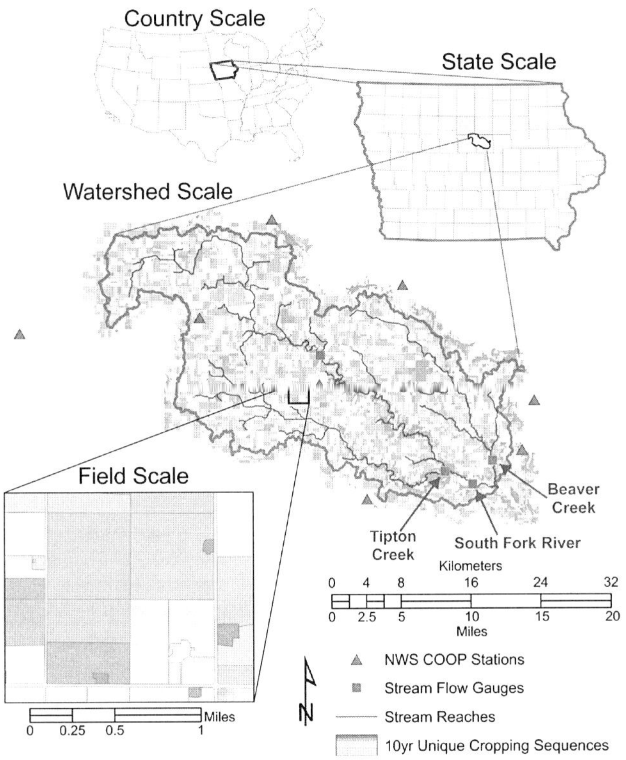

FIGURE 22.5
Map of the South Fork of the Iowa River watershed in Iowa. Please see color plate section at the back of the book.

at the weather stations was spatially distributed using the NEXRAD data (Beeson et al., 2011). Crops grown in each field of the watershed were extracted from the Cropland Data Layer for 2000–2009 (Stern et al., 2011). Soil tillage intensity was determined by field surveys in 2005 (Tomer et al., 2008) and derived using SPOT imagery and roadside surveys in 2008–2010. Fields with <30% residue cover were classified as conventional tillage and fields with ≥30% residue were classified as conservation tillage. Fertilizer application rates and other management practices critical for mapping soil carbon were based on grower surveys or state averages.

Watershed Simulations

The Soil and Water Assessment Tool (SWAT) model is a quasi-physically-based water quality simulation model that predicts the impact of management decisions on water, sediment, and agricultural chemicals in watersheds (Arnold et al., 1998). Gassman et al. (2007) provided a thorough evaluation of the SWAT model. Good results from SWAT are dependent on correct precipitation input and the abundance of observation data to calibrate the water budget.

Simulation results, based on best available inputs (e.g. soil type, land use, crop rotations, precipitation, management practices), showed sediment discharges predicted by SWAT were well correlated with sediment discharges measured near the outlet of the South Fork watershed (Beeson et al., 2011). SWAT estimated annual sediment discharge for six management scenarios for the South Fork watershed (Figure 22.6). Six management scenarios for three soil tillage intensities and two residue removal rates were simulated for continuous corn grown on all fields in the watershed. Tillage intensities were (1) conventional tillage (<30% residue cover), (2) conservation tillage (≥30% cover), and (3) no tillage (except for planting and fertilizer applications, >60% cover). Crop residue removal rates were 0% and 80% for each tillage scenario.

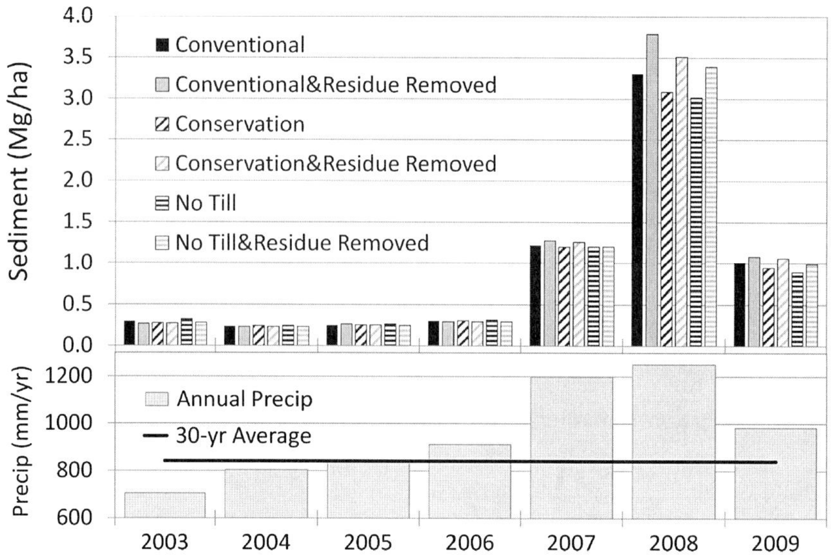

FIGURE 22.6
SWAT simulations of annual sediment discharge at the outlet of the South Fork watershed for six management scenarios. The scenarios assumed all fields were continuous corn with (1) conventional tillage (<30% residue cover), (2) conservation tillage (≥30% cover), or (3) no tillage (>60% cover) and two crop residue removal rates (0% and 80%). Annual rainfall for 2000–2009 and 30-year mean (black line) are shown in the lower graph.

Water quality as measured by sediment discharge in the South Fork river was strongly affected by the amount of precipitation received (Figure 22.6). When precipitation was less than or equal to normal amounts, tillage intensity had little effect on annual amount of sediment delivered to the outlet of the South Fork. However, when precipitation was above normal, such as 2008, sediment discharge was an order of magnitude greater than the sediment load in a normal year. Sediment discharge also increased as soil tillage intensity increased (No-Till - < Conservation < Conventional) and as crop residue cover decreased. Crop residue removal (e.g. for biofuel or feed) also increased the amount of sediment discharge from the watershed for all tillage intensities due to decreased infiltration and reduced cover on the soil surface. Because of the environmental risks associated with above normal precipitation, particularly for intense storm events, proactive management strategies for the South Fork watershed should encourage reducing soil tillage intensity to maintain crop residue cover on the soil, establishing filter strips to retain sediment, and implementing other conservation practices to compensate for the loss of erosion protection provided by crop residues on the soil surface.

Field and Farm Simulations

EPIC is an ecosystem model capable of simulating many processes in agricultural lands, such as crop growth and yield, water balance, and nutrient cycling as affected by weather, soil, and management practices (Izaurralde et al., 2012; Williams, et al., 1984). EPIC is well suited for addressing the effects of crop and soil management practices for homogeneous fields. For the balance of this section, several hypothetical scenarios are presented using the EPIC field model to evaluate the impact of management decisions on soil sustainability over time.

A farmer in the South Fork watershed has two fields with different dominant soil types. Both fields have similar topographic relief (≤2% slope) and have grown corn continuously with conventional tillage for several years. One field is a loam (66.1 Mg ha^{-1} stable soil carbon in the top 20 cm) and the other field is a silty clay loam (92.9 Mg ha^{-1} stable soil carbon). The farmer would like to evaluate the impact of corn residue harvest for biofuel and tillage intensity on the stable soil carbon in the top 20 cm of the soil and sediment loss in each field over the next 10 years. Tillage options are to continue with conventional tillage (<30% crop

TABLE 22.2 EPIC Simulated Changes in Stable Carbon Pools and Annual Sediment Loss from Two Soils in the South Fork Watershed After 10 Years with Different Management Scenarios

		Residue removal		
Soil texture	**Tillage intensity**	0%	40%	80%
		Stable soil carbon, Mg ha^{-1}		
Silty clay loam	Conventional	93.5	91.5	88.5
	No-till	96.1	94.9	93.3
Loam	Conventional	67.8	66.5	63.8
	No-till	70.4	69.5	67.3
		Annual sediment loss, Mg ha^{-1}		
Silty clay loam	Conventional	2.8	3.8	6.1
	No-till	0.5	0.6	1.1
Loam	Conventional	3.0	3.9	6.1
	No-till	0.6	0.6	1.1

residue cover) or switch to no-till. Residue harvesting would remove either 40 or 80% of the corn residue remaining on the soil surface after corn grain harvest.

The EPIC predictions showed that after 10 years, the stable organic carbon in the top 20 cm of both soils increased and the soil loss decreased as soil tillage intensity changed from conventional tillage to no-till (Table 22.2). Although conventional tillage with no residue removal maintained soil carbon, annual sediment losses $\geq$2.8 Mg ha^{-1}. Removing residue and continuing to use conventional tillage reduced SOC and increased soil loss from both soils. The positive effects of reduced soil disturbance, i.e. switching from conventional tillage to no-till, more than offset any losses due to residue removal for both soils. These simulations indicate that removal of up to 80% of the corn residue from these two relatively flat soil areas may be sustainable if tillage intensity is reduced by switching from conventional tillage to no-till management. However, this is not true for all crops, soils, and climatic regions (Wilhelm et al., 2004). For example, corn residue harvest adversely affected SOC of the loamy sand soils (i.e. coarser soil texture) of the coastal plain region (i.e. warmer and wetter climate) of the southeastern U.S. (Gollany et al., 2010). It is also important to recognize that as slope increases the amount of crop residue that can be harvested in a sustainable manner decreases significantly (Andrews, 2006; Johnson et al., 2010; Wilhelm et al., 2004). With the natural variation in topography, soil type, and crop yields typical across most agricultural fields, it is difficult to predict the amount of corn residue that can be removed in a sustainable manner at the field scale (Johnson et al., 2010). In fact, different rates of crop residue harvest may be required for different areas of the same field.

Crop residues on the soil surface are the first line of defense against the erosive forces of water and wind (Johnson and Novak, 2012). Excessive corn residue harvest for bio-energy feedstock or any other use can accelerate soil erosion and loss of soil organic matter. The crop residue harvest decisions must be site specific because the environmental consequences can be minimal in some soil, crop, climate, topography, and management combinations, but much greater in others (Andrews, 2006). Well-validated models can evaluate combinations of management practices for a given soil—climate—crop combination that maximizes biomass production and minimizes environmental impact. Nevertheless, proactive management practices should include frequent monitoring of fields for visual signs of soil erosion and periodic checks of SOC and soil fertility whenever crop residue is removed.

Currently, no single model can adequately address all agronomic and environmental issues. Robust decision support systems that utilize suites of models to address the wide range of agronomic, environmental, and economic questions likely to be posed by farmers, stakeholders, and policy-makers are required and several prototypes are nearing completion (Causarano et al., 2008; Daughtry et al., 2011; Muth et al., 2010). The issues to be addressed and the level of detail (spatial and temporal resolution) required will determine which combination of models, described in Chapters 14–18 of this volume and elsewhere, should be used.

CONCLUSIONS

The overall goal of GRACEnet is to develop strategies for managing soil C sequestration and reducing emissions of greenhouse gases for agricultural lands. The geographical distribution of GRACEnet sites and use of common protocols provide an excellent sampling of U.S. agricultural conditions. Simulation models have been developed or modified to predict the impacts of crop and soil management practices on SOC, greenhouse gas emissions, and water quality. However, in order to assess the overall impact of the adoption of these management practices the data acquired from small plots or fields must be extended to regional and national scales. The spatial scaling task is nontrivial because the mechanisms controlling carbon, water, and energy exchanges are nonlinear and interact with each other.

While *in situ* measurements accurately describe the temporal variability of surface fluxes, remote sensing offers the only practical method to account for the spatial variability inherent across agricultural landscapes. Many of the biophysical characteristics of vegetation and soils, needed by the process models at field or watershed scales, can be derived directly or indirectly from remotely sensed data. However, remote sensing systems are constrained by spatial and temporal resolution compromises. Recently developed data fusion and data assimilation techniques have shown considerable promise for merging data acquired at differing spatial and temporal resolutions and creating synthetic datasets with high spatial and temporal resolutions. Based on the field and watershed scenarios evaluated, process models of varying scales together with these enhanced datasets appear to provide much better descriptions of ecosystem functions at field to national scales.

Acknowledgments

The authors very much appreciate the sustained support of Drs. Jerry Hatfield, Mark Tomer, Doug Karlen, and others from USDA-ARS National Laboratory for Agriculture and the Environment in Ames, Iowa. Finally, we would like to give special recognition to the contributions of one of our own, the late Dr. Paul C. Doraiswamy.

References

Aase, J.K., Tanaka, D.L., 1991. Reflectance from four wheat residue cover densities as influenced by three soil backgrounds. Agron. J. 83, 753–757.

Allen, R., Tasumi, M., Trezza, R., 2007. Satellite-based energy balance for mapping evaportranspiration with internalized calibration (METRIC) model. J. Irrigation Drainage Engineering 133, 380–394.

Anderson, M.C., Kustas, W.P., 2008. Thermal remote sensing of drought and evapotranspiration. EOS Trans. AGU 89, 233–240.

Anderson, M.C., Kustas, W.P., Norman, J.M., 2007. Upscaling flux observations from local to continental scales using thermal remote sensing. Agron. J. 99, 240–254.

Anderson, M.C., Norman, J.M., Diak, G.R., Kustas, W.P., Mecikalski, J.R., 1997. A two-source time-integrated model for estimating surface fluxes from thermal infrared satellite observations. Remote Sens. Environ. 60, 195–216.

Anderson, M.C., Kustas, W.P., Norman, J.M., Hain, C.R., Mecikalski, J.R., Schultz, L., González-Dugo, M.P., Cammalleri, C., d'Urso, G., Pimstein, A., Gao, F., 2011. Mapping daily evaporation at field to continental scales using geostationary and polar orbiting satellite imagery. Hydrology and Earth System Sciences 15, 223–239. DOI: 10.5194/hess-15-123-2011.

Andrews, S.S., 2006. Crop residue removal for biomass energy production: effects on soils and recommendations. Agricultural Outlook Forum 2006. USDA Natural Resource Conservation Service, Arlington, VA. Available at http://soils.usda.gov/sqi/management/files/AgForum_Residue_White_Paper.pdf (verified December 13, 2011).

Arnold, J.G., Srinivasan, R., Muttiah, R.S., Williams, J.R., 1998. Large-area hydrologic modeling and assessment: Part I. Model development. J. American Water Resources Assoc. 34, 73–89.

Asrar, G., Myneni, R.B., Kanemasu, E.T., 1989. Estimation of plant-canopy attributes from spectral reflectance measurements. In: Asrar, G. (Ed.), Theory and Applications of Optical Remote Sensing. John Wiley, New York, pp. 252–296.

Badhwar, G.D., Henderson, K.E., 1981. Estimating development stages of corn from spectral data: an initial model. Agron. J. 73, 748–755.

Baird, F., Baret, F., 1997. Crop residue estimation using multiband reflectance. Remote Sens. Environ. 59, 530–536.

Baldocchi, D.D., Meyers, T., 1998. On using eco-physiological, micrometeorological and biogeochemical theory to evaluate carbon dioxide, water vapor and trace gas fluxes over vegetation: a perspetive. Agric. For. Meteorol. 90, 1–25.

Baldwin, M., Kellogg, C.E., Thorp, J., 1938. Soils and Men, Yearbook of Agriculture, U.S. Department of Agriculture, Washington, DC, pp. 979–1001.

Bauer, M.E., 1985. Spectral inputs to crop identification and condition assessment. Proceedings of the IEEE 73, 1071–1085.

Baumgardner, M.F., Silva, L.F., Biehl, L.L., Stoner, E.R., 1985. Reflectance properties of soils. Advances Agron. 8, 1–44.

Beeson, P.C., Doraiswamy, P.C., Sadeghi, A.M., Di Luzio, M., Tomer, M.D., Arnold, J.G., Daughtry, C.S.T., 2011. Treatments of precipitation inputs to hydrologic models. Trans. ASABE. 54 (6) In press.

Ben-Dor, E., 2002. Quantitative remote sensing of soil properties. Advances Agron. 75, 173–243.

Ben-Dor, E., Chabrillat, S., Dematte, J.A.M., Taylor, G.R., Hill, J., Whiting, M.L., Sommer, S., 2009. Using imaging spectroscopy to study soil properties. Remote Sens. Environ. 113, S38–S55.

Bonan, G.B., Oleson, K.W., Vertenstein, M., Levis, S., Zeng, X., Dai, Y., Dickinson, R.E., Z- Yang, L., 2002. The land surface climatology of the community land model coupled to the NCAR community climate model. J. Climate. 15, 3123–3149.

Brown, D.J., Shepherd, K.D., Walsh, M.G., Mays, M.D., Reinsch, T.G., 2006. Global soil characterization with VNIR diffuse reflectance spectroscopy. Geoderma 132, 273–290.

Brown, D.J., Hunt Jr., E.R., Izaurralde, R.C., Paustian, K.H., Rice, C.W., Schumaker, B.L., West, T.O., 2010. Soil organic carbon change monitored over large areas. EOS Trans. AGU 91, 441–456.

Causarano, H.J., Doraiswamy, P.C., McCarty, G.W., Hatfield, J.L., Milak, S., Stern, A.J., 2008. EPIC modeling of soil organic carbon sequestration in croplands of Iowa. J. Environ. Qual. 37, 1345–1353.

Chang, J., Clay, D.E., Dalsted, K., Clay, S., O'Neill, M.O., 2003. Corn (*Zea mays* L.) yield prediction using multi-spectral and multidate refelctance. Agron. J. 95, 1447–1453.

Chen, F., Kissel, D.E., West, L.T., Adkins, W., Rickman, D., Luvall, J.C., 2008. Mapping soil organic carbon concentration for multiple fields with image similarity analysis. Soil Sci. Soc. Am. J. 72, 186–193.

Corak, S.J., Kaspar, T.C., Meek, D.W., 1993. Evaluating methods for measuring residue cover. J. Soil Water Conserv. 48, 700–704.

Crow, W.T., Kustas, W.P., Prueger, J.H., 2008. Monitoring root-zone soil moisture through the assimilation of a thermal remote sensing-based soil moisture proxy into a water balance model. Remote Sens. Environ. 112, 1268–1281.

Daughtry, C.S.T., Gallo, K.P., Goward, S.N., Prince, S.D., Kustas, W.P., 1992. Spectral estimates of absorbed radiation and phytomass production in corn and soybean canopies. Remote Sens. Environ. 39, 141–152.

Daughtry, C.S.T., Walthall, C.L., Kim, M.S., Brown de Colstoun, E., McMurtrey III, J.E., 2000. Estimating corn leaf chlorophyll concentration from leaf and canopy reflectance. Remote Sens. Environ. 74, 229–239.

Daughtry, C.S.T., Serbin, G., Reeves III, J.B., Doraiswamy, P.C., Hunt Jr., E.R., 2010. Effects of decomposition of remotely sensed estimates of wheat residue cover. Remote Sens. 2, 416–431.

Daughtry, C.S.T., Doraiswamy, P.C., Hunt Jr., E.R., Stern, A.J., McMurtrey III, J.E., Prueger, J.H., 2006. Remote sensing of crop residue cover and soil tillage intensity. Soil Tillage Res. 91, 101–108.

Daughtry, C.S.T., Beeson, P.C., Milak, S., Akhmedov, B., Hunt Jr., E.R., Sadeghi, A.M., Tomer, M.D., 2011. Remote sensing of soil tillage intensity in a CEAP watershed in central Iowa. In: Medley, C.N., et al. (Eds.), Observing, Studying, and Managing for Change. U.S.G.S. Scientific Investigation Report 2011–5169, U.S. Geological Survey, Reston, VA, pp. 168–173.

Delgado, J.A., 2010. Crop residue is a key for sustaining maximum food production and for conservation of our biosphere. J. Soil Water Conserv. 65, 111A–116A.

Doraiswamy, P.C., Hatfield, J.L., Jackson, T.J., Akhmedov, B., Prueger, J., Stern, A., 2004. Crop condition and yield simulations using Landsat and MODIS. Remote Sens. Environ. 92, 548–559.

Doraiswamy, P.C., Sinclair, T.R., Hollinger, S., Akhmedov, B., Stern, A., Prueger, J., 2005. Application of MODIS derived parameters for regional crop yield assessment. Remote Sens. Environ. 97, 192–202.

Dorigo, W.A., Zurita-Milla, R., deWit, A.J.W., Brazile, J., Singh, R., Schaepman, M.E., 2007. A review on reflective remote sensing and data assimilation techniques for enhanced agroecosystem modeling. Int. J. Appl. Earth Obs. Geoinfo. 9, 165–193.

Drolet, G.G., Huemmrich, K.F., Hall, F.G., Middleton, E.M., Black, T.A., Barr, A.G., Margolis, H.A., 2005. A MODIS-derived photochemical reflectance index to detect inter-annual variations in the photosynthetic light-use efficiency of a boreal deciduous forest. Remote Sens. Environ. 98, 212–224.

Duriancik, L.F., Bucks, D., Dobrowolski, J.P., Drewes, T., Eckles, S.D., Jolley, L., Kellogg, R.F., Lund, D., Makuch, J.R., O'Neill, M.P., Rewa, C.A., Walbridge, M.R., Parry, R., Weltz, M.A., 2008. The first five years of the Conservation Effects Assessment Project. J. Soil Water Conserv. 63, 185A–197A. DOI: doi:10.2489/jswc.63.6.185A.

Eitel, J.U., Long, D.S., Gessler, P.E., Hunt Jr., E.R., Brown, D.J., 2009. Sensitivity of ground-based remote sensing estimates of wheat chlorophyll content to variation in soil reflectance. Soil Sci. Soc. Am. J. 73, 1715–1723.

Eitel, J.U.H., Long, D.S., Gessler, P.E., Hunt Jr., E.R., 2008. Combined spectral index to improve ground-based estimates of nitrogen status in dryland wheat. Agron. J. 100, 1694–1702.

Ek, M.B., Mitchell, K.E., Lin, Y., Rogers, E., Grunman, P., Koren, V., Gayno, G., Tarpley, J.D., 2003. Implementation of Noah land surface model advances in the National Centers for Environmental Prediction operational mesoscale Eta model. J. Geophys. Res. (D Atmospheres) 108, 1–16.

Elvidge, C.D., 1990. Visible and near infrared reflectance characteristics of dry plant materials. Int. J. Remote Sens. 11, 1775–1795.

Entekhabi, D., Njoku, E.G., O'Neill, P.E., Kellogg, K.H., Crow, W.T., Edelstein, W.N., Entin, J.K., Goodman, S.D., Jackson, T.J., Johnson, J., Kimball, J., Piepmeier, J.R., Koster, R.D., Martin, N., McDonald, K.C., Moghaddam, M., Moran, S., Riechle, R., Shi, J.C., Spencer, M.W., Thurman, S.W., Tsang, L., Van Zyl, J., 2010. The Soil Moisture Active Passive (SMAP) mission. Proceedings of the IEEE 98, 704–716.

Fitzgerald, G., Rodriguez, D., O'Leary, G., 2010. Measuring and predicting crop canopy nitrogen nutrition in wheat using a spectral index—the crop chlorophyll content index (CCCI). Field Crops Res. 116, 318–324.

Gao, F., Masek, J.G., Hall, J., Schwaller, M., 2006. On the blending of the Landsat and MODIS surface reflectance: predicting daily Landsat surface reflectance. IEEE Trans. Geosci. Remote Sens. 44, 2207–2218.

Garbulsky, M.F., Peñuelas, J., Gamon, J., Inoue, Y., Filella, I., 2011. The photochemical reflectance index (PRI) and the remote sensing of leaf, canopy and ecosystem radiation use efficiencies, a review and meta-analysis. Remote Sens. Environ. 115, 281–297.

Gassman, P.W., Reyes, M., Green, C.H., Arnold, J.G., 2007. The Soil and Water Assessment Tool: historical development, applications, and future directions. Trans. ASABE. 50, 1211–1250.

Gates, D.M., 1980. Biophysical Ecology. Springer-Verlag, New York.

Glenn, E.P., Nagler, P.L., Huete, A., 2010. Vegetation index methods for estimating evapotranspiration by remote sensing. Surv. Geophys. 31, 531–555.

Gollany, H.T., Novak, J.M., Liang, Y., Albrecht, S.L., Rickman, R.W., Follet, R.F., Wilhelm, W.W., Hunt, P.G., 2010. Simulating soil organic carbon dynamics with residue removal using the CQESTR model. Soil Sci. Soc. Am. J. 74, 372–383.

Gomez, C., Viscarra, R.A., McBratney, A.B., 2008. Soil organic carbon prediction by hyperspectral remote sensing and field vis-NIR spectroscopy: an Australian case study. Geoderma 146, 403–411.

Goward, S.N., Huemmrich, K.F., 1992. Vegetation canopy PAR absorptance and the Normalized Difference Vegetation Index: an assessment using the SAIL model. Remote Sens. Environ. 39, 119–140.

Goward, S.N., Arvindson, T., Williams, D.L., Irish, R., Irons, J.R., 2009. Moderate spatial resolution optical sensors. In: Warner, T.A., et al. (Eds.), The Sage Handbook of Remote Sensing. Sage Publications, Thousand Oaks, CA, pp. 123–138.

Graham, R.L., Nelson, R., Sheehan, J., Perlack, R.D., Wright, L.L., 2007. Current and potential U.S. corn stover supplies. Agron. J. 99, 1–11.

Guerschman, J.P., Hill, M.J., Renzullo, L.J., Barrett, D.J., Marks, A.S., Botha, E.J., 2009. Estimating fractional cover of photosynthetic vegetation, non-photosynthetic vegetation and bare soil in the Australian tropical savana region upscaling the EO-1 Hyperion and MODIS sensors. Remote Sens. Environ. 113, 928–945.

Haboudane, D., Miller, J.R., Tremblay, N., Zarco-Tejada, P.J., Dextraze, L., 2002. Integrated narrow-band vegetation indices for prediction of crop chlorophyll content for application to precision agriculture. Remote Sens. Environ. 81, 416–426.

Hatfield, J.L., Prueger, J.H., 2010. Value of using different vegetative indices to quantify agricultural crop characteristics at different growth stages under varying management practices. Remote Sens. Environ. 2, 562–578.

Heinsch, F.A., Zhao, M., Running, S.W., Kimball, J.S., Nemani, R.R., Davis, K.J., Bolstad, P.V., Cook, B.D., Desai, A.R., Ricciuto, D.M., Law, B.E., Oechel, W.C., Kwon, H., Luo, H., Wofsy, S.C., Dunn, A.L., Munger, J.W.,

Baldocchi, D.D., Xu, L., Hollinger, D.Y., Richardson, A.D., Stoy, P.C., Siqueira, M.B.S., Monson, R.K., Burns, S.P., Flanagan, L.B., 2006. Evaluation of remote sensing based terrestrial productivity from MODIS using regional tower eddy flux network observations. IEEE Trans. Geosci. Remote Sens. 44, 1908–1925.

Henderson, T.L., Baumgardner, M.F., Franzmeier, D.P., Stott, D.E., Coster, D.C., 1992. High dimensional reflectance analysis of soil organic mater. Soil Sci. Soc. Am. J. 56, 865–872.

Huete, A.R., Didan, K., Miura, T., Rodriguez, E.P., Gao, X., Ferreira, L.G., 2002. Overview of the radiometric and biophysical performance of the MODIS vegetation indices. Remote Sens. Environ. 83, 195–213.

Hunt Jr., E.R., Rock, B.N., 1989. Detection of changes in leaf water content using near- and middle-infrared reflectances. Remote Sens. Environ. 30, 43–54.

Hunt Jr., E.R., Running, S.W., 1992. Simulated dry matter yields for aspen and spruce stands in the North American boreal forest. Can. J. Remote Sens. 18, 126–133.

Hunt Jr., E.R., Miyake, B.A., 2006. Comparison of stocking rates from remote sensing and geospatial data. Rangeland Ecol. Manage. 59, 11–18.

Hunt Jr., E.R., Daughtry, C.S.T., Eitel, J.U., Long, D.S., 2011. Remote sensing of leaf chlorophyll content using a visible waveband index. Agron. J. 103, 1090–1099. DOI: doi:10.2134/agronj2010.0395.

Hunt Jr., E.R., Kelly, R.D., Smith, W.K., Fahnstock, J.T., Welker, J.M., Reiners, W.A., 2004. Estimation of carbon sequestration by combining remote sensing and net ecosystem exchange data for northern mixed-grass prairie and sagebrush-steppe ecosystems. Environ. Manage. 33, S432–S441.

Izaurralde, R.C., McGill, W.B., Williams, J.R., 2012. Development and application of the EPIC model for carbon cycle, greenhouse-gas mitigation, and biofuel studies. In: Liebig, M.A., et al. (Eds.), Managing Agricultural Greenhouse Gases: Coordinated Agricultural Research through GRACEnet to Address Our Changing Climate. Academic Press, San Diego, CA.

Izaurralde, R.C., Williams, J.R., McGill, W.B., Rosenberg, N.J., Quiroga Jakas, M.C., 2006. Simulating soil C dynamics with EPIC: model description and testing against long-term data. Ecolog. Model. 192, 362–384.

Jackson, T.J., Le Vine, D.M., Hsu, A.Y., Oldak, A., Starks, P.J., Swift, C.T., Isham, J.D., Haken, M., 1999. Soil moisture mapping at regional scales using microwave radiometry: the Southern Great Plains hydrology experiment. IEEE Trans. Geosci. Remote Sens. 37, 2136–2151.

Jackson, T.J., Cosh, M.H., Bindlish, R., Starks, P.J., Bosch, D.D., Seyfried, M., Goodrich, D.C., Moran, M.S., Du, J., 2010. Validation of Advanced Scanning Radiometer soil moisture products. IEEE Trans. Geosci. Remote Sens. 48, 4256–4272.

Jacquemoud, S., Verhoef, W., Baret, F., Bacour, C., Zarco-Tejada, P.J., Asner, G.P., François, C., Ustin, S.L., 2009. PROSPECT + SAIL models: a review of use for vegetation characterization. Remote Sens. Environ. 113, S56–S66.

Jawson, M.D., Shafer, S.R., Franzluebbers, A.J., Parkin, T.B., Follett, R.F., 2005. GRACEnet: greenhouse gas reduction through carbon enhancement network. Soil Tillage Res. 83, 167–172.

Johnson, D.M., Mueller, R., 2010. The 2009 Cropland Data Layer. Photogramm. Eng. Remote Sens. 76, 1201–1205.

Johnson, J.M.F., Novak, J.M., 2012. Sustainable bioenergy feedstock production systems: integrating C dynamics, erosion, water quality, and greenhouse gas production. In: Liebig, M.A., et al. (Eds.), Managing Agricultural Greenhouse Gases: Coordinated Agricultural Rsearch through GRACEnet to Address Our Changing Climate. Academic Press, San Diego, CA.

Johnson, J.M.F., Karlen, D.L., Andrews, S.S., 2010. Conservation considerations for sustainable bioenergy feedstock production: if, what, where, and how much? J. Soil Water Conserv. 65, 88A–91A.

Justice, C.O., Tucker, C.J., 2009. Coarse spatial resolution optical sensors. In: Warner, T.A., et al. (Eds.), The Sage Handbook of Remote Sensing. Sage Publications, Thousand Oaks, CA, pp. 139–150.

Kalma, J.D., McVicar, T., McCabe, M.F., 2008. Estimating land surface evaporation: a review of methods using remotely sensed surface temperature data. Surv. Geophysics. 29, 421–469.

Kimball, J.S., Jones, L.A., Zhang, K., Heinsch, F.A., McDonald, K.C., Oechel, W.C., 2009. A satellite approach to estimate land–atmosphere CO_2 exchange for boreal and arctic biomes using MODIS and AMSR-E. IEEE Trans. Geosci. Remote Sens. 47, 569–587.

Knipling, E.B., 1970. Physical and physiological basis for the reflectance of visible and near-infrared radiation from vegetation. Remote Sens. Environ. 1, 133–144.

Kollenkark, J.C., Daughtry, C.S.T., Bauer, M.E., Housley, T.L., 1982. Effects of cultural practices on agronomic and reflectance characteristics of soybean canopies. Agron. J. 74, 751–758.

Kumar, M., Monteith, J.L., 1981. Remote sensing of crop growth. In: Smith, H. (Ed.), Plants and the Daylight Spectrum. Academic Press, London, pp. 133–144.

Kustas, W.P., Anderson, M.C., 2009. Advances in thermal infrared remote sensing for land surface modeling. Agric. For. Meteorol. 149, 2071–2081.

Lai, C.-T., Katul, G.G., Oren, R., Ellsworth, D., Schafer, K., 2000. Modeling CO_2 and water vapor turbulent flux distributions within a forest canopy. J. Geophys. Res. (D Atmospheres) 105, 26333–26351.

Lal, R., Kimble, J.M., Follett, R.F., Cole, C.V., 1999. The Potential of U.S. Cropland to Sequester Carbon and Mitigate the Greenhouse Effect. Lewis Publishers, Boca Raton, FL.

Lang, M., McCarty, G., 2009. Improved detection of forested wetland hydrology with LiDAR intensity. Wetlands 29, 1166–1178.

Liang, S., Fang, H., Morisette, J.T., Chen, M., Shuey, C.J., Walthall, C.L., Daughtry, C.S.T., 2002. Atmospheric correction of Landsat ETM+ land surface imagery: II. Validation and applications. IEEE Trans. Geosci. Remote Sens. 40, 2736–2746.

MacDonald, R.B., Hall, F.G., 1980. Global crop forecasting. Science 208, 670–679.

MacDonald, R.B., Bauer, M.E., Allen, R.D., Clifton, J.W., Erickson, J.D., Landgrebe, D.A., 1972. Results of the 1971 Corn Blight Watch Experiment, Eighth Int. Symp. Remote Sensing of Enviroment, Ann Arbor, MI, pp. 157–190. Available at http://docs.lib.purdue.edu/cgi/viewcontent.cgi?article=1106&context=larstech (verified December 13, 2011).

Mahecha, M.D., Reichstein, M., Carvalhais, N., Lasslop, G., Lange, H., Seneviriatne, S.I., Vargas, R., Ammann, C., Arain, M.A., Cescatti, A., Janssens, I.A., Migliavacca, M., Montagnani, L., Richardson, A.D., 2010. Global convergence in the temperature sensitivity of respiration at the ecosystem level. Sci. Total Environ. 329, 838–840.

Masek, J.G., Vermote, E.F., Saleous, N., Wolfe, R., Hall, E.F., Huemmrich, K.F., Gao, F., Kutler, J., T- Lim, K., 2006. A Landsat surface reflectance data set for North America, 1990–2000. Geosci. Remote Sens. Letters 3, 68–72.

McBratney, A.B., Pringle, M.J., 1999. Estimating average and proportional variograms of soil properties and their potential use in precision agriculture. Precison Agricult. 1, 125–152.

McCarty, G.W., Hively, W.D., Reeves, J.B., Lang, M.W., Lund, E., Weatherbee, O., 2010. Infrared sensors to map soil carbon in agricultural ecosystems. In: Viscarra-Rossel, R., et al. (Eds.), Proximal Soil Sensing, Progress in Soil Science Volume 1. Springer Science, New York, NY, pp. 165–176.

McKenzie, N.J., Ryan, P.J., 1999. Spatial prediction of soil properties using environmental correlation. Geoderma 89, 67–94.

McNairn, H., Protz, R., 1993. Mapping corn residue cover on agricultural fields in Oxford County, Ontario, using Thematic Mapper. Can. J. Remote Sens. 19, 152–159.

Middleton, E.M., Cheng, Y.B., Hilker, T., Black, T.A., Krishnan, P., Coops, N.C., Huemmrich, K.F., 2009. Linking foliage spectral responses to canopy-level ecosystem photosynthetic light-use efficiency at a Douglas-fir forest in Canada. Can. J. Remote Sens. 35, 166–188.

Monteith, J.L., 1972. Solar radiation and productivity in tropical ecosystems. J. Appl. Ecol. 9, 747–766.

Moran, M.S., 2003. Thermal infrared measurement as an indicator of plant ecosystem health. In: Quattrochi, D.A., Luvall, J. (Eds.), Thermal Remote Sensing in Land Surface Processes. Taylor and Francis, New York, pp. 257–282.

Moran, M.S., Jackson, R.D., Slater, P.N., Teillet, P.M., 1992. Evaluation of simplified procedures for retrieval of land surface reflectance factors from satellite sensor output. Remote Sens. Environ. 41, 169–184.

Morrison, J.E., Huang, C., Lightle, D.T., Daughtry, C.S.T., 1993. Residue cover measurement techniques. J. Soil Water Conserv. 48, 479–483.

Murray, I., Williams, P.C., 1988. Chemical principles of near-infrared technology. In: Williams, P.C., Norris, K. (Eds.), Near-Infrared Technology in the Agricultural and Food Industries. American Association Cereal Chemists, St. Paul, MN.

Muth, D., Anex, R., Karlen, D.L., Nelson, R., 2010. Development of a generalized sustainable agricultural residue assessment tool. Sustainable Feedstocks for Advanced Biofuels: Roadmap Conference, Soil and Water Conservation Society. Centennial Olympic Park, Atlanta, GA. Available at http://www.swcs.org/documents/filelibrary/roadmap/SFAB_Program_and_Abstract_Book__FIN_8607CF9B02A16.pdf (verified December 13, 2011).

Nagler, P.L., Daughtry, C.S.T., Goward, S.N., 2000. Plant litter and soil reflectance. Remote Sens. Environ. 71, 207–215.

Pacheco, A., McNairn, H., 2010. Evaluating multispectral remote sensing and spectral unmixing analysis for crop residue mapping. Remote Sens. Environ. 114, 2219–2228.

Paustain, K., Schuler, J., Killian, K., Chambers, A., DelGrosso, S., Easter, M., Alvaro-Fuentes, J., Gurung, R., Johnson, G., Merwin, M., Ogle, S., Olson, C., Swan, A., Williams, S., Vining, R., 2012. COMET2.0—Decision spport system for agricultural greenhouse gas accounting. In: Liebig, M.A., et al. (Eds.), Managing Agricultural Greenhouse Gases: Coordinated Agricultural Research through GRACEnet to Address Our Changing Climate. Academic Press, San Diego, CA.

Potter, C., Klooster, S., Nemani, R., Genovese, V., Hiatt, S., Field, M., Fladeland, M., Gross, P., 2006. Estimating carbon budgets for U.S. ecosystems. EOS Trans. AGU 87, 85–90.

Prince, S.D., 1991. A model of regional primary production for use with coarse resolution satellite data. Int. J. Remote Sens. 12, 1313–1330.

Reeves III, J.B., McCarty, G.W., Calderon, F., Hively, W.D., 2012. Advances in spectroscopic methods for quantifying soil carbon. In: Liebig, M.A., et al. (Eds.), Managing Agricultural Greenhouse Gases: Coordinated Agricultural Research through GRACEnet to Address Our Changing Climate. Academic Press, San Diego, CA.

Reeves, M.C., Zhao, M., Running, S.W., 2005. Usefulness and limits of MODIS GPP for estimating wheat yield. Int. J. Remote Sens. 26, 1403–1421.

Reichstein, M., Rey, A., Freibauer, A., Tenhunen, J., Valentini, R., Banza, J., Casals, P., Cheng, Y., Grünzweig, J.M., Irvine, J., Joffre, R., Law, B.E., Loustau, D., Miglietta, F., Oechel, W., Ourcival, J.M., Pereira, J.S., Peressotti, A., Ponti, F., Qi, Y., Rambal, S., Rayment, M., Romanya, J., Rossi, F., Tedeschi, V., Tirone, G., Xu, M., Yakir, D., 2003. Modeling temporal and large-scale spatial variability of soil respiration from soil water availability, temperature and vegetation productivity indices. Global Biogeochem. Cycles 17, 1104. doi:10.1029/2003GB002035.

Riaño, D., Chuvieco, E., Salas, J., Aguado, I., 2003. Assessment of different topographic corrections in Landsat-TM data for mapping vegetation types. IEEE Trans. Geosci. Remote Sens. 41, 1056–1061.

Rondeaux, G., Steven, M., Baret, F., 1996. Optimization of soil-adjusted vegetation indices. Remote Sens. Environ. 55, 95–107.

Rouse, J.W., Haas, R.H., Schell, J.A., Deering, D.W., 1974. Monitoring vegetation systems in the Great Plains with ERTS. In: Freden, S.C., et al. (Eds.), Third Earth Resources Technology Satellite-1 Symposium. Vol. 1: Technical presentations, NASA SP-351. U.S. Government Printing Office, Washington, DC, pp. 309–317.

Running, S.W., Nemani, R.R., Heinsch, F.A., Zhao, M., Reeves, M.C., Hashimoto, H., 2004. A continuous satellite-derived measure of global terrestrial primary production. Bioscience 54, 547–560.

Sakamoto, T., Warlow, B.D., Gitelson, A.A., Verma, S.B., Suyker, A.E., Arkebauer, T.J., 2010. A two-step filtering approach for detecting maize and soybean phenology with time-series MODIS data. Remote Sens. Environ. 114, 2146–2159.

Selige, T., Bohner, J., Schmidhalter, U., 2006. High resolution topsoil mapping using hyperspectral image and field data in multivariate regression modeling procedures. Geoderma 136, 235–244.

Serbin, G., Hunt Jr., E.R., Daughtry, C.S.T., McCarty, G.W., Doraiswamy, P.C., 2009a. An improved ASTER index for remote sensing of crop residue. Remote Sens. 1, 971–991.

Serbin, G., Daughtry, C.S.T., Hunt Jr., E.R., Brown, D.J., McCarty, G.W., 2009b. Effect of soil spectral properties on remote sensing of crop residue cover. Soil Sci. Soc. Am. J. 73, 1545–1558.

Serbin, G., Daughtry, C.S.T., Hunt Jr., E.R., Reeves III, J.B., Brown, D.J., 2009c. Effects of soil composition and mineralogy on remote sensing of crop residue cover. Remote Sens. Environ. 113, 224–238.

Simbahan, G.C., Dobermann, A., 2006. Sampling optimization based on secondary information and its utilization in soil carbon mapping. Geoderma 133, 345–362.

Skinner, H., Wagner-Riddle, C., 2012. Micrometeorological methods for assessing greenhouse gas flux. In: Liebig, M.A., et al. (Eds.), Managing Agricultural Greenhouse Gases: Coordinated Agricultural Research through GRACEnet to Address Our Changing Climate. Academic Press, San Diego, CA.

Slaton, M.R., Hunt Jr., E.R., Smith, W.K., 2001. Estimating near-infrared leaf reflectances from leaf structural characteristics. American J. Botany 88, 278–284.

Stern, A.J., Doraiswamy, P.C., Hunt Jr., E.R., 2011. Changes of crop rotations in Iowa determined from the USDA-NASS cropland data layer product. Agricultural Systems in review.

Stoner, E.R., Baumgardner, M.F., 1981. Characteristic variations in reflectance of surface soils. Soil Sci. Soc. Am. J. 45, 1161–1165.

Streck, N.A., Rundquist, D., Connot, J., 2002. Estimating residueal wheat dry matter from remote sensing measurements. Photogramm. Eng. Remote Sensing 68, 1193–1201.

Terra, J.A., Shaw, J.N., Reeves, D.W., Raper, R.L., van Santen, E., Mask, P.L., 2004. Soil carbon relationships with terrain attributes, electrical conductivity, and a soil survey in a coastal plain landscape. Soil Sci. Soc. Am. J. 169, 819–831.

Thoma, D.P., Gupta, S.C., Bauer, M.E., 2004. Evaluation of optical remote sensing methods for crop residue assessment. J. Soil Water Conserv. 59, 224–233.

Thomas, J.R., Gausman, H.W., 1977. Leaf reflectance vs. leaf chlorophyll and carotenoid concentrations for eight crops. Agron. J. 69, 799–802.

Tomer, M.D., James, D.E., 2004. Do soil surveys and terrain analyses identify similar priority sites for conservation? Soil Sci. Soc. Am. J. 68, 1905–1915.

Tomer, M.D., Moorman, T.B., James, D.E., Hadish, G., Green, C.H., 2008. Assessment of the Iowa River's South Fork watershed: Part 2. Conservation practices. J. Soil Water Conserv. 63, 371–379.

Tucker, C.J., 1979. Red and photographic infrared linear combinations for monitoring vegetation. Remote Sens. Environ. 8, 127–150.

USDA-NRCS, 2011. Soil Survey Geographic (SSURGO) Database for Iowa, Natural Resources Conservation Service, United States Department of Agriculture. Available at http://sdmdataaccess.nrcs.usda.gov/WebServiceHelp.aspx (verified December 13, 2011).

van Deventer, A.P., Ward, A.D., Gowda, P.H., Lyon, J.G., 1997. Using Thematic Mapper data to identify contrasting soil plains and tillage practices. Photogramm. Eng. Remote Sens. 63, 87–93.

Verhoef, W., 1984. Light scattering by leaf layers with application to canopy reflectance modeling: the SAIL model. Remote Sens. Environ. 16, 125–141.

Walter-Shea, E.A., Biehl, L.L., 1990. Measuring vegetation spectral properties. Remote Sens. Rev. 1, 179–206.

Walthall, C.L., Roujean, J.L., Morisette, J.T., 2000. Field and landscape BRDF optical wavelength measurements: experience, techniques, and the future. Remote Sens. Rev. 18, 503–531.

Walthall, C.L., Shafer, S.T., Jawson, M.D., 2012. GRACEnet: addressing policy needs through coordinated cross-location research. In: Liebig, M.A., et al. (Eds.), Managing Agricultural Greenhouse Gases: Coordinated Agricultural Research through GRACEnet to Address Our Changing Climate. Academic Press, San Diego, CA.

West, T.O., Brandt, C.C., Baskaran, L.M., Hellwinckel, C.M., Mueller, R., Bernacchi, C.J., Bandaru, V., Yang, B., Wilson, B.S., Marland, G., Nelson, R.G., Ugarte, D.G.D.L.T., Post, W.M., 2010. Cropland carbon fluxes in the United States: increasing geospatial resolution of inventory-based carbon accounting. Ecol. Appl. 20, 1074–1086.

Wilhelm, W.W., Johnson, J.M.F., Hatfield, J.L., Voorhees, W.B., Linden, D.R., 2004. Crop and soil productivity response to corn residue removal: a literature review. Agron. J. 96, 1–17.

Williams, J.R., Jones, C.A., Dyke, P.T., 1984. A modeling approach to determining the relationship between erosion and soil productivity. Trans. ASAE. 27, 129–144.

Xiao, J., Zhuang, Q., Baldocchi, D.D., Law, B.E., Richardson, A.D., Chen, J., Oren, R., Starr, G., Noormets, A., Ma, S., Verma, S.B., Wharton, S., Wofsy, S.C., Bolstad, P.V., Burns, S.P., Cook, D.R., Curtis, P.S., Drake, B.G., Falk, M., Fischer, M.L., Foster, D.R., Gu, L., Hadley, J.L., Hollinger, D.Y., Katul, G.G., Litvak, M., Martin, T.A., Matamala, R., McNulty, S., Meyers, T.P., Monson, R.K., Munger, J.W., Oechel, W.C., Paw U, K.T., Schmid, H.P., Scott, R.L., Sun, G., Suyker, A.E., Torn, M.S., 2008. Estimation of net ecosystem carbon exchange for the conterminous United States by combining MODIS and Ameriflux data. Agric. For. Meteorol. 148, 1827–1847.

Yilmaz, M.T., Hunt Jr., E.R., Jackson, T.J., 2008. Remote sensing of vegetation water content from equivalent water thickness using satellite imagery. Remote Sens. Environ. 112, 2514–2522.

Zheng, B., Campbell, J.B., deBeurs, K.M., 2012. Remote sensing of crop residue cover using multi-temporal Landsat imagery. Remote Sens. Environ. (in press). DOI: doi:10.1016/j.rse.2011.09.016.

Zhu, X., Chen, J., Gao, F., Chen, X., Masek, J.G., 2010. An enhanced spatial and temporal adaptive reflectance fusin model for complex heterogeneous regions. Remote Sens. Environ. 114, 2610–2623.

SECTION 6

Economic and Policy Considerations Associated with Reducing Net Greenhouse Gas Emissions from Agriculture

CHAPTER 23

Economic Outcomes of Greenhouse Gas Mitigation Options

David W. Archer[1], Lyubov A. Kurkalova[2]
[1]USDA-ARS Northern Great Plains Research Laboratory, Mandan, ND
[2]North Carolina A&T State University, Greensboro, NC

CHAPTER OUTLINE

NB: USDA is an equal opportunity provider and employer.

Abbreviations: GHG, greenhouse gas; C, carbon; N, nitrogen; NT, no-till; CT, conventional tillage

INTRODUCTION

Many opportunities for changes in agricultural management to reduce or mitigate GHG emissions have been identified. Opportunities include direct reduction in GHG emissions, sequestering carbon (C), and providing offsets through production of biofuels and biomass energy (McCarl and Schneider, 2001, Weersink et al., 2005). Practices to reduce GHG emissions or sequester C include reducing tillage, eliminating fallow, keeping the soil covered with residue, cover crops, or perennial vegetation, and avoiding over-application of N fertilizer and using split N applications rates (Johnson et al., 2007). While there are many factors influencing the adoption of management practices, a key driver is the profitability of these practices. Conservation practices that are not profitable at the farm level will likely only be adopted by farmers who have strong conservation goals (Weersink et al., 2005). However, profitability of these practices may change with crop and input market conditions. Our goal was to provide an overview of the economic outcomes of cropland management options for mitigating GHGs. We discuss the sensitivity of these outcomes to crop and input market conditions, and specifically on how these conditions are affected by the existing and developing markets for biomass and bioenergy production.

Managing Agricultural Greenhouse Gases. DOI: 10.1016/B978-0-12-386897-8.00023-1
2012, Published by Elsevier Inc.

MANAGEMENT ALTERNATIVES FOR GREENHOUSE GAS MITIGATION

Switching to less intensive tillage has been identified as a method to cost effectively sequester C, and perhaps even provide a win-win-win-win opportunity of simultaneously sequestering C, reducing soil erosion, reducing nutrient runoff, and increasing net returns under many soil and climatic conditions (Manley et al., 2005). However, in a meta-analysis of studies primarily in North America, Manley et al. (2005) showed that the profitability of no-till (NT) relative to conventional tillage (CT) varied by region and within regions, with NT generally less profitable than CT. Within the U.S., regional estimates of reduction in net returns for NT ranged from \$1.20 ha^{-1} for corn/other crops in the South to \$52.58 ha^{-1} for wheat in the Corn Belt/ Prairies. The meta-analysis also indicated that, while NT generally stored C near the soil surface, total C storage over the soil profile may be small or even less than CT when measurements at deeper depths were included. Combining the economic results with the C storage results showed tremendous variation in costs per Mg C stored, ranging from \$1.96 Mg^{-1} C for crops other than wheat in the South to values approaching infinity, due to insignificant C storage with NT for wheat in the Prairies (Manley et al., 2005). This large range in costs per Mg C stored has been observed by others (Pendell et al., 2007) and indicates that relatively small differences in net returns or in C storage rates for different systems can have a large impact on costs per Mg C stored. This may also indicate that producer willingness to adopt GHG mitigation practices in return for C payment may be sensitive to C storage rates (Antle et al., 2002) or changing market conditions.

Important limitations of the meta-analysis were that impacts on soil emissions of other GHG and emissions associated with changes in input use were not included. Even if reducing tillage would not increase soil C storage, reductions in GHG emissions due to lower input use may lead to net GHG benefits (West and Marland, 2002; Meyer-Aurich et al., 2006). Another limitation of the Manley et al. (2005) meta-analysis is that the effects of tillage changes were isolated from the effects of other management changes such as those related to fertilizer applications and rotations. The latter effect may be particularly important in arid areas of the Great Plains where adoption of NT may facilitate reductions in the frequency of fallow, which could increase C storage and economic returns (Liebig et al., 2004; DeVuyst and Halvorson, 2004). Weersink et al. (2005) indicated that the area of conservation tillage and permanent cover has increased and the area of summer fallow has fallen in Canada without any direct C incentive payments. In comparisons of profitability of alternative systems, initial C sequestration may be achieved at a relatively small cost to producers, but rapidly increasing C payments may be needed to induce further adoption of GHG reducing practices (Pautsch et al., 2001; Antle et al., 2002; Kurkalova et al., 2006).

Typically, economic comparisons of alternative tillage systems have not included potential long-term effects of management changes that may improve productivity by building soil quality and potentially enhancing economic returns. In evaluating dynamic effects of management on soil quality in the northern Great Plains, Smith et al. (2000) found that the on-site value of additional soil organic C ranged from \$1 to 4 Mg^{-1} C ha^{-1} yr^{-1}, and that this was enough to influence the optimum tillage system, but not the optimum crop rotation. Better understanding of the long-term impacts of adopting C-conserving tillage practices could potentially lead to greater adoption.

Meyer-Aurich et al. (2006) showed that crop rotation had a greater effect on GHG emissions than tillage; however, there were substantial costs in shifting rotations from the most profitable systems. Rotational benefits could include a change in the mix of crops grown toward less GHG emitting crops or sequencing crops to reduce GHG emissions through synergistic relationships that reduce input needs or enhance productivity. In a model of county-level changes in crop mix and GHG emissions, Popp et al. (2011) showed shifts toward less GHG emitting crops in response to C payments, leading to substantial changes in GHG if N_2O emissions were not included. However, when N_2O emissions were included, production shifts and resulting net GHG changes were minimal. In field research that included both crop mix and potential synergistic relationships, Johnson et al. (2010a, 2011) showed potential reductions in net GHG emissions by diversifying crop rotation, adding a perennial crop, and reducing tillage relative to a corn–soybean rotation under CT (representing typical farm practices). While economic analysis indicated that long-term profitability might be increased by shifting toward less intensive tillage systems (Archer and Reicosky, 2009), analysis of net returns during the transition from typical farm practices indicated that diversifying crop rotation and reducing tillage was not more profitable (Archer et al., 2007). The above studies are among the few that address both the economic and GHG mitigation impacts of crop rotation. Looking at GHG effects alone in a review of tillage and crop rotation alternatives, West and Post (2002) showed greater potential soil C sequestration with a shift from CT to NT than by increasing rotation complexity. This conclusion excluded the effect of tillage changes in wheat–fallow systems, and indicated that changing from continuous corn to corn–soybean rotation may not result in significant accumulation of SOC.

There have been many economic analyses of alternative crop rotation systems, and we provide just a brief overview of findings here. Potential economic advantages of crop rotation are generally associated with higher crop yields, reduced input costs, or farm risk diversification strategies. Generally, crop yields when grown in rotation with other crops tend to be higher than when grown in a continuous sequence (Karlen et al., 1994; Krupinsky et al., 2006; Hennessy, 2006; Stanger et al., 2008). Crop rotations may also reduce input use due to improved N use efficiency and reduced pests (Karlen et al., 1994; Katsvairo and Cox, 2000); although crop drying costs might increase as yields increase (Katsvairo and Cox, 2000). However, rotational crops may not be as profitable as the primary crops. Stanger et al. (2008) showed higher net returns for corn–soybean than for continuous corn or for more diverse rotations that included oats or alfalfa. While Meyer-Aurich et al. (2005) showed highest net returns for more diverse rotations that include wheat and red clover. Also, rotation profitability can be sensitive to relative crop prices, so results based on historical price relationships may not hold if price shifts occur.

Management of nitrogen (N) fertilizer is another option that producers have to mitigate GHG emissions. Higher N fertilizer rates are often associated with higher N_2O emissions (Robertson et al., 2000; Halvorson et al., 2008). However, N fertilizer is needed to achieve high yields and profit in many cropping systems (Archer et al., 2002; DeVuyst and Halvorson, 2004). Complex interactions between tillage, rotation, and fertilizer management on yield, economics, and GHG emissions are not well understood. Wu and Babcock (1998) found that farmers rotating crops are also likely to adopt conservation tillage, and rotating corn with soybeans and/or adopting conservation tillage seem to decrease N fertilizer rates in central Nebraska. De La Torre Ugarte et al. (2004) found that the incentive payments needed to induce conservation tillage adoption in the Corn Belt for continuous corn could be more than three times higher than that for a corn–soybean rotation. Hennessy (2006) estimated a 1.0 Mg ha^{-1} (16.5 bu ac^{-1}) yield increase and 57 kg ha^{-1} (51 lb ac^{-1}) N fertilizer savings for corn after soybean as opposed to corn after corn in north-eastern Iowa. Additionally, soybeans planted after 2 years of corn had yields that were on average 0.8 Mg ha^{-1} (11.6 bu ac^{-1}) higher, and soybeans planted after 1 year of corn had yields that were 0.5 Mg ha^{-1} (7 bu ac^{-1}) higher than continuous soybeans in the same region (Hennessy, 2006).

For irrigated cropping systems in Colorado, Archer and Halvorson (2010) showed that producers could increase profitability by switching from CT continuous corn to an NT corn–bean (soybeans or dry beans) rotation while reducing net GHG emissions. Net GHG emissions generally increased with higher N rates, so avoiding over application of N could increase net returns and reduce net GHG. However, reducing fertilizer N below economic optimum levels would result in trade-offs between profitability and net GHG reductions. Small reductions in N fertilizer and lower yield could be compensated with a relatively small C payment, but C payments would have to increase dramatically with further reductions in N fertilizer. The greatest reductions in net GHG could be obtained by switching to NT continuous corn; however, this would significantly reduce net returns and would require an incentive of \$341 Mg^{-1} CO_2 (\$1250 Mg^{-1} C) to provide equivalent returns to the NT corn–bean system (Archer and Halvorson, 2010). Even though the study results showed potential increases in profitability by switching from CT continuous corn to an NT corn–bean rotation, it is important to consider why producers have not already adopted the more profitable practices. Is it due to lack of awareness about the alternative practices or are there economic or other barriers not obvious that prevent adoption of these practices?

Assessments of the relative profitability of alternative rotation, tillage, and fertilizer application systems can be sensitive to relative crop and input prices. In the CT continuous corn versus NT corn–bean example (Archer and Halvorson, 2010), greater N fertilizer price reduces the economic optimum N rate, providing a condition to reduce net GHG emissions. With greater corn price relative to beans, the gap would narrow between the profitability of the NT corn–bean system and NT continuous corn, reducing the C credit payment needed to induce a shift to the lower net GHG system. With greater diesel fuel price, production costs would be greater for the CT system relative to the two NT systems, since CT uses more fuel than NT. This would provide a greater incentive for producers to switch from CT to NT, even without C credit payments.

Other N fertilizer management practices that may influence GHG emission include changes in not only the quantity, but also the timing of application (Matson et al., 1998; Phillips et al., 2009) and N fertilizer formulation (Halvorson et al., 2008, 2010). There have been few economic analyses of N fertilizer practices aimed specifically at reducing GHG emissions. However, many of these practices are useful for reducing N losses in general, and there is a substantial body of literature on the economics of N fertilizer management. We discuss a few general observations here. Economically optimum fertilizer rates are often determined as a function of the ratio of crop price to fertilizer N price (Bullock and Bullock, 1994; Sawyer et al., 2006). Greater fertilizer N price provides a market incentive for producers to reduce N application rate, while greater crop price provides a market incentive for producers to apply more N. This is based on the assumption that the expected yield response to N rate does not change. Technologies such as soil testing and variable rate applications can change this relationship by refining expectations about yield response and better matching N applications to meet crop needs. It has often been suggested that producers apply more N fertilizer than is agronomically needed (Yadav et al., 1997; Sheriff, 2005). However, the observed applications may be economically rational when considerations of substitutability of other farm inputs, opportunity costs, and uncertainty about soil and weather conditions are included (Sheriff, 2005). Therefore, technologies that reduce uncertainty about soil and weather conditions, and reduce costs of obtaining information regarding soil nutrient status, could potentially reduce N application rates and associated GHG emissions.

In general, continuous biotechnological advances result in development of new practices and crop varieties that enhance yields and/or fertilizer efficiency, and producers have an economic incentive to adopt these practices when the value of benefits (yield changes and N fertilizer cost savings) exceeds information gathering or technology costs (Fuglie and Bosch, 1995). Thus producers are more likely to adopt these practices as the cost of information gathering or technology decreases. For information and technologies that reduce N fertilizer applications or increase crop yield, the economic incentive to adopt these practices increases as crop prices or N prices increase. For some technologies like variable rate fertilizer application, impacts on yields and amount of N fertilizer applied can be ambiguous (Batte, 2000), and thus impact of crop and N fertilizer price on adoption and resulting GHG emissions cannot be generalized.

Risk and uncertainty likewise influence conservation tillage adoption decisions. Estimates of conservation tillage adoption are typically based on comparison of net returns, assuming that producers would be willing to adopt conservation tillage practices if they were more profitable than existing practices. However, there is evidence that producers may demand a premium to adopt conservation tillage practices due to producers' risk preferences and uncertainties about the profitability of conservation tillage (De La Torre Ugarte et al., 2004; Kurkalova, et al., 2006; Kim et al., 2008). Positive uncertainty premiums would further increase the C credit value needed for producers to adopt conservation tillage. If conservation tillage were less risky than CT, the C credit value needed could decrease (Pendell et al., 2007). Because of varying soil, weather, and farmer characteristics, both the uncertainty premiums and overall cost of conservation tillage adoption differ significantly among producers and across the U.S. De La Torre Ugarte et al. (2004) accounted for the uncertainty premiums and reported the estimated subsidies needed to induce the switch from intensive to conservation tillage for representative crop production systems in several regions across the U.S. For the switch from CT to NT, the estimated annual subsidy ranged from \$0.74 ha^{-1} for sorghum–soybean rotation in Central Great Plains to \$99.94 ha^{-1} for continuous corn on poorly drained soils in the Corn Belt. For the switch from CT to ridge till in the Corn Belt, the subsidy ranged from \$3.51 ha^{-1} for continuous corn on poorly drained soils to \$101.22 ha^{-1} for continuous corn on well-drained soils. An uncertainty premium in Iowa was, on average, \$23.22 ha^{-1} for corn and \$85.46 ha^{-1} for soybeans in 1992 to overcome the perceived risk of switching from CT to conservation tillage (Kurkalova et al., 2006). Since net returns under conservation tillage were estimated to be higher than those under CT, the actual subsidy needed to induce the adoption of conservation tillage was \$10.13 ha^{-1} for corn and \$14.82 ha^{-1} for soybeans.

Our review of the economic outcomes of GHG mitigation practices indicates that results are sensitive to crop and input market conditions. The rapidly developing markets for bioenergy represent a significant change in crop and input market conditions, which is likely to impact the economic outcomes of GHG mitigation practices. We discuss these impacts in detail in the following section.

IMPACT OF BIOENERGY MARKETS

Large-scale assessments of C sequestration potential under different biomass-based fuels scenarios have been conducted. Agricultural products could be substituted for fossil fuels in two ways: as feedstock for electrical power plants (direct burning of biomass) or as inputs to liquid fuel production (ethanol or biodiesel) (McCarl and Schneider, 2000, 2001, 2005; Schneider and McCarl, 2003). These studies suggest that dedicated energy crops such as switchgrass or short rotation woody biomass such as hybrid poplar and willow could become important in the overall portfolio of agricultural GHG mitigation options, but only at high C prices, and/or in selected U.S. regions. At lower C prices, management changes, such as switching from CT to NT would be more cost effective than biofuel production. Biofuel-based C sequestration in the Great Lakes states became economically preferred at high C prices (above \$50 Mg^{-1} C equivalent) (McCarl and Schneider, 2001), but other C sequestration options were more profitable at lower prices and/or in other regions. In particular, soil-based strategies such as reduced tillage and crop mix alterations dominated in the Corn Belt (McCarl and Schneider, 2001).

While useful, earlier assessments could not explicitly account for the unprecedented increase in bioenergy production that has occurred over the last several years for reasons only partially related to GHG mitigation. U.S. energy policy to date has promoted production and use of first-generation biofuels, specifically corn ethanol, through economic incentives such as mandates and tax credits (see Secchi et al., 2011a, for a review of the recent bioenergy policy). Energy policy continues to provide incentives for the development of second-generation cellulosic ethanol production that would use crop residues such as corn stover and dedicated energy crops such as perennial grasses as feedstocks. Bioenergy policy and recent energy price spikes have significantly altered the economic landscape of U.S. agriculture through changes in returns to crop producers. Greater demand for corn for ethanol production may have been the most important driver of the recent increase in U.S. corn acreage (Feng and Babcock, 2010). Total U.S. land planted to corn rose by 16.5% between 2001 and 2010 (USDA-NASS, 2011), with major crop production areas having an increase in continuous corn. Based on analysis of NASS land cover satellite images, Stern et al. (2008) reported that land in Iowa under continuous corn increased from 15% in 2001/02 to 21% in 2006/07, although this value declined somewhat in 2008–2010. In addition, corn production has expanded into areas that have not been typically used for corn production. For example, land planted to corn in North Dakota increased 133% from 2001 to 2010 (USDA-NASS, 2011). At the same time, higher energy price has been suggested as a potential driver for greater use of conservation tillage (Werblow, 2005). Change in the relative profitability of both alternative crops and alterative tillage systems need to be accounted for when evaluating the economic potential of agricultural GHG mitigation options.

Another aspect of the link between biomass-based fuels and agricultural GHG mitigation is land use and C footprint consequences of biofuels. Significant changes in spatial cropping patterns could emerge because of the growth in bioenergy. Potential changes have been identified for Illinois (Khanna et al., 2010), Iowa (Kurkalova et al., 2010; Secchi et al., 2011b), Midwest (English et al., 2005; Secchi et al., 2011a), Arkansas (Popp et al., 2008, 2010), U.S. southern states (English et al., 2006), and globally (Searchinger et al., 2008; Melillo et al., 2009; Plevin et al., 2010). Importantly, many assessments have specifically focused on the GHG consequences of indirect land-use change, referring to the changes in cropping patterns caused by crop prices and demand for biomass. Two types of land-use change have been considered, i.e. conversion of land that is currently not cropped (such as land under CRP and pastures) to crop production (including energy crops) (extensive margin) and land that is presently cropped (Secchi et al., 2011b) (intensive margin). Changes on the extensive margin are expected to draw poorer quality land into crop production (Feng and Babcock, 2010). In Iowa, impacts of higher corn price on soil C released into the atmosphere may be substantial when bringing more and more environmentally vulnerable CRP land into production (Secchi et al., 2009). However, this study did not estimate the impact on total GHG emissions. Other studies investigating potential conversion of pasture and marginal cropland to switchgrass with increasing biomass prices (English et al., 2006) suggest potential increases in soil C storage (Bransby et al., 1998).

On the intensive margin, predicted changes due to higher corn prices include substantial shifts in rotations favoring continuous corn over corn–soybean rotation with the corresponding higher N application rates and higher tillage intensity in Iowa (Secchi et al., 2011b) and in the Upper Mississippi River Basin (Secchi et al., 2011a). While there are many potential annual crops that could be grown for energy production, including forage sorghum, sweet sorghum, sugar beets, energy beets, and oilseeds such as canola, industrial rape, and camelina, there have been limited analyses regarding the agronomic feasibility of introducing these energy crops on existing cropland (Hallam et al., 2001; Jaeger, 2008; Bennett and Anex, 2009). Future research is needed to understand potential regional shifts in production and impacts on net GHG emissions for cropping systems that include dedicated energy crops, particularly if these crops displace production of existing crops resulting in indirect land-use impacts. In Arkansas, significant switchgrass and forage sorghum production could displace production of existing crops with important implications on irrigation water availability (Popp et al., 2008, 2010). Impacts on GHG emissions and interactions with C offset policy are being investigated.

Changes in relative profitability of alternative farming practices could have significant implications for GHG mitigation potential. If biofuel production results in wider use of GHG mitigation-friendly farming practices such as reduced tillage, then the overall costs of GHG mitigation through soil-based strategies would decline as many farmers would find it profitable to reduce tillage without additional incentives. However, if the opposite were true, i.e. if biofuels expansion leads to a wider use of CT, then the costs of GHG mitigation would increase. The economic literature is scant and mixed on such predictions. For example, Khanna et al. (2010) investigated the land-use consequences of meeting a predetermined biofuels production target for the state of Illinois through the production of corn and cellulosic ethanol from corn stover, switchgrass, and miscanthus. The study considered several alternative scenarios about future production costs and crop yields and found that the biofuels mandate could result in a significant shift from a corn–soybean rotation to continuous corn with the share of corn–soybean rotation declining from 80 to ~30% and a simultaneous increase in conservation tillage from 27 to 61% under some of the scenarios. The latter finding seems to be driven mostly by the assumption that a higher percentage of corn residues could be removed from the field under NT than under CT. Kurkalova et al. (2010) found a significant shift from corn–soybean rotation to continuous corn in response to a corn stover market in Iowa. However, the more intensive corn production and stover markets were associated with higher not lower tillage intensity. Growing corn after corn as opposed to corn after soybean was assumed to result in a yield decline (drag) that could be partially eliminated with higher N application rates and/or greater tillage intensity. More information on the relative impacts of tillage, rotation, and residue harvest on soil C levels is needed. In a meta-analysis, Johnson et al. (2006) showed that less residue would be needed to maintain soil C levels in NT systems than in moldboard plow systems, and the amount of harvestable corn stover would likely be higher in a continuous corn rotation than in a corn–soybean rotation (Johnson et al., 2007). In a modeling analysis for Iowa, Sheehan et al. (2003) showed that harvesting the maximum amount of corn stover, constrained by erosion limits, while shifting from a predominantly corn–soybean rotation under CT to an NT continuous corn rotation resulted in modest statewide increases in soil C over a 90-year time period. This result indicates the potential for tillage and rotation changes to compensate for SOC losses due to residue removal, but more research is needed to verify and quantify such effects in this and other regions.

Potential changes in N fertilizer use in response to biofuel harvest have not been thoroughly investigated. Harvest of crop residues for bioenergy removes additional nutrients (Johnson et al., 2010b) which could increase fertilizer application rates. In addition, rates could increase if rotations shifted from corn–soybean to continuous corn. Large-scale analysis (De La Torre Ugarte et al., 2010) indicated a decrease in N fertilizer use in the U.S. for two cellulose-to-ethanol scenarios including use of corn grain, wood residues, wheat straw, corn stover, and dedicated energy crops for bioenergy production. Projected N fertilizer reductions for switchgrass and other dedicated energy crops relative to annual crop production were assumed. Tillage shifts from CT to NT were also assumed. Potential increases in N fertilizer use occurred when tillage shifts were not included. There is a need for more detailed analysis of potential impacts on N fertilizer use including N management alternatives, and interactions with tillage and crop rotation decisions.

Another important economic implication of the expansion of biofuels production is the tightening integration between crop and energy markets. Higher crop prices are now more likely to be associated with higher energy prices because of the wider use of agricultural products for energy production (Du and Hayes, 2009). Higher energy prices favor reduced tillage, as CT involves more passes over the field and hence greater fuel expenditures, relative to mulch and NT. Higher energy prices have also been commonly associated with the higher N fertilizer prices, as natural gas accounts for most of the fertilizer cost (Huang, 2009). While higher energy prices could attract more conservation tillage and reduced fertilizer use (Werblow, 2005), only a few recent empirical estimates have been developed. In irrigated corn production in Nebraska's central Platte River valley, Exner et al. (2010) reported a significant decline in average N application rates concurrent with a sharp increase in N fertilizer price in 2001 and 2003. However, other economic factors such as crop prices, agricultural sector wages, and interest rates may have contributed to the observed decline in fertilizer rates. Yet, the need to measure the impact of energy prices on the use of conservation tillage and N fertilizer applications is imperative as both changes in farming practices represent important GHG mitigation options.

Cover crops have been suggested as another bioenergy option that could potentially mitigate GHG emissions. Cover crops could either be used to replace residue removed for bioenergy use or could be harvested as a bioenergy feedstock. Economic analysis of this practice has been limited. For continuous cotton in the southeast U.S., Duzy et al. (2010) showed that removal of cereal rye cover crop for bioenergy was more profitable than either retaining the cover crop as soil cover or a no cover crop treatment assuming a biomass price of \$55 Mg^{-1}. In a review of cropping system impacts on SOC in the southeast, Franzluebbers (2005) identified an interaction effect between tillage and cover crop use, whereby SOC sequestration under NT was about two times greater with cover cropping than without cover cropping. Important questions to be investigated include whether removal of the cover crop for bioenergy negates the benefit of the cover crop on SOC sequestration and the potential for cover crops to retain N or serve as an N source for subsequent crops. Soil moisture impacts, interactions of cover crop management with other cropping activities, and economic factors influencing producer adoption of cover crops are also not well understood.

Overall, a better understanding of the impacts of the growing biofuels market is needed for assessing economic outcomes of GHG mitigation options in a bioenergy world. Some of the poorly understood economic questions and data needs are detailed below.

1. Costs of coupled changes in tillage and rotations need to be better understood, especially with increased use of continuous corn. Costs used in recent studies rarely account for differences in crop rotations. For the purpose of budget analyses that can provide the expected costs of management alternatives, there is a need to better understand soil- and location-specific yield impacts associated with alternative bundles of rotation and tillage systems, including systems that include long-term, occasional, or varying tillage practices.
2. Improved collection of data on tillage practices, including multiple year tillage choice information, and information on rotations is needed. The Conservation Tillage Information Center provided a consistent source of tillage data for the U.S. from 1989 to 2004, but national surveys have not been conducted since 2004, and the survey did not include multiple year tillage choice information for crop producers. Tillage practice information is collected for selected crops and states with the USDA Agricultural Resource Management Survey. However, the survey is not conducted each year for a particular site, and so does not provide multiple year tillage information. However, most recent surveys have included information on whether a field had been in NT over the previous 4 years. Similarly, there are no national data on rotation practices. Using the NASS satellite crop cover imagery, consistent rotation data have presently been published for Iowa only (Stern et al., 2008; Secchi et al., 2009, 2011b).
3. A better understanding of the economic costs of alternative GHG mitigation-friendly farming practices under crop residue harvest scenarios is needed.
4. A better understanding of the interactions between GHG and bioenergy markets in conjunction with other resource constraints (e.g. irrigation water use) is needed at the regional and farm level and should include effects of soil characteristics, and tillage, rotation, and N fertilizer management.
5. Overall, the developing bioenergy markets have clearly demonstrated that important economic drivers such as energy and crop prices could change significantly over short periods of time. To make the results of GHG mitigation analyses more easily transferable

and comparable across alternative market conditions, it would be highly desirable for future studies to be explicit about the crop and energy prices used. Reporting the results of the assessments of agricultural GHG mitigation options as functions of alternative crop–energy price combinations would greatly facilitate such transferability.

CONCLUSION

Producers most likely have already adopted GHG mitigation practices that are profitable for them. Further adoption of GHG mitigation practices requires improving economic outcomes including: identification of new practices that improve profitability and reduce risk, increasing awareness and reducing uncertainty about the economic performance of conservation practices, and use of policy tools such as GHG incentive payments or GHG markets to increase the value of GHG mitigation practices to producers. Research results indicate potential for mitigating GHG at a relatively low cost through changes in agricultural practices. However, quantifying this potential is challenging since economic outcomes of GHG mitigation options vary by site, and may be sensitive to interactions among soil characteristics, and tillage, rotation, and N fertilizer management practices. A further challenge is that market drivers can rapidly change the economic outcomes of GHG mitigation options. This means that adoption of systems may change with changing market conditions, and that producers may demand a premium to adopt conservation practices in the face of uncertainty about the upcoming changes. This is apparent in the rapid growth of the emerging bioenergy market, which will likely have substantial impacts on the economic performance of GHG mitigation alternatives and the resulting adoption of these practices. Overall, better understanding of the impacts of the growing biofuels market is needed for assessing economic outcomes of GHG mitigation options in a bioenergy world. Realizing that markets will likely continue to change, this extends in general to a need for research to better anticipate changes in the relative profitability of GHG mitigation practices in response to changing farming input prices, crop prices, as well as to the constraints and incentives provided by conservation and energy policies.

Acknowledgments

This publication is based on work supported by the USDA-ARS under the GRACEnet Project. L.A.K. gratefully acknowledges partial support from USDA-CSREES grant 2009-10002-05149. The authors thank two anonymous reviewers for their helpful suggestions on an earlier version of this chapter.

References

Antle, J., Capalbo, S., Mooney, S., Elliott, E.T., Paustian, K.H., 2002. Sensitivity of carbon sequestration costs to soil carbon rates. Environ. Pollution. 116, 413–422.

Archer, D.W., Halvorson, A.D., 2010. Greenhouse gas mitigation economics for irrigated cropping systems in northeastern Colorado. Soil Sci. Soc. Am. J. 74, 446–452.

Archer, D.W., Jaradat, A.A., Johnson, J.M.F., Weyers, S.L., Gesch, R.W., Forcella, F., Kludze, H.K., 2007. Crop productivity and economics during the transition to alternative cropping systems. Agron. J. 99, 1538–1547.

Archer, D.W., Pikul Jr., J.L., Riedell, W.E., 2002. Economic risk, returns and input use under ridge and conventional tillage in the northern Corn Belt, USA. Soil Tillage Res. 67, 1–8.

Archer, D.W., Reicosky, D.C., 2009. Economic performance of alternative tillage systems in the northern Corn Belt. Agron. J. 101, 296–304.

Batte, M.T., 2000. Factors influencing the profitability of precision farming. J. Soil Water Conserv. 55, 12–18.

Bennett, A.S., Anex, R.P., 2009. Production, transportation and milling costs of sweet sorghum as a feedstock for centralized bioethanol production in the upper Midwest. Bioresource Technol. 100, 1595–1607.

Bransby, D.I., McLaughlin, S.B., Parrish, D.J., 1998. A review of carbon and nitrogen balances in switchgrass grown for energy. Biomass Bioenergy 14, 379–384.

Bullock, D.G., Bullock, D.S., 1994. Quadratic and quadratic-plus-plateau models for predicting optimal nitrogen rate of corn: a comparison. Agron. J. 86, 191–195.

De La Torre Ugarte, Hellwinckel, C.M., Larson, J.A., 2004. Enhancing agriculture's potential to sequester carbon: a framework to estimate incentive levels for reduced tillage. Eviron. Manage. 33, S229–S237.

De La Torre Ugarte, D.G., Lixia, H., Jensen, K.L., English, B.C., 2010. Expanded ethanol production: implications for agriculture, water demand, and water quality. Biomass Bioenergy 34, 1586–1596.

DeVuyst, E.A., Halvorson, A.D., 2004. Economics of annual cropping versus crop–fallow in the northern Great Plains as influenced by tillage and nitrogen. Agron. J. 96, 148–153.

Du, X., Hayes, D.J., 2009. The impact of ethanol production on US and regional gasoline markets. Energy Policy 37, 3227–3234.

Duzy, L., Arriaga, F., Balkom, K., 2010. In: Endale, D.M., Iversen, K.V. (Eds.), Rye cover crop as a source of biomass feedstock: an economic perspective. Proc. 32nd So. Conserv. Agric. Syst. Conf., Jackson, TN, pp. 1–3. July 20–22, 2010.

English, B.C., Menard, R.J., De La Torre Ugarte, D.G., Walsh, M.E., 2005. Economic impacts of ethanol production from maize stover in selected Midwestern states. In: Outlaw, J., Collins, K., Duffield, J. (Eds.), Agriculture as a Producer and Consumer of Energy. CABI, Cambridge, MA, pp. 218–231.

English, B.C., De La Torre Ugarte, D.G., Walsh, M.E., Hellwinkel, C., Menard, J., 2006. Economic competitiveness of bioenergy production and effects on agriculture of the Southern Region. J. Agric. Appl. Econ. 38 (2), 389–402.

Exner, M.E., Peres-Estrada, H., Spalding, R.F., 2010. Long-term response of groundwater nitrate concentrations to management regulations in Nebraska's central Platte valley. Sci. World J. 10, 286–297.

Feng, H., Babcock, B.A., 2010. Impacts of ethanol on planted acreage in market equilibrium. Amer. J. Agric. Econ. 92, 789–802.

Fuglie, K.O., Bosch, D.J., 1995. Economic and environmental implications of soil nitrogen testing: a switching-regression analysis. Amer. J. Agric. Econ. 77, 891–900.

Franzluebbers, A.J., 2005. Soil organic carbon sequestration and agricultural greenhouse gas emissions in the southeastern USA. Soil Tillage Res. 83, 120–147.

Hallam, A., Anderson, I.C., Buxton, D.R., 2001. Comparative economic analysis of perennial, annual, and inter-crops for biomass production. Biomass Bioenergy 21, 407–424.

Halvorson, A.D., Del Grosso, S.J., Alluvione, F., 2010. Nitrogen source effects on nitrous oxide emissions from irrigated no-till corn. J. Environ. Qual. 39, 1554–1562.

Halvorson, A.D., Del Grosso, S.J., Reule, C.A., 2008. Nitrogen, tillage, and crop rotation effects on nitrous oxide emissions from irrigated cropping systems. J. Environ. Qual. 37, 1337–1344.

Hennessy, D.A., 2006. On monoculture and the structure of crop rotations. Amer. J. Agr. Econ. 88, 900–914.

Huang, W., 2009. Factors contributing to the recent increase in U.S. fertilizer prices, 2002–08. World Agricultural Outlook AR-33. U.S. Dept. Agric., Econ. Res. Serv., Washington, DC.

Jaeger, W., 2008. Economics of oilseed crops and their biodiesel potential in Oregon's Willamette Valley. Special Report 1081. Oregon State University Extension Service, Corvallis, OR.

Johnson, J.M.F., Allmaras, R.R., Reicosky, D.C., 2006. Estimating source carbon from crop residues, roots, and rhizodeposits using the national grain-yield database. Agron. J. 98, 622–636.

Johnson, J.M.F., Archer, D.W., Weyers, S.L., Barbour, N., 2011. Global warming potential of three cropping systems in the United States' northern Corn Belt. Atmosphere. In review.

Johnson, J.M.F., Archer, D., Barbour, N., 2010a. Greenhouse gas emission from contrasting management scenarios in the northern Corn Belt. Soil Sci. Soc. Am. J. 74, 396–406.

Johnson, J.M.F., Franzluebbers, A.J., Weyers, S.L., Reicosky, D.C., 2007. Agricultural opportunities to mitigate greenhouse gas emissions. Environ. Pollut. 150, 107–124.

Johnson, J.M.F., Wilhelm, W.W., Karlen, D.L., Archer, D.W., Wienhold, B., Lightle, D.T., Laird, D., Baker, J., Ochsner, T.E., Novak, J.M., Halvorson, A.D., Arriaga, F., Barbour, N., 2010b. Nutrient removal as a function of corn stover cutting height and cob harvest. Bioenerg. Res. 3, 342–352.

Karlen, D.L., Varvel, G.E., Bullock, D.G., Cruse, R.M., Donald, L.S., 1994. Crop rotations for the 21st century. In: Adv. Agron., vol. 53. Academic Press, pp. 1–45.

Katsvairo, T.W., Cox, W.J., 2000. Economics of cropping systems featuring different rotations, tillage, and management. Agron. J. 92, 485–493.

Khanna, M., Onal, H., Chen, X., Huang, H., 2010. Meeting biofuels targets: implications for land use, greenhouse gas emissions, and nitrogen use in Illinois. In: Khanna, M., et al. (Eds.), Handbook of Bioenergy Economics and Policy. Springer, New York, NY, pp. 287–305.

Kim, M.-K., McCarl, B.A., Murray, B.C., 2008. Permanence discounting for land-based carbon sequestration. Ecol. Econ. 4, 763–769.

Krupinsky, J.M., Tanaka, D.L., Merrill, S.D., Liebig, M.A., Hanson, J.D., 2006. Crop sequence effects of 10 crops in the northern Great Plains. Agric. Sys. 88, 227–254.

Kurkalova, L.A., Kling, C.L., Zhao, J., 2006. Green subsidies in agriculture: estimating the adoption costs of conservation tillage from observed behavior. Can. J. Agric. Econ. 54, 247–267.

Kurkalova, L.A., Secchi, S., Gassman, P.W., 2010. Corn stover harvesting: potential supply and water quality implications. In: Khanna, M., et al. (Eds.), Handbook of Bioenergy Economics and Policy. Springer, New York, NY, pp. 307–326.

Liebig, M.A., Tanaka, D.L., Wienhold, B.J., 2004. Tillage and cropping effects on soil quality indicators in the northern Great Plains. Soil Tillage Res. 78, 131–141.

Manley, J., van Kooten, G., Moeltner, K., Johnson, D., 2005. Creating carbon offsets in agriculture through no-till cultivation: a meta-analysis of costs and carbon benefits. Clim. Change. 68, 41–65.

Matson, P.A., Naylor, R., Ortiz-Monasterio, I., 1998. Integration of environmental, agronomic, and economic aspects of fertilizer management. Science 280, 112–115.

McCarl, B.A., Schneider, U.A., 2000. U.S. agriculture's role in a greenhouse gas emission mitigation world: an economic perspective. Rev. Agric. Econ. 22 (1), 134–159.

McCarl, B.A., Schneider, U.A., 2001. Greenhouse gas mitigation in U.S. agriculture and forestry. Science 294, 2481–2482.

McCarl, B.A., Schneider, U.A., 2005. Implications of a carbon-based energy tax for U.S. agriculture. Agric. Res. Econ. Rev. 3492, 265–279.

Melillo, J.M., Reilly, J.M., Kicklighter, D.W., Gurgel, A.C., Cronin, T.W., Paltsev, S., Felzer, B.S., Wang, X., Sokolov, A.P., Schlosser, C.A., 2009. Indirect emissions from biofuels: how important? Science 326, 1397–1399.

Meyer-Aurich, A., Janovicek, K., Deen, W., Weersink, A., 2005. Impact of tillage and rotation on yield and economic performance in corn-based cropping systems. Agron. J. 98, 1204–1212.

Meyer-Aurich, A., Weersink, A., Janovicek, K., Deen, B., 2006. Cost efficient rotation and tillage options to sequester carbon and mitigate GHG emissions from agriculture in eastern Canada. Agric. Ecosyst. Environ. 117, 119–127.

Pautsch, G.R., Kurkalova, L.A., Babcock, B.A., Kling, C.L., 2001. The efficiency of sequestering carbon in agricultural soils. Contemporary Econ. Policy 19, 123–134.

Pendell, D.L., Williams, J.R., Boyles, S.B., Rice, C.W., Nelson, R.G., 2007. Soil carbon sequestration strategies with alternative tillage and nitrogen sources under risk. Rev. Agric. Econ. 29, 247–268.

Phillips, R.L., Tanaka, D.L., Archer, D.W., Hanson, J.D., 2009. Fertilizer application timing influences greenhouse gas fluxes over a growing season. J. Environ. Qual. 38, 1569–1579.

Plevin, R.J., O'Hare, M., Jones, A.D., Torn, M.S., Gibbs, H.K., 2010. Greenhouse gas emissions from biofuels' indirect land use change are uncertain but may be greater than previously estimated. Eviron. Sci. Technol. 44, 8015–8021.

Popp, M., Nalley, L., Fortin, C., Smith, A., Brye, K., 2011. Estimating net carbon emissions and agricultural response to potential carbon offset policies. Agron. J. 103, 1132–1143.

Popp, M., Nalley, L., Vickery, G., 2008. Expected changes in farm landscape with the introduction of a biomass market. Farm Foundation Proceedings Paper of the Transition to Bioeconomy: Environmental and Rural Development Impacts Conference, Oct. 15–16, St. Louis, 2008.

Popp, M., Nalley, L., Vickery, G., 2010. Irrigation restriction and biomass market interactions: the case of the Alluvial Aquifer. J. Agric. Appl. Econ. 42, 69–86.

Robertson, G.P., Paul, E.A., Harwood, R.R., 2000. Greenhouse gases in intensive agriculture: contributions of individual gases to the radiative forcing of the atmosphere. Science 289, 1922–1925.

Sawyer, J., Nafziger, E., Randall, G., Bundy, L., Rehm, G., Joern, B., 2006. Concepts and rationale for regional nitrogen rate guidelines for corn. Iowa State University extension publication PM 2015. April.

Schneider, U.A., McCarl, B.A., 2003. Economic potential of biomass based fuels for greenhouse gas emission mitigation. Environ. Res. Econ. 24, 291–312.

Searchinger, T., Heimlich, R., Houghton, R.A., Dong, F., Elobeid, A., Fabiosa, J., Tokgoz, S., Hayes, D., T- Yu, H., 2008. Use of U.S. croplands for biofuels increases greenhouse gases through emissions from land-use change. Science 319, 1238–1240.

Secchi, S., Gassman, P.W., Jha, M., Kurkalova, L.A., Kling, C.L., 2011a. Potential water quality changes due to corn expansion in the Upper Mississippi River basin. Ecol. Appl. In press.

Secchi, S., Gassman, P.W., Williams, J.R., Babcock, B.A., 2009. Corn-based ethanol production and environmental quality: a case of Iowa and the Conservation Reserve Program. Environ. Manage. 44, 732–744.

Secchi, S., Kurkalova, L.A., Gassman, P.W., Hart, C., 2011b. Land use change on the intensive and extensive margin in a biofuel hotspot: the case of Iowa, USA. Biomass Bioenergy In press.

Sheehan, J., Aden, A., Paustian, K., Killian, K., Brenner, J., Walsh, M., Nelson, R., 2003. Energy and environmental aspects of using corn stover for fuel ethanol. J. Indust. Ecol. 7, 117–146.

Sheriff, G., 2005. Efficient waste? Why farmers over-apply nutrients and the implications for policy design. Rev. Agric. Econ. 27, 542–555.

Smith, E.G., Lerohl, M., Teklay, M., Janzen, H.H., 2000. Soil quality attribute time paths: optimal levels and values. J. Agric. Resource Econ. 25, 307–324.

Stanger, T.F., Lauer, J.G., Chavas, J.-P., 2008. The profitability and risk of long-term cropping systems featuring different rotations and nitrogen rates. Agron. J. 100, 105–113.

Stern, A.J., Doraiswamy, P.C., Akhmedov, B., 2008. Crop rotation changes in Iowa due to ethanol production. [CDROM] Int. Geoscience and Remote Sens. Symp. Proc., Boston, MA. 6–11 Jul. 2008.

USDA-NASS, 2011. Quick stats 1.0. USDA Natl. Agric. Statistics Serv., Washington, DC. Available at http://www.nass.usda.gov/Data_and_Statistics/Quick_Stats_1.0/index.asp (accessed February 18, 2011).

Weersink, A., Pannell, D., Fulton, M., Meyer-Aulrich, A., 2005. Agriculture's likely role in meeting Canada's Kyoto commitments. Can. J. Agric. Econ. 53, 425–441.

Werblow, S., 2005. "New math": will fuel and fertilizer bills drive adoption of conservation tillage? Partners 23 (3) November 2005, Available at http://partnersarchive.ctic.org/partners/112005/default.asp (accessed February 18, 2011). Conserv. Technol. Inf. Center, West Lafayette, IN.

West, T.O., Marland, G., 2002. A synthesis of carbon sequestration, carbon emissions, and net carbon flux in agriculture: comparing tillage practices in the United States. Agric. Ecosyst. Environ. 91, 217–232.

West, T.O., Post, W.M., 2002. Soil organic carbon sequestration by tillage and crop rotation: a global data analysis. Soil Sci. Soc. Am. J. 66, 1930–1946.

Wu, J., Babcock, B.A., 1998. The choice of tillage, rotation, and soil testing practices: economic and environmental implications. Am. J. Agric. Econ. 80, 494–511.

Yadav, S., Peterson, W., Easter, K., 1997. Do farmers overuse fertilizer to the detriment of the environment? Environ. Res. Econ. 9, 323–340.

CHAPTER 24

Agricultural Greenhouse Gas Trading Markets in North America

Don C. Reicosky[1], Tom W. Goddard[2], Dale Enerson[3], Al.S.K. Chan, Mark A. Liebig[5]
[1]USDA-ARS (retired), Morris, MN
[2]Alberta Agriculture & Rural Development, Edmonton, Alberta, Canada
[3]North Dakota Farmers Union, Jamestown, ND
[4]Collaborator USDA-ARS, National Laboratory for Agriculture and the Environment, Ames, IA
[5]USDA-ARS, Northern Great Plains Research Laboratory, Mandan, ND

CHAPTER OUTLINE

NB: The U.S. Department of Agriculture (USDA) prohibits discrimination in all its programs and activities on the basis of race, color, national origin, age, disability, and, where applicable, sex, marital status, familial status, parental status, religion, sexual orientation, genetic information, political beliefs, reprisal, or because all or part of an individual's income is derived from any public assistance program. Mention of trade names or commercial products in this chapter is solely for the purpose of providing specific information and does not imply recommendation or endorsement by the U.S. Department of Agriculture.

Abbreviations: BAU, business as usual; CO_2, carbon dioxide; CO_2e, carbon dioxide equivalent; CCEMA, Climate Change and Emissions Management Act; CER, Certified Emission Reduction; CCX, Chicago Climate Exchange; GWP, global warming potential; GHG, greenhouse gas; CH_4, methane; MT, metric ton; N_2O, nitrous oxide; NDFU, North Dakota Farmers

Managing Agricultural Greenhouse Gases. DOI: 10.1016/B978-0-12-386897-8.00024-3
2012, Published by Elsevier Inc.

Union; NFU, National Farmers Union; OTC, over-the-counter; RGGI, Regional Greenhouse Gas Initiative

INTRODUCTION

Human activities have strained the natural functions of earth such that the ability of the planet's ecosystems to sustain future generations can no longer be taken for granted. Global environmental change concerns many. Scientists have assembled evidence for climate change and emphasized its anthropogenic causes (Davidson and Ackerman, 1993; Houghton et al., 1999; IPCC, 2000, 2007). In a remarkably short time, scientists have concluded that change of the climate system is "unequivocal" and that there is a "very high confidence that the globally averaged net effect of human negativity since 1750 has been one of warming" (IPCC, 2007). A much greater problem for agricultural producers is destabilization of weather patterns, resulting in a more vigorous hydrological cycle and shifts in vegetation zones towards the poles (Rosenzweig and Hillel, 1998; IPCC, 2007; Hatfield et al., 2008).

Environmental degradation, water shortages, soil erosion, pests, disease, and desertification all pose serious threats to global food supplies, and are exacerbated by climate change (Kossoy and Ambrosi, 2010). Moreover, many of the conventional approaches (conventional tillage, increased use of fertilizers and pesticides, limited rotations) used to overcome these environmental problems further increase the consumption of finite oil and gas reserves. Responses to these challenges will require the development of agricultural production systems that are highly productive, effectively utilize renewable resources, and minimize damage to the environment (Hanson et al., 2007). Encouraging the rapid adoption of these production systems will require social interaction and public policy implementation using easily understood metrics linked to greenhouse gas (GHG) mitigation (Lal et al., 1998; Lal, 2002; Kimble et al., 2007; Follett and Reed, 2010).

Carbon (C) may serve as a useful currency to address concerns about climate change and environmental impacts that relate to hunger and poverty, rural livelihoods, nutrition, food security, renewable/bio-energy, dependence on foreign oil and human health (Smith et al., 2000). These issues may be addressed in an environmentally, socially, and economically sustainable manner with a common thread of improved CO_2 emissions management (Janzen, 2004). An emissions trading system may enhance storage of C in soils using conservation agriculture techniques to help offset GHG emissions while providing numerous environmental benefits such as increasing site productivity, increasing water infiltration, and maintaining soil flora and fauna diversity (Lal et al., 1998; Lal, 2002). Emissions trading may provide financial incentives that make it possible to achieve reductions in net GHG emissions faster than without trading (Dudek et al., 1997; Manly et al. 2005; Ribaudo et al., 2006, 2008; Paustian et al., 2009; Baker et al., 2010; Agricultural Carbon Markets Working Group, 2011).

Soil quality is the fundamental foundation of environmental quality and is largely governed by soil organic matter (SOM) content, which is dynamic and responds to changes in soil management, primarily tillage and C inputs (Lal, 1989, 2003). Maintaining or improving soil quality can be encouraged through an emissions trading system, where management practices sequester atmospheric C in soil, thereby increasing SOM (Lal, 1989, 1995, 2003).

Additional GHGs from agricultural sources include methane (CH_4) and nitrous oxide (N_2O), with respective global warming potentials (GWP) of 25 and 298 (100-year time horizon) (IPCC, 2007). Because of their significant capacity to trap radiant energy in the atmosphere, GHG trading schemes must also address CH_4 and N_2O emissions from agricultural sources (Ravendranath and Otswald, 2008).

Initial targets for GHG reductions are stated in the Kyoto Protocol to the United Nations Framework Convention on Climate Change (UNFCCC, 1997). The protocol allows for the trading of "credits", which represent verified GHG emission reductions and/or removal of GHGs from the atmosphere (UNFCCC, 1997). For example, C stored in soils would reduce atmospheric levels of CO_2, and the stored C could be traded or sold as a "carbon credit" in a market system.

Emissions trading may make it possible to achieve reductions in net GHG emissions offsets for less cost than without trading (Dudek et al., 1997). Additionally, with the introduction of GHG emissions offsets trading, agricultural producers would have a new source of income and incentives to put C into the soil. The objective of this review is to discuss GHG emissions offset trading in North America by providing historical background, current status, and future outlook of the role of C credits as economic incentives for reducing GHG emissions from agriculture. Other reviews have addressed broader aspects of C management/sequestration (Kimble et al., 2007; Follett and Reed, 2010).

HISTORICAL BACKGROUND

SO_2 Emission Trading Allowances

Tradable permits emerged as a cost-effective measure leading to a nationwide sulfur dioxide (SO_2) emission permits market (Burtraw, 2000). One of the first broad scale applications of emissions trading was the trading of SO_2 allowances under the US Acid Rain Program. The Acid Rain Program, regulated under the Clean Air Act 1990, aimed to reduce annual SO_2 emissions by 10 million tons below 1980 levels. The program was implemented in two phases (1995–1999 and 2000–present) and targeted the emissions of SO_2 from fossil-fuel power generators (Burtraw, 2000).

Reductions of SO_2 emissions have been facilitated through a cap-and-trade emissions trading system (U.S.-EPA, 2009). Power generators that reduce their emissions below the number of allowances they hold may trade allowances with other participants in the system or bank them to cover emissions in future years. In this way, power generators regulated under the Acid Rain Program can decide the most cost-effective way to use available resources to comply with the requirements of the Clean Air Act.

The success of the SO_2 emissions trading program does not automatically imply success for GHG emissions trading programs. Differences in the chemical properties of CO_2 and SO_2 present new challenges to any trading scheme. The chemical reactions and substrate material providing the GHGs come in many different forms from simple, readily decomposable, crop residues through chemically complex recalcitrant geochemical sources. The link between SO_2 and acid rain was easily demonstrated and understood by the public. The link between CO_2 and global climate change, however, is less clear and not universally agreed upon (U.S. Senate Environment and Public Works Committee, 2009).

Evolution of the Kyoto Protocol

The Kyoto Protocol was developed as an international system of governance, implemented under the United Nations Framework Convention on Climate Change, for the purpose of regulating levels of GHGs in the earth's atmosphere (Grubb et al., 1999). The Protocol was first adopted in principle at a 1997 United Nations-sponsored meeting held in Kyoto, Japan, and officially came into force in 2005, after being formally ratified by the required number of nations (UNFCCC, 1997). As a system of governance, the Protocol was underwritten by national governments and operated under the sponsorship of the United Nations. Participating nations agreed to meet certain GHG emissions targets addressed in the Protocol, as well as submit to external review and enforcement of these commitments by United Nations-based bodies. The Protocol was meant to serve as a framework by which participating countries could work cooperatively to stabilize concentrations of GHGs in the earth's atmosphere. Greenhouse gases in the protocol included CO_2, CH_4, N_2O, hydrofluorocarbons, perfluorocarbons, and sulfur hexafluoride.

Since the United States did not adopt binding emissions for GHGs through the Kyoto Protocol, national demand for C sequestration and emissions offset services has been limited. The Kyoto Protocol initially did not allow parties to use agricultural C sequestration to meet their emissions commitments, and to engage in credit trading (UNFCCC, 1997), but was later discussed as part of the Clean Development Mechanism (CDM). However, credit trading was limited to ratified parties, so that international demand for U.S. credits has not occurred (Grubb et al., 1999).

Proposed agricultural C sequestration programs would have political appeal for a number of reasons. First, proposed programs would be voluntary, in line with the stance against mandatory programs. Second, agricultural sequestration of C produces environmental benefits in addition to reducing GHGs, including reduced soil erosion and improved soil quality. Third, the new programs could likely be designed to be compatible with U.S. commitments on agriculture, as these programs would meet the requirements of programs that are exempt from limits. The voluntary nature of these programs, the public support for environmental programs, and the continuing pressure to support farm income all combined for the evolution of several volunteer C markets in anticipation of future national and international markets.

CURRENT STATUS

Chicago Climate Exchange

The Chicago Climate Exchange (CCX) was North America's first voluntary, rules-based CO_2 emission registry, reduction and trading system (Walsh, 2002). In this context, rules-based meant that its registered companies agreed to take on targets and achieve them by either taking action internally to reduce emissions or purchase offsets on the trading platform. The CCX was positioned as a pilot initiative to provide its member companies with experience in C markets. Although the CCX was a voluntary system, the offsets were backed by contracts—a legally binding aspect that imposed rigor in the system. To some policymakers, the fact that the CCX was in the business of both defining the rules to create its own credits and offering a trading platform for its member companies was seen as a conflict of interest if it migrated to a regulatory or compliance-based market. At its founding in November 2000, some estimated that the size of CCX's C trading market could reach \$500 billion.

Because CCX was established to service a voluntary market, trading rules did not need to reflect the requirements of regulated initiatives. This gave CCX much more flexibility in how it designed its credit creation side. The low value of the C reflects the voluntary nature of the market—the price fluctuated between \$3 and \$4 MT^{-1} CO_2e since its inception and recently decreased to $<\$0.15\ MT^{-1}$ prior to its dissolution. To support the market, CCX defined standardized guidelines for several project types, including select agricultural, forestry, renewable energy, and fuel switching practices. Specific examples of project types included Forest Carbon Emissions Offsets, Agricultural Soil Carbon Offsets, Agricultural Methane Emissions Offsets, Landfill Methane Emission Offsets, and Renewable Energy Emission Offsets.

Tradable offsets were registered and traded on CCX by both offset providers and offset aggregators. Both of these entities had to be accepted by the CCX administration. An offset provider is an owner of an offset project(s), registered and sold on its own behalf.
An offset aggregator serves as the administrative representative of multiple offset generating projects on behalf of the owners. Members include corporations, utilities, universities, non-governmental organizations (NGO), cities, and several states. All projects were verified by a third party auditing agency, approved by the CCX. Offset projects involving less than

10,000 MT of CO_2e were sold through an aggregator (due to the transaction costs and complexity of management). Overall, the framework of the CCX delivered offsets to their trading platform similar to most of the requirements of project-based systems where the price of C was as high as \$7.40 MT^{-1}.

On October 26, 2010, CCX ended C trading, the very purpose for which it was founded. The end of C trading resulted from broad economic issues, the lack of U.S. legislative activity, no formal compliance market, and other issues to be discussed in more detail below. The CCX was sold in 2010 to the Intercontinental Exchange (ICE), an electronic futures and derivatives platform based in Atlanta and London. The ICE also acquired the European Climate Exchange (ECX) as part of the transaction. The ECX remains open to accommodate the Kyoto Protocol-required C trading among EU nations.

Pacific Northwest Direct Seed Association Carbon Trading

The Pacific Northwest Direct Seed Association (PNDSA) was one of the first farmer-run organizations in the U.S. to develop a protocol to aggregate CO_2 from agricultural soils (PNDSA, 2010a). The association also developed offset emissions guidelines for growers and took a lead role to consolidate a position on C trading with a broad range of commodity groups and conservation farming organizations. In 2002, a contract was signed between the PNDSA and Entergy, an energy producing company based in New Orleans, LA. The contract was for a 10-year lease of CO_2 credits generated through the practice of direct seeding (a type of no-till) crop land in the Pacific Northwest. The "lease" aspect of the contract was unique and minimized the risk to the farmers in the event something unexpected happened. The lease allowed temporary control of the management of the land by the energy company. The sale of a C-credit allowed control in perpetuity, and the sale raised a number of legal issues concerning obligations, measurement, and performance that were not clearly understood by either potential sellers or buyers. The lease allowed the grower to retain ownership of the C-credits at the end of the contract. The lease in the opinion of PNDSA was a win-win for the environment and the contract parties (PNDSA, 2010b).

In 2003, the PNDSA was the first group in North America to compile and register a listing of direct seed areas available for a C offset trade (PNDSA, 2010b). The areas were contracted to be direct seeded for 10 years and the stored C was marketed, in the form of a lease, to an energy company. Growers were paid for their CO_2 as an offset to the energy company's emissions.

An annual trade was contracted between PNDSA and Entergy for the next 10 years for a total of 30,000 MT CO_2e. The association was paid \$75,000 to aggregate a base of growers for this project. The PNDSA then contracted with 77 grower members representing 2618 ha of farmland to meet its obligation with Entergy. The grower was paid a nominal amount to direct seed a designated area for the next 10 years, to sequester 1.36 MT of $CO_2e\ ha^{-1}\ yr^{-1}$. The area was monitored and verified as direct seeded by local NRCS Conservation Districts, which had participation of contracted growers. The contract met the Kyoto Protocol requirements involving additionality, duration, permanence and leakage, but did not require soil C measurement and verification. The ultimate goal of PNDSA in this contract was to stimulate research to develop a whole-farm accounting of C equivalent changes occurring as a result of direct seed cropping systems. While the group had hope for developing a whole-farm carbon accounting system, the lack of a national energy policy leading to potential carbon markets and the voluntary nature of the existing carbon markets offered little incentive to continue. However, there remains a sense of optimism of improved carbon management through direct seeding to enhance environmental quality.

This project highlighted the ability of the private sector to manage an environmental change without federal mandates. Furthermore, PNDSA's relationship with Entergy and Environmental Defense to develop such a program was unprecedented in U.S. agriculture.

North Dakota Farmers Union/National Farmers Union Carbon Trading

Carbon markets managed by the National Farmers Union (NFU) started with the trading system developed by the North Dakota Farmers Union (NDFU) (NDFU/NFU, 2011). The NDFU is the state's largest farm organization, and its member-driven policy emphasizes good stewardship of natural resources and good conservation practices, but always balanced with a goal of enhancing farmers' and ranchers' income with sound innovative production practices. Policy adopted by members of NDFU encouraged the organization to investigate and develop a successful system for helping farmers benefit from environmental markets. Once the NDFU trading system was running smoothly, the NFU solicited applications from other states evolving into a national C trading program.

North Dakota Farmers Union's C trading program began formally in May 2006 with the announcement that NDFU had been approved as an aggregator for the CCX coupled with the establishment of a new online enrollment system for C sequestration project areas. While there were other farm organizations active with agricultural offsets in the Corn Belt states, NDFU established the first web-based enrollment system and the first integration with satellite mapping applications for enrollment.

Prior to 2006, the pilot project area for C sequestration offsets as established by the CCX included only Corn Belt states centered around Iowa. The NDFU, in addition to applying for membership as an aggregator for CCX, also submitted soils research data for states in the northern Plains to the CCX offsets technical committee. Carbon sequestration rates were available only for conservation tillage and long-term grass establishment (including CRP) in the region. North Dakota Farmers Union also worked with USGS scientists to submit data for C sequestration rates for restored or created wetlands in the Prairie pothole region, but those sequestration rates and approvals were tabled by the offsets committee and never became a part of the CCX offsets protocols.

The closing date of the first pool of enrollments in North Dakota was in September 2006, and about 323,756 ha of no-till and grass establishment contracts were initially enrolled. Due to the interest in the program, other states as part of NFU wanted to join the project. In October of that same year, enrollment was opened for other states where there were C sequestration rates established, and by the November 2006 deadline, a total of another 121,408 ha were enrolled. The NDFU continued as the fiscal agent for the C sequestration program, and managed the enrollments and contracts for all others states in the program.

Farmers were asked to agree to 5 consecutive years of no-till or grass practices, with additional pools of enrollments established in subsequent years. If producers enrolled by the deadline each year, and if they had been doing the practice prior to the contract year, they could receive offsets for the 1 previous year. If the tract of land was sold or rented prior to the 5-year contract period, then previous credits had to be returned and penalties were assessed. Otherwise, the general trading rules of CCX and their definition of no-till were employed as part of the aggregation process.

As an aggregator, NDFU arranged for third party verification each year for a portion of the contracts. Private firms such as SES, Inc. (Salina, KS), the Association of Seed Certifying Agencies, North Dakota Association of Soil Conservation Districts, and others were hired as verifiers to spot-check agricultural practices applied to contracted land. The aggregator had to cover costs of verification as well as the cost of registering the offsets on the CCX registry.

The first NFU credit pool began selling in January 2007, and the pool was sold out by May 2007, with \$2 million paid to farmers for their registered offsets. The first year pool averaged about \$3.30 MT^{-1} CO_2e. In subsequent years, the program grew to about 2 Mha including sustainable rangeland management areas. Protocols for sustainable rangeland management required documentation of rotational grazing, controlled stocking rates, and grazing plans consistent with NRCS grazing system recommendations. Enrollments for range management offset contracts were very popular in the Sand Hills region of Nebraska and other Great Plains states.

The NDFU/NFU continued improvements with its online mapping and enrollment system using Google maps (Google, Inc., Mountain View, CA) and gave enrollees options of selecting and/or drawing land tracts for enrollments. This mapping enrollment system was combined with the database and accounting system to allow automated calculation and check printing for producer payments for offsets sold. Verifiers in many states could upload verification reports, digital photos, and other documentation online into the system. In 2008, NDFU/NFU partnered with a remote sensing and satellite imagery company to implement verification practices. However, the C market decline in 2009 led to program termination.

For the pool of credits sold in 2008, the average price on the CCX market averaged about \$4.40 MT^{-1} CO_2e. After 2009 offset prices and trading volume declined as policy discussions related to GHGs and environmental regulation changed as a result of legislative inactivity. After 2008, very few credits were sold on the CCX auction, and only limited volumes were traded on the over-the-counter (OTC) market through brokers. In all, nearly \$10 million were paid to farmers, ranchers, and foresters for all C credits sold through NFU.

In the spring of 2008, increased optimism in environmental markets was voiced by U.S. presidential candidates through the development of a "cap and trade" mechanism. In response, C prices on the CCX spiked above \$7 MT^{-1} CO_2e briefly, and then began a steady decline to the spring of 2010, when prices reached \$0.10 MT^{-1} CO_2e. Given the cost of verification and registration, sales proceeds paid the costs of the sequestration program alone, leaving nothing for the enrolled producer or aggregator.

Enrollment in NFU from several states continued through 2010, but on January 1, 2011, NFU suspended enrollment in new contracts, and was left with millions of MT of C credits "in the bin" to sell. All participants in the NFU program were affected by the low market value of carbon. The NFU was liable for all contracts established before January 1, 2011, but suspended activity for new contracts. Presently, a U.S. cap and trade or compliance market appears unlikely in the near future, though there is still hope of state or regional emission reduction markets requiring verifiable offsets.

Producers, who enrolled in C sequestration contracts through NFU/NDFU, never paid enrollment, verification, or registry fees, and accordingly, never incurred costs of the program. The contract language included a clause stating that if trading was suspended on the CCX, obligations to continue sequestration practices ended.

Both NDFU and NFU still have policies favoring cap and trade emission reduction systems as part of any forthcoming U.S. federal regulations. Members of NFU still support a well-designed cap and trade system allowing agricultural offsets as potential income opportunities for producers to incentivize good environmental and production practices. This system is much preferred to a C tax, which would penalize agriculture as an intensive energy user.

Despite its brief tenure, the NFU C Credit program was a very successful project for demonstrating an innovative mechanism for accruing environmental benefits on agricultural land while providing an additional source of farm/ranch income. Positive lessons learned included the feasibility of the system, as well as the development of the technology for contract enrollment and maintenance, verification, and marketing of C credits. Additionally, an educational component of the program was highly successful involving innovative agricultural producers and policymakers about environmental markets. Unfortunately, with more vocal climate change skeptics and without a regulated and managed national or international market for GHG emissions offset reductions, it was pointless to continue to contract areas, verify practices, and hold unmarketable offsets.

Carbon Markets and Trading in Canada

Alberta is natural resource rich and contributes significantly to Canada's GHG emission profile. The long-term agriculture industry and rapid growth of the energy extraction industry have underlined the need to address GHG emissions. Alberta was the first province in Canada with a climate change action plan in 2002 and an enabling act to address emissions management (Climate Change and Emissions Management Act; Alberta Environment, 2002). The act required large final emitters to report their emission intensity on an annual basis. Since Alberta was the first jurisdiction in North America to implement a constraint in its economy and, since it relies heavily on export markets for agriculture and energy products, an emission intensity [GHG output per unit of gross domestic product (GDP)] was chosen as an economically favorable method of accounting. Emission intensities show companies their efficiencies, and if they have to reduce their emissions they become more efficient and thus more competitive. From 2005 to 2008, the GHG/GDP intensity in Alberta decreased 20% while Canada as a whole decreased 15% (calculated from Environment Canada, 2010c and Statistics Canada, 2011). Various methods to provide incentives for GHG emission reductions included financial incentives, taxes, and compliance programs. Alberta chose a regulation compliance program.

Alberta amended the Climate Change Emissions Management Act in 2007 so companies with annual emissions of $\geq$100,000 MT CO_2e had to reduce their emissions below their baseline (Alberta Government, 2007). The baseline was the 3-year average of emissions from 2003 to 2005, which included the first 3 years of the Act. If the >100 regulated companies could not meet their obligations, they had three options to reconcile their accounts: (1) pay into a fund at C\$15 MT^{-1} CO_2e, (2) use performance credits for facility reductions beyond requirements (permits), or (3) purchase GHG offsets from Alberta sources using approved protocols. Among the three options, performance credits were perhaps the most limiting, as facilities have limited ability to generate them on a continuing basis. For example, extra credits may be generated when a facility undergoes a retrofit or mechanical upgrade that reduces emissions. Performance credits may be carried over to be used in a later year.

The Climate Change Emissions Management Fund ("Tech Fund") was created specifically by GHG compliance legislation. Facilities may pay into the fund at a set price of C\$15 MT^{-1} CO_2e for shortfalls in their annual reduction targets. The Tech Fund is administered by a board of directors from industry, and funds are intended to be reinvested in research and technologies that reduce GHG emissions and enhance adaptation to climate change. Biologic sequestration or emission reduction research proposals are allowed under the fund's priorities. Fund contributions up to July 2007 were nearly C\$200 million.

For the third option, legislation stipulated offset credits must be produced from Alberta sources using provincial government approved protocols. Alberta has built a Registry and all offsets must be registered once they are approved. Each MT is serialized so that it is used only once for meeting compliance targets.

No restrictions were placed on how much of any of the three options could be used by regulated facilities. The annual proportion that offsets comprise of the total compliance liability has increased each year from an initial 18% in 2007 to 36% in 2009, indicating regulated industries have confidence in the offsets and an increasing willingness to buy. The initial year (2007) saw a total of seven projects registered from four protocols. In 2009, 58 projects from 15 protocol types were registered, with soil tillage projects the most popular, accounting for just less than half of the total registered projects.

Canadian agriculture experts have been involved in protocol development since 2002 at various scales. The Canadian federal government stopped work on a potential offset market in 2006, but Alberta continued. When Alberta went on alone, initial protocol development was perhaps more provincially focused but grew to accommodate more national or international aspects with the long-term goal of harmonization across future emerging offset markets. Protocol development has followed the International Standards Organization (ISO) 14064-2 specification. The ISO process was embedded in a six-step process that could easily take 12 months or more to complete. An "intent to develop a protocol" is reviewed by the regulator and if approved is followed by a technical seed document reviewing the science, in turn followed by a technical science review, stakeholder review, and finally a public review posting (Alberta Government, 2011b). Following the ISO 14064-2 specification, all sources and sinks are accounted using a partial comparative life-cycle analysis approach evaluating GHG emissions that are controlled, related, and affected. All 26 GHGs listed in the Act must be considered, but agricultural activities focus only on CO_2, N_2O, and CH_4. Currently, 10 of the 23 approved protocols are agriculture related: afforestation, beef feeding edible oils, beef days on feed, beef life-cycle, biofuel, biogas, nitrous oxide emission reduction, pork, and tillage management (Alberta Government, 2011b).

Eligible offsets in Alberta must adhere to the following rules:

- Actions must have been taken on or after 1 January 2002 (where applicable due to the nature of the protocol). This acknowledges the signals sent in 2002 with the Climate Change and Emissions Management Act and gives credit for early actions. In 2012 this will no longer apply;
- Real, demonstrable, quantifiable, and directly measurable—offsets must be net of all relevant GHG sources and sinks stated in the Act. Suppliers, buyers and the public must be confident in what is being created and sold;
- Occur at a place other than a regulated facility and from actions not otherwise required by law;
- Ownership is established and clear;
- Accounted only once for compliance purposes (they are unique);
- Verified by a qualified third party (engineer or an accountant);
- Credits occur from Alberta-only projects.

Additionality, a measure of quality and eligibility of a project or practice to qualify as an emissions offset, is dealt with in two ways. First, federal, provincial, or municipal laws, regulations, or bylaws that require certain practices or modifies emissions cannot be included. Second, emissions must be shown to be below the "business as usual" (BAU) baseline. There is uncertainty internationally as to what additionality means and what threshold level of adoption of a practice must pass to be called BAU. Alberta has taken a guide of 40% or greater adoption of a practice to be no longer considered as additional (Alberta Government, 2011a). Additionality filters for protocol development also include considerations of technical, financial, and social factors (Alberta Government, 2011b).

Baseline methodology varies, as do protocol types, but each protocol development explicitly identifies the baseline approach and conditions that must be met. Alberta protocol allows for any of five methodologies to be used: historic benchmark, performance standard, comparison approach, projection based, and adjusted baseline (Alberta Government, 2011b). The Nitrous Oxide Emission Reduction Protocol (NERP) uses a historic benchmark approach. Historic survey data establish background information and farms must provide 3 years of site-specific data for each crop type prior to the start of the project. The Tillage System Management protocol (no-till) uses an adjusted baseline approach which adopts the same baseline for all projects in the same agro-climatic area. Early adopters are not encouraged to revert back to tillage and farms do not have to keep site specific climate, soil, and yield records. In turn everyone gets a lower, static coefficient. The no-till coefficient is further discounted to account for risk of tillage reversion over the pooled project.

Leakage and permanence are also recognized and addressed in the Alberta system. Leakage could occur if incentives in one area cause emissions to increase in another. Leakage must be identified and dealt with in protocol developments using various management tools such as minimum project boundaries, monitoring plans or conservative emission factors.

Permanence is a concern especially in bio-sequestration offsets such as afforestation and tillage management. A reversal (carbon losses) could occur accidentally (forest fire) or intentionally (revert to conventional cultivation). Protocol development is a new area and must take the risk of reversals, potential political issues, historic data, and future projections into consideration across regions. A coefficient is then established as a "risk-based assurance factor" (Alberta Government, 2011b). The assurance factor allocates "buffer credits" into a government account where they can be used to adjust future losses. Permanence factors, like other protocol components, can be reviewed when a protocol is reviewed (every 5 years) and adjusted as necessary.

As protocols were implemented, two new offset facilitators emerged, the Aggregator and the Verifier. Agricultural offsets do not often generate enough tonnes of credit on a farm to attract interest of a large final emitter. Aggregator companies went through a learning exercise of developing data management systems including sub-tasks of quality assurance and reporting. This in turn helped keep verification costs down and was useful for other farm service businesses requiring data systems. Verifier companies also learned integrated disciplinary approaches to verification. Accountants, engineers, and agricultural professionals worked together to ensure technical integration.

The fourth year of compliance in the Alberta GHG offset system was completed in 2010. Companies had until the following March (e.g. March 31, 2011) to settle their accounts for the previous year. Experience has shown that no matter how much time is spent designing a project, fine-tuning is needed to help a new system run well on the ground. Over time, protocol developers, regulators, aggregators, and verifiers have refined their approaches and have become more efficient implementing the offset system.

As of March 2011, there were 71 projects registered with >11 million MT CO_2e (Alberta GHG Offset-System Registry, 2011). The last completed year summarized to date (2009) showed that agriculture offsets provided ~3.2 million MT CO_2e and contributed 15% of the targeted reductions in emissions for the province. These agricultural offsets generated ~C$35 million in revenue to farmers, aggregators, and verifiers. While the results of the Alberta Carbon Offset Market are encouraging, the long-term market resiliency needs to be determined. These uncertain results should not deter the development of other emission offset platforms that are developing, such as the Verified Carbon Standard (http://www.v-c-s.org/www.v-c-s.org) and Climate Action Reserve (http://www.climateactionreserve.org/ http://www.climateactionreserve.org). However, in the absence of a legislated national or international standard, these groups will likely be susceptible to some of the risks described above for CCX and NFU.

FUTURE OUTLOOK

Lessons Learned from U.S. Greenhouse Gas Markets

Despite the lack of U.S. legislative initiatives and no formal compliance market, there are still active voluntary markets for C credits. Unfortunately, these are private transactions between businesses with no publicly available tracking system. Within the U.S., the existing C credit verification business may eventually expand as climate change concerns are better understood and economic solutions are tested. A no-compliance market with less stringent rules translates to a lower value per MT of C (Figure 24.1). Daily closing prices of Certified Emission Reduction (CER) prices (from the CDM) and historical CCX prices show substantial and consistent differences in value. Both trading systems have been market driven, but because each system has had a different set of regulations, the value given to C can be different. Low price often leads to low potential capital income, and therefore little financial interest from potential project developers and investors.

There were several issues contributing to the "failure/limited acceptance" of a thriving compliance or voluntary U.S. C market. The "voluntary" nature of the market limited acceptance, because without some form of compliance, prices were depressed and C emitting companies were not required to reduce emissions. An alternative model to consider might be when C would not be the major financial portion of a project, but rather energy production, crop production or another revenue engine would be the main financial driver and C would be a secondary benefit. This alternative system would require rules and regulations that may not be as stringent as legislative compliance. Additionally, C markets were not encouraged because there was the absence of political will for the U.S. to sign the Kyoto Protocol. Concerns with the protocol in the U.S. were based on anticipated negative economic impacts and a strong resistance to regulations and set emissions limits.

In addition to fractionated mechanisms, well-known internationally accepted concepts were sometimes ignored (e.g. additionality). Additionality is one of the criteria used to measure quality and eligibility of a project or practice to qualify as an emissions offset. In the early days of the voluntary U.S. C market, this concept was often not considered in protocols. As the market evolved and matured and credit buyers became more educated, this concept slowly reappeared in many voluntary protocols.

The uncertain baseline date, i.e. the time to start GHG accounting, was another unfortunate feature of immature systems. There was much confusion on project baseline dates, which defined when a project would start earning credits. Baselines were different not only across industries but even within the same industry (e.g. agriculture, methane digester versus soil C sequestration). Different baseline approaches can be warranted if practices are quite different such as afforestation versus feedlot or dairy operations. Where controversies arose was when

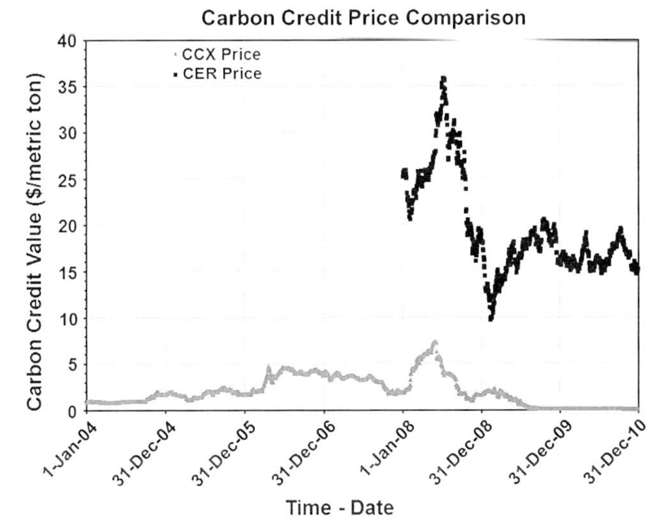

FIGURE 24.1
Daily closing C prices (US$) of Certified Emission Reduction (CER) (from the UNFCCC Clean Development Mechanism) and Chicago Climate eXchange (CCX) trading registries, 2004–2010.

different baseline approaches were used for the same practice. There were instances where emissions credits were produced and sold originating from projects with an unknown baseline.

The emergence of multiple mechanisms and inconsistent standards led to a situation where an MT of CO_2e in one system was not equal to an MT of CO_2e in a different system. In other words, an MT of CO_2e created in a specific platform (e.g. CCX) could be different in another platform (i.e. Regional Greenhouse Gas Initiative). This led to confusion and the creation of millions of C credits under easy, less stringent methodologies.

Complicated issues arose for soil C sequestration due to computational processes and prediction methods, verification issues, and the temporary nature of the C credit. Simple factors were used in many occasions to predict C credits, which were then verified by a simple observation of farm tillage practice. In other occasions, model-based approaches were used (e.g. Zimmerman et al., 2005, Easter et al., 2007) followed by verification of data entry to the model resulting in C credits. Either of these approaches could result in vast over- or underestimation of the actual C sequestered. Neither of these approaches verified the actual C sequestered in the soil, but gave a relative estimate or trend of expected results. These observations suggest a need for more research to agree on a practical compromise to be reached between direct measurement of soil C and an accepted and validated model prediction of soil C.

A troublesome issue associated with soil C sequestration in no-till systems is the unintended effects on the N_2O and CH_4 fluxes within the managed ecosystem. Rochette et al. (2008) found differences in the response of N_2O emissions when converting to a no-till practice between clay and loam soils in eastern Canada. While emissions were similar in both tillage treatments in the well-aerated loam, the emissions more than doubled under no-till in the clay soil. While some research groups have reported no change in CO_2, N_2O and CH_4 flux intensities between tillage systems, others have observed different ecosystem behaviors (Six et al., 2004), including potential loss of biodiversity (Putz, 2009), which further complicates an already complex system.

Early presentation of optimistic monetary returns for soil C credits with subsequent low returns eventually discouraged farmer involvement. Initial projections of C credits ranged from \$20 to 25 MT^{-1} CO_2e, which raised farmer expectations for large paybacks from simple practice changes. When contrasted with the reality of <\$5 MT^{-1} CO_2e in voluntary markets, confidence in offset programs plummeted. Additionally, estimated values did not include costs of transaction, verification, and aggregation. As a result, farmers often became frustrated and disillusioned with the low financial expectations of voluntary C markets.

Recommendations for Greenhouse Gas Markets

Carbon trading/accounting systems must be transparent, consistent, comparable, accurate, and verifiable (IPCC, 2000). Trading systems must be economically and environmentally sustainable and should meet social expectations. Some further recommendations aimed at highlighting co-benefits and strengthening GHG markets for agriculture include:

- Develop educational strategy on urgency of climate change issues;
- Provide a solid science foundation for issues related to GHG emissions and allow science and policymakers to come together to address both technical and social issues;
- Develop local and regional aggregation networks addressing economic and environmental issues to decrease GHG emissions;
- Develop national policies based on international standards for GHG emissions to include CO_2, CH_4, and N_2O;
- Enable farmers to develop better record systems for improved C management, including issues of permanence, additionality, and environmental policy;

- Develop national measurement standards for the creation, verification, and issuance of GHG emissions offsets that are economical, transparent, consistent, comparable, complete, accurate, verifiable, and environmentally friendly;
- Aggregators and verifiers should learn to work collaboratively using an interdisciplinary approach;
- Provide for interdisciplinary collaboration among scientists to develop a sophisticated prediction system/model accounting for all GHGs and their interactions with sequestered soil C;
- Develop new non-intrusive and/or remote sensing technologies for verifying sequestered soil C (Chatterjee et al., 2009; Wielopolski et al., 2008);
- Develop procedures for placing true economic value on currently unpriced ecosystem services and environmental benefits of C management;
- Clarify that ecosystem services from soil carbon management may be more beneficial to farmers than just C credits.

CONCLUSIONS

Currently, prices for C credits are below their social opportunity cost ($<\$0.15\ MT^{-1}\ CO_2e$) due to uncertainty over future policies and a lack of national markets in the U.S. and Canada. However, some businesses are reducing emissions and buying C credits at low prices as a hedge against the possibility that emission limits will be enacted in the future. Firms may be motivated to purchase C credits with the intent of publicizing this action and enhancing their reputation as a good environmental citizen. Innovative companies are strategically moving to reduce emissions, as it makes economic sense and moves in the right direction in a global environment of climate change consciousness. Multinational firms will recognize that their facilities will have to meet limits in countries that have ratified the Kyoto Protocol, and many anticipate GHG limits will eventually be enacted in the U.S. While provincial markets continue in Canada with optimism for a national market, the Canadian federal government stopped work on a potential offset market in 2006.

The U.S. is presently involved in intense political debate, industry discussion and market formation to deal with GHG reductions. The lack of clarity/transparency in voluntary markets, consistency in regulations, and the lack of legislative direction/mandate have contributed to the slow acceptance of C trading markets. The future of C trading in the U.S. likely remains unclear until there is some type of energy efficiency legislation that will encourage the reduction of GHG emissions, especially C emissions. Recent political discussions and uncertainty have dimmed hopes of energy efficiency legislation in the near term as major economic recession, unemployment, mortgage foreclosures, energy security, health care, and unrest in the Middle East have risen to a higher priority in Congress.

Acknowledgments

We thank Audrey Myers and Jill Gunderson for assistance with manuscript formatting and Rattan Lal and John Kimble whose comments improved an earlier draft of the manuscript.

References

Agricultural Carbon Markets Working Group, 2011. Climate change and national policies: Implications for corn in the United States Accessed at http://www.agcarbonmarkets.com/documents/ACMWG_Corn_Climate.pdf (Verified: April 11, 2011).

Alberta Environment, 2002. Climate Change and Emissions Management Act. Alberta Queen's Printer, Edmonton, Alberta. Accessed at http://www.qp.gov.ab.ca/documents/Acts/C16P7.cfm?frm_isbn=077974697X (Verified: March 22, 2011).

Alberta GHG Offset-System Registry. 2011. Accessed at http://carbonoffsetsolutions.climatechangecentral.com/offset-registry (Verified: March 22, 2011).

Alberta Government, 2007. Climate Change and Emissions Management Amendment Act, 2007. Accessed at http://www.assembly.ab.ca/ISYS/LADDAR_files/docs/bills/bill/legislature_26/session_3/20070307_bill-003.pdf (Verified: March 22, 2011).

Alberta Government, 2011a. Technical guidance for offset project developers, version 2.0, p. 69. Accessed at http://environment.gov.ab.ca/info/library/7915.pdf (Verified: March 22, 2011).

Alberta Government, 2011b. Technical guidance for offset protocol developers, version 1.0, p. 92. Accessed at http://environment.gov.ab.ca/info/library/8331.pdf (Verified: March 22, 2011).

Baker, J.S., McCarl, B.A., Murray, B.C., Rose, S.K., Alig, R.J., Adams, D., Latta, G., Beach, R., Daigneault, A., 2010. Net-farm income and land use under a U.S. greenhouse gas cap-and-trade Policy Issues, P17 (April).

Burtraw, D., 2000. Innovation under the tradable sulfur dioxide emission permits program in the U.S. Electricity sector. Discussion Papers dp-00-38, Resources for the Future. EconPapers. Örebro University, Örebro, Sweden.

Chatterjee, A., Lal, R., Wielopolski, L., Martin, M.Z., Ebinger, M.H., 2009. Evaluation of different soil carbon determination methods. Critical Rev. Plant Sci. 28, 164–178.

Climate Change Emissions Management Fund ("Tech Fund"), 2010. Accessed at http://ccemc.ca (Verified: March 22, 2011).

Davidson, E.A., Ackerman, I.L., 1993. Changes in soil carbon inventories following cultivation of previously untilled soils. Biogeochemistry 20, 161–193.

Dudek, D.J., Goffman, J., Wade, S.M., 1997. Emissions trading in nonattainment areas: potential, requirements, and existing programs. In: Kosobud, R.F., Zimmermann, J.M. (Eds.), Market-based Approaches to Environmental Policy: Regulatory Innovations to the Fore. Van Nostrand Reinhold, New York, pp. 151–185.

Easter, M., Paustian, K., Killian, K., Williams, S., Feng, T., Al Adamat, R., Batjes, N.H., Bernoux, M., Bhattacharyya, T., Cerri, C.C., Cerri, C.E.P., Coleman, K., Falloon, P., Feller, C., Gicheru, P., Kamoni, P., Milne, E., Pal, D.K., Powlson, D.S., Rawajfih, Z., Sessay, M., Wokabi, S., 2007. The GEFSOC soil carbon modeling system: a tool for conducting regional-scale soil carbon inventories and assessing the impacts of land use change on soil carbon. Agric. Ecosyst. Environ. 122, 13–25.

Environment Canada, 2010. National inventory report 1990–2008, Part 1. 2010, p. 221. Accessed at http://www.ec.gc.ca/Publications/default.asp?lang=En&xml=492D914C-2EAB-47AB-A045-C62B2CDACC29 (Verified: March 22, 2011).

Follett, R.F., Reed, D.A., 2010. Soil carbon sequestration in grazing lands: societal benefits and policy implications. Rangeland Ecol. Manage. 63, 4–15.

Grubb, M., Vrolijk, C., Brack, D., 1999. The Kyoto Protocol—a guide and assessment. Earthscan, London.

Hanson, J.D., Liebig, M.A., Merrill, S.D., Tanaka, D.L., Krupinsky, J.M., Stott, D.E., 2007. Dynamic cropping systems: increasing adaptability amid an uncertain future. Agron. J. 99, 939–943.

Hatfield, J.L., Boote, K., Fay, P., Hahn, L., Izaurralde, C., Kimball, B.A., Mader, T., Morgan, J., Ort, D., Polley, W., Thomson, A., Wolfe, D., 2008. Agriculture. In: The Effects of Climate Change on Agriculture, Land Resources, Water Resources, and Biodiversity. U.S. Climate Change Science Program and the Subcommittee on Global Change Research, Washington, DC, pp. 21–74.

Houghton, R.A., Hackler, J.L., Lawrence, K.T., 1999. The U.S. carbon budget: contributions from land-use change. Science 285, 574–577.

Intergovernmental Panel on Climate Change (IPCC), 2000. Land use, land-use change, and forestry. Special report of the Intergovernmental Panel on Climate Change. Cambridge University Press, Cambridge, U.K., p. 377.

IPCC, 2007. Climate change 2007: The physical science basis. In: Solomon, S., Qin, D., Manning, M., Chen, Z., Marquis, M., Averyt, K.B., Tignor, M., Miller, H.L. (Eds.), Contribution of Working Group I to the Fourth Assessment Report of the Intergovernmental Panel on Climate Change. Cambridge University Press, Cambridge, United Kingdom and New York, NY, USA.

Janzen, H.H., 2004. Carbon cycling in earth systems—a soil science perspective. Agric. Ecosyst. Environ. 104, 399–417.

Kimble, J.M., Rice, C.W., Reed, D., Mooney, S., Follett, R.F., Lal, R., 2007. Soil carbon management: economic, environmental and societal benefits. CRC Press, Boca Raton, Florida, p. 288.

Kossoy, A., Ambrosi, P., 2010. State and trends of the carbon market 2010 CF-Assist Program (www.Carbonfinance.org). Accessed at http://siteresources.worldbank.org/INTCARBONFINANCE/Resources/State_and_Trends_ofthe_Carbon_Market_2010_low_res.pdf (Verified: April 11, 2011).

Lal, R., 1989. Land degradation and its impact on food and other resources. In: Pimentel, D., Hall, C.W. (Eds.), Food and Natural Resources. Academic Press, San Diego, CA, pp. 65–68.

Lal, R., et al., 1995. Global soil erosion by water and carbon dynamics. In: Lal, R. (Ed.), Soils and Global Change. Lewis Publishers, Boca Raton, FL, pp. 131–140.

Lal, R., 2002. Why C sequestration in agricultural soils? In: Kimble, J.M., Lal, R., Follett, R.F. (Eds.), Agricultural Practices and Policies for C Sequestration in Soil. Lewis Publishers, CRC Press, Boca Raton, FL, pp. 21–30.

Lal, R., 2003. Soil erosion and the global carbon budget. Environ. Int. 29 (4), 437–450.

Lal, R., Kimble, J.M., Follett, R.F., Cole, V., 1998. Potential of U.S. Cropland for Carbon Sequestration and Greenhouse Effect Mitigation. Ann Arbor Press, Chelsea, MI, p. 128.

Manley, J., Van Kooten, G.C., Moeltner, K., Johnson, D.W., 2005. Creating carbon offsets in agriculture through no-till cultivation: a meta-analysis of costs and carbon benefits. Climate Change 68, 41–65.

North Dakota Farmers Union (NDFU)/National Farmers Union (NFU), 2011. Accessed at www.carboncredit.ndfu.org (Verified: March 29, 2011).

Pacific Northwest Direct Seed Association (PNDSA), 2010a. Carbon market history. Accessed at http://directseed.org/carbonhistory.html (Verified: July 29, 2010).

Paustian, K., Brenner, J., Easter, M., Killian, K., Ogle, S., Olson, C., Schuler, J., Vining, R., Williams, S., 2009. Counting carbon on the farm: reaping the benefits of carbon offset programs. J. Soil Water Conserv. 64, 36A–40A.

PNDSA, 2010b. Carbon trading. Accessed at http://www.directseed.org/carbon%20trading.html (Verified: July 29, 2010).

Putz, F.E., 2009. Dangers of carbon-based conservation. Global Environmental Change. doi:10.1016/j.gloenvcha.2009.07.005.

Ravindranath, N.H., Ostwald, M., 2008. Carbon inventory methods—Handbook for Greenhouse Gas Inventory, Carbon Mitigation and Roundwood Production Projects. Advances in Global Change Research 29. Springer, Heidelberg, Germany.

Ribaudo, M., Hansen, L., Hellerstein, D., Greene, C., 2008. The use of markets to increase private investment in environmental stewardship. ERR-64. USDA, Economic Research Service. September 2008.

Ribaudo, M., Johansson, R., Jones, C., 2006. Environmental credit trading: can farming benefit? In: Amber Waves, vol. 4, Special Issue USDA, Economic Research Service. July 2006.

Rochette, P., Angers, D.A., Chantigny, M.H., Bertrand, N., 2008. Nitrous oxide emissions respond differently to no-till in a loam and a heavy clay soil. Soil Sci. Soc. Am J. 72, 1363–1369.

Rosenzweig, C., Hillel, D., 1998. Climate Change and the Global Harvest: Potential Impacts of the Greenhouse Effect on Agriculture. Oxford Univ. Press, Oxford, U.K.

Six, J., Ogle, S.M., Breidt, F.J., Conant, R.T., Mosier, A.R., Paustian, K., 2004. The potential to mitigate global warming with no-tillage is only realized when practiced in the long-term. Global Change Biol. 10, 155–160.

Smith, O.H., Petersen, G.W., Needelman, B.A., 2000. Environmental indicators of agroecosystems. Adv. Agron. 69, 75–97.

Statistics Canada, 2011. Gross domestic product, expenditure-based, by province and territory—Table 384-0002. Accessed at http://www40.statcan.ca/l01/cst01/econ15-eng.htm (Verified: March 22, 2011).

U.S. Environmental Protection Agency (EPA), 2009. SO_2 reductions and allowance trading under the acid rain program. Accessed at http://www.epa.gov/airmarkets/progsregs/arp/s02.html (Verified: March 22, 2011).

U.S. Senate Environment and Public Works Committee, 2009. U.S. senate minority report: more than 700 international scientists dissent over man-made global warming claims scientists continue to debunk "consensus" in 2008 & 2009, Minority report. 11 Dec 08. Updated 16 Mar 09, p. 7. Accessed at http://www.scribd.com/doc/25998234/U-S-Senate-Minority-Report (Verified: April 26, 2011).

United Nations Framework Convention on Climate Change (UNFCCC), 1997. Kyoto Protocol to the United Nations Framework Convention on Climate Change. United Nations Framework Convention on Climate Change Secretariat, Bonn, Germany.

Walsh, M.J., 2002. Growing the market: recent developments in agricultural sector C trading. In: Kimble, J.M., Lal, R., Follett, R.R. (Eds.), Agricultural Practices and Policies for C Sequestration in Soil. Lewis Publishers, CRC Press, Boca Raton, FL, pp. 375–385.

Wielopolski, L., Hendrey, G., Johnsen, K.H., Mitra, S., Prior, S.A., Rogers, H.H., Torbert, H.A., 2008. Nondestructive system for analyzing carbon in the soil. Soil Sci. Soc. Am. J. 72, 1269–1277.

Zimmerman, P.R., Price, M., Peng, C., Capehart, W.J., Updegraff, K., Kozak, P., Vierling, L., Baker, E., Kopp, F., Duke, G., Das, C., 2005. C-lock (patent pending): a system for estimating and certifying carbon emission reduction credits for the sequestration of soil carbon on agricultural land. Mitigation and Adaptation Strategies for Global Change 10 (2), 307–331.

CHAPTER 25

Eligibility Criteria Affecting Landowner Participation in Greenhouse Gas Programs

Robert Johansson[1], Greg Latta[2], Eric White[2], Jan Lewandrowski[1], Ralph Alig[3]
[1]USDA Office of the Chief Economist, Climate Change Program Office, Washington, DC
[2]Department of Forest Engineering, Resources and Management, Oregon State University Corvallis, OR
[3]USDA Forest Service, Pacific Northwest Research Station, Corvallis, OR

CHAPTER OUTLINE

INTRODUCTION

Farmers, ranchers, and forest landowners have a wide variety of production and land management practices that could either lower the greenhouse gas (GHG) emissions of their operations or increase the quantity of carbon (C) sequestered in soils, biomass, and products. Over the last 20 years, numerous agronomic and economic studies have concluded that if accomplished on a large enough scale, adoption of these production and land management practices could result in significant and cost-effective GHG mitigation (e.g. Johnson et al., 2010a; Eagle et al., 2010; Horowitz and Gottlieb, 2010; Smith et al., 2008).

State and national policymakers have also recognized the benefits of incentivizing GHG mitigation in the agriculture and forest sectors in the context of pursuing a broad set of climate and other environmental goals. A variety of program designs have been proposed to

Managing Agricultural Greenhouse Gases. DOI: 10.1016/B978-0-12-386897-8.00025-5
2012, Published by Elsevier Inc.

promote the adoption of GHG sequestration and less GHG intensive production and land management practices on the nation's farms, ranches, and forest lands (see Table 25.1). For example, during the 111th Congress, the House of Representatives passed the American Clean Energy and Security Act (H.R. 2454) on June 26, 2009 and the Senate passed out of committee the Clean Energy Jobs and American Power Act (S. 1733) on November 5, 2009. If enacted, either would have gradually reduced U.S. GHG emissions by implementing a cap-and-trade framework for limiting greenhouse gas emissions from several energy-intensive sectors. While agriculture and forestry were not included as regulated sectors in either bill, each contained provisions that allowed farmers, ranchers, and forest landowners to be awarded offsets for voluntarily taking specific actions that reduced their emissions or sequestered additional C.

Neither bill became law. However, they generated considerable debate about the potential role of agriculture and forestry in a national GHG mitigation framework. Much of the analyses (AFPC, 2009; APAC, 2009; U.S. EPA, 2009, 2010) and discussion that accompanied the debate focused on how the various analyses of the legislation had handled a number of key issues to affect the participation rates of farmers and forest landowners, and the quantity of GHG mitigation that would result from their actions. Proposals to incentivize GHG mitigation activities in the farm and forest sectors have typically included eligibility criteria to limit the scope of participation, help control total cost, and/or increase cost effectiveness of different mitigation efforts (Table 25.1). In general, such criteria target land management and production practices that result in verifiable GHG emissions reductions or C sequestration that are additional, (relatively) permanent, and not prone to leakage (see Chapter 24 for a discussion of these concepts).

In this chapter we use one model, whose simulations and results were central to the debate over H.R. 2454 and S. 1733 to investigate how alternative formulations of five key eligibility criteria (identified below) affect farm and forest sector GHG mitigation potential.

ELIGIBILITY CRITERIA FOR GHG MITIGATION PROGRAMS

We examine five incentive designs that address eligibility criteria likely to be important in future GHG mitigation programs (Ramseur, 2010). Specifically, we consider:

- allowing the included forest and agriculture C abatement activities to supply offsets with few constraints;
- reducing landowner participation in another conservation program;
- making emissions reductions related to bioelectricity ineligible for offset supply to address additionality concerns;
- constraining land-use change to limit potential C leakage; and
- providing enhanced levels of incentives for certain types of GHG mitigation efforts that might require additional support to ensure permanence.

We believe those issues would play a key role in the design of any future framework to incentivize GHG mitigation in the farm and forestry sectors, whether that framework takes the form of a cap-and-trade, direct government payments, private sector contracts, or another legally binding approach (Ramseur, 2010).

For each of the above incentive design issues, we compared a number of policy scenarios to a business-as-usual (BAU) state of the world that does not include specific programs to incentivize GHG mitigation (see Table 25.2). Simulations were conducted to examine agriculture and forest sector adjustments in response to C payments for reducing emissions of GHGs or increasing sequestration of C in forests or soils over a broad range of farm and forest mitigation activities. Simulations were conducted over a 2000 through 2080 period to accommodate forest rotations. The policy was assumed enacted in 2010 and remained in effect through the end of the 2080 period; however, results were only reported through 2050 in order

TABLE 25.1 Elements of Eligibility Criteria for Agricultural and Forestry Programs

	Status	Coverage	Eligibility criteria
H.R. 2454—American Clean Energy and Security Act and S. 1733—Clean Energy Jobs and American Power Act	Passed the U.S. House of Representatives on June 26, 2009 [1]; Reported by the Senate Committee on Environment and Public Works on November 5, 2009 [2]	Most economic sectors included in the cap. Domestic agricultural and forestry sectors can supply up to 1 billion C offsets in H.R. 2454 and 1.5 billion C offsets in S. 1733.	Additional and permanent afforestation, reforestation, management of peatland or wetland, conservation of grassland and forested land, improved forest management, including accounting for C stored in wood products, reduced deforestation or avoided forest conversion, urban tree-planting and maintenance, agroforestry are allowed. Term offset credits used to address permanence. [3]
Conservation Reserve Program (CRP)	Land rental program [4] and an informational tool to track C sequestered on CRP contracts [5]	Retires approximately 13 Mha of environmentally sensitive cropland.	Long-term contracts used to maximize environmental benefits including C sequestration; CarbonNet is expected to provide data to facilitate verification of C sequestration from tree planting activities on CRP contracts in Arkansas, Georgia, and Minnesota.
Environmental Quality Incentives Program (EQIP)	Cost-share conservation program [6] and Conservation Innovation Grant Program [7]	Working farms and ranches	Limits contract payments to 3 years and eligible efforts to a list of pre-approved management practices to ensure that investments in conservation have a high probability of generating environmental benefits in a cost-effective manner. Targeted grants to encourage landowner implementation of systems and practices designed to demonstrate effective techniques to mitigate GHG emissions.
Renewable Fuels Standard	Authorized by the Energy Independence and Security Act of 2007 [8]	Mandates the sale of 136 billion liters of biomass-derived liquid fuels per year by 2022	Explicit criteria on the GHG emissions from biofuel production used to determine eligibility for the program.
Renewable Portfolio Standards	State programs that require electricity generation to be from renewable energy [9]	All but five states have some binding target for the provision of renewable electricity generation	Most state programs allow the use of certain types of biomass combustion to qualify as a renewable energy for the mandates. See, for example, California Energy Commission (2011) "Renewables Portfolio Standard Eligibility," Commission Guidebook CEC-300-2010-007-CMF (January).
Regional Greenhouse Gas Initiative (RGGI)	Regional Cap-and-Trade program beginning in 2008 [10]	Covers the power sector in 10 northeastern and Mid-Atlantic states through 2018	Offsets allowed for afforestation (enforced with a permanent easement) and for avoided methane emissions from agricultural manure management operations through anaerobic digesters.

Continued

TABLE 25.1 Elements of Eligibility Criteria for Agricultural and Forestry Programs—continued

	Status	Coverage	Eligibility criteria
California Climate Program (AB 32)	Regional Cap-and-Trade program beginning in 2012 [11]	Covers the electricity, fuel, and transportation sectors in California through 2020	Offsets allowed for reforestation, improved forest management, urban tree planting, and biogas control using anaerobic digesters.

[1] Details of the bill are available online at http://www.govtrack.us/congress/bill.xpd?bill=h111-2454.
[2] Details of the bill are available online at http://www.govtrack.us/congress/billtext.xpd?bill=s111-1733.
[3] "Term offset credits" are offset activities that may expire after 5 years. These were included in the bill to address activities such as soil carbon sequestration, which may be difficult to make permanent. Upon expiration of term offsets, the owner of the credit was required to replace it with another term offset, with another permanent offset or allowance. For additional discussion see Brent Yacobucci, Jonathan L. Ramseur, and Larry Parker (2009) Climate Change: Comparison of the Cap-and-Trade Provisions in H.R. 2454 and S. 1733. Congressional Research Service Report No. R40896 (November 5).
[4] Heimlich, Ralph (2008) USDA's Conservation Reserve Program: Is it Time to Ease into Easements? RFF Weekly Policy Commentary (September 8).
[5] U.S. Department of Agriculture (2010) Farm Service Agency's Voluntary Carbon Data Project. Fact sheet available online at http: www.fsa.usda.gov/Internet/FSA_File/voluntary_carbon_data.pdf (last accessed March 21, 2011).
[6] Stubbs, Megan (2010) Environmental Quality Incentives Program (EQIP): Status and Issues. Congressional Research Service Report No. R40197 (August 13).
[7] U.S. Department of Agriculture (2010) CIG FY 2011 Announcement for Program Funding-GHG. Available online at http://www.nrcs.usda.gov/technical/cig/index.html (last accessed March 21, 2011).
[8] For additional information on the greenhouse gas benefits of biofuel use see EPA (2010) Renewable Fuel Standard Program (RFS2) Regulatory Impact Analysis. EPA-420-R-10-006 (February).
[9] See Energy Information Administration (2009) Annual Energy Outlook 2009 with Projections to 2030. Table 3. State renewable portfolio standards. EIA Report No. DOE/EIA-0383 (March). Some national programs have been proposed in recent legislation (see, for example, S. 1462 American Clean Energy Leadership Act of 2009, reported by Senate Committee on Energy and Natural Resources, July 16, 2009).
[10] Information on eligible RGGI offsets can be found online at http://www.rggi.org/market/offsets.
[11] Information on eligible AB 32 offsets can be found online at http://www.arb.ca.gov/cc/capandtrade/capandtrade.htm.

TABLE 25.2 Scenarios for Examining Eligibility Criteria

Scenario	Carbon prices [1]	Description
Business-as-Usual (BAU)	$0	Baseline agricultural and forestry production with no carbon policy. Liquid biofuels meet RFS mandates and bioelectricity generation meets the EIA AEO 2010 projections extending out to 2035 and carry forward with that level as a floor. 13 million acres of CRP are maintained.
Carbon Price Scenario (CPS)	$5 and $15 (static and rising at 4% annual)	Same as above, but with C prices placed on the additional annual carbon flux generated through agriculture, bioenergy, and forestry management options relative to BAU conditions. Liquid biofuels do not receive the carbon price.
CRP reduction	$15 (rising at 4% annual)	Same assumptions as under CPS, but with a floor on CRP allowed to fall to 8.7 Mha.
Bioenergy limitation (NBIO)	$15 (rising at 4% annual)	Same assumptions under CPS, but flux generated from bioelectricity not credited with a C price.
Sector Specific Scenario (SSS)	$15 (rising at 4% annual)	Same assumptions under CPS, but land not allowed to move between agriculture and forestry; bioelectricity subject to the C price payment.
Sector Specific no Bioelectricity (SSB)	$15 (rising at 4% annual)	Same assumptions under SSS, but bioelectricity not subject to the C price payment.
Payments for tillage reduction (NTILL)	$15 (rising at 4% annual)	Same assumptions as under SSB, but no-till management provided an extra adoption incentive equal to 2000 EQIP cost share payments

[1] Carbon prices are provided for eligible sectors or activities that generate reductions in net GHG flux relative to a business-as-usual simulation of future production (i.e. relative to the baseline).

to minimize the influence of terminal effects on policy conclusions. The simulations provided insights into how different eligibility criteria for incentive payments might impact landowners' decisions regarding agricultural and forest management.

In the first set of simulations, with all GHG accounts eligible to supply offsets, we compared two alternative time path structures that C price could follow in a national GHG mitigation framework. Regardless of framework, a key element of any incentive to mitigate GHG emissions is that they explicitly or implicitly create a value for reducing emissions of CO_2, N_2O, and CH_4 and/or increasing C sequestration. This value is typically called a C price and is usually stated in CO_2 equivalents to put removals of C, and emissions of CO_2, N_2O, and CH_4 on a comparable scale. We assumed that the U.S. government adopted a range of values representing the social cost of C (USG, 2010) to represent the value of mitigating a tonne (t) of GHG emissions. Those ranged from \$5 to \$21 per t CO_2e in 2010 and rose to \$15 to \$45 by 2050. The simulations used C prices that bounded those values by selecting a C price of \$5 and \$15 in 2010. One path kept those values static in real terms throughout the simulations. In the other price path, C prices rose at 4% per year (reaching \$24 and \$72 in 2050 for the initial \$5 and \$15 levels, respectively). Static real prices might be more reflective of a stand-alone conservation program that seeks to encourage GHG mitigation within a fixed budget. C prices that increase in real terms over time are more reflective of a C tax program or a cap-and-trade program that seeks increasing levels of emissions reductions over time.

The second set of simulations addressed implications of allowing some lands now enrolled in other USDA conservation programs to also qualify for GHG mitigation offsets. The Conservation Reserve Program (CRP) provides farmers with relatively long-term contracts to retire environmentally sensitive cropland from active production. By doing so, a number of environmental benefits can be realized, including water and air quality benefits, as well as mitigating GHGs. However, when commodity prices increase, farmers have less incentive to keep potential cropland out of production. Increasing demands for land to grow energy crops could also be a factor in determining the willingness of landowners to retire lands. Furthermore, if budgetary restrictions were to reduce federal appropriations for CRP, the amount of land that could be enrolled could be lower than the current cap of 13 Mha (see, for example, a 2010 USDA-FSA Environmental Assessment of CRP that includes a policy reducing CRP by 25%). We simulated the effects on GHG mitigation efforts of reducing the cap on CRP by 38% to 8.7 Mha.

A third set of simulations assessed the implications of allowing electrical utilities to supply offsets from substituting biomass feedstocks for fossil fuels in the generation of electricity. In some programs, such substitution would be directly rewarded such as if fuel refineries or electrical utilities were either charged a C tax or subject to a cap for emissions related to the use of fossil fuels but were not charged for emissions related to biomass fuels (since an equivalent amount of CO_2e would be removed from the atmosphere in subsequent production of the biomass feedstock).[a] Conversely, use of biomass as a substitute for fossil fuels might not be considered additional and therefore would not be counted as an offset. This might be the case if bioenergy use were mandated under a renewable electricity standard (Congressional Budget Office, 2009b).

Any national GHG mitigation framework has the potential to affect numerous interactions between the agriculture and forest sectors. Interactions would likely occur across regions (e.g. Adams et al., 2012) and over time through changes in commodity prices and land uses.

[a] Controversy regarding GHG accounting of energy generated from biomass feedstocks as discussed in Searchinger et al. (2010) was avoided in FASOMGHG due to full accounting of soil, fuel, fertilizer, livestock emissions, and tree-derived GHGs in both the forest and agricultural sectors.

Policymakers may also want to limit potential problems with leakage associated with impacts spilling over from one sector to another.

The fourth set of scenarios looked at the implications of effectively addressing leakage by limiting land transfers among sectors. Tillage management entails the use of reduced or conservation tillage practices such as no-till, ridge-till, or mulch-till (USDA-ERS, 2005). Manipulating crop residue levels through tillage management has been advocated as a means to reduce nutrient and sediment runoff for many years. For some farms, reductions in tillage make sense from a strictly economic sense due to the reduced input costs of production associated with no-till farming (Aigner et al., 2003). Another benefit of tillage management is to minimize the release of soil C into the atmosphere. As such, tillage management has been pointed to as a farm management technique that could allow farmers to generate C offsets (Sperow et al., 2003; Horowitz et al., 2010).[b]

However, C stored in soils in one year could easily be released in another year if the field were tilled. No-till management of a field could require periodic reduced tillage operations in certain years to prevent soil compaction. Therefore, creating the correct eligibility criteria for no-till management would be a complicated task. It may be that credits for additional, permanent, and verifiable soil C storage require an additional payment to cover higher management costs. In the final set of simulations we considered several payment designs for tillage management.

SIMULATING ELIGIBILITY CRITERIA

To examine eligibility criteria we used the Forest and Agriculture Sector Optimization Model—Greenhouse Gases, or FASOMGHG (Beach et al., 2010). FASOMGHG combines U.S. agriculture, forest, and bioenergy sectors and uses an inter-temporal dynamic optimization approach to simulate markets for numerous agriculture and forest products (Alig et al., 1998), including biofuels and bioelectricity markets (Alig et al., 2010). Because they are linked, sectors compete for the use of private lands that can be used to produce either agriculture or forest products or the supply of substitutable products including sequestered C and biomass energy feedstocks. It is for this reason that FASOMGHG was used as a basis for EPA's analysis of offset provisions in H.R. 2454 and S. 1733 (U.S. EPA, 2009, 2010).

FASOMGHG includes agricultural GHG emissions from livestock production and manure management, soil disturbance, fertilizer application, and use of fossil fuels in agricultural production. Emissions of N_2O from agricultural lands under specific cropping practices were accounted using parameters estimated from the DAYCENT model (Del Grosso et al., 2001). CH_4 emissions associated with the handling of livestock manure and from enteric fermentation were also represented based on factors established per head of livestock (U.S. EPA, 2011, Annex L). Fossil fuel emissions included those from use of gasoline and diesel in planting, management, and harvesting of agricultural products. Sequestration of C in agricultural soils was based on the CENTURY agroecosystem model (Parton, 1996) and was influenced by tillage practices, use of irrigation, land use (e.g. land in pasture), and planting of perennials (e.g. switchgrass).

Bioelectricity generation was assessed from the higher heating value of combustion of biomass feedstocks in either direct fired boilers or by co-firing with coal. Agricultural feedstocks included residues from crops such as corn or wheat, dedicated energy crops such as switchgrass

[b] Note that the model was updated since initial analyses of H.R. 2454 and S. 1733 were conducted. For example, agricultural productivity assumptions were updated to make them closer to recent USDA baseline projections (http://www.ers.usda.gov/Briefing/Baseline/). In addition, sequestration estimates from afforestation efforts were updated to better reflect current observations of timber stands in the FIA data and model documentation was updated.

or willow, and manure. Forest feedstocks included logging and milling residues, along with whole trees harvested for biomass feedstock. Lignin produced as a byproduct of cellulosic ethanol was also available for combustion. Emission abatement attributed to bioelectricity generation was derived from avoidance of coal emissions required to produce the same amount of electricity.

Within the forest sector, GHG emissions occurred from use of fossil fuels in management, application of fertilizers, deforestation, and burning of residues. Forest ecosystem C included standing live and dead tree biomass (Smith et al., 2006), down dead material and forest understory vegetation (U.S. EPA, 2003, Annex O), organic litter on the forest floor (Smith and Heath, 2002), and forest soils (Birdsey, 1996). Wood product C was based on manufacturing levels of a variety of products, their end uses, and how long they remained in use (Skog and Nicholson, 2000).

The model is a dynamic optimization model; therefore agriculture, bioelectricity, and forestry decisions that allocate resources in the most effective manner possible to achieve the greatest benefits over time were simulated. This assumption of intertemporal optimal resource allocation, based on an assumed perfect knowledge of current and future conditions and subject to the confines of constraints or limits in the model, was necessary for simulating production and mitigation choices into the future. For example, considering one of the scenarios, simulation indicted how agriculture would adjust to a C price of \$15 per t CO_2e in 2010 and that increased to \$72 per t CO_2e in 2050. In such an adjustment, the agriculture sector would generate economic benefits from reallocating resources and planting trees immediately to take advantage of the higher C prices in the future.

SIMULATION RESULTS

Simulations reflected potential changes in complex systems based on many simplifying assumptions. Actual management decisions would change over time in response to conditions that were not built into the model. For example, projections of management changes in response to C prices over time reflect, in part, assumptions about future crop yields. Lower crop yields would likely lead to more cropland under agriculture production to meet growing world demands for food commodities. However, crop yields higher than model assumptions would result in more land available for afforestation, all else equal.

We categorized mitigation results into four basic groups and two time horizons. Afforestation was mitigation that occurred on agricultural land planted to trees. Agriculture included crop and livestock production. Bioenergy included avoided fossil fuel emissions from bioelectricity, ethanol, and biodiesel due to utilization of feedstocks from both agricultural land and forests. Forestry included current and future forests and wood products manufacturing that could utilize agricultural feedstocks as an energy source. To simplify the analysis of temporal differences in C price response, we generated annuitized short-term GHG mitigation values for the first 5-year period post-policy (2010–2014) and long-term values for 40 years (2010–2049).

Business as Usual (BAU) and C Price Scenarios

Under the BAU scenario, there were no programs that provided C payments for mitigation activities. Under such conditions, the combined agriculture and forest sectors were a net GHG source through 2020, a net GHG sink between 2020 and 2040, and then a net source again (Figure 25.1). Net GHG emissions from agricultural activities increased from a current level of 224 Mt CO_2e per year, but plateaued at 320 beginning in about 2030, owing to increased use of agricultural feedstocks for production of biofuels and bioelectricity. Afforestation of 7.4 Mha of agricultural land (4.3 Mha in the 2000 to 2010 period) resulted in up to 151 Mt CO_2e per year removed from the atmosphere in 2015, but annual

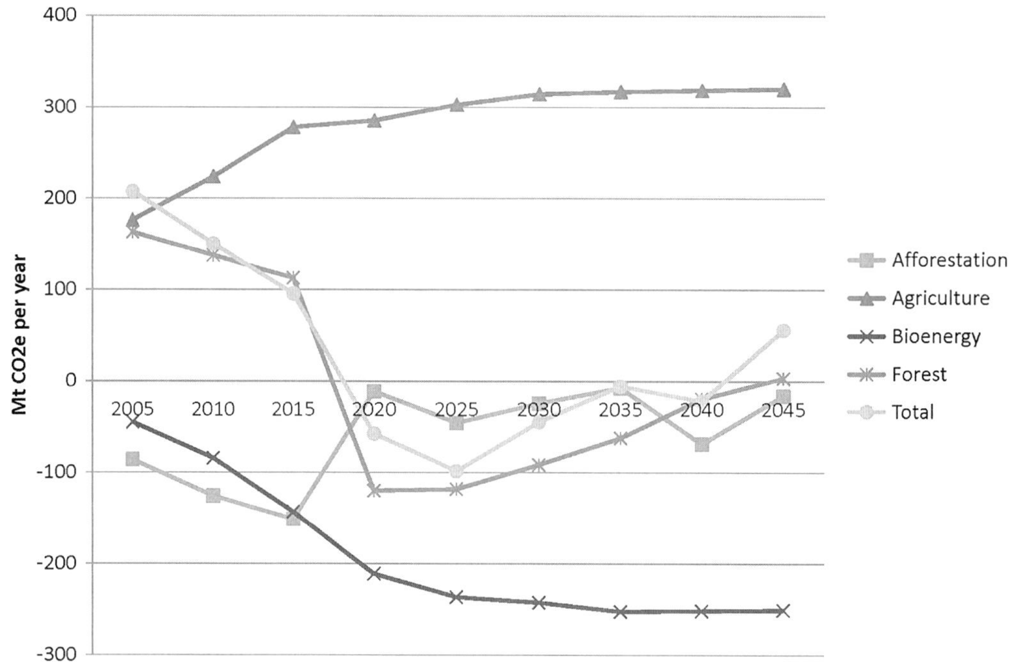

FIGURE 25.1
Simulated annual CO_2e flux values (Mt CO_2e/year) for the business-as-usual case.

C sequestration ranged from 10 to 50 Mt CO_2e per year. Increasing bioenergy production offset 45 to 250 Mt CO_2e per year of GHG emissions from burning of fossil fuels as agriculture and forestry combined to meet the bioenergy levels forecasted by the U.S. Energy Information Administration (U.S. EIA, 2010). Conversion of timberland for agricultural use resulted in a reduction of 6.3 Mha of timberland through the end of the 2020 period. The land-use change led to net GHG emissions in the near term; however, improved growth coupled with harvesting of the first afforested stands planted in 2000 led to a sequestration rate of 120 Mt CO_2e per year in 2020, and subsequently the rate gradually declined until 2050.

Carbon prices were applied each period to the difference between the with-policy C flux and the BAU C flux (Table 25.3). Under rising C prices, simulations showed agriculture GHG emissions to be a net source in both the short and long term under both the low and high price paths. In the short term, forestry was the dominant mitigation option yet both afforestation and bioenergy had potential for sharp gains as C price moved from a low to high price path. In the long term at the lower C price, afforestation was the dominant mitigation technique; however, on the high price path bioenergy took over. For total net GHG mitigation, the higher C price resulted in four times the sequestration rate of the lower C price in the short run, but only just over three times the rate in the long run.

Relative to rising price scenarios, simulations with static C prices had higher GHG sequestration in the short term, but were less of a C sink in the long term (Table 25.3). At the low price in the short term, forestry was the only mitigation activity to change substantially and its value increased three-fold. Forestry responds by changing forest management and afforestation more quickly relative to the case of rising C prices, with over 5 Mha more afforestation at \$15 than BAU in the short term and 1.7 Mha more than the \$15 rising C price case. However, in the long term, afforestation in the static scenario was only 26 and 70%, respectively, of afforestation in the low and high C rising price scenarios. Production of bioelectricity remained a significant GHG mitigation strategy under

TABLE 25.3 Reductions in Greenhouse Gas (GHG) Emissions (Mt CO_2e/year). [1] See Definitions for Abbreviations of Scenarios in Table 25.2

	Short Term					Long Term				
Scenario and CO_2 Price	*Afforestation*	*Agriculture*	*BioEnergy*	*Forestry*	*Total*	*Afforestation*	*Agriculture*	*BioEnergy*	*Forestry*	*Total*
	discount annualized flux value (Mt CO_2 e/year)									
BAU										
$0	−128	228	−86	140	153	−74	287	−189	3	27
	change from BAU discount annualized flux value (Mt CO2 e/year)									
CPS										
$5@4%	−11	14	−6	−35	−39	−80	28	−55	−48	−155
$15@4%	−83	81	−69	−87	−159	−239	74	−250	−78	−493
$5	−18	12	−8	−101	−115	−52	7	−11	−67	−123
$15	−133	122	−60	−161	−231	−294	60	−47	−139	−420
CRP										
$5@4%	−11	26	−9	−35	−29	−85	36	−57	−49	−155
$15@4%	−82	92	−68	−92	−150	−257	83	−255	−79	−509
NBIO										
$5@4%	−5	7	0	−25	−22	−87	26	−1	−42	−104
$15@4%	−63	58	−1	−74	−81	−280	47	−2	−70	−301
SSS										
$5@4%	63	−40	−7	−84	−69	35	−12	−57	−78	−111
$15@4%	63	−37	−73	−106	−153	36	−9	−268	−103	−344
SSB										
$5@4%	63	−48	0	−83	−68	36	−15	−1	−73	−53
$15@4%	63	−54	−2	−98	−91	36	−19	−3	−100	−86
NTILL										
$5@4%	63	−47	2	−75	−57	36	−14	6	−74	−46
$15@4%	63	−53	1	−96	−85	36	−19	4	−101	−79

[1] Positive values indicate increased GHG emissions and negative values indicate net reduction in GHG emissions. Reductions are given in discounted annuitized changes in flux as measured by millions of metric tons of CO_2e over either the first 5-year period post-policy ("Short Term") or over the first 40-year period post-policy ("Long Term").

a static C price scenario, particularly in the short term, but less so in the long term relative to a rising C price scenario.

In the remainder of this chapter, discussion of results was limited to the rising C price scenario, which better reflected the future price path of a more stringent emission cap than static C price, as suggested in recent policy.

Reductions in the Conservation Reserve Program (CRP)

Land within the agriculture sector was constrained by enrollment in CRP. Currently, about 15 Mha are idled in CRP, maintained in perennial grasses or planted to trees. In the BAU scenario, CRP enrollment fell to 13 Mha by 2030 by assumption. However, with increased demand for bioenergy or new markets for C sequestration, agriculture operators were assumed to convert this land to production when C incentive payments were available. In the CRP scenarios, the cap on CRP enrollment was reduced to 8.1 Mha by 2030. Under the high price path, CRP land fell to the lower cap of 8.1 Mha; however, in the low price path, CRP land fell to only 10.1 Mha.

Simulations suggested that allowing more CRP to revert to production did not appreciably change C flux in the two sectors in aggregate or in the agriculture sector in isolation. Of the 4.9 Mha land that reverted from CRP, roughly half of that was afforested, leading to further reduction in emissions in that activity. Another 0.8 Mha planted with a high C price to hard red winter wheat had more than two times the land change as any other crop. The other land that remained in crops was primarily planted to switchgrass, hay, corn, soybeans, and sorghum and that cultivation led to additional emissions from agriculture. GHG emissions associated with bioelectricity offsets had minor gains from the C price scenarios with the cap set at 13 Mha. Less land deforested for agriculture led to reductions of forestry emissions of $<5\%$.

Limitations on Bioelectricity Inclusion

Policies for reduction in GHG emissions from fossil fuel-based electricity create market opportunities for bioenergy feedstocks. Production of bioelectricity was assumed to be at least that projected in the Annual Energy Outlook (EIA, 2010) until 2030 and to remain at the 2030 level thereafter. In simulations with increasing C prices, bioelectricity production increased above that baseline. Under the low C price path, significant increases in bioelectricity production occurred after 2030 and rose to almost three times BAU levels by 2050, with over 90% of biomass feedstock coming from dedicated energy crops on agricultural land and the rest from forestry (Figure 25.2).

Under the high C price path, bioelectricity production immediately increased above the EIA projection. Simulations suggested that such payments would result in twice as much bioenergy production as the BAU case by as early as 2025. By 2035 crop residues exceeded forestry as the second largest bioelectricity feedstock, with dedicated energy crops, crop residues, and forestry comprising 85, 9, and 6%, respectively, of the total bioelectricity production by 2050. Bioelectricity offsets were the largest mitigation option available to the agriculture and forest sectors in the long term, amounting to 250 Mt CO_2e per year (Table 25.3).

In simulations that did not allow bioelectricity offsets, the reduction in need for biomass feedstocks resulted in an increase in agricultural net GHG emissions under both the low and high C price paths. Other mitigation options available to agriculture were not large enough to overcome the lost offset potential associated with increasing bioelectricity production. In those cases, afforestation on agricultural lands became more attractive due to reduced pressure on agricultural land to produce bioenergy feedstocks. Under the high C price scenario, long-term afforestation increased by 6.9 Mha compared to a scenario with bioelectricity offsets. Greater

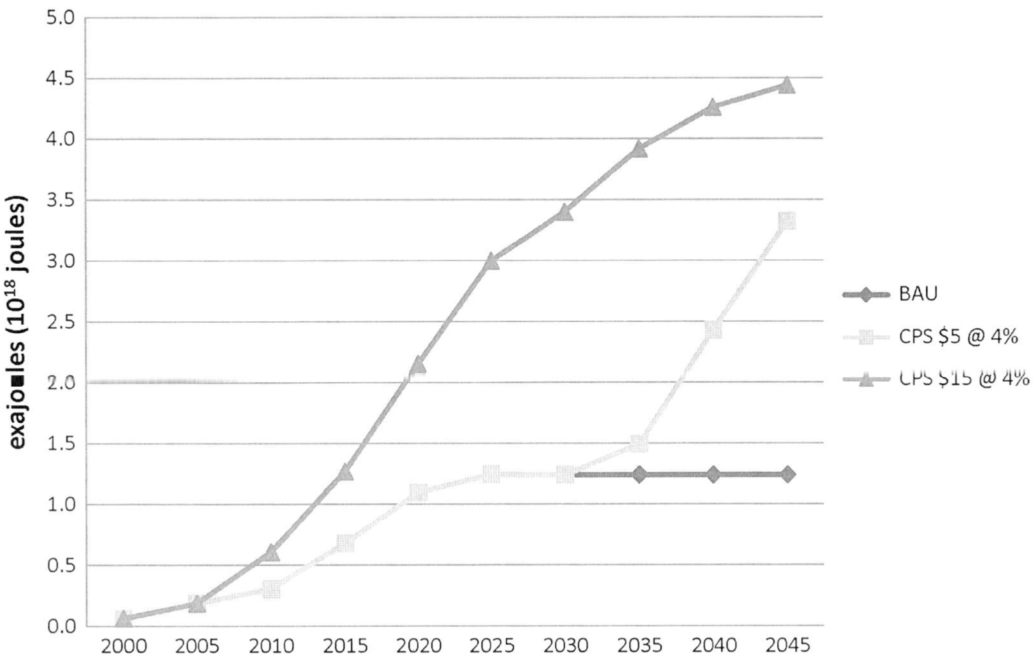

FIGURE 25.2
Simulated bioelectricity energy production from biomass harvested from the agriculture and forestry sector under a baseline case (Annual Energy Outlook Projection; \$0/t CO_2e) and two other CO_2e price scenarios (initial price per t CO_2e/year and increasing at 4%/year).

afforestation led to C sequestration in trees of 42 Mt CO_2e per year greater in the long term than when bioelectricity offsets were eligible for payments (Table 25.3).

Limitation on Land-use Change with and without Bioelectricity Inclusion

When C incentives were available, a significant amount of land shifted between agriculture and forestry mainly due to landowner incentives to increase revenues from afforestation, which rose more quickly than incentives to produce crops due to the assumption of rising C prices and rising crop yields. However, it is possible that future GHG mitigation programs may limit land exchange between sectors. With no land exchange under rising C prices, long-term GHG sequestration from both sectors was improved (Table 25.3) compared to the BAU case and in the short run the combined total C flux value improved relative to the less restrictive scenario for the lower C price. Improvements in agricultural C flux values were largely a result of soil C that would have been transferred to afforestation accounts in the BAU case remaining in agriculture. Lacking afforestation as an option, most GHG mitigation was achieved through bioelectricity production.

Alternatively, some mitigation programs might focus solely on one sector in isolation and not allow offset payments via production of bioelectricity. Under such a policy and for both C price paths combined, agriculture and forestry C sequestration was higher in the short term than the long term. Similar to the limitation on land exchange scenario, short-term C flux values were larger than with the scenario at the lower C price. Despite projected short-term success, C sequestration improvement in the long term due to C valuation in the market was 34 and 17% of that occurring in the low and high C price scenarios, respectively (where all practices were allowed), and 51 and 29% of the long-term no bioenergy C flux values.

Additional Payments for Reduction in Tillage

A rising C price could facilitate changes in tillage systems as farm operators would likely attempt to take advantage of incentives to reduce GHG emissions from their operations. In the

BAU scenario, a general decline in conventional tillage over time, from 63% in 2010 to 61% by 2050, and a countervailing increase in no-till systems, from 17 to 24% over the same period occurred. The scenario with the high C price path produced an extra 12.7 Mha of no-tillage cropping by 2045; however, this increase came from dedicated energy crops such as switchgrass for which there existed only no-till options. We simulated an additional scenario in which land use and bioenergy were limited as in the no bioelectricity scenario and cropland that adopted no-till practices would qualify for an additional land payment equal to the 2000 Environmental Quality Incentives Program (EQIP) cost share value. Such a program could be implemented in hopes of further increasing tillage changes to reduce GHG emissions and achieve other environmental benefits. A C market might also be developed in which agriculture operators receive payments for C sequestration resulting from tillage changes, as well as assistance payments for implementing tillage changes.

To examine the impacts of the C and cost share payment on tillage practices, we focused on two primary crops, corn and soybeans, that together accounted for over 47% of the nation's cropland. At low C prices and supporting reductions in tillage, little difference arose in tillage practices for these two crops until 2040 (Figure 25.3). When assistance payments and C payments were both included, cropland planted using no-till increased for corn (Figure 25.3A) and was roughly offset by reductions in no-till acres of soybeans (Figure 25.3B), resulting in an aggregate acreage for the two crops equivalent to BAU. Although the change in GHG mitigation from BAU (Table 25.3) showed that short-term gains in additional C flux were greater than at the low C price point, at the high price point and in the long term the results were for less mitigation than C price scenarios and even less compared to the sector-specific no bioelectricity scenario. This suggested that to optimize farm revenue, agricultural operators shifted toward tillage practices that captured additional income from the payment and converted to more GHG intensive cropping practices when afforestation and bioelectricity options were limited.

SUMMARY

Defining a policy framework that incentivizes GHG mitigation and C sequestration in the U.S. is a complex task. Neither H.R. 2454 nor S. 1733 became law; however, analyses of the two bills on the various provisions from various government agencies, academic institutions, and other research organizations revealed how program designs might encourage mitigation efforts in the agriculture and forest sectors (Johnson et al., 2010b). Mitigation programs at the state and regional levels have also been developed, similar to the design options in H.R. 2454 and S. 1733. Those efforts include the Regional Greenhouse Gas Initiative in the northeast and Mid-Atlantic states and the cap-and-trade program of the California Air Resources Board.

Some analyses have concluded that the agriculture and forest sectors could play an important role in GHG mitigation, either through reduced emissions or increased C sequestration in soils and forests (Parker and Yacobucci, 2009). Since agricultural producers and forest owners might adjust production and land management practices when incentives would be provided, these regulated entities could provide lower-cost alternatives by reducing their own emissions. Between 2010 and 2050, FASOMGHG simulations showed that the agriculture and forest sectors could reduce GHG emissions as much as 550 Mt CO_2e per year (with upper values of 1200 Mt CO_2e in several years) on a discounted annualized basis, depending on eligibility criteria and payment structure. Even facing a modest, static C payment of \$5 per metric ton CO_2e over time, agriculture and forest landowners could increase their incomes and reduce emissions in the U.S. by an average of 100 million metric tons per year over the next 40 years—equivalent to the annual GHG emissions from 17 million mid-sized conventional gasoline cars (Chew and Nguyen, 2011).

However, reaching consensus on what constitutes an appropriate mitigation program for eligible landowners has also proven to be a difficult process, due in large part to differing

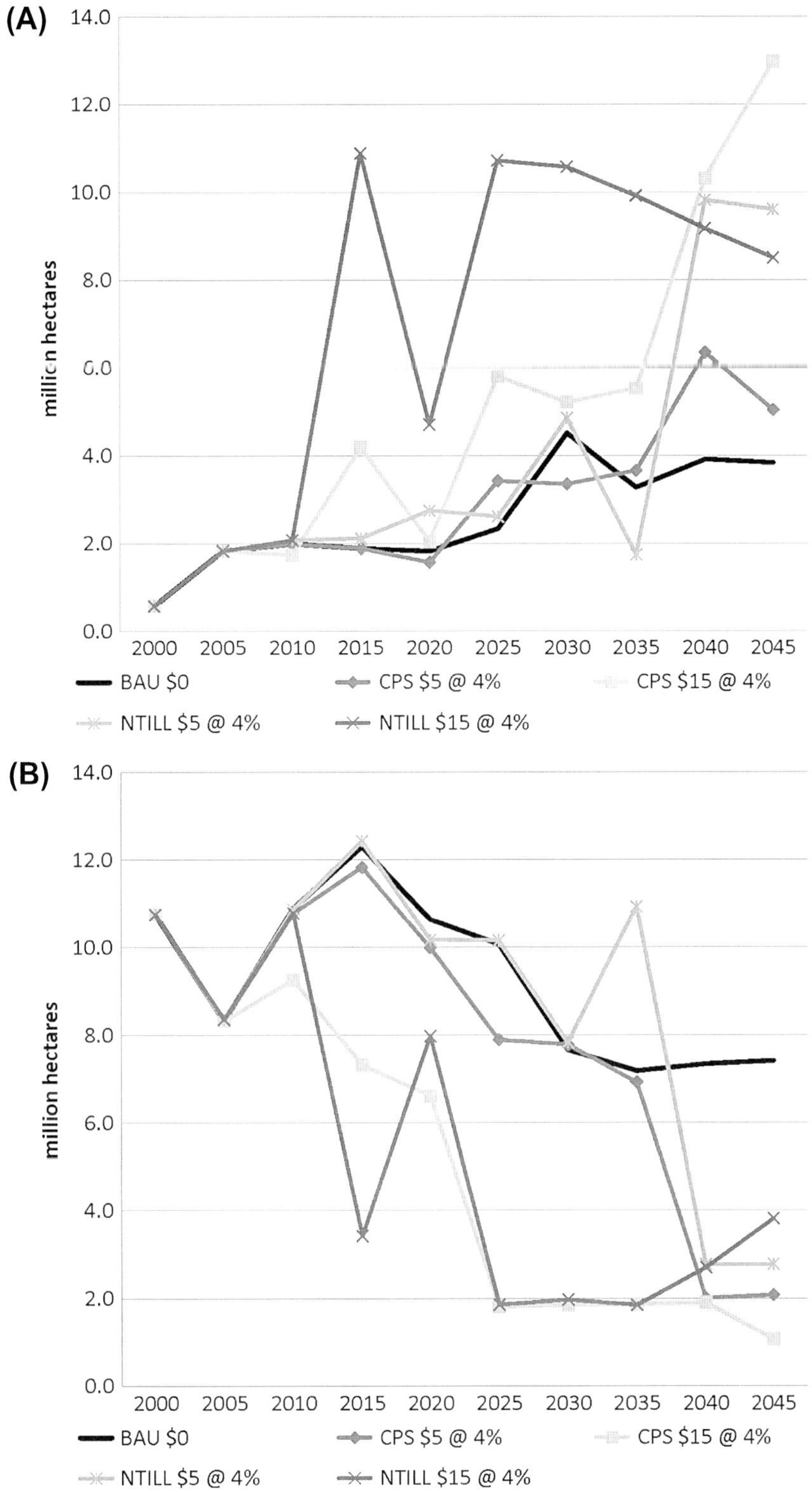

FIGURE 25.3
Simulated changes in corn and soybean land utilizing no-tillage production relative to business as usual (BAU) under two CO_2e prices (CPS) and with CO_2e prices and an additional payment for no-tillage and no land exchanges between agriculture and forestry (NTILL).

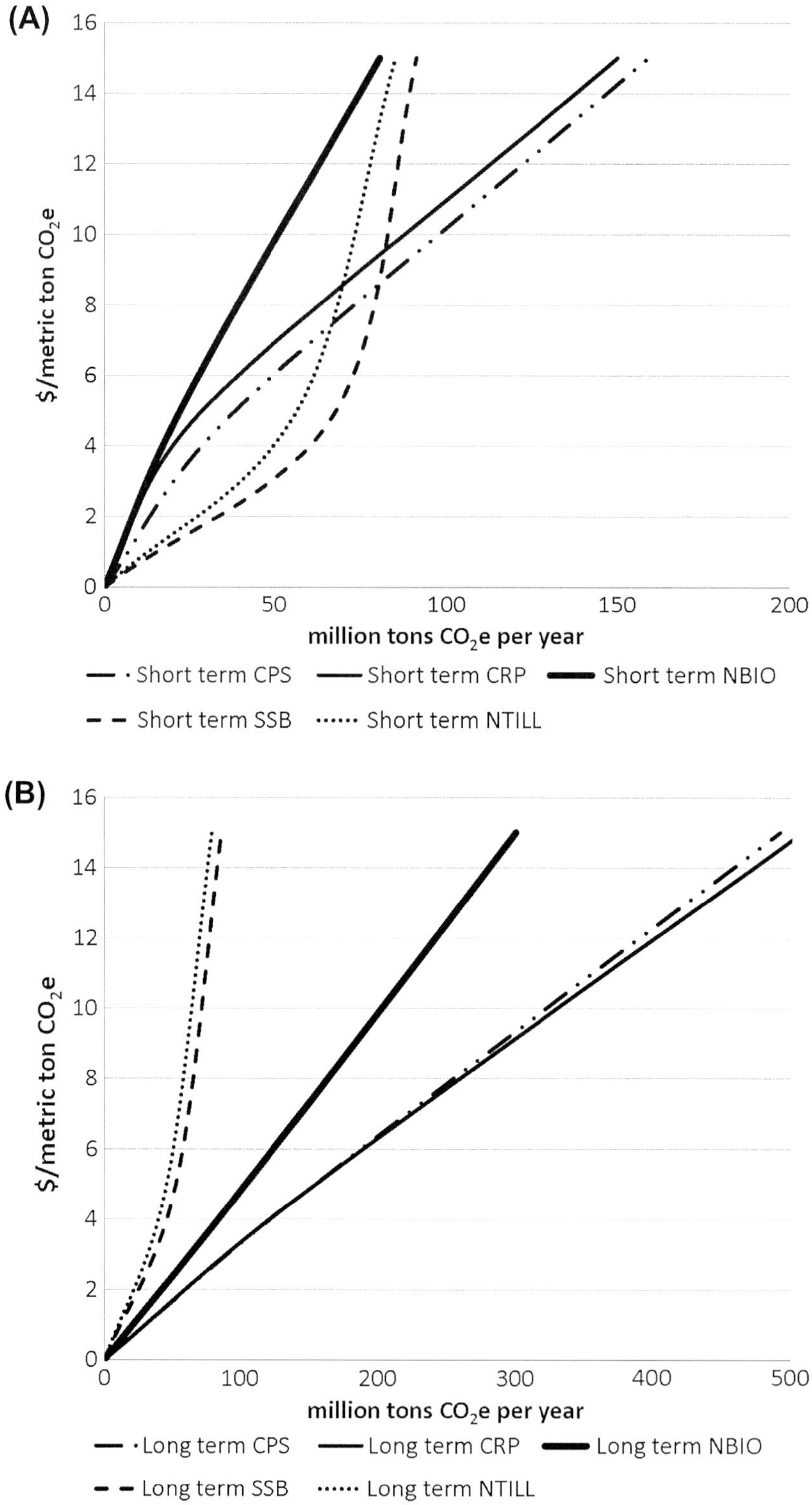

FIGURE 25.4
Mitigation supply curves for GHG mitigation for carbon price scenario (CPS), conservation reserve program reduction (CRP), bioenergy limitation (NBIO), sector specific (SSS), sector specific no bioenergy (SSB), and payment for tillage reduction (NTILL) policy options.

viewpoints on additionality, permanence, and verification of mitigation benefits. When modeled for U.S. agriculture and forests, more flexible program designs would generate greater mitigation at lower costs. Simulation analysis highlighted two activities that accounted for the majority of GHG mitigation potential from agriculture and forest sectors: using biomass feedstocks to produce electricity and sequestering C in woody material in afforested stands. Since those activities accounted for so much of the GHG mitigation potential, programs seeking to incentivize landowners to generate GHG reductions at low cost should consider including these cost-effective bioenergy and forest sector efforts.

As eligibility criteria are built into a GHG mitigation program, it should be recognized that the cost of potential mitigation for a given C price falls with time. This would occur for a number of reasons, most notably technological improvements, which is why most mitigation from policy scenarios in early years would be more costly than mitigation achieved later. The amount of mitigation supplied at a rising C price of $15 per t CO_2e was three times larger in the 2050 period relative to 2015 for the C price scenarios. In some cases that result did not hold. The most restrictive of mitigation programs for agriculture and forestry could exclude many potential activities or sectors and thereby inflate the overall cost per unit of GHG emissions reduced or C sequestered. Our analysis indicated that more restrictive policies would reduce supplies of available mitigation efforts at any given price (Figure 25.4). Under a program that would have greater flexibility, such as when CRP reversion to production were unchecked, total GHG mitigation might improve only slightly because more land would be under crop production. However, greater mitigation might be available under different programs for a given price in earlier than in later years. In the short term, policies that target reductions in certain sectors could generate cost-effective actions, but such policies might not be cost effective in the long run.

References

Adams, D.M., Alig, R.J., Latta, G.S., White, E.M., 2011. Regional impacts of a program for forest carbon offset sales. J. Forestry. 109 (8), 444–453.

Agricultural and Food Policy Center (AFPC), 2009. Economic Implications of the EPA Analysis of the CAP and Trade Provisions of H.R. 2454 for U.S. Representative Farms. AFPC Research Paper 09-2. Available online at http://www.farmland.org/documents/A-Comparison-of-Select-Cost-Benefit-Studies-HR2454-Impacts-On-Agriculture-Sector.pdf (last accessed October 1, 2010).

Agricultural Policy Analysis Center (APAC), 2009. Analysis of the implications of climate change and energy legislation to the agriculture sector. Available online at http://www.25x25.org/storage/25x25/documents/ut_climate_energy_report_25x25_november11.pdf (last accessed October 1, 2010).

Aigner, D., Hopkins, J., Johansson, R., 2003. Beyond compliance: sustainable business practices and the bottom line. Am. J. Agric. Econ 85 (5), 1126–1139.

Alig, R., Adams, D., McCarl, B., 1998. Impacts of incorporating land exchanges between forestry and agriculture in sector models. J. Agric. Appl. Econ. 30 (2), 389–401.

Alig, R., Latta, G., Adams, D., McCarl, B., 2010. Mitigating greenhouse gases: the importance of land base interactions between forests, agriculture, and residential development in the face of changes in bioenergy and carbon prices. Forest Policy Econ. 12 (1), 67–75.

Beach, R.H., Adams, D., Alig, R., Baker, J., Latta, G.S., McCarl, B.A., Murray, B.C., Rose, S.K., White, E., 2010. Model documentation for the Forest and Agricultural Sector Optimization Model with Greenhouse Gases (FASOMGHG). RTI International. Available online at http://www.cof.orst.edu/cof/fr/research/tamm/FASOMGHG_Model_Documentation_Aug2010.pdf (last accessed October 1, 2010).

Birdsey, R.A., 1996. Carbon storage for major forest types and regions in the conterminous United States. In: Hair, D., Sampson, Neil R. (Eds.), Forests and Global Change. Forest Management Opportunities for Mitigating Carbon Emissions, vol. II. American Forests, Washington, DC, pp. 1–25.

Chew, A., Nguyen, T., 2011. Number of cars equivalent to 100 metric tons of avoided greenhouse gases per year. DOE Hydrogen Program Record No. 11002. Available online at www.hydrogen.energy.gov/pdfs/11002_cars_equivalent_ghg.pdf (last accessed October 1, 2010).

Congressional Budget Office, 2009b. S. 1733 Clean Energy Jobs and American Power Act. Cost Estimate (December 16).

Del Grosso, S.J., Parton, W.J., Mosier, A.R., Hartman, M.D., Brenner, J., Ojima, D.S., Schimel, D.S., 2001. Simulated interaction of carbon dynamics and nitrogen trace gas fluxes using the DAYCENT model. In: Schaffer, M., Ma, L.,

Hansen, L.S. (Eds.), Modeling Carbon and Nitrogen Dynamics for Soil Management. CRC Press, Boca Raton, Florida, pp. 303–332.

Eagle, A., Henry, L.R., Olander, L.P., Haugen-Kozra, K., Millar, N., Robertson, G.P., 2010. Greenhouse gas mitigation potential of agricultural land management in the United States Nicholas Institute for Environmental Policy Solutions Report No. NI-R-10–04 (October).

Horowitz, J., Gottlieb, J., 2010. The role of agriculture in reducing greenhouse gas emissions USDA Economic Research Service, Economic Brief No. 15 (September).

Horowitz, J., Ebel, R., Ueda, K., 2010. No-till farming is a growing practice Economic Information Bulletin No. 70 (November).

Johnson, R., Ramseur, J.L., Gorte, R.W., 2010a. Estimates of carbon mitigation potential from agricultural and forestry activities Congressional Research Service Report No. R40236 (January).

Johnson, R., Ramseur, J.L., Gorte, R.W., Stubbs, M., 2010b. Potential implications of a carbon offset program to farmers and landowners Congressional Research Service Report No. R41086 (February 26).

Parker, L., Yacobucci, B.D., 2009. Climate change: costs and benefits of cap-and-trade provisions of H.R. 2454 Congressional Research Service Report No. R40809 (September 14).

Parton, W.J., 1996. The CENTURY Model. In: Powlson, D.S., Smith, P., Smith, J.U. (Eds.), Evaluation of Soil Organic Matter Models Using Existing Long-Term Datasets. Springer-Verlag, Berlin, Germany, pp. 283–293.

Ramseur, J.L., 2010. Estimating offset supply in a cap-and-trade program Congressional Research Service Report No. RL 34705 (April 16).

Searchinger, T.D., 2010. Biofuels and the need for additional carbon. Environ. Res. Lett. 5 (2010) 024007.

Skog, K., Nicholson, G., 2000. Carbon sequestration in wood and paper products. In: Joyce, L., Birdsey, R. (Eds.), The Impact of Climate Change on American Forests. USDA Forest Service, General Technical Report RMRS-GTR-59. pp. 79–88.

Smith, J., Heath, L.S., 2002. A Model of Forest Floor Carbon Mass for United States Forest Types. USDA Forest Service, Newton Square, PA.

Smith, J.E., Heath, L.S., Skog, K.E., Birdsey, R.A., 2006. Methods for calculating forest ecosystem and harvested carbon with standard estimates for forest types of the United States. Gen. Tech. Rep., NE-343. Newtown Square, PA: U.S. Department of Agriculture, Forest Service, Northeastern Research Station. p. 216.

Smith, P., Martino, D., Cai, Z., Gwary, D., Janzen, H., Kumar, P., McCarl, B., Ogle, S., O'Mara, F., Rice, C., Scholes, B., Sirotenko, O., Howden, M., McAllister, T., Pan, G., Romanenkov, V., Schneider, U., Towprayoon, S., Wattenbach, M., Smith, J., 2008. Greenhouse gas mitigation in agriculture. Phil. Trans. Royal Soc. B: Biological Sciences 363 (1492), 789–813.

Sperow, M., Eve, M., Paustian, K., 2003. Potential soil C sequestration on U.S. agricultural soils. Climatic Change 57, 319–339.

U.S. Department of Agriculture, Economic Research Service, 2005. Agricultural Chemicals and Production Technology: Soil Management Briefing Room. Available online at http://www.ers.usda.gov/briefing/agchemicals/soilmangement.htm#croplandacreage (last accessed October 1, 2011).

U.S. Energy Information Administration, 2010. Annual Energy Outlook 2010 Report No. DOE/EIA-0383–2010 (May 11).

U.S. Environmental Protection Agency (EPA), 2003. Inventory of U.S. Greenhouse Gas Emissions and Sinks: 1990–2001, All Annexes EPA 430-R-03-004. Available online at http://www.epa.gov/climatechange/emissions/downloads/US_GHG_Inv_Annexes_1990-2001.pdf (last accessed October 1, 2011).

U.S. Environmental Protection Agency (EPA), 2003. Inventory of U.S. Greenhouse Gas Emissions and Sinks: 1990–2001, Annex O.

U.S. EPA, 2009. EPA Analysis of the American Clean Energy and Security Act of 2009 H.R. 2454 in the 111th Congress June 23. Available online at http://www.epa.gov/climatechange/economics/pdfs/HR2454_Analysis.pdf (last accessed October 1, 2010).

U.S. EPA, 2010. EPA Analysis of the American Power Act in the 111th Congress (June 14), Available online at http://www.epa.gov/climatechange/economics/pdfs/EPA_APA_Analysis_6-14-10.pdf (last accessed October 1, 2010).

United States Government (USG), 2010. Technical Support Document: Social Cost of Carbon for Regulatory Impact Analysis under Executive Order 12866. The Interagency Working Group on Social Cost of Carbon (February).

SECTION 7

Looking Ahead: Opportunities for Future Research and Collaboration

CHAPTER 26

Potential GRACEnet Linkages with Other Greenhouse Gas and Soil Carbon Research and Monitoring Programs

John M. Baker[1], Ronald F. Follett[2]
[1]USDA Agricultural Research Service, Soil & Water Management Unit, St. Paul MN
[2]USDA Agricultural Research Service, Soil Plant Nutrient Research Unit, Ft. Collins CO

CHAPTER OUTLINE

Abbreviations: GHG, greenhouse gas; GRA, Global Research Alliance

INTRODUCTION

One of the original goals of the GRACEnet program was to coordinate ARS research on the potential role of agricultural practices in the mitigation of global climate change (Walthall

Managing Agricultural Greenhouse Gases. DOI: 10.1016/B978-0-12-386897-8.00026-7
2012, Published by Elsevier Inc.

et al., 2012). Because relevant research was already underway at a number of locations, the GRACEnet organizers believed that synergistic benefits would result if these scientists became part of a group that met regularly, planned research in a coordinated way, and cooperated in the sharing and dissemination of data. The same logic has motivated the formation of other groups of scientists who conduct climate change-related research, and also argues for development of higher-level connections among these various climate science collectives—groups of groups, so to speak. In some cases connections already exist through the multiple affiliations of individual scientists, but too often organizations with similar interests labor along in ignorance of each other's efforts. This chapter is intended to improve awareness of group and network activities that are relevant to the interests and goals of GRACEnet.

The subject areas, carbon sequestration and greenhouse gas (GHG) mitigation, can particularly benefit from coordinated research activity among multiple scientists, for a number of reasons. Perhaps the most obvious is the complexity of the topic. It is simply not possible for one scientist, one laboratory, or one university to develop the expertise necessary to unravel all aspects of soil carbon and GHG dynamics. This is due in part to regional differences in the key drivers of these processes—vegetation type, management practices, weather, climate, and soil—all of which can vary tremendously from one ecoregion to another. Taking the example of N_2O emission, the factors affecting episodic release from northern farm fields during spring thaw (Wagner-Riddle et al., 2008) may be quite different from those that control emission in tropical forests (Kiese and Butterbach-Bahl, 2002). And yet, while the methodologies and skill sets required to conduct research in these two cases may be quite different, there are almost certainly sufficient commonalities that the groups working on each will benefit from sharing their ideas and results in a coordinated way. The long-term nature of soil carbon and GHG research is another factor that argues for a coordinated, networked approach. Management-induced changes in soil carbon can be substantial, but they are spatially variable, susceptible to inter-annual weather variability, and in many locations must be detected against a large background signal. Consequently, treatment comparisons often must be conducted over periods of 5 years or more to obtain significant, meaningful results. Organized networks can play a crucial role in this regard, not only through data archival, but also through archival of soil samples, and by providing a framework for collaborative continuity as key scientists retire and younger counterparts join the field. Other benefits that can flow from such activity are standardization of methods and systematic archival of data. The latter provides a platform and resource for the development, testing, and application of models, which are critical tools for predicting the impact of human activities on climate change, and informing policy decisions to moderate those impacts. Finally, there is the value of redundancy. Networks facilitate the systematic testing of theories by multiple researchers, a key component in the progression of scientific understanding.

Presentation of the various networks with relevance to GRACEnet has been organized according to their primary focus: soil-based networks, GHG networks, and general ecological networks. These distinctions are necessarily somewhat arbitrary, so in some cases the fit is imperfect. For convenience, basic information for all listed networks is summarized in Table 26.1.

SOIL-BASED NETWORKS

National Soil Carbon Network

This group was organized by an ad hoc steering committee that includes members from the U.S. Forest Service, USDA-NRCS, USDA-ARS, USGS, EPA, Lawrence Berkeley Laboratory, a number of universities, and the private sector. Though relatively new, the NSCN (http://www.fluxdata.org/nscn/SitePages/Home.aspx) has obtained sufficient funding to hire a coordinator, sponsored symposia at National Society meetings, and is currently spearheading an effort to develop a national archive of soil samples that would serve as a resource for carbon cycle scientists and facilitate regional- and continental-scale synthesis studies. A related focus

TABLE 26.1 Summary of Networks with Relevance to Soil Carbon and Greenhouse Gas Research

Network	Primary location	Primary focus	Currently active?	Data available?	Website
National Soil Carbon Network	U.S.	SOC	Y	Y	http://www.fluxdata.org/nscn/SitePages/Home.aspx
LASCANET	Latin America	SOC	N	N	http://cirit.osu.edu/clusterone/LASCANET/
European Soil Bureau	Europe	SOC	Y	Y	http://eusoils.jrc.ec.europa.eu/
Tragnet	U.S.	GHG	N	Y	http://www.nrel.colostate.edu/projects/tragnet/
N_2O Network	Australia	N_2O	Y	Y	http://www.n2o.net.au/index.jsp
Global Research Alliance	Global	GHG	Y	TBD	http://www.globalresearchalliance.org/
CarboEurope	Europe	GHG & SOC	N	Y	http://www.carboeurope.org/
FLUXNET-Canada	Canada	GHG & SOC	N	Y	http://www.fluxnet-canada.ca/home.php
Green Crop Network	Canada	GHG & SOC	N	N	www.greencropnetwork.com
CarboAfrica	Africa	GHG & SOC	N	Y	http://www.carboafrica.net/index_en.asp
AsiaFlux	Asia	GHG & SOC	Y	Y	http://www.asiaflux.net/index.html
ChinaFlux	China	GHG & SOC	Y	Y	http://www.chinaflux.org/en/index/index.asp
Ozflux	Australia and NZ	GHG	Y	Y	http://www.ozflux.org.au/index.html
Ameriflux	U.S.	GHG	Y	Y	http://public.ornl.gov/ameriflux/index.html
FLUXNET	Global	GHG	Y	Y	http://www.fluxnet.ornl.gov/fluxnet/index.cfm
BASIN	Global	Stable isotopes	Y	Y	http://basinisotopes.org.
LTER	U.S.	General ecology	Y	Y	http://www.lternet.edu/
NEON	U.S.	General ecology	Y	TBD	http://www.neoninc.org/science/overview

has been the development of a database and a fair use policy for data. In this effort they are following the approach used by the USGS research team at Menlo Park, CA, with soil carbon data from Alaska (http://carbon.wr.usgs.gov/deepsoil.html).

Latin American Soil Carbon Network (LASCANet)

This group was organized at a workshop in Brazil in 2004 (Lal et al., 2006). The Ohio State University was a primary organizer, with sponsorship from the U.S. State Department, USDA, and a broad range of Central and South American government agencies. The group, which included participants from Costa Rica, Colombia, Brazil, Argentina, Mexico, France, and the U.S., developed the following mission statement:

> The LASCANet is developing a research, extension, outreach and training program in Latin America on soil carbon sequestration to improve soil quality, increase agronomic productivity, decrease non-point source pollution, and reduce the rate of enrichment of atmospheric concentration of CO_2. The LASCANet facilitates exchange of scientists and students, collates and disseminates scientific information, creates awareness about the importance of soil carbon sequestration to

> sustainable management of natural resources, liaises with land managers and policy makers, and establishes cross-linkages among stakeholders in the Latin America region. The LASCANet is identifying ways to minimize carbon loss and maximize retaining carbon in land, reducing the effects of CO_2 and other trace gases (CH_4, N_2O, NO_x) on global climate change, and improving environment quality.

The group has not met since the original workshop, and is not currently active, though a website is still online (http://cirit.osu.edu/clusterone/LASCANET/).

European Soil Bureau Network

The European Soil Bureau Network (http://eusoils.jrc.ec.europa.eu/) was created in 1996 as an umbrella organization of European national soil science institutions. It coordinates a broad array of activities related to all phases of soil science within the European Union, and is headquartered at the Joint Research Centre of the European Community in Ispra, Italy. Of specific relevance to GRACEnet is the mapping of soil organic carbon in the surface horizon in Europe, data that are freely available in raster format through their web portal at http://eusoils.jrc.ec.europa.eu/ESDB_Archive/octop/octop_data.html. The ESBN has also played an active role in biochar research, through the co-sponsorship of symposia and a comprehensive literature review (Verheijen et al., 2010).

GREENHOUSE GAS NETWORKS

Tragnet

The United States Trace Gas Network (Ojima et al., 2000) was established to provide a repository for CO_2, CH_4, and N_2O flux data in important ecosystems, to improve understanding of the factors controlling those fluxes, and to improve prediction of future fluxes. While data are no longer being added, the database has been maintained, and can be accessed at http://www.nrel.colostate.edu/projects/tragnet/.

N_2O Network

This is an Australian research program (http://www.n2o.net.au/index.jsp) created to study and document N_2O emissions from agricultural soils in Australia. It provides a repository and source of data collected with automated chambers in a variety of agricultural settings. The data are publicly available (http://isrrepository.n2o.net.au/knb/) and other researchers are encouraged to take part. A major goal is the development of N_2O emissions models that can be useful in the development of national policies.

Green Crop Network

The Green Crop Network (www.greencropnetwork.com) is a Canadian nationwide network of scientists engaged in research on GHG management in agricultural systems, headquartered in the Plant Science Department at McGill University. Dr. Donald Smith is the Scientific Director, assisted by a board of directors with representation from universities, industry and government, and a Science Advisory Board. The Network has received sponsorship from a variety of university, industry, and government sources, and has in turn funded a range of research activities, with extensive involvement of graduate students. The network was organized around four theme areas: N_2O emissions, soil carbon stocks, plant response to rising CO_2, and biofuel crops. The network has conducted annual workshops where research findings are disseminated, and has funded inter-project travel to allow scientists to collaborate across locations. However, funding for the project has recently expired, so the future of the Network is uncertain (Henry Janzen, Xiaomin Zhou, personal communications).

CarboEurope

CarboEurope (http://www.carboeurope.org/) was a massive 5-year project funded by the European Union in 2004, building on previous EU projects dating from 2000, with the charge of answering the question "What is the role of the European continent in the global carbon cycle?" It included 75 primary partner institutions and 30 associate partner institutions, governed by a "Scientific Management" executive board, assisted by an advisory panel. Much attention was paid to both measurement protocols and data management, and a comprehensive data policy was developed. A primary goal throughout the program was "to apply a single comprehensive experimental strategy...[integrated] into a comprehensive carbon data assimilation framework...[to] allow for the first time a consistent match of bottom-up and top-down estimates of the regional variation in carbon sources and sinks." The project has generated at least 21 theses and dissertations, and hundreds of peer-reviewed scientific publications, in addition to numerous technical reports and policy papers. It officially concluded at the end of 2008, but two successors have begun, albeit with a somewhat different focus: GHG Europe and CarboExtreme (Angelika Thuille, personal communication). The former, http://www.ghg-europe.eu/, is similar in many respects to CarboEurope (Annette Freibauer, personal communication). It links together sites throughout Europe where flux measurements and GHG-related research are conducted, with the following scientific objectives:

1. Quantify the annual to decadal variability of CO_2, N_2O, and CH_4 budgets for European ecosystems.
2. Obtain a better and more comprehensive understanding of the terrestrial carbon cycle and responses of GHG fluxes to variability in natural and anthropogenic drivers.
3. Identify the carbon pools and GHG processes most sensitive and vulnerable to changes.
4. Assess the impact of possible strategies/policies on future carbon pools and GHG fluxes in Europe.

CarboExtreme, as its name implies, is primarily concerned with weather extremes and climate variability, although the response of soil carbon to increasing temperature is also a key interest. Among the sponsored projects are an ecosystem manipulation experiment at 14 European sites along a latitudinal gradient from Italy to northern Norway and a comparison of generic ecosystem models. More information is available from the network website (http://www.carbo-extreme.eu/index.php?n=Main.HomePage).

CarboAfrica

As one might suspect from the name, this program is in some respects an outgrowth of CarboEurope; many of the investigators involved with it are also affiliated with CarboEurope, and it was funded by a grant from the European Union. It is coordinated by Riccardo Valentini at the University of Tuscia in Viterbo, Italy. The original stated goal of the project was the quantification, understanding, and prediction of the carbon cycle and other GHGs in sub-Saharan Africa (Bombelli et al., 2009), and it was organized around six thematic areas: long-term observation, ecosystem process understanding, model—data integration, fires, capacity building, and clean development mechanisms. The project began in 2006, and though it officially terminated in 2010 (Antonio Bombelli, personal communication), several sites are still operating, as is the database (http://gaia.agraria.unitus.it/newtcdc2/CarboAfrica_home.aspx). The network website (http://www.carboafrica.net/index_en.asp) contains a list of relevant publications and links.

AsiaFlux

Asiaflux was established in 1999 and remains active today (Kim, 2009; Mizoguchi, et al., 2009). Its primary purpose has been to develop collaborative research and data sets on the carbon, water, and energy cycles in key Asian ecosystems. Database development has been a critical component of the project, and contributed data are made available to registered users through AsiaFluxDB (http://www.asiaflux.net/datapolicy.html). AsiaFlux includes more than 80 sites, encompassing a broad range of land covers and land uses, ranging from Siberian tundra to tropical forests. The network conducts regularly scheduled workshops and training sessions, and has generated a substantial collection of network-associated publications that can be found on their website (http://www.asiaflux.net/index.html).

ChinaFlux

This is a long-term flux measurement network that is affiliated with the Chinese Ecosystem Research Network (Yu et al., 2006; Leuning et al., 2006). It uses eddy covariance and chamber systems to measure exchanges of CO_2, H_2O, and energy throughout China, covering a broad range of vegetation types. There are currently 24 sites; the characteristics of each are described on the network website (http://www.chinaflux.org/en/index/index.asp). Data are quality controlled and archived in a central database. The network regularly conducts science meetings and training courses, and is a participant in FLUXNET.

OzFlux

OzFlux (http://www.ozflux.org.au/index.html) is a network of sites in Australia and New Zealand encompassing a range of vegetation types, where fluxes of CO_2, water, and energy are measured. The goals of the network are to: understand mechanisms controlling exchanges of CO_2, H_2O, and energy in terrestrial ecosystems; provide data on carbon and water balances for ecosystem testing; provide data for validation of remotely sensed estimates of net primary productivity, evaporation, and energy absorption; validate new developments in micrometeorological theory; and provide high precision CO_2 concentration measurements for inverse studies of the carbon cycle. Ultimately the network will include 21 sites; currently 15 are operational. Data are contributed to the Global FLUXNET site.

FLUXNET-Canada/Canadian Carbon Program

FLUXNET-Canada and its successor the Canadian Carbon Program (CCP) were coordinated nationwide research efforts to quantify carbon and GHG budgets at multiple time scales across Canada (Margolis et al., 2006). The system has a board of directors, a network management office, and a program director, Dr. Hank Margolis, at Laval University. Ecosystem fluxes of CO_2, H_2O, and sensible heat have been measured continuously at multiple sites representative of all major ecosystems in the country, and both CH_4 and N_2O fluxes have also been measured at a subset of sites. A broad range of additional physical and physiological data has been collected, and the data have been archived and made available through the FLUXNET-Canada Data Information Center, under the direction of Dr. Alan Barr of Environment Canada, MSC Climate Research Branch. These data have supported modeling efforts that were a key component of the network goals and objectives (e.g. Grant et al., 2010; Ju et al., 2010). An extensive list of publications and products can be accessed from the network website (http://www.fluxnet-canada.ca/home.php). This monitoring program, which commenced operation in 2003, ceased to exist in September 2011, although some sites will continue to function on an individual basis if the respective investigators are successful in funding them from alternate sources (Hank Margolis, personal communication).

Ameriflux

The Ameriflux network (http://public.ornl.gov/ameriflux/index.html) has been in existence since 1996 (Kaiser, 1998). It is funded by five different federal agencies under the aegis of the U.S. Climate Change Science Program. Overall objectives of the network are to: (1) quantify carbon storage and exchanges of carbon, water, and energy in major ecosystems subjected to a range of climatic conditions and disturbances; (2) advance understanding of carbon assimilation, respiration, and storage through measurements and modeling; and (3) produce high quality data for analyses, synthesis activities, and data archival. Participants agree to adhere to measurement protocols and timely submission of data to the central data archive, which is maintained by personnel at Oak Ridge National Laboratory. A key feature of the network is its QA/QC laboratory, situated at Oregon State University, which maintains a roving measurement system that travels to the individual sites to allow comparison of measurement systems. The Ameriflux network does not fund the individual sites; rather, each site has its own funding source(s), although most sites have been funded by one or more of the participating agencies. An annual science meeting is held, and a number of training sessions and workshops have also been sponsored. The data archived by the Ameriflux network is readily available under fair use standards, and has been widely used for a broad range of synthesis activities (e.g. Xiao et al., 2008; Law, 2005).

FLUXNET

FLUXNET is a global "network of networks" that links many of the previously discussed programs (Baldocchi et al., 2001). It provides a common repository for archiving flux data from sites throughout the world and for distributing it to users. It also provides a mechanism for calibration and intercomparison to facilitate use of data from different networks in modeling and data synthesis activities. Currently there are 24 different networks that contribute data to FLUXNET, which relies on the individual networks or measurement locations for QA/QC, gap filling, and other preprocessing. FLUXNET is maintained by the Oak Ridge National Laboratory Distributed Active Archive Center (DAAC) at http://www.fluxnet.ornl.gov/fluxnet/index.cfm.

FLUXNET also provides links to ancillary data that are frequently useful in synthesis and modeling efforts, including MODIS data, maps, and vegetation data.

The Global Research Alliance on Agricultural Greenhouse Gases

The Global Research Alliance (GRA) was established in December 2009 at a climate change conference in Copenhagen. Conceived by scientists and leaders in New Zealand earlier in 2009, the vision for an international alliance of researchers focusing on mitigating GHG emissions from agricultural systems gathered support, culminating in an event during the Copenhagen meetings where agricultural and environmental ministers from 21 countries (including U.S. Agriculture Secretary Thomas Vilsack) endorsed a ministerial statement establishing the GRA. The GRA on Agricultural Greenhouse Gases brings together more than 30 governments to pool funding allocations and coordinate research. The alliance's goal is to improve the agricultural system's ability to grow more food without increasing GHG emissions.

The GRA will seek to increase international cooperation, collaboration, and investment in both public and private research activities to:

- Understand GHG emissions from agriculture
- Improve measurement and estimation of GHG emissions and carbon sequestration
- Develop ways to reduce emissions
- Develop ways to increase C sequestration
- Mitigate GHGs while sustaining or enhancing productivity and resilience as climate changes
- Transfer new knowledge and technology to farmers/land managers worldwide
- Build scientific capacity in developing countries via partnerships.

An organizational meeting was held in Wellington, NZ, in early April 2010 and subsequent meetings have been held in San Diego, CA, U.S.; Paris, FR; and Rome, Italy. Thus far, a charter to guide the GRA's work and operations has been drafted, organizing research groups have met, research needs and priorities have been identified, and a partnership network of collaborating non-government organizations has been established.

Each country's participation is based on current and developing interests and existing resources and capacity. As a whole, the GRA will encourage and seek opportunities to expand funding for research and technology transfer. Scientific exchanges and partnerships among participating countries are intended to leverage resources and expertise in the world's scientific community working on mitigating GHG emissions from agricultural systems. Research groups have been established around three major categories of agricultural production systems, and countries were delegated coordination roles for these groups: croplands, including agroforestry (coordinated by the United States); paddy rice (Japan); and livestock systems (New Zealand and the Netherlands). GRACEnet is a key component of the U.S. contribution to the GRA. In addition, two key cross-cutting issues were also identified: soil carbon and nitrogen (France and Australia), and inventories and measurement (Canada and the Netherlands).

The responsibilities of the coordinating country or countries for each research group or cross-cutting issue include:

- Organize meetings and information exchanges to identify research gaps and needs
- Coordinate the development of research plans
- Facilitate identification of appropriate partnerships
- Disseminate information among collaborators.

A secretariat has been established to guide overall operations of the GRA, with New Zealand providing the staffing and support for the first year, which is responsible for:

- Reporting and meeting documentation
- Providing logistical support for research groups, including collating and analyzing cross-group information
- Maintaining a website and contact database
- Managing Alliance-level outreach to partners and other entities.

GENERAL ECOLOGICAL NETWORKS

BASIN

A complementary effort to FLUXNET is the Biosphere-Atmosphere Stable Isotope Network, BASIN, which coordinates stable isotope research related to the carbon and water cycles. The network has received funding over the past ten years from the National Science Foundation and the European Science Foundation. Data are archived in a database that is publicly available at http://basinisotopes.org. The network has sponsored symposia at the American Geophysical Union meetings, as well as a number of stand-alone, isotope-specific conferences.

LTER

The U.S. Long Term Ecological Research Network, LTER, has been in existence for nearly 30 years (Robertson, 2008). It is currently comprised of 26 sites that represent a wide range of biome types, from urban locales to a tundra site on the north slope of Alaska. The stated mission of the LTER network is "To provide the scientific community, policy makers, and society with the knowledge and predictive understanding necessary to conserve, protect, and manage the nation's ecosystems, their biodiversity, and the services they provide" (http://www.lternet.edu/). Historically, there has been more emphasis on natural rather than managed (agricultural) systems, although Michigan State University's Kellogg Biological Station has been a network member since 1987, contributing a broad range of agriculturally-based research. LTER is governed by a chair and an executive board, advised by a science council.

A network office, funded by the National Science Foundation, provides administrative support. There are five research themes at the core of LTER:

1. Pattern and control of primary production.
2. Spatial and temporal distribution of populations selected to represent trophic structure.
3. Pattern and control of organic matter accumulation in surface layers and sediments.
4. Patterns of inorganic inputs and movements of nutrients through soils, groundwater, and surface waters.
5. Patterns and frequency of site disturbances.

NEON

The National Ecological Observatory Network (NEON) is a massive NSF funded effort to develop a coordinated network of core sites in each major *ecoclimatic zone* across the entire United States, including Hawaii, Alaska, and Puerto Rico (Keller et al., 2008). The number of such regions has been set at 20, although the establishment of the individual sites will take place incrementally over a number of years. Each site will have its own administrator and technical staff. The permanent core sites will be supplemented by re-locatable sites that will be moved on a 3- to 5-year time scale to address specific questions and to attempt to capture ecosystem variability within each ecoregion. Airborne remote sensing platforms will also be employed. A central organizational structure is already in place, with a permanent staff and headquarters in Boulder, CO. The planned scope and intended duration of investigation are unprecedented:

> In those domains, NEON will collect site-based data about climate and atmosphere, soils and streams and ponds, and a variety of organisms. Additionally, NEON will provide a wealth of regional and national-scale data from airborne observations and geographical data collected by Federal agencies and processed by NEON to be accessible and useful to the ecological research community. NEON will also manage a long-term multi-site stream experiment and provide a platform for future observations and experiments proposed by the scientific community. The data collected and generated across NEON's network—all day, every day, over a period of 30 years—will be synthesized into information products that can be used to describe changes in the nation's ecosystem through space and time. It will be readily available in many formats to scientists, educators, students, decision makers and the general public. (http://www.neoninc.org/science/overview)

NEON is not a funding agency; rather, it will provide basic data and infrastructure to support research at its sites funded through conventional granting agencies and other sources. Data collected in NEON research projects will be publicly available, and a database development staff is already in place.

CONCLUDING REMARKS

One of the lessons learned from a review of carbon and GHG-related networks is that, as with organisms, conceiving new offspring is more exciting than tending to those already present. And yet, to carry the analogy further, the latter activity generally produces more lasting rewards. Henry Janzen (2009) has eloquently made the case for the importance of long-term research sites—"listening places" as he calls them. Baldocchi (2008) has shown how valuable the data from these networks are in detecting responses to climatic change, and in deciphering the role of land management practices, both in causing these changes and in responding to them. However, the forces that drive decisions on funding, promotion, and tenure emphasize short-term, hypothesis-driven research. Particularly in the agricultural sciences, proposals that carry even a scent of monitoring tend to fare poorly in the peer review process. Ecologists have been more effective than agricultural scientists in communicating to funding sources the value of multi-site, long-term networks. Perhaps the best strategy for the future is to build linkages that will increase the representation of managed lands, and the participation of agricultural scientists, in these broader, interdisciplinary programs.

References

Baldocchi, D.D., 26 others, 2001. FLUXNET: a new tool to study the temporal and spatial variability of ecosystem-scale carbon dioxide, water vapor, and energy flux densities. Bull. Amer. Meteorol. Soc. 82, 2415–2434.

Baldocchi, D.D., 2008. "Breathing" of the terrestrial biosphere: lessons learned from a global network of carbon dioxide flux measurement systems. Aust. J. Bot. 56, 1–26.

Bombelli, A., Henry, M., Castaldi, S., Adu-Bredu, S., Arneth, A., de Grandcourt, A., Grieco, E., Kutsch, W.L., Lehsten, V., Rasile, A., and others, 2009. An outlook on the Sub-Saharan Africa carbon balance RID A-7494-2011. Biogeosciences. 6, 2193–2205.

Grant, R.F., Black, T.A., Jassal, R.S., Bruemmer, C., 2010. Changes in net ecosystem productivity and greenhouse gas exchange with fertilization of Douglas fir: mathematical modeling in ecosys. J. Geophys. Res. 115, G04009. doi:10.1029/2009JG001094.

Janzen, H.H., 2009. Long-term ecological sites: musings on the future, as seen (dimly) from the past. Glob. Change Biol. 15, 2770–2778.

Ju, W., Chen, J.M., Black, T.A., Barr, A.G., McCaughey, H., Roulet, N.T., 2010. Spatially explicit modelling of coupled water and carbon cycles in Canada's forests and wetlands. Tellus 62 (3), 140–159.

Kaiser, J., 1998. New network aims to take the world's CO_2 pulse. Science 281, 506–507.

Keller, M., Schimel, D.S., Hargrove, W.H., Hoffman, F.M., 2008. A continental strategy for the national ecological observatory network. Frontiers Ecol. Environ. 6, 282–284.

Kiese, R., Butterbach-Bahl, K., 2002. N_2O and CO_2 emissions from three different tropical forest sites in the wet tropics of Queensland, Australia. 2002. Soil Biol. Biochem. 34, 975–987.

Kim, J., Guiui, Y., Miyata, A., 2009. AsiaFlux—sustaining ecosystems and people through resilience thinking. In: Climate Sense, WMO. Tudor Rose, Leicester, U.K, pp. 165–168.

Lal, R., Cerri, C.C., Bernoux, M., Etchevers, J., Cerri, E. (Eds.), 2006. Carbon Sequestration in Soils of Latin America. Food Products Press, NY, p. 553.

Law, B.E., 2005. Carbon dynamics in response to climate and disturbance: recent progress from multi-scale measurements and modeling in Ameriflux. In: Omasa, K., Nouchi, I., Kok, L.J. (Eds.), Plant Response to Air Pollution and Global Change. Springer, Japan, pp. 205–213.

Leuning, R., G- Yu, R., 2006. Carbon exchange research in ChinaFLUX. 2006. Agric. Forest Meteorol. 137, 123–124.

Margolis, H.A., Flanagan, L.B., Amiro, B.D., 2006. The Fluxnet-Canada research network: influence of climate and disturbance on carbon cycling in forests and peatlands. Agric. Forest Meteorol. 140, 1–5.

Mizoguchi, Y., Miyata, A., Ohtani, Y., Hirata, R., Yuta, S., 2009. A review of tower flux observation sites in Asia. J. Forest Res. 14, 1–9.

Ojima, D., Mosier, A., Del Grosso, S., Parton, W., 2000. TRAGNET analysis and synthesis of trace gas fluxes. Glob. Biogeochem. Cycles. 14, 995–997.

Robertson, G.P., 2008. Long-term ecological research: reinventing network science. Frontiers Ecol. Environ. 6, 281.

Verheijen, F., Jeffery, S., Bastos, A.C., van der Velde, M., Diafas, I., 2010. Biochar application to soils: a critical scientific review of effects on soil properties, processes and functions. JRC Tech. Rept. 55799 EU, Luxembourg.

Wagner-Riddle, C., Hu, Q.C., van Buchove, E., Jayasundara, S., 2008. Linking nitrous oxide flux during spring thaw to nitrate denitrification in the soil profile. Soil Sci. Soc. Am. J. 72, 908–916.

Walthall, C.L., Shafer, S.R., Jawson, M.D., 2012. Clear objectives, communication, organization: elements for successful science. In: Liebig, M.A., Franzluebbers, A.J., Follett, R.F. (Eds.), Managing Agricultural Greenhouse Gases: Coordinated Agricultural Research through GRACEnet to Address our Changing Climate. Academic Press, San Diego, CA.

Xiao, J., 39 others, 2008. Estimation of the net ecosystem exchange for the conterminous United States by combining MODIS and Ameriflux data. Agric. Forest Meteorol. 148, 1827–1847.

Yu, G.-R., Wen, X.-F., X-Sun, M., Tanner, B.D., Lee, X., J- Chen, Y., 2006. Overview of ChinaFLUX and evaluation of its eddy covariance measurement. Agric. Forest Meteorol. 137, 125–137.

CHAPTER 27

Elevated CO_2 and Warming Effects on Soil Carbon Sequestration and Greenhouse Gas Exchange in Agroecosystems: A Review

Feike A. Dijkstra[1], Jack A. Morgan[2]
[1]Faculty of Agriculture, Food and Natural Resources, the University of Sydney, Eveleigh NSW, Australia
[2]USDA Agricultural Research Service, Rangeland Resources Research Unit, Fort Collins, CO

CHAPTER OUTLINE

Abbreviations: C, carbon; CH_4, methane; CO_2, carbon dioxide; eCO_2, elevated atmospheric CO_2 concentration; FACE, free air CO_2 enrichment; GHG, greenhouse gas; HC, heating cable; IPCC, Intergovernmental Panel on Climate Change; IRH, infrared heater; MAP, mean annual precipitation; MAT, mean annual temperature; N, nitrogen; N_2O, nitrous oxide; OTC, open top chamber; PT, plastic tunnel; SACC, screen-aided CO_2 control

Concentrations of CO_2 and other greenhouse gases (GHGs) have been increasing dramatically in earth's atmosphere since the industrial revolution, and are expected to continue increasing from ~385 ppmv today to more than 600 ppmv by the end of this century (IPCC, 2007). Global surface temperatures are expected to rise between 1.1 to 5.4°C by 2100, depending on how fast greenhouse gas concentrations increase. Precipitation dynamics are also predicted to

Managing Agricultural Greenhouse Gases. DOI: 10.1016/B978-0-12-386897-8.00027-9
2012, Published by Elsevier Inc.

TABLE 27.1 Summary of Studies Reporting Elevated CO_2 Effects On Soil C, Soil Respiration, N_2O Emission, and CH_4 Exchange

Agroecosystem	Site	Facility[1]	Other treatment	Reference Soil C	Respiration	N_2O	CH_4
N Fertilized Studies							
Sorghum (Sorghum AL)	Auburn AL, U.S.	OTC		Prior et al. 2004			
Sorghum (Sorghum AZ)	Maricopa AZ, U.S.	FACE	Low and high irrigation	Cheng et al. 2007		Welzmiller et al. 2008	
Cotton (Cotton AZ)	Maricopa AZ, U.S.	FACE	Low and high irrigation	Wood et al. 1994	Nakayama et al. 1994		
Wheat (Wheat AZ)	Maricopa AZ, U.S.	FACE		Leavitt et al. 1996, Prior et al. 1997			
Wheat (Wheat AZ)	Maricopa AZ, U.S.	FACE	Low and high N fertilization	Leavitt et al. 2001	Pendall et al. 2001		
Wheat (Wheat GE)	Hohenheim, Germany	FACE		Marhan et al. 2010			
Wheat (Wheat CH)	Beijing, China	FACE			Lam et al. 2011	Lam et al. 2011	Lam et al. 2011
Wheat (Wheat SE)	Ostad, Sweden	OTC				Pleijel et al. 1998	
Soybean (Soybean AL)	Auburn AL, U.S.	OTC		Prior et al. 2004			
Rice (Rice JA)	Iwate, Japan	FACE	Unwarmed and warmed				Inubushi et al. 2003, Tokida et al. 2010
Rice (Rice FL)	Gainesville FL, U.S.	PT	Unwarmed and warmed				Schrope et al. 1999
Rice (Rice PH)	Los Baños, Phillipines	OTC	Unwarmed and warmed, dry and wet season				Ziska et al. 1998
Wheat/sugar beet (Wheat/sugar GE)	Braunschweig, Germany	FACE		Giesemann and Weigel 2008			
Wheat/soybean (Wheat/soy MD)	Beltsville MD, U.S.	OTC	Low and high irrigation, low and high ozone	Islam et al. 2000			
Corn/soybean (Corn/soy IL)	Urbana IL, U.S.	FACE		Peralta and Wander 2008	Peralta and Wander 2008		
Sorghum/soybean (Sorghum/soy AL)	Auburn AL, U.S.	OTC	Conventional and Conservation tillage	Prior et al. 2005	Smith et al. 2010	Smith et al. 2010	Smith et al. 2010

Rice/wheat (Rice/wheat JI)	Jiangdu, Jiangsu, China	FACE	Low and high N fertilization	Zhong et al. 2009	Kou et al. 2007		
Rice/wheat (Rice/wheat WU)	Wuxi, Jiangsu, China	FACE	Low and high N fertilization	Juan et al. 2007			Xu et al. 2004, Zheng et al. 2006
White clover (White clover SW White clover bin SW)	Eschikon, Switzerland	FACE	Low and high N fertilization	Van Kessel et al. 2006, 2000	Baggs et al. 2003	Baggs et al. 2003	
Rye grass (Rye grass SW Rye grass bin SW)	Eschikon, Switzerland	FACE	Low and high N fertilization	Van Kessel et al. 2006, 2000	Ineson et al. 1998, Baggs et al. 2003	Ineson et al. 1998, Baggs et al. 2003	Ineson et al. 1998
White clover/rye grass (White clover/rye SW)	Eschikon, Switzerland	FACE	Low and high N fertilization	Van Kessel et al. 2006	Baggs et al. 2003	Baggs et al. 2003	
Rye grass (Rye grass FR)	France	PT	Low and high N fertilization		Casella et al. 1997		
Temperate grassland (Temperate grass GE)	Giessen, Germany	FACE		Jäger et al. 2003		Kammann et al. 2008	
Alpine grassland (Alpine grass SW)	Switzerland	OTC		Niklaus and Körner 1996			
Non-N Fertilized Studies							
Alpine grassland (Alpine grass SW)	Switzerland	OTC		Niklaus and Körner 1996			
Temperate grassland (Temperate grass SW)	Switzerland	SACC		Niklaus et al. 2001			
Temperate grassland (Temperate grass NZ)	Manawatu, New Zealand	FACE		Ross et al. 2004			
Temperate grassland (Temperate grass AU)	Tasmania, Australia	FACE	Unwarmed and warmed; C_3 and C_4 grasses	Pendall et al. 2011			
Annual grassland (Annual grass CA)	Stanford CA, U.S.	OTC	Two soil types	Hungate et al. 1997a	Luo et al. 1996, Hu et al. 2001	Hungate et al. 1997b	
Tall grass prairie (Tall grass prairie KS)	Manhattan KS, U.S.	OTC		Williams et al. 2000			
Short grass steppe (Short grass steppe CO)	Nunn CO, U.S.	OTC		Pendall et al. 2004	Mosier et al. 2002	Mosier et al. 2002	Mosier et al. 2002
Northern mixed grassland (Northern mixed grass WY)	Cheyenne WY, U.S.	FACE	Unwarmed and warmed	Unpublished results	Unpublished results	Unpublished results	Dijkstra et al. 2011 Unpublished results

[1]*FACE: Free Air CO_2 Enrichment; OTC: Open Top Chamber; PT: Plastic Tunnel; SACC: Screen-Aided CO_2 Control.*

change, although there is still considerable uncertainty in these projections. While some of the details of these events are unclear, most agree climate change has already affected agro-ecosystems worldwide, and will have even more profound effects as climate change accelerates (Solomon et al., 2009). Important feedback exists between the atmosphere and the soil (Heimann and Reichstein, 2008), and a clear understanding of how climate change and rising atmospheric CO_2 might affect soil C sequestration and greenhouse gas exchange in agro-ecosystems is urgently needed.

Our review will address the effects of warming and rising CO_2 on the GHG balance. Although precipitation can have strong effects on C sequestration and greenhouse gas exchange in agro-ecosystems, current projections about precipitation responses remain highly uncertain. Our review will focus on manipulative field experiments in which researchers alter the environment to evaluate ecosystem responses. These experiments include manipulations of atmospheric CO_2 through Open Top Chambers (OTC), Free Air Carbon dioxide Enrichment (FACE), or Screen-Aided CO_2 Control (SACC), manipulations of temperature using heating cables (HC) or infrared heaters (IRH), or a combination of atmospheric CO_2 and temperature. We assess important mechanisms and identify critical knowledge gaps regarding the effects of elevated CO_2 (eCO_2) and warming on C sequestration and greenhouse gas exchange in agroecosystems.

METHODS

In our review we focused on manipulative field experiments, while we excluded growth chamber and greenhouse studies, studies conducted in arctic and subarctic environments, and studies conducted in systems with no direct agronomic benefit (e.g. forests). In most experiments, CO_2 concentrations were manipulated above present-day ambient concentrations (~375–385 ppmv) to enriched levels (470–720 ppmv). Temperature increases ranged between 1 and 5°C above ambient, consistent with IPCC projections for the end of the 21st century (IPCC, 2007). We reviewed a total of 32 eCO_2 and 13 warming studies (Tables 27.1 and 27.2).

The eCO_2 and warming effects on soil C, soil respiration, and N_2O emission were separated in N fertilized and non-N fertilized studies with the expectation that eCO_2 and warming effects on these properties largely depend on soil N availability. For instance, in other meta-analyses a significant increase in soil C under eCO_2 required N fertilization (Van Groenigen et al., 2006; Hungate et al., 2009). We further separated eCO_2 and warming effects on CH_4 exchange conducted in dry land sites (non-rice) where the net CH_4 efflux is predominantly negative (i.e. net CH_4 uptake in soils), and in rice paddy field studies where the net CH_4 efflux is much larger and always positive (i.e. net CH_4 production in soils). When other treatments were included, eCO_2 and warming effects were averaged across those other treatments (e.g. irrigation, ozone).

We calculated the effect of eCO_2 and warming on soil C as the absolute change in soil C (in $g\,C\,kg^{-1}$ soil) divided by the number of years of treatment. We used absolute changes rather than relative changes because absolute changes provide more biogeochemical significance (Hungate et al., 2009). The absolute changes were calculated for the shallowest soil depths reported, which ranged between 0–5 and 0–26 cm among studies. The number of years of treatment effects on soil C ranged between 2 and 10 years. We calculated the effect of CO_2 and warming on soil respiration, N_2O, and CH_4 flux rates as the absolute change in the average flux rates measured during the growing season (in $kg\,C\,ha^{-1}\,d^{-1}$, $g\,N\,ha^{-1}\,d^{-1}$, and $g\,C\,ha^{-1}\,d^{-1}$ for CO_2, N_2O, and CH_4, respectively). When flux rates were measured in multiple years, we averaged the flux rates across years. All flux rates were measured using static chambers.

Because the effect of eCO_2 on soil C, soil respiration, N_2O, and CH_4 flux rates were highly variable among studies, we tested whether this variability could be explained by climate factors or soil properties of the study site. For the climate factors we chose mean annual temperature

TABLE 27.2 Summary of Studies Reporting Warming Effects On Soil C, Soil Respiration, N_2O Emission, and CH_4 Exchange

Agroecosystem	Site	Facility[1]	Other treatment	Reference Soil C	Respiration	N_2O	CH_4
N Fertilized Studies							
Wheat (Wheat GE)	München, Germany	HC				Kamp et al. 1998	
Wheat (Wheat UK)	York, U.K.	HC			Hartley et al. 2007		
Corn (Corn UK)	York, U.K.	HC			Hartley et al. 2007		
Rice (Rice JA)	Iwate, Japan	HC	Ambient and elevated CO_2				Tokida et al. 2010
Rice (Rice FL)	Gainesville FL, U.S.	TGC	Ambient and elevated CO_2				Schrope et al. 1999
Rice (Rice PH)	Los Baños, Phillipines	OTC	Ambient and elevated CO_2				Ziska et al. 1998
Rye grass (Rye grass FR)	France	PT	Low and high N fertilization		Casella et al. 1997		
Non-N Fertilized Studies							
Alpine grassland (Alpine grass TI)	Tibet, China	IRH	No grazing and grazing			Hu et al. 2010	
Temperate grassland (Temperate grass CH)	Inner Mongolia, China	IRH	No irrigation and irrigation; day, night, and diurnal warming		Liu et al. 2009, Xia et al. 2009		
Temperate grassland (Temperate grass UK)	Scotland, U.K.	HC			Briones et al. 2009		
Temperate grassland (Temperate grass AU)	Tasmania, Australia	IRH	Ambient and elevated CO_2; C_3 and C_4 grasses	Pendall et al. 2011			
Tall grass prairie (Tall grass OK)	OK, U.S.	IRH	No clipping and clipping	Luo et al. 2009	Zhou et al. 2007		
Northern mixed grassland (Northern mixed grass WY)	Cheyenne WY, U.S.	IRH	Ambient and elevated CO_2	Unpublished results	Unpublished results	Unpublished results	Unpublished results

[1]*HC: Heating Cable; IRH: Infrared Heater; PT: Plastic Tunnel.*

(MAT) and mean annual precipitation (MAP) of the site where the studies were conducted and for soil properties we chose %clay and pH. We chose these climate and soil factors because they can have significant effects on plant growth and biological activity in the soil (Epstein et al., 1997; Guo et al., 2006; Fierer et al., 2009) and therefore we expected that they could significantly influence eCO_2 and warming effects on soil C and GHG flux rates among sites. These parameters are also frequently reported in the literature. We used %clay when reported in the study, but often only the textural class was reported. In that case we used the average %clay of the two boundaries of the textural class according to the textural triangle. For instance, if it was reported that the study was conducted in a sandy clay loam with a clay content between 20 and 35% according to the textural triangle, we designated that soil with the average clay content of 27.5%. We related CO_2 effects on soil C and GHG flux rates to each of MAT, MAP, %clay, and pH using linear regression. With the linear regressions, we put more weight on studies that were conducted over a longer time period, because we assumed that studies over longer time periods provide more reliable data. We weighted the absolute rate of change in soil C by the treatment length (in years) after which soil C was measured and weighted the absolute change GHG flux rates by the duration of the measurements (in years; Wu et al., 2011). Some studies were conducted at the same location and soil type, but in different years (e.g. the wheat, sorghum, and cotton studies at Maricopa, AZ). In those cases eCO_2 effects and treatment length/duration of measurements were averaged across the different studies conducted at the same site. We only constructed relationships when there were data available for four or more sites. All linear regressions were performed with JMP (version 8.0.1; SAS Institute, Cary, NC, USA).

THE EFFECT OF eCO_2 ON SOIL C

We found 27 studies (19 N fertilized and 8 non-N fertilized studies) where the effect of eCO_2 on soil C was reported (Figure 27.1). In 74% of the studies (79% of the N fertilized and 63% of

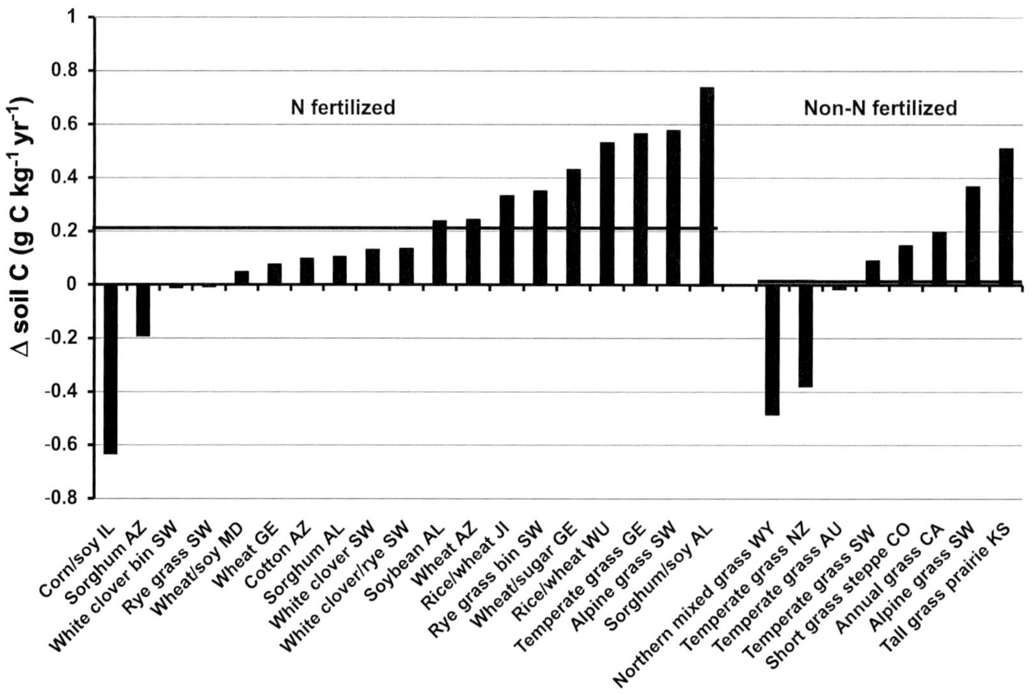

FIGURE 27.1
The rate of change in soil C in response to eCO_2 among different studies. Horizontal bold lines represent averaged values for N fertilized and non-N fertilized studies.

the non-N fertilized studies) a positive effect of eCO_2 on soil C was found, although only in a few occasions were these positive effects statistically significant (e.g. Williams et al., 2000; Prior et al., 2004, 2005; Zhong et al., 2009). While the positive effects of CO_2 enrichment on plant production are generally observed in the initial treatment year (Kimball et al., 2002), detection of significant changes in soil C may take many years of CO_2 enrichment due to the high amount of C present in soils and the relatively small amounts that accrue on an annual basis (Conant and Paustian, 2002; Smith, 2004). On average soil C increased by $0.205\ g\ kg^{-1}\ yr^{-1}$ in the N fertilized studies and by $0.008\ g\ kg^{-1}\ yr^{-1}$ in the non-N fertilized studies. If we assume that in all studies the soil had a bulk density of $1.3\ g\ cm^{-3}$ and that the change in soil C occurred in the top 20 cm, this would correspond to an average rate of soil C increase of 1460 and 57 $g\ ha^{-1}\ d^{-1}$ in the N fertilized and non-N fertilized studies, respectively. A greater response in N fertilized studies was also found by Van Groenigen et al. (2006) in their meta-analysis, where they included greenhouse and growth chamber studies, and non-agronomic sites. Our results confirm the notion that soil C sequestration under eCO_2 is generally constrained by the availability of N and that N fertilization enhances the capacity to increase soil C under eCO_2 (Reich et al., 2006a; Van Groenigen et al., 2006).

Nitrogen fixation by legumes is often enhanced under eCO_2, especially with the addition of non-N nutrients (van Groenigen et al., 2006), and legume responses to CO_2 tend to be greater than non-legumes under conditions of low soil N (van Kessel et al., 2006). Yet no strong evidence was found of greater C sequestration under eCO_2 in studies with legumes. Soil C only slightly increased in plots of *Trifolium repens* (white clover) in the FACE experiment at Eschikon, Switzerland, the most extensive evaluation yet of legume CO_2 responses in a field setting (van Kessel et al., 2006), while soil C decreased under eCO_2 in a temperate pasture with legumes in New Zealand (Ross et al., 2004).

Although on average soil C sequestration in response to eCO_2 was higher in N fertilized than in non-N fertilized studies, within the N fertilized and non-N fertilized studies eCO_2 effects on soil C showed large variation (Figure 27.1). For example, the most negative response to eCO_2 was observed in an N fertilized study with a corn–soybean rotation in Illinois (Peralta and Wander, 2008) and the most positive response was observed in an N fertilized study with a sorghum–soybean rotation in Alabama (Prior et al., 2005). Both these extreme responses were larger than any of the responses observed in the non-N fertilized studies. We tested whether this variability in soil C response among sites could be explained by site differences in climate and soil type. When we related the absolute rate of change in soil C in response to eCO_2 to climate and soil parameters, only %clay in the fertilized sites exhibited a relationship, with marginal significance ($P = 0.08$, Table 27.3). Soil C sequestration in response to eCO_2 tended to decrease with increased clay content (Figure 27.2). Although only marginally significant, this result is remarkable (and any marginal or significant relationships discussed further on) given that all these studies were done using different methods under a variety of conditions. A possible explanation for the decrease in soil C sequestration with increased clay content in response to eCO_2 is that rhizosphere priming effects on soil organic matter decomposition under eCO_2 may be stronger in more clayey soils. Rhizosphere priming, where microbial decomposition of relative recalcitrant soil organic matter is enhanced because of microbial stimulation by energy-rich root exudates, may increase under eCO_2 (Cheng, 1999), particularly in soils with greater clay content (Dijkstra and Cheng, 2007). Thus, an eCO_2-induced increase in rhizosphere priming in more clayey soils may result in less C sequestration, or even cause a net loss of soil C as was observed in the silty clay loam in Illinois, U.S. (Peralta and Wander, 2008).

THE EFFECT OF eCO_2 ON SOIL RESPIRATION

Soil respiration increased with eCO_2 in 12 of the 13 studies reviewed (Figure 27.3). A decrease in soil respiration under eCO_2 was observed in a study with rye grass in Switzerland (Ineson

TABLE 27.3 Summary of Regression Analyses Explaining Variation in Soil C, Soil Respiration, N_2O Emission and CH_4 Exchange in Response to eCO_2

A. Soil C

	All sites			Fertilized sites			Non-fertilized sites		
	*n**	*Corr. C.*	*P*	*n*	*Corr. C.*	*P*	*n*	*Corr. C.*	*P*
MAT	19	0.26	0.26	11	0.22	0.51	8	0.30	0.46
MAP	20	0.28	0.23	11	0.12	0.73	9	0.36	0.34
pH	20	−0.21	0.38	11	−0.20	0.56	9	−0.18	0.65
% Clay	22	−0.19	0.39	13	−0.51	*0.08*	9	0.11	0.79

B. Soil Respiration

	All sites			Fertilized sites		
	n	*Corr. C.*	*P*	*n*	*Corr. C.*	*P*
MAT	9	0.56	0.13	6	0.29	0.60
MAP	8	0.46	0.22	5	0.15	0.80
pH	8	−0.71	*0.06*	5	−0.82	0.06
% Clay	10	0.71	*0.03*	7	0.50	0.33

C. N_2O Emission

	All sites			Fertilized sites		
	n	*Corr. C.*	*P*	*n*	*Corr. C.*	*P*
MAT	8	0.63	*0.09*	6	0.51	0.24
MAP	7	0.06	0.86	5	−0.16	0.85
pH	7	−0.10	0.90	5	0.36	0.50
% Clay	9	0.62	*0.09*	7	0.60	0.21

D. CH_4 Exchange

	Non-rice		
	n	*Corr. C.*	*P*
MAT	5	−0.15	0.85
MAP	4	0.66	0.32
pH	4	0.24	0.76
% Clay	5	0.87	0.07

**n: number of sites included in the regression; Corr. C.: Pearson's correlation coefficient; P: P-value of linear regression.*

et al., 1998). Because root respiration was included in all studies, it is not surprising that soil respiration increased under eCO_2 in most studies. Elevated CO_2 generally increases plant productivity in agroecosystems (Kimball et al., 2002), and thus the increase in soil respiration under eCO_2 may largely have been driven by an increase in root production. The effect of eCO_2 on soil respiration was on average 6.4 and 1.9 $kg\,C\,ha^{-1}\,d^{-1}$ in N fertilized and non-N fertilized studies respectively. Again, the greater eCO_2 effect on soil respiration in N fertilized studies may have been caused by an increase in root productivity and respiration that often occurs under eCO_2 with N additions (Van Groenigen et al., 2006).

Of the two climate and two soil parameters that we tested, both soil pH and %clay explained most of the variability in eCO_2 effects on soil respiration, although only the relationship with % clay was significant ($P = 0.03$, Table 27.3). When we included both N fertilized and non-N fertilized sites in the regression, the increase in soil respiration in response to eCO_2 increased with increased clay content, explaining 45% of the variability, and decreased with increased soil pH, explaining 48% of the variability (Figure 27.4). As was argued for the relationship between

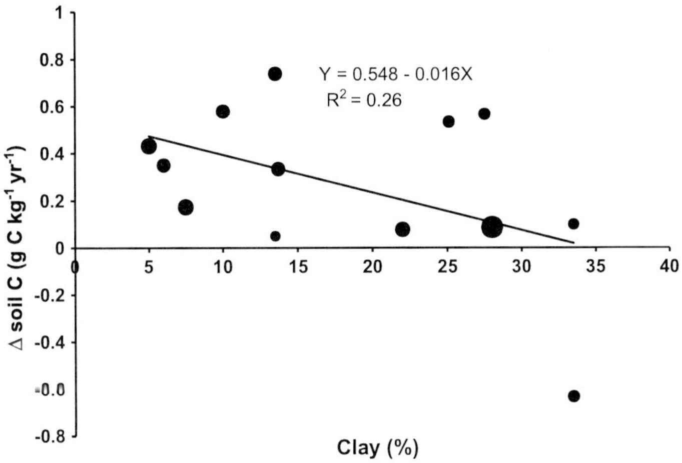

FIGURE 27.2
The rate of change in soil C in response to eCO_2 in the N fertilized studies as a function of the soil clay content. The size of the dots indicate the weight used in the regression (i.e. bigger dots have more weight).

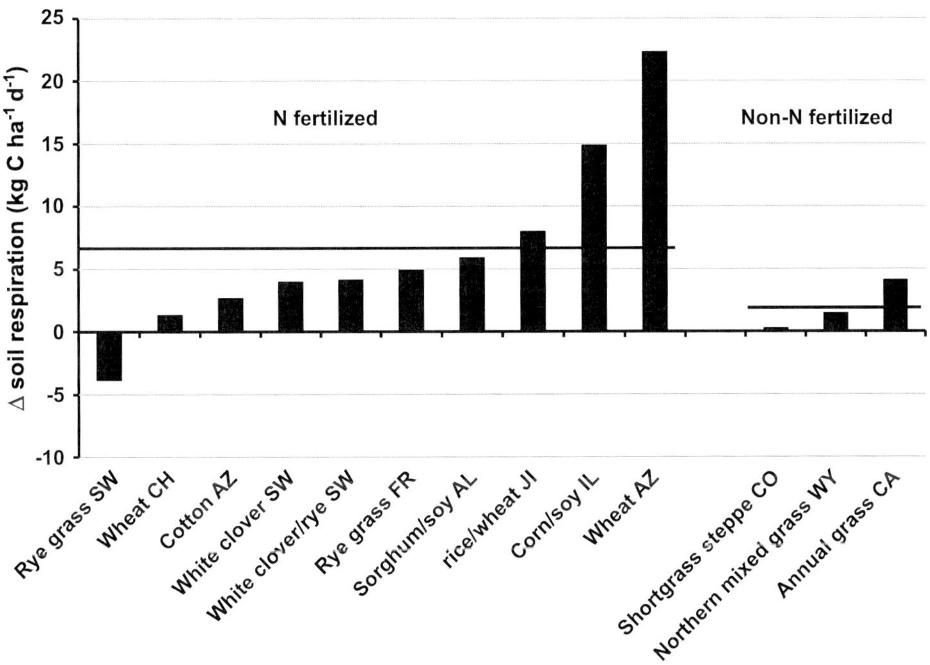

FIGURE 27.3
The change in soil respiration in response to eCO_2 among different studies. Horizontal bold lines represent averaged values for N fertilized and non-N fertilized studies.

clay content and eCO_2 effects on soil C, an increase in rhizosphere priming under eCO_2 may have resulted in larger eCO_2 effects on soil respiration with increased clay content. The negative relationship with soil pH is less clear. Microbial community composition and enzyme activity are often strongly affected by soil pH (Sinsabaugh et al., 2008; Fierer et al., 2009) that can be altered by changes in substrate inputs (Aciego Pietri and Brookes, 2009). It is, however, unclear to what degree the negative relationship that we observed between soil pH and soil respiration in response to eCO_2 was caused by changes in microbial or plant respiration.

THE EFFECT OF eCO_2 ON N_2O EMISSION

Of the 8 N fertilized studies that we found, the N_2O emission increased under eCO_2 in 6 studies. The average increase in N fertilized studies was 9.3 g N ha^{-1} d^{-1}, while in the 3 non-N fertilized studies eCO_2 had hardly any effect on N_2O emission with an average decrease of

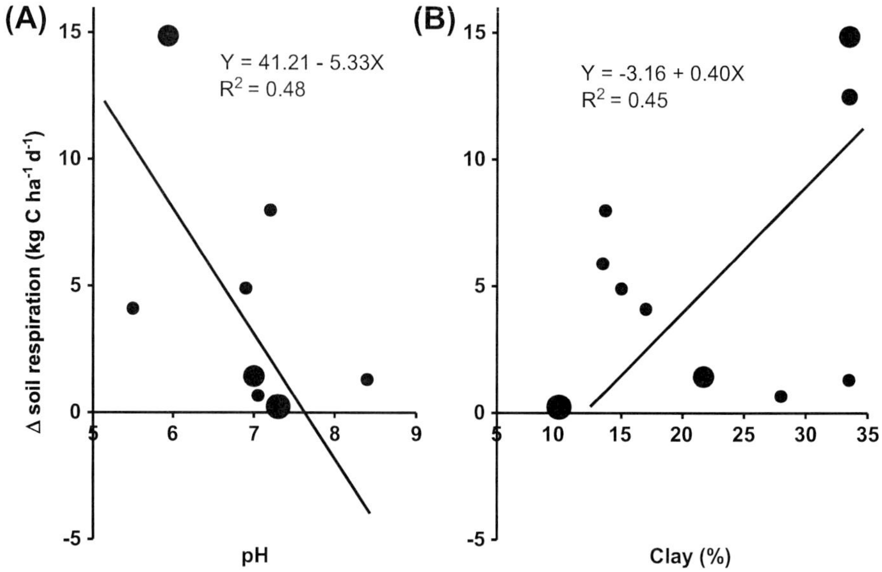

FIGURE 27.4
The change in soil respiration in response to eCO_2 as a function of (A) soil pH and (B) clay content. The size of the dots indicate the weight used in the regression (i.e. bigger dots have more weight).

0.5 g N ha^{-1} d^{-1} (Figure 27.5). With a global warming potential 298 times greater than CO_2 (IPCC, 2007), N_2O emission in N fertilized studies correspond on average to 1188 g C-CO_2 equivalents ha^{-1} d^{-1}. This is slightly less than the average rate that we calculated for C sequestration in the top 20 cm of the soil in N fertilized studies in response to eCO_2 (see above). This suggests that, although N fertilization has the potential to increase soil C under eCO_2 (Van Groenigen et al., 2006), these soil C gains can potentially be almost completely offset by increased N_2O emissions under eCO_2 when N fertilizer is applied.

In the N fertilized studies N_2O emission in response to eCO_2 showed large variation between a decrease of 0.2 g N ha^{-1} d^{-1} in white clover in Switzerland (Baggs et al., 2003) and an increase of 38 g N ha^{-1} d^{-1} in rye grass at the same site in Switzerland (Ineson et al., 1998). This large variation is to a great extent caused by the timing and frequency of measurements after the N fertilizer application. For instance, Ineson et al. (1998) observed some of the

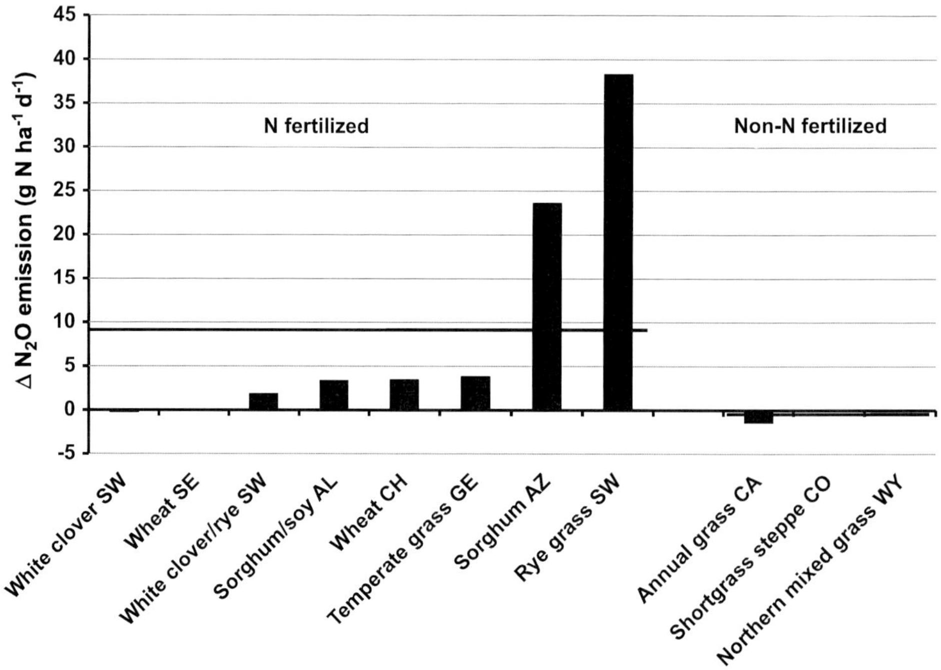

FIGURE 27.5
The change in N_2O emission in response to eCO_2 among different studies. Horizontal bold lines represent averaged values for N fertilized and non-N fertilized studies.

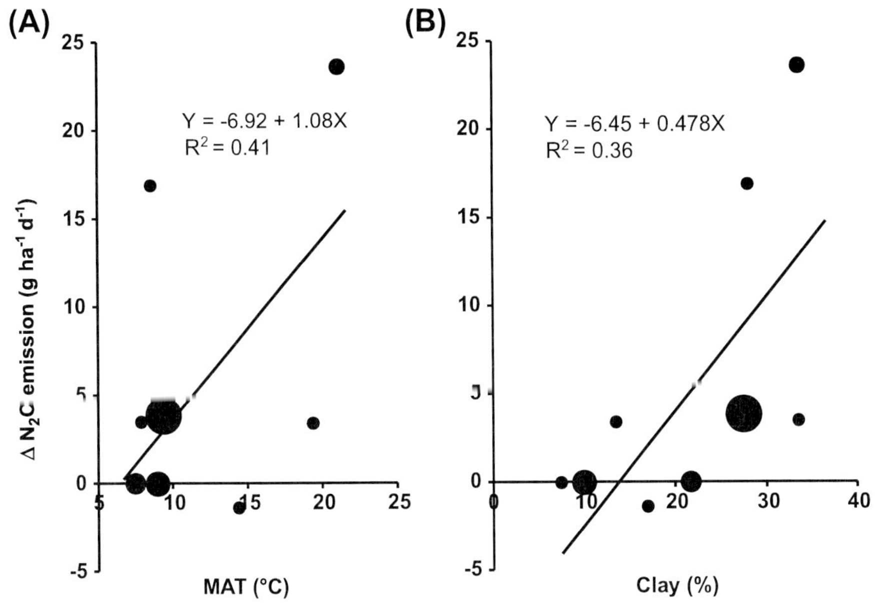

FIGURE 27.6
The change in N_2O emission in response to eCO_2 as a function of (A) mean annual temperature (MAT) and (B) clay content. The size of the dots indicate the weight used in the regression (i.e. bigger dots have more weight).

highest N_2O emissions ever recorded directly after N fertilizer application in grassland systems. Because of these high rates of N_2O emission, eCO_2 effects on N_2O emissions can also be high directly after N application (Ineson et al., 1998; Welzmiller et al., 2008). On the other hand, Kammann et al. (2008) measured the effect of eCO_2 on N_2O emissions during 9 years in a temperate grassland in Germany (the longest study conducted on the effect of eCO_2 on N_2O emission) and observed the greatest eCO_2 effects during vegetative growth periods in the summer when soil mineral N concentrations were low, while eCO_2 had no effect on N_2O emission directly after the N application in the spring. Regardless of the timing and frequency of measurements in relation to N fertilizer application, the majority of N fertilized studies showed an increase in N_2O emission in response to eCO_2. It has been suggested that the increase in N_2O emission under eCO_2 in some of these N fertilized studies was caused by an increase in labile C substrates fueling denitrification (Ineson et al., 1998; Kammann et al., 2008).

In contrast, no eCO_2 effect, or even a slight reduction in N_2O emissions, was observed in the 3 non-N fertilized studies. Possibly, eCO_2 increased plant N uptake and reduced soil N availability in these unfertilized systems where available soil N was already low, causing no or reduced effects on N_2O emission (Hungate et al., 1997b; Mosier et al., 2002).

Both MAT and %clay showed a positive relationship with N_2O emission in response to eCO_2, although both relationships were only marginally significant (Table 27.3, Figure 27.6). N_2O emissions are highly sensitive to temperature (Grant and Pattey, 2008), which could explain why N_2O emissions respond more to eCO_2 in combination with higher MAT. Further, an increase in soil moisture, because of decreased stomatal conductance under eCO_2 (Kimball and Idso, 1983; Morgan et al., 2004; Wand et al., 1999), can increase anaerobic conditions in the soil conducive to denitrification, particularly in clayey soils that have relatively more small pores than sandy soils. These results suggest that, apart from N fertilization, MAT and soil texture play important roles in the large variability in N_2O emission in response to eCO_2 among different sites.

THE EFFECT OF eCO_2 ON CH_4 EXCHANGE

In studies with rice, eCO_2 resulted in large increases in CH_4 emission in 3 out of 4 studies (Figure 27.7). The average increase in CH_4 emission in rice studies in response to eCO_2 was

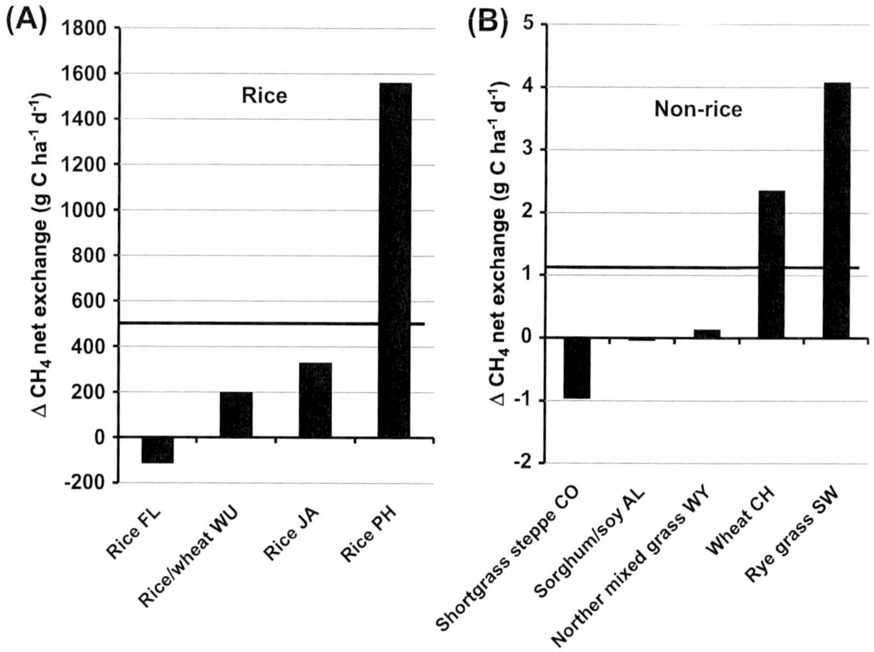

FIGURE 27.7
The change in CH_4 exchange in response to eCO_2 among different studies. Horizontal bold lines represent averaged values for (A) rice and (B) non-rice studies.

$493\ g\ C\ ha^{-1}\ d^{-1}$. With a global warming potential 25 times greater than CO_2 (IPCC, 2007), this increase in CH_4 emission corresponds to an increase of 4482 g C-CO_2 equivalents $ha^{-1}\ d^{-1}$ in response to eCO_2. Evidently, rice paddy fields show some of the greatest responses to eCO_2 in terms of GHG emissions (e.g. the average rate of C–CO_2 equivalents associated with CH_4 emission in rice is 3.8 times higher than the average rate associated with N_2O emission in N fertilized studies).

In the rice studies, we found no relationships between MAT, MAP, soil pH, or %clay with the rate of CH_4 emission in response to eCO_2. CH_4 emission in rice fields is to a large degree controlled by inputs of C substrates, and increased CH_4 emission under eCO_2 has been associated with increased plant residues, root productivity, and exudation (Inubushi et al., 2003; Xu et al., 2004; Tokida et al., 2010). The increase in CH_4 emission in response to eCO_2 through increased inputs of C substrates may simply have overwhelmed any soil or external climate effect. We should note that with only 4 studies, we had limited statistical power to do the regressions.

In most non-rice studies, soil is a net sink for CH_4, where it is oxidized by methanotrophic bacteria [an exception was the study by Smith et al. (2010) where soil was sometimes a CH_4 source in a sorghum–soybean rotation]. The effect of eCO_2 on CH_4 fluxes (where we used the same convention as in the rice studies, i.e. a positive flux indicates CH_4 emission, while a negative flux indicates CH_4 uptake) was mixed where both increases and decreases were observed with an average increase of $1.1\ g\ C\ ha^{-1}\ d^{-1}$ among the 5 studies we evaluated (or on average a *reduction* in CH_4 uptake in response to eCO_2 because in most studies there was an overall net CH_4 uptake; Figure 27.7). The reduced transpiration and consequent higher soil water content that often occurs under eCO_2 can have opposite effects on CH_4 fluxes depending on whether methanotroph activity is limited by soil moisture (in most arid and semiarid environments) or by CH_4 diffusivity into the soil (in most mesic environments; Dijkstra et al., 2011). Note that responses of CH_4 fluxes to eCO_2 in non-rice or dry land systems are orders of magnitude smaller than in rice systems. Nevertheless, because a much larger proportion of the global area is covered by dry land systems than by rice paddy fields, small changes in CH_4 fluxes in dry land systems can still have a significant impact on the global CH_4 flux (Mosier et al., 1991).

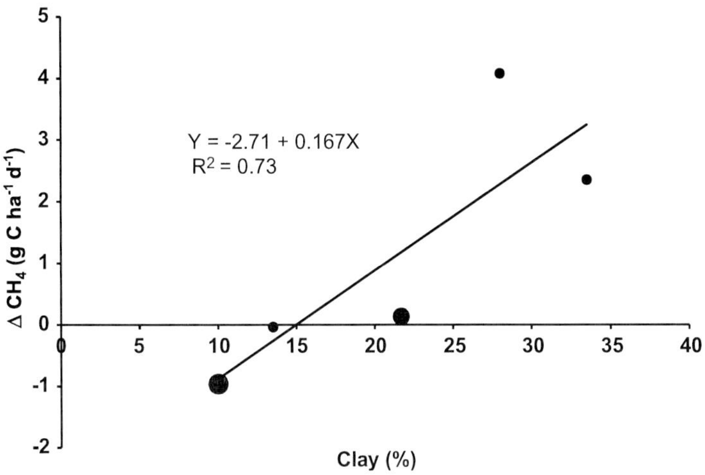

FIGURE 27.8
The change in CH_4 exchange in response to eCO_2 in non rice studies as a function of soil clay content. The size of the dots indicate the weight used in the regression (i.e. bigger dots have more weight).

As with the greenhouse gases CO_2 and N_2O, the CH_4 flux in response to eCO_2 was positively related to the clay content of the soil, explaining 73% of the variation (Table 27.3, Figure 27.8). Soil texture should influence CH_4 fluxes to the extent that methanotroph activity is limited by soil moisture or by CH_4 diffusivity. Since sandy soils have better aeration than clayey soils, CH_4 uptake is more likely to be limited by direct effects of soil moisture on methanotroph activity than by CH_4 diffusivity. Thus, a CO_2-induced increase in soil moisture would tend to increase CH_4 uptake (thus *decrease* the CH_4 flux in response to eCO_2) more in sandy soils (e.g. as observed by Mosier et al., 2002, in a sandy loam with only 10% clay). On the other hand, in clayey soils with poorer aeration, CH_4 uptake is more likely to be limited by CH_4 diffusivity, and a CO_2-induced increase in soil moisture might therefore decrease CH_4 uptake (or increase the CO_2 response) as observed in a clay loam by Lam et al. (2011). The positive relationship of CH_4 fluxes with clay content is based on only five observations, and it remains to be seen if this relationship will hold with more observations.

WARMING EFFECTS ON SOIL C, SOIL RESPIRATION, N_2O EMISSION AND CH_4 EXCHANGE

Little work has been done on the effects of warming on soil C sequestration and GHG fluxes in agroecosystems. We found only two published studies (Luo et al., 2009; Pendall et al., 2011) and one unpublished study (northern mixed grass prairie in Wyoming, U.S.), all in non-N fertilized grassland systems, reporting warming effects on soil C. Results are not consistent among those three studies, with an average decrease in soil C by 0.026 g C kg^{-1} soil yr^{-1} (Figure 27.9A). The effect of warming on soil C in all three studies is relatively small compared to the effect of eCO_2 in many studies (Figure 27.1). It is noteworthy that warming induced C loss only in Wyoming northern mixed-grass prairie, the driest of these three grasslands. Although warming has the potential to enhance biological activity and extend the length of growing season, it also desiccates, and in dry grasslands, such desiccation can lead to C loss (Zhang et al., 2010).

A little more work has been done evaluating the effects of warming on soil respiration, with more consistent results. In 6 of 7 studies, soil respiration increased with warming (Figure 27.9B). The exception was a study with ryegrass in France, where no change in soil respiration was observed (Casella and Soussana, 1997). On average soil respiration increased more in N fertilized (6.8 kg C ha^{-1} d^{-1}) than in non-N fertilized studies (4.4 kg C ha^{-1} d^{-1}), although the highest increase was observed in a non-N fertilized study (Briones et al., 2009). Warming often leads to increased rates of SOM decomposition, and likely led to increased soil respiration (Rustad et al., 2001), particularly when N is not limited. We related soil respiration

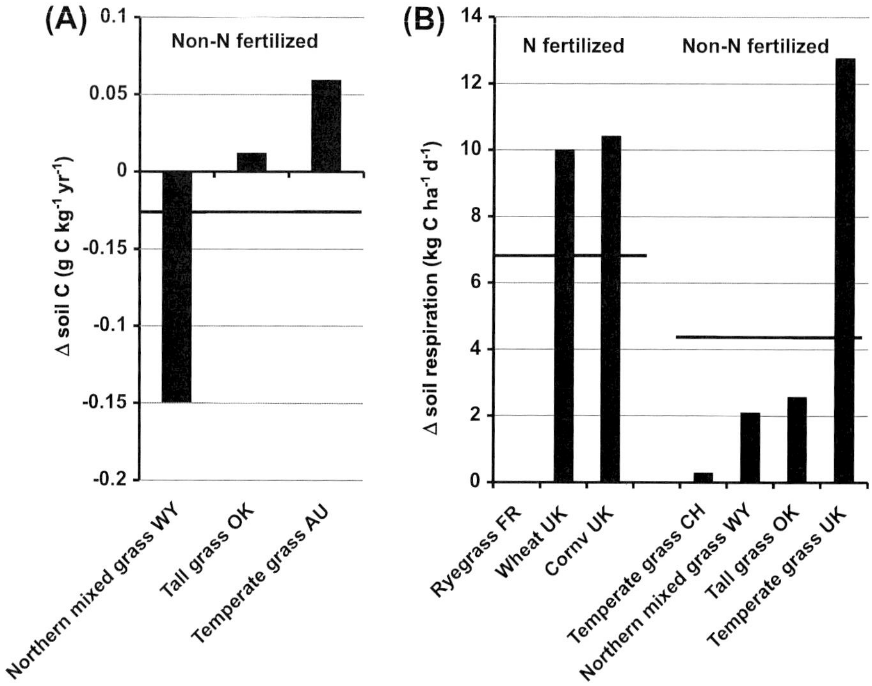

FIGURE 27.9
The rate of change in (A) soil C and (B) soil respiration in response to warming among different studies. Horizontal bold lines represent the averaged values for N fertilized and non-N fertilized studies.

rates in response to warming to climate and soil parameters, but observed no significant relationships (data not shown).

Little information is available on how warming affects N_2O and CH_4 fluxes. In the only N fertilized study we found, the N_2O emission decreased in response to warming (Kamp et al., 1998), while in the two non-N fertilized studies warming had very little effect on N_2O emission (Hu et al., 2010, and unpublished results from northern mixed grassland, Wyoming, U.S., Figure 27.10A). Warming had mixed effects on CH_4 emission in rice paddy fields where both decreased (Ziska et al., 1998) and increased emission rates (Tokida et al., 2010) were reported (Figure 27.10B). The only non-rice study conducted in a semiarid grassland showed that warming decreased CH_4 uptake (Figure 27.10C). It was argued that in this semiarid climate, methanotroph activity was mostly directly limited by a soil moisture and that the drying effect of warming therefore directly reduced methanotroph activity (Dijkstra et al., 2011). Because of the limited number of studies, we did not perform regressions with climate and soil parameters.

INTERACTIVE $eCO_2 \times$ WARMING EFFECTS ON SOIL C, SOIL RESPIRATION, N_2O EMISSION, AND CH_4 EXCHANGE

Few studies included both atmospheric CO_2 and temperature manipulations (Ziska et al., 1998; Schrope et al., 1999; Tokida et al., 2010; Dijkstra et al., 2011; Pendall et al., 2011). In only two studies were $CO_2 \times$ warming interactive effects on soil C investigated, both in non-N fertilized grassland systems. In both studies, soil C under eCO_2 decreased more in combination with warming than without warming (Table 27.4). In the northern mixed grassland soil respiration under eCO_2 also increased in combination with warming but slightly decreased without warming. These results suggest that eCO_2 effects on SOM decomposition rates may accelerate with increased temperature. However, we found little evidence for $CO_2 \times$ warming interactions on soil C from our regressions with MAT. Despite the relatively large numbers of studies included in this regression (19 studies, Table 27.3), we observed no significant relationship between soil C in response to eCO_2 and MAT, suggesting that eCO_2 effects on soil C sequestration did not depend on the temperature regime that the experiment was conducted

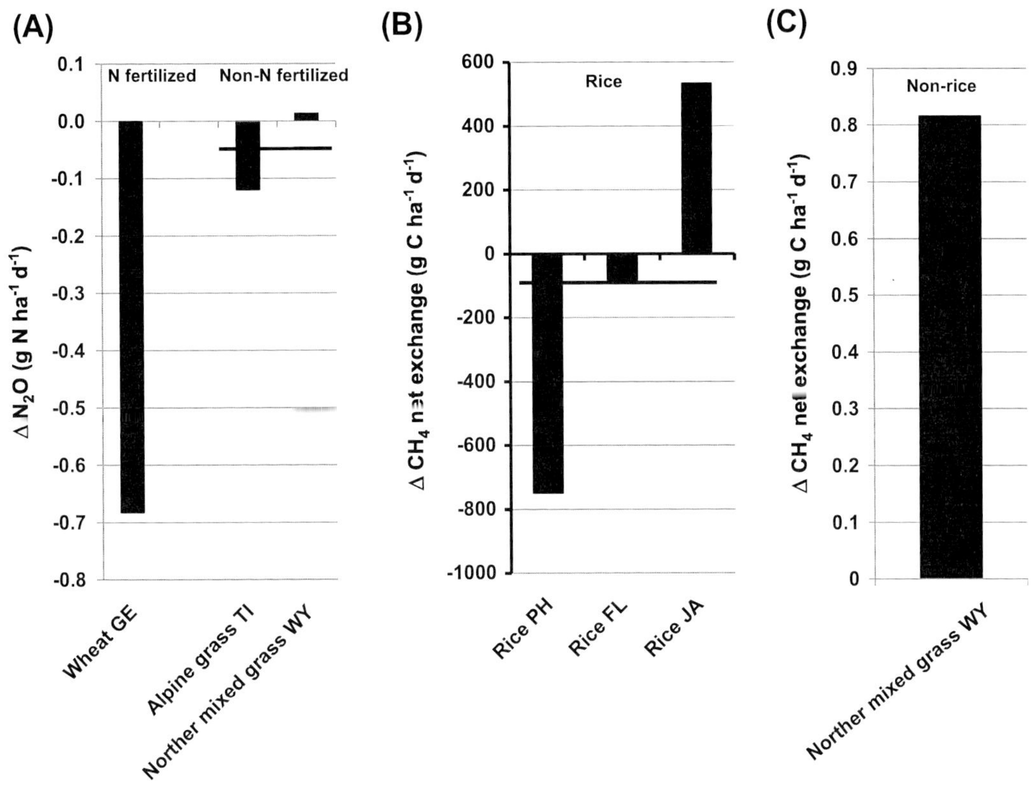

FIGURE 27.10
The change in (A) N_2O emission in N fertilized and non-N fertilized studies, (B) CH_4 exchange in rice studies, and (C) CH_4 exchange in non-rice studies in response to warming. Horizontal bold lines represent averaged values for non-N fertilized and rice studies.

TABLE 27.4 Interactive Effects of eCO_2 and Warming on Soil C, Soil Respiration, N_2O Emission, and CH_4 Exchange

	eCO_2 effect (% change from aCO_2)		
Agroecosystem	*Low temperature*	*High temperature*	**Reference**
Soil C—Non-fertilized			
Temperate grass AU	12.5	−3.4	Pendall et al. 2011
Northern mixed grass WY	−2.8	−14.1	Unpublished results
Respiration—Non-fertilized			
Northern mixed grass WY	−2.0	13.6	Unpublished results
N_2O—Non-fertilized			
Northern mixed grass WY	−94.9	−21.6	Unpublished results
CH_4—Rice			
Rice FL	−84.2	−90.0	Schrope et al. 1999
Rice JA	22.1	29.1	Tokida et al. 2010
Rice PH	48.3	214.5	Ziska et al. 1998
CH_4—Non-rice			
Northern mixed grass WY	0.1	5.0	Dijkstra et al. 2011 Unpublished results

in. Although studies were done at locations that occupied a relatively large range of MAT between 7.5 and 21.1°C, we note that reported MAT values do not always reflect the actual temperature that occurred during the time-frame of the experiment. This may have contributed to finding no significant relationship with eCO_2 effects on soil C.

Only N_2O emission in response to eCO_2 showed a marginally significant relationship with MAT (Figure 27.6), suggesting a $CO_2 \times$ temperature interactive effect where eCO_2 effects are stronger with increased temperature. In the non-N fertilized northern mixed grassland study in Wyoming, the N_2O emission under eCO_2 decreased less with warming than without warming (Table 27.4). These results suggest that $CO_2 \times$ warming interactive effects may be important for N_2O emissions.

The increase in CH_4 emission under eCO_2 was much greater in combination with warming than without warming in a study with rice in the Philippines (Ziska et al., 1998), but no $CO_2 \times$ warming interactive effects on CH_4 exchange were observed in three other studies (Table 27.4). Clearly, more research is needed to identify clear patterns of $eCO_2 \times$ warming interactive effects on soil C sequestration and GHG emissions.

CONCLUSIONS

We reviewed studies conducted in agroecosystems to determine sensitivity of C cycling and GHG emissions to the effects of eCO_2 and warming. We found that eCO_2 had the potential to increase soil C, particularly in combination with N fertilization. Similar results were found in other reviews (Reich et al., 2006b; Van Groenigen et al., 2006; Hungate et al., 2009). However, we also found that the increase in soil C in combination with N fertilization did not come without a price. N_2O emissions also increased under eCO_2 with N fertilization, and indeed on average the CO_2-induced increase in N_2O emission in terms of its Global Warming Potential almost completely offset the average increase in soil C in N fertilized studies. Thus, particularly in N fertilized agroecosystems, one can come to the wrong conclusion about the effect of eCO_2 on the GHG balance expressed in CO_2 equivalents if N_2O emissions are not accounted for. A similar conclusion was reached by van Groenigen et al. (2011), who estimated that the Global Warming Potential caused by a CO_2-induced increase in N_2O and CH_4 emission in agricultural and non-agricultural lands could offset as much as 16.6% of the global increase in terrestrial C storage in response to eCO_2 by 2050.

Research is needed to determine whether or not practices like precision application of N fertilizers or use of nitrification inhibitors can be used to minimize N_2O emission but that can help capitalize on the potential for rising CO_2 to enhance C sequestration. Legumes have been suggested as a possible remedy to the N-limitation problem of plants exposed to eCO_2 (van Groenigen et al., 2006) and legumes tend to respond positively to eCO_2 (Newton et al., 1994; Teyssonneyre et al., 2002). However, lack of a significant positive effect of CO_2 on C sequestration in pastures with legumes (Ross et al., 2004; van Kessel et al., 2006) suggests that simply enhanced N fixation under increasingly higher CO_2 concentrations may not necessarily lead to greater C sequestration. More research is needed to evaluate how and under what conditions various mixtures of legumes and forage grasses in combination with non-N fertilization practices might lead to increased C sequestration under future CO_2-enriched atmospheres. Such research should consider appropriate combinations of legumes and non-legumes whose morphology and development might optimize the capture and cycling of N so as to minimize the release of N_2O.

We found that soil clay content is an important factor in explaining the large variability in GHG exchange among sites in response to eCO_2. We observed marginally significant to significant positive relationships between %clay and all three GHGs in response to eCO_2. Our results suggest that GHG exchange from clayey soils is more sensitive to eCO_2 than from sandy soils. The relationships we found between %clay and GHG exchange should be explored further to better understand the mechanisms involved and how widely applicable this relationship might be to scale up the effect of eCO_2 on GHG exchange in agroecosystems to regional or even global levels.

More research is needed about warming effects on soil C sequestration and GHG exchange. The limited studies that we found from agroecosystems often showed mixed warming effects, and no clear strong patterns emerged on how soil C and GHG exchange is affected by warming. Similarly, it remains unclear what the interactive effects of eCO_2 and warming are on soil C sequestration and GHG exchange in agroecosystems. While important challenges remain for agriculture to identify systems and practices that can mitigate global warming, future research with the objective to enhance C sequestration and mitigate GHG emissions will need to remain vigilant as climate change continues to increase in coming decades.

Acknowledgments

We would like to thank Jeffrey Amthor and Carina Moeller for helpful comments on a previous version of the manuscript.

References

Aciego Pietri, J.C., Brookes, P.C., 2009. Substrate inputs and pH as factors controlling microbial biomass, activity and community structure in an arable soil. Soil Biol. Biochem. 41, 1396–1405.

Baggs, E.M., Richter, M., Hartwig, U.A., Cadisch, G., 2003. Nitrous oxide emissions from grass swards during the eighth year of elevated atmospheric pCO_2 (Swiss FACE). Glob. Change Biol. 9, 1214–1222.

Briones, M.J.I., Ostle, N.J., McNamara, N.P., Poskitt, J., 2009. Functional shifts of grassland soil communities in response to soil warming. Soil Biol. Biochem. 41, 315–322.

Casella, E., Soussana, J.F., 1997. Long-term effects of CO_2 enrichment and temperature increase on the carbon balance of a temperate grass sward. J. Exp. Bot. 48, 1309–1321.

Cheng, L., Leavitt, S.W., Kimball, B.A., Pinter Jr., P.J., Ottman, M.J., Matthias, A., Wall, G.W., Brooks, T., Williams, D.G., Thompson, T.L., 2007. Dynamics of labile and recalcitrant soil carbon pools in a sorghum free-air CO_2 enrichment (FACE) agroecosystem. Soil Biol. Biochem. 39, 2250–2263.

Cheng, W., 1999. Rhizosphere feedbacks in elevated CO_2. Tree Physiol. 19, 313–320.

Conant, R.T., Paustian, K., 2002. Spatial variability of soil organic carbon in grasslands: implications for detecting change at different scales. Environ. Poll. 116, 127–135.

Dijkstra, F.A., Cheng, W., 2007. Moisture modulates rhizosphere effects on C decomposition in two different soil types. Soil Biol. Biochem. 39, 2264–2274.

Dijkstra, F.A., Morgan, J.A., Von Fischer, J.C., Follett, R.F., 2011. Elevated CO_2 and warming effects on CH_4 uptake in a semiarid grassland below optimum soil moisture. J. Geophys. Res.—Biogeosci. 116, G01007. doi: 10.1029/2010JG001288.

Epstein, H.E., Lauenroth, W.K., Burke, I.C., Coffin, D.P., 1997. Productivity patterns of C_3 and C_4 functional types in the U.S. Great Plains. Ecology 78, 722–731.

Fierer, N., Grandy, A.S., Six, J., Paul, E.A., 2009. Searching for unifying principles in soil ecology. Soil Biol. Biochem. 41, 2249–2256.

Giesemann, A., Weigel, H.J., 2008. Soil carbon isotopic composition and soil carbon content in an agroecosystem during six years of Free Air Carbon dioxide Enrichment (FACE). Isotopes Environ. Health Studies. 44, 349–363.

Grant, R.F., Pattey, E., 2008. Temperature sensitivity of N_2O emissions from fertilized agricultural soils: mathematical modeling in ecosys. Glob. Biogeochem. Cycles. 22, GB4019. doi:4010.1029/2008GB003273.

Guo, Y., Gong, P., Amundson, R., Yu, Q., 2006. Analysis of factors controlling soil carbon in the conterminous United States. Soil Sci. Soc. Am. J. 70, 601–612.

Hartley, I.P., Heinemeyer, A., Evans, S.P., Ineson, P., 2007. The effect of soil warming on bulk soil vs. rhizosphere respiration. Glob. Change Biol. 13, 2654–2667.

Heimann, M., Reichstein, M., 2008. Terrestrial carbon dynamics and climate feedbacks. Nature 451, 289–292.

Hu, S., Chapin III, F.S., Firestone, M.K., Field, C.B., Chiariello, N.R., 2001. Nitrogen limitation of microbial decomposition in a grassland under elevated CO_2. Nature 409, 188–191.

Hu, Y., Chang, X., Lin, X., Wang, Y., Wang, S., Duan, J., Zhang, Z., Yang, X., Luo, C., Xu, G., Zhao, X., 2010. Effects of warming and grazing on N_2O fluxes in an alpine meadow ecosystem on the Tibetan plateau. Soil Biol. Biochem. 42, 944–952.

Hungate, B.A., Holland, E.A., Jackson, R.B., Chapin III, F.S., Mooney, H.A., Field, C.B., 1997a. The fate of carbon in grassland under carbon dioxide enrichment. Nature 388, 576–579.

Hungate, B.A., Lund, C.P., Pearson, H.L., Chapin III, F.S., 1997b. Elevated CO_2 and nutrient addition alter soil N cycling and N trace gas fluxes with early season wet-up in a California annual grassland. Biogeochem. 37, 89–109.

Hungate, B.A., van Groenigen, K.J., Six, J., Jastrow, J.D., Luo, Y., de Graaff, M.A., van Kessel, C., Osenberg, C.W., 2009. Assessing the effect of elevated carbon dioxide on soil carbon: a comparison of four meta-analyses. Glob. Change Biol. 15, 2020–2034.

Ineson, P., Coward, P.A., Hartwig, U.A., 1998. Soil gas fluxes of N_2O, CH_4 and CO_2 beneath Lolium perenne under elevated CO_2: the Swiss free air carbon dioxide enrichment experiment. Plant Soil 198, 89–95.

Inubushi, K., Cheng, W., Aonuma, S., Hoque, M.M., Kobayashi, K., Miura, S., Kim, H.Y., Okada, M., 2003. Effects of free-air CO_2 enrichment (FACE) on CH_4 emission from a rice paddy field. Glob. Change Biol. 9, 1458–1464.

Intergovernmental Panel on Climate Change, 2007. Working Group I Report. The Physical Science Basis. Technical Summary Available online at. http://www.ipcc.ch (verified January 17, 2011).

Islam, K.R., Mulchi, C.L., Ali, A.A., 2000. Interactions of tropospheric CO_2 and O_3 enrichments and moisture variations on microbial biomass and respiration in soil. Glob. Change Biol. 6, 255–265.

Jäger, H.J., Schmidt, S.W., Kammann, C., Grünhage, L., Müller, C., Hanewald, K., 2003. The University of Giessen Free-Air Carbon dioxide Enrichment study: description of the experimental site and of a new enrichment system. J. Appl. Bot. 77, 117–127.

Juan, L., Yong, H., Zucong, C., Huilin, L., 2007. Changes in CH_4 and CO_2 emissions from soils under flooded conditions after being exposed to FACE (free-air CO_2 enrichment) for three years. Acta Ecologica Sinica 27, 2184–2190.

Kammann, C., Müller, C., Grünhage, L., H- Jäger, J., 2008. Elevated CO_2 stimulates N_2O emissions in permanent grassland. Soil Biol. Biochem. 40, 2194–2205.

Kamp, T., Steindl, H., Hantschel, R.E., Beese, F., Munch, J.C., 1998. Nitrous oxide emissions from a fallow and wheat field as affected by increased soil temperatures. Biol. Fert. Soils. 27, 307–314.

Kimball, B.A., Idso, S.B., 1983. Increasing atmospheric CO_2: effects on crop yield, water use, and climate. Agric. Water Manage 7, 55–72.

Kimball, B.A., Kobayashi, K., Bindi, M., 2002. Responses of Agricultural Crops to Free-air CO_2 Enrichment. Adv. Agron. vol. 77 Academic Press.

Kou, T., Zhu, J., Xie, Z., Hasegawa, T., Heiduk, K., 2007. Effect of elevated atmospheric CO_2 concentration on soil and root respiration in winter wheat by using a respiration partitioning chamber. Plant Soil 299, 237–249.

Lam, S.K., Lin, E., Norton, R., Chen, D., 2011. The effect of increased atmospheric carbon dioxide concentration on emissions of nitrous oxide, carbon dioxide and methane from a wheat field in a semi-arid environment in northern China. Soil Biol. Biochem. 43, 458–461.

Leavitt, S.W., Paul, E.A., Galadima, A., Nakayama, F.S., Danzer, S.R., Johnson, H., Kimball, B.A., 1996. Carbon isotopes and carbon turnover in cotton and wheat FACE experiments. Plant Soil 187, 147–155.

Leavitt, S.W., Pendall, E., Paul, E.A., Brooks, T., Kimball, B.A., Pinter Jr., P.J., Johnson, H.B., Matthias, A., Wall, G.W., LaMorte, R.L., 2001. Stable-carbon isotopes and soil organic carbon in wheat under CO_2 enrichment. New Phytol 150, 305–314.

Liu, W., Zhang, Z., Wan, S., 2009. Predominant role of water in regulating soil and microbial respiration and their responses to climate change in a semiarid grassland. Glob. Change Biol. 15, 184–195.

Luo, Y., Jackson, R.B., Field, C.B., Mooney, H.A., 1996. Elevated CO_2 increases belowground respiration in California grasslands. Oecologia 108, 130–137.

Luo, Y., Sherry, R., Zhou, X., Wan, S., 2009. Terrestrial carbon-cycle feedback to climate warming: experimental evidence on plant regulation and impacts of biofuel feedstock harvest. GCB Bioenergy 1, 62–74.

Marhan, S., Kandeler, E., Rein, S., Fangmeier, A., Niklaus, P.A., 2010. Indirect effects of soil moisture reverse soil C sequestration responses of a spring wheat agroecosystem to elevated CO_2. Glob. Change Biol. 16, 469–483.

Morgan, J.A., Pataki, D.E., Korner, C., Clark, H., DelGrosso, S.J., Grunzewig, J.M., Knapp, A.K., Mosier, A.R., Newton, P.C.D., Niklaus, P.A., Nippert, J.B., Nowak, R.S., Parton, W.J., Polley, H.W., Shaw, M.R., 2004. Water relations in grassland and desert ecosystems to elevated atmospheric CO_2. Oecologia 140, 11–25.

Mosier, A., Schimel, D., Valentine, D., Bronson, K., Parton, W., 1991. Methane and nitrous oxide fluxes in native, fertilized and cultivated grasslands. Nature 350, 330–332.

Mosier, A.R., Morgan, J.A., King, J.Y., LeCain, D., Milchunas, D.G., 2002. Soil-atmosphere exchange of CH_4, CO_2, NO_x, and N_2O in the Colorado shortgrass steppe under elevated CO_2. Plant Soil 240, 201–211.

Nakayama, F.S., Huluka, G., Kimball, B.A., Lewin, K.F., Nagy, J., Hendrey, G.R., 1994. Soil carbon dioxide fluxes in natural and CO_2-enriched systems. Agric. For. Meteorol. 70, 131–140.

Newton, P.C.D., Clark, H., Bell, C.C., Glasgow, E.M., Campbell, B.D., 1994. Effects of elevated CO_2 and simulated seasonal chagnes in temperature on the species composition and growth rate of pasture turves. Ann. Bot. 73, 53–59.

Niklaus, P.A., Körner, C., 1996. Responses of soil microbiota of a late successional alpine grassland to long term CO_2 enrichment. Plant Soil 184, 219–229.

Niklaus, P.A., Wohlfender, M., Siegwolf, R., Körner, C., 2001. Effects of six years atmospheric CO_2 enrichment on plant, soil, and soil microbial C of a calcareous grassland. Plant Soil 233, 189–202.

Pendall, E., Leavitt, S.W., Brooks, T., Kimball, B.A., Pinter Jr., P.J., Wall, G.W., LaMorte, R.L., Wechsung, G., Wechsung, F., Adamsen, F., Matthias, A.D., Thompson, T.L., 2001. Elevated CO_2 stimulates soil respiration in a FACE wheat field. Basic Appl. Ecol. 2, 193–201.

Pendall, E., Mosier, A.R., Morgan, J.A., 2004. Rhizodeposition stimulated by elevated CO_2 in a semiarid grassland. New Phytol 162, 447–458.

Pendall, E., Osanai, Y.U.I., Williams, A.L., Hovenden, M.J., 2011. Soil carbon storage under simulated climate change is mediated by plant functional type. Glob. Change Biol. 17, 505–514.

Peralta, A.L., Wander, M.M., 2008. Soil organic matter dynamics under soybean exposed to elevated [CO_2]. Plant Soil 303, 69–81.

Pleijel, H., Sild, J., Danielsson, H., Klemedtsson, L., 1998. Nitrous oxide emissions from a wheat field in response to elevated carbon dioxide concentration and open-top chamber enclosure. Environ. Poll. 102, 167–171.

Prior, S.A., Runion, G.B., Rogers, H.H., Torbert, H.A., Reeves, D.W., 2005. Elevated atmospheric CO_2 effects on biomass production and soil carbon in conventional and conservation cropping systems. Glob. Change Biol. 11, 657–665.

Prior, S.A., Runion, G.B., Torbert, H.A., Rogers, H.H., 2004. Elevated atmospheric CO_2 in agroecosystems: soil physical properties. Soil Science 169, 434–439.

Prior, S.A., Torbert, H.A., Runion, G.B., Rogers, H.H., Wood, C.W., Kimball, B.A., LaMorte, R.L., Pinter, P.J., Wall, G.W., 1997. Free-air carbon dioxide enrichment of wheat: soil carbon and nitrogen dynamics. J. Environ. Qual. 26, 1161–1166.

Reich, P.B., Hobbie, S.E., Lee, T., Ellsworth, D.S., West, J.B., Tilman, D., Knops, J.M.H., Naeem, S., Trost, J., 2006a. Nitrogen limitation constrains sustainability of ecosystem response to CO_2. Nature 440, 922–925.

Reich, P.B., Hungate, B.A., Luo, Y., 2006b. Carbon-nitrogen interactions in terrestrial ecosystems in response to rising atmospheric carbon dioxide. Ann. Rev. Ecol. Evol. System. 37, 611–636.

Ross, D.J., Newton, P.C.D., Tate, K.R., 2004. Elevated [CO_2] effects on herbage production and soil carbon and nitrogen pools and mineralization in a species-rich, grazed pasture on a seasonally dry sand. Plant Soil 260, 183–196.

Rustad, L., Campbell, J., Marion, G., Norby, R., Mitchell, M., Hartley, A., Cornelissen, J., Gurevitch, J., GCTE-NEWS, 2001. A meta-analysis of the response of soil respiration, net nitrogen mineralization, and aboveground plant growth to experimental ecosystem warming. Oecologia 126, 543–562.

Schrope, M.K., Chanton, J.P., Allen, L.H., Baker, J.T., 1999. Effect of CO_2 enrichment and elevated temperature on methane emissions from rice, *Oryza sativa*. Glob. Change Biol. 5, 587–599.

Sinsabaugh, R.L., Lauber, C.L., Weintraub, M.N., Ahmed, B., Allison, S.D., Crenshaw, C., Contosta, A.R., Cusack, D., Frey, S., Gallo, M.E., Gartner, T.B., Hobbie, S.E., Holland, K., Keeler, B.L., Powers, J.S., Stursova, M., Takacs-Vesbach, C., Waldrop, M.P., Wallenstein, M.D., Zak, D.R., Zeglin, L.H., 2008. Stoichiometry of soil enzyme activity at global scale. Ecol. Lett. 11, 1252–1264.

Smith, P., 2004. How long before a change in soil organic carbon can be detected? Glob. Change Biol. 10, 1878–1883.

Smith, K.E., Runion, G.B., Prior, S.A., Rogers, H.H., Torbert, H.A., 2010. Effects of elevated CO_2 and agricultural management on flux of greenhouse gases from soil. Soil Sci. 175, 349–356.

Solomon, S., Plattner, G.K., Knutti, R., Friedlingstein, P., 2009. Irreversible climate change due to carbon dioxide emissions. Proc. Nat. Acad. Sci. 106, 1704–1709.

Teyssonneyre, F., Picon-Cochard, C., Falcimagne, R., J- Sousanna, R., 2002. Effects of elevated CO2 and cutting frequency on plant community structure in a temperate grassland. Glob. Change Biol. 8, 1034–1046.

Tokida, T., Fumoto, T., Cheng, W., Matsunami, T., Adachi, M., Katayanagi, N., Matsushima, M., Okawara, Y., Nakamura, H., Okada, M., Sameshima, R., Hasegawa, T., 2010. Effects of free-air CO_2 enrichment (FACE) and soil warming on CH_4 emission from a rice paddy field: impact assessment and stoichiometric evaluation. Biogeosciences 7, 2639–2653.

Van Groenigen, K.J., Six, J., Hungate, B.A., De Graaff, M.A., Van Breemen, N., Van Kessel, C., 2006. Element interactions limit soil carbon storage. Proc. Nat. Acad. Sci. 103, 6571–6574.

Van Groenigen, K.J., Osenberg, C.W., Hungate, B.A., 2011. Increased soil emissions of potent greenhouse gases under increased atmospheric CO_2. Nature 475, 214–216.

Van Kessel, C., Boots, B., de Graaff, M.A., Harris, D., Blum, H., Six, J., 2006. Total soil C and N sequestration in a grassland following 10 years of free air CO_2 enrichment. Glob. Change Biol. 12, 2187–2199.

Van Kessel, C., Horwath, W.R., Hartwig, U., Harris, D., Lüscher, A., 2000. Net soil carbon input under ambient and elevated CO_2 concentrations: isotopic evidence after 4 years. Glob. Change Biol. 6, 435–444.

Wand, S.J.E., Midgley, G.F., Jones, M.H., Curtis, P.S., 1999. Responses of wild C_4 and C_3 grasses (Poaceae) species to elevated atmospheric CO_2 concentrations: a meta-analytic test of current theories and perceptions. Glob. Change Biol. 5, 723–741.

Welzmiller, J.T., Matthias, A.D., White, S., Thompson, T.L., 2008. Elevated carbon dioxide and irrigation effects on soil nitrogen gas exchange in irrigated sorghum. Soil Sci. Soc. Am. J. 72, 393–401.

Williams, M.A., Rice, C.W., Owensby, C.E., 2000. Carbon dynamics and microbial activity in tallgrass prairie exposed to elevated CO_2 for 8 years. Plant Soil 227, 127–137.

Wood, C.W., Torbert, H.A., Rogers, H.H., Runion, G.B., Prior, S.A., 1994. Free-air CO_2 enrichment effects on soil carbon and nitrogen. Agric. For. Meteor. 70, 103–116.

Wu, Z., Dijkstra, P., Koch, G.W., Peñuelas, J., Hungate, B.A., 2011. Responses of terrestrial ecosystems to temperature and precipitation change: a meta-analysis of experimental manipulation. Glob. Change Biol. 17, 927–942.

Xia, J., Han, Y., Zhang, Z., Wan, S., 2009. Effects of diurnal warming on soil respiration are not equal to the summed effects of day and night warming in a temperate steppe. Biogeosciences 6, 1361–1370.

Xu, Z., Zheng, X., Wang, Y., Han, S., Huang, Y., Zhu, J., Butterbach-Bahl, K., 2004. Effects of elevated CO_2 and N fertilization on CH_4 emissions from paddy rice fields. Glob. Biogeochem. Cycles. 18, GB3009.

Zhang, L., Wylie, B.K., Ji, L., Gilmanov, T.G., Tieszen, L.L., 2010. Climate-driven interannual variability in net ecosystem exchange in the Northern Great Plains grasslands. Rangeland Ecol. Manag. 63, 40–50.

Zheng, X., Zhou, Z., Wang, Y.Y., Zhu, J., Wang, Y., Yue, J., Shi, Y., Kobayashi, K., Inubushi, K., Huang, Y., Han, S., Xu, Z., Xie, B., Butterbach-Bahl, K., Yang, L., 2006. Nitrogen-regulated effects of free-air CO_2 enrichment on methane emissions from paddy rice fields. Glob. Change Biol. 12, 1717–1732.

Zhong, S., Liang, W., Lou, Y., Li, Q., Zhu, J., 2009. Four years of free-air CO_2 enrichment enhance soil C concentrations in a Chinese wheat field. J. Env. Sci. 21, 1221–1224.

Zhou, X., Wan, S., Luo, Y., 2007. Source components and interannual variability of soil CO_2 efflux under experimental warming and clipping in a grassland ecosystem. Glob. Change Biol. 13, 761–775.

Ziska, L.H., Moya, T.B., Wassmann, R., Namuco, O.S., Lantin, R.S., Aduna, J.B., Abao Jr., E., Bronson, K.F., Neue, H.U., Olszyk, D., 1998. Long-term growth at elevated carbon dioxide stimulates methane emission in tropical paddy rice. Glob. Change Biol. 4, 657–665.

CHAPTER 28

Mitigation Opportunities from Land Management Practices in a Warming World: Increasing Potential Sinks

J.L. Hatfield, T.B. Parkin, T.J. Sauer, J.H. Prueger
USDA Agricultural Research Service, National Laboratory for Agriculture and the Environment, Ames, IA

CHAPTER OUTLINE

Abbreviations: CO_2, carbon dioxide; N_2O, nitrous oxide; C, carbon; N, nitrogen; SOM, soil organic matter; SOC, soil organic carbon; GHG, greenhouse gas

CHANGING CLIMATE

Climate across the Northern Hemisphere is changing and will continue to change into the future, and as a result will affect temperature, precipitation and atmospheric carbon dioxide (CO_2) concentration. The rising CO_2 concentrations have led to scenarios of increased temperatures around the world. Within the United States, Karl et al. (2009) presented an analysis of the recent changes and projected changes over the next century. Patterns of temperature and precipitation across the United States for the next 30 years show a warming trend of 1.5 to 2°C and a slight increase in precipitation over most of the country (Tebaldi et al., 2006; Karl et al., 2009). The changes in temperature are projected to be more uniform than precipitation in which the projections are for increased within-season variation and more spatial variation across the United States. Across the United States, spring precipitation will increase while summer precipitation is projected to decrease. Karl et al. (2009) showed that

Managing Agricultural Greenhouse Gases. DOI: 10.1016/B978-0-12-386897-8.00028-0
2012, Published by Elsevier Inc.

precipitation events would change in frequency and intensity with a projected increase in spring precipitation, particularly in the northeast and Midwest U.S., and a decline in the southwestern U.S. Analysis of climatological data from Iowa has revealed that these changes are already evident in the data record with an increase in springtime precipitation and more days with rainfall above 30 mm per day (Hatfield, 2011, unpublished data).

Both Tebaldi et al. (2006) and Karl et al. (2009) projected an increase in the number of days when the temperature will be higher than the climatic normals by 5°C (heat waves), which will impact agricultural systems. They also project an increase in warm nights, defined as occurring when the minimum temperature is above the 90th percentile of the historical climatological temperature distribution for the day (Tebaldi et al., 2006; Karl et al., 2009). Observations from Iowa have shown that minimum temperatures have steadily increased over the past 15 years. Coupled with these changes is the decrease in a number of frost days by 10% in the eastern half of the U.S., increasing the growing season by over 10 days. Agricultural production will be impacted by these increases in extreme temperature events, warm nights, and more variable precipitation. The potential impact of climate change on crop production has been reviewed by Hatfield et al. (2011) and on forage and rangeland production by Izaurralde et al. (2011). The changes in climate will impact crop production and will also affect how various management practices will impact greenhouse gas (GHG) emissions. The efficacy of mitigation practices on greenhouse gas emissions will be tempered by the increased variation in climate. In this chapter we will address how these factors potentially interact.

Changes in climate over the next 50 years will be evident in the atmospheric CO_2 concentrations leading to a positive impact on plant growth. In general, the effect of increased CO_2 to nearly 450 μmol mol^{-1} will be to increase plant productivity. The effect of the increase in CO_2 concentration differs among species. Hatfield et al. (2011) provided a summary of the current synthesis of increased CO_2 concentrations on different agricultural species. In general, the changes in productivity are positive; however, they are sometimes offset by changes in temperature and precipitation. The interactions among the effects of CO_2, temperature, and precipitation complicate our ability to provide a precise estimate of the impacts of future climate on production systems.

Over the next 50 years we can expect temperatures to increase by 0.5 to 0.8°C. This increase in mean temperatures will vary across the United States leading to regional and seasonal differences. It is important to realize that changes in temperature will be coupled with increased temporal and spatial variation in precipitation. These changes will interact with the backdrop of a rising CO_2 affecting plant growth. These components of a changing climate will interact with soil management practices to affect the efficacy of mitigation practices and emissions of greenhouse gases. Effects of climate change on the efficacy of soil management practices will be expressed in two ways, direct effect on the processes affecting the emission of GHG and the indirect effect of climate change on the growth of crops or forage affecting the water, carbon (C), or nitrogen (N) dynamics within the soil profile. Our goal is to summarize the current information on how the mitigation potential of soil management practices will respond to climate change and how climate change will affect our ability to utilize crop and soil management practices to mitigate GHG emissions.

SOIL MANAGEMENT PRACTICES

Soil management practices can be classified as those affecting tillage, placement, and incorporation of residue and nutrients. It is important to understand that changes in soil management practices affect soil water balance, temperature, biological activity, and gas exchange between the soil and the atmosphere. Follett (2001) summarized the effect of soil management on soil organic carbon (SOC) as being attributed to: tillage and soil management systems; management to increase the amount of crop cover; and increased efficiency in the use of inputs

(N and water) by the cropping system. To be able to assess how soil management practices may be changed to better respond to a changing climate requires that we understand how each of these factors affect the processes governing the emission of GHG. There have been several reviews in the past decade on the role of soil management practices on mitigation of GHG (e.g. Martens et al., 2005). An interesting conclusion from their work was that agricultural impacts on C cycling in soil needed greater understanding before the full potential of soil management practices on C mitigation strategies could be better quantified. Earlier, West and Marland (2002) summarized the effect of changing tillage practices that could result in a reduction in CO_2 emissions from agriculture equivalent to 368 kg C ha^{-1} yr^{-1} but would vary by cropping system and climate regime. Similar conclusions were reached by Franzluebbers (2005) in which he found that reductions in tillage intensity would increase the SOC content but the impact on other GHG emissions was less well defined. These statements suggest that if we are to understand the role of soil management practices on mitigation potential we need first to understand the soil processes affected by changes in soil management practices.

SOIL WATER CHANGES

Dynamics of soil C and N cycling within the soil are affected by soil water distribution within the profile and over time. In general, a reduction in tillage intensity increases the soil water content in the soil profile because of the reduced rate of soil water evaporation caused by the presence of the crop residue on the surface and the reduction in episodic bursts in soil water evaporation caused by the disturbance of the soil and exposing wet soil to the atmosphere. Maintenance of crop residues on the soil surface increases infiltration into the soil and decreases soil water evaporation (Steiner, 1994). Tilling the surface disrupts the soil surface which often leads to an initial increase in soil water storage by increasing infiltration into soil for the first precipitation event while in subsequent precipitation events infiltration rates in tilled soil may be limited by surface sealing due to lack of stability of soil aggregates. Disruption of the soil surface increases soil water evaporation compared to residue covered surfaces or undisturbed surfaces because tillage moves moist soil up to the surface where losses from drying may offset increased infiltration rates (Burns et al., 1971; Papendick et al., 1973). Soil water evaporation is affected by soil water content at the surface and the degree of plant cover on the surface (Ritchie, 1971). Sauer et al. (1996) observed that the presence of surface residue reduced soil water evaporation by 34 to 50% and manipulating the weathered residue to create a 15 cm bare strip increased soil water evaporation by only 7% as compared to the fall residue cover. Deibert et al. (1986) found precipitation storage efficiency in the northern Great Plains was similar among tillage systems but varied among years and locations during the non-growing season under continuous wheat (*Triticum aestivum* L.). Steiner (1994) summarized the impacts of crop residue on water conservation in agricultural systems and her review serves as a point of reference of how tillage and residue management affects the water balance in soils. She concluded from the studies throughout the world that maintaining crop residue on the soil surface reduced soil water evaporation and increased soil water storage in the soil profile leading to increased plant growth because of the greater amount of available soil water. The variation in precipitation potentially induced by climate change will affect the soil water balance and methods to increase soil water availability, and will have a positive impact on plant growth and yield. Tillage and residue management affects the soil water balance and with the increasing variability in precipitation will increase the need for improved soil water management and water savings. Improved soil water management will occur as a result of maintaining crop residues, limiting tillage operations, and potentially altering the crop rotations to avoid high water use crops in water limited areas.

CO_2 FLUXES FROM SOIL

Extending the concepts of soil water dynamics to CO_2 flux from the soil reveals tillage management practices when coupled with variation in precipitation does affect soil–atmospheric CO_2 dynamics. However, there is an additional impact induced by N dynamics and management because of the effect on soil microbiological activity. Variation in CO_2 flux has been associated with cultivation type and tillage history with variation across soil types along with soil water content and presence and age of crop residue when tillage is implemented (Wood and Edwards, 1992; Prior et al., 1997; Kessavalou et al., 1998; Calderón et al., 2000). Additionally, tilling previously untilled soil also increased CO_2 emissions. There have been a number of mechanisms proposed for this response including aggregate disruption or exposure of protected organic matter to decomposition (La Scala et al., 2008). In addition to the short-term tillage effects, some of the long-term effects of tillage include modification of soil environmental conditions, SOC content, and stratification (Franzluebbers et al., 1995; Bono et al., 2008). Reicosky et al. (1997) and Reicosky and Archer (2007) observed in cultivated soils that tillage operations caused a release of gases entrapped in the soil pores with soil CO_2 fluxes showing an initial burst followed by stabilization within hours. There is variation among systems in the CO_2 loss depending on the characteristics of each system (Alvarez et al., 2001). There is a positive effect of rainfall on soil CO_2 fluxes because of the improved transfer of gases through the soil pores (Rochette et al., 1991). A seasonal trend in soil CO_2 flux occurs in relation to soil environmental conditions and crop phenology related to the changes in soil temperature throughout the year and the modification of the soil temperature caused by the shading from the plant canopy as it develops throughout the growing season. During drought periods, soil CO_2 flux is negatively related to soil water content (Akinremi et al., 1999). Precipitation events cause increased CO_2 emissions from heterotrophic respiration (Inglima et al., 2009), and recovery of microbial activity (Borken and Matzner, 2009). Precipitation events are offset by temperature effects which limit CO_2 emission (Almagro et al., 2009). However, the seasonal trends in CO_2 fluxes from the soil are affected by tillage and precipitation events (Sanchez et al., 2003).

Observations of CO_2 emissions from agricultural systems reveal no uniform response to management practices. Morell et al. (2010) observed CO_2 fluxes from three tillage systems: conventional, minimum, and no-tillage in a barley (*Hordeum vulgare* L.) monoculture system. Their results showed a burst of CO_2 flux with tillage where the magnitude was related to the CO_2 flux from the previous day. The CO_2 fluxes increased from 1 to 8 $\mu mol\ m^{-2}\ s^{-1}$ following tillage with the differences among years caused by variations in the soil water content with the wetter years creating a larger flux of CO_2 from the soil. As the soil dried, there was a decrease in CO_2 loss from the soil. In their study, rainfall events caused soil CO_2 fluxes to increase to an even a larger magnitude in minimum tillage systems than fluxes under conventional tillage operations (Morell et al., 2010). They also observed that N fertilization led to higher soil CO_2 fluxes particularly under the conventional tillage system. The observations by Morell et al. (2010) and Lal (2009) on the effect of N fertilization on soil CO_2 fluxes suggested N management would affect the soil organic matter (SOM) dynamics within the soil and by inference of soil CO_2 fluxes. Morell et al. (2010) observed the response of soil CO_2 flux to N application and the interaction with tillage systems is similar to the interaction with residue production. There was an increase in residue production by 10–20% under medium (60 $kg\ ha^{-1}$) and high (120 $kg\ ha^{-1}$) N fertilization levels under conservation tillage systems, but not under conventional tillage. The positive impact on biomass production was a result of increased available water under conservation tillage systems during the growing seasons. As a further effect, the increased residue production may have increased substrate availability and soil microbial activity that would have increased soil CO_2 flux (Morell et al., 2010). Similar results were observed by Sainju et al. (2008a, b) in which management of N with both manure and commercial fertilizer caused a positive response on plant growth which in turn affected the soil C dynamics through the addition of more root material into the soil leading to an increase in the soil carbon pools.

Soil management practices affect soil water balance, loss of C through enhanced CO_2 release via tillage, and changes in the soil C through the enhancement effects on plant growth induced by improved N management. Changes in soil management affect the dynamics of soil water, C, and temperature. These changes interact with the future climate variability in precipitation and temperature which will directly impact both the soil environment and plant response. The interactions among climate variability, crop management, and soil management are complex, but need to be understood in the context of how these interactions will affect the C and N dynamics at the soil–plant–atmosphere interface.

CLIMATE IMPACTS ON SOIL AND PLANT

To fully understand the potential of various mitigation practices it is also important to understand the impact of climate on crop growth and development. The impacts of climate change on agricultural crops were summarized by Hatfield et al. (2011) and on forage and rangeland crops by Izaurralde et al. (2011). We live in a warming world with increasing CO_2 concentrations. Current CO_2 concentration is 387 $\mu mol\ mol^{-1}$ and is expected to increase to nearly 450 $\mu mol\ mol^{-1}$ by 2050. Evaluation of the impacts of increasing CO_2 on plant growth has been investigated through the use of a variety of controlled chambers. Kimball (2010) summarized the results of free air CO_2 enrichment (FACE) studies as: increasing CO_2 to 550 $\mu mol\ mol^{-1}$ stimulated photosynthesis by 10 to 45% in C_3 plants but only up to 10% in C_4 plants. This effect was decreased by low soil N availability and the stimulation increased at air temperatures above 25°C compared to less than 25°C. Increasing CO_2 also increases water use efficiency in crops by decreasing the evapotranspiration because there is a greater uptake of CO_2 compared to transpiration losses; however, under limited soil N the increases in water use efficiency were diminished (Kimball, 2010). Hatfield et al. (2011) in their summary showed the interacting effects of increasing temperatures along with CO_2 and showed that many of the positive benefits of rising CO_2 would be offset by rising temperatures. In their analysis, the inclusion of temperature revealed that warming temperature effects on crops was dependent upon the temperature response of the individual species because some crops will experience temperature stresses with a warming environment. Temperature stresses are induced throughout the growing period and during the vegetative growth period cause an increase in the rate of growth and smaller plant size along with an increase in water use because of the increased evaporative demand. During the reproductive period (grain-filling), exposure to warmer temperatures causes more rapid grain-fill and increased rate of senescence leading to lower yields. One of the most sensitive stages to temperature is the pollination stage and failures in pollination will negatively impact grain yields (Hatfield et al., 2011). In summary, exposure of temperate crops to warmer temperatures causes plants to grow more quickly and hence tend to be smaller plants with reduced yield. Coupled with the warmer temperatures is an increase in evapotranspiration resulting in increased crop water use.

One complication in the climate change scenarios is the projected increase in climate variation, e.g. temperature and precipitation, during the year and even more so during the growing season. The increase in temperature extremes places a potential stress on plants which may limit growth. Temperature extremes during pollination of grain crops would severely impact yield and even more so when coupled with limited soil water (Hatfield et al., 2011). Variation in temperature and precipitation can limit the growth and yield of crops and exposure to warming temperatures during the grain-filling period of most grain crops will reduce yield. Hatfield (2010) observed from yield records across the Midwest (Illinois, Indiana, Iowa, and Kansas) that precipitation was the dominant factor in preventing crop yields from reaching their maximum potential. If we assume the harvest index (ratio of grain to total biomass production) remains relatively constant then a reduction in growth induced by decreases in precipitation would cause less biomass to be returned to the soil. In small grains (wheat and barley), water stress from limited precipitation decreases the grain yield significantly and Hatfield (2010) showed for winter wheat in Kansas and South Dakota yield losses up to 70% of the potential yield occurred in years with dry periods during grain filling. A reduction in biomass returned to the soil would affect the soil organic matter changes and also alter the N uptake patterns by the crop. Additional variation in temperature and precipitation as evidenced through extreme events, e.g. heat waves or intense storms, will have a negative impact on plant growth and our current production systems may not be able to effectively withstand these events. The result will be a decrease in biomass production or even total crop failure. Grassini et al. (2009) evaluated the response of corn (*Zea mays* L.) across the Midwest U.S. at 18 different locations under both irrigated and rainfed environments using the Hybrid-Maize model. They found the Hybird-Maize model accurately simulated corn yields for both rainfed and irrigated environments with radiation, temperature, and potential evapotranspiration gradients impacting corn productivity. They also found potential grain yields were related to cumulative incident solar radiation and temperature during the post-silking period for irrigated conditions, while under rainfed conditions grain yields were closely related to available soil water supply (Grassini et al., 2009). Lobell et al. (2006) studied the impact of climate on crop yields in California and found that yields for six perennial crops [almonds (*Prunus dulcis*), walnuts (*Juglans regla* L.), avocados (*Persea Americana* P. Mill.), wine grapes (*Vitis vinifera* L.), table grapes (*Vitis vinifera* L.), and oranges (*Citrus sinensis* L. Osbeck)] had projected yield losses of 0 to 40% primarily due to limited water availability. These examples demonstrate the impact that variable rainfall can have on crop production and all crops will be affected to some degree by the increased variability in climate. This is typified in the results reported by Haim et al. (2008) for wheat and cotton (*Gossypium hirsutum* L.) crops in Israel that scenarios of climate change would cause yield reductions in both crops with water limitation being a dominant factor. They also found that moderate changes in climate (temperature and precipitation) could be offset with improved management practices, e.g. N management.

Corn has been one of the most studied crops in terms of response to climate and serves as an example of how changes in climate can affect crop production. An original study by Runge (1968) provides a foundation for demonstrating how corn yields were affected by an interaction of daily maximum temperature and rainfall in the period 25 days prior to and 15 days after anthesis. If rainfall was low (0 to 44 mm per 8 days) yield was reduced by 1.2 to 3.2% per 1°C rise in temperature from the normal temperatures for the grain-filling period. Conversely, if temperature was warm (T_{max} of 35°C), yield declined 9% per 25.4 mm rainfall decline. Using corn yields across a number of locations, Muchow et al. (1990) found the highest observed and simulated grain yields occurred with relatively cool temperature (growing season mean of 18.0 to 19.8°C at Grand Junction, CO), compared to warmer sites, e.g. Champaign, IL (21.5 to 24.0°C), or warm tropical sites (26.3 to 28.9°C). Warming temperatures and variable rainfall will have major impacts on growth and yield of crops and will result in increasing variation in crop production.

An interesting observation from climate change and CO_2 increase is from studies on rangeland ecosystems where production is often limited by N, and where changing climate affects cycling between organic and inorganic N compounds. In these systems, plant material deposited on the soil surface or accumulated in the root zone is decomposed by soil fauna and micro flora and ultimately becomes part of the SOM pool. As the SOM decomposes there is a release of CO_2 as well as minerals and other plant-available forms of N. As temperatures increase, especially in the colder regions, SOM decomposition rates and therefore soil CO_2 emission rates will increase (Raich and Schlesinger, 1992; Reich et al., 2006b; Rustad et al., 2001). At temperatures below the biological optimum temperature, exponential increases in biological activity can occur in response to increases in temperature. The biological optimum temperature is defined as the temperature at which there is a maximum growth rate of the plant with no evidence of thermal stress. Raich and Schlesinger (1992) summarized the Q10 temperate response factors for soil CO_2 production and reported a median factor of 2.4. Using this factor it is calculated that global temperature increases of 0.5 to 0.8°C may increase soil CO_2 production by 4.5 to 7.2% (Raich and Schlesinger, 1992).

Whereas warming may stimulate microbial activity resulting in greater soil CO_2 emissions, more rapid soil drying resulting from enhanced plant growth and increased rate of soil water use may reduce microbial activity (Wan et al., 2005). Warming may lead to an increase in the length of the growing season causing a further increase in the rate of decomposition processes (Wan et al., 2005). In water-limited rangeland systems, soil water content will become the major factor leading to changes in decomposition rates (Epstein et al., 2002; Wan et al., 2005).

Rising CO_2 does not directly affect soil microbial processes. Luo et al. (2004) observed that CO_2 enrichment reduced plant-available N through a stimulation of plant growth that increased plant N uptake and N sequestration in long-lived plant biomass and SOM pools. Luo et al. (2004) developed the Progressive Nitrogen Limitation (PNL) hypothesis to link CO_2 enrichment to reduced plant-available N by increasing plant demand for N and enhancing the sequestration of N in long-lived plant biomass and SOM pools. The greater plant demand for N is caused by effects of elevated CO_2 on plant growth. The continual accumulation of N in organic compounds may eventually reduce soil N availability and limit plant growth response to CO_2 or other climate changes (Reich et al., 2006a, b; Van Groenigen et al., 2006; Parton et al., 2007). On the other hand, increased C input may stimulate N accumulation and may be linked to a number of processes, including increased biological fixation of N, greater retention of atmospheric N deposition, reduced losses of N in gaseous or soluble forms, and more complete exploration of soil by expanded root systems (Luo et al., 2006). N concentration of leaves or aboveground tissues has declined on shortgrass steppe, tallgrass prairie, and mesic grassland at elevated CO_2, and on tallgrass prairie with warming, but total N content of aboveground tissues increased with plant biomass in these ecosystems and on annual grasslands (Owensby et al., 1993; Hungate et al., 1997; King et al., 2004; Wan et al., 2005; Gill et al., 2006). The degree to which N responds to increasing atmospheric CO_2 is not well defined and varies among ecosystems (Luo et al., 2006). These interactions between forage quality and soil C storage provide some insights to how changing climate will affect the N and C dynamics under a changing climate. These effects are especially critical in cultivated agroecosystems with more N management options as part of the production system and the interactions of elevated CO_2 and N are as equally complex as in native species. Ghannoum et al. (2007) found the critical N concentration for adequate growth was reduced for a number of species with increasing CO_2 concentrations. They suggested increasing CO_2 concentrations affects the nutrient concentration in plants and defines when plant organs stop accumulating mass. Poorter and Navas (2003) suggested that the amount of N fertilizer required in C_3 crops to achieve maximum productivity under increasing CO_2 will not increase. However, in C_4 plants the N required for maximum production may increase because a decrease in N concentration with elevated CO_2 has not been observed (Poorter and Navas, 2003). These observations present new insights into how rising CO_2 will affect the N balance while changes in temperature and precipitation will affect the soil decomposition rates of organic material from cropping systems.

As further evidence of the interactions of N with climate, Hunsaker et al. (2000), using a FACE study on wheat in Arizona, observed that different N application rates (70 vs. 350 kg ha^{-1}) applied four times during the growing season, affected evapotranspiration and water use efficiency. Evapotranspiration was reduced by 3.8% over the 2 years with enriched CO_2 at 570 $\mu mol\ mol^{-1}$ with the high N treatment and only 0.9% with the low N treatment (Hunsaker et al., 2000). Both N treatments caused a yield increase under enriched CO_2 with 16% increase with high N and 8% with low N treatments. In their experiment there was an increase in water-use efficiency induced by the increase in CO_2 rather than an effect from N management. In their study, they did not find a CO_2 by N interaction showing the observations by Poorter and Navas (2003) to be reasonable; however, future studies in crop response to climate should consider N levels as part of the experimental protocols to ensure potential N effects are part of the study (Norby et al., 2010).

MITIGATION OPPORTUNITIES

In agricultural systems there are three potential opportunities for mitigation of GHG emissions from implementation of different practices that reduce CO_2 emissions, CH_4 emissions, or N_2O emissions. Methane emissions are more closely linked with animal production systems and in the remainder of this chapter we will focus primarily on CO_2 and N_2O because these offer the greatest potential impact to mitigating GHG emissions. Additionally, the implementation of soil management practices also afford the opportunity to increase the resilience of both irrigated and rainfed cropping systems to climate change and climate variation.

MITIGATION OF CO_2 FROM SOIL

Tillage induces CO_2 loss from soil and the short CO_2 bursts observed during tillage operations were attributed to changes in mass flow processes rather than alterations in microbial activity (Reicosky et al., 1997). In their study a comparison was made of tillage systems including (untilled, moldboard plow, and chisel plow) in Bermudagrass [*Cyndon dactylon* (L.) Pers.], continuously cultivated grain sorghum (*Sorghum vulgare* L. Moench.), and no-till grain sorghum. Short-term measurements obtained following tillage operations in each of the systems revealed large episodic emissions of CO_2 from the soil. The results of these comparisons are shown in Table 28.1 along with summaries from other similar experiments. In another experiment, Reicosky et al. (1999) observed CO_2 emissions from tilled soil was affected by irrigation with an increase in emission rates from 0.2 to 2.7 g CO_2 m^{-2} h^{-1} in no-till systems and from 0.6 to 3.8 g CO_2 m^{-2} h^{-1} in the conventional tillage systems. They attributed this increase to a limitation of soil gas fluxes in a dry soil caused by a lack of microbial activity; however, they also observed that CO_2 fluxes from no-till soils began to decline more rapidly. These observations suggest that an increased variability in precipitation also increases the variation in CO_2 fluxes. These studies have been conducted as short-term experiments; however, extension of the measurement of CO_2 fluxes throughout the year showed that conventional tillage still had higher fluxes than no-till (Almaraz et al., 2009). In their observations they found the emission pattern for CO_2 fluxes to be closely and positively linked with soil temperature patterns. Calderón et al. (2000) observed short responses of CO_2 flux may not be as large in soil with a long tillage history because of the lower initial microbial biomass and nutrient pools. Duiker and Lal (2000) compared CO_2 fluxes from a no-till system with varying amounts of wheat residue applied and found no significant differences in the CO_2 fluxes from the soil but a major impact on soil temperature from differing amounts of residue.

There are temporal dynamics in the CO_2 fluxes that add to the complexity of being able to draw simple conclusions about the effect of tillage systems. Franzluebbers (2005) suggested that an effective mitigation strategy for C in the southeast U.S. would be to decrease tillage and increase the complexity of the crop rotation with further increases in SOC created by the addition of manure into the system. Martens et al. (2005) offered a similar solution for the more arid southwestern U.S. with the major reduction in emissions and increase in SOC occurring under reduced tillage. There have been few long-term comparisons of the effect of different management practices on the changes in the C or N balance in soils. Dalal et al. (2011) examined these changes after 40 years of no-tillage, crop residue management, and N fertilizer for a wheat cropping system on a Vertisol in Australia. In their comparison, the effects of tillage on soil C and N were small and there was an interaction between crop residue retention and N fertilization. They found SOC and N increased under crop residue retention only when N was applied and under N fertilization when the residue was retained. They also observed that the positive effects of tillage, crop residue retention, and N fertilization occurring during the early years of the study (Dalal et al., 2011). These observations would suggest that with the increased variation in temperature and precipitation patterns expected in a warming climate, management of tillage, crop residue, and N fertilizer will produce some interacting

TABLE 28.1 Summary of Cropping Experiments with CO_2 Mitigation Analysis

Crop	Location	Soil	Comparison	Response	Reference
Bermudagrass	Temple, TX	Fine montmorillinitic, thermic Udic Pellusterts	Untilled (u), moldboard plow (m), Chisel plow (c)	115 g CO_2 m^{-2} d^{-1} in tilled vs 10 g CO_2 m^{-2} d^{-1} in untilled	Reicosky et al., 1997
Cultivated grain sorghum	Temple, TX	Fine montmorillinitic, thermic Udic Pellusterts	Untilled, moldboard plow, Chisel plow	45 (m), 30 (c), and 5 (u) g CO_2 m^{-2} d^{-1}	Reicosky et al., 1997
No-till grain sorghum	Temple, TX	Fine montmorillinitic, thermic Udic Pellusterts	Untilled, moldboard plow, Chisel plow	45 (m), 70 (c), 3 (u) g CO_2 m^{-2} d^{-1}	Reicosky et al., 1997
Soybean	Anne de Bellevue, Quebec	fine clay, mixed, Nonacid, frigid Humaquept	Conventional tillage (cn) vs No-till (nt)	160 g CO_2 m^{-2} d^{-1} (ct) vs 68 g CO_2 m^{-2} d^{-1} for peak summer emissions	Almaraz et al. 2009
Corn/soybean rotation with crimson clover cover crop	Auburn, AL	Fine loamy siliceous thermic Typic Kandiudults	Conventional tillage (cn) vs No-till (nt)	150 g CO_2 m^{-2} (cn) vs 50 g CO_2 m^{-2} (nt) accumulated for 80 hrs	Reicosky et al. 1999
Wheat	Columbus, OH	Fine, mixed mesic Aeric Ochraqualf	No-till with varying amounts of wheat residue added	No significant impact of residue on CO_2 fluxes	Duiker and Lal (2000)
Sorghum-wheat-soybean rotation	Southcentral Texas	Westfield silty clay loam	Conventional tillage (cn) vs no-till (nt)	Fallow period was not different with fluxes of 1.87 g CO_2 m^{-2} d^{-1} nt was higher in sorghum growing period, 2.15 vs 1.95 g CO_2 m^{-2} d^{-1}	Franzluebbers et al. (1995)
Wheat-soybean rotation	Southcentral Texas	Westfield silty clay loam	Conventional tillage (cn) vs no-till (nt)	nt was higher in fallow period 1.06 vs 0.32 g CO_2 m^{-2} d^{-1}	Franzluebbers et al. (1995)

effects on plant growth and consequently on the C and N dynamics in the soil. Grant et al. (2004) used the DNDC model to simulate the effect of changing agricultural practices on both CO_2 and N_2O fluxes. The results from this work show that converting cultivated land to grassland, conventional tillage to no-tillage, and reducing summer fallow in crop rotations could substantially increase C sequestration and decrease net GHG emissions. Alberti et al. (2010) evaluated the effect of converting continuous corn to alfalfa (*Medicago sativa* L.) but found that a short-term comparison of 3 years was not sufficient to provide an adequate assessment of impacts of land conversion.

MITIGATION OF N_2O FLUXES FROM SOIL

Emissions of N_2O from the soil are more complex than CO_2 emissions because of the interacting dynamics of the N cycle within the soil. The summary of results from several studies is shown in Table 28.2. Almaraz et al. (2009) observed seasonal increases in N_2O emissions with precipitation being the primary variable affecting the seasonal pattern. Ball et al. (2008) observed N_2O fluxes in a barley production system comparing plowing vs. no-till and found fluxes from no-till were 1.5 to 35 times higher compared to plowed fields. They attributed this difference to the surface soil water dynamics and observed peaks of N_2O following wet conditions and found these fluxes increased when the water-filled pore space exceeded 80%. This study is relatively short (one growing season) and may not adequately capture the dynamics of N_2O emissions among seasons with different precipitation patterns. However, the response of N_2O emissions to no-tillage is variable. For example, in a 2-year study of N_2O emissions from a corn–soybean rotation, Parkin and Kaspar (2006) observed no significant differences in annual N_2O emissions in a chisel plow system vs. a no-till system. Venterea et al. (2005) observed that N_2O emissions were higher under no-till and conservation tillage than conventional tillage with broadcast urea; however, when anhydrous ammonia was applied, N_2O emissions were higher under conventional and conservation tillage than no-till. Tillage systems had no effect on N_2O emissions when the urea ammonium nitrate was surface applied. In a later study, Venterea et al. (2010) found that N_2O emissions were twice those observed with urea under a continuous corn and a corn–soybean rotation when anhydrous ammonia was used compared to urea as a nitrogen form. They also observed that converting from a corn–soybean rotation to continuous corn would further increase the annual N_2O emissions. In a summary of 25 field studies comparing N_2O emissions from conventional and no-till cropping systems, Rochette (2008) concluded that no-till generally increased N_2O emissions in soils with poor aeration, but in soils with good aeration N_2O emissions are not stimulated by no-tillage. The projected increase in springtime precipitation across the Midwest may lead to increases in N_2O emissions because of the potential for wetter soils. Clay et al. (1995) observed that peaks of C and N mineralization did not occur at the same time in the soil with peak N values in the early spring and peak C occurring during the summer. This is consistent with the observations by Almaraz et al. (2009) linking these changes to patterns of soil water for N and soil temperature for C and reinforces the concept that any mitigation strategy will have to account for the temporal dynamics in the N and C cycles.

Mitigation strategies for N_2O were reviewed by Snyder et al. (2009) and they proposed that changes in N fertilizer management (rate, source, and timing) along with the use of controlled release or nitrification inhibitors provide reductions in potential N_2O emissions. Delgado and Mosier (1996) observed N_2O fluxes from a polymer-coated urea were initially less than with the uncoated material; however, the emissions persisted longer in the growing season. Hyatt et al. (2010) found in potato (*Solanum tuberosum* L.) that changing timing with conventional urea had higher N_2O emissions than applications of polymer-coated urea and there was no effect on potato yield. Halvorson et al. (2008) compared different rates of N application across different cropping systems in an irrigated study in Colorado with polymer-coated urea applied at sidedress periods and observed a reduction in N_2O emissions with differences among cropping systems primarily caused by the amount of N applied and differences among the growing

TABLE 28.2 Summary of Cropping Experiments with N_2O Mitigation Analysis

Crop	Location	Soil	Comparison	Response	Reference
Soybean	Anne de Bellevue, Quebec	fine clay, mixed, Nonacid, frigid Humaquept	Conventional tillage (cn) vs No-till (nt)	18.1 mg N_2O m^{-2} d^{-1} (cn) vs 7.4 mg N_2O m^{-2} d^{-1} (nt) for peak emissions	Almaraz et al. 2009
Barley	Midlothian, Scotland	Eutric Gleysol	Plowed (p) vs no-till (nt)	4.44 μg N m^{-2} h^{-1} (nt) vs 2.88 μg N m^{-2} h^{-1}	Ball et al. 2008
Barley	Ft. Collins, CO	Fine montmorillonitic, mesic Aridic Argiustoll	Control vs Urea vs urea with dicyandiamide vs polyolefin coated urea	4.5 g N ha^{-1} c^{-1} 8.2 g N ha^{-1} c^{-1} 5.2 g N ha^{-1} c^{-1} 6.9 g N ha^{-1} c^{-1}	Delgado and Mosier (1996)
Conv. Tillage cont. corn No-Till Cont. corn No-Till corn-drybean No-Till corn-barley	Ft. Collins, CO	Fine loamy mixed superacitve, mesic Aridic Haplustaffs	N rates of 0, 67, 134, and 246 kg ha^{-1} with urea vs polymer −coated urea	Reduction in cumulative N_2O emissions from polymer-coated urea and urea with inhibitors	Halvorson et al. (2008, 2010a)
Annual vs Perennial crops	Potsdam, Germany	Sandy loam soil	Annual vs perennial crops with 0, 75, and 150 kg N ha^{-1} applied	Annual (4.3 kg ha^{-1}) vs perennial (1.9 kg ha^{-1})	Kavidr et al. (2008)
Barley, wheat, canola rotation	Star City, Saskatchewan, Canada	Gray Luvisol	Conventional tillage vs no-tillage under differ residue removal	Rates of N_2O emission ranged from 30 to 1300 g N ha^{-1}	Malhi and Lemke (2007)
Oilseed rape-wheat rotation	Research Center Foulum, Demark	Typic Hapludult	Conventional tillage, reduced tillage, and direct seed with manure application	Conventional tillage (3.81 kg N ha^{-1} yr^{-1}), reduced tillage (3.32 kg N ha^{-1} yr^{-1}), direct drill (2.4 kg N ha^{-1} yr^{-1})	Chatskikh et al. (2008)
Greenhouse gas emission study	Ames, IA	Storden fine-loamy, mixed, mesic Typic Udorthents Webster fine-loamy, mixed, mesic Typic Haplaquolls	N sources and manure application to containers Control (C), UAN, (U), Manure (M)	mg N_2O-N m^{-2} Storden emissions are listed first then Webster C, 85.1, 29.6 U, 746, 76.6 M, 628, 397	Jarecki et al. (2008)
Wheat-corn rotation	Jiangsu Province, China	Hydromorphic soil	Plow vs No-tillage following each crop	No effect of tillage on emissions	Chen and Huang (2009)
Forage crop with mixed species	LaBroquerie, Mantobia, Canada	Coarse textured soil	Varying rates of swine manure, control (0), split (72 kg N ha^{-1} at 2 times) and single (148 kg N ha^{-1})	Control (60 g N ha^{-1}), split (372 g N ha^{-1}) and single (785 g N ha^{-1})	Tenuta et al. (2010)

Continued

TABLE 28.2 Summary of Cropping Experiments with N_2O Mitigation Analysis—continued

Crop	Location	Soil	Comparison	Response	Reference
Wheat (pot study) Corn (field study)	Nanjing University Experimental Farm, Nanjing, China	Anthrosol	Urea, urea coated with C-Ma-P (U-Ca), urea coated with polymer (Up), urea coated with sulfur(Us), urea formaldehyde (Uf), urea with dicyandiamide with hydrochinone (Ud)	All units as kg N_2O-N ha^{-1}) first number is 2007 and second is 2008 results U, 2.26, 2.82; U-Ca, 3.70; Up, 4.49, 2.85; Us, 3.63, 2.71; Uf, 1.93, 2.10; Ud, 1.52, 1.04	Jiang et al. (2010)
Wheat	New Dehli, India	Typic Ustochrept	Urea (U), urea with S-benzylisothiouronium butanoate (U-B), urea with benzylisothiouronium furoate (U-F), urea with dicyandiamide (U-D)under conventional and no-tillage systems	g N_2O-N ha^{-1} Conv till is the first value and No-tillage is the second value Control 313, 355 U, 778, 873 U-B, 708, 737 U-F, 673, 705 U-D 704, 733	Bhatia et al. (2010)
Chisel Plow Corn	Ames, IA USA	Silty clay loam	Nitrification inhibitor vs. no inhibitor with fall-applied anhydrous ammonia	Inhibitor changed timing of N_2O emissions but annual emissions not affected	Parkin and Hatfield (2010)
Corn/soybean	Ames, IA USA	Fine-loamy mixed mesic Typic Hapludolls	No tillage vs. chisel plow	Tillage did not affect annual N_2O emissions, but crop did	Parkin and Kaspar (2006)
Barley	Quebec City Canada	Heavy clay and Gravelly loam	No-tillage vs. moldboard plow	Higher N_2O emissions in NT vs MP in the clay soil. Tillage did not influence N_2O in the coarse textured soil	Rochette et al. (2008)

seasons. Halvorson et al. (2010a) observed that the polymer-coated urea reduced N_2O emission rate in no-till corn production systems in Colorado. In their study, they found differences among forms of N fertilizer and inhibitors on the reduction of N_2O emissions. For example, under conventional tillage continuous corn there was no difference between urea and ESN (environmentally stable nitrogen, a controlled-release polymer-coated urea) on the cumulative N_2O emissions; however, under no-till with continuous corn there was a 49% reduction with the ESN vs. urea. They found that under no-till systems for other crop rotations there was a decrease emission of N_2O with the enhanced efficiency fertilizers (Halvorson et al. (2010a). Jiang et al. (2010) only found a significant effect on the cumulative values from a corn field study from different forms of urea when the inhibitor was applied. Halvorson et al. (2010b) also observed reductions in N_2O emissions with the use of inhibitors in no-till cropping systems with no effect on corn grain yields. Bhatia et al. (2010) found nitrification inhibitors to reduce N_2O emissions by 9 to 20% compared to untreated urea on wheat. However, Parkin and Hatfield (2010) observed that the nitrification inhibitor nitrapyrin reduced late fall/early spring N_2O emissions, but annual emissions were not significantly affected. Seasonality of N_2O fluxes are caused by the addition of N fertilizer to the soil as observed by Halvorson et al. (2008), Jarecki et al. (2008), Kavdir et al. (2008), and Parkin and Kaspar (2006). Kavdir et al. (2008) observed increased N_2O emissions after N fertilization, after plant harvest in the summer, and during freeze–thaw cycles in the winter and early spring with twice the emissions from annual (4.3 kg ha^{-1}) compared to perennial (1.9 kg ha^{-1}) crops. They found plant type and fertilizer rates to be significant factors in their study but no interactions among the variables across the years. Jarecki et al. (2008) observed greater N_2O emission from the sandy loam soil than the clay soil at all N application rates and types and was attributed to the differences in the NH_4^+ fixation between the two soils. In their study there was no significant differences detected due to water-filled pore space. Jarekci et al. (2008) concluded that soil factors controlling N availability may be a stronger factor in N_2O emissions than soil water. Malhi and Lemke (2007) did observe large differences among years with below normal precipitation years having cumulative N_2O emissions of less than 300 g N ha^{-1} and above normal precipitation years exceeding 1300 g N ha^{-1}. In only one year of their study when the precipitation was high was there a significant difference caused by tillage systems (conventional vs. no-till). There was an effect of N rate on N_2O emission but the increase in plant growth and yield was greater than the rate of increase in emissions. Chatskikh et al. (2008) observed from their study on oilseed rape (*Brassica napus* L.) and wheat rotations that soil water was the primary factor affecting N_2O emissions. Chen and Huang (2009) found there was no effect of tillage systems; however, the rates were dependent upon the soil water regime during the study period.

Reduction of N_2O emissions is not consistent among all practices and the degree of effectiveness of any management practice is dependent upon the weather. Application of advanced analytical methods similar to the discussion of wavelet analysis described by Furon et al. (2008) may enhance our understanding of the interactions among CO_2 or N_2O emissions and different variables in the soil–plant–atmosphere system. Soil water dynamics are the most consistent factor affecting emissions and with the changing climate regime with increasing precipitation events in the spring coupled with more intense storms during the growing season. The mitigation of N_2O with improved N management and altered forms of N with inhibitors and coatings appears to offer promise; however, the degree of effectiveness will vary among years. Climate (semiarid vs. humid) appears also to result in different responses to N sources (ARS N source project).

CONCLUSIONS

Changing climate components of rising atmospheric CO_2, increasing temperature, and temporal and spatial variation in precipitation will affect plant growth and management practices. These effects have the potential of inducing negative impacts on growth and yield because of the additional stresses on the plant; however, increases in CO_2 with favorable temperatures could potentially induce more growth. If we are to achieve reductions in the GHG emissions from soil management practices then we have to consider the impacts from the implementation of practices primarily affecting CO_2 or N_2O emissions. Observations from a range of experiments reveal that CO_2 fluxes are affected by changes in soil temperature and tillage management. The potential exist from the rising atmospheric CO_2 to increase plant growth and add crop residues to the soil which in turn would reduce CO_2 emissions. However, there is an offsetting effect on plant growth caused by warmer temperatures which will increase the rate of water use from the crop and when coupled with variable precipitation and the increased variation in soil water availability, creates a negative feedback on rate of plant material added to the soil and a positive feedback on the rate of decomposition. The end result is increase in CO_2 emissions. A consistent observation is that adoption of decreased-tillage systems maintains crop residues which will reduce the emission of CO_2 from the soil (Desjardins et al., 2005). The recent approach by Huth et al. (2010) to utilize simulation models to evaluate these dynamics will further enhance insights into these processes.

Conversely, N_2O emissions are more closely related to the dynamics of the soil water regime and improved N management under more variable climates so that reductions in N_2O emissions will occur. However, large variations in the amounts of N_2O emissions will continue across years. A consistent fact is that increasing N application rates above the optimum for plant growth or yield contributes to increased N_2O emissions. Under a changing climate more emphasis will have to be given to improved N management strategies. The coupling between the C and N cycles in soils under increased CO_2, along with the variable soil water regimes, will alter the ability of the plant to utilize N and thus affect the N-uptake patterns. An increased emphasis on understanding linkages between C and N under a changing climate will enhance our management capabilities. Millar et al. (2010) has proposed an approach for producers wherein they would adopt N management practices, e.g. rate reductions and timing, to reduce N_2O emissions; however, the effect of the changing climate on the effectiveness of these practices still needs to be determined.

There are promising agronomic practices, e.g. reduced tillage and increased crop residue, improved N management including alternate forms of N, which will decrease CO_2 and N_2O emissions but the consistency and magnitude of these responses will be tempered by the increased climate variation over the next 30 to 50 years. Understanding the dynamics of CO_2 and N_2O emissions coupled with improved crop management strategies will enhance our ability to produce feed, forage, and food as well as to decrease the environmental impact of crop production systems. However, the strategies associated with CO_2 and N_2O mitigation in agricultural systems, e.g. increasing soil carbon and changing the forms and rates of N along with reduced tillage systems, are those potentially providing an enhanced soil resource and one capable of withstanding the weather variation associated with climate change. Development of strategies for mitigation will have far-reaching benefits beyond reductions in greenhouse gas emissions. The development of a comprehensive extension network similar to the NitroEurope described by Skiba et al. (2009) that extends beyond the GRACEnet program in scope and complexity may be necessary to refine our understanding of the complexities of GHG emissions to changes in soil management practices while enhancing our mitigation strategies. Although there is a large amount of literature on CO_2 and N_2O emissions from agricultural practices, these studies are best characterized by short-term assessments of the range of practices available to producers. Development of more research efforts to quantify the complexities will benefit agriculture and the environment through an improved understanding of the mitigation potential of a wider range of agricultural practices.

References

Akinremi, O.O., McGinn, S.M., McLean, H.D.J., 1999. Effects of soil temperature and moisture on soil respiration in barley and fallow plots. Can. J. Soil Sci. 79, 5–13.

Alberti, G., Vedove, G.D., Zuliani, M., Peressotti, A., Castaldi, S., Zerbi, G., 2010. Changes in CO_2 emissions after crop conversion from continuous maize to alfalfa. Agric. Ecosys. Environ. 136, 139–147.

Almagro, M., Lopez, J., Querejeta, J.I., Martinez-Mena, M., 2009. Temperature dependence of soil CO_2 efflux is strongly modulated by seasonal patterns of moisture availability in a Mediterranean ecosystem. Soil Biol. Biochem. 41, 594–605.

Almaraz, J.J., Zhou, X., Mabood, F., Madramootoo, C., Rochette, P., Ma, B., Smith, D.L., 2009. Greenhouse gas fluxes associated with soybean production under two tillage systems in southwestern Quebec. Soil Tillage Res. 104, 134–139.

Alvarez, R., Alvarez, C., Lorenzo, G., 2001. Carbon dioxide fluxes following tillage from a Mollisol in the Argentine Rolling Pampa. Eur. J. Soil Biol. 37, 161–166.

Ball, B.C., Crichton, I., Horgan, G.W., 2008. Dynamics of upward and downward N_2O and CO_2 fluxes in ploughed or no-tilled soils in relation to water-filled pore space, compaction and crop presence. Soil Tillage Res. 101, 20–30.

Bhatia, A., Sasmal, S., Jain, N., Pathak, H., Kumar, R., Singh, A., 2010. Mitigating nitrous oxide emission from soil under conventional and no-tillage in wheat using nitrification inhibitors. Agric. Ecosys. Environ. 136, 247–253.

Bono, A., Alvarez, R., Buschiazzo, D.E., Cantet, R.J.C., 2008. Tillage effects on soil carbon balance in a semiarid agroecosystem. Soil Sci. Soc. Am. J. 72, 1140–1149.

Borken, W., Matzner, E., 2009. Reappraisal of drying and wetting effects on C and N mineralization and fluxes in soils. Glob. Change Biol. 15, 808–824.

Burns, R.L., Cook, D.J., Phillips, R.E., 1971. Influence of no tillage on soil moisture. Agron. J. 73, 593–596.

Calderón, F.J., Jackson, L.E., Scow, K.M., Rolston, D.E., 2000. Microbial responses to simulated tillage in cultivated and uncultivated soils. Soil Biol. Biochem. 32, 1547–1559.

Chatskikh, D., Olesen, J.E., Hansen, E.M., Elsgaard, L., Petersen, B.M., 2008. Effects of reduced tillage on net greenhouse gas fluxes from loamy sand soil under winter crops in Denmark. Agric. Ecosys. Environ. 128, 117–126.

Chen, S., Huang, Y., 2009. Soil respiration and N_2O emission in croplands under different ploughing practices: a case study in south-east China. Aust. J. Soil Res. 47, 198–205.

Clay, D.E., Schumacher, T.E., Brix-Davis, K.A., 1995. Carbon and nitrogen mineralization in row and interrow areas of chisel and ridge tillage systems. Soil Tillage Res. 35, 167–174.

Dalal, R.C., Allen, D.E., Wang, W.J., Reeves, S., Gibson, I., 2011. Organic carbon and total nitrogen stocks in a Vertisol following 40 years of no-tillage, crop residue retention, and nitrogen fertilization. Soil Tillage Res. In press.

Deibert, E.J., French, E., Hoag, B., 1986. Water storage and use by spring wheat under conventional tillage and no-till in continuous and alternate crop-fallow systems in the northern Great Plains. J. Soil Water Conserv. 41, 53–58.

Delgado, J.A., Mosier, A.R., 1996. Mitigation alternatives to decrease nitrous oxides emissions and urea-nitrogen loss and their effect on methane flux. J. Environ. Qual. 25, 1105–1111.

Desjardins, R.L., Smith, W., Grant, B., Campbell, C., Riznek, R., 2005. Management strategies to sequester carbon in agricultural soils and to mitigate greenhouse gas emissions. Climatic Change 70, 283–297.

Duiker, S.W., Lal, R., 2000. Carbon budget study using CO_2 flux measurements from a no-till system in central Ohio. Soil Tillage Res. 54, 21–30.

Epstein, H.E., Burke, I.C., Lauenroth, W.K., 2002. Regional patterns of decomposition and primary production rates in the U.S. Great Plains Ecol. 83, 320–327.

Follett, R.F., 2001. Soil management concepts and carbon sequestration in cropland soils. Soil Tillage Res. 61, 77–92.

Franzluebbers, A.J., 2005. Soil organic carbon sequestration and agricultural greenhouse gas emissions in the southeastern USA. Soil Tillage Res. 83, 120–147.

Franzluebbers, A.J., Hons, F.M., Zuberer, D.A., 1995. Tillage and crop effects on seasonal dynamics of soil CO_2 evolution, water-content, temperature and bulk density. Appl. Soil Ecol. 2, 95–109.

Furon, A.C., Wagner-Riddle, C., Smith, C.R., Warland, J.S., 2008. Wavelet analysis of wintertime and spring thaw CO_2 and N_2O fluxes from agricultural fields. Agric. Forest Meteorol. 148, 1305–1317.

Ghannoum, O., Searson, M.J., Conroy, J.P., 2007. Nutrient and water demands of plants under global climate change. In: Newton, P.C.D., Carran, R.A., Edwards, G.R., Niklaus, P.A. (Eds.), Agroecosystems in Changing Climate. Taylor & Francis, New York, NY, pp. 53–83.

Gill, R.A., Anderson, L.J., Polley, H.W., Johnson, H.B., Jackson, R.B., 2006. Potential nitrogen constraints on soil carbon sequestration under low and elevated atmospheric CO_2. Ecology 87, 41–52.

Grant, B., Smith, W.N., Desjardins, R., Lemke, R., Li, C., 2004. Estimated N_2O and CO_2 emissions as influenced by agricultural practices in Canada. Climatic Change 65, 315–332.

Grassini, P., Yang, H., Cassman, K.G., 2009. Limits to maize productivity in western Corn-Belt: a simulation analysis for fully irrigated and rainfed conditions. Agric. Forest Meteorol. 149, 1254–1565.

Haim, D., Shechter, M., Berliner, P., 2008. Assessing the impact of climate change on representative field crops in Israeli agriculture: a case study of wheat and cotton. Climate Change 86, 425–440.

Halvorson, A.D., Del Grosso, S.J., Ruele, C.A., 2008. Nitrogen, tillage, and crop rotation effects on nitrous oxide emissions from irrigated cropping systems. J. Environ. Qual. 37, 1337–1344.

Halvorson, A.D., Del Grosso, S.J., Alluvione, F., 2010a. Tillage and inorganic nitrogen source effects on nitrous oxide emissions from irrigated cropping systems. Soil Sci. Soc. Am. J. 74, 436–445.

Halvorson, A.D., Del Grosso, S.J., Alluvione, F., 2010b. Nitrogen source effects on nitrous oxide emissions from irrigated no-till corn. J. Environ. Qual. 39, 1554–1562.

Hatfield, J.L., 2010. Climate impacts on agriculture in the United States: the value of past observations. In: Hillel, D., Rosenzwieg, C. (Eds.), Handbook of Climate Change and Agroecosystems: Impact, Adaptation and Mitigation. Imperial College Press, London, U.K, pp. 239–253.

Hatfield, J.L., Boote, K.J., Kimball, B.A., Ziska, L.H., Izaurralde, R.C., Ort, D., Thomson, A.M., Wolfe, D.W., 2011. Climate impacts on agriculture: implications for crop production. Agron. J. 103, 351–370.

Hungate, B.A., Chapin III, F.S., Zhong, H., Holland, E.A., Field, C.B., 1997. Stimulation of grassland nitrogen cycling under carbon dioxide enrichment. Oecologia 109, 149–153.

Hunsaker, D.J., Kimball, B.A., Pinter Jr., P.J., Wall, G.W., LaMorte, R.L., Adamsen, F.J., Leavitt, S.W., Thompson, T.L., Matthias, A.D., Brooks, T.J., 2000. CO_2 enrichment and soil nitrogen effects on wheat evapotranspiration and water use efficiency. Agric. Forest Meteorol. 104, 85–105.

Huth, N.I., Thorburn, P.J., Radford, B.J., Thornton, C.M., 2010. Impacts of fertilizers and legumes on N_2O and CO_2 emissions from sols in subtropical agricultural systems: a simulation study. Agric. Ecosys. Environ. 136, 351–357.

Hyatt, C.R., Venterea, R.T., Rosen, C.J., McNearney, M., Wilson, M.L., Dolan, M.S., 2010. Polymer-coated urea maintains potato yields and reduces nitrous oxide emissions in a Minnesota loamy sand. Soil Sci. Soc. Amer. J. 74, 419–428.

Inglima, I., Alberti, G., Bertolini, T., Vaccaris, F.P., Giolis, B., Migliettas, F., Cotrufo, M.F., Peressotti, A., 2009. Precipitation pulses enhance respiration of Mediterranean ecosystems: the balance between organic and inorganic components of increased soil CO_2 efflux. Glob. Change Biol. 15, 1289–1301.

Izaurralde, R.C., Thomson, A.M., Morgan, J.A., Fay, P.A., Polley, H.W., Hatfield, J.L., 2011. Climate impacts on agriculture: implications for forage and rangeland production. Agron. J. 103, 371–381.

Jarecki, M.K., Parkin, T.B., Chan, A.S.K., Hatfield, J.L., Jones, R., 2008. Greenhouse gas emissions from two soils receiving nitrogen fertilizer and swine manure slurry. J. Environ. Qual. 37, 1432–1438.

Jiang, J., Hu, Z., Sun, W., Huang, Y., 2010. Nitrous oxide emissions from Chinese cropland fertilized with a range of slow-release nitrogen compounds. Agric. Ecosys. Environ. 135, 216–225.

Karl, T.R., Melillo, J.M., Peterson, T.C. (Eds.), 2009. Global climate change impacts in the United States. Cambridge University Press, New York, NY, p. 188.

Kavdir, Y., Hellebrand, H.J., Kern, J., 2008. Seasonal variation of nitrous oxide emission in relation to nitrogen fertilization and energy crop types in sandy soil. Soil Tillage Res. 98, 175–186.

Kessavalou, A., Doran, J.W., Mosier, A.R., Drijber, R.A., 1998. Greenhouse gas fluxes following tillage and wetting in a wheat-fallow cropping system. J. Environ. Qual. 27, 1105–1116.

Kimball, B.A., 2010. Lessons from FACE: CO_2 effects and interactions with water, nitrogen, and temperature. In: Hillel, D., Rosenzweig, C. (Eds.), Handbook of Climate Change and Agroecosystems: Impacts, Adaptation, and Mitigation. ICP Series on Climate Change Impacts, Adaptation, and Mitigation, vol. 1. Imperial College Press, London, pp. 87–107.

King, J.Y., Mosier, A.R., Morgan, J.A., LeCain, D.R., Milchunas, D.G., Parton, W.J., 2004. Plant nitrogen dynamics in shortgrass steppe under elevated atmospheric carbon dioxide. Ecosystems 7, 147–160.

La Scala Jr., N., Lopes, A., Spokas, K., Bolonhezi, D., Archer, D.W., Reicosky, D.C., 2008. Short-term temporal changes of soil carbon losses after tillage described by a first-order decay model. Soil Tillage Res. 99, 108–118.

Lal, R., 2009. Sequestration of atmospheric CO_2 in global C pools. Energy Environ. Sci. 1, 86–100.

Lobell, D.B., Field, C.B., Cahill, K.N., Bonfils, C., 2006. Impacts of future climate change on California perennial crop yields: model projections with climate and crop uncertainties. Agric. Forest Meterol. 141, 208–218.

Luo, Y., Su, B., Currie, W.S., Dukes, J.S., Finzi, A., Hartwig, U., Hungate, B., McMurtrie, R.E., Oren, R., Parton, W.J., Pataki, D.E., Shaw, M.R., Zak, D.R., Field, C.B., 2004. Progressive nitrogen limitation of ecosystem responses to rising atmospheric carbon dioxide. BioScience 54, 731–739.

Luo, Y., Hui, D., Zhang, D., 2006. Elevated CO_2 stimulate net accumulations of carbon and nitrogen in land ecosystems: a meta-analysis. Ecology 87, 53–63.

Malhi, S.S., Lemke, R., 2007. Tillage, crop residue and N fertilizer effects on crop yield, nutrient uptake, soil quality and nitrous oxide gas emissions in a second 4-yr rotation cycle. Soil Tillage Res. 96, 269–283.

Martens, D.A., Emmerich, W., McLain, J.E.T., Johnsen, T.N., 2005. Atmospheric carbon mitigation potential of agricultural management in the southwestern USA. Soil Tillage Res. 83, 95–119.

Millar, N., Robertson, G.P., Grace, P.R., Gehl, R.J., Hoben, J.P., 2010. Nitrogen fertilizer management for nitrous oxide (N_2O) mitigation in intensive corn (maize) production: an emissions reduction protocol for US Midwest agriculture. Mitig. Adapt. Strateg. Glob. Change 15, 185–204.

Morell, F.J., Alvaro-Fuentes, J., Lampurlanes, J., Cantero-Martinez, C., 2010. Soil CO_2 fluxes following tillage and rainfall events in a semiarid Mediterranean agroecosystem: effects of tillage systems and nitrogen fertilization. Agro. Ecosys. Environ. 139, 167–173.

Muchow, R.C., Sinclair, T.R., Bennett, J.M., 1990. Temperature and solar-radiation effects on potential maize yield across locations. Agron. J. 82, 338–343.

Norby, R.J., Warren, J.M., Iverson, C.M., Medlyn, B.E., McMurtrie, R.E., 2010. CO_2 enhancement of forest productivity constrained by limited nitrogen availability. PNAS 107, 19368–19373.

Owensby, C.E., Coyne, P.I., Auen, L.M., 1993. Nitrogen and phosphorus dynamics of a tallgrass prairie ecosystem exposed to elevated carbon dioxide. Plant Cell Environ. 16, 843–850.

Papendick, R.I., Lindstrom, M.J., Cochran, V.L., 1973. Soil mulch effect on seedbed temperature and water during fallow in eastern Washington. Soil Sci. Soc. Am. Proc. 37, 307–314.

Parkin, T.B., Hatfield, J.L., 2010. Influence of nitrapyrin on N_2O losses from soil receiving fall-applied anhydrous ammonia. Ag. Ecosys. Environ. 136, 81–86.

Parkin, T.B., Kaspar, T.C., 2006. Nitrous oxide emissions from corn-soybean systems in the Midwest. J. Environ. Qual. 35, 1496–1506.

Parton, W., Silver, W.L., Burke, I.C., Grassens, L., Harmon, M.E., Currie, W.S., King, J.Y., Adair, E.C., Brandt, L.A., Hart, S.C., Fasth, B., 2007. Global-scale similarities in nitrogen release patterns during long-term decomposition. Science 315, 361–364.

Poorter, H., Navas, M.L., 2003. Plant growth and competition at elevated CO_2: on winners, losers and functional groups. New Phytologist 157, 175–198.

Prior, S.A., Rogers, H.H., Runion, G.B., Torbert, H.A., Reicosky, D.C., 1997. Carbon dioxide-enriched agro-ecosystems: influence of tillage on short-term soil carbon dioxide efflux. J. Environ. Qual. 26, 244–252.

Raich, J.W., Schlesinger, W.H., 1992. The global carbon dioxide flux in soil respiration and its relationship to vegetation and climate. Tellus 44b, 81–99.

Reich, P.B., Hungate, B.A., Luo, Y., 2006b. Carbon-nitrogen interactions in terrestrial ecosystems in response to rising atmospheric carbon dioxide. Ann. Rev. Ecol. Syst. 37, 611–636.

Reich, P.B., Hobbie, S.E., Lee, T., Ellsworth, D.S., West, J.B., Tilman, D., Knops, J.M.H., Naeem, S., Trost, J., 2006a. Nitrogen limitation constrains sustainability of ecosystem response to CO_2. Nature 440, 922–924.

Reicosky, D.C., Archer, D.W., 2007. Moldboard plow tillage depth and short-term carbon dioxide release. Soil Tillage Res. 94, 109–121.

Reicosky, D.C., Reeves, D.W., Prior, S.A., Runion, G.W., Rogers, H.H., Raper, R.L., 1999. Effects of residue management and controlled traffic on carbon dioxide and water vapor loss. Soil Tillage Res. 51, 153–165.

Reicosky, D.C., Dugas, W.A., Torbert, H.A., 1997. Tillage-induced soil carbon dioxide loss from different cropping systems. Soil Tillage Res. 4, 105–118.

Ritchie, J.T., 1971. Dryland evaporative flux in a subhumid climate. 1. Micrometeorological influences. Agron. J. 70, 723–728.

Rochette, P., 2008. No-till only increases N_2O emissions in poorly-aerated soils. Soil Tillage Res. 101, 97–100.

Rochette, P., Angers, D.A., Chantigny, M.H., Bertrand, N., 2008. Nitrous oxide emissions respond differently to no-till in a loam and heavy clay soil. Soil Sci. Soc. Am. J. 72, 1363–1369.

Rochette, P., Desjardins, R.L., Pattey, E., 1991. Spatial and temporal variability of soil respiration in agricultural fields. Can. J. Soil Sci. 71, 189–196.

Runge, E.C.A., 1968. Effect of rainfall and temperature interactions during the growing season on corn yield. Agron. J. 60, 503–507.

Rustad, L.E., Campbell, J.L., Marion, G.M., Norby, R.J., Mitchell, M.J., Hartley, A.E., Cornelissen, J.H.C., Gurevitch, J., 2001. A meta-analysis of the response of soil respiration, net nitrogen mineralization, and aboveground plant growth to experimental ecosystem warming. Oecologia 126, 543–562.

Sainju, U.M., Jabro, J.D., Stevens, W.B., 2008a. Soil carbon dioxide emission and carbon content as affected by irrigation, tillage, cropping system, and nitrogen fertilization. J. Environ. Qual. 37, 98–106.

Sainju, U.M., Senwo, Z.N., Nyakatawa, E.Z., Tazisong, I.A., Reddy, K.C., 2008b. Soil carbon and nitrogen sequestration as affected by long-term tillage, cropping systems, and nitrogen sources. Agric. Ecosys. Environ. 127, 234–240.

Sanchez, M.L., Ozores, M.I., Lopez, M.J., Colle, R., De Torre, B., Garcia, M.A., Perez, I., 2003. Soil CO_2 fluxes beneath barley on the central Spanish plateau. Agric. For. Meteorol. 118, 85–95.

Sauer, T.J., Hatfield, J.L., Prueger, J.H., 1996. Corn residue age and placement effects on evaporation and soil thermal regime. Soil Sci. Soc. Am. J. 60, 1558–1564.

Skiba, U., Drewer, J., Tang, Y.S., van Dijk, N., Helfter, C., Nemitz, E., Famulari, D., Cape, J.N., Jones, S.K., Twigg, M., Pihlatie, M., Vesla, T., Larsen, K.S., Carter, M.S., Ambus, P., Ibrom, A., Beier, C., Hensen, A., Frumau, A., Erisman, J.W., Bruggemann, N., Gasche, R., Butterbach-Bahl, K., Neftel, A., Spirig, C., Horvath, L., Freibauer, A., Cellier, P., Laville, P., Loubet, B., Magliulo, E., Bertolini, T., Seufert, G., Andersson, M., Manca, G., Laurila, T., Aurela, M., Lohila, A., Zechmeister-Bolenstern, S., Kitzler, B., Schaufler, G., Siemens, J., Kindler, R., Flechard, C., Sutton, M.A., 2009. Biosphere-atmosphere exchange of reactive nitrogen and greenhouse gases at the Nitro-Europe core flux measurement sites: measurement strategy and first data sets. Agric. Ecosys. Environ. 133, 139–149.

Snyder, C.S., Bruulsema, T.W., Jensen, T.L., Fixen, P.E., 2009. Review of greenhouse gas emissions from crop production systems and fertilizer management effects. Agric. Ecosys. Environ. 133, 247–266.

Steiner, J.L., 1994. Crop residue effects on water conservation. In: Unger, P.W. (Ed.), Managing Agricultural Residues. Lewis Publishers, Boca Raton, FL, pp. 41–76.

Tebaldi, C., Hayhoe, K., Arblaster, J.M., Meehl, G.E., 2006. Climate change, Going to the extremes, An intercomparison of model simulated historical and future changes in extreme events. Clim. Change 79, 185–211.

Tenuta, M., Mkhabela, M., Tremorin, D., Coppi, L., Phipps, G., Flaten, D., Ominski, K., 2010. Nitrous oxide and methane emission from a coarse-textured grassland soil receiving hog slurry. Agric. Ecosys. Environ. 138, 35–43.

Van Groenigen, K.-J., Six, J., Hungate, B.A., Graaff, M.-A., Van Breemen, N., van Kessel, C., 2006. Element interactions limit soil carbon storage. Proc. Nat. Acad. Sci. USA 103, 6571–6574.

Venterea, R.T., Burger, M., Spokas, K.A., 2005. Nitrogen oxide and methane emissions under varying tillage and fertilizer management. J. Environ. Qual. 34, 1467–1477.

Venterea, R.T., Dolan, M.S., Ochsner, T.E., 2010. Urea decreases nitrous oxide emissions compared with anhydrous ammonia in a Minnesota corn cropping system. Soil Sci. Soc. Amer. J. 74, 407–418.

Wan, S., Hui, D., Wallace, L., Luo, Y., 2005. Direct and indirect effects of experimental warming on ecosystem carbon processes in a tallgrass prairie. Global Biogeochem. Cycles 19, 2014 (doi:10.1029/2004GB002315).

West, T.O., Marland, G., 2002. A synthesis of carbon sequestration, carbon emissions.

Wood, C.W., Edwards, J.H., 1992. Agroecosystem management effects on soil carbon and nitrogen. Agric. Ecosys. Environ. 39, 123–138.

CHAPTER 29

Beyond Mitigation: Adaptation of Agricultural Strategies to Overcome Projected Climate Change

Ronald F. Follett
USDA-Agricultural Research Service, Soil Plant Nutrient Research Unit, Fort Collins, CO

CHAPTER OUTLINE

Abbreviations: GCC, global climate change; GHG, greenhouse gases; CO_2, carbon dioxide; N_2O, nitrous oxide; CH_4, methane; SOM, soil organic matter; SOC, soil organic carbon; GDP, gross domestic product; Nr, reactive nitrogen; FTA, free trade agreement

CLIMATE AND CLIMATE CHANGE

Current science indicates an association between climate change over the last century and human activity, specifically land-use change, burning of fossil fuels, and regional changes in temperature and weather patterns. More broadly, global climate change (GCC) is a significant alteration from one climatic condition to another, including those caused by both non-anthropogenic and anthropogenic causes (Le Treut et al., 2007). While global warming refers to the overall warming of the planet, GCC refers to the fact that, if warming continues,

Managing Agricultural Greenhouse Gases. DOI: 10.1016/B978-0-12-386897-8.00029-2

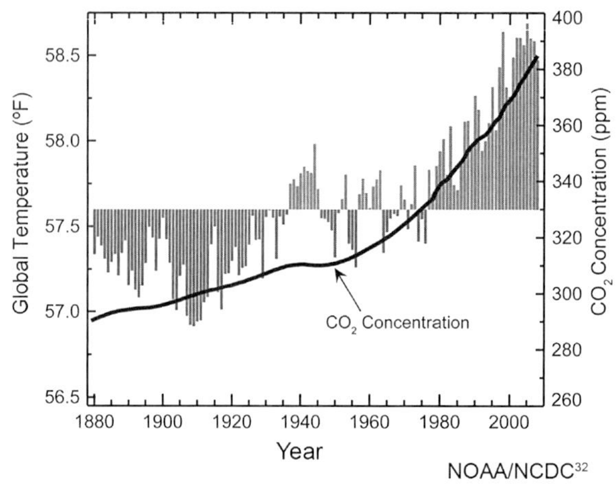

FIGURE 29.1

Increases in annual global surface temperature (over both oceans and land) since 1880. Red bars indicate temperatures above and blue bars represent temperatures below the average temperature period 1901–2000. The black line is atmospheric CO_2 concentration in parts per million. Global Climate Change Impacts in the United States. *USGCRP (2009)*. Please see color plate section at the back of the book.

it probably will not be uniform across the planet. Some regions on earth may become warmer while others become cooler. Although, the earth's temperature has fluctuated considerably over millions of years, the study of change in the earth's surface temperature and alterations of weather patterns associated with the "greenhouse effect" addresses the part of GCC associated with human activity (see Figure 29.1).

Future Temperatures

Fortunately, the greenhouse effect has allowed the earth to provide a hospitable environment for life. Naturally occurring GHGs result in a warming effect of about +33°C. The earth's mean global temperature is thereby increased from about −19°C to +14°C.[a] Their contributions and the fraction of the greenhouse effect they cause include water vapor (36–70%), CO_2 (9–26%), CH_4 (4–9%), and ozone (3–7%) (Kiehl and Trenberth, 1997; Schmidt, 2005; and Russell, 2007). Human activity since the beginning of the industrial revolution has increased GHGs in the atmosphere and their associated radiative climate forcing. Since 1830, concentrations of CO_2, N_2O, and CH_4 have increased by 36, 18, and 148%, respectively (U.S.-EPA, 2007).

Global warming changes can be observed in the instrumental temperature record, rising sea levels, and observed decreases in snow and ice extent (Le Treut et al., 2007). The most common measure of global warming is the trend in mean global temperature near the earth's surface, which increased by 0.74 ± 0.18°C from 1906 to 2005. In the future, and even with current GHG reduction efforts, global emissions are expected to grow during the coming decades (Pachauri and Reisinger, 2007). During the 21st century, increases in emissions at or above their current rate are expected to induce changes in the climate system larger than those observed in the 20th century (IPCC, 2007). A recent analysis of future temperature trends projects that under a high GHG emissions scenario mean temperatures in the U.S. (Figure 29.2) may increase by approximately 4–6°C (7–11°F) by the end of the century (USGCRP, 2009). Even if CO_2 emissions were to completely cease, changes in earth's climate

[a] "To emit 240 W m^{-2}, a surface would have to have a temperature of around −19°C. This is much colder than the conditions that actually exist at the Earth's surface (the global mean surface temperature is about 14 °C). Instead, the necessary −19°C is found at an altitude about 5 km above the surface" http://www.ipcc.ch/pdf/assessment-report/ar4/wg1/ar4-wg1-chapter1.pdf. Retrieved April 21, 2009

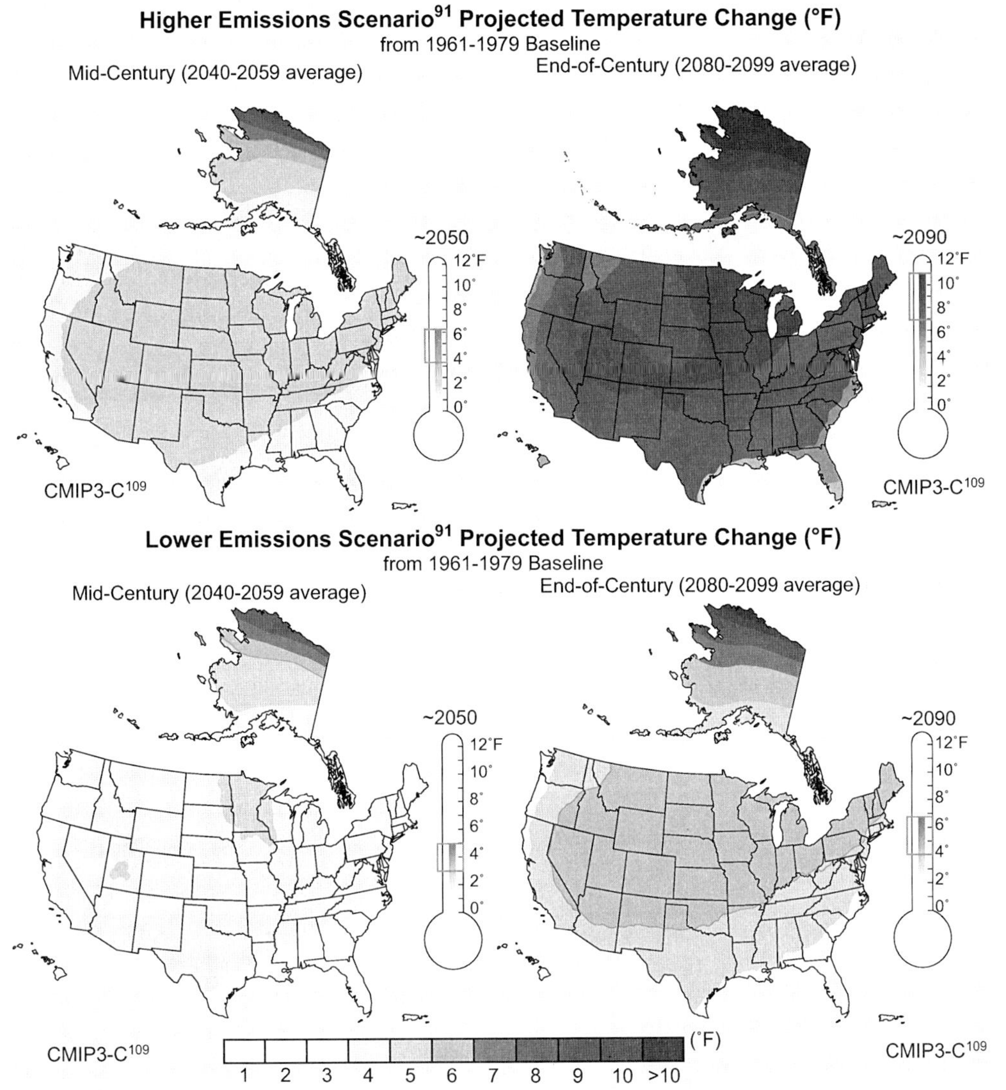

FIGURE 29.2

Projections of future temperature from 16 of the Coupled Model Inter-comparison Project. The maps feature a higher and lower GHG emission scenario. Brackets on the thermometers represent likely ranges of model predictions. Global Climate Change Impacts in the United States.*USGCRP (2009).* Please see color plate section at the back of the book.

system due to the rise in atmospheric GHGs over the past 200 years likely would sustain a warmer climate for another thousand years (Solomon et al., 2009).

Climate and GCC have many interacting features that are exceedingly complex and dominated by the balance of incoming energy from the sun. Earth's energy balance can be changed in three basic ways: (1) by the amount of incoming radiation from a change in solar activity or earth's orbit, (2) by the amount of radiation reflected back into space from changes in vegetation or cloud cover, and (3) by the amount of re-radiation of long-wave radiation from atmospheric GHG concentrations (Somerville et al., 2007). Many uncertainties exist about climate change with the severity differing regionally (Somers, 2010). Increased incidences of violent weather, warmer temperatures, retreating glaciers, thawing permafrost, and increased pests are already having a significant impact on water, energy, transportation, agriculture, and ecosystems (IPCC, 2007).

Future Precipitation

In general, increases in global temperatures will increase humidity and alter atmospheric circulation patterns. Observations confirm the impact of increasing temperatures on earth's hydraulic cycle through increased precipitation intensity, with varying effects for different regions. Precipitation amounts have increased about 5% in the past 50 years in the U.S., especially in the northeast (USGCRP, 2009). Records show a 10–20% decline in summer precipitation between 1901 and 2007 in most of the southeastern U.S. Deeper incursions of warm, humid air are expected to lead to increased precipitation in the northern U.S. (Figure 29.3). Furthermore, greater drought is expected for the southwest quadrant of the U.S. with increased temperatures and declining annual precipitation in the southern U.S. during spring and summer (Seager et al., 2010; Wang et al., 2005). Consequently, as the above described climatic patterns change, so do spatial distribution of agroecological zones, habitats, plant diseases, and pests.

Two types of climate change impacts specifically identified by the FAO (2007) were divided into *biophysical* and *socio-economic*, with their application to agriculture as follows:

Biophysical impacts include:

- physiological effects on crops, pasture, forests, and livestock
- changes in land, soil, and water resources
- increased weeds and pests
- shifts in spatial and temporal distribution of impacts.

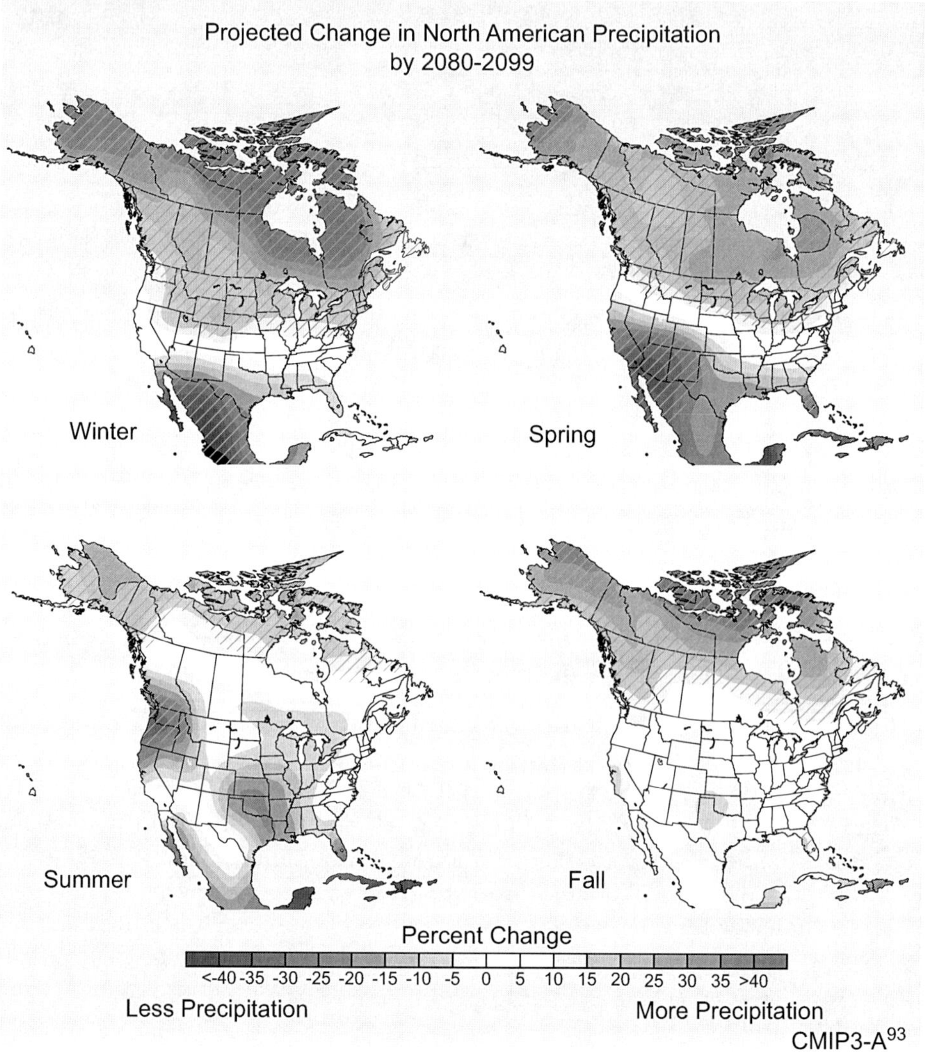

FIGURE 29.3
Projected future changes in precipitation relative to the recent past as simulated by 15 climate models. Simulations are the late 21st century, under a higher emissions scenario. Global Climate Change Impacts in the United States. *USGCRP (2009).* Please see color plate section at the back of the book.

Socio/economic impacts include:

- decline in yields and production
- reduced gross domestic product (GDP) from agriculture
- fluctuations in world market prices
- changes in geographical distribution of trade regimes
- increased number of people at risk of hunger and food insecurity
- increased human migration and civil unrest.

Biophysical Impacts

For the *biophysical impacts* a key concern is water. Climate change adaptation for agricultural cropping systems will require higher resilience against both excess (high intensity rainfall) and deficient water (extended drought periods). The soil's organic matter (SOM) content is a key factor to respond to both problems because it improves and stabilizes soil structure and allows soils to absorb more water without causing surface runoff. Soil organic matter improves water absorption capacity to help buffer against drought. Intensive tillage decreases SOM through aerobic mineralization while conservation tillage maintains SOM through permanent soil cover (crop residues or cover crops and use of diversified crop rotations). Conservation agriculture is an adaptation that combines zero or low tillage and permanent soil cover to (1) increase and maintain SOM and soil organic carbon (SOC), its major component, (2) improve nitrogen- and energy-use efficiency, and (3) lower input costs. Conservation agriculture protects soil macrospores and channels that can aid in the drainage of excess water. Surface cover modulates surface soil temperature and reduces evaporation. Accordingly, conservation agriculture is a climate change "adaptation" technology, but also an effective GCC "mitigation" strategy. However, adaptation strategies will need to become increasingly advanced, more effective, and include additional approaches to address the *biophysical impacts* of GCC. In doing so, such strategies also help decrease the severity of the associated *socio/economic impacts* of GCC.

Socio-Economic Impacts

The most negative impacts from future GCC will likely be in countries that are most stressed in their GDP contribution from agriculture and in their dietary energy supply per capita. Those countries are reported to occur in areas already experiencing severe poverty (Figure 29.4) (World Bank, 2005). World GDP from Oceania, Africa, Asia, North and South America, and Europe is 2, 3, 27, 33, and 35%, respectively, with world agricultural GDP projected to decrease by 16% by 2020 due to impacts from GCC (FAO, 2009). Output in developing countries is projected to decline by 20%, while output in industrial countries is projected to decline by 6% (Cline, 2007). Carbon fertilization, with higher atmospheric CO_2 levels increasing photosynthesis, might limit the severity of climate change effects to only 3% (Cline, 2007). However, technology is not expected to adequately alleviate output losses or increase yields sufficiently to keep up with growing food demand (Cline, 2007), while global mean temperature increases of 5.5°C or more could lead to an average increase in food prices of 30% (Easterling et al., 2007).

Risk-coping, resilient production systems require adaptable structures such as diversified crop rotations, agroforestry, crop–livestock integration, possibly crop–fish systems, use of vegetative buffer strips or hedges, and other on-farm practices. Such strategies can have great impact for adaptation to drought, heavy rains, and strong winds. Land and water management techniques to mitigate GHG emissions are currently being researched and adopted by many land managers. However, better programs that further these efforts are still needed. The best adapted and most workable practices, under specific biophysical and socio-economic conditions, need to be fully identified and implemented on a continuing basis. Regional and national land-use planning approaches can potentially work best, but need to stress participatory approaches that identify and prioritize where investments are most needed for future GCC.

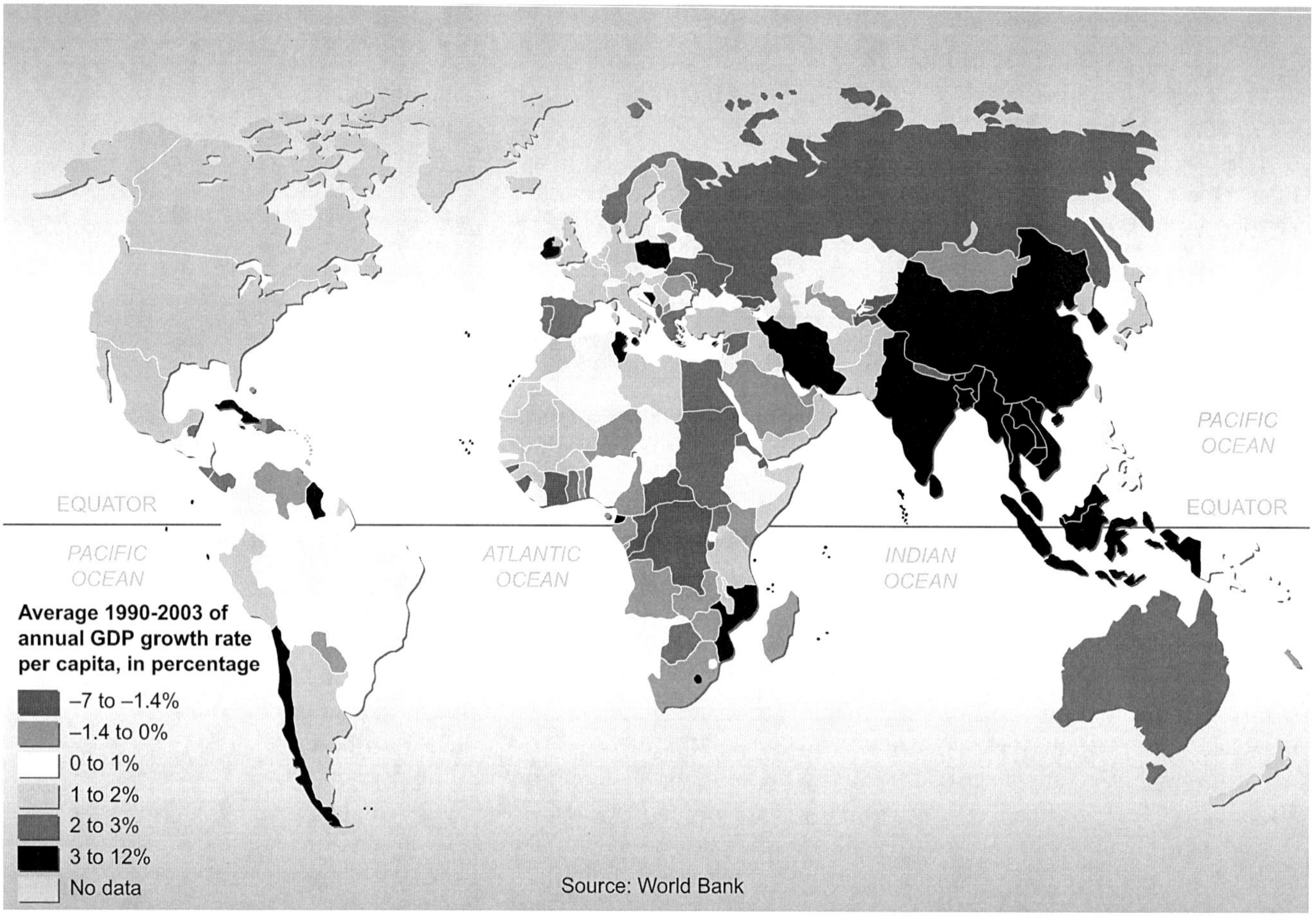

FIGURE 29.4

The global economic importance of agriculture. Per capita gross domestic product (GDP) average annual growth, 1990–2003. Average annual percentage growth rate of GDP per capita at market prices based on constant local currency. Dollar figures for GDP are converted from domestic currencies using 1995 official exchange rates. GDP is the sum of gross value added by all resident producers in the economy plus any product taxes and minus any subsidies not included in the value of the products (World Bank, 2005). Cartographer/Designer Philippe Rekacewicz, and Emmanuelle Bournay, UNEP/GRID-Arendal. http://maps.grida.no/go/graphic/per-capita-gross-domestic-product-gdp-average-annual-growth-1990-2003 (verified May 23, 2011). Please see color plate section at the back of the book.

Agricultural sustainability and the ability to feed a growing world population without degrading the environmental and natural resource base will be affected by climate change and is critical to future production systems. In addition to GCC, many forces will shape world economies and the supply and demand for agricultural products. Changes in population numbers, economic activity, and technology will be major driving forces. The response of producers and governments will have major implications to natural resources and the environment.

Feedbacks from changes in resource quality and availability will affect agricultural production. A particular concern is water, including from heavy downpours, drought, and water quality as they are manifested in different regions. In regions around the world where snowpack is relied on as a primary water resource, including in the U.S., both agricultural and municipal water supplies will be challenged. Sea level rise and storm surges are a particular concern for coastal areas, with consequences for transportation and energy infrastructure. Human health concerns include increasing heat stress, waterborne diseases, poor air quality, extreme weather events, and disease (IPCC, 2007).

What are the limits to adaptation? How much can adaptation deal with, what can be accomplished, and what types of tools need to be available? These questions are daunting and beyond the scope of this chapter. Even though the agricultural sector is the focus here, the above questions are worthy of consideration by all the economic sectors including: transportation, power, industry, commercial, and residential.

THE ROLE OF AGRICULTURE

Mitigation of Greenhouse Gas Emissions

Future *adaptations* to climate change by agriculture will be based upon improvement to current *mitigation* strategies. Improved practices can reduce net GHG emission; however, the effectiveness of the agricultural practices used depends on climate, soil type, and farming system. The largest total mitigation opportunity is from sink enhancement (SOC sequestration) followed by emissions reduction. However, maintaining and possibly increasing amounts of sequestered SOC may be critical to future climate change adaptations. Technologies and practices critical to the mitigation of GHG emissions from the agriculture sector (Pachauri and Reisinger, 2007, Morgan et al., 2010, CAST, 2011) include:

1. Improved crop and grazing land management to increase soil carbon storage
2. Restoration of cultivated peaty soils and degraded lands
3. Improved rice cultivation techniques
4. Improved livestock and manure management to reduce CH_4 emissions
5. Improved nitrogen fertilizer application techniques to reduce N_2O emissions
6. Dedicated energy crops to replace fossil fuel use
7. Improved energy efficiency in agricultural operations.

Greenhouse gas emissions can be decreased by substituting energy production from agricultural feed stocks (e.g. bioenergy sources) for fossil fuels. There is a potential to use agricultural products and residues as feedstocks, but concerns exist about potential competition for other land uses. In response to these concerns, growing and harvesting diverse mixtures of long-lived, deep-rooted native plants for biofuels on underutilized grazing lands or less productive farmland may be an alternative that can allow biomass to be produced without competing with the production of food or higher-value crops.

Agriculture is not only affected by the climate, but contributes to GCC through its own exchanges of GHG with the atmosphere. The management of agricultural systems to sequester atmospheric CO_2 as SOC and to minimize GHG emissions has potential to help mitigate climate change. Thus, even as new strategies and technologies emerge, it will remain important for agriculture to continue to develop and implement successful mitigation practices which maintain or enhance soil C pools while minimizing CH_4 and N_2O emissions.

Agricultural Adaptations to Climate Change

CROPLAND SYSTEMS

Many available GCC mitigation options can be further developed as adaptation options. Agricultural technologies are rapidly improving, but also need to be adopted rapidly enough to begin addressing existing and emerging socio-economic impacts of degraded natural resources in many developing countries. Future adaptation includes responding to: (1) changes in long-term mean temperature or precipitation conditions, (2) year-to-year variability of climate, (3) increasing challenges of invasive or better adapted pests and diseases, (4) increasing competition for natural resources, and (5) greater demands for food, feed, and fiber from an ever-growing human population. Consequently it becomes important to define the issues of "adaptation to what", "who or what is adapting", and "how does adaptation occur". Adaptation is an ongoing process whereby improvements are already leading to cost reductions, improved varieties, better adapted livestock, improved management, and enhanced marketing strategies. Many of the current adaptations are already those that address either "changes in long-term mean climate conditions" or adaptation to "year-to-year variability". There is considerable need and opportunity to improve the science of adaptation and its potential application to GCC impacts and for policy improvement.

Management Systems

Currently, cropland soils in the U.S. are a net source of GHG emissions of around 150 Tg CO_2-eq year, almost entirely due to N_2O. Soil C has increased on most cropped soils by 46 Tg CO_2 yr^{-1} due to a long-term trend of increased crop-residue production, reductions in tillage intensity, cropland conversion to pasture/hay, and set-aside of annual cropland in the Conservation Reserve Program. However, there is a soil C offset by emissions from cultivated cropland on organic (e.g. peat) soils of around 30 Tg CO_2-eq and net emissions (ca. 6 Tg CO_2-eq) from land recently converted to cropland (Ogle et al., 2010; U.S.-EPA, 2010). Already known are a number of practices that contribute to increases soil C stocks. The classes of practices and their primary mode of action, i.e. increased soil C inputs (I) or reduced decay rates of SOM (R), are:

- High residue yielding crops and residue retention (I)
- Conversion of annual crops to perennial grasses and legumes (I, R)
- No-tillage and other conservation tillage practices (R)
- Cover crops (I, R)
- Manure additions (I)
- Reduced frequency of bare fallow (I, R)
- Practices to increase C additions to soils (e.g. irrigation, improved fertility) (I)
- Rewetting (flooding) of organic (i.e. peat and muck) soils (R)
- Tree planting on annual cropland (I, R).

Changes in soil carbon stocks from adoption of the above "carbon-friendly" management practices depend on interactions of several management variables (e.g. crop type and rotation, tillage, nutrient and water management), soil type, climate regime, and land-use history (Paustian et al., 1997). Recently, the concept of precision farming, or precision conservation, is becoming increasingly recognized (Khosla et al., 2002; Delgado et al., 2008).

The largest proportion of N from mineral fertilizers and manure in the U.S. is applied to annual cropland. Additional N additions result from N_2 fixation by grain legumes (e.g. soybeans) and legume hay (e.g. alfalfa and clover) grown in rotation with annual crops. This input of reactive N (Nr) is the primary driver of increased N_2O emissions from cropland with emission rates typically amounting to 0.5–3% of N input (Pachauri and Reisinger, 2007). The Nr, based upon its form, can move by processes of *atmospheric transport* and by *water and sediment associated transport*. The transport of Nr can be between the atmosphere, terrestrial, and aquatic ecosystems and the fraction of lost N may be higher still when applied N inputs exceed crop demand (Follett et al., 2010). Emission of N_2O for a location and/or management system is highly variable, but the direct relationship in space and time between N_2O emissions and N input and the importance of N supply in determining crop yield are primary challenges to increase plant N-uptake efficiency (NUE) and to reduce Nr losses.

Management practices to increase NUE of productive cropping systems has four main options (Robertson and Vitousek, 2009): (1) more diverse and continuous crop rotations that increase NUE; (2) providing farmers with the decision support tools that allow better crop N predictions to avoid over-fertilizing; (3) managing timing, placement, and formulation of fertilizer N to better ensure N is available where and when needed; and (4) managing watersheds to mitigate indirect Nr losses and N_2O emissions, including those from downstream denitrification of nitrate leached from farm fields (Beaulieu et al., 2011).

Adaptation of annual crops to GCC will be a continuing process of evaluation, selection, and testing of germplasm in important environments and to a range of climatic conditions. Information will be required about germplasm susceptibility pests and diseases. Depending upon how rapidly climate changes, a range of both cultivar and suitable crop species need to be identified and tested over time across anticipated temperature and moisture conditions, against potential pests and diseases, and on a continuing basis that includes field tests and possible genomic approaches.

GRAZING LANDS

This vast global land area has high potential to sequester SOC, produce animal protein, and provide important environmental services across the land and climate resource found throughout the world (Follett and Reed, 2010). Lal (2004) suggested SOC sequestration on grazing lands (range and pasture) might represent about 15% of the potential for U.S. soils. Follett and Schuman (2005) estimated that the world's SOC sequestration rate on grazing lands was about 0.06 Mt C ha^{-1} yr^{-1}, resulting in about 0.2 Gt SOC yr^{-1} on 3.5 billion ha of land. Lal (2004) estimated the world's grazing lands could sequester between 0.01 and 0.3 Gt SOC yr^{-1} on 3.7 billion ha. In the U.S., grazing is the predominant use on the 237 Mha of pastureland and rangeland, occurs on the 54 Mha of forested grazing lands, and 25 Mha of fall and winter grazing of small grains and after-harvest grazing of haylands (USDA-ERS, 2007). Numerous ecoregions and climates around the world are occupied by grazing lands, each with unique characteristics that affect rates of SOC sequestration and with different potential to respond to management. The SOC sequestration potential of grazing lands worldwide is nearly 30 Pg CO_2 over a 40-year period (Lal, 2004; Follett and Schuman, 2005; Conant et al., 2001). Whether this C remains in storage, and for how long, and whether amounts can be increased through improved management are important questions.

Land Management

Key considerations for sequestration and subsequent retention of SOC in grazing lands are: (1) aboveground C is a minor component of total ecosystem C storage and mean residence time of this C pool is only a few years, so yearly variations in aboveground biomass only minimally affects C storage; (2) most SOC is recalcitrant and well protected from minor natural disturbances and resistant organic C and soil carbonates are least sensitive (Allen et al., 2010); microbial biomass and particulate or light-fraction organic C are most sensitive to management or land-use change; (3) major pathways of SOC input are through decomposition of belowground root biomass, surface deposition of animal feces, and decaying litter from aboveground vegetation (Pinerio et al., 2010); and (4) protection of grazing lands requires avoiding large perturbations in the SOC pool that can occur with major soil disturbances, such as surface denudation with overgrazing, wind and water erosion, and tillage, these effects occur naturally with extreme weather conditions or through human-induced management decisions that cause poor vigor of plant communities (Follett, 2001). Important management factors controlling fate of SOC in grazing lands are: (1) long-term changes in production and quality of above- and belowground biomass which can alter quantity of available N and C:N ratios of SOM (Derner and Hart, 2007; Pineiro et al., 2010); and (2) grazing-induced effects on vegetation composition, which can directly impact grazing intensity (Derner and Schuman, 2007; Bagchi and Ritchie, 2010).

ECOSYSTEM SERVICES

Among the many *societal and environmental benefits* recognized for grazing lands (Follett and Reed, 2010) is to offset atmospheric CO_2 emissions by its long-term storage in SOM, thus helping to improve the soil's productivity. Adaptation to GCC will require a greater awareness of the importance of practices and policies to encourage maintaining and improving grazing land condition. Policies to support protecting and enhancing terrestrial C stocks are beneficial for soil erosion control. Enhancing grazing land soil C sequestration consistently can be associated with improved soil and water quality; reductions in silt loads and sediments into streams, lakes, and rivers; and air quality improvement (Lal et al., 2003). Risk of soil loss from grazing lands is usually less than from cropland, and magnitude can depend on conditions such as drought, high winds, intense rainfall, and extreme runoff events resulting in wind and water erosion. Management to enhance SOC generally results in increased soil protection from erosion. Adaptation research is needed on national and regional scales to overcome poor management such as abusive grazing, intensive burning, or similar factors that can result in SOC losses that cause land vulnerability to soil erosion, especially on steeper slopes. Lal (2007) estimates approximately 0.8–1.2 Pg yr^{-1} of SOC displaced by erosion is mineralized and emitted into the atmosphere as CO_2. Poor or potentially abusive grazing management may decrease vegetative cover, increase soil erosion risks by wind and water, and heighten the potential to degrade water quality (Thurow et al., 1988; Lal, 2001). Changes in available C inputs are caused by a decline or increase in aboveground biomass production and change in plant residue amount returned to the soil. Soil moisture and temperature regimes, soil erosion, leaching, and water-holding capacity are all important to SOC increase (or loss), degree of SOM protection from oxidation, and the potential for SOC to provide societal benefits (Follett and Reed, 2010).

Encouraging Successful Adaptation

Agriculture and the agricultural industry in developed countries have many resources available to respond and adapt to GCC. The U.S. has a strong resilient agriculture based upon abundant land, genetic resources, research capacity, institutions, and local and global market access. Less developed countries, as they evaluate their resources, may find soil and water resources are compromised, as well as issues of poverty, lack of suitable infrastructure, flawed or lack of suitable institutions or in-country policies, and limited market availability. Climate change will impose several human, economic, and environmental costs that are country or region specific. Use of an asset portfolio (CAST, 1992) may be an important future strategy to better allow adaptation resources to be considered across a range of scales depending upon needs and anticipated climate change conditions.

An Asset Portfolio to Prepare for Climate Change

An asset portfolio to address climate change requires planners to evaluate risk and to conduct an inventory of available or needed resources (assets) within the geographic area for which they are planning. Identified assets need to be quantifiable to construct a portfolio and to optimize or maximize expected returns based on given levels of risk. Portfolio theory has been shown to fit the climate change problem well, because it is concerned with decisions involving outcomes (CAST, 1992). Modern portfolio theory, as it has been referred to, was pioneered by Harry Markowitz (1952) and included four basic steps for construction of a portfolio: (a) security valuation, (b) asset allocation, (c) portfolio optimization, and (d) performance measurement. One of the most important attributes to construct the portfolio is reduction in risk in terms of quantifying the benefits of diversification. In theory, the risk in a portfolio of diverse assets will be less than that inherent in holding only a single asset (provided the risks of the various assets are not directly related). Critical to portfolio theory is asset allocation, which refers to the strategy of dividing the total portfolio among various asset classes and where asset allocation is an organized and effective method for diversifying. Effective portfolio management as described recently by Cooper et al. (1999) is vital to successful innovation requiring strategic choices about investments in products and technologies. A key component identified by Cooper et al. (1999) is recognition of the importance of considering resource allocation; engineering, research and development, and marketing resources. Also needed is to have the right balance among available resources and/or capabilities. This can be illustrated by climate change resulting in some farm land becoming unproductive, but greater investment in research might result in technology that can restore some of the lost productivity. Such research can also enhance productivity for other parts of the total land base. Important in an asset portfolio is to recognize that assets are valuable, can contribute to an effective adaptation

TABLE 29.1 Portfolio of Assets to Prepare for Climate Change

Asset	Value for adapting to climate change	Steps for increasing flexibility
Land	Extensive cropland across diverse climates.	Encourage flexible land uses, diversifying, and adaptations.
Water	Water may already limit farming in some regions and is crucial to adaptation.	Encouragement of prudent use of water. Raise the value of crop produced per volume of water consumed.
Energy	Reliable energy supply essential for many adaptations to new climate.	Improve the efficiency of energy to food production. Explore new biological fuels, solar, and other sources of energy.
Physical infrastructure	Facilitates input flows, trade, and markets.	Maintain and improve input supply and export delivery infrastructure to respond to market signals.
Genetic diversity	Source of genes to adapt crops and animals to new climates.	Assemble, preserve, and characterize plant and animal genes. Conduct research on alternative crops and animals.
Research	Source of knowledge and adaptation technology.	Broaden research agenda on adaptation strategies. Improve sustainable and alternative farming and food systems.
Information	Provide information needed to track climate.	Enhance national/international systems to exchange information about climate change and how to adapt.
Human resources	Provide pool of skills enabling farmers and researchers to adapt to climate change.	Make adaptation skills the hallmark of agriculture's human resources. Strengthen rural and continuing education.
Political Institutions	Determine the policies and rules to facilitate and remove those that hinder adaptation.	Harmonize agricultural institutions and policies.
World markets	Enables trade to mediate shifts in farm production and sends price signals that eventually adjust production to new climates.	Promote freer trade and avoid protectionism.

(Adapted from CAST, 1992)

strategy, but also must be flexible in asset use. For example, a rapid change in weather in one region, even if temporary, may idle some land, but allow improved conditions in another region. Table 29.1 is a generalized portfolio of assets for possible use at regional, national, or larger scales and can be adapted to planning for GCC.

LAND

Land in the U.S. includes large spatial and temporal differences and the large expanses available for agricultural uses, along with appropriate inputs, knowledgeable land managers, and effective use of advanced cultural practices allows the U.S. to produce abundant food, feed, and fiber. The overall agricultural area of the world is 4968 Mha, while cropland area is 1562 Mha, with 688 Mha in cereals (FAOstat, 2009). Large areas of agricultural land exist around the world (Figure 29.4), but many exist on degraded soils. Although there are differences around the world, estimates of soil degradation causes on agricultural lands include overgrazing (35%), non-conserving agricultural practices (28%), deforestation (30%), over-exploitation to produce fuelwood (7%), and industrialization (4%) (FAOstat, 2009). In North

America, agriculture is considered responsible for 66% of the soil loss, while in Africa, overgrazing is responsible for about half of the soil degradation (FAOstat, 2009). Lightly degraded soils can be improved by crop rotation, minimum tillage, and other improved practices, but more severely degraded soils are difficult to restore. Highly worrisome is that continued loss of arable land will jeopardize the world's ability to feed its growing populations.

Climate change adaptation for world agricultural lands will require drawing upon not only current technologies, but new technology development. Assuming adequate recognition is given to the potential impacts of climate change, developed and developing countries need to implement restorative and land protective measures for one of their most valuable assets. Underdeveloped countries, with their limited resources and economic/societal issues, may continue to abandon land as it becomes more severely damaged while less severely damaged lands will continue to degrade. Changes in soil conservation practices can slow land degradation, but often cannot restore soil fertility in the short term. National/international programs are needed for such lands and can require major structural change (e.g. draining, contour terraces, reseeding, etc.).

WATER

Water is crucial to adaptation and already limits production in many regions. Water conservation practices that contribute to greater water-use-efficiency (WUE) are critical to adapting to climate change for both irrigated and dryland cropping systems (Table 29.1). Emphasis to improve WUE will be essential to adaptation to climate change. Conservation practices and adaptive strategies will be required that increase the availability of water for food and feed production at critical plant growth stages by increasing soil water storage in the soil profile, decreasing evaporative demand, and synchronizing timing availability with plant water requirements for both irrigated and dryland agriculture. Such practices help increase the amount of either irrigation water or precipitation that can be available for food, feed, and fiber production. However, excess water will also need to be managed. Conservation practices will need to increase infiltration under high precipitation as well as control drainage to improve excess water removal and prevent salinization. Prevention of soil and water degradation to maintain the productivity and efficient use of both will become increasingly important adaptive strategies. In addition, excess water, where appropriate, requires enhanced means for its storage and recycling to subsequent beneficial use.

Developed countries are beginning to address their hydroclimatic vulnerability by expanding economic development and improving the management of their water resources, but continue to face limitations from flooding and drought. Less developed countries remain hampered by water-related challenges such as flooding, drought, and severe water pollution (Falkenmark, 2009) with the poorest economies, often in semiarid climates, still remaining hostage to water problems and suffering large-scale poverty, disease, and uncurbed population growth. Causes of water insecurity include overpumped aquifers, excessive water consumption by irrigation, pollutants in return flows that contaminate clean water sources, and neglect of critical maintenance of urban and other important water-supply systems. A fundamental tool to master the variability of hydroclimatic water delivery systems is water storage. Statistics show North America has access to over 6000 cubic meters per person per year stored in reservoirs, while the poorest African countries have less than 700. Ethiopia, for example, has less than 50 cubic meters per person per year of water storage (Falkenmark, 2009).

Climate change is expected to alter atmospheric circulation patterns and is likely to expand aridity, change rainfall patterns and seasonality of river flows, and increase melting rates of winter snow (Figures 29.2 and 29.3). Planning is required to allow safe water and development of new or modified infrastructures that are not based upon hydrologic averages from the past. It will need to be recognized that the water asset includes precipitation, recharge of the soil moisture, and water that flows through rivers and aquifers. Sustainable water management will require including water withdrawal and its return use, which often contains pollutants. Treatment and reuse of water can be done repeatedly, but unlimited consumption of water can potentially lead to human and ecological shortages. Increased water scarcity will require improved planning and management to reconcile upstream and downstream uses of water in an integrated manner.

ENERGY

Reliable energy supplies are essential to adapting to climate change. Even though agriculture uses only a small proportion of a nation's total energy, direct and indirect energy inputs are critical to agricultural production. In the U.S. in 2002, 1.7 quadrillion Btu were used by the agricultural sector; of that amount, 65% (1.1 quadrillion Btu) was consumed as direct energy (electricity, gasoline, diesel, LP gas, and natural gas), compared with 35% (0.6 quadrillion Btu) consumed as indirect energy (fertilizers and pesticides) (Schnepf, 2004). Energy is a key asset because high or unstable energy prices can make agriculture unprofitable. Agriculture and other economic sectors must find ways to become more energy independent. Various strategies are emerging whereby agriculture is being asked to address not only its own energy future, but also that of the U.S. Perhaps most apparent is the increased use of biofuels. Specifically, U.S. research is focusing on ways that agriculture can efficiently increase biomass production, increase grower profit, reduce biorefinery transaction costs, and optimally incorporate biomass and other dedicated feedstocks into existing agricultural and forestry-based systems. Adaptation will require uncertainties of expanded production be addressed early in the adaptation process. Use of biofuels will require that their use minimizes negative impacts on existing ecosystems, markets, and services while finding new ways to utilize value-added co-products to help enable commercial biorefining technologies.

Prices of commodities used in biofuel production are increasingly linked with energy prices (von Braun, 2007). In Brazil, the price of sugar is very closely connected to the price of ethanol. A worrisome implication of the link between energy and food prices is that high energy-price fluctuations are increasingly translated into high food-price fluctuations. Price fluctuations for oilseeds, wheat, and corn have increased dramatically compared to previous decades. Feed prices for livestock products, such as meat and dairy products, have also increased. Effects of price increase on consumption are different among countries and consumer groups. Consumers in low-income countries are much more affected by price changes than those in high-income countries with the demand for meat, dairy, fruits, and vegetables more sensitive to price than cereals (von Braun, 2007).

Adaptation to GCC will include the need to produce more food from each unit of energy in agricultural systems. Greater efficiency will lessen the emission of CO_2 and the other GHGs. Moreover, rising energy prices will increase the importance of greater energy efficiency so that such costs will be less of a limiting factor in food production. Use of biological materials produced by agriculture can offer a new revenue stream to producers while offsetting domestic expenditures on fossil fuels.

PHYSICAL INFRASTRUCTURE

An important asset for adaptation to climate change is a nation's physical infrastructure to support its agricultural production. Physical infrastructure is critical to facilitate input flows, trade, and markets. Efficient systems distribute inputs to agricultural producers, provide delivery points for storing and handling output, process food, and transport and distribute products. When infrastructure functions efficiently, markets serve as a change agent in the regional/rural sector functions so that farmers and suppliers are assured free access to improved technologies. Four categories of infrastructure needing major attention are: (1) a source of farm inputs, (2) access to markets for agricultural products, (3) farm product processing and/or distribution resources, and (4) technology transfer services such as those provided by consultants, extension service agents, or government action agencies.

Infrastructure will play a critical supporting role in adaptation. An efficient system to allow transport of inputs and farm output facilitates existing flows of trade. Adaptation to future GCC may result in greater infrastructure efficiency through consolidation and/or elimination of various supply and marketing segments. Though difficult to predict, some of the pieces lost in some countries or regions may be necessary under a different climate. This suggests the need for countries to review infrastructure components and recognize the current and potential future roles for each to beneficially contribute to adapting to climate change.

GENETIC DIVERSITY

As an overall theme, genetic diversity creates flexibility for adaptation to environmental change. However, although a wide range of climate change conditions may be anticipated, the actual change may be sudden and unpredictable in timing and intensity. It is clear that even if immediate and concerted action is taken to adapt, dramatic and significant changes appear inevitable in the next 20–30 years (IPCC, 2007). Rapid climate change makes both the scientific and technical issues more difficult and the response of agriculture more complicated. Whatever means are developed to maintain food and agricultural production it will be difficult to disregard threatened species and ecosystems.

Predicted climate change scenarios will necessarily present numerous challenges to national legislative frameworks designed to protect habitat types and important species. As reviewed by (Harris et al., 2006), there is an increasing need to recognize goods and services that accrue from the natural capital of the world's combined ecosystems. As discussed by Follett et al. (2010), agricultural production systems are an integral part within the terrestrial ecosystem and "agro effects" generated by human systems impact ecosystem processes and components during the production of food and fiber. Potential impacts affect not only terrestrial ecosystems but also aquatic ecosystems and the atmosphere. A near-term, and possibly unsatisfactory, adaptation strategy for agricultural genetic resources and their diversity will likely be to utilize current cultivars, varieties, and species in new locations or regions. Regionally adapted alternative crops and animal species will be incorporated into mainstream production and their suitability tested as the climate conditions change. Should there be a rapid change of climate, translocation of both plant and animal genetic resources may occur under "trial and error" conditions, thus potentially introducing the wrong genetic material in the wrong place at the wrong time, with associated unintended environmental consequences.

A diverse portfolio of genetic resources is an asset that will be needed to adapt to climate change. A large library of agricultural plant and animal genetic materials currently exists that, if properly understood, could be a major asset for adaptation. However, action is needed to strengthen the asset, to sufficiently understand how the asset would respond to various possible climate change scenarios, and finally to know how to bring this asset into play. Important efforts have been ongoing for many years to assemble, preserve, and characterize plant and animal genes. A major constraint in developing a cost-effective strategy for collecting, preserving, and using genetic resources must be the timely updating and adequate characterization of the materials in both *in situ* and *ex situ* collections. Genetic resources have also been preserved from many locations from around the world at the National Center for Genetic Resources Preservation (NCGRP, 2010). The NCGRP preserves more than 500,000 samples of genetic materials of crops, livestock, and agriculturally important microbes. Seed in the plant base collection comes from USDA-ARS National Plant Germplasm System (NPGS) regional or crop-specific field sites. A sample of seed from each accession is retained at the field site for regeneration, multiplication, distribution, characterization, and evaluation and comprises an active collection. The NPGS base collection contains all the inventories of each accession including the original sample and earlier regenerations (NPGS, 2010). Testing seed

TABLE 29.2 Ecosystem Functions with Examples of Processes, Goods, and Services

Ecosystem function	Ecosystem process and components	Goods and services
Regulation functions	Maintenance of essential ecological processes and life support processes	
Gas regulation	Biogeochemical cycling	CO_2, CH_4, and N_2O recycling between atmosphere and terrestrial systems. UVb protection by ozone
Climate regulation	Influence of land-cover vegetation type	Maintenance of a favorable climate
Water supply	Filtering, retention, and storage of water	Provision of water for consumption
Habitat functions	Providing habitat for plant and animal species	
Refugium function	Niche availability and genetic diversity	Maintenance of biological and genetic diversity (and hence most other functions)
Production functions	Provision of food and fiber	
Raw materials	Conversion of solar energy into edible plants and animals	Fuel, structural materials
Information functions	Conversion of solar energy into edible plants and animals	Fuel, structural materials
Cultural functions	Cultural and artistic information	Use of nature as motive in books, films, and paintings

Adapted from Harris et al., 2006

inventories under GCC scenarios can identify suitable germplasm and guide successful implementation strategies so that alternatives can emerge quickly to replace existing species now in production.

Maintenance of genetic richness of natural and less managed ecosystems is also critically important for continuation of important ecosystems services. A needed goal is to minimize the potential for degradation of and to enhance ecosystem services provided wherever possible while agriculture adapts it genetic assets to climate change. Though less holistic than at present the importance of ecosystems was recognized by L.D. Harris (1984) in his book covering fragmented forests in terms of island biogeography theory and preserving biotic diversity. Harris (1984) recognized the role of both genetic diversity and the dangers of local extinctions that could result from landscape fragmentation. More recently, J.A. Harris et al. (2006) addressed the goods and services, including climate regulation and the provision of food and fiber, and several other important ecosystem functions that could be provided when ecosystem processes and components are maintained (Table 29.2). As each region and country must address climate change, it will be important that adapting genetic material for food, feed, and fiber and also an appreciation of the important relationship of agriculture as one part of entire ecosystems are both considered.

RESEARCH CAPACITY

A nation's research capacity is both a valuable and versatile asset to prepare for GCC. Although much of the world's research capacity is located in developed countries, it is still a resource for new discoveries, training, and technology transfer across national boundaries. Researchers will be asked to address the adaptation of agriculture and other ecosystems to climate change. Moreover, research must not only consider agricultural production during GCC but also its role in minimizing its own contributions to GHG emissions. The following are some recommended issues that research might address (CAST, 1992):

- Agricultural research systems will need to become more flexible. If climate change is rapid, social costs could mount rapidly before researchers can provide technological solutions. It is important to consider ways to shorten the time between discovery and application. Collaborations must be encouraged and multidisciplinary teams considered critical because of the need to develop new technologies to enable food and fiber producers to adapt to climate change in an environmentally sustainable manner. Funding structures for research will need to enable effective responses to uncertain regional or local effects. Team approaches are needed to encourage and support effective research responses to climate change.

- Regulations and government policies must not inhibit private sector research. Consistent regulatory regimes across nations are an increasingly important factor in international trade. Many technical and institutional innovations are possible. It will be necessary to devote substantial research effort to identifying and quantifying scientific, technical, economic, and societal information needed to rationalize regulatory information in the future. Much of agriculture's research on climate change will be conducted by publicly funded researchers; examples being environmental quality, sustainable agriculture, and conservation practices.
- Agriculture effects on the environment may intensify and lead to major impacts in conjunction with the effects of climate change. The major impacts on the environment must be expected to differ across regions (Figure 29.4). Research must enhance sustainable use of an increasing number of technical possibilities in advanced farming systems, possibly including advanced germplasm, more environmentally friendly pesticides and farm chemicals, and soil management technologies. Such technologies contribute to preserving other portfolio assets, such as reducing soil degradation or water pollution. Research to improve water, energy, and input efficiencies and enhancing genetic assets will be critical. Research can enhance and encourage adoption of better technologies.
- If climate change is severe in various regions or nations, then entirely new food systems may be needed. A food-system perspective, along with infrastructure development, may need to become the organizing principle for improving existing and/or designing new systems. Some of the alternatives will involve possible radical changes in food sources and use of other plant species alone or in combination with grain crops, both for animal production and human consumption.

INFORMATION

Information and information systems are vital to managing modern agriculture and adapting to GCC. Information technology will be important to maintain, share, compile, and interpret research and other data. Proprietary claims may become increasingly attached to new technologies as private sector entities develop and distribute technologies and information internally across regional and national boundaries. Information about climate and weather will need to be maintained, updated, and utilized for predictions as well as to identify progress in adaptation. Continued upgrading and availability of meteorological data will be needed to improve predictions made by climate models. Weather data availability is often weak for developing countries and climate change will result in such data being increasingly important for improved global weather monitoring. Whole season weather predictions will be especially important to producers, suppliers, market planning, and for food distribution networks. Providing, sharing, and utilizing weather data will largely remain the responsibility of governments even as private weather services continue to innovate. Attempts to respond effectively to GCC will require multiple mechanisms to exchange both public and private information.

HUMAN RESOURCES

People manage land used for food, feed, and fiber production and need access to technology and training to successfully adapt to GCC. Successful food producers, plant breeders, and actions agencies in the U.S. and in other countries have shown their ability to adapt crops and cropping systems beyond the original climate zones and conditions, wherein various crops were grown. However, future climate change will need to encourage improved and flexible skills for agriculture's human resources. Climate change will cause new and different demands.

Many different climate possibilities will exist in the future and the need will exist to alter existing practices. Maintaining and expanding education to rural areas can facilitate and assist in improving the flexibility of producers to adapt to GCC.

POLITICAL INSTITUTIONS

How will political institutions respond to GCC? Can there be concerted or even coordinated approaches at national and international levels? If institutions remain passive or only react to a changing climate then inventiveness, application, and the flexibility that might facilitate adaptation to GCC will all be deterred. For political institutions to successfully adapt to climate change they will need to adopt a proactive response, improve efforts to be informed, reprioritize agendas, and increase resourcefulness. Institutional response will be required on numerous issues, often region or country specific. However, disharmonies among various agricultural priorities can seriously deter effective adaptation responses to climate change and major efforts need to be made to identify and correct conflicting policies. Institutions need to encourage compatibility among individuals, organizations, and among nations. Policy changes will be needed in many areas, but more flexible commodity programs and improved water allocation will be among those most needed to adapt to climate change.

WORLD MARKETS

Access to global markets allows U.S. agriculture to sell its abundant production abroad, thereby increasing foreign exchange while distributing food where it is needed. A USDA Foreign Agriculture Service (FAS) report on international trade (August 2010) indicates that 2011 agricultural exports were at their second highest level ever at $114 billion, or $5.5 billion greater than the 2010 forecast and nearly $17 billion above 2009 (USDA FAS, 2010). Record exports to China are forecast as it surpasses those to Mexico to become the second largest U.S. market for agricultural commodities (USDA FAS, 2010). According to USDA FAS (2010), "Though imports are forecast at a record $81.5 billion, the agricultural trade surplus is forecast to be the second highest ever at $31.5 billion as agriculture remains one of the few industry sectors to consistently post a trade surplus."

Free trade, avoidance of protectionism, and removal of trade barriers and subsidies are assets to help address climate variability. The Administrator of FAS recently estimated that for every $1 billion in additional agricultural exports, there are 8000 jobs created and $1.4 billion in economic activity generated. An important contributor to foreign exchange of U.S. agricultural products has been the implementation of free trade agreements (FTAs) (i.e. international agreements between two or more countries to eliminate tariffs). The above export amounts have been most beneficial with those of several countries that are fully participating FTA partners with the U.S. Modern FTAs generally go beyond eliminating tariffs to include commitments on services, customs cooperation, intellectual property, and foreign investment. Eliminating restrictions between FTA partners leads to greater integration of economies and mutual benefits, including more export opportunities and product choices for importing countries. Trade barriers can curtail development of climate-change adaptation approaches globally because distortive subsidies and other barriers result in restriction to the required flexibility, and distribution of needed products and knowledge (CAST, 1992).

References

Allen, D.E., Pringle, M.J., Page, K.L., Dalal, R.C., 2010. A review of sampling designs for the measurement of soil organic carbon in Australian grazing lands. Rangeland J. 32, 227–246.

Bagchi, S., Ritchie, M.E., 2010. Introduced grazes can restrict potential soil carbon sequestration through impacts on plant community composition. Ecol. Lett. 13, 959–968.

Beaulieu, J., Tank, J., Hamilton, S., Wollheim, W., Hall, R., Mulholland, P., Peterson, B., Ashkenas, L., Cooper, L., Dahm, C., 2011. Nitrous oxide emission from denitrification in stream and river networks. Proc. Nat. Acad. Sci. 108, 214–220.

Cline, W.R., 2007. Global warming and agriculture: impact estimates by country. Center for Global Development and Peterson Institute for International Economics, Washington, DC.

Conant, R.T., Paustian, K., Elliott, E.T., 2001. Grassland Management and Conversion into Grassland: Effects on Soil Carbon. Ecolog. Appl. 11, 343–355.

Cooper, R.G., Edgett, S.J., Kleinschmidt, E.J., 1999. New product portfolio management: practices and performance. J. Prod. Innov. Mgmt. 16, 333–351.

CAST, 1992. Preparing agriculture of global climate change. Council of Agricultural Science and Technology. Report 119, 93.

CAST, 2011. Carbon sequestration and greenhouse gas fluxes in agriculture: challenges and opportunities. Task Force Report No. 142, p. 106.

Delgado, J.A., Berry, J.K., Khosla, R., 2008. New advances and practices for precision conservation. Paper presented at 9th International Conference on Precision Agriculture, Denver, Colorado. July 21–23, 2008.

Derner, J.D., Hart, R.H., 2007. Grazing-induced modifications to peak standing crop in northern mixed-grass prairie. Rangeland Ecol. Mgmt. 60, 270–276.

Derner, J.D., Schuman, G.E., 2007. Carbon sequestration and rangelands: a synthesis of land management and precipitation effects. J. Soil Water Conserv. 62, 77–85.

Easterling, W.E., Aggarwal, P.K., Batima, P., Brander, K.M., Erda, L., Howden, S.M., Kirilenko, A., Morton, J., J-Soussana, F., Schmidhuber, J., Tubiello, F.N., 2007. Food, fibre and forest products. In: Parry, M.L., Canziani, O.F., Palutikof, J.P., van der Linden, P.J., Hanson, C.E. (Eds.), Climate Change 2007: Impacts, Adaptation and Vulnerability. Contribution of Working Group II to the Fourth Assessment Report of the Intergovernmental Panel on Climate Change. Cambridge University Press, Cambridge, U.K., pp. 275–303.

Falkenmark, M., 2009. Foreword, pp. xi–xii. In: Gleick, P.H., Cooley, H., Cohen, M.J., Morikawa, M., Morrison, J., Palaniappan, M. (Eds.), The World's Water 2008–2009. The Biennial Report on Freshwater Resources. Island Press, Washington, DC, p. 373.

FAO, 2007. Adaptation to Climate Change in Agriculture, Forestry and Fisheries: Perspective, Framework and Priorities. Food and Agriculture Organization of the United Nations, Rome, p. 24.

FAO, 2009. Profile for Climate Change. Food and Agriculture Organization of the United Nations, Rome, p. 25.

FAOstat, 2009. Food and Agriculture Organization of the United Nations. Rome, 2009 <http://www-958.ibm.com/software/data/cognos/manyeyes/visualizations/cropland-area-in-hectares-relative-t> (retrieved May 19, 2011).

Follett, R.F., 2001. Organic carbon pools in grazing land soils. In: Follett, R.F., Kimble, J.M., Lal, R. (Eds.), The Potential of U.S. Grazing Lands to Sequester Carbon and Mitigate the Greenhouse Effect. CRC Press, Boca Raton, FL, pp. 65–86.

Follett, J.R., Follett, R.F., Herz, W.R., 2010. Environmental and human impacts of reactive nitrogen, pp. 1–37. In: Delgado, J.A., Follett, R.F. (Eds.), Advances in Nitrogen Management for Water Quality. Soil and Water Conservation Society, Ankeny, IA, p. 429.

Follett, R.F., Reed, D.A., 2010. Soil carbon sequestration in grazing lands: societal benefits and policy implications. Rangeland Ecol. Manage. 63, 4–15.

Follett, R.F., Schuman, G.E., 2005. Grazing land contributions to carbon sequestration (invited Keynote paper for the 2005 International Grassland Congress, Belfast, Ireland), pp. 266–277. In: McGilloway, D.A. (Ed.), Grazingland: A Global Resource. Wageningen Academic Publishers.

Harris, J.A., Hobbs, R.J., Higgs, E., Aronson, J., 2006. Ecological restoration and global climate change. Restor. Ecol. 14, 170–176.

Harris, L.D., 1984. The Fragmented Forest: Island Biogeography Theory and the Preservation of Biotic Diversity. The Univ. of Chicago Press, Chicago, IL, p. 211.

IPCC, 2007. International Panel on Climate Change. Fourth Assessment Report: Climate Change (AR4). Synthesis Report. Projected climate change and its impacts. <http://www.ipcc.ch/publications_and_data/ar4/syr/en/spms3.html> (retrieved May 19, 2011).

Kiehl, J.T., Trenberth, K.E., 1997. Earth's Annual Global Mean Energy Budget (PDF). Bull. Am. Meteorol. Soc. 78 (2), 197–208.

Khosla, R., Fleming, K., Delgado, J., Shaver, T., Westfall, D., 2002. Use of site-specific management zones to improve nitrogen management for precision agriculture. J. Soil Water Cons. 57, 513–518.

Lal, R., 2001. The physical quality of soil on grazing lands and its effects on sequestrating carbon, pp. 249–266. In: Follett, R.F., Kimble, J.M., Lal, R. (Eds.), The Potential of US Grazing Lands to Sequester Carbon and Mitigate the Greenhouse Effect. Lewis Publishers, A CRC Company, Boca Raton, FL, p. 442.

Lal, R., 2007. Carbon management in agricultural soils. Mitig. Adapt. Strat. Glob. Change 12, 303–322.

Lal, R., 2004. Soil carbon sequestration impacts on global climate change and food security. Science 304, 1623–1627.

Lal, R., Follett, R.F., Kimble, J.M., 2003. Achieving soil carbon sequestration in the United States: a challenge to the policy makers. Soil Scie. 168, 827–845.

Le Treut, H., Somerville, R., Cubasch, U., Ding, Y., Mauritzen, C., Mokssit, A., Peterson, T., Prather, M., 2007. Historical overview of climate change. In: Solomon, S., Qin, D., Manning, M., Chen, Z., Marquis, M.,

Averyt, K.B., Tignor, M., Miller, H.L. (Eds.), Climate Change 2007: The Physical Science Basis. Contribution of Working Group I to the Fourth Assessment Report of the Intergovernmental Panel on Climate Change. Cambridge University Press, Cambridge, United Kingdom and New York, NY, U.S .

Markowitz, H., 1952. Portfolio selection. J. Finance 7, 77–91.

Morgan, J.A., Follett, R.F., Allen Jr., L.H., Delgrosso, S., Derner, J.D., Dijkstra, F., Franzluebbers, A., Fry, R., Paustian, K., Shoeneberger, M.M., 2010. Carbon sequestration in agricultural lands of the United States. J. Soil Water Cons. 65, 6A–13A.

NCGRP, 2010. National Center for Genetic Resources and Preservation. <http://www.ars.usda.gov/main/site_main.htm?modecode=54-02-05-00> (retrieved May 19, 2011).

NPGS, 2010. The National Genetic Resources Program. <http://www.ars-grin.gov/> (retrieved May 19, 2011).

Ogle, S.M., Breidt, F.J., Easter, M., Williams, S., Killian, K., Paustian, K., 2010. Scale and uncertainty in modeled soil organic carbon stock changes for US croplands using a process-based model. Glob. Change Biol. 16, 810–822.

Pachauri, R.K., and Reisinger, A. (Eds.). 2007 . Contribution of working Groups I, II and III to the Fourth Assessement Report of the Intergovernment Panel on Climate Change. IPCC, Geneva, Switzerland, p. 104.

Paustian, K., Collins, H.P., Paul, E.A., 1997. Management controls on soil carbon. In: Paul, E.A., Paustian, K., Elliott, E.T., Cole, C.V. (Eds.), Soil Organic Matter in Temperate Agroecosystems: Long-term Experiments in North America. CRC Press, Boca Raton, FL, U.S, pp. 15–49.

Pineiro, G., Paruelo, J.M., Oesterheld, M., Jobbagy, E.G., 2010. Pathways of grazing effects on soil organic carbon and nitrogen. Range. Ecol. Manage. 63, 109–119.

Robertson, G.P., Vitousek, P.M., 2009. Nitrogen in agriculture: balancing the cost of an essential resource. Ann. Rev. Environ. Res. 34, 97–125.

Russell, R., May 16, 2007. The Greenhouse Effect & Greenhouse Gases. University Corporation for Atmospheric Research Windows to the Universe. <http://www.windows.ucar.edu/tour/link=/earth/climate/greenhouse_effect_gases.html&edu=high> (retrieved May 19, 2011).

Schnepf, R., 2004. Energy Use in Agriculture: Background and Issues. CRS Report for Congress. Congressional Research Service, Order Code RL32677, (November 19), p. 40.

Schmidt, G., April 6, 2005. Water vapour: feedback or forcing? RealClimate. <http://www.realclimate.org/index.php?p=142> (retrieved May 19, 2011).

Solomon, S., Plattner, G.K., Knutti, R., Friedlingstein, P., 2009. Irreversible climate change due to carbon dioxide emissions. Proc. Nat. Acad. Sci. 106, 1704–1709.

Somers, B., 2010. Uncertainty should be powerful motivator on climate, expert says. Science 30, 586.

Somerville, R., Le Treut, H., Cubasch, U., Ding, Y., Mauritzen, C., Mokssit, A., Peterson, T., Prather, M., 2007. Historical overview of climate change. In: Solomon, S., Qin, D., Manning, M., Chen, Z., Marquis, M., Averyt, K.B., Tignor, M., Miller, H.L. (Eds.), Climate Change 2007: The Physical Science Basis. Contribution of Working Group I to the Fourth Assessment Report of the Intergovernmental Panel on Climate Change. Cambridge University Press, Cambridge, United Kingdom and New York, NY, U.S.

Thurow, T.L., Blackburn, W.H., Taylor Jr., C.A., 1988. Infiltration and inter-rill erosion responses to selected livestock grazing strategies, Edwards Plateau, Texas. J. Range Manage. 41, 296–302.

U.S.-EPA, 2007. Recent Climate Change: Atmosphere Changes. Climate Change Science Program. United States Environmental Protection Agency. <http://www.epa.gov/climatechange/science/recentac.html> (retrieved May 19, 2011).

U.S.-EPA, 2010. Inventory of U.S. Greenhouse Gas Emissions and Sinks: 1990–2008. Office of Atmospheric Programs. EPA 430-R-10-006, Washington, DC. <www.epa.gov/climatechange/emissions/usinventoryreport.html> (retrieved May 19, 2011).

USDA ERS (USDA Economic Research Service), 2007. Conservation Policy: Farmland and Grazing Land Protection Programs. USDA ERS, Washington, DC. <http://www.ers.usda.gov/Briefing/ConservationPolicy/farmland.htm> (retrieved May 19, 2011).

USDA FAS, 2010. Foreign Agricultural Service. 2010 Trade News. FY 2010 & 2011 U.S. Agricultural Trade Outlook (08/31/10). <http://www.fas.usda.gov/itp/Policy/newsdata_2010.asp> (retrieved May 19, 2011).

USGCRP, 2009. U.S. Global Change Research Program. Global Climate Change Impacts in the United States 2009. <http://www.globalchange.gov/publications/reports/scientific-assessments> (retrieved May 19, 2011).

von Braun, J., 2007. The World Food Situation: New Driving Forces and Required Actions. International Food Policy Research Institute, Washington, DC.

Wang, G., 2005. Agricultural drought in a future climate: results from 15 global climate models participating in the IPCC 4th assessment. Clim. Dyn. 25, 739–753.

World Bank, 2005. Millennium Ecosystem Assessment. <http://www.maweb.org/en/About.aspx> (retrieved May 19, 2011).

INDEX

Page numbers followed by "*f*" indicate figures, "*t*", tables.

B

C

D

E

H

I

J

K

L

M

N

O

Q

R

S

T

Y

Z

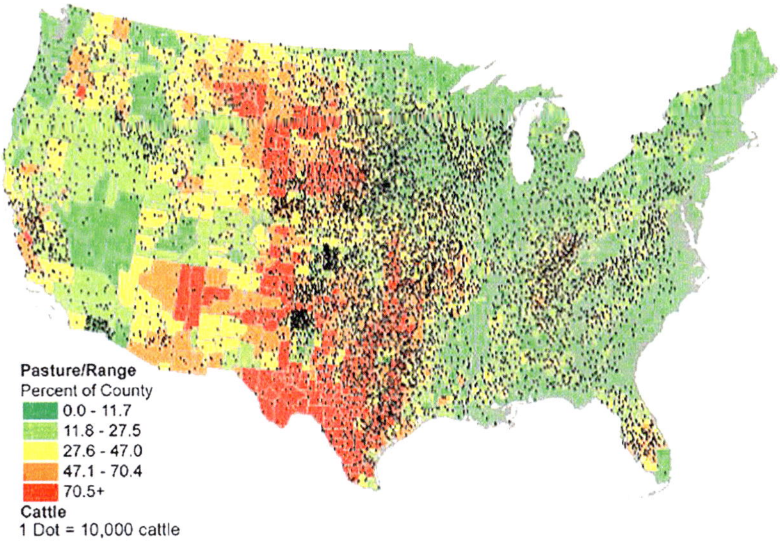

FIGURE 6.2

Extent of grazing land as a proportion of total county area and livestock production cattle inventory in the U.S. (one dot = 10,000 cattle; USGCRP, 2009).

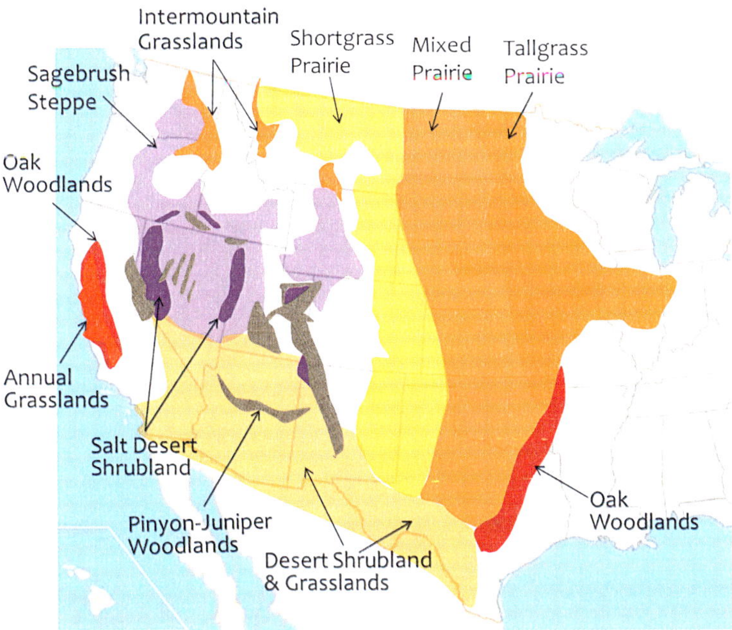

FIGURE 6.3

Major rangelands ecosystems of the U.S. *(from Kuchler, 1964; courtesy of K. Launchbaugh).*

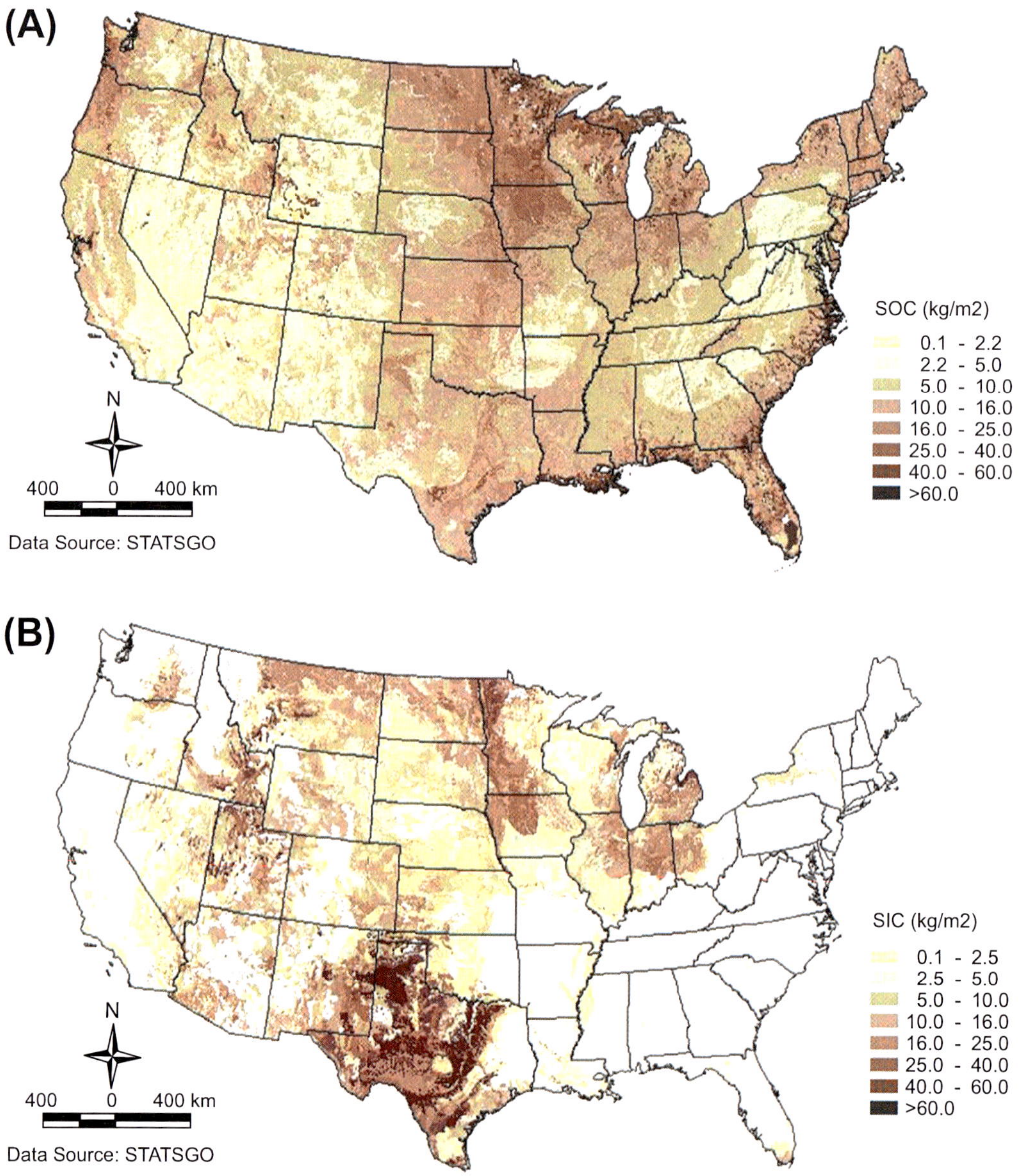

FIGURE 6.6
Distribution of (A) soil organic C (SOC) and (B) soil inorganic C (SIC) to 2 m soil depth in the conterminous U.S. *(from Guo et al., 2006).*

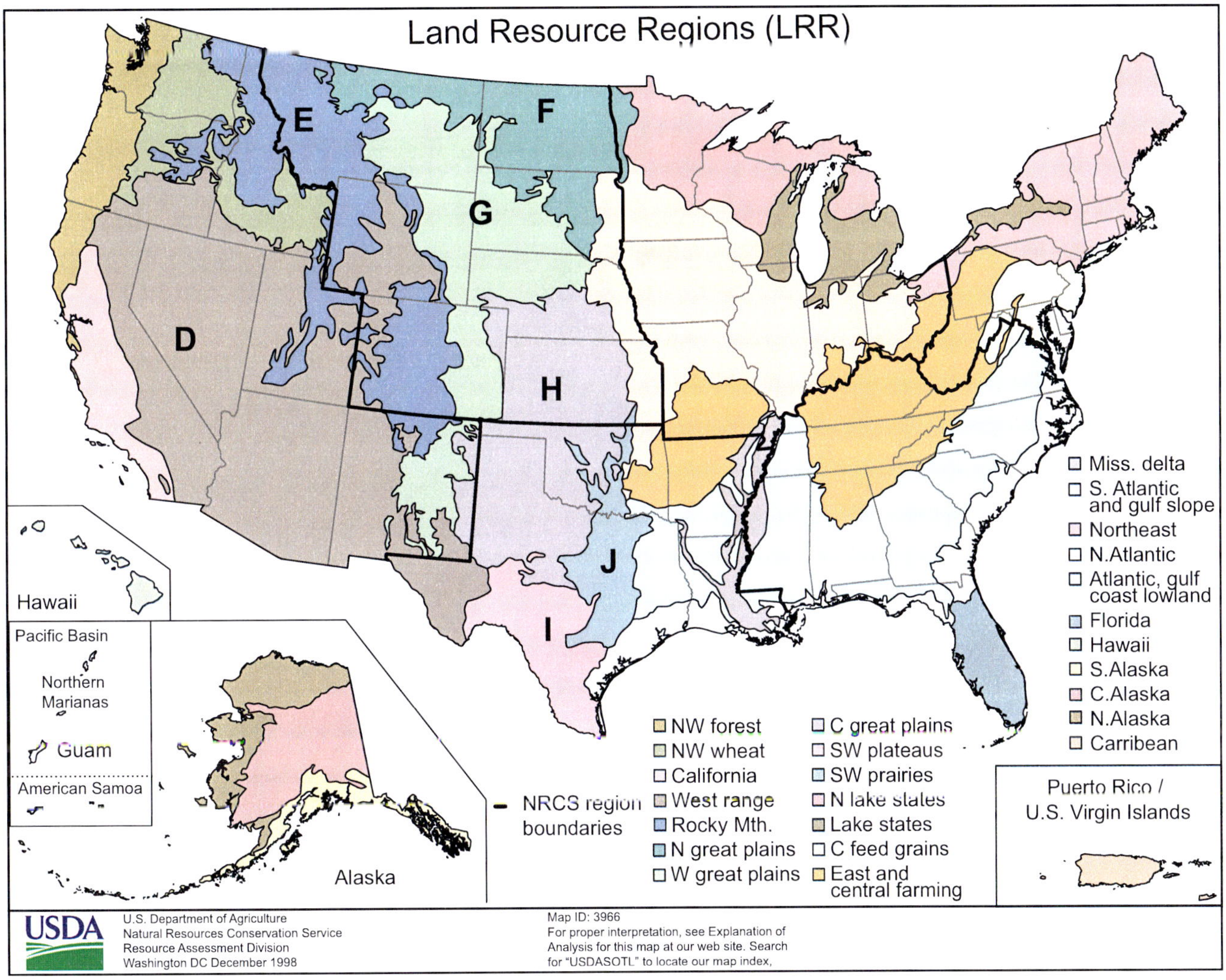

FIGURE 6.7
Land Resource Regions (LRRs) in the U.S. (USDA-NRCS). LRRs in Table 6.2 are labeled.

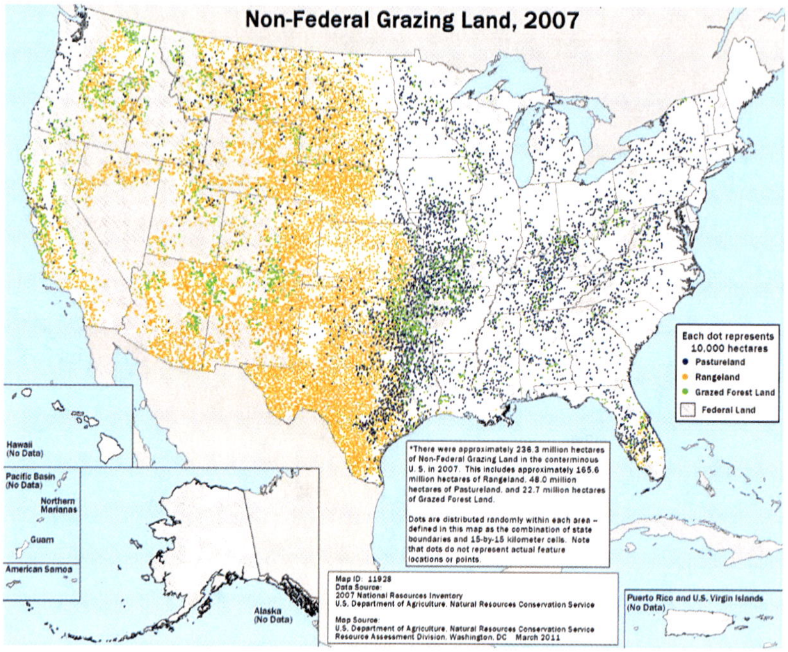

FIGURE 11.1

Distribution of private pastureland, rangeland, and grazed forest land in the 48 contiguous United States, 2007. *[adapted from USDA (2009)].*

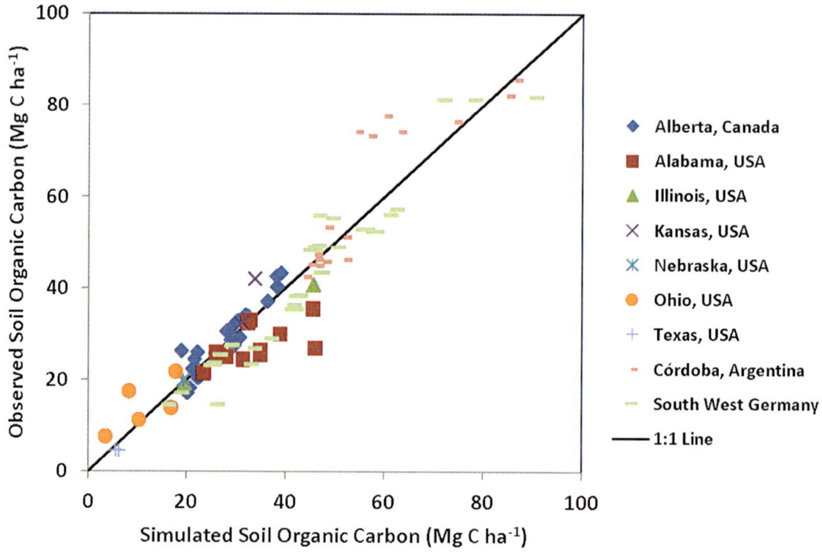

FIGURE 17.2

Comparison between simulated and observed soil organic C for a range treatment × site combinations ranging from 10 to 90 Mg C ha^{-1}. Fitted linear regression: OBS = 0.9SIM + 4.16; $R^2 = 0.907$, $p < 0.01$, n = 109).

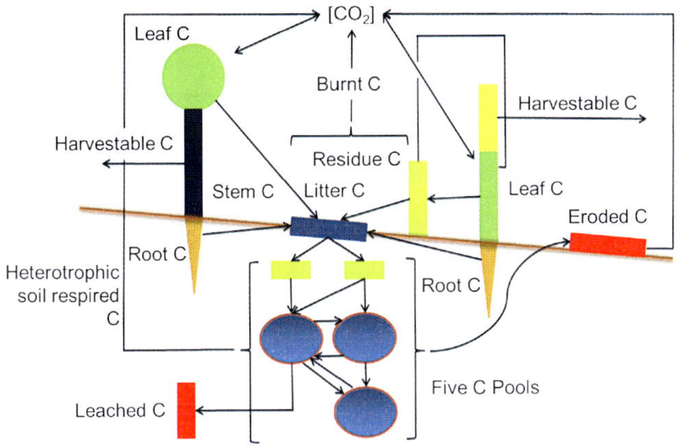

FIGURE 17.3
Diagram representing the ecosystem carbon balance in EPIC.

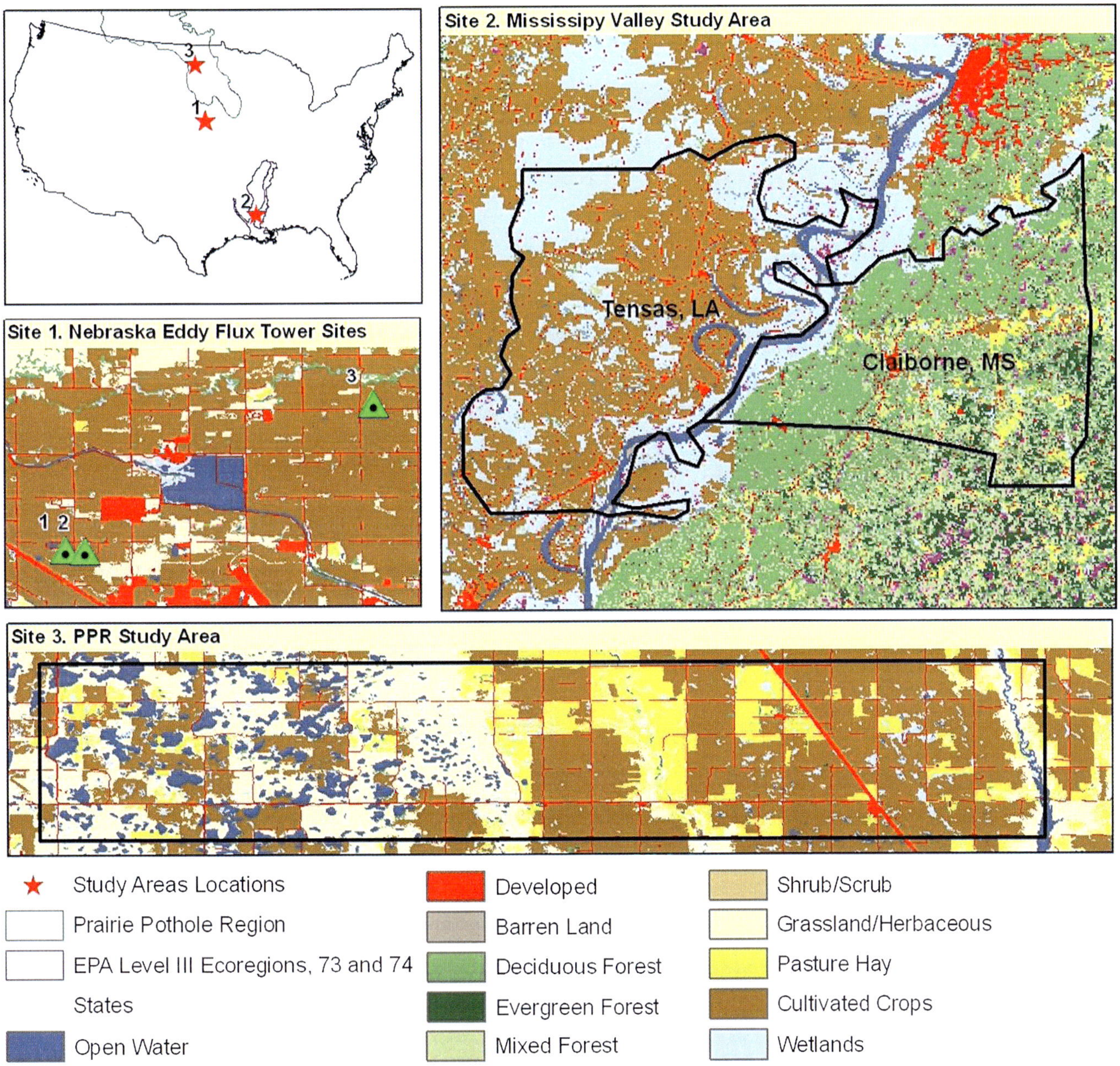

FIGURE 18.2
The locations of the study areas.

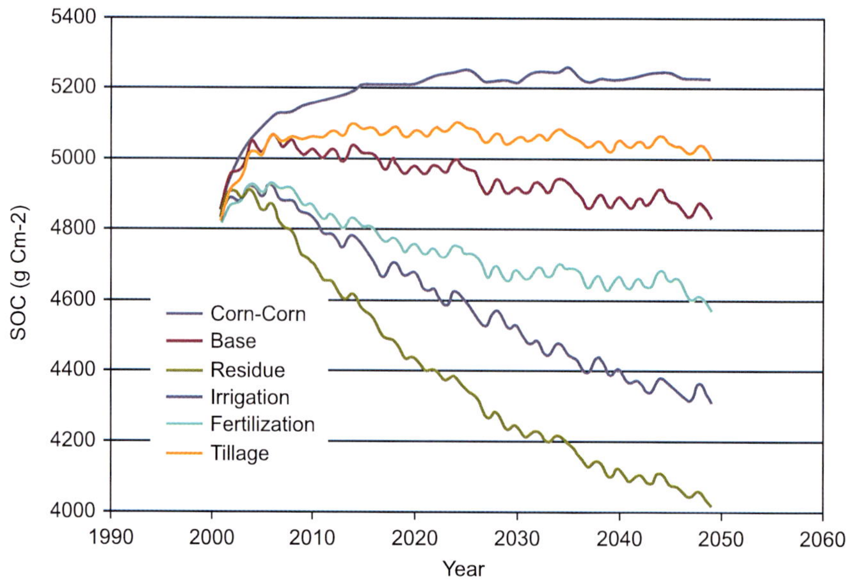

FIGURE 18.3
Soil carbon dynamics under various management practices. The scenarios are specified in Table 18.1.

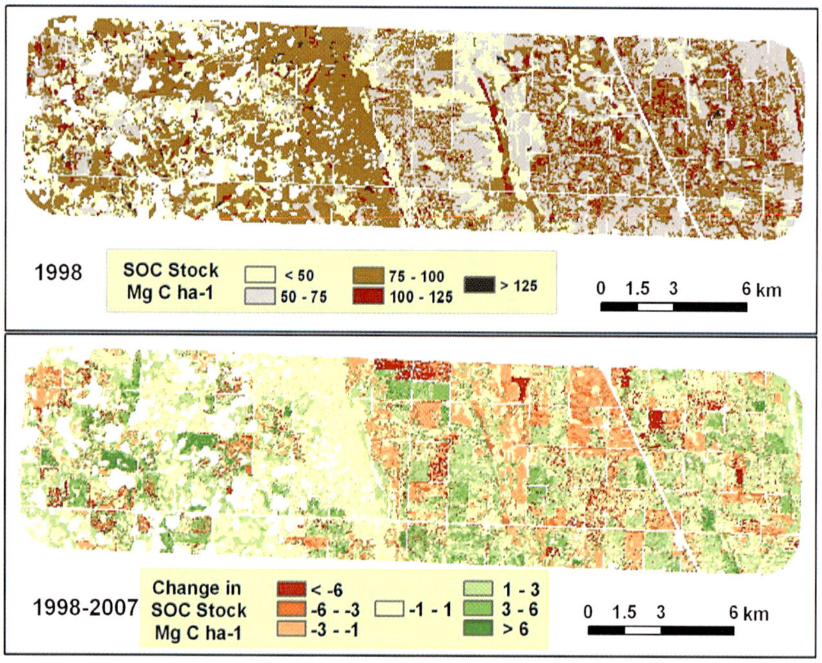

FIGURE 18.4
GEMS simulated changes in soil organic carbon in the 0–100 cm depth within the study area of the Prairie Pothole region.

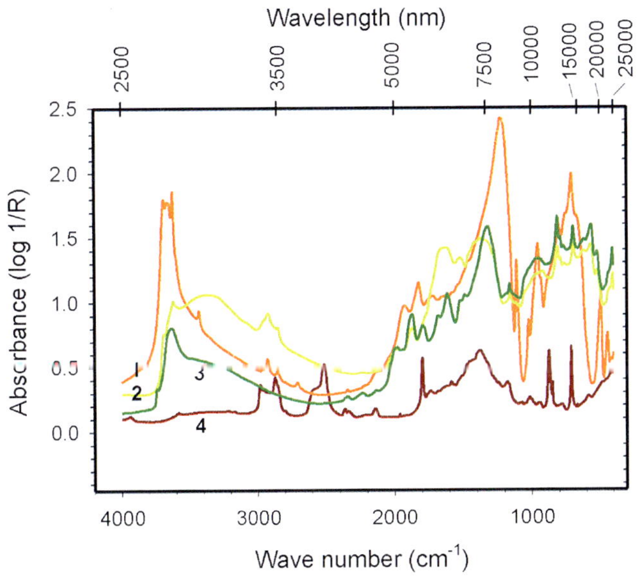

FIGURE 20.4
Mid-infrared spectra of: kaolinite clay (1); soil with 9.80% organic carbon (0.59% carbonate) before (2) and after (3) ashing at 550°C, and 5% calcium carbonate diluted in KBr (4).

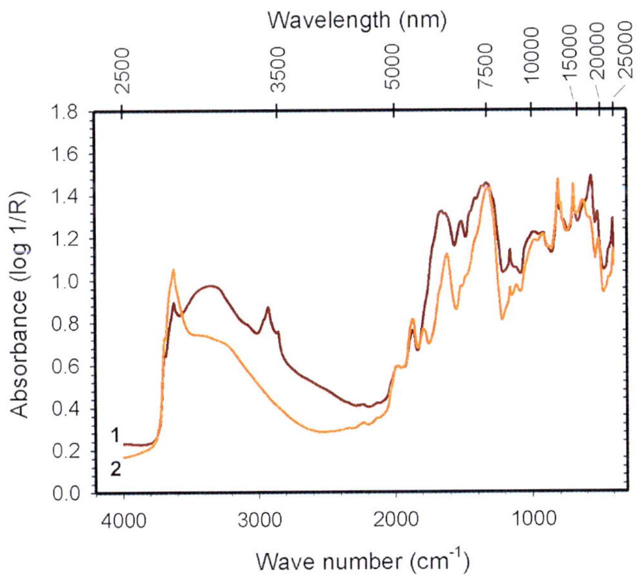

FIGURE 20.6
Mid-IR spectra of soils with low carbonate content carbonates (acidified) and respective organic carbon contents at 9.80% (1) and 0.09% (2), respectively.

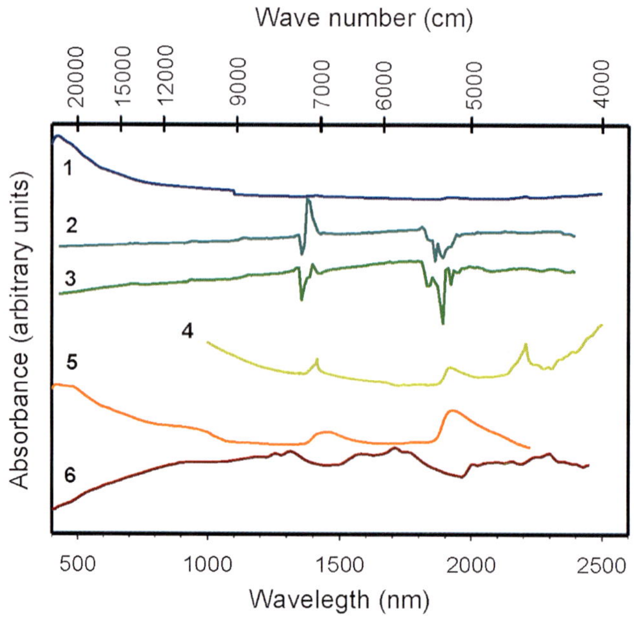

FIGURE 20.8

Near-infrared spectra of soils including: (1) pseudo-absorbance (PA) NIRSystems 6500 scanning monochromator; (2) reflectance (R) from Hyperion satellite at 4.5% and (3) 80% ground cover; (4) PA from an FT-NIR laboratory instrument; (5) PA from a tillage-shank VIS-NIR Veris scanning system; and (6) airborne reflectance collected using a hyperspectral VIS-NIR imaging spectrometer (HyperSpecTIR®). All X units converted to nm and all spectra normalized to between 0 and 1 and shifted for display.

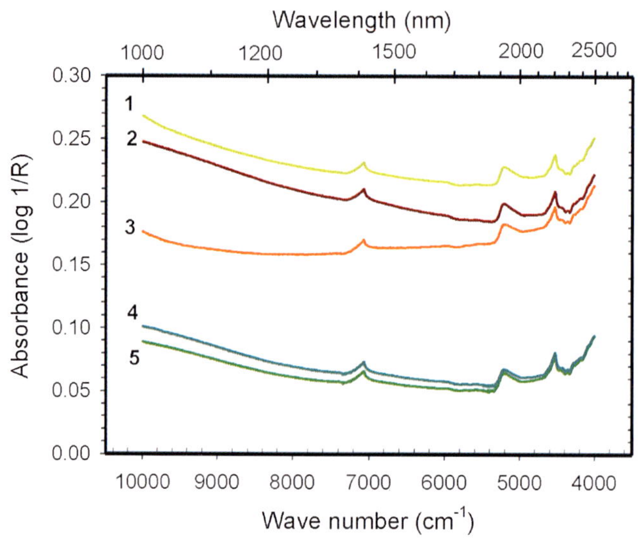

FIGURE 20.9

Near-infrared spectra of five soil samples from the eastern shore of Maryland demonstrating differences in overall soil brightness as reflected by overall absorbance differences, and scattering effects as reflected by overall slope differences, due to particle size differences among soils.

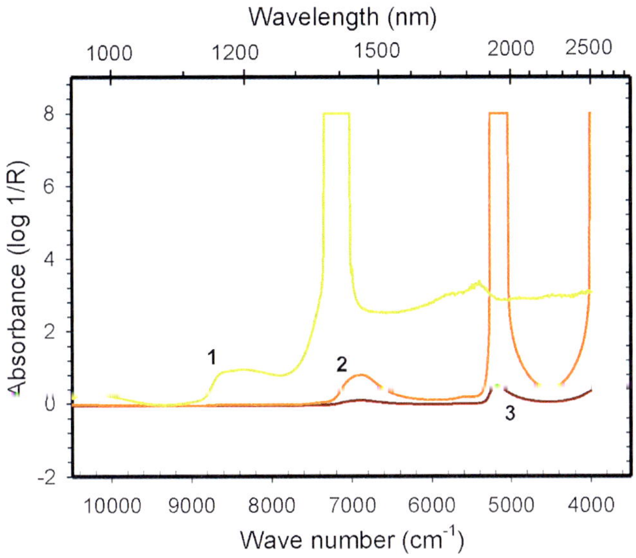

FIGURE 20.10
Near-infrared spectrum of water with a path length of 2 mm (1), 0.5 mm (2), and 0.1 mm (3); all spectra zeroed to baseline.

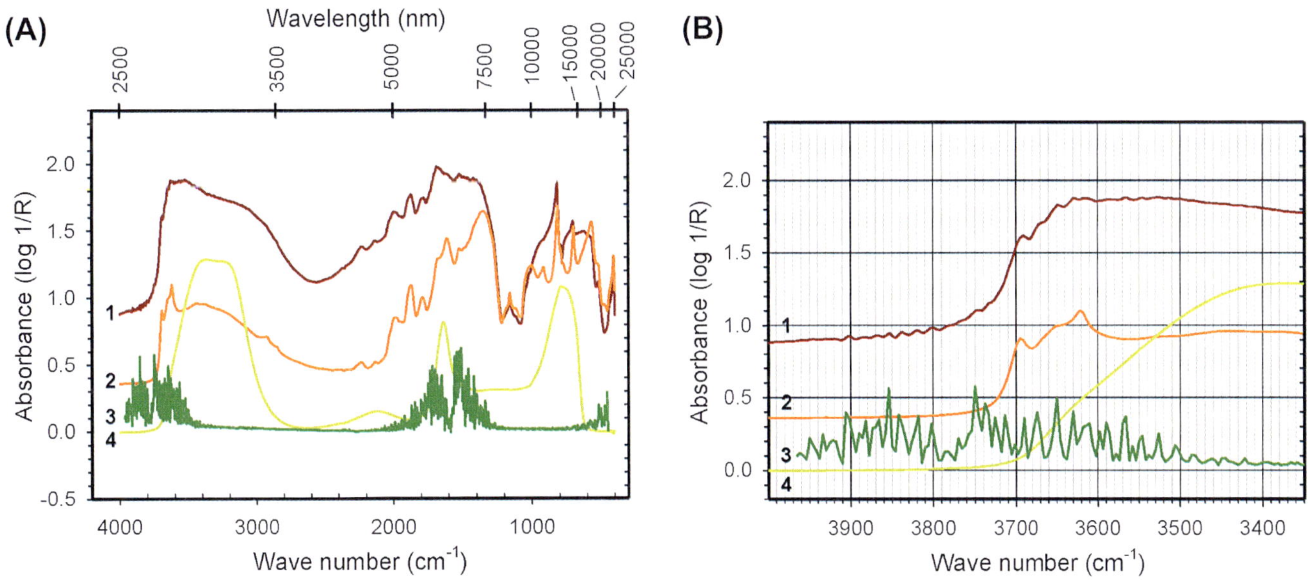

FIGURE 20.11
(A) Average infrared (IR) spectra of: 201 soils scanned field moist in open cells (1), scanned dried (2), water vapor by pyrolysis of $CuSO_4 \cdot (H_2O)_5$ (3), and liquid water by attenuated total reflection using 9 bounce ZnSe crystal (4). (B) Average near infra-red (NIR) of: 201 soils scanned field moist in open cells (1), scanned dried (2) water vapor by pyrolysis of $CuSO_4 \cdot (H_2O)_5$ (3), liquid water by attenuated total reflection using 9 bounce ZnSe crystal (4).

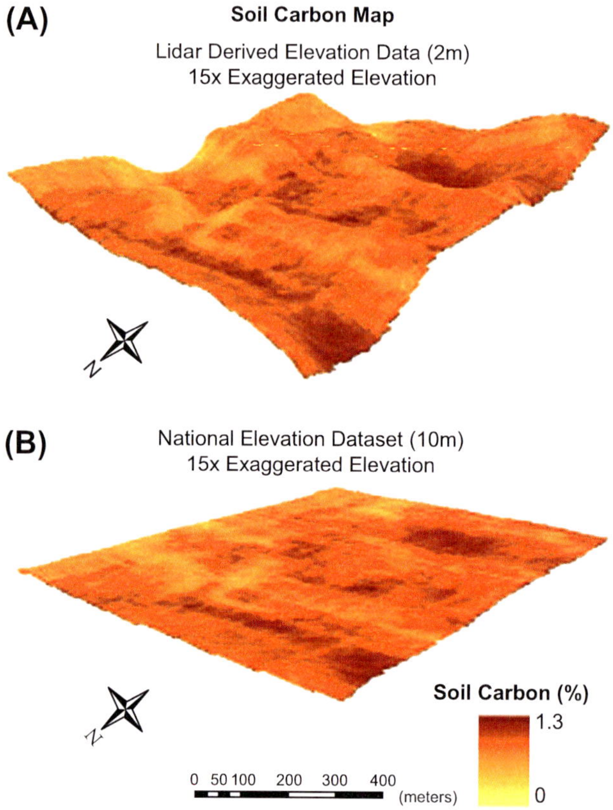

FIGURE 22.3

Soil organic carbon (SOC) map produced from aircraft hyperspectral data over a field in Maryland, draped on DEM products generated by: (A) high resolution LiDAR (2 m resolution) and (B) National Elevation Dataset (NED, 10 m resolution). Elevation within the field associated with the LiDAR data ranged from 11.1 to 17.3 m above sea level (ASL) and the NED data ranged from 11.9 to 14.2 m ASL. Although the general slope of the field is represented to some degree by the NED data, it contains essentially no information related to SOC distributions in the field as a function of topographic relief. By contrast, visual inspection shows that the LiDAR product is rich in information related to field-scale distribution of SOC from analysis of hyperspectral data.

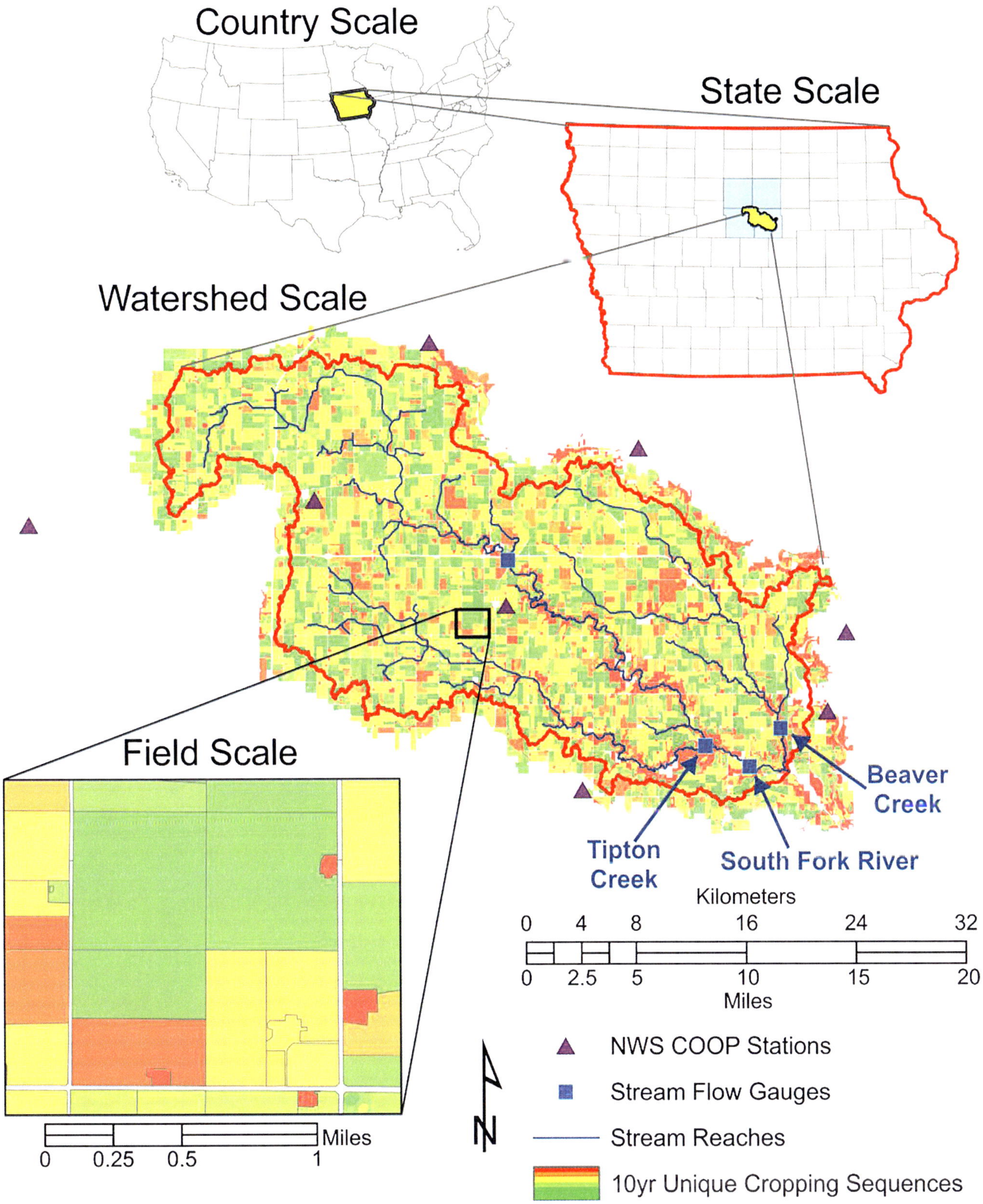

FIGURE 22.5
Map of the South Fork of the Iowa River watershed in Iowa.

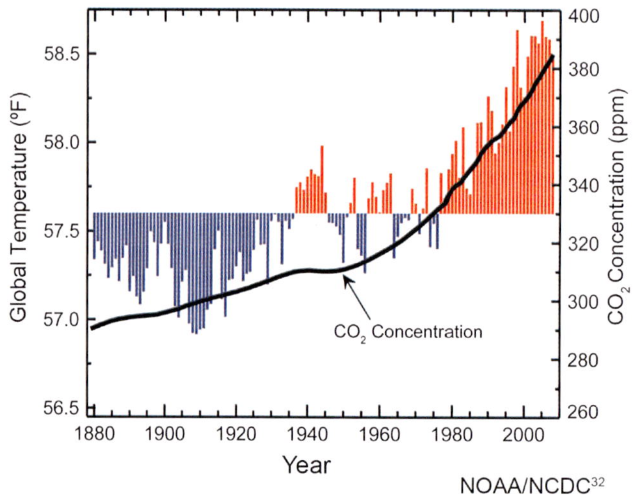

FIGURE 29.1

Increases in annual global surface temperature (over both oceans and land) since 1880. Red bars indicate temperatures above and blue bars represent temperatures below the average temperature period 1901–2000. The black line is atmospheric CO_2 concentration in parts per million. Global Climate Change Impacts in the United States. *USGCRP (2009).*

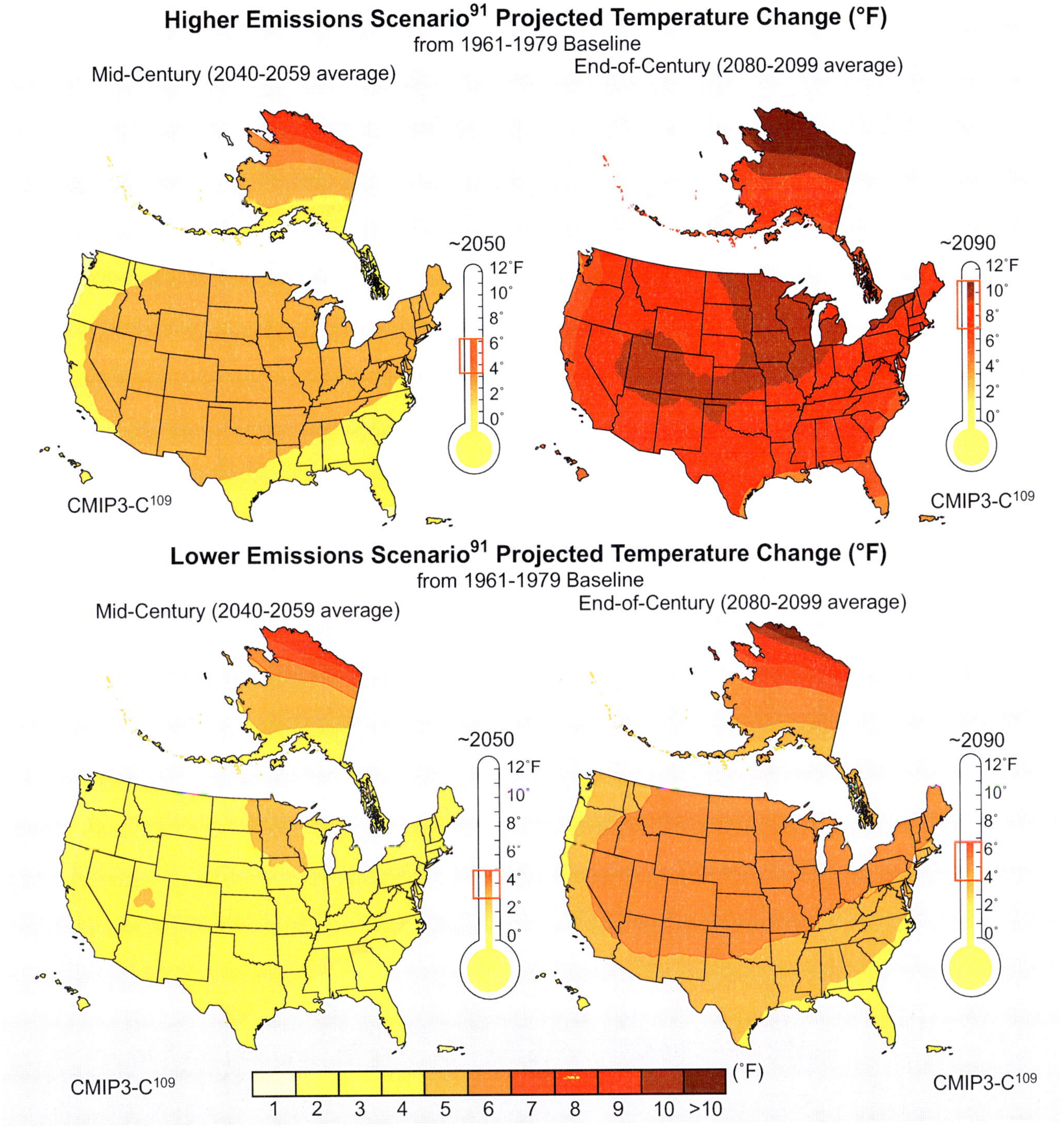

FIGURE 29.2

Projections of future temperature from 16 of the Coupled Model Inter-comparison Project. The maps feature a higher and lower GHG emission scenario. Brackets on the thermometers represent likely ranges of model predictions. Global Climate Change Impacts in the United States. *USGCRP (2009).*

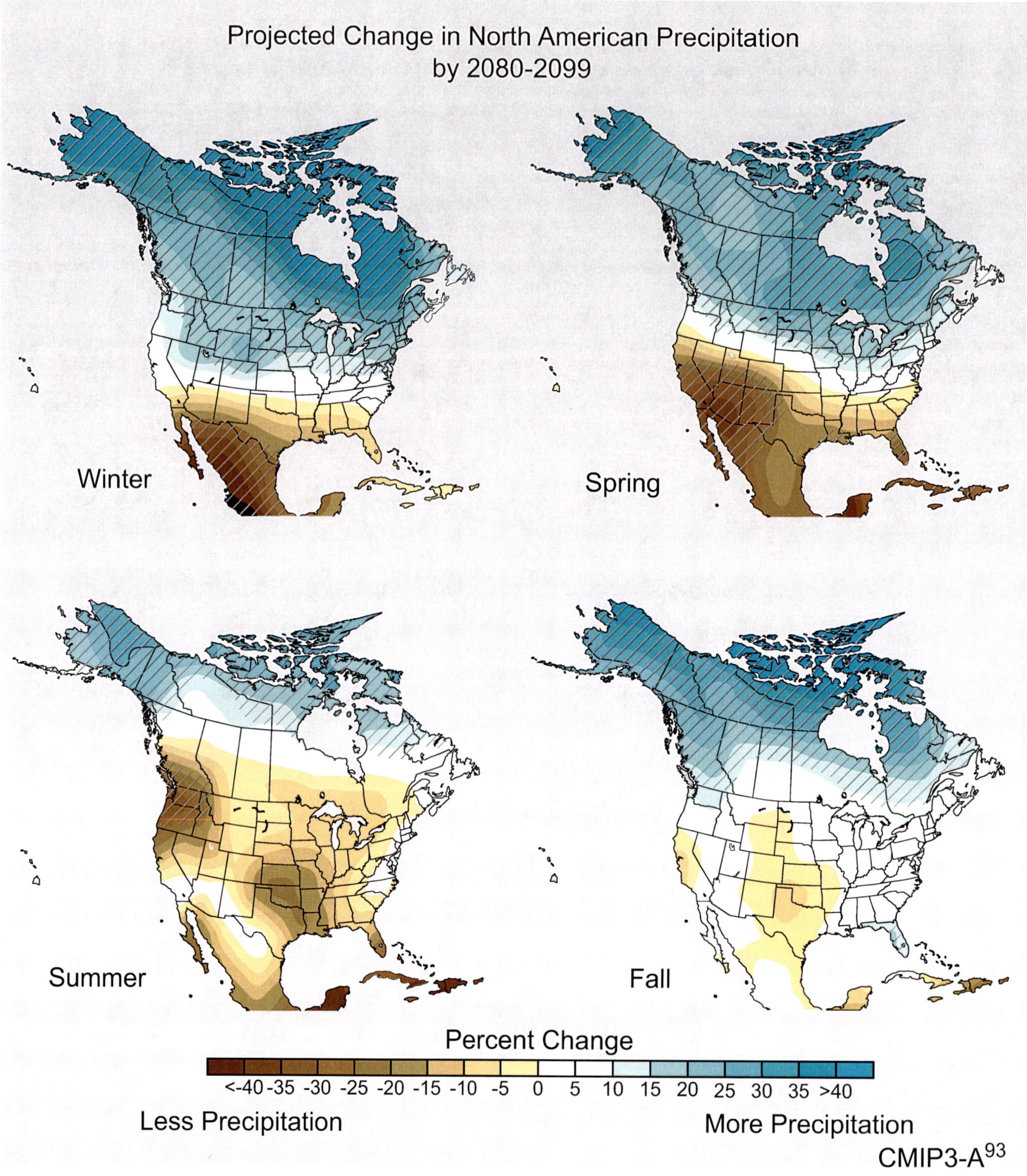

FIGURE 29.3

Projected future changes in precipitation relative to the recent past as simulated by 15 climate models. Simulations are the late 21st century, under a higher emissions scenario. Global Climate Change Impacts in the United States. *USGCRP (2009).*

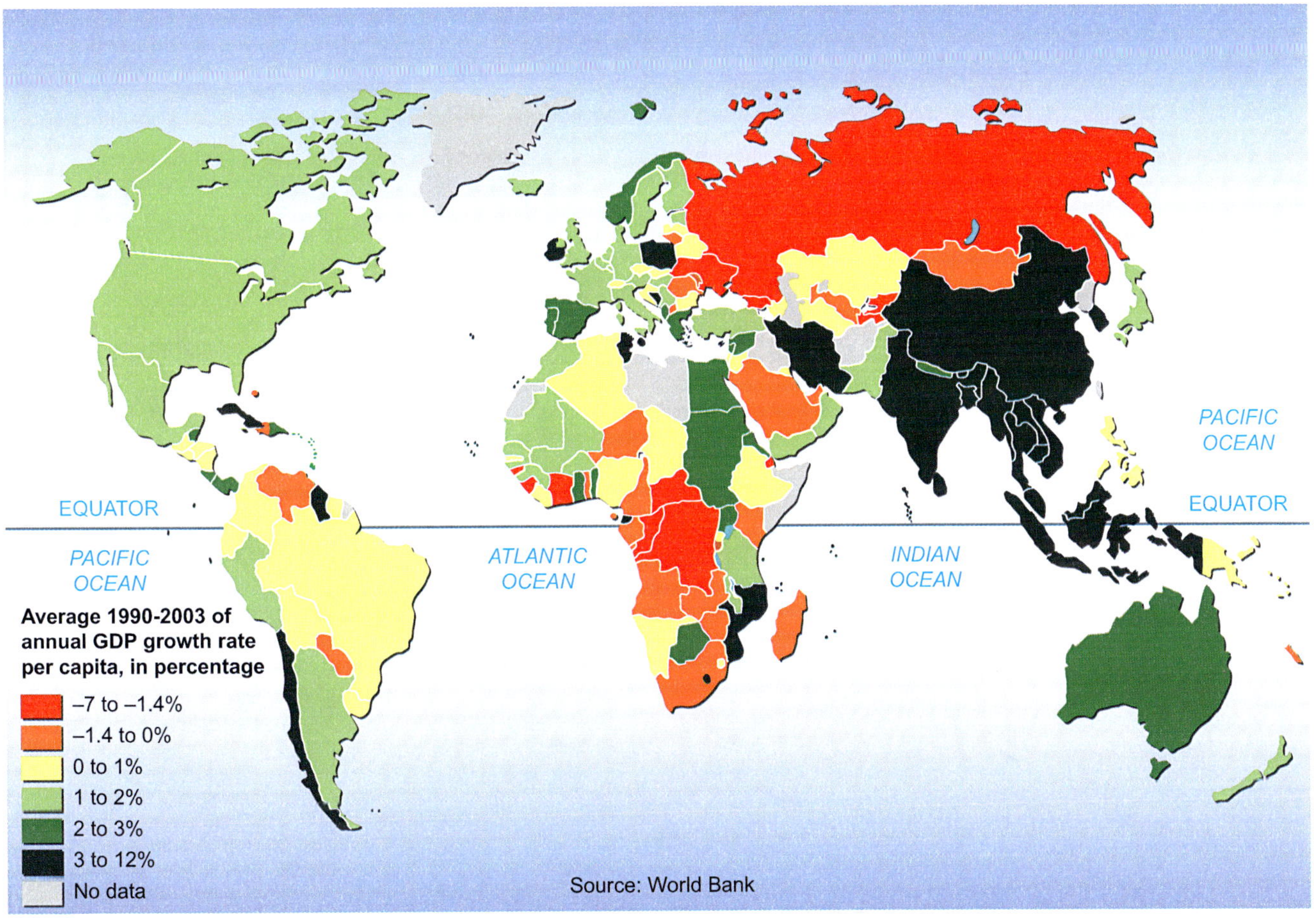

FIGURE 29.4

The global economic importance of agriculture. Per capita gross domestic product (GDP) average annual growth, 1990–2003. Average annual percentage growth rate of GDP per capita at market prices based on constant local currency. Dollar figures for GDP are converted from domestic currencies using 1995 official exchange rates. GDP is the sum of gross value added by all resident producers in the economy plus any product taxes and minus any subsidies not included in the value of the products (World Bank, 2005). *Cartographer/Designer Philippe Rekacewicz, and Emmanuelle Bournay, UNEP/GRID-Arendal. http://maps.grida.no/go/graphic/per-capita-gross-domestic-product-gdp-average-annual-growth-1990-2003 (verified May 23, 2011).*

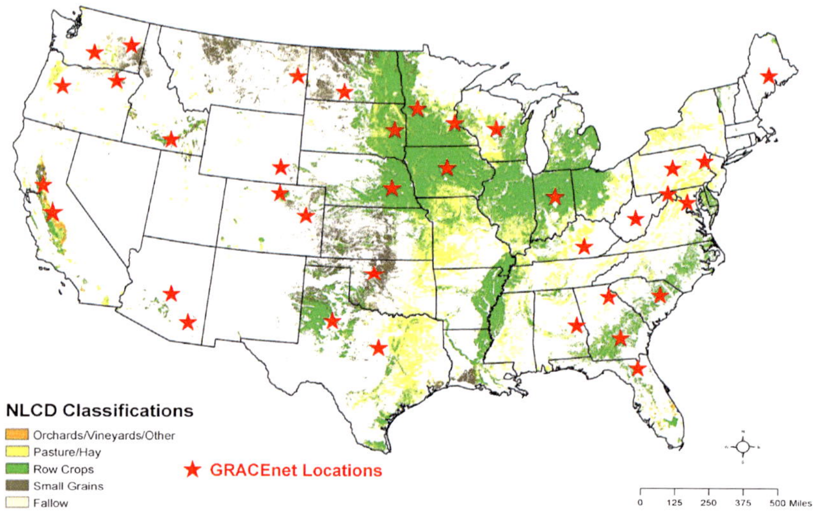
NLCD Classifications
Orchards/Vineyards/Other
Pasture/Hay
Row Crops
Small Grains
Fallow
GRACEnet Locations
0 125 250 375 500 Miles

Edwards Brothers Malloy
Ann Arbor MI. USA
November 27, 2012